TECHNIQUES OF CHEMISTRY

WILLIAM H. SAUNDERS, JR., *Series Editor*
ARNOLD WEISSBERGER, *Founding Editor*

VOLUME XX

**TECHNIQUES FOR
THE STUDY OF
ION–MOLECULE
REACTIONS**

TECHNIQUES OF CHEMISTRY

VOLUME XX

TECHNIQUES FOR THE STUDY OF ION–MOLECULE REACTIONS

Edited by

JAMES M. FARRAR

and

WILLIAM H. SAUNDERS, JR.

Department of Chemistry
University of Rochester
Rochester, New York

QD61
T4
v.20
1988

WILEY

A WILEY-INTERSCIENCE PUBLICATION
JOHN WILEY & SONS
New York · Chichester · Brisbane · Toronto · Singapore

Copyright © 1988 by John Wiley & Sons, Inc.

All rights reserved. Published simultaneously in Canada.

Reproduction or translation of any part of this work
beyond that permitted by Section 107 or 108 of the
1976 United States Copyright Act without the permission
of the copyright owner is unlawful. Requests for
permission or further information should be addressed to
the Permission Department, John Wiley & Sons, Inc.

The Library of Congress has cataloged this serial Publication as follows:
Techniques of chemistry.—   —New York:
    Wiley,
       v.
    Irregular.
    Began in 1971.
    Description based on: Vol. 4, pt. 1.
    Photo-offset reprint separately cataloged and classified in LC.
    Editor: A. Weissberger.
    Merger of: Technique of inorganic chemistry; and: Technique of organic
chemistry.
      ISSN 0082-2531 = Techniques of chemistry.
      1. Chemistry—Manipulation—Collected works.   I. Weissberger, Arnold,
1898–   . II. Title: Techniques of chemistry

    [DNLM: W1 TE197K]
    QD61.T4   542—dc19   85-649508

    ISBN 0-471-84812-3   AACR 2 MARC-S
    Library of Congress   [8610]

    Printed in the United States of America

  10 9 8 7 6 5 4 3 2 1

To the Memory of Bruce H. Mahan

CONTRIBUTORS

NIGEL G. ADAMS, *Department of Space Research, University of Birmingham, Birmingham, England*

MICHAEL T. BOWERS, *Department of Chemistry, University of California, Santa Barbara, California*

JOHN I. BRAUMAN, *Department of Chemistry, Stanford University, Stanford, California*

FULVIO CACACE, *Department of Chemistry, University of Rome, Rome, Italy*

JAMES M. FARRAR, *Department of Chemistry, University of Rochester, Rochester, New York*

BEN S. FREISER, *Department of Chemistry, Purdue University, West Lafayette, Indiana*

CRIS E. JOHNSON, *Department of Chemistry, Stanford University, Stanford, California*

MARK A. JOHNSON, *Department of Chemistry, Yale University, New Haven, Connecticut*

PAUL KEBARLE, *Department of Chemistry, University of Alberta, Edmonton, Alberta, Canada*

PAUL R. KEMPER, *Department of Chemistry, University of California, Santa Barbara, California*

W. CARL LINEBERGER, *Department of Chemistry, University of Colorado, Boulder, Colorado*

THOMAS H. MORTON, *Department of Chemistry, University of California, Riverside, California*

CHEUK-YIU NG, *Department of Chemistry, Iowa State University, Ames, Iowa*

DAVID SMITH, *Department of Space Research, University of Birmingham, Birmingham, England*

MAURIZIO SPERANZA, *Department of Chemical Science, University of Camerino, Camerino, Italy*

MASAHARU TSUJI, *Research Institute of Industrial Science, Kyushu University, Fuku-oka, Japan*

INTRODUCTION TO THE SERIES

Techniques of Chemistry is the successor to Technique of Organic Chemistry and its companion, Technique of Inorganic Chemistry. The newer series reflects the fact that many modern techniques are applicable over a wide area of chemical science. All of these series were originated by Arnold Weissberger and edited by him for many years.

Following in Dr. Weissberger's footsteps is no easy task, but every effort will be made to uphold the high standards he set. The aim remains the same: the comprehensive presentation of important techniques. At the same time, authors will be encouraged to illustrate what can be done with a technique rather than cataloging all known applications. It is hoped in this way to keep individual volumes to a reasonable size. Readers can help with advice and comments. Suggestions of topics for new volumes will be particularly welcome.

WILLIAM H. SAUNDERS, JR.

Department of Chemistry
University of Rochester
Rochester, New York

PREFACE

New techniques are usually developed for very specific and limited purposes. In a few cases, such as magnetic resonance, a new technique turns out to have a range of applications far beyond anything envisioned by the original discoverers. The techniques employed to study ion–molecule reactions in the gas phase are now finding a similarly rapidly expanding array of uses.

Chemical physicists interested in the detailed dynamics of chemical reactions are discovering many new ways of approaching previously unanswerable questions. The importance of ion–molecule reactions in unraveling the complex chemistry of planetary atmospheres and interstellar space has become apparent in recent years. The role of such reactions in electrical discharges, plasma chemistry, and combustion systems has also provided great impetus for the development of new probes of gas phase ion chemistry.

Organic chemists have long discussed the role of solvation in organic reactions. The recently acquired ability to observe in the gas phase reactions that formerly could be carried out only in solution is providing for the first time a sound experimental foundation for theories of solvation. Many long-accepted ideas have already required revision as a result of gas-phase scales of acidity and basicity which often differ strikingly from scales determined in water or other solvents. A new appreciation is emerging of the profound effects of solvation on such important ionic organic reactions as substitution, addition, and elimination. It is not too much to expect that even synthetic applications of gas-phase ion–molecule reactions may soon appear.

The aim of this volume is to provide descriptions of the various techniques presently available or under development for the study of ion–molecule reactions in the gas phase. While we have tried to include all important techniques, we have not aimed for encyclopedic coverage. We have asked each author to show the range of applicability of a technique by means of carefully chosen examples. What we have tried to do is produce a volume that will be useful to a new research student in the field, or to an established researcher who wants to find a technique capable of answering the questions that concern him or her. A major part of the volume is devoted to descriptions of the construction and operation of apparatus so as to provide help in practice as well as in theory. At the same time, we hope to provide the more general reader with a useful overview of this exciting and rapidly developing field.

The order in which the chapters are presented is to a considerable extent arbitrary. We have, however, put first the six chapters describing methods used primarily to probe "bulk" properties or behavior such as thermodynamic quantities and rate constants. The remaining five chapters cover "microscopic" techniques capable of giving more specific or more highly resolved information.

We wish to thank our authors for their excellent contributions to this project. It is only through their efforts that we have been able to bring this volume to fruition.

JAMES M. FARRAR
WILLIAM H. SAUNDERS, JR.

Rochester, New York
July 1988

CONTENTS

Chapter I
Ion Cyclotron Resonance Spectrometry 1
               Paul R. Kemper and Michael T. Bowers

Chapter II
Fourier Transform Mass Spectrometry 61
               Ben S. Freiser

Chapter III
Neutral Products from Electron Bombardment Flow
Studies 119
               Thomas H. Morton

Chapter IV
Flowing Afterglow and SIFT 165
               Nigel G. Adams and David Smith

Chapter V
Pulsed Electron High Pressure Mass Spectrometer 221
               Paul Kebarle

Chapter VI
Nuclear-Decay Techniques 287
               Fulvio Cacace and Maurizio Speranza

Chapter VII
Ion-Beam Methods 325
               James M. Farrar

Chapter VIII
State-Selected and State-to-State Ion–Molecule
Reaction Dynamics by Photoionization Methods 417
               Cheuk-Yiu Ng

Chapter IX
Spectroscopic Probes 489
               Masaharu Tsuji

Chapter X
Infrared Laser Photolysis 563
               Cris E. Johnson and John I. Brauman

Chapter XI
Pulsed Methods for Cluster Ion Spectroscopy. 591
Mark A. Johnson and W. Carl Lineberger

Index. 637

Chapter 1

ION CYCLOTRON RESONANCE SPECTROMETRY

Paul R. Kemper and Michael T. Bowers

---

1 **Introduction**
2 **Equations of Motion**
  2.1 Introduction
  2.2 Motion in Static Fields
  2.3 Power Absorption
  2.4 Oscillation in the Trapping Field
3 **Drift ICR**
  3.1 Introduction
  3.2 Experimental Apparatus
      3.2.1 Cells
      3.2.2 Vacuum System
      3.2.3 Electronics
  3.3 Types of Experiments
      3.3.1 Double Resonance
      3.3.2 Ion–Neutral Reactions
      3.3.3 Collision Frequencies
      3.3.4 Temperature Variable ICR
      3.3.5 Equilibrium Measurements
      3.3.6 Miscellaneous Experiments
4 **Trapped ICR**
  4.1 Introduction
  4.2 Experimental Apparatus
      4.2.1 Cells
      4.2.2 Electronics
      4.2.3 Detection
  4.3 Data Analysis
  4.4 Double Resonance
  4.5 Performance
  4.6 Applications
5 **Tandem ICR**
  5.1 Introduction
  5.2 Design
      5.2.1 Vacuum Systems
      5.2.2 Ion Sources

        5.2.3   Ion Optics
        5.2.4   ICR Reaction Cell
    5.3 Performance
    5.4 Applications
        5.4.1   Product Distributions
        5.4.2   Total Rate Constants
        5.4.3   Reactant Ion Kinetic Energy Studies
        5.4.4   Effect of Reactant-Ion Internal Energy
    5.5 Tricyclotron Resonance Spectrometer
6 **Kinetic Energy ICR**
    6.1 Introduction
    6.2 Theory
    6.3 Experimental Technique
    6.4 Applications
7 **Detectors**
    7.1 Introduction
    7.2 Marginal Oscillator
    7.3 Sensitivity Calibration
    7.4 Capacitance Bridge Detector
    7.5 Comparison of MO and CBD Sensitivities
    7.6 The Q Meter
    7.7 Electrometer Detection
References

1 INTRODUCTION

The beginnings of ion cyclotron resonance (ICR) spectrometry probably lie in the 1940s with Sommer, Thomas, and Hipple [1] and their low resolution mass spectrometer called the omegatron. In the early 1960s, Wobschall and Wobschall et al. [2, 3] used the ICR technique to measure collision frequencies and other phenomena at high pressures. Probably the "present age" of ICR dates from Beauchamp and co-workers' article on ICR in 1967 [4] and Baldeschwieler's article in 1968 [5]. Since that time several reviews have appeared, including those by Beauchamp [6] (1971), Bowers and Su [7] (1973), and Lehman and Bursey [8] (1976). The first two reviews discuss experimental techniques but concentrate mainly on the chemistry studied. Lehman and Bursey's book divides about equally between technique and results and includes a good discussion of both drift and trapped cell experiments. In the 10 years since their book, several new techniques have emerged such as kinetic energy analyzed ICR and tandem ICR. Advances have also been made in detector technology. This article presents an overview of ICR experimentation in 1986.

Several topics have been excluded: most notably, photoexcitation/ionization experiments along with any mention of FTICR. These topics are covered elsewhere in this volume. We have also cut discussion of ion–neutral chemistry to an absolute minimum. Examples of experiments unique to a specific technique are given, but a general discussion of the chemistry studied using ICR would require its own book. Specifically, we include sections on (1) ion motion in the ICR cell (in both static and varying fields); (2) drift ICR, including a discussion of kinetic energy studies using semiquantitative double resonance; (3) trapped ICR; (4) tandem ICR, in which reactant ions are mass selected before reaction; (5) kinetic energy ICR, in which reaction translational energy release is measured; and (6) a section on detectors. We hope that the discussion will be of use.

2 EQUATIONS OF MOTION

2.1 Introduction

An understanding of ICR spectroscopy depends on an understanding of the ion motion in the ICR cell. In this section we state the phenomenological equations for a charged particle in crossed electric and magnetic fields, we describe the shape of the electric field, and we outline the solution of the equations giving the ion position in the cell as a function of time and initial conditions. The rate of power absorption from an alternating electric field is also derived. Almost all of the work presented here has appeared in the literature. The coordinate system used here is shown in Fig. 1; MKS units are used throughout [B in tesla (10^4 gauss) and E in volts/meter].

2.2 Motion in Static Fields

The motion of a charged particle in a magnetic field is governed by Eq. 2.1

$$m\dot{\bar{v}} = q\bar{v} \times \bar{B} \tag{2.1}$$

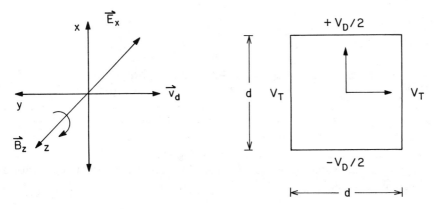

Fig. 1. Coordinate system of ion motion and ICR cell. Positive ions rotate in direction shown.

which describes radial motion by the ion perpendicular to the magnetic field and has solution [9]

$$\omega_c = \frac{qB}{m} \tag{2.2}$$

In 2.1 and 2.2 v is the ion velocity, q is the ion charge, B the magnetic field strength, and ω_c the cyclotron frequency of the particle. Since ω_c is actually a vector quantity, positive and negative ions rotate in opposite directions; the direction of rotation is shown in Fig. 1. In the presence of both electric and magnetic fields, the governing equation becomes

$$m\dot{\bar{v}} = q[\bar{E} + \bar{v} \times \bar{B}]. \tag{2.3}$$

If $\bar{E}$ and $\bar{B}$ are perpendicular, Eq. 2.3 is solved [9] by changing to a frame of reference moving at velocity $\bar{v}_d$ and substituting

$$\bar{v} = \bar{u} + \bar{v}_d \tag{2.4a}$$

where

$$\bar{v}_d = \frac{\bar{E}_x \times \bar{B}}{B^2} = \text{constant} \tag{2.4b}$$

an $\bar{u}$ is the velocity in the moving reference frame.

Substituting Eq. 2.4b in Eq. 2.3, the effect of the electric field vanishes and

$$m\dot{\bar{v}} = m\dot{\bar{u}} = q\bar{u} \times \bar{B} \tag{2.5}$$

The ion motion then consists of a constant drift velocity, $\bar{v}_d$, and the previously derived cyclotron motion. The drift velocity is in the $-y$ direction and has magnitude

$$v_d = \frac{E_x}{B} \tag{2.6}$$

It is clear that we must determine the functional form of the electric field in the cell to specify the ion motion. In the absence of other potentials, E_x is approximately equal to V_D/h, where V_D is the voltage between the top and bottom plates of the cell and h is the plate spacing; however, to constrain motion of positive ions in the z direction, a pair of positively biased side plates at potential V_T is added to the cell. The resulting field is due to both V_T and V_D.

The field due to the trapping potential alone is often approximated with a quadrupolar potential [10, 11]

$$V = V_o + Az^2 - Bx^2 \tag{2.7}$$

where V_o (the potential in the center of the cell) and the constants A and B depend on the cell geometry. For the square cell (width = height = d) with trapping plate potential equal V_T [10],

$$V = V_T\left(0.5 + \frac{2}{d^2}(z^2 - x^2)\right). \tag{2.8a}$$

For the flat configuration (width = $d = 2 \times$ height) [11],

$$V = V_T\left[0.11 + \left(\frac{1.76}{d^2}\right)(2z^2 - x^2)\right] \tag{2.8b}$$

Equation 2.8b is fairly accurate near the cell center but fails badly near the boundaries; it is not a solution of Laplace's equation.

For a square cell with equal potentials on the top and bottom plates (V_x), the resulting field is given by [12]

$$V = \frac{(V_T + V_x)}{2} + (V_T - V_x)\left(\frac{2}{d^2}\right)(z^2 - x^2) \tag{2.8c}$$

In trapped ICR cells, the electric field is due entirely to the difference between the trapping potential and that on the top, bottom, and end plates (usually all at the same potential). In this case Eq. 2.8c is a good approximation to the actual cell potentials [12]. In the drift ICR experiment, as described above, a voltage V_D is also applied between the top and bottom cell plates, and the resulting field is due to the sum of the trapping and drift potentials. For reasons discussed later, V_D is split equally between top and bottom plates ($\pm V_D/2$). Knott and Riggin [13] have obtained an exact formulation of the electrostatic field in such a drift ICR cell and the corresponding ion equations of motion. They expressed the fields in terms of power series involving the cell dimensions, plate potentials, etc. Sharp et al. [12] have done the same for the trap cell case. Although a considerable gain in accuracy is achieved with these solutions, a corresponding loss in intuitive understanding also occurs. For this reason we have elected to present the potential in the cell as a sum of the quadrupolar trapping field (Eqs. 2.7 and 2.8) and a drift potential approximated by

$$V \cong -\frac{V_D}{h}x \tag{2.9}$$

where h is the spacing between the plates. This approximation breaks down near the sides of the cell [11, 13]. The electric field is derived from the sum of the potentials given by Eqs. 2.8a and 2.9:

$$\bar{E} = -\nabla V = -\left(\frac{4V_T x}{d^2} + \frac{V_D}{d}\right)\bar{i} - \frac{4V_T z}{d^2}\bar{k} \tag{2.10}$$

The x and z coordinates are shown in Fig. 1. In the trapped ICR cell, a very small $\bar{E}_y$ is also present due to the potential on the (widely spaced) end plates [12]. Equation 2.10 and the subsequent discussion apply to the square cell geometry; equations appropriate to other geometries are easily derived.

Substituting $\bar{E}$ from Eq. 2.10 into the force Eq. 2.3 gives

$$\dot{\bar{v}}_x = \frac{qV_D}{md} + \omega_T^2 x + \omega_c \bar{v}_y \quad (2.11a)$$

$$\dot{\bar{v}}_y = -\omega_c \bar{v}_x \quad (2.11b)$$

$$\dot{\bar{v}}_z = -\omega_T^2 z \quad (2.11c)$$

where

$$\omega_T^2 = \frac{4qV_T}{md^2} \quad (2.11d)$$

This set of equations has the following solutions:

$$x(t) = x_0 + \left(\frac{v_{x0}}{\omega_e}\right) \sin \omega_e t + \left(\frac{v_{y0} \omega_c^2}{\omega_e}\right)(1 - \cos \omega_e t) \quad (2.12a)$$

$$y(t) = y_0 - \left(\frac{v_{x0} \omega_c}{\omega_e^2}\right)(1 - \cos \omega_e t) + \left(\frac{v_{y0} \omega_c^2}{\omega_e^3}\right) \sin \omega_e t - v_d t \quad (2.12b)$$

$$z(t) = z_0 \cos \omega_T t + \frac{v_{z0}}{\omega_T} \sin \omega_T t \quad (2.12c)$$

where x_0, y_0, v_{y0} and v_{x0} give the initial position and velocity, and

$$\omega_e = \omega_{\text{effective}} = (\omega_c^2 - \omega_T^2)^{1/2} \quad (2.12d)$$

$$v_d = \frac{V_D}{Bd} + \frac{\omega_T^2}{\omega_c}\left(x_0 + \frac{v_{y0}}{\omega_e}\right) \quad (2.12e)$$

is the drift velocity of the ion. We can simplify Eq. 2.12 by assuming $\omega_e \simeq \omega_c$ (since $\omega_T \ll \omega_c$), yielding

$$x(t) \simeq \frac{v_{x0}}{\omega_e} \sin \omega_e t - \frac{v_{y0}}{\omega_e} \cos \omega_e t + \left(x_0 + \frac{v_{y0}}{\omega_e}\right) \quad (2.13a)$$

$$y(t) \simeq \frac{v_{y0}}{\omega_e} \sin \omega_e t + \frac{v_{x0}}{\omega_e} \cos \omega_e t + \left(y_0 - \frac{v_{x0}}{\omega_e}\right) - v_d t \quad (2.13b)$$

The first two terms in each equation describe circular motion in the x–y plane with angular frequency equal to ω_e and radius equal to $(v_{y0}^2 + v_{x0}^2)^{1/2}/\omega_c$. The third term gives the x (or y) coordinates of the origin of the circle at $t = 0$. The $-v_d t$ term in Eq. 2.13b describes the drift motion in the $-y$ direction. Equation 2.12c shows the ions undergo simple harmonic motion in the z direction.

Note the dependence of v_d on V_T (Eq. 2.12e). For ions near the top or bottom of the cell, the contribution from V_T to the drift field can be comparable to that from V_D. Normally ions are formed in the center of the cell, making the effect of V_T small; however, since the ions follow equipotential lines as they move through the cell, drift potentials that are unbalanced in one region of the cell can cause the ions to move away from $x = 0$ as they enter or leave a region. To avoid this, the drift potentials are split evenly about ground. The effect of electron space charge in the source on ion position has been discussed by Woods et al. [14], Riggin and Woods [15], and Beauchamp [16].

The effect of momentum transfer collisions on ion motion in static fields has been considered by Beauchamp [11, 16, 17]. Equation 2.3 is assumed to have a new form:

$$m\dot{\bar{v}} = q(\bar{E} + \bar{v} \times \bar{B}) - \xi \bar{v} \tag{2.14}$$

where ξ is the collision frequency for momentum transfer. Beauchamp [17] has discussed some of the assumptions implicit in Eq. 2.14; the fact that ξ does not depend on $\bar{v}$ is the most important. To the extent that the Langevin potential [18] governs the ion–neutral collisions, this assumption is a good one. The time averaged solutions to Eq. 2.14 are [11, 16, 19]

$$\langle v_x \rangle = \frac{E_x}{B} \cdot \frac{\omega_e \xi}{(\omega_e^2 + \xi^2)} \simeq v_d \cdot \frac{\xi}{\omega_e} \tag{2.15a}$$

$$\langle v_y \rangle = -\frac{E_x}{B} \cdot \frac{\omega_e^2}{(\omega_e^2 + \xi^2)} \simeq -v_d \left(1 - \frac{\xi^2}{\omega_e^2}\right) \tag{2.15b}$$

The effect of collisions is to give the ions a net velocity in the x direction, proportional to the collision frequency. At very high pressures or very long times, this can result in ion loss. The effect of collisions on $\langle v_y \rangle$ is negligible.

A few sample calculations may serve to show the size of the quantities discussed and the degree of error in the approximations. For a $^{40}Ar^+$ ion with $B = 1$ T (10^4 G), $V_T = 1.0$ V, $V_D = 0.5$ V, flat cell geometry (width = 2.54 cm, height = 1.27 cm), we have $\omega_c/2\pi = 381.0$ kHz, $\omega_T/2\pi = 25.8$ kHz, $\omega_e/2\pi = 380.2$ kHz ($\sim 0.2\%$ different than $\omega_c/2\pi$). Assuming an average thermal velocity of ~ 420 m/s and $x_0 = y_0 = z_0 = 0$, we find the cyclotron radius to be ~ 0.14 mm and the z axis oscillation to be $\sim \pm 1.5$ mm. The ion is confined near the center of the cell. The drift velocity $v_d = 40$ m/s, adding $\sim 20\%$ to the average ion kinetic energy. To gauge the effects of momentum transfer collisions, we use the

Langevin orbiting collision frequency [18]:

$$k_L = 2\pi e \left(\frac{\alpha}{\mu}\right)^{1/2} \quad (2.16)$$

where (departing from MKS units) e is the ion charge in esu, α is the average polarizability in cm^3 and μ is the ion–neutral reduced mass in grams. The collision frequency for momentum transfer is given by [18]

$$\xi = 1.105 n k_L \left(\frac{M}{m+M}\right) \quad (2.17)$$

where n is the neutral gas density, and M and m are the neutral and ion masses, respectively. At a pressure of 10^{-3} torr (about the maximum possible) $\xi \sim 1.2 \times 10^4 \, \text{s}^{-1}$, $\xi/\omega \sim 5 \times 10^{-3}$, and $\langle v_x \rangle \sim 0.2 \, \text{m/s}$. This results in a net motion toward the negative drift plate of ~ 0.2 mm in 10^{-3} s and no ion loss should be observed. The effect on the drift velocity (Eq. 2.15b) is completely negligible.

2.3 Power Absorption

In almost every application of ICR, ion detection is accomplished by exciting the ions' cyclotron motion using an alternating electric field. In the presence of such a field, Eq. 2.14 still governs the ion motion; however, the electric field now becomes

$$\bar{E} = [E_{\text{RF}} \sin(\omega t + \phi) + E_D]\bar{i} + E_T \bar{k} \quad (2.18)$$

where $E_{\text{RF}} = V_{\text{peak}}/h$, ϕ is the phase angle between the oscillating field and the ion's initial cyclotron motion and the drift and trapping fields are those in Eq. 2.10. Since these fields do not directly affect the power absorption, we shall ignore them, remembering that the observed cyclotron frequency will be ω_e rather than ω_c. As noted before, approximating E_{RF} as V_p/h (where V_p is the base to peak RF voltage) ignores the rather severe edge effects present [13]. This point is discussed later (Section 3.3.1), but the main effects are (1) a reduction in the actual value of E_{RF} in the center of the cell and (2) a coupling between E_{RF} and the z axis oscillation.

Proceeding, we again have coupled equations

$$\dot{\bar{v}}_x = \frac{qE_{\text{RF}}}{m} \sin(\omega t + \phi) + \omega_e \bar{v}_y - \xi \bar{v}_x \quad (2.19a)$$

$$\dot{\bar{v}}_y = -\omega_e \bar{v}_x - \xi \bar{v}_y \quad (2.19b)$$

Exact solutions for $\bar{v}_x$ and $\bar{v}_y$ are given by Comisarow [20] and Dunbar

[21]. In the limits of zero collisions and $\omega = \omega_e$ the solutions are

$$\bar{v}_x = \frac{qE_{RF}}{2m} t \sin \omega_e t \qquad (2.19c)$$

$$\bar{v}_y = \frac{qE_{RF}}{2m} t \cos \omega_e t \qquad (2.19d)$$

The instantaneous power absorption (joule/second/ion) is the dot product of the electric force and the ion velocity:

$$A = \bar{F} \cdot \bar{v} = q\bar{E}(t) \cdot (\bar{v}_x + \bar{v}_y) \qquad (2.20)$$

Both Comisarow [20] and Dunbar [21] have solved the above using the simplifying assumptions that $\xi \ll \omega_c$ and $\omega_c + \omega \simeq 2\omega_c$. Also, since all values of ϕ are equally probable, it is permissible to average over ϕ. The result from Comisarow's derivation is given in Eq. 2.21:

$$A(t, \omega_c, \omega, \xi) = \frac{q^2 E_{RF}^2}{4m} \left(\frac{1}{\Delta\omega^2 + \xi^2} \right) \times [\xi + (\Delta\omega \sin \Delta\omega t - \xi \cos \Delta\omega t) \exp(-\xi t)] \qquad (2.21)$$

where $\Delta\omega = \omega_c - \omega$. With additional simplifying, Dunbar's expression is equivalent to the above; however, as presented in Ref. [21], as $\xi \to 0$, the sign of the power absorption expression depends on the sign of $\Delta\omega$. Equation 2.21 has two limiting forms:

$$\xi \to 0 \qquad A(t, \Delta\omega) = \frac{K}{m} \frac{\sin \Delta\omega t}{\Delta\omega} \qquad (2.22a)$$

$$\xi t \gg 1 \qquad A(\xi, \Delta\omega) = \frac{K}{m} \frac{\xi}{(\Delta\omega^2 + \xi^2)} \qquad (2.22b)$$

where $K = q^2 E_{RF}^2/4$. The low-pressure ("zero collision") form has been derived by Baldeschwieler [5] and Buttrill [22]; the high-pressure ("collision limited") form was first derived by Wobschall et al. [3] and has been discussed by Beauchamp [17]. Comisarow [20] also discusses these limiting expressions.

At resonance ($\Delta\omega = 0$) Eq. 2.21 reduces to

$$A(t, \xi) = \frac{K}{m} \frac{[1 - \exp(-\xi t)]}{\xi} \qquad (2.23)$$

In the limits of collision frequency discussed above,

$$\xi \to 0 \qquad A(t) = \frac{K}{m} t \qquad (2.24a)$$

$$\xi t \gg 1 \qquad A(\xi) = \frac{K}{m\xi} \qquad (2.24b)$$

The above formulas calculate the instantaneous power absorption (joule/second/ion); Eq. 2.23 is referred to as the general power absorption equation. The quantity usually measured in ICR is the power absorption integrated over the time the ion is in resonance. If the ion concentration is constant, Eq. 2.21 is simply integrated to give the observed intensity as a function of ξ, $\Delta\omega$ and t:

$$\text{Int.} = \frac{K}{m}\left(\frac{1}{\xi^2 + \Delta\omega^2}\right)\left[\left(\frac{(\xi^2 - \Delta\omega^2)\cos\Delta\omega\tau - 2\xi\Delta\omega\sin\Delta\omega\tau}{\xi^2 + \Delta\omega^2}\right)\right.$$
$$\left. \times \exp(-\xi\tau) + \frac{\Delta\omega^2 - \xi^2}{\Delta\omega^2 + \xi^2} + \xi\tau\right] \qquad (2.25)$$

where τ is the time spent absorbing power. Comisarow [20] refers to Eq. 2.25 as the general line shape equation. Again we have the limiting cases:

$$\xi \to 0 \qquad \text{Int.} = \frac{K}{m}\frac{(1 - \cos\Delta\omega\tau)}{\Delta\omega^2} \qquad (2.26a)$$

$$\xi t \gg 1 \qquad \text{Int.} = \frac{K}{m}\frac{\xi\tau}{\xi^2 + \Delta\omega^2} \qquad (2.26b)$$

which describe the peak shape in the limits of very low and very high pressure. At resonance ($\Delta\omega = 0$) the observed peak height is

$$\text{Int.}_{\Delta\omega=0} = \frac{K}{m\xi}\left(\tau - \frac{[1 - \exp(-\xi\tau)]}{\xi}\right) \qquad (2.27)$$

with the usual high- and low-pressure limits:

$$\xi \to 0 \qquad \text{Int.}_{\Delta\omega=0} = \frac{K}{m}\frac{\tau^2}{2} \qquad (2.28a)$$

$$\xi\tau \gg 1 \qquad \text{Int.}_{\Delta\omega=0} = \frac{K}{m}\frac{\tau}{\xi} \qquad (2.28b)$$

Equations 2.27 and 2.28 describe the observed peak heights of nonreactive ions.

The case of reacting ions is discussed in Section 3.3.2; in general, the power absorption from these ions is calculated according to

$$\text{Int.}(\tau, k, \xi) = \int_0^\tau I^+(t, k)\, A(t, \xi)\, dt \qquad (2.29)$$

where $I^+(t,k)$ is the ion concentration as a function of time and reaction rate, and $A(t, \xi)$ is the instantaneous power absorption (Eq. 2.23). Beauchamp [17] has theoretically investigated systems of ions that are chemically coupled by resonant charge transfer and discussed the effects on the equations of motion. The rather difficult formalism and restrictive assumption of an infinite homogeneous medium limit the application of this method, however.

Some interesting conclusions follow from the line shape and resonant intensity expressions above. At low pressure Buttrill [22] has shown that the peak width (half-width at half-height) $\Delta_{1/2} \simeq 2.783/\tau$ (from Eqs. 2.26a and 2.28a). This result has little use in determining the ions' time in resonance (due to other peak broadening effects) but strongly affects the available resolution. At high pressure the line shape becomes Lorentzian (Eq. 2.26b) and $\Delta\omega_{1/2} = \xi$. This result allows determination of momentum transfer collision frequencies using ICR line widths.

2.4 Oscillation in the Trapping Field

The ion oscillation along the z axis can also be excited [10]. If an alternating electric field is applied to the side plates of the ICR cell, Eq. 2.11c becomes

$$\dot{v}_z = -\omega_T^2 z + \frac{qE_{RF}}{m} \sin \omega t \tag{2.30}$$

where we assume $E_{RF} = V_{peak}/d$. If we assume $v_{z0} = z_0 = 0$, Eq. 2.30 has the solution

$$z(t) = \frac{qE_{RF}}{m\omega_T} \left(\frac{\omega_T \sin \omega t - \omega \sin \omega_T t}{\omega_T^2 - \omega^2} \right) \tag{2.31}$$

At resonance ($\omega = \omega_T$)

$$z(t) = \frac{qE_{RF}}{m\omega_T} \left(t \cos \omega_T t - \frac{\sin \omega_T t}{\omega_T} \right) \tag{2.32}$$

and the amplitude of the harmonic motion increases with time. Eventually $z > d/2$ and the ion is lost to one of the side plates. This occurs at time

$$t_{eject} = \frac{m\omega_T d}{2qE_{RF}} \tag{2.33}$$

This ejection technique has been used to remove unwanted ions from the ICR cell; however, the resolution is poor ($M/\Delta M \sim 4$) since $\omega_T \propto 1/\sqrt{m}$. Trapping ejection is most commonly used to remove trapped electrons in negative ion experiments. The trapping field oscillation can also be excited by a low-frequency field on the top or bottom plate of the cell [13, 23] due to the distortion of the field near the side plates. This distortion gives the nominal $\bar{E}_x$ field an $\bar{E}_z$ component that allows coupling to the z axis motion.

3 DRIFT ICR

3.1 Introduction

Drift ICR experiments are distinguished by continuous ion formation and continuous movement of the ions through the cell. Experiments are generally done as a function of neutral reactant pressure at some reaction time (typically ~ 1 ms) that is fixed by the ion's drift velocity and the length of the ICR cell. This is in contrast to the various pulsed ICR experiments in which ion formation and detection are pulsed and the reaction time is easily varied. Resolution in drift ICR experiments is inferior to that in pulsed experiments for two reasons. First, as shown in Section 2, the ICR linewidth depends inversely on the ion observation time:

$$\Delta\omega \simeq \frac{5.57}{\tau} \tag{3.1}$$

where $\Delta\omega$ is the full width at half-height of the peak and τ is the length of time the ion is observed. In pulsed experiments, the detection period, and thus resolution, can be increased considerably. The second contribution to linewidth is collisional broadening at the high pressures used in some drift ICR experiments. In pulsed experiments, much lower pressures are used and collisional broadening is not important. The pressure range used in drift ICR experiments is typically $\sim 5 \times 10^{-7} \rightarrow 1 \times 10^{-3}$ torr. This wide range, and in particular the relatively high pressures available, make drift ICR the technique of choice for high-pressure applications such as three-body association reactions, ion–neutral collision frequencies, and collisional relaxation processes.

In the remainder of this section, we will describe first the experimental apparatus and second some typical experiments, including data analysis.

3.2 Experimental Apparatus

3.2.1 Cells

The ICR cell provides the electric fields used to contain and analyze the ions. Drift ICR cells have metal plates on top, bottom, and sides mounted on insulating rails (Fig. 2). Various plate materials have been used: electropolished stainless steel; rhodium flashed, silver-plated copper; and pure molybdenum. The insulating rails are either Vespel [24], Macor ceramic [25], or occasionally alumina. Screens may be used instead of solid plate. The top and bottom plates (carrying the drift potential) are divided into two or three sections: the source region where ions are formed, an optional "reaction" region, and the analyzer or "resonance" region where ions are detected. Including the intermediate

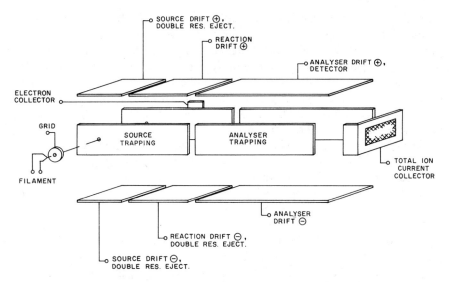

Fig. 2. Exploded view of typical four-section drift ICR cell, showing split trapping plates (to allow drift time measurement) and total ion current collector.

reaction region allows the source section to be kept small, and the ions spend more time in a region of well-defined electric fields (away from the electron beam) [26]. The side or trapping plates of the cell are split into two sections to allow measurement of ion drift times. The division is at the beginning of the analyzer region. Both rectangular and square cross-section cells have been used with usual values of height × width = 1.27 × 2.54 cm or 2.54 × 2.54 cm. The rectangular cross section provides superior drift field homogeneity and is more commonly used. The square cross section does allow for better trapping ejection (Section 3.3.6). Cell lengths are limited by the diameter of the homogeneous magnetic field. With a 9-in. pole face diameter and a 2-in. air gap, cell lengths of about 15 cm are used. Source and analyzer lengths are about equal. A longer analyzer section gives more signal but results in greater ion heating from the analyzer field.

Top and bottom plate potentials are typically 0.2–0.5 V (with the 1.27 cm cell height). This corresponds to an ion drift velocity (at 1 T) of ~ 30–80 m/s compared with the ion thermal velocity of ~ 400 m/s. Top and bottom plate voltages are balanced to allow a smooth match up of equipotential lines between sections and to keep the ions near the center of the cell [11]. Some deviation in the source region may be required to compensate for the perturbation due to the electron beam. Trapping plate voltages of 0.5–1.0 V are typical. Passive RC filtering of the DC drift voltages outside the vacuum chamber may be useful.

Ions are collected at the end of the cell and read out as a total ion current (typically 10^{-13}–10^{-10} A). The current is used to check tuning, determine ion densities, check for ion ejection, etc. Originally, a set of small (0.5 cm) top, bottom, and side plates were all connected to the electrometer [5]. Ions entering the region were not constrained in the z direction and were theoretically collected on one of the side plates. We have found improved collection with the design shown in Fig. 2. One side plate extends completely to the end of the cell, providing a repeller voltage to move the ions to the collector opposite. The top and bottom analyzer plates extend into the collector, eliminating the sharp drift field transition at the beginning of the collector. This also ensures that the ions are in the analyzer field until the time they are collected, which is the assumption made when measuring the drift times.

Various filament assemblies have been used in ICR cells. One that has worked well for general use consists of a rhenium ribbon (0.0012 × 0.032 in. cross section × ∼0.25 in. length [27] spot welded to 0.032-in. stainless-steel rods. The rods are mounted in a Macor [25] ceramic block and retained with set screws. The filament requires ∼4 A of current. The filament bias voltage (electron energy) is variable between 0 and 100 V. A grid is used to modulate the electron beam [28]. Electrons exiting the cell are collected and the current (0.1–30×10^{-9} A, typically) is read on a picoammeter. The picoammeter output can be used to control the emission current [29, 30], giving better long-term stability. The electron entrance and exit holes (∼3 mm diameter) in the cell are usually screened to prevent field penetration. Attaching the filament assembly directly to the cell results in significant local heating (perhaps 30–40°C rise) [31]. This can cause shorting between cell plates due to nonuniform expansion. A better method is to mount the filament and grid on the structure that supports the cell.

3.2.2 Vacuum System

The vacuum system is constructed of nonmagnetic stainless steel using the usual high vacuum, all metal, and ceramic techniques [32]. The chamber enclosing the cell is bakable to 150–200°C. Pumping is usually with a 2-in. diffusion pump or diffusion pump/ion pump combination. The diffusion pump is necessary if experimental pressures approaching 10^{-3} torr are used, and a throttling valve may be necessary at very high pressures. A base pressure of 10^{-8}–10^{-7} torr is adequate due to the relatively high experimental pressures and short reaction times. The pressure is usually measured with an ion gauge calibrated against a capacitance manometer. Some systems continue the manometer inlet tube inside the vacuum chamber to a point near the cell to ensure that the true cell pressure is sampled. This is probably not necessary in chambers with a high conductance to pumping speed ratio.

A separate inlet vacuum system is necessary, including bakable leak valves if condensables are to be used. A mechanical pump plus small diffusion pump is probably the best pumping combination, but a trapped mechanical pump will work nearly as well.

3.2.3 Electronics

With the exception of the ion detectors (Section 7), very few specialized electronic circuits are used in drift ICR experiments. Phase sensitive detection is used, requiring that some parameter be modulated. Electron beam modulation (using a grid) [28] is probably most common today; other possibilities include modulating the electron energy [33]; modulating one trapping plate between $\pm V_T$ [11] (switching the static trapping plate voltage from $+V_T$ to $-V_T$ allows observation of negative ions); modulating the double resonance ejection amplitude [34]; however, better ways of doing these experiments are described later; gating a trapping plate ejection field [35] (see Section 3.3.6); finally, modulating the ion detection field (Section 7). Magnetic field modulation [5] is no longer common due to the resulting derivative line shape. Source region drift potential modulation is also in disrepute because of the higher source ion densities present. Whatever modulation scheme is used, the period must be much greater than the drift time through the analyzer region.

Drift times in the source and analyzer regions must be measured if quantitative work is to be done. McMahon and Beauchamp [11] first described the pulsed electron beam/pulsed trapping plate method commonly used. A point to note is that ion transit times are very dependent on the ion/electron density in the cell. Thus if the electron beam is pulsed, the average electron current must be reduced to reflect the change in duty factor. Van der Hart and van Sprang [36] described a method of determining ion transit times by analyzing the phase delay between the modulating square wave and the detector output signal. The method requires a very high-quality output waveform, however, and has not found wide application.

3.3 Types of Experiments

3.3.1 Double Resonance

Double resonance experiments are unique to ICR mass spectrometry [23, 34, 37–39]. The basic experiment consists of simultaneously observing one ion at its cyclotron frequency while translationally exciting a second ion with an electric field oscillating at the second ion's cyclotron frequency. This excitation increases the angular velocity of the irradiated ion and its cyclotron orbital radius according to

$$\omega_c = \frac{v_c}{r_c} = \frac{qB}{m} \qquad (3.2)$$

If the ions are chemically coupled such that the observed ion is a product of a reaction involving the translationally excited ion, then a change in the product ion intensity will almost always occur. Such changes are due to two effects. First the cyclotron radius of the ion can be increased to the point where the ion is ejected from the cell [37]. The average ion kinetic energy (KE) in the absence of collisions is given by the time integral of the instantaneous power absorption (Eq. 2.24a)

$$\mathrm{KE}_{\mathrm{ion}} = \frac{q^2 E_{\mathrm{RF}}^2 t^2}{8m} \qquad (3.3)$$

where q is the ion charge in coulombs, E_{RF} is the peak irradiating field in volts/meter, t is the time in resonance, and m is the ion mass in kilograms. Ejection occurs when $r \geqslant h/2$ (h = cell height). The ejection time in a cell of height "h" equals

$$t_{ej} = \frac{\omega h^2 m}{q V_{RF}} = \frac{B h^2}{V_{RF}} \tag{3.4}$$

where V_{RF} is the peak irradiating voltage. Typical values of t_{ej} are 10–50 μs. This technique is used to remove unwanted ions, which may complicate the reaction kinetics, and to sort out reaction sequences by removing possible precursor ions one at a time. Care must be taken to avoid exciting adjacent mass ions as well as coupling of the double resonance excitation with the trapping field oscillation. This results in indiscriminant ejection of ions at frequencies $\lesssim 30$ kHz [23].

The second double resonance effect involves observing changes in reaction dynamics with translational energy. If the reaction is not strictly governed by the Langevin interaction potential [18], changes in rate coefficient will occur with reactant ion energy [38, 39]. Endothermic reactions and reactions involving barriers can be qualitatively studied as well.

The double resonance excitation is usually applied to either the source or reaction regions. This avoids coupling with the ion detector and allows unwanted ions to be removed immediately after formation. The original double resonance experiments [34] detected the double resonance effects by modulating the amplitude of the excitation. This is inconvenient, and now the product ion is observed as a normal signal while the excitation frequency is swept. The simple relation between observed and irradiated masses

$$m_1 \omega_1 = m_2 \omega_2 \tag{3.5}$$

follows from Eq. 3.2. Monitoring the total ion current shows the presence or absence of ion ejection.

Semiquantitative studies of reaction kinetic energy dependence are possible using Eq. 3.3 to calculate values of the ion energy. The irradiation time is usually fixed by either pulsing the double resonance oscillator for a known time [40] or by continuous irradiation in the source region and measuring the source residence time. Two major errors arise. First, the value of E_{RF} is very approximate. Constant resistance paper plots [11, 41] show that if the excitation voltage is applied to only one plate, the resulting field in the center of the cell is $\sim 60\%$ of that calculated. The resulting kinetic energy would be $\sim 1/3$ of that calculated. Furthermore, the field strength varies greatly through the cell. A major improvement is obtained by transformer coupling the excitation voltage to both top and bottom plates. With this arrangement, the field in the cell center is $> 95\%$ of that calculated (flat cell geometry) and is constant over a large fraction of the cell. End effects (i.e., transitions from one region to the next) can also be important. As the end of the excitation region is approached, the field strength drops off. For a cell height of 1.27 cm this is important in the last 0.5 cm where the field strength decreases from ~90% to 0.

The second major source of error is due to nonreactive ion–neutral collisions that reduce the actual ion kinetic energy. The power absorbed by the ions (Eq. 2.27) is partitioned between the ions and the bath gas. In the "high-pressure" limit where the collision frequency $\xi > 10/t$ (t = irradiation time) the ion kinetic energy does not increase with time [17, 42] and

$$KE = \frac{q^2(m+M)E_{RF}^2}{8m^2\xi^2} + \tfrac{3}{2}kT \tag{3.6}$$

where k is Boltzmann's constant, M is the neutral mass in kilograms and T is the temperature in Kelvins. Between the high- and zero-pressure limits, no simple expression exists for the ion energy imparted by the excitation field. Nonreactive collisions that occur after the ion leaves the excitation field result in an exponential decay of the average kinetic energy [43]. Reactive collisions during excitation will occur at lower than calculated kinetic energies. All these factors conspire to produce ions with less kinetic energy than the experimenter calculates. This trend is supported by comparison of ICR double resonance studies with flowing afterglow DRIFT experiments in which the fields and ion energies are better characterized [44].

One additional factor enters at low double resonance energies, namely, the phase between the exciting field and the initial cyclotron motion. A more complete version of Eq. 3.3 is

$$KE = \frac{q^2 E_{RF}^2 t^2}{8m} + \frac{qE_{RF}v_0 t}{4}\cos\phi + \tfrac{1}{2}mv_0^2 \tag{3.7}$$

where v_0 is the initial ion cyclotron velocity and ϕ is the phase angle [20, 21]. The $\cos\phi$ term averages to zero but with a maximum error of $\pm qE_{RF}v_0 t/4$. The percent uncertainty decreases with increasing double resonance energy. As an example, consider an ion of mass 50, with thermal KE = 0.04 eV and an irradiation time of 5×10^{-4} s: If $E_{RF} = 1.26$ V/m, $KE_{ion} = 0.14 \pm 0.047$ eV; if $E_{RF} = 2.83$ V/m, $KE_{ion} = 0.54 \pm 0.1$ eV; if $E_{RF} = 4$ V/m, $KE_{ion} = 1.04 \pm 0.15$ eV, and if $E_{RF} = 8.95$ V/m, $KE_{ion} = 5.04 \pm 0.35$ eV. Thus at low ion energies the uncertainty can be substantial. The ion thermal energy (third term in Eq. 3.4) adds a constant amount to the total energy (~0.04 eV) and is easily included in the calculated energies.

Lest the above sound too pessimistic, we should take note of the ease with which these experiments can be done. Using a high-frequency rectifier [45] we can obtain an experimental plot of product ion intensity vs. reactant ion irradiation amplitude ($\propto \sqrt{KE}$) in a few minutes. Despite the limitations noted, no other technique allows such an easy examination of reaction kinetic energy dependence.

3.3.2 Ion–Neutral Reactions

Probably the most common ICR experiments involve measuring rate coefficients and product distributions of ion–neutral reactions. If we have a series of primary, secondary, tertiary, etc., ions governed by the reaction scheme

$$A^+ \xrightarrow{k_1} B_1^+ \xrightarrow{k_{11}} C_{11}^+ \quad \xrightarrow{k_{12}} C_{12}^+ \quad \xrightarrow{k_2} B_2^+ \tag{3.8}$$

the ion abundances are given by

$$A^+ = A_0^+ \exp(-It) \tag{3.9a}$$

$$B_1^+ = A_0^+ [n] \frac{k_1}{J-I} [\exp(-It) - \exp(-Jt)] \tag{3.9b}$$

$$C_{11}^+ = A_0^+ [n]^2 \frac{k_1 \cdot k_{11}}{J-I} \left(\frac{[1 - \exp(-It)]}{I} - \frac{[1 - \exp(-Jt)]}{J} \right) \tag{3.9c}$$

where

$$I = [n](k_1 + k_2 + \cdots) \tag{3.10a}$$

$$J = [n](k_{11} + k_{12} + \cdots) \tag{3.10b}$$

$[n]$ is the neutral concentration, and t is the reaction time. Product distributions are given by ratios of the rate coefficients.

The observed drift ICR intensities are not completely governed by the above kinetic equations, however, because of the time-dependent nature of the power absorption (see Section 2). Several papers have been presented describing the deconvolution of ICR intensities to give the underlying rate coefficients [20, 22, 46–48]. The analysis we present here is a combination of several approaches.

In general, the observed intensity at resonance is given by

$$\text{Int.}(I^+) = \int_0^\tau [I^+(t)] \cdot \text{power absorption } (t) \, dt \tag{3.11}$$

where $[I^+(t)]$ is the time dependent concentration of I^+ (Eq. 3.9), the power absorption (PA) is the instantaneous PA (Eq. 2.23 or 2.24), and the integral is over the time spent in resonance.* This is the basis for calculating the observed

---

*Equation 3.11 is actually a time average. The $1/\tau$ normalizing term usually present exactly cancels the time dependence of the ion flux to concentration conversion; i.e., since a constant flux of ions/second are produced by the filament, ion concentration is $\propto$ flux $\times \tau$.

intensity of any ion. As an example, we apply Eq. 3.11 to a reactive parent ion and a single, nonreactive secondary ion:

$$\text{Int.}(A^+) = \frac{A_0^+ q^2 E_{RF}^2}{4m\xi} \int_0^\tau \exp[-I(\tau'+t)] \cdot [1 - \exp(-\xi t)] \, dt \quad (3.12)$$

where τ' and τ are the source and resonance region drift times, respectively (i.e., the total drift time is $\tau' + \tau$). Evaluating:

$$\text{Int.}(A^+) = K \cdot \frac{A_0^+}{m\xi} \exp(-I\tau') \cdot \left(\frac{(1 - \exp(-I\tau))}{I} - \frac{[1 - \exp(-(I+\xi)\tau)]}{1+\xi} \right) \quad (3.13)$$

where $K = q^2 E_{RF}^2 / 4$. The signal from secondary ions is evaluated in two parts: First, those B^+ ions formed in the source region proceed unchanged through the analyzer and are evaluated as unreactive primary ions:

$$\text{Int.}(B_{\text{source}}^+) = \frac{K A_0^+}{m\xi} \cdot [1 - \exp(-I\tau')] \int_0^\tau [1 - \exp(-\xi t)] \, dt \quad (3.14a)$$

$$= \frac{K A_0^+}{m\xi} [(1 - \exp(-I\tau'))] \left(\tau - \frac{[1 - \exp(-\xi\tau)]}{\xi} \right) \quad (3.14b)$$

The signal from those B^+ ions formed in the resonance region requires a double integral to evaluate, since the length of time they absorb power varies from 0 to τ. We diagram the time periods in Scheme I:

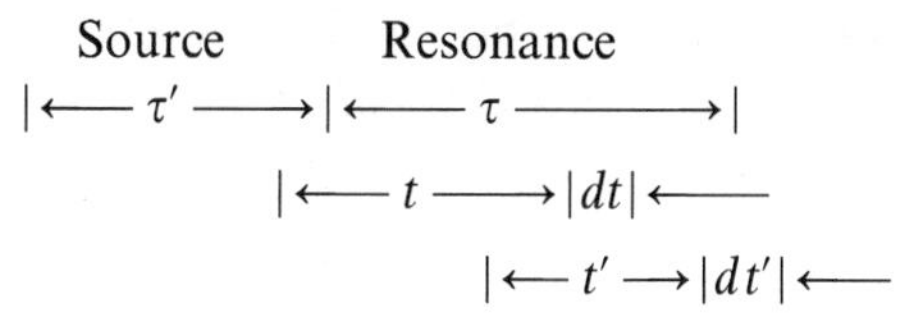

```
         Source        Resonance
      |←── τ' ──→|←──── τ ────→|
                 |←── t ──→|dt|←
                 |←── t' ──→|dt'|←
```

Scheme 1

Consider those B^+ ions formed during time interval dt:

$$dB^+ = IA^+(t) \, dt \quad (3.15a)$$

$$= A_0^+ I \exp[-I(\tau'+t)] \, dt \quad (3.15b)$$

The power absorption due to this group of ions is

$$PA = \frac{K}{m\xi} dB^+ \int_0^{\tau-t} [1 - \exp(-\xi t')] \, dt' \quad (3.16)$$

since $[dB^+]$ is constant from time t onwards. To find the signal from all B^+ produced in the analyzer, we integrate over t:

$$\text{Int.}(B^+_{\text{RESONANCE}}) = \frac{K}{m\xi} \int_0^\tau dB^+ \int_0^{\tau-t} [1 - \exp(-\xi t')] \, dt' \, dt \tag{3.17}$$

Evaluating and adding the source term (Eq. 3.14b) yields

$$\text{Int.}(B^+) = \frac{K}{m\xi} A_0^+ \left(\tau - \frac{[1-\exp(-\xi\tau)]}{\xi} + \frac{\exp(-I\tau')}{I(I-\xi)} \cdot \{\xi[1-\exp(-I\tau)] \right.$$

$$\left. - I[1-\exp(-\xi\tau)]\} \right) \tag{3.18}$$

Equations 3.13 and 3.18 give the intensities of reactive primary ions and nonreactive secondary ions as a function of time, reaction rate, nonreactive collision frequency, and mass.* They have been derived by several authors [20, 22, 30, 46–48]. Comisarow [20], Marshall and Buttrill [46], and Kemper [30] have derived corresponding equations for reactive secondary, tertiary, and nonreactive quaternary product ions. Reactive product ions have an additional $\exp(-Jt')$ term in Eq. 3.17 to account for reaction of the $[dB^+]$ packet during t', otherwise the analysis is the same as that outlined above.

Equations 3.13 and 3.18 represent a "best attempt" to account for the factors noted above. They are too complex, however, to permit an analytical solution for the rate coefficient I. It is possible to solve for I if we restrict the experimental conditions such that only small amounts of reaction occur and such that only very low or very high pressures are used. In these limits, exponentials can be expanded and simpler expressions produced. Anicich and Bowers [47] and Goode et al. [48] have derived appropriate equations that we summarize here for a single primary ion (with observed peak height P^+) and multiple secondary ions (with observed peak heights S_i^+) in the low-conversion, low-pressure limit:

$$k_i = \frac{S_i^+}{P^+ \cdot C_{psi} + D\left(\sum_j S_j^+ \cdot C_{sisj}\right)} \cdot \frac{1}{[n]} \cdot \frac{1}{(\tau_p' + \tau_p/3)} \tag{3.19a}$$

where

$$C_{psi} = \frac{m_p}{m_{si}} \cdot \left(\frac{\tau_{si}}{\tau_p}\right)^2 \cdot \frac{\tau_{si}' + \tau_{si}/3}{\tau_p' + \tau_p/3} \tag{3.19b}$$

$$C_{sisj} = \frac{m_{sj}}{m_{si}} \cdot \left(\frac{\tau_{si}}{\tau_{sj}}\right)^2 \cdot \frac{\tau_{si}' + \tau_{si}/3}{\tau_{sj}' + \tau_{sj}/3} \tag{3.19c}$$

---

*It should be obvious that although m, ξ, τ, and τ' are not subscripted in Eqs. 3.12–3.18, they relate to the ion in resonance.

$$D = \frac{\tau'_p + 2\tau_p/3}{\tau'_p + \tau_p/3} \tag{3.19d}$$

The rate coefficient k_i is usually obtained from the slope of a plot of the corrected extent of reaction (first term in 3.19a) vs. neutral concentration. Equation 3.19 is accurate for extents of conversion $\lesssim 25\%$. The "correction" terms C_{ps} and C_{ss} take explicit note of the different drift times for each different ion. This is necessary if the detector frequency is fixed and the magnetic field scanned. A major reduction in complication results if the magnetic field is kept constant and the detector frequency scanned, since drift times are then equal for all ions. Frequency scans also avoid problems of changing ion densities and differential losses. It is definitely the technique of choice. Scanning detectors are discussed in Section 7.

Rate coefficients $\gtrsim 5 \times 10^{-10}\,\mathrm{cm^3/s}$ in some systems can be measured easily by observing the decrease in parent ion as a function of neutral pressure. This is possible if the parent reactant ion is generated from one neutral while the observed reaction is with a second neutral. Neutral 1 is introduced at a fixed pressure, and the resulting P^+ is observed while the pressure of neutral 2 is swept. It is possible to show that

$$\mathrm{P}^+ = K \cdot \exp(-[n\#2]k\tau_{rxn}) \tag{3.20}$$

The reaction time (τ_{rxn}) depends on the initial pressure of neutral 1 and varies from $\tau' + 2\tau/3$ (low pressure) to $\tau' + \tau/2$ (high pressure). This technique is confined to large rate coefficients since nonreactive collisions with neutral 2 will also reduce P^+ (although with an apparent rate coefficient less than the collision frequency).

If experimental conditions are such that simplified expressions such as Eq. 3.19 cannot be used; e.g., data taken over a wide pressure range, tertiary reaction products etc., then the only choice is to begin with Eq. 3.11 and use the complete expressions (such as Eqs. 3.13 and 3.18) for the ion intensities. Knowing drift times, masses, and momentum transfer collision frequencies (either calculated or measured), the different rate coefficients are iterated until the calculated ratios of ion intensities match the experimental. This is a tedious procedure and prone to error with very small rate coefficients due to the "single point" nature of the analysis (i.e., k's are calculated from intensities at a single pressure rather than from intensities vs. pressure).

The sensitivity of the drift ICR technique is good. Rate coefficients of $\gtrsim 5 \times 10^{-13}\,\mathrm{cm^3/s}$ can be measured, and coefficients of $\sim 5 \times 10^{-14}\,\mathrm{cm^3/s}$ can be detected. Whatever experimental techniques are used, for accurate results it is mandatory to measure the ion drift times and calibrate the neutral gas pressures.

3.3.3 Collision Frequencies

Collision frequencies are used in ICR data analysis, testing ion–neutral interaction theories, determining ion–neutral reaction probabilities, etc. They are usually determined in mobility experiments using flow tubes, but they are also easily determined using high-pressure ICR techniques. Three approaches have been presented: Collisional line broadening (Beauchamp [17]), transient power absorption level (Huntress [42]), and transient heterodyne (Dunbar [21]).

Some definitions follow: The ion–neutral collision frequency is the sum of reactive and nonreactive collisions. This is usually approximated by the Langevin [18] expression 3.21:

$$k_{coll} = 2\pi e \left(\frac{\alpha}{\mu}\right)^{1/2} \tag{3.21}$$

where e is the ion charge in electrostatic units, α is the average polarizability in cubic centimeters, and μ is the reduced mass. The rate coefficient for momentum transfer (without reaction) includes a grazing collision correction and mass term [18]:

$$\frac{\xi}{n} = 1.105 \frac{M}{M+m} k_{coll} \tag{3.22}$$

where M and m are the neutral and ion masses, respectively. If reactive collisions occur,

$$\frac{\xi}{n} = (1.105 - f) \frac{M}{M+m} \cdot k_{coll} \tag{3.23}$$

where f is the fraction of reactive collisions. We also define the total collision frequency

$$C = \xi + [n]k_{rxn} \tag{3.24}$$

where k_{rxn} is the rate coefficient for reactive collisions.

Measurement of collision frequencies using line broadening is possible due to the Lorentzian line shape of the ICR signal at high pressure $C \gtrsim 10/\tau$, where τ is the time in resonance). As first shown by Beauchamp [17] and Marshall [49],

$$A(\Delta\omega, \xi, \tau) = \frac{q^2 E_{RF}^2 \tau}{4m} \cdot \left(\frac{C}{(\Delta\omega)^2 + C^2}\right) \tag{3.25}$$

The peak width at 1/2 maximum ($\Delta\omega_{1/2}$) equals the *total* collision frequency (reactive and nonreactive). Usually $\Delta\omega_{1/2}$ (or its equivalent in gauss) is plotted vs. neutral pressure and $C/[n]$ is taken from the slope. This avoids static contributions to the linewidth due to ions exiting the cell, inhomogeneous fields, etc. Collision frequencies for both primary and product ions can be determined, and in general they compare well with mobility data.

Huntress [42] has developed a transient ICR technique to determine collision frequencies. An ion packet of width $\Delta t (\sim 150\,\mu s)$ is created by pulsing the electron beam, and its power absorption signal is monitored as a function of time. The average power absorption of the packet (at time $t > \Delta t$) is given by

$$\bar{A}(t,\xi) = N^+ \frac{q^2 E_{RF}^2}{4m\xi}\left(1 + \frac{[1-\exp(\xi \Delta t)]}{\xi \Delta t}\cdot \exp(-\xi t)\right) \quad (3.26)$$

The transient signal climbs linearly and then saturates at long times. This saturated level is $\propto 1/\xi$; thus by studying the saturated intensity as a function of pressure, ξ/n can be determined. These experiments are difficult for several reasons. First, the number of ions must remain constant as the pressure is changed; second, there can be no pressure dependent ion loss; third, systems with reactive collisions cannot be studied. Good agreement was obtained in the N_2^+/N_2 and CO^+/CO systems with the corresponding mobility data.

Dunbar's heterodyne technique [21] also involves monitoring a transient signal from a narrow ion packet (width $\sim 50\,\mu s$). Here, however, the detector is tuned slightly off resonance ($\omega - \omega_c \neq 0$). The resulting signal is a beat pattern (with frequency equal to $\omega_c - \omega$) that decays with time constant C (the total collision frequency). The beat pattern is superimposed on a base line that decays to 0 with time constant equal to $[n]k_{rxn}$. Values of C are determined by studying the heterodyne pattern as a function of pressure. Although this is possible in principle, the experimental results are not good enough to allow separate determinations of k_{rxn} and ξ/n. Again, these experiments are difficult. Dunbar notes that the method is very sensitive to instrument tuning; measurement of secondary ion collision frequencies will also be difficult.

3.3.4 Temperature Variable ICR

Various temperature variable ICR systems have been reported [31, 50]. Usually either the vacuum chamber surrounding the cell or the cell itself is temperature controlled. In the former method [50], the vacuum chamber is jacketed and a heating/cooling fluid circulates in the innerspace. This method relies on radiation and convection to heat the cell and although easier to implement, experiments indicate this will result in significant temperature gradients. With the heated cell method [31], the heating/cooling fluid line is brought into the vacuum and attached to copper plates, which are in turn attached to the cell plate mounting rails. The rest of the cell is heated by conduction. The cell is surrounded with a temperature-controlled radiation shield to reduce heat loss and to heat or cool the neutral molecules before they enter the cell. This is necessary due to the poor thermal conductivity of the mechanical cell joints in vacuum. The filament/grid assembly must be thermally isolated from the cell to prevent localized heating. Correction for thermal transpiration must be made in calibrating pressures [31, 50].

With either method, dry N_2 gas is the preferred heating/cooling fluid. Cooling is accomplished with a copper heat exchanger coil immersed in liquid N_2; the temperature is varied by controlling the N_2 gas flow rate. Heating is done with an electric in-line heater; either the heater voltage or the N_2 flow rate can be varied.

An important consideration is the effect of the various fields in the ICR cell and resulting ion velocities on the ion energy distribution. Lehman and Bursey [8] and Chesnavich et al. [51] discuss this point. Three perturbations are present: the ion's drift velocity, cyclotron heating of the ion by the detector field, and acceleration due to the trapping field. Consider a specific example: ion mass = 50 amu; drift field = 0.5 V/cm, magnetic field = 1.0 T, detector field = 10 mV/cm (low, but not unreasonable), source and resonance drift times = 1×10^{-3} s and trapping potential = 1 V. The resulting drift velocity = 50 m/s. If we take $E_{trans} = 3/2 T$, the drift velocity causes a 20% increase over thermal energy at 100 K and a 10% increase at 500 K, probably negligible. The extent of cyclotron heating in the absence of collisions is ~ 0.25 eV—clearly not negligible. This is mitigated somewhat by noting that this is the energy at the end of the cell; through the source the ion energy is thermal, through the first half of the analyzer the heating is less than 0.06 eV. The only real answer to this, however, is to raise the pressure. At a collision frequency of $\sim 10^4$/s (6×10^{-4} torr pressure and $\xi/n = 5 \times 10^{-10}$ cm^3/s) the ion heating is only 1% of that at low pressure. By a happy coincidence, many temperature-dependent experiments are high-pressure experiments as well: for example, equilibrium experiments [52, 53] three-body association reactions [54], and ion–neutral collision rates [55].

The ion kinetic energy due to the trapping potential depends on the ion's point of formation. It varies from 0.0 in the center of the cell to $\sim V_T/2$ eV near the sides. This motion is relaxed quickly both by collisions and by resistive heating of the control circuitry due to the ion image currents [56, 57]. It is thus usually not a problem, although using minimal trapping potentials to reduce the initial energy distribution is wise.

3.3.5 Equilibrium Measurements

Considerable thermochemical information can be obtained from equilibrium measurements. Free energies of reaction (ΔG^0) are obtained directly from the equilibrium constants (K_{eq}). A variable temperature experiment is theoretically required to determine ΔS^0 and ΔH^0; however, in many cases the ΔS term is small and can be estimated without large error. This is true of most proton transfer equilibria [52].

Several specific guidelines for the attainment of equilibrium are given in Refs. [52] and [53]. They help ensure that sufficient ion–neutral collisions occur to establish equilibrium in a time significantly shorter than the total drift time in the cell. Other problems that can occur include higher-order reactions (such as dimerization) that can differentially deplete the equilibrium and nonreactive fragment ions that can interfere with peak measurement. Experiments must be done with different neutral gas ratios and with different total pressures to check for equilibrium. Since experiments are done at high pressures ($1–10 \times 10^{-4}$ torr), internal excitation should be quickly removed by resonant collisions. Pressures must be measured on a capacitance manometer. A constant magnetic field is almost a necessity.

Data analysis is trivial since the ion concentration is constant. The observed intensity (at high pressure and constant magnetic field) is

$$\text{Int.}(A^+) \propto \frac{[A^+]}{m_A \xi_A} \tag{3.27}$$

(see Eq. 2.28b). The collision freqency (ξ_A) is obtained from the ICR linewidth (see Section 3.3.3) allowing calculation of the ion concentration. Accuracies better than $\sim \pm 0.2$ kcal/mol for $\Delta G^0 \lesssim 2$–3 kcal/mol are obtained. Lias and Ausloos [58] have proposed that the ICR intensities must be corrected for differing ion–neutral collision frequencies to yield true equilibrium data. This correction has been widely disputed [59].

3.3.6 Miscellaneous Experiments

A variety of specialized experiments have been done on drift ICR spectrometers. They include:

1. Reactions of ions and hydrogen atoms [60]. A differentially pumped microwave cavity was used to produce a beam of partially dissociated H_2 or D_2 that was directed into a normal drift ICR cell. The ions present in the cell react with the molecules and atoms in the beam. By recording spectra with and without the discharge and using ion ejection, rate coefficients and product distributions were obtained.
2. Measurement of inelastic excitation by low-energy electrons [61]. Here, zero energy electrons were produced in resonant excitation of vibronic states in neutral N_2. These low kinetic energy electrons are quickly scavenged by dissociative attachment with CCl_4 to produce Cl^-. By observing the Cl^- signal as a function of electron energy, the inelastic excitation spectrum of N_2 was obtained.
3. The study of ionizing reactions of metastable molecules [62]. Metastable N_2^* was produced by electron impact on a beam of N_2 gas. The beam was directed into the ICR cell where the ionizing reaction with C_6H_6 was observed. The cell was biased to prevent N_2^+ ions from entering.
4. Trapping ejection experiments. As mentioned in Section 2, excitation of the resonant ion motion in the trapping well is possible. This method is used to eject unwanted electrons in negative ion experiments ($\omega_{ej}/2\pi \sim 5$–10×10^6 Hz) since cyclotron ejection is not feasible ($\omega/2\pi \sim 30 \times 10^9$ Hz). It is a poor technique for ion ejection due to low resolution ($M/\Delta M \sim 4$ with a square cell, much less with the flat cell geometry). It was used to study the reactions of low mass ions (H_2^+, H_3^+, etc.) by modulated ejection [35]. This was necessary at the time because magnetic field sweeps were used, and the ejection frequency could not depend on magnetic field.

4 TRAPPED ICR

4.1 Introduction

Trapped ICR (TICR) spectrometry is characterized by the study of ion–neutral reactions as a function of time at a fixed pressure. This is in contrast to drift ICR, where the reaction time is fixed and the pressure varied. In TICR, the ion formation and detection times are pulsed. The time between the pulses is varied (from $\sim 1 \times 10^{-3}$ to 10 s), thereby sampling different reaction times. Pressures between 10^{-7} and 10^{-5} torr are commonly used. The technique has several advantages over drift ICR: higher resolution, easier data analysis, and no ion heating by the detection field during reaction. There are disadvantages also; namely, the near impossibility of studying three-body processes and generally lower signal levels due to the reduced duty factor. Trapped ICR has, to some extent, been eclipsed by FTICR in recent years; however, TICR remains a versatile technique for studying ion–neutral reactions.

4.2 Experimental Apparatus

In this section we discuss cell design, pulsed electronics, ion detection, data analysis, and double resonance experiments. Much of this has been reviewed by McIver [63].

4.2.1 Cells

Two different types of ICR cells have been used in TICR: the single section cell of McIver [64] and the modified drift cell of McMahon and Beauchamp

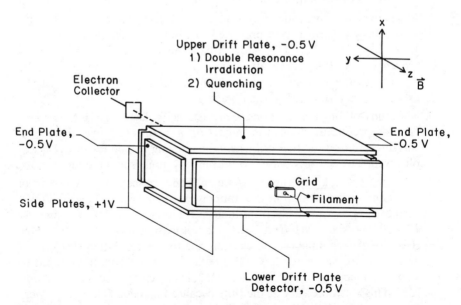

Fig. 3. Schematic view of trapped ICR cell. The dc potentials shown are suitable for trapping positive ions. [From *Rev. Sci. Instrum.*, **41**, 555 (1970).]

[65]. The single section cell (Fig. 3) is a rectangular box (dimensions ~ 2.54 × 2.54 × 8.9 cm). Plate material is plated copper [63] or stainless steel. The solid cell plates can be replaced with high line density screen without adverse effect [63, 66] (see Ref. [66] for source). It is necessary to screen the electron entrance and exit holes. The electron beam must be pulsed with a grid between the filament and cell. Cell plates are mounted on either Vespel [24] or Macor [25] insulating rails. Typical potentials are +1 V on the side (trapping) plates and 0.0–0.5 V on the top, bottom, and end plates. A version of this cell with hyperbolic-shaped top, bottom, and side plates has also been described [66]. An elongated version of the cell has been built for use in a superconducting solenoid [67]. The advantages of this arrangement are (1) operation at higher magnetic field for better ion trapping and resolution, (2) less perturbation of the cyclotron frequency due to the trapping field, and (3) greater signal intensity due to the ability to store a larger number of ions. Potentials similar to those listed above are applied to the cell plates.

The McMahon–Beauchamp cell [65] is a modified drift cell (Fig. 4). An end screen is attached to either the top or bottom source drift plate, and the electron entrance and exit holes are screened. Ion trapping in the source region occurs when the top and bottom source potentials are set equal to 0 V and the top and bottom analyzer to −1.0 V. This closes the equipotential lines in the source. Drift operation resumes when the normal +/− drift potentials are applied to the top and bottom plates.

Trapped cells have also been built [68, 69] in which the top and bottom plates of a drift cell are concentric cylinders. Ions drift in a continuous circle

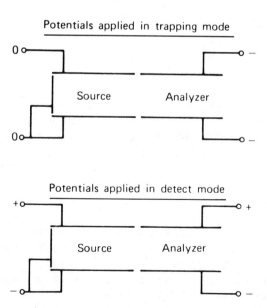

Fig. 4. Potentials applied to McMahon–Beauchamp cell in trapping and drift (detect) modes. [From *Rev. Sci. Instrum.*, **43**, 509 (1972).]

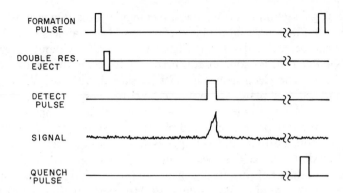

Fig. 5. Timing diagram for trapped cell ICR. The detect pulse is usually scanned between the formation and quench pulses. Multiple RF ejection pulses can be used.

about the center electrode (bottom drift plate). A pulsing scheme similar to that of the single section trap cell is used (Fig. 5).

A general analysis of the electric fields present in the ICR cell, based on a quadrupolar approximation, is given in Section 2. Sharp et al. [12] have examined the potentials in the one-section trapped cell and have obtained exact solutions for the electric fields as a function of cell geometry and plate voltages. Hunter et al. [67] have used these results to calculate ion motion in cubic, "normal," and elongated one-section cells. Hunter et al. show the total ion motion to be a fourfold composite of (1) normal cyclotron motion, (2) harmonic motion in the trapping potential well, (3) a radial motion, and (4) a slow motion along the equipotential lines in the cell ("magnetron" motion). All these movements affect the observed frequency of cyclotron motion to a small extent (typically reducing it $\sim 0.5\%$).

4.2.2 Electronics

Both electromagnets [63] and high field superconducting solenoids [67] have been used in TICR. Both ion trapping and resolution improve with magnetic field strength, making high fields desirable. The single section TICR experiment requires three pulses: grid, detect, and quench (Fig. 5). The cycle is initiated with a 1–5-ms grid pulse to form the ions. Pulsing the electron energy with a continuous electron beam allows control of the filament emission but results in gross perturbation of the cell potentials and loss of trapping [63]. Pulsed double resonance ejection is used after formation to remove unwanted ions. After a variable reaction time, the detect pulse brings the ions into resonance. The resulting power absorption is detected with a marginal oscillator or capacitance bridge detector. The sequence ends with a $+10\,\text{V}$ quench pulse on the top plate to remove all ions. The grid-detect delay (reaction time) is usually scanned from time zero to the grid-quench delay of between 2 ms and 10 s. Dugan et al. [70] have described a digitally timed control system for TICR.

A number of methods have been used to bring the ions into resonance during the detect pulse in the single section TICR cell. Originally the magnetic field was pulsed [64]. This is inconvenient and not usually done. Since the effective cyclotron frequency depends on the trapping potential (see Section 2), it is possible to pulse V_T and effect resonance [63, 71]. The dependence of ω_{eff} on V_T is given by [10]

$$\omega_{\text{eff}} \cong \omega_c \left(1 - \frac{2qV_T}{\omega_c^2 d^2 m}\right) \tag{4.1}$$

where ω_c is the unperturbed cyclotron frequency, q is the ion charge in coulombs, d is the spacing between the trapping plates, and m is the ion mass. A change in V_T of ~ 1 V changes the mass in resonance by ~ 0.5 amu ($\omega_c/2\pi = 153$ kHz, $V_T = 1$ V). At higher pressures this method is less successful due to collisional broadening of the resonance. Discrimination and ion heating during the reaction period can result. One solution is to pulse the detector oscillation, either with a pulsed marginal oscillator [72] or bridge detector [73, 74] (see Section 7).

The sequence is somewhat simpler in the store drift cell [65]. A 1–5-ms grid pulse forms the ion packet that is stored in the source region. After the desired reaction time, the top and bottom plate voltages are switched to drift mode. The ion's passage through the analyzer region then produces a power absorption transient on the continuously oscillating detector. The transient is integrated with a boxcar integrator. The detection time is limited to the analyzer region drift time. Thus the resolution of the store/drift cell is essentially the same as the drift cell.

4.2.3 Detection

Detection of ions in TICR is usually done either with marginal oscillator (MO) or capacitance bridge detectors (CBD). Electrometers have also been used. Detectors are discussed in Section 7. Two solid-state MO designs adapted for use in TICR have been described [66, 72]. The marginal oscillator is a sensitive detector and can be operated at different, fixed frequencies with a constant magnetic field for intensity vs. time plots of different masses. The sensitivity at the different frequencies must be measured (Section 7). Mass scans can be done either by scanning the magnetic field or with a frequency scanning detector. Marginal oscillators are not amenable to external frequency control and must be scanned slowly. As noted above, they can be used in a pulsed mode. Higher detector oscillation levels can be used in TICR (vs. drift ICR) due to the lack of reaction during the detect period. The only real constraint is minimal ion ejection from the cell.

Bridge detectors are becoming more common, partly due to the ease with which they are pulsed and frequency programmed. Operation at constant magnetic field is natural. The CBD is somewhat less sensitive than the MO (see Section 7), but it is possible to observe both the excitation and relaxation of the coherent ion motion with the CBD. Two CBD designs suitable for TICR use have appeared [73, 74]. Transient signals from either the MO or CBD detector are integrated and sampled, either with a boxcar integrator or a separate integrator and sample and hold.

An alternate method of signal acquisition is the "rapid scan" of Hunter and McIver [75, 76]. Mass spectra are acquired by rapidly scanning the excitation frequency of a CBD. Scan rates of ~ 60 kHz/s are used (~ 15 s for 20–1000 amu). The output of the CBD is a convolution of ion power absorption and emission at frequency ω_e and the scanning excitation frequency. This output is digitized, stored, and deconvoluted by correlation with a known peak shape observed under rapid scan conditions. The technique is very similar to FTICR; the difference is the lower frequency components in the transient signal with the rapid scan method ($\omega_{ex} - \omega_{eff}$ beat frequencies, rather than ω_{eff}) which are easier to digitize accurately. The deconvolution of the transient is not as straightforward as in FTICR, however, and depends on accurate heterodyning of the ion power absorption signal to lower frequencies.

Electrometer detection has also been used in TICR [77]. The ions are irradiated with a low-level ejection field. Ions in resonance are driven to the top or bottom plates of the cell which are connected to the electrometer. By scanning the ejection frequency, a mass spectrum results. Reasonable resolution is obtained ($M/\Delta M \sim 500$). The sensitivity is lower than either the CBD or MO.

4.3 Data Analysis

Simplified data analysis is one of the nice features of TICR. Since the detect time is relatively short (~ 5 ms) and the neutral gas pressure low, collisions during the detect period can usually be neglected. The resulting instantaneous power absorption per ion at resonance (Eq. 2.24a) is then

$$\mathrm{PA} = \frac{q^2 E_{RF}^2 t}{4m} \tag{4.2}$$

where E_{RF} is the peak excitation field (volt/meter), m is the ion mass (kilograms) and t is the time in resonance. Since the transients are usually integrated, the observed signal is

$$\mathrm{Int.} = \frac{q^2 E_{RF}^2 t^2}{8m} \tag{4.3}$$

Since the detection time is equal for all ions, the observed intensities need only be multiplied by ion mass to give true abundances. If different MO observing frequencies are used, the MO sensitivity must be considered also. The CBD sensitivity is proportional to mass, thus the observed CBD response is theoretically flat [73]. If the relaxation period is included in the CBD signal, the detection time may no longer be short with respect to the collision period, and collisional effects should be considered.

4.4 Double Resonance

The use of double resonance to eject ions and study reaction kinetic energy dependencies is discussed in Section 3.3.1. In TICR, pulsed double resonance ejection is used extensively to remove unwanted ions after the formation pulse. Continuous ejection can be used to remove product ions. Semiquantitative reaction energy dependencies can also be measured by irradiating a reactant ion at its resonant frequency for a known period of time [71]. The resulting average kinetic energy is

$$KE = \frac{q^2 E_{RF}^2 t^2}{8m} + \tfrac{1}{2} m v_0^2 \qquad (4.4)$$

There are a number of limitations on the accuracy of these calculated energies that are discussed in Section 3.3.1. Double resonance irradiation may perturb the output of a MO detector. Several compensation circuits have been published to reduce the problem [71, 78]. All will reduce the sensitivity of the MO to some extent. Bridge detectors are essentially immune to this interference.

4.5 Performance

We will consider three aspects of TICR performance here: resolution, ion trapping efficiency, and sensitivity. Both McIver and Baranyi [66] and Comisarow [57] have discussed the factors affecting resolution in TICR. The major consideration is allowing uninterrupted power absorption for the longest possible time. Collisions interrupt power absorption and limit resolution according to [66]

$$\frac{M}{\Delta M} \leqslant \frac{\omega}{\xi} \qquad (4.5)$$

where ΔM is the peak width at half-maximum and ξ is the momentum transfer collision frequency. The length of time in resonance, t, also influences the linewidth [22, 66]

$$\frac{M}{\Delta M} \leqslant \frac{\omega t}{5.57} \qquad (4.6)$$

Excitation levels that result in ion ejection will reduce resolution; below that level there should be no effect. Note that signal intensity is proportional to E_{RF}^2 (Eq. 4.3), and thus for a given length of detection time the largest value of E_{RF}, commensurate with insignificant ion loss, should be used. The variation in the homogeneity of the magnetic field over the volume of the TICR cell is approximately ± 1–$2\,G$ [79]. Since minimum ICR linewidths are $\sim 2\,G$, this may be a significant cause of line broadening [57, 66]. Resolution is directly proportional to magnetic field strength independent of homogeneity, hence the push to higher field magnets. Electric field inhomogeneities are also present in the TICR cell resulting in a dependence of ω_{eff} on ion position. In an attempt to gauge the magnitude of this effect, McIver and Baranyi [66] constructed a cell with hyperbolic cell plates. This change in cell design had no appreciable effect on resolution or ion-trapping efficiency. Electric field penetration from the grid or electron collector does perturb both the parameters [63], and, as noted, the electron entrance and exit holes must be screened.

Resolution of 5700 ($M/\Delta M_{1/2}$) at mass 28 and 1690 at mass 204 have been reported [66]. The resolution decreases with increasing mass theoretically as $1/m$ [57], experimentally [66] as roughly $1/\sqrt{m}$. The limit of unit mass resolution is ~ 500 amu. Mass, frequency, magnetic field, and excitation level interact in a complex way to determine the actual resolution.

The subject of ion loss from TICR cells was first analyzed theoretically by Sharp et al. [12] as a random walk phenomenon caused by ion–neutral collisions. They concluded that retention times should be proportional to $B^2 \cdot V^2/\xi$, where B is the magnetic field strength and V is the volume of the TICR cell. Their conclusion was qualitatively correct, but the predicted retention times were ~ 10 times larger than actual values. Ridge and Beauchamp [19] pointed out that ion loss results from nonrandom motion due to collisions in the presence of a drift field (Eq. 2.15a) Francl et al. [80] analyzed ion motion as diffusion in the presence of electric and magnetic fields. They found the same general dependencies of trapping efficiency as Sharp et al. [12], with the addition of a $1/V_T$ dependence. The resulting theoretical predictions of trapping efficiency is given by [80]

$$t_{1/2} = \frac{1.57 \times 10^{-28}(aB)^2}{P(V + 0.81)(\alpha\mu)^{1/2}} \quad (4.7)$$

where $t_{1/2}$ is the time for half the ions to escape, a is the spacing between upper and lower cell plates (centimeters), B is the magnetic field strength (tesla), P is the pressure (torr), V is the trapping potential ($V_T - V_0$, the center potential), μ is the ion–neutral reduced mass (grams), and α is the neutral polarizability (cm^3). Calculated retention times are within a factor of 2 of those observed experimentally. In practice, at pressures of $\sim 1 \times 10^{-6}$ torr and magnetic fields of ~ 1 T, more than 80% of the ions are retained for 1–2 s.

The sensitivity of the TICR technique is governed largely by the minimum detectable signal. This, in turn, depends on detector sensitivity (discussed in Section 7). With the CBD, McIver et al. [73] state that ~ 7000 ions are required to give a signal to noise ratio = 1, with a single 5-ms detect period under typical conditions. Sensitivity with an MO would be somewhat better. The space charge storage limit for a "typical" TICR cell of $2.54 \times 2.54 \times 8.9$ cm dimensions is $\sim 350,000$ ions [73], giving a dynamic range of ~ 50. The "elongated" TICR cell has a greater volume and a storage capacity of $\sim 20 \times 10^6$ ions [67]. The effect of space charge on TICR frequencies has been discussed by Francl et al. [81].

4.6 Applications

The TICR can obviously be used as a mass spectrometer; however, the limited mass range ($\lesssim 500$ amu) and relatively long acquisition times limit its usefulness. The FTICR is far better suited to analytical mass spectrometry. Trapped ICR spectrometers are well suited to study ion–neutral reactions, however [63]. Knowing the neutral gas pressure, rate coefficients are obtained from semilog plots of reactant ion loss as a function of reaction time. Account must be taken of nonreactive ion loss, especially with slow reactions. Rate coefficients less than $\sim 0.1\%$ of the collision rate are difficult to measure for this reason [63].

A great many measurements have been made of ion–ion equilibria using TICR [63, 82]. Often interfering product ions or termolecular reactions are absent in the low-pressure TICR experiments. The equilibrium is measured by observing the ion intensities over a period of time to ensure a constant ratio. Different neutral gas pressure ratios are also used to check for equilibrium. The forward and backward rate coefficients can also be measured directly by continuously ejecting one product ion and observing the decrease in the other.

Rate coefficients for momentum transfer collisions can be measured by observing the exponential decay of the CBD transient signal. This represents the loss of coherent ion motion by several mechanisms: reaction, loss from cell, ohmic heating of resistive elements in the detector and nonreactive collisions. If the first three effects are small, then a good estimate of the collision frequency can be made [57, 83].

5 TANDEM ICR

5.1 Introduction

Drift ICR experiments (and to a lesser extent trapped ICR) commonly suffer from interfering reactions involving unwanted ionic or neutral species. The tandem ICR is one solution to the problem. By separating the ion formation and reaction regions and mass selecting the reactant ion, unwanted species are removed. The tandem ICR has an obvious analog in the flowing afterglow SIFT technique of Smith and Adams [84]. The singular advantage of the SIFT technique is that it provides translationally and rotationally thermal energy ions. The corresponding disadvantage is the difficulty of studying the effects of vibrational/rotational excitation. The tandem ICR has the advantage of being able to study such reactant ion excitation and of using ICR double resonance to both eject unwanted secondary ions and study reaction kinetic energy dependencies. The price for this is a nonthermal reactant ion translational energy distribution. Ions entering the reaction cell may have kinetic energies up to ~ 0.5 eV (lab).

In this section we outline the design of the tandem ICR vacuum system, electronics, and sources, pointing out differences between the original design of Smith and Futrell [85] and that of Kemper and Bowers [86].

5.2 Design

5.2.1 Vacuum Systems

The general layout of the tandem ICR is shown in Fig. 6. The vacuum system consists of (1) a main chamber containing the ion source and ion optics and (2) a differentially pumped ICR chamber housing the reaction cell. Pumping in the main chamber is with either diffusion pump [85] or turbomolecular pump [86]. The ICR chamber is differentially pumped with a 2-in. trapped diffusion pump. Background pressure in the main chamber is $\sim 5 \times 10^{-8}$ torr, in the ICR chamber $\sim 5 \times 10^{-7}$ torr. Communication between the chambers is only via the ICR entrance slit. The section of the vacuum system housing the source, ion optics, and ICR reaction cell fits between the pole faces of a 12-in. electromagnet (2-in. air gap). A separate vacuum system is used for the two inlets that control pressures in the source and reaction chamber.

5.2.2 Ion Sources

Two types of sources have been used in the tandem ICR. Originally a high-pressure ($\sim 10^{-4} \to 0.5$ torr) mass spectrometer type source was used [85–87] (Fig. 7). Ionization was by electron impact. A split repeller must be fitted to control the ion motion in the strong magnetic field. The ion exit slit may also be

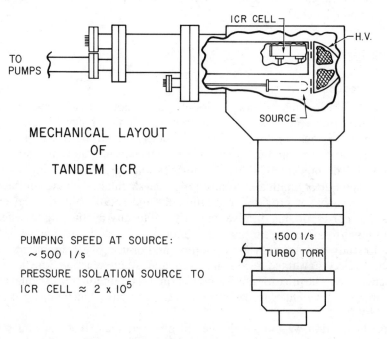

Fig. 6. Cutaway view of UCSB tandem ICR, showing the mechanical layout. [From *Int. J. Spectrom. Ion Phys.*, **52**, 1 (1983).]

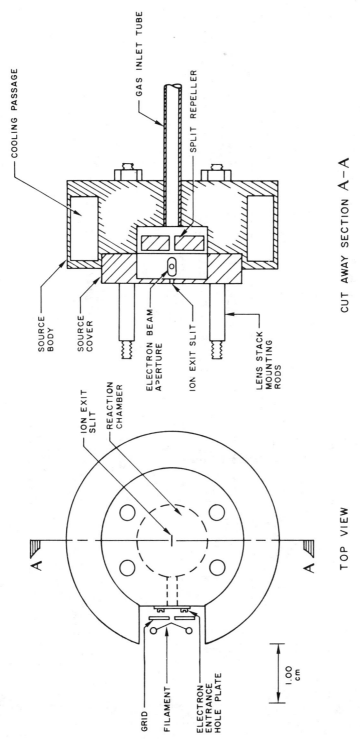

Fig. 7. Modified high-pressure mass spectrometer ion source for tandem ICR. [From *Int. J. Mass Spectrom. Ion Phys.*, **52**, 1 (1983).].

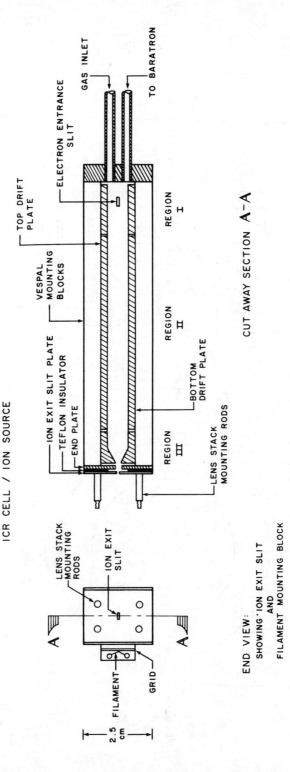

Fig. 8. Drift ICR ion source for tandem ICR. [From *Int. J. Mass Spectrom. Ion Phys.*, **52**, 1 (1983).]

biased. Source pressures were either sampled directly using a capacitance manometer [86] or indirectly from the chamber pressure [85]. Pressure isolation between the source and the ICR reaction chamber is between $\sim 2 \times 10^4$ [85] and 10^5 [86], and mixing of the source and reaction gases can usually be neglected. A temperature variable version of this source has also been described [86]. The maximum ion current from the source is $\sim 5 \times 10^{-8}$ A. The source body voltage is variable ± 10 V about ground to allow fine tuning of the ion energy as it enters the reaction cell. The repellers, electron energy, and trap voltages are referenced to the source body and move with it.

In addition to the mass spectrometer source, an ICR cell source has also been described [86] (Fig. 8). This source is a modified drift ICR source consisting of three regions. Ionization occurs by electron impact in region 1, which is kept short to minimize reaction. Region 2 is a long reaction region. In region 3 the top and bottom plates are bent to provide an increasing drift field and a focusing effect. The ion exit slit (0.05 × 0.38 cm or 0.02 × 0.15 in.) can also be biased for additional focusing. The cell plates were originally silver plated copper [86]; however, pure molybdenum plates appear to require far less cleaning. Pressure isolation between this source and the ICR chamber is $\sim 10^4$; however, the lower pressures used (10^{-4}–10^{-2} torr) actually result in less interference. Beam currents of $\sim 3 \times 10^{-8}$ A are typical. Again the mean source voltage is variable about ground, and the various plate voltages follow this reference voltage.

The advantage of the ICR cell source is the ability to use ICR double resonance to separate electron impact (EI) and chemical ionization (CI) formed ions. If ion A^+ is formed both by electron impact and reaction (chemical ionization), the EI-formed A^+ can be ejected in region 1 immediately after formation. Since the CI-formed A^+ is produced throughout the cell, the A^+ exiting the source will be largely from reaction. An example is given later.

5.2.3 Ion Optics

Mass selection in the tandem ICR is accomplished with a 180° symmetric magnetic sector or Dempster mass filter (Fig. 9). Ions leaving the source are accelerated by a three-element lens system through a high-voltage slit into a high-voltage field free region. Here they move in a circular orbit of radius

$$r = \left(\frac{2Vm}{qB^2}\right)^{1/2} \tag{5.1}$$

where V is the accelerating voltage, m the ion mass, q the ion charge, and B the magnetic field strength (all in MKS units). Ions of a particular mass pass through a second high-voltage slit (deceleration slit) and are decelerated to near 0 eV as they pass through the ICR entrance slit (at ground potential) and into the ICR reaction cell.

Typical accelerating voltages are between 0.8 and 4 kV. The high-voltage field free region is screened for better pumping. The lenses outside the high-voltage slits provide both focusing and steering effects. Their mean voltage

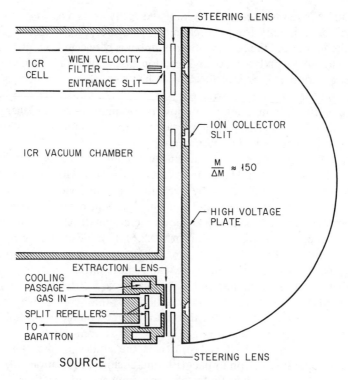

Fig. 9. Dempster mass filter in tandem ICR, showing ion lenses. [From *Int. J. Mass Spectrom. Ion Phys.*, **52**, 1 (1983).]

ranges from ∼100 to 600 V and is derived from the high-voltage supply. The total differential voltage (between lenses) ranges from ∼0 to ±300 V and is supplied by four 150-V power supplies. Modulation of the ion beam (for phase-sensitive detection) can be accomplished with electron beam modulation; however, a better method is to modulate the deceleration steering lens. A floatable, bipolar, ×5 amplifier supplies complementary 50-V square waves to the two lens halves. When measuring drift times, a pulse is applied to the amplifier instead of the square wave. This gating method has the advantages of maintaining a constant, stable source environment and of maintaining the proper ion density in the reaction cell during drift time measurements.

Two designs have been used for the ICR entrance slit. The Utah tandem ICR [85] uses a "slot" of 0.035 cm height × 0.110 cm thickness × 0.38 cm length (0.014 × 0.045 × 0.15 in.). The Santa Barbara instrument [86] uses a relatively large slit 0.075 cm height × 0.38 cm width (0.03 × 0.15 in.) followed by a Wien velocity filter [9] that admits only a narrow bandwidth of ion velocities. This filter consists of two parallel plates (0.51 × 1.27 cm; 0.2 × 0.5 in.) separated by 0.043 cm (0.017 in.). The plates must be coated with Aquadag. These parallel plates, rather than the entrance slit, form the image-defining slit. Performance of the two designs is discussed in Section 5.3.

An intermediate ion collector is fitted to help in source tuning. The collector is biased with a 300-V battery and the current read on a picoammeter. The bottom plate of the velocity filter is also used as an intermediate collector to locate the beam at the entrance slit of the reaction cell. The source body voltage is raised to $\sim +5\,\text{V}$ to drive the ions to the plate (at 0 V). This allows both the magnetic field and the steering lens voltage to be set easily. This procedure helps greatly in "finding the beam," since usually only the source body must be adjusted to put the beam in the cell.

5.2.4 ICR Reaction Cell

Various tandem ICR reaction cells have been described [85, 86]. The cell currently used in the Santa Barbara tandem ICR is a two-section drift ICR cell. Dimensions are 1.27 cm height × 3.30 cm wide (0.50 × 1.30 in.); source length 3.81 cm, analyzer length 7.62 cm (1.5 and 3 in.) A short cell is necessary since both the cell and the mass filter must fit in the magnetic field. A total ion collector similar to that in Fig. 2 is used. The plate material is molybdenum; the insulating rails are Macor ceramic. The cell essentially never needs cleaning.

Double resonance excitation is applied to both top and bottom source plates as described in Section 3.3.1. This greatly improves the field homogeneity for quantitative double resonance but causes greater detector coupling. The problem could be solved with an intermediate region held at ac ground.

Since the magnetic field must remain fixed, a scanning detector must be used (either marginal oscillator or capacitance bridge detector, see Section 7). A frequency to voltage converter provides the x-axis drive [88]. Detector sensitivity at different frequencies is usually determined with a Q-spoiler type calibration [89]. Calibration has also been done by comparison of double resonance ejection spectra with MO signals [85].

The ion beam is gated for drift time measurement by pulsing the deceleration steering lens to admit a 100-μs-wide ion packet. Times are then measured as described by McMahon and Beauchamp [11]. No change in source conditions is necessary (Section 3.2.3).

5.3 Performance

The resolution of the mass selector in the tandem ICR ranges from $M/\Delta M \sim 150$ [86] to 500 [85]. The significantly higher resolution in the Utah instrument may be due to the slightly larger Dempster radius (5.7 vs. 4.35 cm), or to the use of a commercial ion source [87], or to ions cyclotroning in the thick slit (effectively narrowing the slit) [85, 87].

Beam currents out of the source are typically $1-5 \times 10^{-8}$ A. Currents into the ICR reaction cell are typically 1×10^{-10} A. Larger currents can be obtained, but they are usually unstable, probably due to the space charging at the low-voltage slits.

The question of reactant ion kinetic energy is obviously critical. Put another way, what range of ion energies is admitted to the reaction cell? According to Smith and Futrell [85], ions pass through the thick slit in two ways: either by executing several cyclotron orbits of $\lesssim 0.017$ cm radius at low energy ($\lesssim 0.1$ eV) or in a single pass at high energy ($\gtrsim 1.0$ eV). Ions with translational energies between 0.1 and 1.0 eV should be filtered out. This is supported by scans of percent transmission vs. source body voltage [85, 87] that show a sharp maximum around 0 V. Studies of endothermic reactions tended to confirm the low ion energy, but not unambiguously [85].

Kemper and Bowers [86] have compared the product distribution obtained for the $N^+ + O_2$ reaction using the thick slit geometry with that obtained from SIFT-DRIFT experiments [90]. Kinetic energies in the DRIFT experiments range from ~ 0.04 to 5 eV and are presumably well characterized. The results they obtained agreed well with the previous tandem ICR results of Anicich et al. [91] and indicated the N^+ kinetic energy to be ~ 0.6 eV (c.m.). Thus, there appears to be some question as to the actual ion energy.

Including the Wien velocity filter in the ion optics should greatly reduce transmission of high-energy ions (> 1 eV) and better characterize the energy of the ions that are transmitted. It is possible to show that the range of ion energies that can pass through the filter is 0 to $\sim V_F$ eV [9], where V_F is the potential across the filter (divided as $\pm V_F/2$). Variation of the $N^+ + O_2$ product distribution with filter voltage showed the direct dependence of the ion energy on filter voltage. Using the velocity filter, minimum N^+ energies of 0.3 eV (c.m.) were obtained, compared with ~ 0.6 eV using the thick slit [91]. Kemper and Bowers [86] thus concluded that the reactant ion energies were ~ 0.5 eV (lab) with the velocity filter.

Whatever excess kinetic energy is present can be removed by use of a buffer gas (usually Ne or He). This is necessary for systems that are sensitive to reactant kinetic energy. Such systems can be easily identified using the ICR double resonance experiments described below.

5.4 APPLICATIONS

5.4.1 Product Distributions

The tandem ICR is extremely well suited for measuring product distributions. The reactant ion of choice is focused into the ICR reaction cell where it reacts with the desired neutral species. The different product–ion intensities are then measured by scanning the detector frequency. Product distributions are recorded at four or five different neutral-species pressures (usually between 2×10^{-6} and 2×10^{-5} torr) and extrapolated to zero pressure to minimize the effects of subsequent reactions on the measured product distribution. The relative sensitivities are measured [89] and, together with the ion mass m, are used to calculate the relative sensitivity, $S = $ MO sensitivity/m* The apparent ion intensities are

---

*This assumes that the low-pressure (single-collision) form of the ICR power-absorption equation applies. In the absence of momentum-transfer collisions, the observed intensity is proportional to

divided by their relative sensitivities to yield the true intensities. Changes in relative sensitivity are not large with a MO detector, since sensitivity is approximately proportional to mass. A change in ion mass by a factor of 2 will produce a maximum change in sensitivity of $\sim 30\%$.

5.4.2 Total Rate Constants

Total rate constants are measured in two ways using the tandem ICR: First, a decrease in the primary-ion intensity can be observed as reactant neutral gas is added. This decrease is governed by the pseudo-first-order rate law

$$A^+ = A_0^+ \exp(-[n]kt) \qquad (5.2)$$

where $[n]$ is the neutral-species concentration and t the reaction time. The rate constant is then derived from the slope of a semilogarithmic plot of intensity vs. pressure (see Section 3.3.2, Eq. 3.20). This is an approximate method, but calculations and experiments show that it applied well up to $\sim 60\%$ conversion. Reaction times are measured and ionization-gauge pressures are calibrated against a capacitance manometer. The disadvantage of this method is that primary-ion scattering and reactions that occur before the ions enter the ICR cell can at times give rise to an apparent rate constant as large as $1-2 \times 10^{-10}$ cm^3 s^{-1}. This effect can be compensated for by measuring the decrease in total ion current in the ICR cell with added neutral species and subtracting the resulting ion-scattering rate. Even with this correction, however, large uncertainties result when rates smaller than 2×10^{-10} cm^3 s^{-1} are measured in this way.

For slower reactions, both the primary- and secondary-ion intensities are measured at a series of pressures. The rate constant is then obtained from a plot of the extent of reaction vs. neutral-species concentration (Section 3.3.2, Eq. 3.19).

5.4.3 Reactant Ion Kinetic Energy Studies

Double resonance can be easily used to determine (semiquantitatively) the energy dependence of a reaction. A given product ion is monitored in the analyzer region while the parent ion is irradiated in the source. Different product ions may be examined while keeping the parent irradiation unchanged. Use of a high-frequency rectifier [92] allows x-y plots of intensity vs. E_{RF} to be made directly. Measurement of the source drift time ($=$ irradiation time) allows KE_{ion}

$[I^+] \times t^2/m \times$ (detector sensitivity), where $[I^+]$ is the true ion current in the cell, m is the ion mass, and t is the time in the detection region. Since t is equal for all ions, it follows that $[I^+] = m/$(detector sensitivity) $\times$ (observed intensity). Intensities measured at high pressure must also be multiplied by the ion–neutral collision frequency.

to be calculated from

$$\mathrm{KE}_{\mathrm{ion}} = \frac{q^2 V_{\mathrm{RF}}^2 t^2}{8md^2} + \tfrac{1}{2} m v_0^2 \qquad (5.3)$$

where q is the ion charge (coulomb), V_{RF} is the peak irradiating voltage (volt), t is the irradiation time, m is the ion mass (kilogram) and d is the cell spacing (m). The available energy range is usually $0.5 - \sim 5\,\mathrm{eV}$ (lab). The maximum energy is limited by the onset of ion ejection, the minimum by the ion injection energy. The calculated ion energies are only semiquantitative due to ion–neutral collisions and a distribution of irradiation times. See Section 3.3.1 for a complete discussion.

5.4.4 Effect of Reactant–Ion Internal Energy

The tandem ICR can be used to gauge the average internal energy present in a reactant ion as in a recent study of energy partitioning in a series of charge-transfer reactions that produce excited $(NH_3^+)^*$ [93]. The method used can be diagrammed as follows:

Tandem source $X^+ + NH_3 \rightarrow (NH_3^+)^* + X^*$

Mass-select

ICR cell $(NH_3^+)^* + H_2O \longrightarrow \begin{array}{l} \rightarrow NH_4^+ + OH \quad \Delta H = -0.20\,\mathrm{eV} \\ \rightarrow H_3O^+ + NH_2 \quad \Delta H = +0.47\,\mathrm{eV} \end{array}$

$$(5.4)$$

The ions X^+ used included Xe^+, Kr^+, Ar^+, N_2^+, O_2^+, CO^+, and CO_2^+. The product distribution in the reaction with H_2O is sensitive to $(NH_3^+)^*$ internal energy. Thus, when different charge-transfer reactions are used to form NH_3^+ in the tandem source, different product distributions are observed in the $NH_3^+ + H_2O$ reaction in the ICR cell, thus gauging the $(NH_3^+)^*$ internal energy. By using complementary kinetic-energy-release experiments (Section 6), it is possible to derive a calibrated product-distribution vs. $(NH_3^+)^*$ internal-energy curve. This curve may then be used to calculate the complete energy-partitioning between $E_{\mathrm{int}}(NH_3^+)^*$, $E_{\mathrm{int}}(X)^*$, and $E_{\mathrm{trans}}(NH_3^+ \cdots X)$ in the $X^+ + NH_3$ reactions.

A second example involves electronic-state selection in Kr^+. Adams et al. [94] have shown that the two electronic states of Kr^+ react differently with CH_4:

$$Kr^+(^2P_{3/2}) + CH_4 \xrightarrow{1.00} CH_4^+ + Kr \qquad (5.5a)$$

$$Kr^+(^2P_{1/2}) + CH_4 \begin{array}{l} \xrightarrow{0.10} CH_4^+ + Kr \\ \xrightarrow{0.90} CH_3^+ + H + Kr \end{array} \qquad (5.5b)$$

The total rate constants for reactions 5.5a and 5.5b are equal ($1 \times 10^{-9}\,\text{cm}^3\,\text{s}^{-1}$). Thus, the product distribution in reaction 5.5 can be used to gauge the relative amounts of $\text{Kr}^{+\,2}P_{3/2}$ and $^2P_{1/2}$. If Kr^+ is formed by 30-eV electron ionization, a statistical ratio $^2P_{3/2}/^2P_{1/2}$ of 2:1 is expected initially. If this ratio does not change significantly in the time required for drift through the source and mass selection ($\sim 5 \times 10^{-4}\,\text{s}$), then a $\text{CH}_4^+ : \text{CH}_3^+$ ratio of $\sim 0.7 : 0.3$ is expected when the Kr^+ reacts with the CH_4 in the ICR cell. If the $^2P_{1/2}$ state is deactivated, then 100% CH_4^+ product should be observed.

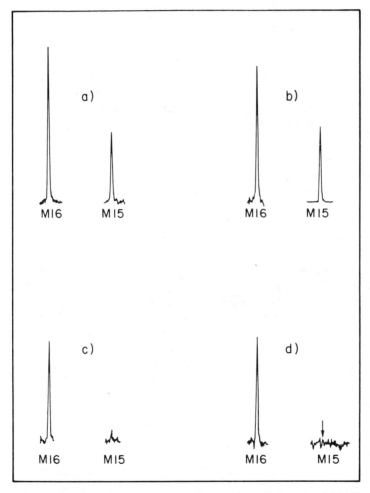

Fig. 10. CH_4^+ (m/e 16) and CH_3^+ (m/e 15) products from $\text{Kr}^+ + \text{CH}_4$ (reaction 5.5). Differences in $16^+/15^+$ ratio are due to changes in the relative amounts of $\text{Kr}^+(^2P_{3/2})$ and $(^2P_{1/2})$ electronic states. (a) Kr^+ formed by 30-eV electron impact; (b) CO added to ion source; (c) same conditions as (b) but with Kr^+ ejected in the first region of ion source at medium power; (d) same conditions as (b) but with Kr^+ ejected at high power. [From *Int. J. Mass Spectrom. Ion Phys.*, **52**, 1 (1983).]

The experimental result indicates a ratio of 0.7:0.3, suggesting no deactivation of the $Kr^{+\,2}P_{1/2}$ state. However, it is possible that translationally hot $Kr^{+\,2}P_{3/2}$ ions produce the observed CH_3^+ signal. To check for this possibility, a beam of pure $Kr^{+\,2}P_{3/2}$ was produced by adding CO to the Kr already present in the ICR ion source and ejecting the electron-beam-generated Kr^+ ions in the first region of the source (immediately after ionization). $Kr^{+\,2}P_{3/2}$ is then produced by near-resonant charge transfer between CO^+ and Kr:

$$CO^+ + Kr = Kr^{+\,2}P_{3/2} + CO \qquad (5.6)$$

If the Kr^+ mixture formed in the electron beam is completely ejected, then any Kr^+ leaving the source must be produced by charge transfer and hence be $Kr^{+\,2}P_{3/2}$. The results of this experiment are given in Fig. 10, which shows the CH_4^+ and CH_3^+ peak heights (a) with only Kr in the source, (b) with Kr and CO present but without ejection, (c) with Kr and CO and some ejection of Kr^+, and (d) with increased Kr^+ ejection. The CH_3^+ product vanishes with increased ejection efficiency, indicating that when the Kr^+ is ejected in the first region of the source, only $Kr^{+\,2}P_{3/2}$ is present in the Kr^+ beam leaving the source.

5.5 Tricyclotron Spectrometer

Mauclaire and Marx have very recently reported the construction of a three-section ICR spectrometer [95]. The first and last sections (the ion source and analyzer) are straight; the center section connecting them is C shaped. Each section is differentially pumped. "Slits" are formed by compressing the side plates at the transitions between sections. Each section operates as a McMahon–Beauchamp trapped/drift cell (Section 4). Thus ions can be stored for varying times before being moved to the next section. There is no inherent mass selection, and unwanted ions must be removed by double resonance ejection. There is no high-voltage acceleration either, so the ion energy remains essentially thermal. Ion detection is done with a capacitance bridge detector (Section 7).

The spectrometer is obviously well suited to the study of time-dependent processes such as radiative decay of excited states. The analyzer is also designed to function as a kinetic energy ICR (see Section 6).

6 KINETIC ENERGY ICR

6.1 Introduction

ICR spectrometry can be used to determine the kinetic energy produced in a simple thermal energy charge transfer or atom transfer reaction

$$A^+ + BC \longrightarrow \begin{matrix} BC^+ + A & (6.1a) \\ AB^+ + C & (6.1b) \end{matrix}$$

Such measurements allow calculation of the product ion internal energy [IE

(BC^+)] using a simple energy balance equation. For example, for charge transfer reactions

$$RE(A^+) - IP(BC) = IE(BC^+) + KE(A - BC^+) \tag{6.2}$$

where RE (A^+) is the recombination energy of A^+, IP (BC) is the ionization potential of BC, and KE $(A—BC^+)$ is the kinetic energy release. We assume IE $(A) = 0$. If the charge-transfer reaction is dissociative

$$A^+ + BC \to B^+ + C + A \tag{6.3}$$

then

$$RE(A^+) - IP(B^+) = IE(B^+, C) + KE(A, B^+, C) \tag{6.4}$$

Finally, for atom interchange reactions 6.1b,

$$IE(AB^+) = \Delta H_{react} - KE(AB^+ - C) - IE(C) \tag{6.5}$$

The kinetic energy determination is based on the fact that ions can escape the cell if their kinetic energy is greater than the trapping potential. Only that ion velocity parallel with the magnetic field is useful in promoting such escape, since velocities perpendicular to the field only result in cyclotron motion. The possibility of such kinetic energy spectroscopy was first suggested by Riggin and Woods [15] and Orth et al. [96] who reported the first application in a photodissociation study. Since those initial experiments, the great majority of the experimental development has been done by Mauclaire et al. [97, 98]. See also Ref. [99] for experimental details.

6.2 Theory

The quantitative expression for trapping in the ICR cell is

$$qV_0 > \tfrac{1}{2}mv_\parallel^2 = E_k \cos^2\theta \tag{6.6}$$

where q is the charge on the ion, V_0 is the trapping potential seen by the ion in the center of the cell,* m is the ion mass, $v_\parallel$ is the ion velocity parallel to the magnetic field, E_k is the total ion kinetic energy, and θ is the angle between the ion velocity and the magnetic field. Since we will consider only singly charged ions and work in electron volt units, the q in Eq. 6.6 will be dropped. For $V_0 > E_k$, all ions will be trapped. If $V_0 < E_k$, ions are trapped when $V_0/E_k > \cos^2\theta$; i.e., for θ such that $\pi/2 \geqslant \theta \geqslant \cos^{-1}(V_0/E_k)^{1/2}$. The velocity distribution in the ICR cell is isotropic and thus $v_\parallel$ has a $\sin\theta$ distribution. The fraction of

---

*V_0 is slightly less than V_T (the trapping plate potential). The difference is small (5–10%) and can be calculated from the cell geometry. The experimental procedure keeps the ions near the center of the cell and thus ensures that all ions see the same potential.

product ions trapped (f) is then

$$f = 1 \qquad\qquad V_0 > E_k \qquad (6.7a)$$

$$f = \int_{\cos^{-1}(V_0/E_k)^{1/2}}^{\pi/2} \sin\theta\, d\theta \qquad V_0 \leqslant E_k \qquad (6.7b)$$

$$= \left(\frac{V_0}{E_k}\right)^{1/2}$$

A plot of f vs. $(V_T)^{1/2}$ will thus increase linearly until $V_0 = E_k$, then remain constant at $f = 1$. A sharp break in an experimental curve indicates a single value of E_k. Curvature about the break point in the experimental curve indicates a distribution of translational energies (and hence product internal energies). Any curvature present is compressed at high V_T due to the $(V_T)^{1/2}$ plot. Multiple break points are also observed.

The greatest amount of information is obtained with atomic ion reactants due to the lack of accessible excited states in the corresponding neutral atom. In nondissociative charge transfer reactions (6.1a), a complete determination of energy partitioning can be obtained. In the case of a dissociating diatomic product (6.3), the kinetic energies of the neutral products are unknown; however, limits on their values can be obtained by assuming a reaction mechanism. For example, a common assumption is that the reaction occurs in two steps:

$$\begin{array}{l} A^+ + BC \to (BC^+)^* + A \\ \hookrightarrow B^+ + C \end{array} \qquad (6.8)$$

Limits on the relative kinetic energy of the $(BC^+)^*$–A products vary from zero for a resonant process with no momentum transfer to an upper limit that just leaves enough internal energy in $(BC^+)^*$ to dissociate to $B^+ + C$. Comparison of the predictions of the various models with experiment often yields insight into the reaction mechanism.

6.3 Experimental Technique

All experiments to date use a combination Trap-Drift cell of the McMahon–Beauchamp type [65] with electrometer detection. A drift mode spectrometer cannot be used due to the influence of V_T on the drift potential and hence on the reaction time (see Section 2). The drift field may also perturb the trapping potential. The cell developed by Mauclaire and Marx [97] has five sections: source, eject 1, reaction, eject 2, and collector. The cell is initially in the "drift" mode. Ions are formed by a 100-μs electron beam pulse; a low V_T ($\lesssim 0.1$ V) is used in the source to avoid forming translationally excited reactant ions. The ions drift into the first ejection region where unwanted ions are ejected; V_T is raised (~ 4.0 V) to center the ion beam. Usually a small pressure of unreactive neutral ($1-5 \times 10^{-5}$ torr of He or Ne) is maintained to relax the ions collisionally. When the ion packet reaches the center of the reaction region, the cell is switched to "trap" mode and the desired V_T is set. After reaction (1–20 ms), "drift" mode is restored and the ions drift through the second ejection region to the collector. High trapping potentials are used to avoid further ion loss. Selective ejection before the collector allows determination of the relative abundances. Three ejection measurements at each V_T are required. The product ion abundance is normalized to that of the unreacted parent ion to account for cell losses other than those due to reaction kinetic energy. This ratio is the fraction (f) of product ions trapped. The entire experimental sequence is repeated at many V_T values until a smooth f vs $(V_T)^{1/2}$ curve is obtained. The curve must be corrected for the small difference between V_0 and V_T [99]. Also, curves may not pass through the origin due to offsets in V_T caused by contact potentials or cell contamination [99]. Such offsets must be subtracted from V_0. The experimental sequence is controlled by a microcomputer [99].

A second cell configuration that has been used [99] has a single region for use as ion source, first ejection, and reaction region. A timing sequence is used to vary the trapping and drift voltages in this portion of the cell. The experimental sequence is identical to the above. The experiment should also be feasible with a single section trapped cell and bridge detector (see Section 7). Marginal oscillator detectors are effectively excluded due to the difficulty of computer control.

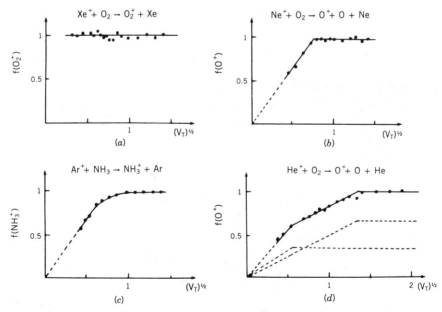

Fig. 11. Four examples of KE ICR measurements made by Mauclaire et. al. at Orsay: (a) no kinetic energy release, (b) only one discrete value of kinetic energy, (c) a distribution of kinetic energies, (d) two discrete values of kinetic energy. (From Ref. [100].)

6.4 Applications

Four examples of kinetic energy determinations are shown in Fig. 11 [100]. The lack of any break in the f vs. $(V_T)^{1/2}$ curve for the $Xe^+ + O_2$ reaction (Fig. 11a) indicates the charge transfer is resonant with no kinetic energy release. This result is perhaps obvious since RE (Xe^+) = IP (O_2), but it gives a positive check on the technique. The sharp break in the curve for the dissociative $Ne^+ + O_2$ system (Fig. 11b) indicates that a single discrete value of kinetic energy is released. Consideration of the O_2^+ potential curves led Mauclaire et al. [97] to conclude that the reaction involved resonant population of an O_2^+ state at 21.6 eV followed by separation to the second dissociation asymptote. The data for the $Ar^+ + NH_3$ reaction (Fig. 11c) shows a pronounced curvature, indicating a spread in NH_3^+ internal energy. By modeling the expected kinetic energy release for different NH_3^+ vibrational populations, Derai et al. [101] were able to derive the vibrational population produced in the charge transfer. Finally, in Fig. 11d, the $He^+ + O_2$ dissociative system is seen to produce two discrete values of kinetic energy release. These correspond to two different O^+ product electronic states [102].

7 DETECTORS

7.1 Introduction

A surprising number of detection methods have been used in ICR experiments; all depend on ion power absorption from an oscillating electric field. We discuss four techniques here: marginal oscillator detectors, capacitance bridge detectors, the Q meter, and electrometer detection. We also include a comparison of marginal oscillator and capacitance bridge sensitivity.

7.2 Marginal Oscillator

The diagram in Fig. 12 shows the various components that make up the marginal oscillator (MO) detector. The ICR cell is included in a resonant L–C "tank" circuit. A portion of the voltage (V_T) present across the tank is amplified,

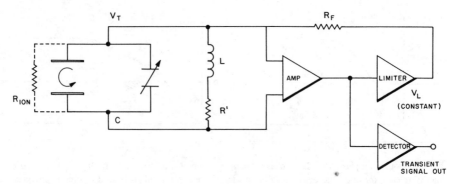

Fig. 12. Schematic diagram of marginal oscillator detector.

limited to a constant value V_L, and returned to the circuit via the feedback resistor R_F. The frequency of free oscillation, ω

$$\omega \cong \frac{1}{\sqrt{LC}} \tag{7.1}$$

depends slightly on the resistance (R') of the inductor (L), but with high-quality inductors this can be ignored.

The theory of MO response to ICR signals has been discussed by Anders [103], McIver [72], Comisarow [56, 104] and Kemper and Bowers [88]; all achieve the same results. The following is an outline of the theory.

The voltage across the L–C circuit is determined by the feedback resistor (R_F) and the resistive losses in the tank circuit (R_T):

$$V_T = \frac{R_T}{R_T + R_F} \cdot V_L = \frac{1}{1 + R_F/R_T} \cdot V_L \tag{7.2}$$

These losses are almost entirely due to R', and it can be shown that

$$R_T \cong \frac{\omega^2 L^2}{R'} = Q\omega L \tag{7.3}$$

where Q is a "quality" factor equal to X_L/R' (X_L = inductive impedance = ωL). When ions absorb power an additional resistance appears across the L–C circuit

$$R_{\text{ION}} = \frac{PA}{V_T'^2} \tag{7.4}$$

where PA is the instantaneous power absorption (in whatever form) and V_T' is the new (lower) tank circuit voltage. Then

$$V_T' = V_L \frac{1}{1 + R_F(1/R_T + 1/R_{\text{ION}})} \tag{7.5}$$

and the resulting signal voltage is

$$\Delta V_T = V_T - V_T' = \frac{PA}{V_T'} \cdot \frac{R_F}{1 + R_F/R_T} \tag{7.6}$$

If $R_{\text{ion}} \gg R_T$, then $V_T' \sim V_T$; if in addition $R_F \gg R_T$, then

$$\Delta V \cong \frac{PA}{V_T} \cdot Q\omega L = \frac{PA}{V_T} \cdot Q\left(\frac{L}{C}\right)^{1/2} \tag{7.7}$$

Equations 7.6 and 7.7 point out the desirability of maximizing R_F, L, and Q while minimizing C. The maximum value of Q may be limited by the modulation bandwidth necessary [105]. The 3-dB bandwidth is $\sim \omega/2\pi Q$. Thus for $\omega/2\pi =$ 100 kHz and $Q = 300$ (for example), the maximum modulation frequency is ~ 160 Hz (one-half the 3-dB band width). The value of R_F must be increased with caution due to parasitic capacitance in R_F. This leads to phase shifts in the feedback as well as frequency-dependent feedback. These effects can be avoided by using a split capacitor current drive [105] or by combining several smaller resistors to form R_F [88].

There have been numerous calculations of the minimum detectable signal using an MO. [56, 103, 106–108] The usual procedure is to calculate the minimum theoretical noise voltage according to

$$V_N = (4kTFBR_T)^{1/2} = \left[4kTFBQ\left(\frac{L}{C}\right)^{1/2} \right]^{1/2} \qquad (7.8)$$

where k is Boltzmann's constant, T is the temperature in Kelvins, F is the noise figure (usually of the preamp [105]), and B is the noise bandwidth. This noise voltage is then equated with $n \times (\Delta V_T/\text{ion})$, which allows determination of the minimum detectable number of ions, n. The usual numbers are about 8–15 ions. These values are best case estimates since most sources of experimental noise are ignored (noisy grounds, radiative pickup, ground loops, etc.). A detection limit of ~ 50 ions [88, 109] has been demonstrated, however.

The ion power absorption actually adds both a resistive and a capacitive (dispersive) component to the MO L–C circuit. This dispersion causes a shift in the circuit's resonant frequency. These shifts are proportional to the power absorption and are typically a few hertz. Anders [103] reports shifts of up to 15 Hz.

Many different MO designs appear in the literature. Robinson [106] discusses design, noise, etc., in general, and for vacuum tube MO's in particular. Viswanathan et al. [110] and Adler et al. [107, 108] theoretically model sensitivity, noise, and response of the MO. McIver [72] gives a bibliography of solid-state designs (mostly for NMR applications). Warnick et al. [105] describe a solid-state MO for ICR use that has a number of interesting aspects. They discuss points such as impedance matching, the split capacitor current drive mentioned earlier, and sensitivity calibration. McIver has published several solid-state MO designs [66, 72], including a pulsed version [72] in which the oscillation level is gated on by an external pulse. The output is the difference between the tank circuit voltage and that of a compensation circuit designed to mimic the tank circuit response when no ions are present. This method of gating the ion detection has advantages at high pressure where ICR linewidths are so broad that power absorption occurs even when the ion is nominally off resonance. The high Q tank circuit does have a slow rise time (300 μs) [72], leading to possible errors in power absorption intensities measured at short times. Also the difficulty of matching the compensation and tank circuit responses leads to small spikes at the start and stop of oscillation [70, 72].

The above marginal oscillators cannot be frequency swept. There are great advantages to maintaining a constant magnetic field in both trapped and drift ICR [77b, 88]. Two designs for a frequency swept MO have appeared. Amano et al. [111] designed a solid-state, constant sensitivity circuit in which the frequency sweep is done by a variable capacitance (varactor) diode. The varactor has a range of 25–900 pf and a $Q \sim 1000$. Constant sensitivity is achieved by keeping R_T constant through use of a variable shunt resistance across the L–C circuit. The resistance is controlled by an error amplifier that senses the oscillation level. The MO is limited to its lowest sensitivity with this approach.

Kemper and Bowers [88] also describe a frequency scanning MO in which a constant oscillation amplitude is maintained by varying the limiter gain. There is no degradation of sensitivity, but the gain varies with frequency and must be measured (see Section 7.3). The detector is shown to be sensitive to ~ 75 ions. Frequency scans are done mechanically with a variable speed motor. The authors also describe a frequency to voltage converter suitable for driving an x–y recorder.

7.3 Sensitivity Calibration

In order to operate a marginal oscillator at different frequencies, some means of measuring the sensitivity as a function of frequency is needed. This is usually done with a standard signal of the "Q-spoiler" type shown in Fig. 13, which is attached in place of the cell. The device is discussed by Warnick et al. [105], Kemper and Bowers [89], and Anicich and Huntress [122]. With the switch open, it appears as a capacitor C_T of value $nC_1/(n+1)$. This is selected to approximate the ICR cell capacitance. With the switch closed, an effective resistance $R_T = (n+1)^2 R$ is placed across the L–C circuit. Typical values of C_1, n, and R are 30 pF, 100 and 20 kΩ giving $R_T \sim 200$ MΩ. Direct switching of such high-value resistors is impossible due to capacitance across the switch contacts. With large resistor values and at ICR frequencies, the switch would appear essentially closed all the time. Small, frequency dependent calibration signals would result.

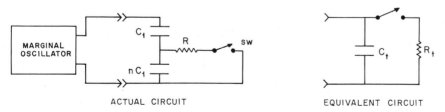

Fig. 13. Schematic diagram of Q-spoiler standard signal, showing actual and equivalent circuits. Typical component values are $C_1 \sim 30$ pF, $R \sim 20$ kΩ, and $n = 100$. A low-capacitance reed relay can be used as the switch.

7.4 Capacitance Bridge Detector

A capacitance bridge detector (CBD) was first used in ICR by Wobschall in 1965 [2]. The need for a detector capable of fast automated frequency scans and pulsed operation led to its revival by McIver et al. [73, 75] in 1977 and Wronka and Ridge [74] in 1982. The basic circuit is shown in Fig. 14.

The excitation voltage and double resonance voltage are independently gated and summed. The output of the summing amplifier drives an L–C bridge with the ICR cell as one capacitive leg. The balance capacitor and null potentiometer adjust the in-phase and quadrature signals to produce a null at the preamp input when no ions are present. When power absorption occurs, the resulting imbalance is amplified, rectified (in the phase-sensitive detector), and filtered to give the output transient that is then usually integrated. Once excited, coherent ion motion in the cell continues, even in the absence of excitation. This effect greatly enhances the CBD sensitivity; there is no corresponding effect with the MO. Relaxation of this motion occurs mainly through collisions and heating of resistive circuit components by the resulting image currents [56, 57]

The theory of CBD response has been investigated by Wobschall [2, 3] and McIver et al. [73, 83]. McIver's image current analysis allows modeling of both excitation and relaxation. The following is a condensation of the analysis in Ref. [73].

The image current I produced by n ions of charge q positioned between two flat plates of spacing l moving with velocity v toward one of the electrodes is given by

$$I = \frac{nqv}{l} \tag{7.9}$$

This current will flow according to

$$I = (C_1 + C_2 + C_3)\frac{dV_0}{dt} + \frac{V_0}{R} \tag{7.10}$$

where V_0 is the bridge imbalance voltage, C_1 is the cell capacitance, C_2 is the value of the balance capacitor, C_3 is the preamp input capacitance, and R is the preamp input resistance. We define $C = C_1 + C_2 + C_3$. Generally $R \gg X_c$ (the capacitive impedance $= 1/\omega C$), and current flow through R may be neglected. Then

$$C\frac{dV_0}{dt} = \frac{nqv}{l} \tag{7.11}$$

The velocity v of an ion absorbing power is given in Section 2. In the absence of collisions and with resonant excitation,

$$v_x = \frac{qV_p t}{2ml} \sin \omega t \tag{7.12}$$

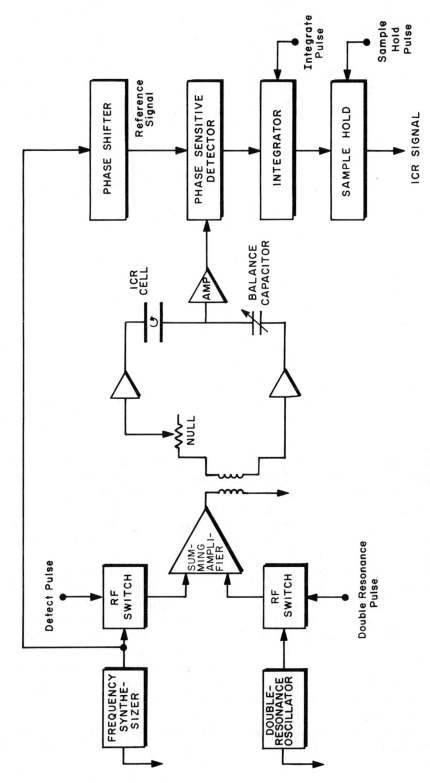

Fig. 14. Block diagram of capacitance bridge detector. [From *Int. J. Mass Spectrom. Ion Phys.*, **39**, 65 (1981).]

where V_p is the base-to-peak excitation voltage, t is the time in resonance, and ω is the resonant frequency. The bridge imbalance is then

$$V_0 = \frac{nq^2 V_p}{2ml^2 C\omega^2} (\sin \omega t - \omega t \cos \omega t) \tag{7.13}$$

After rectification and low pass filtering the resulting transient signal for one ion is

$$V_{\text{SIG}} = \frac{q^2 V_p t}{2ml^2} \cdot \frac{1}{\omega C} = \frac{2\text{PA}}{V_p} \cdot \frac{1}{\omega C} \tag{7.14}$$

where PA is the instantaneous power absorption per ion. Remembering that $\omega = qB/m$, we have

$$V_{\text{SIG}} = \frac{qV_p t}{2l^2 \cdot C \cdot B} \tag{7.15}$$

which indicates the response of the CBD should be independent of mass at constant B.

A more general treatment including ion–neutral collisions, reactions, and the relaxation signal is given in Ref. [83]. The dispersion signal is easily detected using the CBD by shifting the phase-sensitive detector reference phase by 90°.

Three ICR CBD circuits have appeared in the literature. Wobschall's [2] original design is somewhat outdated (except for the actual excitation bridge) and will not be discussed. In McIver's [73] design, the bridge imbalance voltage is amplified and then rectified by a (sin) × (sin) multiplication. The multiplier reference is phase shifted 90° (adjustable) with respect to the excitation voltage. The reference amplitude and phase must remain constant as the frequency is swept. This is difficult, and for exact comparison of two ions, two phase shifter boards may be used [113].

Wronka and Ridge [74] use qualitatively the same design for their CBD; however, they use a balanced modulator/demodulator as the phase-sensitive detector. The reference must also be shifted 90°, but they note that the reference output on many commercial oscillators is already shifted 90° and can be used directly. Fine phase adjustments are done with a constant time delay line (phase shift $\propto \omega$). They argue that this compensates better for phase shifts over a wide frequency range than the constant phase shift of McIver. Although both approaches are certainly adequate, neither appears perfect.

A third alternative is to use a commercial lock-in amplifier as the phase-sensitive detector. The limitations are lower frequency (usually $\gtrsim 200\,\text{kHz}$), higher cost, and slower scan rates; the advantages are ease and (probably) higher-quality output signals.

7.5 Comparison of MO and CBD Sensitivities

Since marginal oscillators and bridge detectors are the most common ICR detectors, it is useful to compare them; see Comisarow [56, 57], and Kemper and Bowers [88]. We define sensitivity as the ratio of signal and excitation voltages (V_{SIG}/V_{EX}). The MO sensitivity is taken from Eq. 7.7 ($V_{SIG} = \Delta V$, $V_{EX} = V_T$, $\omega L = 1/\omega c$);

$$\frac{V_{SIG}}{V_{EX}} \cong \frac{PA}{V_{EX}^2} \frac{Q}{\omega C_{MO}} = \frac{R_{TANK}}{R_{ION}} \quad (7.16)$$

From equation 7.14 we obtain the corresponding CBD sensitivity,

$$\frac{V_{SIG}}{V_{EX}} = \frac{PA}{V_{EX}^2} \frac{2}{\omega C_{BD}} = \frac{2Z_{BD}}{R_{ION}} \quad (7.17)$$

where Z_{BD} is the impedance (largely capacitive) of the bridge detector. Equations 7.16 and 7.17 are intuitively obvious: Since a resistance (R_{ION}) is being measured, the detector's sensitivity depends on its impedance. For typical values of Q (100–400) C_{MO} (150–650 pF) and C_{BD} ($\sim$ 30 pF) [88], the MO should be 2–40 times as sensitive as the CBD [114].

This sensitivity calculation ignores the contribution of the relaxation signal that (in the absence of collisions) can greatly increase signal to noise in the CBD. Such emission signals may not be strictly proportional to ion intensities, however. The calculation also ignores noise effects. Comisarow [57] has done signal to noise calculations for both MO and CBD and agrees that the ratio of MO and CBD sensitivities should be similar to that calculated above.

7.6 The Q Meter

Huntress and Simms [115] have designed a Q-meter detector for use at high frequencies (1 – 15 MHz). The circuit (Fig. 15) is similar to a MO with the amplifier and limiter replaced by an external oscillator. The resonant frequency of the $L-C$ circuit is tuned to match the oscillator frequency. Since no feedback is used, very stable components and oscillator are necessary to maintain resonance. Ion detection is identical to the MO: The power absorption causes

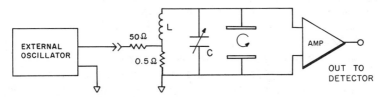

Fig. 15. Schematic diagram of Q-meter detector. Note method used to couple driving oscillator. This is an alternative to the feedback resistor method shown in Fig. 12. [From *Rev. Sci. Instrum.*, **44**, 1274 (1973).]

a decrease in oscillation level that is amplified and filtered. The advantage of the Q meter is its ability to operate with good sensitivity at both high frequencies and low levels. One novel experiment was done in which free electrons in the cell were detected by their resonant motion in the trapping field ($\omega_T = 6$–8 MHz).

7.7 Electrometer Detection

With the electrometer detection technique, the total ion current (TIC) is monitored as a function of double resonance ejection frequency. Usually ejection is done in a short region immediately before the TIC collector. If perfect ejection is achieved, then the decrease in TIC equals the ion abundance. The method has the advantages of (1) no mass discrimination, (2) no complex power absorption deconvolution in drift ICR, as the ion abundances are measured at a single specific time, and (3) simplicity. The disadvantages are (1) difficulty in ensuring perfect ejection, (2) restriction to fairly low pressures, (3) lower resolution due to high excitation levels necessary, (4) lower sensitivity, and (5) problems at low frequency due to coupling with trapping motion [23]. The disadvantages have outweighed the advantages, and the technique is not commonly used.

A variation of this technique was used by McIver et al. [77] in a trapped cell experiment in which individual ion currents were collected on the top and bottom plates of the cell. The frequency of a low-level double resonance field was swept generating a mass spectrum. This is similar to the detection method used in the omegatron of Hipple et al. [1].

REFERENCES

1. (a) J. A. Hipple, H. Sommer, and H. A. Thomas, *Phys. Rev.*, **76**, 1877 (1949); (b) H. Sommer, H. A. Thomas, and J. A. Hipple, *Phys. Rev.*, **82** 697 (1951).
2. D. Wobschall, *Rev. Sci. Instrum.*, **36**, 466 (1965).
3. D. Wobschall, J. R. Graham, and D. P. Malone, *Phys. Rev.*, **131**, 1565 (1963).
4. J. L. Beauchamp, L. R. Anders, and J. D. Baldeschwieler, *J. Am. Chem. Soc.*, **89**, 4569 (1967).
5. J. D. Baldeschwieler, *Science*, **159**, 263 (1968).
6. J. L. Beauchamp, *Annu. Rev. Phys. Chem.*, **22**, 527 (1971).
7. M. T. Bowers and T. Su, *Adv. Electronics Electron Phys.*, **34**, 223 (1973).
8. T. A. Lehman and M. M. Bursey, *Ion Cyclotron Resonance Spectrometry*, Wiley, New York, 1976.
9. See any text on electricity and magnetism, e.g., L. Page and N. I. Adams, *Principles of Electricity*, D. Van Nostrand, Princeton, New Jersey, 1931.
10. J. L. Beauchamp and J. T. Armstrong, *Rev. Sci. Instrum.*, **40**, 123 (1969).
11. T. B. McMahon and J. L. Beauchamp, *Rev. Sci. Instrum.*, **42**, 1632 (1971).
12. T. E. Sharp, J. R. Eyler, and E. Li, *Int. J. Mass Spectron. Ion Phys.*, **9**, 421 (1972).
13. T. F. Knott and M. Riggin, *Can. J. Phys.*, **52**, 426 (1974).
14. I. B. Woods, M. Riggin, T. F. Knott, and M. Bloom, *Int. J. Mass Spectrom. Ion Phys.*, **12**, 341 (1973).
15. M. Riggin and I. B. Woods, *Can. J. Phys.*, **52**, 456 (1974).
16. J. L. Beauchamp, Ph.D. Thesis, Harvard University, Cambridge, Mass. (1967).

17. J. L. Beauchamp. *J. Chem. Phys.*, **46**, 1231 (1967).
18. E. W. McDaniel, *Collision Phenomenon is Ionized Gases*, Wiley, New York, 1964.
19. D. P. Ridge and J. L. Beauchamp, *J. Chem. Phys.*, **64**, 2735 (1976).
20. M. B. Comisarow, *J. Chem. Phys.*, **55**, 205 (1971).
21. R. C. Dunbar, *J. Chem. Phys.*, **54**, 711 (1971).
22. S. E. Buttrill, Jr., *J. Chem. Phys.*, **50**, 4125 (1969).
23. B. S. Freiser, T. B. McMahon, and J. L. Beauchamp, *Int. J. Mass Spectron. Ion Phys.*, **12**, 249 (1973).
24. Vespel is a high temperature polyimide resin manufactured by E. I. Dupont, Inc.
25. Macor machinable glass ceramic. Corning Glass Works, Corning, New York 14830.
26. R. P. Clow and J. H. Futrell, *Int. J. Mass Spectrom. Ion Phys.*, **4**, 165 (1970).
27. Rhenium ribbon from H. Cross Co., 363 Park Avenue, Weehawken, New Jersey 07087.
28. R. T. McIver Jr., *Rev. Sci. Instrum.*, **41**, 126 (1970).
29. F. P. Clay Jr., F. J. Brock, and L. T. Melfi Jr., *Rev. Sci. Instrum.*, **46**, 528 (1975).
30. P. R. Kemper, Ph.D. Thesis, University of California, Santa Barbara, Calif. (1977).
31. A. G. Wren, P. T. Gilbert, and M. T. Bowers, *Rev. Sci. Instrum.*, **49**, 531 (1978).
32. J. H. Moore, C. C. Davis, and M. A. Coplan, *Building Scientific Apparatus*, Addison-Wesley, London, 1983.
33. J. M. S. Henis and W. Frasure, *Rev. Sci. Instrum.*, **39**, 1772 (1968).
34. L. R. Anders, J. L. Beauchamp. R. C. Dunbar, and J. D. Baldeschwieler, *J. Chem. Phys.*, **45**, 1062 (1966).
35. M. T. Bowers and D. D. Elleman, *J. Chem. Phys.*, **51**, 4606 (1969).
36. W. J. van der Hart and H. A. van Sprang, *Chem. Phys. Lett.*, **36**, 215 (1975).
37. J. L. Beauchamp and R. C. Dunbar, *J. Am. Chem. Soc.*, **92**, 1477 (1970).
38. J. L. Beauchamp and S. E. Buttrill, *J. Chem. Phys.*, **48**, 1783 (1968).
39. R. C. Dunbar, *J. Chem. Phys.*, **47**, 5445 (1967).
40. L. R. Anders, *J. Phys. Chem.*, **73**, 469 (1969).
41. P. R. Kemper, unpublished results.
42. W. T. Huntress, Jr., *J. Chem. Phys.*, **55**, 2146 (1971).
43. C. A. Lieder, R. W. Wien, and R. T. McIver, *J. Chem. Phys.*, **56**, 5184 (1972).
44. P. R. Kemper and M. T. Bowers, *Int. J. Chem. Kinetics*, **16**, 707 (1984).
45. P. R. Kemper, unpublished design.
46. A. G. Marshall and S. E. Buttrill, Jr., *J. Chem. Phys.*, **52**, 2752 (1970).
47. V. G. Anicich and M. T. Bowers, *Int. J. Mass Spectrom. Ion Phys.*, **11**, 329 (1973).
48. G. C. Goode, R. M. O'Malley, A. J. Ferrer-Correia, R. L. Massey, K. R. Jennings, J. H. Futrell, and P. M. Llewellyn, *Int. J. Mass Spectrom. Ion Phys.*, **5**, 393 (1970).
49. A. G. Marshall, *J. Chem. Phys.*, **55**, 1343 (1971).
50. S. E. Buttrill, Jr., *J. Chem. Phys.*, **58**, 656 (1973).
51. W. J. Chesnavich, T. Su, and M. T. Bowers, *J. Chem. Phys.*, **65**, 990 (1976).
52. D. H. Aue, H. M. Webb, and M. T. Bowers, *J. Am. Chem. Soc.*, **98**, 311 (1976).
53. W. R. Davidson, M. T. Bowers, T. Su, and D. H. Aue, *Int. J. Mass Spectrom. Ion Phys.*, **24**, 83 (1977).
54. L. M. Bass, P. R. Kemper, V. G. Anicich, and M. T. Bowers, *J. Am. Chem. Soc.*, **103**, 5283 (1981).
55. D. C. Parent and M. T. Bowers, *Chem. Phys.*, **60**, 257 (1981).
56. M. B. Comisarow, *J. Chem. Phys.*, **69**, 4097 (1978).

57. M. B. Comisarow, in H. Hartmann and K-P. Wanczek, Eds., *Lecture Notes in Chemistry: Ion Cyclotron Resonance Spectrometry, II*, Springer-Verlag, New York, 1982.
58. See S. G. Lias in *Kinetics of Ion-Molecule Reactions*, P. Ausloos, Ed., Plenum Press, New York, 1979, and references therein.
59. See W. J. Chesnavich, H. Metiu, and M. T. Bowers, *Int. J. Mass Spectrom. Ion Phys.*, **41**, 143 (1982) and references therein.
60. Z. Karpas, V. G. Anicich, and W. T. Huntress, Jr., *J. Chem. Phys.*, **70**, 2877 (1979).
61. D. P. Ridge and J. L. Beauchamp, *Chem. Phys.*, **51**, 470 (1969).
62. W. T. Huntress and J. L. Beauchamp, *Int. J. Mass Spectrom. Ion Phys.*, **3**, 149 (1969).
63. R. T. McIver, Jr., *Rev. Sci. Instrum.*, **49**, 111 (1978).
64. R. T. McIver, Jr., *Rev. Sci. Instrum.*, **41**, 555 (1970).
65. T. B. McMahon and J. L. Beauchamp, *Rev. Sci. Instrum.*, **43**, 509 (1972).
66. R. T. McIver and A. D. Baranyi, *Int. J. Mass Spectrom. Ion Phys.*, **14**, 449 (1974).
67. R. L. Hunter, M. G. Sherman, and R. T. McIver, Jr., *Int. J. Mass Spectrom. Ion Phys.*, **50**, 259 (1983).
68. P. R. Kemper and M. T. Bowers, unpublished results.
69. S-H. Lee, K-P. Wanczek. and H. Hartmann, *Adv. Mass Spectrom.*, **8B**, 1645 (1979).
70. R. J. Dugan, L. M. Morganthaler, R. O. Daubach, and J. R. Eyler, *Rev. Sci. Instrum.*, **50**, 691 (1979).
71. R. T. McIver, Jr., and R. C. Dunbar, *Int. J. Mass Spectrom. Ion Phys.*, **7**, 471 (1971).
72. R. T. McIver, Jr., *Rev. Sci. Instrum.*, **44**, 1071 (1973).
73. R. T. McIver, Jr., R. L. Hunter, E. B. Ledford, Jr., M. J. Locke, and T. J. Francl. *Int. J. Mass Spectrom. Ion Phys.*, **39**, 65 (1981).
74. J. Wronka and D. P. Ridge, *Rev. Sci. Instrum.*, **53**, 491 (1982).
75. R. L. Hunter and R. T. McIver, *Chem. Phys. Lett.*, **49**, 577 (1977).
76. R. L. Hunter and R. T. McIver in *Lecture Notes in Chemistry: Ion Cyclotron Resonance Spectrometry II*, H. Hartmann and K.-P. Wanczek, Eds., Springer-Verlag, New York, 1982.
77. (a) R. T. McIver, Jr., E. B. Ledford, Jr., and J. S. Miller, *Anal. Chem.*, **47**, 692 (1975); (b) E. B. Ledford, Jr., and R. T. McIver, Jr., *Int. J. Mass Spectrom. Ion Phys.*, **22**, 399 (1976).
78. D. J. Defrees and R. T. McIver, Jr., *Rev. Sci. Instrum.*, **48**, 574 (1977).
79. Manufacturer's specification for a Varian V3400 electromagnet.
80. T. J. Francl. E. K. Fukuda, and R. T. McIver, Jr., *Int. J. Mass Spectrom. Ion Phys.*, **50**, 151 (1983).
81. T. J. Francl, M. G. Sherman, R. L. Hunter, M. J. Locke, W. D. Bowers, and R. T. McIver, Jr., *Int. J. Mass Spectrom. Ion Phys.*, **54**, 189 (1983).
82. M. T. Bowers, D. H. Aue, H. H. Webb, and R. T. McIver, *J. Am. Chem. Soc.*, **93**, 4314 (1971).
83. R. T. McIver, Jr., E. B. Ledford, Jr., and R. L. Hunter, *J. Chem. Phys.*, **72**, 2535 (1980).
84. D. Smith and N. G. Adams in *Gas Phase Ion Chemistry*, M. T. Bowers, Ed., Academic Press, New York, 1979, vol. 1.
85. D. L. Smith and J. H. Futrell, *Int. J. Mass Spectrom. Ion Phys.*, **14**, 171 (1974).
86. P. R. Kemper and M. T. Bowers, *Int. J. Mass Spectrom. Ion Phys.*, **52**, 1 (1983).
87. J. H. Futrell and R. G. Orth in *Lecture Notes in Chemistry: Ion Cyclotron Resonance II*, H. Hartmann and K. P. Wanczek, Eds., Springer-Verlag, New York, 1982.
88. P. R. Kemper and M. T. Bowers, *Rev. Sci. Instrum.*, **53**, 989 (1982).
89. P. R. Kemper and M. T. Bowers, *Rev. Sci. Instrum.*, **48**, 1477 (1977).
90. F. Howorka, I. Dotan, F. Fehsenfeld, and D. L. Albritton, *J. Chem. Phys.*, **73**, 758 (1980).
91. V. G. Anicich, W. T. Huntress, and J. H. Futrell, *Chem. Phys. Lett.*, **47**, 488 (1977).

92. P. R. Kemper, unpublished design.
93. P. R. Kemper, M. T. Bowers, D. C. Parent, G. Mauclaire, R. Derai, and R. Marx, *J. Chem. Phys.*, **79**, 160 (1983).
94. N. G. Adams, D. Smith, and E. Alge, *J. Phys. B.*, **13**, 3235 (1980).
95. G. Mauclaire and R. Marx, 10th International Mass Spectrometry Conference, Swansea, U. K., (1986).
96. R. Orth, R. C. Dunbar, and M. Riggin, *Chem. Phys.*, **19**, 279 (1977).
97. G. Mauclaire, R. Derai, S. Fenistein, and R. Marx, *J. Chem. Phys.*, **70**, 4017 (1979).
98. G. Mauclaire, R. Derai, and R. Marx, *Int. J. Mass Spectrom. Ion Phys.*, **26**, 289 (1978).
99. D. C. Parent, Ph.D. Thesis, University of California at Santa Barbara, 1983.
100. R. Derai, M. Mencik, G. Mauclaire, and R. Marx in *Lecture Notes in Chemistry: Ion Cyclotron Resonance II*, Hartmann, H. and K. P. Wanczek, Eds., Springer Verlag, New York, 1982.
101. R. Derai, G. Mauclaire, and R. Marx, *Chem. Phys. Lett.*, **86** 275 (1982).
102. G. Mauclaire, R. Derai, S. Fenistein, and R. Marx, *J. Chem. Phys.*, **70**, 4023 (1979).
103. L. R. Anders, Ph.D. Thesis, Harvard University, Cambridge, Mass., 1966.
104. M. B. Comisarow, *Int. J. Mass Spectrum. Ion Phys.*, **26**, 369 (1978).
105. A. Warnick, L. R. Anders, and T. E. Sharp, *Rev. Sci. Instrum.*, **45**, 929 (1974).
106. F. N. H. Robinson, *J. Sci. Instrum.*, **36**, 481 (1959).
107. M. S. Adler and S. D. Senturia, *Rev. Sci. Instrum.*, **40**, 1481 (1969).
108. M. S. Adler, S. D. Senturia, and C. R. Hewes, *Rev. Sci. Instrum.*, **42**, 704 (1971).
109. Specified for the Varian 5900 series ICR spectrometers, Varian Associates, Palo Alto, California.
110. T. L. Viswanathan, T. R. Viswanathan, and K. V. Sane, *IEEE Trans.* **IM-24**, 55 (1975).
111. C. Amano, Y. Goto, and M. Inoue, *Int. J. Mass Spectrum. Ion Phys.*, **32**, 67 (1979).
112. V. G. Anicich and W. T. Huntress, Jr., *Rev. Sci. Instrum.*, **48**, 542 (1977).
113. R. T. McIver, private communication.
114. McIver et al. [73] quotes a CBD detection limit of ~ 7000 ions/5-ms pulse. Kemper and Bowers [88] quote an MO limit of ~ 75 ions when ~ 20 10-ms pulses are averaged (50 Hz modulation, 0.3 s time constant). The corresponding, single, 5-ms pulse detection limit should be ~ 750 ions with the M.O.
115. W. T. Huntress and W. T. Simms, *Rev. Sci. Instrum.*, **44**, 1274 (1973).

Chapter **II**

FOURIER TRANSFORM MASS SPECTROMETRY

Ben S. Freiser

---

1 **Introduction**
2 **Brief History**
3 **Principles of Operation**
   3.1 Signal Generation
   3.2 Resolution
   3.3 Mass Accuracy
4 **Applications**
   4.1 Mass Spectrometry/Mass Spectrometry
   4.2 Photodissociation
   4.3 Gas Chromatography
5 **Ionization Methods**
   5.1 Electron Ionization and Chemical Ionization
   5.2 Multiphoton Ionization
   5.3 Laser Desorption
   5.4 Cesium Ion–Secondary Ion Mass Spectrometry (Cs^+ – SIMS)
   5.5 Plasma Desorption
6 **New Excitation Methods**
7 **Tomorrow's Instrumentation Here Today**
8 **Conclusions**
Acknowledgments
References

---

1 INTRODUCTION

The field of mass spectrometry has exploded over the last 20 years, permeating just about every area of science in which analytical methodology is used. From biochemistry to catalysis, from energy research to astrochemistry, mass spectrometry is being applied. Not only is mass spectrometry being called into service because of improved instrumentation permitting ever higher masses to be observed, but it is increasingly becoming recognized as a tool for studying the details of fundamental chemical problems. In this setting, one of the most promising techniques to appear on the scene is Fourier transform mass spectrometry (FTMS). As will be discussed, this mass spectrometer, perhaps more than any other, has evolved primarily as a method for studying fundamental ion–molecule processes as opposed to a method for direct analytical applications. However, we can hardly ignore the analytical potential of a mass spectrometer with a virtually limitless mass range and a proven resolution capable of observing the gain in mass of an ion following absorption of a single photon! In this review, a brief history of the ancestry of the technique is presented, together with the fundamental principles, applications, and future directions of FTMS. As an indication of the rising interest in FTMS, a number of other review articles has appeared recently [1–8], and two journals have dedicated an entire issue each to the topic [9, 10].

2 BRIEF HISTORY

As with many modern analytical spectrometers, the early roots of the instrumentation for FTMS can be traced back in the physics journals. A series of articles about 1950 by Sommer, Thomas, and Hipple introduce the omegatron, the "grandparent" precursor instrument to FTMS [11, 12]. In these papers a method is described for measuring the cyclotron frequencies of ions constrained in a magnetic field and using these results to obtain accurate charge-to-mass ratios. Interestingly, the group's first studies were done on the proton and were aimed at yielding the magnetic moment of the proton as well as an accurate measure of the faraday. The heart of the omegatron is the probe, which can be described simply as a "box" consisting of flat metal plates or electrodes to which radio frequency (rf) and direct current (dc) voltages can be applied. Whereas the ion motion is constrained to circular orbits perpendicular to the magnetic field, there is no constraint by the magnetic field along the parallel. The ions are prevented from escaping axially by applying a dc electric field onto the "trapping plates" of the probe. As discussed below, this basic design has not changed significantly in the modern day FTMS. It is interesting to note that these workers had the foresight to predict the analytical use of the omegatron. Specifically, they pointed out that one of the most important applications of the omegatron would be in precise mass measurement. As an impressive example, they demonstrated that the isobars H_2^+ and D^+ could be readily separated, but resolution rapidly degraded going to higher mass. Nevertheless, they stated that it should be possible to build a device capable of measurements at higher masses. They also stated that its simplicity and high sensitivity indicated promise for analytical applications, and that the absence of slits defining the ion beam and the variable travel times held promise for studying fundamental ion processes.

Although the omegatron itself never achieved the promise nor prominence the creators had envisioned and, in fact, has been relegated to the "lowly" role of a partial pressure measuring device, it did give birth in the early sixties to ion cyclotron resonance (ICR) spectroscopy [13-15]. Unlike the omegatron, the ICR spectrometer became available commercially from Varian. This instrument had an improved mass range and resolution compared to the original omegatron and used a marginal oscillator detector. Ions were generated by electron impact and were drifted through the probe, or cell as it is now termed, by virtue of crossed electric and magnetic fields. This instrument caught the attention of Baldeschwieler and he, together with his scientific progeny, began what remains today an exceedingly productive and exciting area of research revolving around ICR spectrometry [16-18] and setting the stage for the introduction of FTMS. A review of this literature is beyond the scope of the present article but can be found elsewhere [5, 19]. In keeping with their interests, Baldeschwieler, and subsequently his students, demonstrated the versatility of ICR spectrometry for monitoring fundamental ion–molecule reactions. The millisecond time scale required for the ions to drift through the cell is sufficient to permit reactive and unreactive collisions with the neutral gas present as the pressure is increased above about 10^{-6} torr. A particularly important breakthrough in ICR methodology was the discovery of the double resonance technique that permits unambiguous determination of reactant and product ion relationships in a way superior to other mass spectrometric methods [17, 20]. This method opened the door to unraveling the very extensive chemistry observed even for relatively simple ion–molecule systems and has yielded a wealth of information [21-25]. Other highlights from the ICR literature include the determination of intrinsic acid base properties of molecules [26-33], the determination of isomeric ion structures by selective ion–molecule reactions [34-37], the study of the reactivity of transition metal complexes [38-41], the photochemistry of gas phase cations [42, 43] and anions [44], theoretical and experimental treatments of the kinematics of reactive and unreactive collisions [45-49], and the report of a tandem Dempster-ICR apparatus [50, 51]. Although this is not an exhaustive list, it provides an indication of the scope of this work and, as we shall discuss below, is being pursued using FTMS. Finally, the single most important breakthrough in ICR spectrometry, which was essential for the development of FTMS, was the successful demonstration of a trapped ion cell by McIver in 1970 [52] and an alternative design by McMahon and Beauchamp in 1972 [53]. Interestingly, this breakthrough did not really involve developing a new type of cell (i.e., it was still a simple "box") but simply was a change in the way the dc and rf potentials were applied. In short, the ions were no longer made to drift through the cell in milliseconds but were permitted to maintain their enclosed trajectories on the order of seconds. Pulsed detection permitted the accurate determination of residence times. Not only did the trapping cell greatly simplify the determination of reaction kinetics, but the lower pressures required in the instrument allowed improved resolution. Photochemical experiments were also facilitated by longer trapping times, as well as the accurate determination of acid–base equilibrium constants.

Despite these breakthroughs, in the early seventies ICR spectroscopy pretty well remained a physical organic or inorganic chemist's tool. Although it was unquestionably one of the most powerful methods for studying ion–molecule reactions, it was still perceived as a low-mass, low-resolution, and slow instrument of limited analytical use. Improved resolution and speed were demonstrated by McIver and co-workers using the trapped ion cell [54], but this was not sufficient to start "the revolution"—although the seeds were there. At a time when the computer was finding increased use in the laboratory and facilitating the application of Fourier transform methods in nuclear magnetic resonance (NMR) and infrared (IR), and with the development of the trapped ion cell, the idea of applying Fourier transform (FT) methods to ICR naturally emerged. But the details of the mechanics of how such an instrument could be constructed provided a sufficient barrier to prevent anyone from trying until, in 1974, Comisarow and Marshall reported the first successful demonstration of FT–ICR [55]. In this landmark paper Comisarow and Marshall pointed out the key advantages of FT–ICR including (1) very high speed, (2) high-signal-to-noise (S/N) and signal averaging, (3) easy operation (i.e., under computer control), (4) easy mass scale calibration, (5) high to ultrahigh resolution, (6) choice of electron impact or chemical ionization by computer control, and (7) retention of the important capabilities of conventional ICR spectrometry including the double resonance technique. Each of these topics will be discussed in greater detail below.

Inertia and the lack of a commercially available instrument prevented what might have been a rapid expansion of the FT–ICR technique in the late seventies, although progress was made on a few homebuilt instruments. In 1981, however, Nicolet introduced the first commercial instrument, the FTMS-1000, at the Pittsburgh conference. The use of the acronym FTMS, as opposed to FT–ICR, was done with purpose to underscore the notion that this was a high-performance analytical instrument not to be confused with the low-performance ICR instrument. Today Spectrospin, a subsidiary of Bruker, Finnigan, and several small companies, are developing or have introduced FTMS instruments or related equipment, and Nicolet has introduced its next generation spectrometer, the FTMS-2000. Thus, it is a very exciting time to be in the area of FTMS and a particularly good time to review where the field is now and where it is going.

3 PRINCIPLES OF OPERATION

3.1 Signal Generation

Detailed treatments of the major principles of operation associated with FTMS have been described by Comisarow, Marshall, and others [55–72]. These studies address such complicated issues as ion motion, sources of signal and noise, and the theoretical limits of the technique. Although such information is of critical importance in the optimization and development of FTMS, fortunately the basis of FTMS can still be described in relatively simple terms [73].

As with ICR spectroscopy, the heart of the FTMS instrument is the cubic

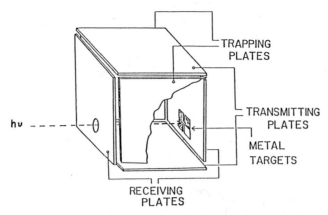

Fig. 1. Schematic view of a cubic trapping cell modified for laser desorption from metal targets.

trapping cell shown schematically in Fig. 1. This simple geometry, used by Comisarow [64] and by Nicolet, has yet to be replaced by anything of superior capability, although a cylindrical cell is used in the Bruker instrument [5] and there is reason to believe a hyperbolic cell may prove advantageous [74]. The cell is placed in a vacuum chamber operated at 10^{-5} torr and lower and positioned in a high magnetic field B (from 1 to 7 T is typical). Once an ion (or any free charged particle) is generated in the cell, it is constrained to move in a circular orbit of angular frequency ω perpendicular to B with no constraints on the ion parallel to B. As in the original omegatron, ions are prevented from escaping axially by applying a small dc electric field (e.g., 1 V/cm) across the trapping plates. Positive or negative ion detection is selected simply by applying positive or negative potentials to the trapping plates. The frequency of the cyclotron motion is independent of velocity and is given in practical units (hertz) by Eq. 1, where z is the unit charge of the ion, B is the magnetic field strength in tesla, and m is the ion mass in atomic mass units:

$$\omega = \frac{1.537 \times 10^7 zB}{m} \quad (1)$$

If an rf field having the same frequency as the cyclotron frequency of the ion is applied differentially across the "transmitter" plates of the cell, there is a resonance effect and the ion absorbs energy, causing its orbital radius and velocity to increase but without a change in frequency. Ions of the same mass-to-charge all orbit at the same frequency before excitation, but are in different spatial positions in their orbits, and are thus out of phase with each other. After an rf excitation pulse, not only have the ions absorbed energy, increasing their kinetic energies and orbital radii, but the ions after only a few rotations are all moving together coherently. It is this coherent motion that gives rise to a signal. The "packet" of ions attracts electrons to whichever "receiving" plate it is

Fig. 2. Schematic depiction of the basic principles of FTMS. Coherent motion of ions in the cell produces an image current signal that is digitized and then Fourier transformed. The result is a frequency or mass spectrum.

approaching, establishing an image current on the two plates. One very interesting consequence of the physics, as pointed out by McIver, is that a packet of 10^5 ions (typical of the number stored in the trapping cell) traveling between two flat plates 2 cm apart at 10^4 m/s can induce an image current of $\sim 10^{-8}$ A corresponding to about 5×10^{10} electrons. Thus, an amplification of 5×10^5 occurs! This signal would normally last only until the ions collided with the plate and were neutralized (2 μs). However, the orbital motion of the ions arising from the magnetic field not only prevents this from occurring but causes a "long-lasting" oscillating signal. As shown in Fig. 2, this image current provides the time domain information that is then digitized and fast Fourier transformed to yield the frequency domain spectrum. A complete mass spectrum is obtained by applying a 0–2 MHz rf pulse, referred to as an rf "chirp", for about 1 ms. This excites all of the ions in a coherent fashion, producing an image current consisting of all of the superimposed frequencies. The Fourier transform yields the frequency spectrum and thus the mass spectrum. Fast Fourier programs [75] can perform the transform at about 1 s per 1 K of data points collected, and the use of array processors can increase the speed 10 fold or more. The intensities of the peaks, as with a conventional mass spectrum, are directly related to the concentrations of the ions present.

3.2 Resolution

One of the truly remarkable features of FTMS is its ultrahigh resolution capability first noted by Comisarow and Marshall [55, 58]. As a direct consequence of the uncertainty principle, the longer the image current transient is sampled, the higher the resolution. This also means sensitivity goes up with resolution as the signal is sampled for a longer time [63, 69], in contrast to conventional mass spectrometry where resolution is achieved by narrowing slits and losing sensitivity! In addition to sampling time t (seconds), Eq. 2 shows that resolution also increases with unit electronic charge z and magnetic field B (tesla) but is inversely related to mass m (atomic mass units):

$$R = \frac{m}{\Delta m} = \frac{\omega}{\Delta \omega} \qquad (\Delta\omega : \text{FWHH}) \leqslant 1.7 \times 10^7 \frac{zBt}{m} \qquad (2)$$

Although many factors can affect the image current decay rate, the single most important mechanism is collisional damping. A collision of one of the ions in the packet causes that ion to no longer be moving coherently with the remainder of the ions in the packet. Thus, with high pressure in the cell ($> 10^{-5}$ torr), the signal decays rapidly (on the order of milliseconds), producing a low-resolution spectrum, while at low pressures ($< 10^{-8}$ torr) the transient may last for tens of seconds producing an ultrahigh-resolution spectrum. The specific effect of pressure on resolution is shown in Eq. 3, where ξ/n is the collisional dampening frequency (typically 1×10^{-9} cm^3 molecule^{-1} s^{-1}) and P is the pressure (torr):

$$R = \frac{m}{\Delta m} = \frac{\omega}{\Delta \omega} \leqslant 8.65 \times 10^{-10} z \frac{B}{mP}\left(\frac{\xi}{n}\right) \tag{3}$$

Thus, collisional dampening renders the experiment useless above $\sim 10^{-3}$ torr. It is interesting to note that unlike FTMS, there is no gain in resolution on going from NMR to FTNMR.

One other interesting feature that FTMS and FTNMR do share in common is that the time-domain data, once collected, can be mathematically manipulated or "apodized" to achieve a variety of different objectives [4, 76]. For example, as shown in Fig. 3, multiplying the time-domain data with a decreasing

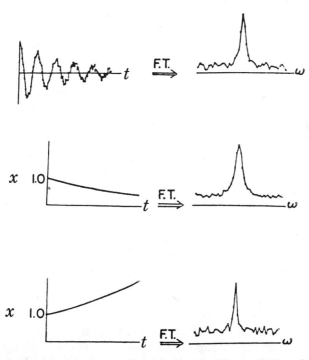

Fig. 3. Signal-to-noise ratio or resolution enhancement produced by apodization of a time-domain transient signal. (Reprinted from Ref. [250] with permission.)

exponential gives an FT spectrum with improved signal-to-noise ratio since the most intense time-domain data have been weighted most strongly. Similarly, multiplying the digitized transient by an increasing exponential effectively increases the length of the time-domain signal while weighting more heavily the low-intensity data points, resulting in an FT spectrum of improved resolution, but worse signal-to-noise ratio. Apodization has also proved to be an effective method for narrowing the "wings" (i.e. baseline peak width) resulting from the Lorentzian peak shape that often obscures the presence of a nearby low-intensity ion.

Assuming a low pressure is used and a long signal transient is achieved, the resolution is no longer limited by the fundamental physics but rather by the instrumentation, in particular the computer memory size. Say, for example, a mass spectrum is to be obtained by applying the rf chirp from 0 to 2 MHz (at $B = 3T$, this would correspond to a mass range of $23-\infty$ amu.) The Nyquist criterion requires that the resulting image current signal be digitized at a rate twice the frequency of the highest frequency to be observed; in this case, an analog-to-digital conversion (ADC) rate of 4 MHz would, therefore, be necessary. Given a 256 K fast memory capacity, data could only be collected for 64 ms despite the fact that the transient may have barely begun to decay at that point! A clever solution to this problem is to perform a heterodyne experiment, in analogy to what is done in laser spectroscopy, by mixing the transient response with a reference frequency. This results in the production of a high-frequency signal, which is the sum of the transient and reference frequencies, and a low-frequency signal, which is the difference. The high-frequency signal is filtered out, and the resultant difference or low-frequency signal is digitized, but now requiring a substantially reduced A to D sampling rate ($2 \times$ highest difference frequency). The disadvantages of this heterodyne mode (also referred to as narrow band or mixer mode) are that only a small portion of the mass spectrum centered around the reference frequency can be observed at a time and that the higher the resolution desired, the narrower the mass range observed. Thus, under these conditions the technique begins to lose its multichannel advantage [55, 77]. Solutions to this problem today include simultaneously mixing in several reference frequencies to observe several portions of the spectrum at high resolution at once [78a] and the use of multiplet foldover techniques [78b], but in the future the development of ever-increasing fast memory storage, as well as computing speed, should correct this difficulty.

In a recent new development, Rahbee has successfully demonstrated the use of a maximum entropy method (MEM) of spectral analysis that appears to have a number of significant advantages [79, 80]. In what would appear to defy the uncertainty principle, higher resolution and signal-to-noise ratio can be achieved using the MEM method on the first 100 points of the digitized transient than using fast Fourier transform (FFT) on the first 16 K data points. Figure 4 provides a convincing example. One possible drawback is that in general MEM takes roughly twice as much computation time as does FFT. In addition, this method results in a limited precision imposed by the short acquisition

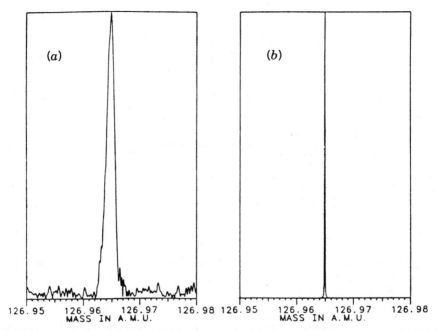

Fig. 4. Spectrum of the SF_5^+ ion. (a) FFT spectrum using 8 K data points with a resolution (FWHH) of 8.3×10^4; (b) MEM spectrum using starting data point $S = 13$, number of data points $N = 32$, and $M = 16$ filter elements. Resolution is 1.035×10^6. (Reprinted from Ref. [80] with permission.)

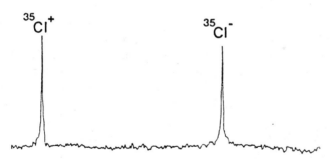

Fig. 5. Mass spectrum of $^{35}Cl^+$ and $^{35}Cl^-$ at a pressure of 10^{-8} torr and a mass resolution of $\sim 1.5 \times 10^6$. The Cl^+ and Cl^- ion signals were alternately acquired, stored, and transformed. The FTMS cannot simultaneously store positive and negative ions. (Reprinted from Ref. [81] with permission.)

period. Nevertheless, it is evident that this new approach could provide some exciting new capabilities.

Two of the most impressive examples of high resolution come from Karl-Peter Wanczek's laboratory in Bremen using the Bruker CMS-47 instrument at 4.7 T. Figure 5 shows a mass spectrum of $^{35}Cl^+$ and $^{35}Cl^-$ at a pressure of 10^{-8} torr and a mass resolution of $\sim 1.5 \times 10^6$, with half the distance between the two

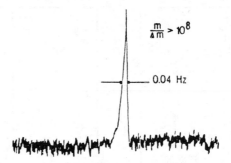

Fig. 6. Mass spectrum of H_2O^+ (m/z 18) taken at a pressure of 8×10^{-11} torr from one scan digitized for 51 s. The resolution, $m/\Delta m > 10^8$, is currently the world's record on any mass spectrometer. (Reprinted from Ref. [72] with permission.)

peaks providing an accurate measure of the mass of an electron [81]. Figure 6 shows a spectrum of H_2O^+ (m/z 18) taken at a pressure of 8×10^{-11} torr from one scan digitized for 51 s at a resolution $m/\Delta m > 10^8$ [72]. On the scale shown in the figure, m/z 19 would be about 2 miles away! In addition, a resolution of 10^8 is sufficient in theory to resolve He^+ in its ground state from He^+ in its first excited state; i.e., the resolution is equivalent to the rest mass of a 21-eV photon. Conceivably, the gain of another order of magnitude in resolution could allow bond energies to be determined directly from mass-defect measurements. Currently in progress in at least three laboratories is an effort to determine by FTMS whether the neutrino does or does not have a finite rest mass by measuring the mass difference between $^3He^+$ and $^3H^+$, requiring a resolution of $> 10^9$. Two Soviet groups were the first to perform this experiment using FTMS, but their results are controversial [82,83]. These types of experiments require extraordinary care and expertise and at the present time are by no means routine, although resolutions of the order of 10^4–10^6 are easily achievable.

3.3 Mass Accuracy

Whereas it is definitely possible to measure the cyclotron frequency of an ion to one part in 10^9, unfortunately this does not directly translate into mass accuracy. Although the basic cyclotron Eq. 1 is suitable for most routine measurements at unit mass, this equation does not take into account space charge effects (i.e. ion–ion repulsion) and nonideal electric field effects, which become important when pushing the limits of mass accuracy. A number of calibration procedures have been proposed [77,84,85]. One of the most recent and popular is given in Eq. 4 where m is mass, ω is the observed ion frequency, and a and b are calibration constants [71]:

$$m = \frac{a}{\omega_{obs}} + \frac{b}{\omega_{obs}^2} \qquad (4)$$

This method was derived taking into account the space charge effects of the model ion distributions of Jeffries et al. [86] and assumes that the ion space charge potential is time invariant and quadratic. To test this model, six reference masses from 1,1,1,2-tetrachloroethane over a mass range from 117 to 135 amu were used to obtain empirical values of a and b [71]. Next, observed ion frequencies were used to calculate theoretical masses with Eq. 4. The results were very good with 15,000 ions in the cell, where the average error was +0.4 ppm and the precision was ±4 ppm, but worsened substantially with 120,000 ions in the cell, where the range of residual errors changed from 11 to 36 ppm. Clearly, this model does not fully account for space charge effects. However, one feature of the model is that errors go down with magnetic field and, for example, an error of 36 ppm at 1.2 T should shrink to 1 ppm at 7.2 T. Efforts are underway to reduce space charge effects by construction of larger cells [85] with concomitant development of more sensitive ion image current amplifiers. In addition Ledford and Rempel [74] have developed an interesting hyperbolic cell that yields good results with 650,000 ions that are comparable to the cubic cell with 15,000 ions. Wang and Marshall have suggested that a greater understanding of the space charge effect could be obtained by calculating ion trajectories [67]. By following the trajectories of two ions excited to larger orbits by rf excitation, they found that the mass resolution increases linearly as the ion orbit increases, until the Lorentzian force overwhelms the coulomb repulsive force. Besides confirming the desirability of a larger cell, their work suggested that nonflat excitation power to excite ions of different mass to different radii might improve resolution.

Needless to say, this important issue remains the subject of further intense investigation in a number of laboratories, some of which will be summarized here. White et al. demonstrated the earlier prediction that the stability of FTMS would permit accurate mass assignment in the absence of a calibration gas [87]. In this work the heterodyne mode was used to monitor very narrow regions of the mass spectrum (e.g., a 2 amu range at m/z 100 corresponding to 6 kHz). An average error of 6 ppm was obtained on nine compounds, and optimum results were achieved by acquiring spectra of the calibrant gas and the unknowns at as nearly identical conditions as possible, in particular by trying to match ion densities in the cell. However, corrections can be applied to obtain accurate mass measurements when unknowns are run under different conditions than the calibrant [87]. A good deal of work has been done in conjunction with gas chromatography (GC)–FTMS, and in general 5–10 ppm errors have been demonstrated in wide-band experiments ranging from about 40 to 300 amu [88, 89].

Interfacing desorption methods (described below) with FTMS to obtain spectra of nonvolatile high molecular weight compounds is currently an active area of research for which mass accuracy is also being evaluated. The results are very encouraging, again mostly in the low ppm error range. Marshall and co-workers, for example, reported a calibration table produced from laser desorbed molecular ions of four clinical drugs ranging from 404 to 819 amu, with an average mass error of 2.4 ppm [90]. An average error of 3 ppm for several compounds ranging up to 1180 amu has been obtained by workers at Nicolet using Cs^+ and laser desorption techniques [91].

Allemann et al. noted that sidebands resulting from the coupling of ion cyclotron and drift frequencies could be used with appropriate corrections to achieve errors in the 1–2 ppm range from 18 to 170 amu, good for several days [92]. Similar stabilities have been noted by White et al. [87].

One other important consideration in obtaining accurate mass measurements at the ppm or better level is the digital resolution. The same number of data points acquired for the transient is used in plotting the frequency or mass spectrum. In a wideband experiment covering a full mass range, this may mean that a peak in the spectrum is defined by only a few points, and there is a likelihood that there is no point exactly at the true peak maximum. Once again, using the narrow band or heterodyne mode yields the expected improved results, but over a small mass range. In addition the technique of zero filling provides a means for circumventing this problem but requires increased computing time [93, 94].

In summary, the present state-of-the-art mass accuracy is generally in the low ppm range using a variety of different ionization techniques and, thus, quite comparable with conventional high-performance sector instruments. Some of the limitations such as digital resolution, space-charge effects, and nonideal electric fields are under active investigation, and future improvements can be anticipated.

4 APPLICATIONS

4.1 Mass Spectrometry/Mass Spectrometry

One of the techniques currently on the forefront of mass spectrometry research is mass spectrometry/mass spectrometry, MS/MS by analogy to GC–MS [95–97]. Figure 7 illustrates one of the basic differences between performing the MS/MS experiment with the more conventional tandem instrument and by using FTMS. In a tandem MS/MS experiment the ions are manipulated in space. An ion generated in the source is selected by using the first mass spectrometer (sector or quadrupole), MS_1, and undergoes a collision in the reaction chamber. The product ions from that collision are analyzed by using a second mass spectrometer, MS_2. This is also the basis of the popular triple quadrupole instrument [97], as well as for those instruments using magnetic and electrostatic analyzers [95, 96]. Clearly, additional sectors or quadrupoles would be required for MS/MS/MS or further combinations. In contrast, in FTMS the ions are constrained in space and the experiments are carried out in time. A typical pulse-timing sequence is also shown in Fig. 7. Following ion formation (by whatever means), a series of rf pulses (often referred to as double resonance pulses from the original ICR technique) of varied amplitude, frequency, and duration occur to manipulate the ions (as discussed below), and finally a detection pulse permits a "snapshot" of the products

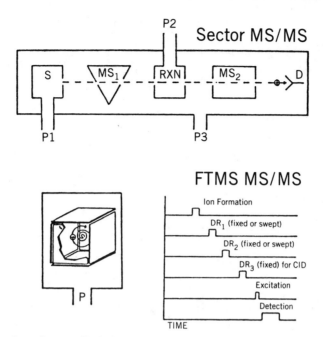

Fig. 7. Comparison of a generalized tandem mass spectrometer and FTMS. Also shown is a typical FTMS pulse timing sequence.

generated. The key advantage of this methodology versus conventional tandem instrumentation is that the pulse sequence can be readily modified by the user at the computer keyboard and, therefore, different experiments (MS/MS, MS/MS/MS, etc.) require changes only in *software* as opposed to *hardware* [6,98].

All of the rf pulses are generated from a frequency synthesizer that is under computer control. This device literally calculates voltage as a function of time to achieve the desired frequency and sends the signal to the cell. The radio frequency from the synthesizer is used for three main functions: (1) for detection as described above, (2) for ejection, and (3) to drive endothermic reaction processes such as collision-induced dissociation (CID). Ejection is achieved when a high-amplitude rf signal is applied at the cyclotron resonance frequency of an ion, causing its orbit to expand rapidly and resulting in the annihilation (neutralization) of the ion against one of the cell plates. Ion ejection was first used in conjunction with conventional ICR, where it was also referred to as double resonance ejection [20]. A single frequency can be applied to eject one ion selectively, or a swept frequency can be applied to eject a range of ions. As was the case with ICR, FTMS ion ejection is a particularly powerful tool for determining reactant ion/product ion relationships in ion–molecule reaction chemistry. Comisarow et al. [99] pointed out that FTMS ejection has the advantage over the technique as applied originally to ICR in that all of the

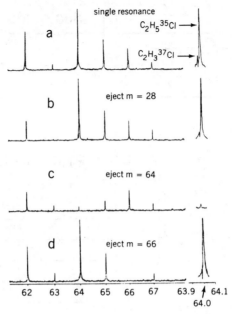

Fig. 8. Single resonance (a) and double resonance (b–d) FTICR spectra for ethyl chloride. All spectra were obtained with an ethyl chloride pressure of 5.8×10^{-7} torr and a reaction time of 60 ms. The inserts show the region around $m = 64$ with the horizontal scale expanded five times to show clearly the mass separation between $C_2H_3^{37}Cl^+$ ($m = 63.989378$) and $C_2H_5^{35}Cl^+$ ($m = 64.007979$). (a) The normal FTMS mass spectrum; (b) The FTMS mass spectrum with $C_2H_4^+$ ejected before the reaction time period of 60 ms; (c) The FTMS mass spectrum with $C_2H_5^{35}Cl^+$ ejected before the reaction time period. $C_2H_3^{37}Cl^+$ is not ejected in this experiment because it is a product ion that is formed after the ejection time period. The intensity of $C_2H_3^{37}Cl^+$ is reduced from that in (a) because some of this ion is formed from $C_2H_5^{35}Cl^+$. (d) The FTMS mass spectrum with $C_3H_5^{37}Cl^+$ ejected before the reaction time period of 60 ms. (Reprinted from Ref. [99] with permission.)

daughter and subsequent granddaughter ions arising from a precursor ion can be determined by a single experiment, as opposed to the multiple experiments necessary using the older technique. Figure 8, taken from Comisarow's original paper, provides an excellent example of the use of ion ejection. By comparing the spectra in Figs. 8a and 8b with and without ejection, missing peaks in the latter must arise from the ejected precursor. Additional experiments on other ions permit us to chart out in detail the ion–molecule reaction "family tree." A simple variation on this experiment is to eject all but one ion, as opposed to ejecting only one ion, and then monitor its chemistry [100, 101]. By performing this selected ion isolation, the reactant ion/product ion relationship is even more unambiguous with regard to initial product ion ratios and low-intensity product ions that might be overlooked by single ion ejection experiments. Figure 9 provides an example of this experiment, and additional examples will be given below. Finally, ejection of more abundant ions in order to enhance

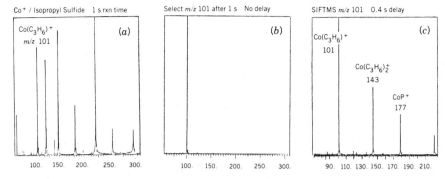

Fig. 9. Selected ion reaction capabilities in FTMS. Spectra shown result from the reaction of laser desorbed Co$^+$ with isopropylsulfide after (a) 1 s reaction time; (b) mass selection of the product ion at m/z 101, Co(propene)$^+$; (c) further reaction of mass selected Co(propene)$^+$ with isopropylsulfide. Note that the spectra are normalized to the biggest peak. (Taken from Ref. [206].)

the detection of less abundant ions can be used to extend FTMS dynamic range as described below [102, 103].

Collision-induced dissociation, whereby an ion is accelerated into a target gas resulting in the ion's fragmentation, remains one of the most powerful and certainly most widely used tools in mass spectrometry for ion structure determination and as an integral part of the MS/MS technique. Collision-induced dissociation is readily accomplished in a Fourier transform mass spectrometer by, once again, applying an rf pulse at the frequency appropriate for an ion of a particular mass [104, 105]. The amount of kinetic energy imparted to an ion in FTMS is clearly limited to being less than that required to eject the ion from the cell totally. An upper limit for the translational energy of the ion undergoing CID may, therefore, be calculated. The maximum time that an ion may be irradiated, assuming that it is located in the center of the cell, is given by Eq. 5, where t_{max} is the time in seconds, r the cell radius in meters, B the magnetic field in tesla, and E_{RF} the electric field in volts per meter:

$$t_{max} = \frac{2rB}{E_{RF}} \qquad (5)$$

For a magnetic field of 3 T, an rf amplitude of 40 V, and a distance of 0.0508 m (i.e., 2-in. cell) between the cell plates, which are typical parameters, the maximum irradiation time is calculated to be 0.19 ms. The maximum translational energy acquired by an ion in excess of thermal energy is given by Eq. 6, where $E_{tr(max)}$ is the translational energy in electron volts, e is the electronic charge, and m is the mass in kilograms:

$$E_{tr(max)} = \frac{E_{RF}^2 e^2 t_{max}^2}{8m} = \frac{e^2 r^2 B^2}{2m}. \qquad (6)$$

Table 1. Maximum Translational Energies Calculated from Eq. 6 for a 2-in. Cubic Cell

	Energy (eV)		
$B\,(T)$	1	3	7
m (atomic mass units)			
10	3.0×10^3	2.7×10^4	1.5×10^5
100	3.0×10^2	2.7×10^3	1.5×10^4
1,000	30	2.7×10^2	1.5×10^3
10,000	3.0	27	1.5×10^2

From this relation, it is evident that the maximum attainable translational energy is proportional to both the square of the cell radius and the square of the magnetic field, inversely proportional to the ionic mass, and independent of the rf amplitude used. Table 1 illustrates the range of kinetic energies attainable at various magnetic fields and masses for a 2-in. cell. For an ion at m/z 100 and using a 7-T superconducting magnet, as shown in the table, the FTMS experiment is capable of spanning from the low-energy CID regime ($< 100\,\text{eV}$) to the high-energy regime ($> 1\,\text{keV}$). Note that an ion of mass 10,000 amu would only attain a translational energy of $\sim 3\,\text{eV}$ before ejection using a 1-T magnet and, thus, the advantage of a larger cell and a larger magnetic field for studies involving high mass ions is evident.

Interestingly, CID was observed early on using ICR spectroscopy [106, 107] but, once again, was too cumbersome to be useful. McIver was the first to go on record to state that the CID technique was possible using FTMS [108], and shortly thereafter Cody et al. provided the first convincing results on the parent molecular ions of α,α,α-trifluorotoluene and acetone [104, 105]. The spectra obtained from trifluorotoluene are shown in Fig. 10. In this experiment the sample served as its own collision gas at a pressure of $\sim 2 \times 10^{-5}$ torr. The major fragments observed in Fig. 10b include the loss of H(m/z 145), F(m/z 127), and CF$_2$(m/z 96) and correspond to the major low-energy fragmentation pathways observed by a number of other techniques [109]. A more extensive comparison of FTMS–CID to a hybrid magnetic-sector/quadrupole (BQQ) instrument at low energies ($< 100\,\text{eV}$) by McLuckey et al. showed good qualitative agreement for the fragmentation of several compounds [110].

Following the introduction of FTMS–CID, it was not difficult conceptually to realize that MS/MS, MS/MS/MS, and so forth, would be possible by simple modifications in software as opposed to hardware [98, 108, 111, 112]. Figure 11 shows the first MS/MS/MS spectra reported using FTMS. Kleingeld [111] has demonstrated (MS)^5 at 1.4 T by sequentially dissociating the m/z 120 ion of acetophenone and its fragments in the following order: $120 \rightarrow 105 \rightarrow 77 \rightarrow 51 \rightarrow 50$. Although the number of MS/MS steps is ultimately limited by the eventual ejection of ions as they are excited closer and closer to the cell plates, at high magnetic fields one is likely to run out of ideas before one runs out of MS/MS

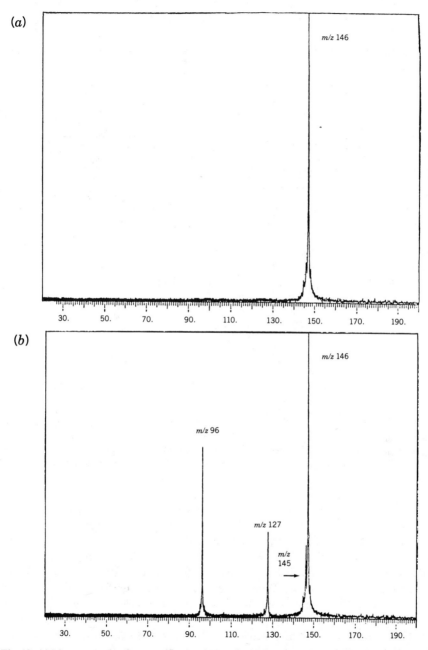

Fig. 10. (a) Mass spectrum of α,α,α-trifluorotoluene obtained at a nominal ionizing electron energy of 13 eV and at a total gas pressure of 2×10^{-5} torr. (b) Same conditions as Fig. 10a except that $C_6H_5CF_3^{+\cdot}$ (m/z 146) is irradiated near its cyclotron frequency for 10 ms before detection. The fragments that are observed arise from the collision-induced dissociation of $C_6H_5CF_3^{+}$. (Reprinted from Ref. [105] with permission.)

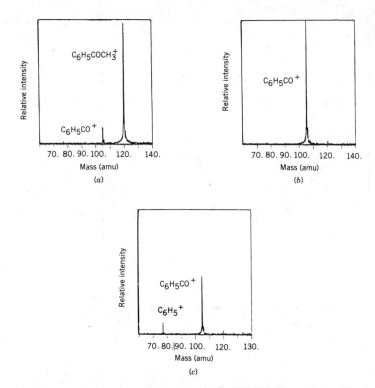

Fig. 11. (a) Background spectrum for the acetophenone MS/MS/MS experiment. The DR_1 pulse is swept to eject fragment ions produced during the 10.6-eV formation pulse. The peak at m/z 105 arises from "uncontrolled" CID resulting from inadvertent excitation of the molecular ion. (b) Same conditions as (a), with the exception that the molecular ion is irradiated with the DR_2 pulse and undergoes CID to increase the relative abundance of the m/z 105 ion. (c) Same conditions as (b), but now the ions of m/z 105 produced by the first CID stage are accelerated by the DR_3 pulse and undergo CID to produce a peak at m/z 77. (Reprinted from Ref. [98] with permission.)

steps! For example, more than $(MS)^{20}$ should be possible at 7 T! Certainly the capability of FTMS to perform $(MS)^n$ has helped the technique gain a higher visibility in the mass spectrometry community.

Applications of FTMS–CID and MS/MS have, as with the general mass spectrometry literature, been expanding rapidly over the past few years [1–10, 113, 114]. White and Wilkins demonstrated its use for mixture analysis by analyzing a five component mixture that had previously been studied by Yost and Enke on a triple quadrupole apparatus [115]. Some differences in the observed fragmentations were noted between the two techniques. A mass resolution of 4000 FWHH, achieved in the direct mode, was sufficient to distinguish the two isobars at m/z 43, CH_3CO^+ and $C_3H_7^+$, and provided additional information for the characterization of the components in the mixture. In another application the same workers were able to identify a major component in a commercial lacquer thinner by MS/MS [115].

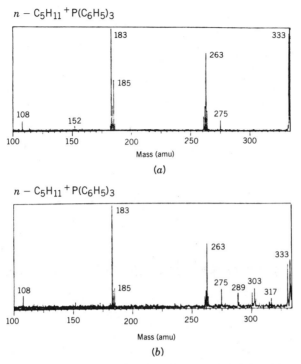

Fig. 12. Low-energy CID spectra of daughter ions from n-pentyltriphenylphosphonium cations generated by laser desorption. Spectra were obtained using (a) He and (b) Ar as the collision gas. Mass given in amu. (Reprinted from Ref. [116] with permission.)

Many of the methods for desorbing high molecular weight species such as laser desorption and Cs^+ desorption produce simple spectra containing little else besides the molecular ion or pseudomolecular ion [i.e. $(M + H)^+$ or $(M + C)^+$ where C = metal such as Li, Na, K, Ag], although this is dependent on the nature of the sample. Thus, MS/MS provides an important means for obtaining structural information. Despite the already large number of FTMS desorption spectra reported, however, surprisingly few workers have included CID results, but this trend will certainly change. McCrery et al. [116] obtained good results on the CID of a variety of laser-desorbed alkyltriphenyl phosphonium ions (Fig. 12), and Cody et al. reported impressive results on laser-desorbed gramicidin S, a cyclic peptide of molecular weight 1141 amu, and on gramicidin D [117]. For gramicidin D, the actual ion dissociated was the K^+ adduct at 1920 amu (Fig. 13).

By far and away the most frequent use of FTMS–CID has been for elucidating the structures of ions resulting from ion–molecule reactions in an effort to obtain mechanistic information. Organometallic ion chemistry has been the focus of a good deal of attention over the last few years [118]. As one of the first examples of FTMS–CID, Cody et al. [105] reported that the Cu^+-

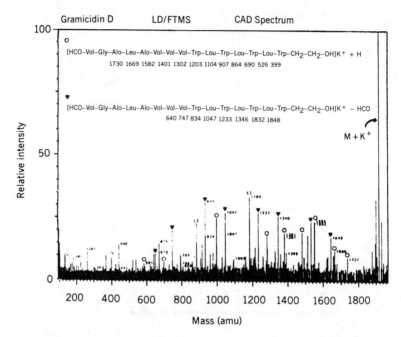

Fig. 13. The CID spectrum of gramicidin D is obtained by accelerating the $[M + K]^+$ adduct ion (m/z 1920) produced by laser desorption FTMS. Amino acid sequence information for the peptide is included. Mass given in amu. (Reprinted from Ref. [117] with permission.)

methylacetate adduct ion undergoes the simple cleavage reaction 7, suggesting an intact methylacetate structure:

$$Cu(CH_3CO_2CH_3)^+ \xrightarrow[Ar]{CID} Cu^+ + CH_3CO_2CH_3 \qquad (7)$$

CID was used to distinguish four isomeric $NiC_4H_8^+$ ions generated from the reactions of Ni^+ with four different neutral precursors [119, 120]. As shown in Table 2, the bis-ethene ion loses one and two ethenes upon CID, while the butene complex loses H_2 to form a stable butadiene complex, and so on. The particular significance of these results was that they provided additional direct evidence regarding the reaction mechanism of bare atomic metal ions. FTMS–CID has been used extensively to assign structures from the reactions of a variety of transition metal ions, metal ion complexes, and metal cluster ions with various organic molecules, particularly branched, linear, and cyclic alkanes and alkenes [119–140].

The study of organic ion structures has also been extensive [115, 141–146]. Quasibreakdown curves (i.e., product ion distributions versus parent ion translational energy) were obtained for a series of proton-bound alcohol dimer ions [141]. The ease of dehydration of the dimers following collisional activation

Table 2. Neutral Losses from CID of $NiC_4H_8^+$ Complexes[a,b]

Structure	H_2	C_2H_4	C_4H_8
$\|\|-\overset{+}{Ni}-\|\|$		X	X
$\overset{+}{Ni}-\|\|\diagup\!\!\diagdown$	X		X
$\overset{+}{Ni}\diagup\text{(cyclobutane)}$	X	X	X
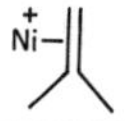 $\overset{+}{Ni}-\diagup\!\!\!\diagdown\diagup$			X

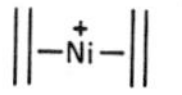

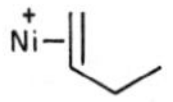

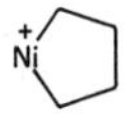

[a]CID fragments observed at 15-eV kinetic energy.
[b]Argon added for total pressure of 1×10^{-5} torr.

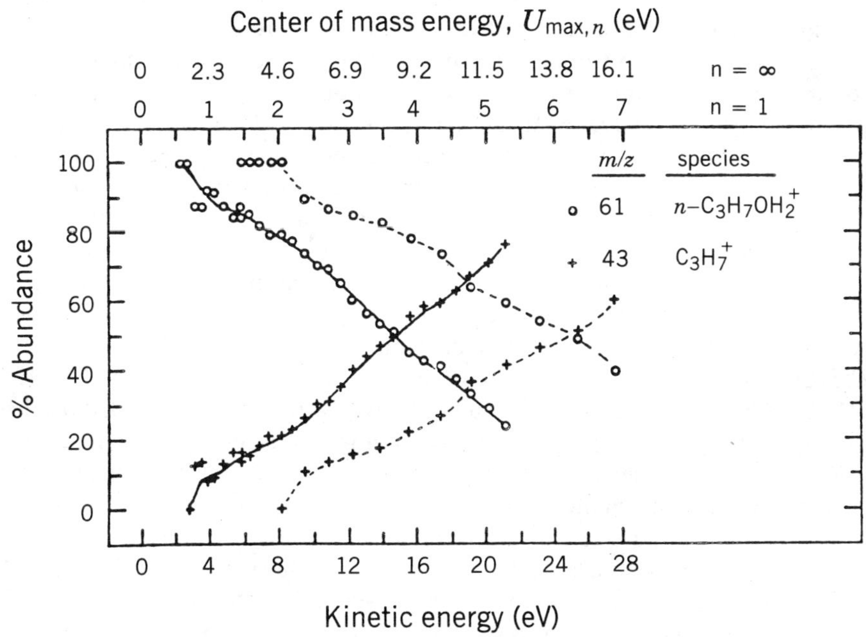

Center of mass energy, $U_{max,n}$ (eV)

m/z species
○ 61 $n\text{-}C_3H_7OH_2^+$
+ 43 $C_3H_7^+$

Kinetic energy (eV)

Fig. 14. Product ion distribution from the proton-bound dimer of n-propyl alcohol, $(n\text{-}C_3H_7OH)_2H^+$, as a function of the reactant ion kinetic energy. Solid curve obtained with a 40-ms CID interaction time. Dashed curve obtained with a 4-ms interaction time. (Reprinted from Ref. [141] with permission.)

was found to follow the same order as their acid-catalyzed dehydration in solution with tertiary alcohols > secondary > primary. In this study the effect of multiple collisions in FTMS–CID was also addressed. As shown in Fig. 14, the product ion distribution from the proton-bound dimer of n-propyl alcohol as a function of the reactant ion kinetic energy shifts dramatically to lower energy when a longer CID interaction time is used, suggesting that multiple collisions can in fact place more energy into an ion than a single collision. These workers felt that although multiple collisions can shift the breakdown curves to lower energy, they do not radically alter the shapes of the curves. Froelicher and co-workers conducted a systematic study of the effect of structural changes on the low-energy CID of ester–enolate ions [143].

One limitation noticeable for most MS/MS instrumentation is the resolution obtainable for the detection of CID fragment ions. This is often termed MS-II or "back-end" resolution in order to distinguish it from MS-I or "front-end" resolution, which is the resolution available for selecting the parent ion to undergo CID. Reverse geometry instruments suffer from poor (often less than unit mass) back-end resolution due to kinetic energy release. Quadrupole instruments are capable of unit resolution in most cases.

Linked-scanning techniques have been used to improve the back-end mass resolution of double-focusing instruments [147, 148]. Improved resolution of the daughter ions is provided at the expense of the resolution of the parent ions. Boerboom and co-workers have improved resolution in their tandem mass spectrometer by using post-collision acceleration of fragments to reduce the relative energy spread due to kinetic energy release [149]. Of particular note are the double–double focusing instruments first developed by McLafferty and co-workers combining two double-focusing mass spectrometers in series and capable of both high front-end and back-end resolution [150, 151]. Again, it was conceptually simple to realize that high resolution on FTMS–CID product ions could be obtained by maintaining low collision gas pressures and operating in the heterodyne mode. Cody and Freiser reported the first examples of high resolution and exact mass measurement on the CID-produced isobaric ions CH_3CO^+ and $C_3H_7^+$ (Fig. 15) achieving a resolution of ca. 3000 FWHH [152]. Wilkins et al. in a similar study achieved a resolution of 10,000 on the same isobars by using about half the collision pressure and twice the magnetic field used in the earlier study [115]. It is clear that the collision gas pressure cannot be lowered indefinitely to improve resolution without adversely affecting collision efficiency. An important development circumventing this problem was the incorporation of a pulsed inlet valve [153]. By pulsing in the collision gas, CID efficiency can be maximized; and by delaying the detection of the trapped CID fragment ions until the lower ambient pressure is restored in the cell, resolution is also maximized. Carlin and Freiser reported that resolutions of $> 60,000$ could be achieved at m/z 43 by pulsed valve CID (Fig. 15). As realized by several laboratories, the pulsed-valve method is not limited to CID studies but can be applied to achieve high resolution in GC–FTMS experiments [88, 154, 155], in chemical-ionization experiments [88, 153, 154, 156], and in ion–molecule reaction studies [123, 126, 137, 157]. Finally, new instrument geometries in which the ion source and the analyzer are separated and differentially pumped, as discussed below, hold promise for even higher

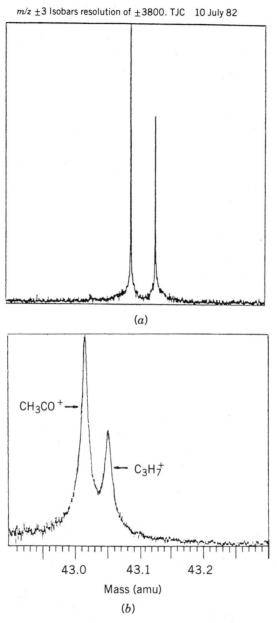

Fig. 15. Comparison of mass resolution on two isobaric ions at m/z 43 generated from CID where (a) the Ar target gas was pulsed permitting detection at 1×10^{-7} torr ($m/\Delta m \sim 44{,}000$) (taken from Ref. [206]) and (b) the Ar target gas was held constant at 6×10^{-5} torr ($m/\Delta m \sim 3000$) (reprinted from Ref. [152] with permission).

resolution. Also discussed below are new excitation methods that will greatly enhance the front-end resolution in CID experiments.

In addition to structure elucidation, FTMS–CID has been found to be a powerful and convenient tool for synthesizing unusual ions in situ. This technique has been used by Freiser and co-workers extensively to generate and study interesting organometallic ions [123, 132, 133, 137, 158, 159], as well as organic ions [144]. Figure 16, taken from Ref. [123], provides a striking example of CID synthesis, as well as the multistep methodology inherent in the FTMS software. In spectrum a, CD_3ONO is pulsed into the instrument concurrently with laser generation of Ni^+. The $^{58}Ni^+$ isotope has been selected, and a variety of primary and secondary reaction products are observed. Next, in spectrum b, all but a selected group of ions have been ejected from the cell. In particular, one of these ions corresponds to $NiOCD_3^+$, generated by reaction 8.

$$Ni^+ + CD_3ONO \longrightarrow NiOCD_3^+ + NO \qquad (8)$$

Selective CID on $NiOCD_3^+$ is evidenced by its absence in spectrum c as well

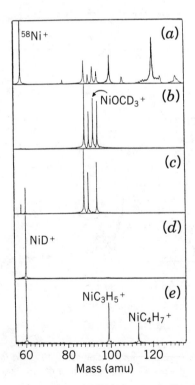

Fig. 16. Multistep synthesis and reaction of NiD^+ (see text). (Reprinted from Ref. [123] with permission.)

as the appearance of two product ions from reactions 9 and 10:

$$\text{NiOCD}_3^+ \xrightarrow[\text{Ar}]{\text{CID}} \text{NiD}^+ + \text{CD}_2\text{O} \tag{9}$$

$$\hookrightarrow \text{Ni}^+ + (\text{CD}_3\text{O}) \tag{10}$$

The NiD$^+$ is isolated, spectrum d, and permitted to react with a background of 2-methylpropane, spectrum e. Interestingly, the products observed from this

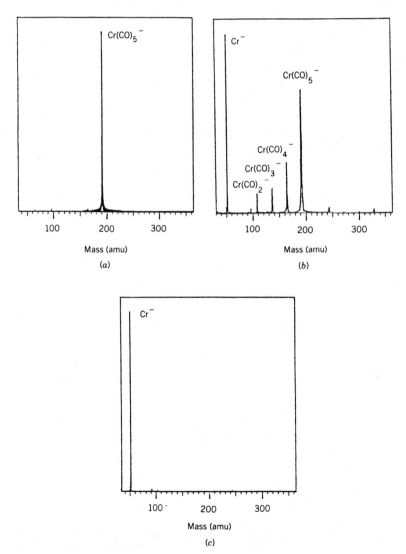

Fig. 17. Multistep synthesis of Cr$^-$ (see text). (Reprinted from Ref. [158] with permission.)

reaction show a complete loss of label. Reactions 8 and 9 were also used to generate and study FeD$^+$ and CoD$^+$ [123]. Sallans et al. [158,159] used CID synthesis to generate and study the transition metal anions V$^-$, Cr$^-$, Fe$^-$, Co$^-$, Mo$^-$, and W$^-$ from the collisional activation of the corresponding metal carbonyl negative ions. As an example, Fig. 17 shows the results obtained from Cr(CO)$_6$, which, like most metal carbonyls, has a large cross section for dissociative electron capture and produces a large signal for Cr(CO)$_5^-$. Next, CID results in the sequential loss of CO ligands producing a variety of products including Cr$^-$. The Cr$^-$ is isolated by ejecting all of the other species present, and its subsequent chemistry can readily be monitored. In particular important thermodynamic data were obtained in this study on the homolytic and heterolytic bond energies of the corresponding metal hydrides. Another interesting example is the formation of metal dimer species by CID as shown for CuFe$^+$ in Fig. 18. The stacked plot shows the sequential loss of carbonyls from CuFe(CO)$_4^+$ as its translational energy is increased, resulting ultimately in the formation of the bare CuFe$^+$. Once generated, the chemistry of the metal dimers can be studied, as exemplified for CoFe$^+$ in Fig. 19 [135]. A variety of metal dimer [132, 135, 137, 160] and trimer [133] ions have been generated and studied in this way. Finally, Froelicher et al. [144] used FTMS–CID to generate the dimethylsilanone enolate ion from trimethylsiloxide anion by reaction 11

$$H_2O \xrightarrow{e^-} OH^- \xrightarrow{(CH_3)_4Si} (CH_3)_3SiO^- \qquad (11)$$
$$\xrightarrow[Ar]{CID} CH_3\overset{\overset{\displaystyle O}{\|}}{Si}{-}CH_2^- + CH_4$$

and found it to exhibit a markedly different chemistry than its carbon analog acetone enolate, CH$_3$COCH$_2^-$.

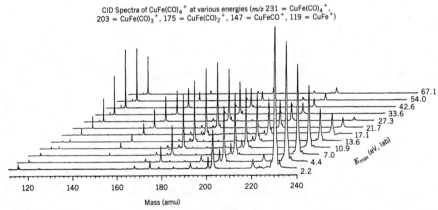

Fig. 18. Stacked plot showing the sequential loss of carbonyls from CuFe(CO)$_4^+$ as its translational energy is increased, resulting ultimately in the formation of bare CuFe$^+$. The CuFe(CO)$_4^+$ was generated by reacting Cu$^+$ with Fe(CO)$_5$.

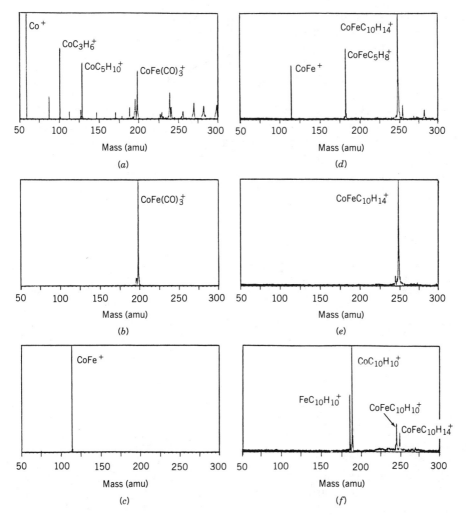

Fig. 19. Multistep synthesis and reaction of CoFe$^+$. (*a*) Reaction of Co$^+$ with a mixture of 1-pentene and Fe(CO)$_5$. The CoFe(CO)$_3^+$ product is isolated, spectrum (*b*), and then its subsequent CID and isolation of the product CoFe$^+$ results in spectrum (*c*). The CoFe$^+$ then reacts with the neutral gas mixture, but it is evident in spectrum (*d*) that it reacts more readily with 1-pentene by loss of H$_2$ than with Fe(CO)$_5$. One of the products, CoFeC$_{10}$H$_{14}^+$, is isolated, spectrum (*e*), and then collisionally activated, spectrum (*f*). The products observed correspond to metallocenes. (Reprinted from Ref. [135] with permission.)

One recently reported variation on FTMS–CID that looks very promising involves exchanging the inert target gas (e.g., Ar) with one that will react with a translationally excited ion in what would normally be an endothermic reaction [161, 162]. Since the kinetic energy of the ion can be varied, reaction thresholds can be determined that yield thermodynamic information about the endothermicity of the process. These types of experiments have been primarily the domain of the ion-beam instruments and have yielded a wealth of thermodynamic data [163, 164]. More recently, triple–quadrupole instruments have been demonstrated to be useful for these energy-resolved experiments. For example, Kinter and Bursey [165] reported a study in which two isomers of $C_2H_5O^+$ were accelerated from about 4 to 35 eV into an unreactive gas, N_2, and a reactive gas, NH_3. Thresholds for all endothermic processes were determined and used as indicators of product ion and product neutral structures. In exact analogy to these experiments, FTMS can be used to determine the onset energy of endothermic processes. The data shown in Fig. 14 provide one of the earliest examples. In this paper [141] the authors reported a threshold value for reaction 12 of 1.4 ± 0.2 eV, which compared favorably to an earlier reported value of 1.3 eV:

$$(i\text{-}C_3H_7OH)_2H^+ \xrightarrow[\text{Ar}]{\text{CID}} i\text{-}C_3H_7OH_2^+ + C_3H_7OH \qquad (12)$$

Bensimon and Houriet [161] have published the first examples of the application of FTMS for determining threshold values using a reactive collision gas, CH_4. In particular they chose to study reactions 13–15, since they had been studied previously by Clow and by Futrell [166] and the thermochemistry was well established:

$$CH_3^+ + CH_4 \rightarrow C_2H_5^+ + H_2, \qquad \Delta H_r = -1.0\,\text{eV} \qquad (13)$$

$$CH_3^+ + CH_4 \rightarrow C_2H_3^+ + 2H_2, \qquad \Delta H_r = +1.1\,\text{eV} \qquad (14)$$

$$C_2H_5^+ \xrightarrow{M} C_2H_3^+ + H_2, \qquad \Delta H_r = 2.1\,\text{eV} \qquad (15)$$

They concluded from this study that thresholds for endothermic processes could be determined with an accuracy of about ± 0.2 eV. Although this error is worse than that obtained by ion-beam experiments, the authors suggest that the inherently larger mass range of FTMS opens the door for studying considerably more complex chemical systems. Additional convincing evidence for the use of FTMS for studying endothermic reactions comes from a more extensive study by Forbes et al. [162] involving transition metal ion chemistry. Thresholds for a variety of reactions were found to be in good agreement with ion-beam results. One of the most interesting examples is the determination of the onset of Fe_2H^+ from reaction 16 for which no previous data exist:

$$Fe_2^+ + C_2H_6 \xrightarrow{\text{KE}} Fe_2H^+ + C_2H_5 \qquad (16)$$

The threshold data, shown in Fig. 20, yield a value for ΔH_r of reaction 16 of ~ 35 kcal/mol, which can be used with other known thermochemistry to yield

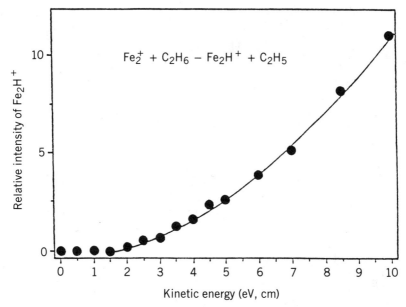

Fig. 20. Relative intensity of Fe_2H^+ generated from reaction 16 as a function of increasing Fe_2^+ kinetic energy.

$D(Fe_2^+—H)$ ~ 63 kcal/mol, $D(Fe^+—FeH)$ ~ 95 kcal/mol, $D(FeH^+—Fe)$ ~ 72 kcal/mol, and $PA(Fe_2)$ ~ 231 kcal/mol. It is evident from the results at this early stage that endothermic reaction processes, in addition to simple CID, will find increasing use both for ion structure determination and ion thermochemistry.

4.2 Photodissociation

The ion storage capability of the ICR spectrometer makes it ideally suited for the study of the photochemistry of ions in the gas phase. In the early seventies this methodology was demonstrated and exploited, principally by Dunbar [42], Beauchamp [167–171], and Brauman [44], and marked the beginning of what has become an extensive literature on simple organic ions. A wide variety of other techniques and instruments have also been brought to bear on this problem, with large bodies of results now available from ion-beams [172], drift tubes [173], tandem quadrupoles [174], and rf quadrupole traps [175]. Like ICR, FTMS enjoys the advantage of long ion storage times but, in addition, it clearly opens up new horizons as to the variety of ions that can be studied (e.g., those synthesized by complex multistep processes and high molecular weight species), and permits us to monitor the precursor ion and all of the photoproducts simultaneously.

In order to observe photodissociation, process 17, three criteria must be satisfied: first, the ion AB^+ must absorb a photon; second, the photon energy

Table 3. Bond Energy Determinations

A^+—B	$D^0(A^+$—$B)$ (kcal/mol) Photodissociation	Other
Fe^+—CH_2	82 ± 5^a	96 ± 5^b
Fe^+—CH	101 ± 5^a	115 ± 20^c
Fe^+—C	94 ± 5^a	89^c
Co^+—CH_2	84 ± 5^a	85 ± 7^d
Co^+—CH	100 ± 5^a	
Co^+—C	90 ± 5^a	98^c
Nb^+—CH_2	109 ± 7^a	112^e
Nb^+—CH	145 ± 8^a	
Nb^+—C	$> 138^d$	
Rh^+—CH_2	91 ± 5^a	94 ± 5^f
Rh^+—CH	102 ± 7^a	
Rh^+—C	$> 120^a$	164 ± 16^g
La^+—CH_2	106 ± 5^a	106 ± 5^a
La^+—CH	125 ± 8^a	
La^+—C	102 ± 8^a	
Fe^+—CH_3	65 ± 5^h	69 ± 5^i
Co^+—CH_3	57 ± 7^h	61 ± 4^j
Fe^+—O	68 ± 5^h	68 ± 3^j
Fe^+—S	65 ± 5^h	$74 > x > 59^k$
Co^+—S	62 ± 5^h	$74 > x > 59^k$
Ni^+—S	60 ± 5^h	
V^+—C_6H_6	62 ± 5^h	
$C_6H_6V^+$—C_6H_6	57 ± 5^h	
Fe^+—C_6H_6	55 ± 5^h	59 ± 5^l
Co^+—C_6H_6	68 ± 5^h	$71 > x > 61^m$
Sc^+—Fe	48 ± 5^n	49 ± 6°
Ti^+—Fe	60 ± 6^n	$> 49^n$
V^+—Fe	75 ± 5^p	$> 62^h$
Cr^+—Fe	50 ± 7^n	
Fe^+—Fe	62 ± 5^n	58 ± 7^q
Co^+—Fe	62 ± 5^n	62 ± 6^q
Ni^+—Fe	64 ± 5^n	$< 68 \pm 5^r$
Cu^+—Fe	53 ± 7^n	$> 52^n$
Nb^+—Fe	68 ± 5^n	$> 60^s$
Ta^+—Fe	72 ± 5^n	$> 60^s$

[a] Reference [176].
[b] P. B. Armentrout, L. F. Halle, and J. L. Beauchamp, *J. Am. Chem. Soc.*, **103**, 6501 (1981).
[c] J. L. Beauchamp, private communication.
[d] P. B. Armentrout and J. L. Beauchamp, *J. Chem. Phys.*, **74**, 2819 (1981).
[e] S. W. Buckner and B. S. Freiser, *J. Am. Chem. Soc.*, **109**, 1247 (1987).
[f] D. B. Jacobson and B. S. Freiser, *J. Am. Chem. Soc.*, **107**, 5870 (1985).
[g] D. B. Jacobson, G. D. Byrd, and B. S. Freiser, *Inorg. Chem.*, **23**, 553 (1984).
[h] Reference [177].
[i] L. F. Halle, P. B. Armentrout, and J. L. Beauchamp, *Organometallics*, **1**, 963 (1982).

must exceed the positive enthalpy required to generate the products (assuming a one-photon process); and third, the quantum yield for photodissociation must be nonzero:

$$AB^+ + h\nu \rightarrow A^+ + B \qquad (17)$$

Thus, the information inherent in a photodissociation experiment includes an ion's absorption characteristics, its electronic states, its structure, its photophysics, and some information on the energetics of process 17.

Cassady et al. [142] reported that $C_7H_7O^+$ derived from benzaldehyde photodissociates using the frequency tripled output of a pulsed Nd: YAG laser at 355 nm, whereas an ion of identical composition formed from tropone does not. These results were in agreement with experiments performed on an ICR instrument using an arc lamp source suggesting two stable $C_7H_7O^+$ isomers. More recently, Freiser and co-workers have been carrying out extensive studies on organometallic ions using FTMS [139, 176, 177]. These studies are performed with a 2.5 kW Hg–Xe arc lamp used in conjunction with a 0.25 m monochromator set for 10 nm resolution. The initial studies have focused on three general categories of metal-containing ions:

1. ML^+ (L = O, S, CH_2, CH_3, C_4H_6, C_4H_8, C_6H_6, etc.)
2. $L_1ML_2^+$ (both $L_1 = L_2$ and $L_1 \neq L_2$: L = C_2H_4, C_3H_6, C_4H_8, C_6H_6, etc.)
3. MFe^+ (M = Sc, Ti, V, Cr, Fe, Co, Ni, Cu, Nb, Ta)

One of the major findings of these studies is that, in general, these metal-containing species absorb broadly in the ultraviolet and visible spectral regions, apparently due to the high density of low-lying electronic states associated with the metal. Because of this broad absorption, photodissociation thresholds are attributed to thermodynamic factors and yield absolute bond energies. With few exceptions, this is not the case for organic ions where photodissociation thresholds yield only upper limits to the endothermicity of process 17. Table 3 lists the metal–ligand bond energies obtained to date by photodissociation and, wherever available, other results for comparison. Isomer differentiation has also been demonstrated in a number of cases by monitoring differences in cross

---

[j]P. B. Armentrout and J. L. Beauchamp, J. Am. Chem. Soc., **103**, 784 (1981).
[k]Reference [206].
[l]Reference [128].
[m]Reference [132].
[n]R. L. Hettich and B. S. Freiser, J. Am. Chem. Soc., **109**, 3537 (1987).
[o]L. M. Lech and B. S. Freiser, unpublished results.
[p]Reference [137].
[q]Reference [132].
[r]D. B. Jacobson and B. S. Freiser, unpublished results.
[s]S. Buckner, T. MacMahon, and B. S. Freiser, unpublished results.

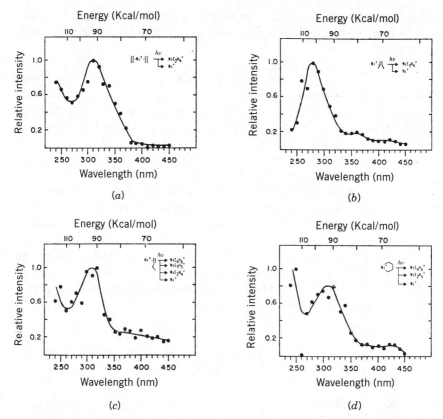

Fig. 21. Photodissociation spectra of four $NiC_4H_8^+$ isomers generated by reacting Ni^+ with (a) n-butane, (b) neo-pentane, (c) n-hexane, and (d) cyclopentanone. (Reprinted from Ref. [177] with permission.)

sections, spectral band positions, and fragmentation pathways (Fig. 21). Finally, these workers have reported that product ions generated by photodissociation are found in some cases to differ significantly from those produced by CID for reasons that are, as yet, not well understood.

Dunbar and co-workers, using a home-built FTMS, have begun to extend their earlier ICR studies on the collisional and radiative relaxation processes of vibrationally excited ions [178]. These studies use a chopped-laser source and monitor multiple-photon absorption processes. Freiser and Beauchamp [168] using ICR first reported a novel sequential two-photon process in which benzene radical cation, $C_6H_6^+$, was observed to photodissociate to $C_6H_5^+$ and H upon irradiation from an intense argon ion laser at energies significantly below thermodynamic threshold. To explain these results, a sequential two-photon excitation was proposed (reaction 18), where I is the intensity of the incident radiation, K_1 and K_2 are the rate constants for the absorption of the first and second photons, respectively, k_3 is the rate constant for collisional deactivation, and [C_6H_6] is the number density of benzene neutrals:

$$C_6H_6^+ \underset{k_3[C_6H_6]}{\overset{IK_1}{\rightleftarrows}} [C_6H_6^+] \overset{IK_2}{\longrightarrow} C_6H_5^+ + H \qquad (18)$$

This work was subsequently extended by Orlowski and co-workers to cyanobenzene radical cation [171] and, in an important conceptual extrapolation, to the study of multiphoton photodissociation in the infrared [43]. Dunbar and co-workers have studied these processes in detail by developing a number of new experimental methodologies [179–183]. Their experiments using FTMS involved exciting parent molecular ions with laser light at 515 nm and yielded radiative rates that ranged from $2 s^{-1}$ for bromobenzene to $75 s^{-1}$ for pentafluorobromobenzene [178]. Baykut and co-workers [184] also report using FTMS for isomer differentiation by multiphoton photodissociation in the infrared using a continuous wave CO_2 laser. They found the multiphoton results to be in accord with CID experiments. In addition, they recently reported multiphoton photodissociation of laser-desorbed ions using one laser to perform both functions, as well as separate lasers for desorption and photodissociation [185].

Finally, there is some controversy over whether collision-induced dissociation will be possible for high molecular weight ions since these species have a large number of degrees of freedom that can "absorb the shock." In addition, the center-of-mass energy decreases as the ion mass increases. Surface-induced dissociation [186] in which, as the name implies, a surface instead of a gas is used as the target, holds promise in this regard and can in theory be applied to FTMS. Bowers et al. [187] proposed that photodissociation could be a useful alternative to CID for these species since multiphoton absorption could add a substantial amount of internal energy to the ion. In this same paper a series of simple oligopeptide ions was photodissociated with a pulsed excimer laser at 193 nm, which produced fragments containing useful sequence information. The equivalent to MS/MS/MS in which daughter ions produced by photodissociation are themselves photodissociated has also recently been reported [112].

4.3 Gas Chromatography

Considering the importance and widespread use of GC–MS, it was critical that a major effort be made to demonstrate its application to FTMS. Ledford et al. [188] at Nebraska spearheaded this effort in the late seventies using a home-built apparatus. The challenge was to develop an interface that went from atmospheric pressure on the GC side to high vacuum ($< 10^{-6}$ torr) on the FTMS side. Two methods that accomplished this were reported in the first article on GC–FTMS from the Nebraska laboratory and included a direct-coupled open-split interface and one using a jet separator [188]. A high-speed turbomolecular pump (500l/s) was also used. Benzene was used as a test case and yielded a resolution (FWHH) of 8000 for the direct-coupled interface and

23,000 when the jet separator was used. In addition both electron impact ionization (EI) and chemical ionization (CI) spectra were obtained. Subsequently, White and Wilkins [189] using a jet separator in conjunction with a SCOT GC column obtained a resolution of 40,000 on benzene molecular ion and attributed it to the lower pressure (4×10^{-8} torr) that was achieved. In addition this experiment was performed by peak switching back and forth every 285 ms between m/z 78 and m/z 156 in narrow band or mixer mode. Broad-band measurements (from m/z 38 to m/z 200) were also obtained on a multicomponent mixture of aromatic molecules with a resolution of 1000 (FWHH) at m/z 78 and at an operating pressure of 5×10^{-8} torr.

One of the clear trade-offs in reducing the gas load into the FTMS using an open-split or jet separator interface is that a high fraction of the sample (> 99%) is "rerouted" away from the cell. Although Wilkins and co-workers have used this to advantage in a number of elegant papers on linked GC–FTIR–FTMS [88, 190, 191] (Fig. 22), the resulting loss of sensitivity rendered the technique inferior to conventional GC–MS for trace analysis. Gross and Sack [154, 155] reported that a pulsed valve interface provided a viable solution to this problem by admitting the GC effluent directly into the cell in short bursts during the ionization step, followed by a delay to permit pump down and then detection under optimal low-pressure conditions (i.e., the same strategy that was applied to obtain high-resolution CID and CI with a pulsed valve described earlier). Using this methodology, Sack et al. [155] reported a resolution of 87,000 for naphthalene parent cation at m/z 128, using a 1.2 T magnetic field, and a detection limit of 400 pg compared to an earlier reported value of 20 ng. This sensitivity was improved even further to a detection limit of 20 pg in a demonstration of GC–MPI–FTMS [192] in which resolutions over 20,000 were achieved. An example from their work demonstrating the selectivity of multiphoton ionization (MPI) in conjunction with GC is shown in Fig. 23. Despite these promising results, it has been pointed out [7] that the pulsed valve is still not the ultimate answer. For example, because of the relatively long pump-down times required between pulses, the valve is only open a few milliseconds during the elution of a GC peak preventing both (a) maximum sensitivity and (b) reproducibility of peak maxima and detection limits (in analogy to digital resolution discussed earlier) from being achieved. New developments in FTMS instrumentation (discussed below) may provide a general solution for the successful coupling of high-pressure/low-pressure regimes. One fundamental limitation of GC–FTMS, however, is that the cycle time required for maximum resolution will exceed the time resolution for capillary GC [54]. Thus, with the exception of raising the magnetic field, a trade-off between mass and time resolution is unavoidable. Finally, Mullen and Marshall [193] have recently demonstrated that the large amounts of digitized data resulting from a GC–FTMS experiment can be reduced by a factor of 20 by storing the FTMS transients in representations that have been "clipped" to one bit per word.

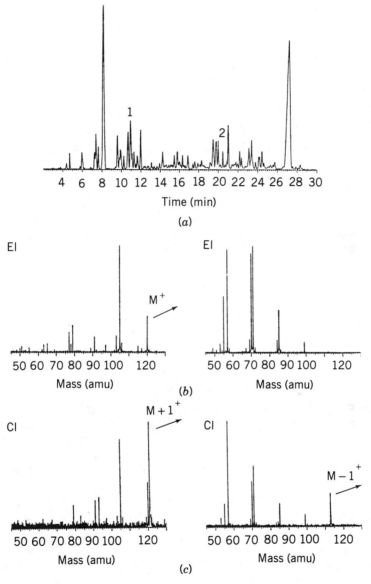

Fig. 22. (*a*) Reconstruction of an EI–GC–FTMS run from a pulsed-valve EI/CI analysis of a commercial fuel additive. Time resolution of the reconstruction was 2 s; (*b*) Mass spectrum of the first labeled peak in (*a*) corresponding to 3-methyl-2-ethylpentane; (*c*) mass spectrum of the second labeled peak in (*a*) corresponding to 1,3,4-trimethylbenzene. (Reprinted from Ref. [88] with permission.)

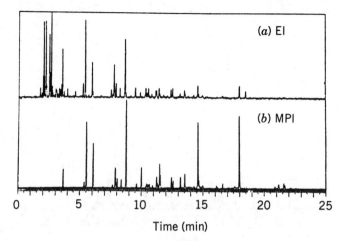

Fig. 23. Reconstructed chromatograms for a neat injection of 0.1 μL of commercial gasoline under (*a*) EI and (*b*) MPI conditions. Because all the aliphatic components of gasoline have ionization potentials greater than 9.3 eV (corresponding to absorption of two photons of 266-nm light), they do not appear in the MPI trace. The column temperature was programmed from 50 to 250°C at 5°C/min. Files were stored at a rate of 1 per second. (Reprinted from Ref. [155] with permission.)

5 IONIZATION METHODS

5.1 Electron Ionization and Chemical Ionization

Dating back to the omegatron, electron impact ionization (EI) is still "standard equipment" on commercial FTMS instruments. The spectra obtained in EI may differ somewhat from those obtained on conventional mass spectrometers (e.g., the millisecond time regime of FTMS versus the microsecond regime on other instruments eliminates observation of metastable ions), but are similar enough to be able to take advantage of the extensive library spectra for positive identification of unknowns [194].

A key advantage of FTMS is that the sample pressures required to obtain good signal-to-noise ratio are 10^{-9}–10^{-8} torr compared to the $\sim 10^{-6}$–10^{-5} torr required for most conventional mass spectrometer sources. Thus, samples introduced on the solids probe can be run at considerably lower temperatures, reducing the risk of pyrolysis. Shore et al. [195] and Hsu et al. [196], for example, have published several impressive examples where FTMS EI and CI spectra were obtained for organometallic compounds that could not be run at higher temperatures by conventional mass spectrometry.

Chemical ionization (CI), as the name implies, involves "indirect" ionization of a sample by an ion–molecule reaction. Considering the extensive literature on ion chemistry, it is evident that the list of possible CI reagent ions has barely been explored. In this context it is significant that both ICR and FTMS first rose to prominence as methods of choice for studying ion–molecule reactions. Thus, the unusual capability and versatility of FTMS to generate a virtually endless list of interesting reagent ions will continue to emerge as one of its most useful analytical features. Two other features distinguish FTMS CI from its use with conventional mass spectrometers. First, the reagent gas pressure used in FTMS is typically 10^{-6} torr and that of the sample is about 10^{-8} torr, compared to 1 torr and 10^{-5} torr, respectively, used in conventional mass spectrometers [197, 198]. One consequence of these lower pressures is that adduct ions, involving the attachment of reagent gas molecules and often seen in high pressure CI, are not observed. In addition operation at lower pressures not only increases the range of thermally unstable compounds that can be analyzed, as discussed for EI above, but also multiplies the list of viable reagent gases orders of magnitude. The concept of "self-chemical ionization" or self-CI, introduced by Ghaderi et al. [198], emphasizes this point by demonstrating that any sample gas can even serve as its own CI reagent gas (i.e., ions generated from the sample gas by electron impact become the reagent ions for chemical ionization of the sample gas). Second, unlike conventional mass spectrometers, the same source can be used for both EI and CI simply by changing the time sequence for the experiment. If EI spectra are desired, detection occurs within a few milliseconds after the electron beam pulse. If CI spectra are desired, following the electron beam pulse with a variable delay (seconds are typical but hours are possible [199]) permits ion–molecule reactions to occur before detection.

High-resolution CI spectra can be achieved, once again, by minimizing system pressure. There are several ways in which this can be accomplished. Literally, the system pressure (reagent and sample or just sample for self-CI) can be lowered, and the delay or trapping time can be increased accordingly to permit the desired chemical ionization to occur [198]. Pulsed-valve addition of reagent gas, as described above for MS/MS, provides the optimum conditions of reasonable pressures for reagent ion formation and low background during detection [153]. Alternatively, "reagentless CI" (originally used as a synonym for self-CI [198] but redefined here), in which reagent ions are introduced into the cell *without* any reagent gas being present, is also ideal. Although at first this might sound contradictory, there are actually several ways of accomplishing this. For example, transition metal ions are produced when a high-powered pulsed laser is focused onto a metal target (termed laser desorption and discussed in greater depth below) and can be used as reagent ions, clearly in the absence of any reagent gas [100, 118, 200, 201]. The feasibility of metal ions as selective chemical ionization reagents has been demonstrated by Freiser and co-workers on extensive studies of Cu^+ with esters and ketones [201], Fe^+ with ethers [202], ketones [202], and hydrocarbons [121, 203], Ti^+ [203], Rh^+ [204], V^+ [138], Y^+ [205], and La^+ [205] with hydrocarbons, and of Fe^+ and Co^+ with sulfur-containing compounds [206]. Pattern recognition methods have been found to be a particularly useful way to evaluate the information content from transition metal ion CI [207]. Results have also been reported from a number of other laboratories [118]. Chemical ionization with Co^+, for example, has been applied to bifunctional compounds [208], and Fe^+ has been found to be useful for locating double bonds in a variety of long-chain compounds [209]. Generalizing this method, any one of the different desorption methods compatible with FTMS (discussed below) can be used to generate ions of all sorts (organic and inorganic) for reagentless CI. Finally, CI reagent ions can be generated externally to the cell and injected, once again, in the absence of reagent gas. The external source approach represents a recent and exciting development in FTMS instrumentation, as discussed below.

5.2 Multiphoton Ionization

Multiphoton ionization (MPI) has proven to be an invaluable tool in molecular spectroscopy and is often used in concert with a mass spectrometer [210–212]. From an analytical point of view, MPI has several advantages: it alleviates many of the problems encountered in photoionization mass spectrometry with vacuum ultraviolet sources, the ionization efficiency during the light pulse can be very high, the "softness" of the ionization can be controlled by varying laser power and wavelength, and it can be very selective. Combining MPI, therefore, with FTMS would enjoy the strengths of both techniques. Irion et al. [213] reported the first examples, followed shortly by Carlin and Freiser [214], both confirming the power of FTMS in tandem with MPI (e.g., Fig. 24). Most previous MPI–MS studies have involved the use of time-of-flight (TOF) mass spectrometers because, unlike sector instruments and quadrupoles, a complete mass spectrum can be obtained with a single laser pulse. This advantage is also inherent in FTMS but with the added advantages of high resolution, exact mass, and MS/MS capabilities. Carlin and Freiser [214] also suggested that MPI at 266 nm, used in conjunction with GC–FTMS, would be an effective approach to selective detection of aromatic compounds. Sack and co-workers [155] provided experimental confirmation of this in an analysis of aromatic constituents in unleaded gasoline (Fig. 23). They also reported a resolution of 90,000 (at 1.2 T) with a detection limit of 10 pg and a linear dynamic range for quantification of 2.5 orders of magnitude. More recently, Watson and co-workers [215] also applied MPI to generate and study isomeric ions.

5.3 Laser Desorption

Many of the advantages described above for combining MPI with FTMS hold for laser desorption (LD) as well. An extensive literature exists on the combination of LD with a variety of mass detectors for analyzing nonvolatile and thermally unstable molecules [216–219] and, once again, it was apparent that the pulsed nature of FTMS would be ideal for LD experiments. The first application of LD–FTMS was reported by Cody et al. [200] for the generation of transition metal ions from the pure metal target. This observation has greatly accelerated the study of bare atomic metal ions, metal–ion complexes, and metal–ion clusters in the gas phase [6]. Of particular interest is the recent report by Reents et al. of the generation and study of positive and negative silicon clusters, Si_{2-8}^+ and Si_{2-6}^-, by direct laser desorption from bulk silicon [220]. McCrery et al. [221] demonstrated the first use of LD–FTMS to generate ions from organic samples, which has led subsequently to a flurry of activity

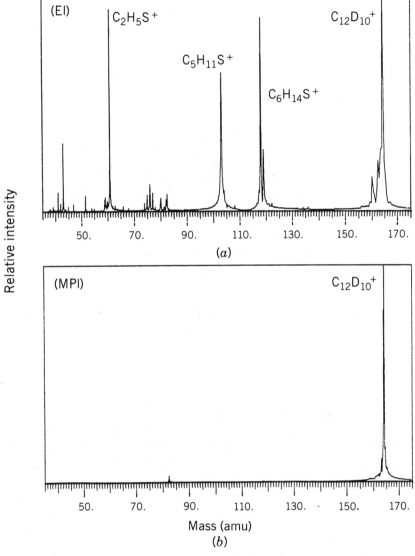

Fig. 24. Mass spectra of a mixture of 2×10^{-7} torr perdeuterodiphenyl and 1×10^{-6} torr isopropyl sulfide. Ionization was accomplished by (a) a 70-eV 20-ms electron beam and (b) a 266-nm laser pulse. Both spectra are averages of 50 scans. (Reprinted from Ref. [214] with permission.)

in a number of laboratories around the world as well as at Nicolet and Bruker. To date a wide variety of compound classes have been examined, including biologically interesting molecules such as carboxylic acids, dipeptides, oligopeptides, porphyrins, oligosaccharides, glycosides, and nucleosides, as well as organometallics and intractable polymers [90, 91, 222–229]. Figure 25

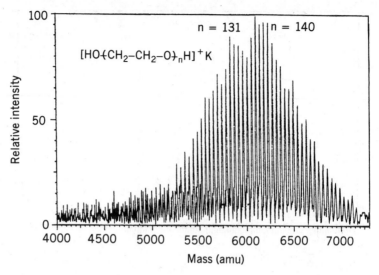

Fig. 25. LD–FTMS spectrum of PEG 6000 doped with KBr. (Reprinted from Ref. [226] with permission.)

demonstrates the high mass capability of LD–FTMS with a spectrum obtained in Wilkins' laboratory on a polyethylene glycol sample having an average molecular weight of 6000 amu [227]. The molecular weight average determined from this data was in excellent agreement both with the manufacturer's information and with the average determined by an ASTM end group titration method. High resolution has been demonstrated (e.g., 125,000 at m/z 1241) [7], as well as collision-induced dissociation of LD generated ions [116, 117]. McCrery et al. [221] noted that neutral compounds produce more abundant negative ions than positive ions, with $(M - H)^-$ yielding molecular weight information. The addition of salts quite often is found to be useful in promoting the cationization of neutral, polar molecules [225]. Shomo and co-workers [90] reported a comparison of spectra obtained on erythromycin by laser desorption/ionization FTMS and fast-atom-bombardment (FAB) using a double-focusing instrument. In their work a single laser shot on a 100 ng sample was found to produce a more prominent $(M + H)^+$ signal at m/z 734 than the FAB spectrum obtained with about 100 times more sample. They also pointed out that four classes of mass spectra can be obtained from a single sample and instrumental configuration, including positive or negative ion spectra produced directly by the laser or from electron impact on neutrals "blown off" by the laser. In this latter regard, Sherman and co-workers [230] have recently reported an exciting study in which detailed information about the nature of simple molecules adsorbed on crystals was obtained. Once again, this experiment involved "gentle" desorption of the neutral species by laser desorption, followed by electron impact. Finally, LD-FTMS holds promise for mixture analysis with little or no sample preparation. For example, Hsu and Marshall (228b) recently reported a study where they detected and identified (by chemical formula) dyes in solid poly-(methyl methacrylate) commercial plastics at concentrations of at least an order of magnitude lower (0.1% vs 1–2% by weight) than those obtained by the best currently available alternative method (attenuated total reflectance FTIR).

5.4 Cesium Ion–Secondary Ion Mass Spectrometry

Before FAB, secondary ion mass spectrometry (SIMS) had gained popularity as a method of analyzing not only inorganic samples but organic samples as well. Although keV Ar^+ is typically used to bombard a solid sample, a variety of metal ions has also been used. Of these, Cs^+ is commonly used because of the relatively simple filament sources required to produce beams of good intensity. Castro and Russell [231] reported the first experiments interfacing a keV Cs^+ source to an FTMS to produce spectra from vitamin B_{12}, and have subsequently reported a study on CsI [232]. They pointed out that the advantage of using Cs^+ is not only in its instrumental simplicity but that, unlike neutral Ar from an Ar^+ source, no increase of pressure above background occurs. This same problem (i.e., introduction of excessive gas loads) precludes the use of FAB (or equivalently, liquid SIMS unless it is frozen [225]) directly in the FTMS. However, even this problem can be overcome with the use of an external source as described below. The Cs^+–SIMS source is provided as an option on both Nicolet and Bruker instruments. Finally, Amster and co-workers [233] reported the most impressive demonstration of mass range by FTMS to date in their work on positive and negative ion clusters generated by Cs^+–SIMS on a CsI sample (Fig. 26).

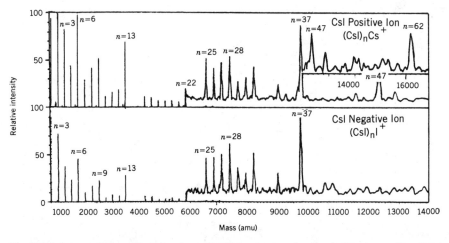

Fig. 26. Cs^+ desorption positive (*a*) and negative (*b*) ion spectra of cesium iodide clusters. (Reprinted from Ref. [233] with permission.)

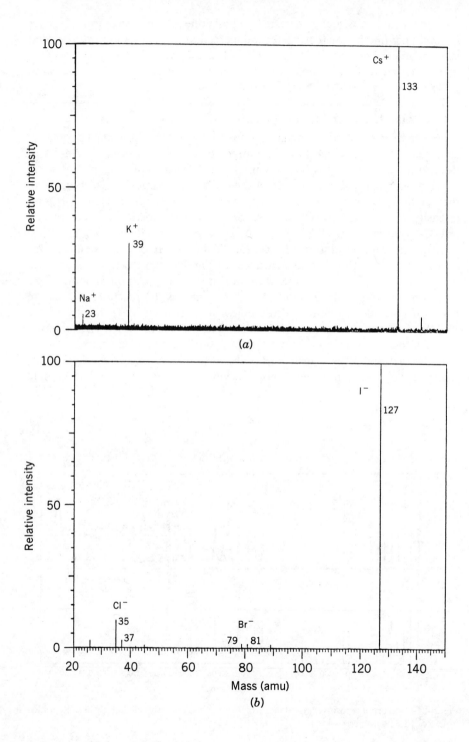

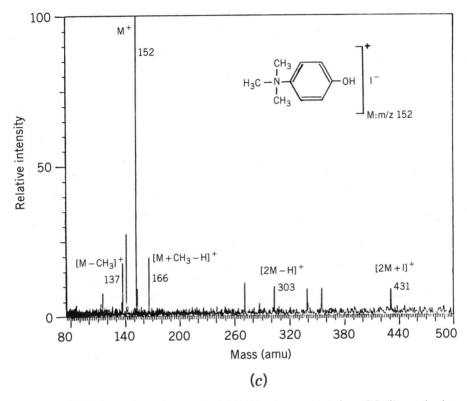

Fig. 27. Cf-252 plasma desorption spectra: (*a*) Positive ion spectrum from CsI; (*b*) negative ion spectrum from CsI; (*c*) positive ion spectrum from N, N, N-d_9-trimethyl-*p*-aminophenolate iodide. (Reprinted from Ref. [237] with permission.)

5.5 Plasma Desorption

Following the pioneering work of MacFarlane and co-workers on ^{252}Cf plasma desorption mass spectrometry for desorbing very high molecular weight molecules [234–236], several FTMS groups have been working on interfacing this important technique to FTMS. The first successful, but preliminary, results have recently been reported by Viswanandham et al. [237]. Because of the low fluxes, trapping times of as much as 30 minutes were required to obtain good signal-to-noise ratios (Fig. 27). Clearly, additional work will be required to demonstrate its advantages over the other desorption techniques.

6 NEW EXCITATION METHODS

In their first experiments, Comisarow and Marshall realized that while a rectangular rf pulse (Fig. 28*a*) could readily be used to excite a narrow-frequency or mass range, in order to excite a wide-frequency range on the order of 2 MHz, a short pulse ($\sim 10^{-7}$ s) having an amplitude $> 10^4$ V was required [55]. Instead,

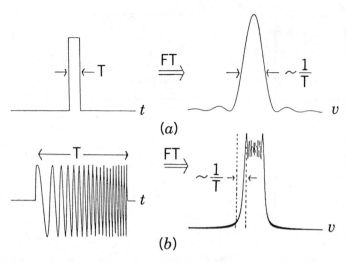

Fig. 28. Time-domain excitation waveforms (left) and their corresponding frequency-domain magnitude-mode spectra (right). FTMS experiments were first performed with a rectangular rf pulse (a), are currently done with a fast frequency sweep (chirp) (b), and may be performed in the future with tailored excitation discussed in the text. (Reprinted from Ref. [4] with permission.)

they developed "chirp" excitation, whereby the range of frequencies is scanned in about 1 ms and requires an amplitude only on the order of 1–100 V [56]. Chirp excitation followed by detection is still the standard used in current FTMS instruments. It has long been realized [70, 238], however, that a serious problem exists with chirp excitation in that its nonuniform power spectrum (Fig. 28b) directly affects relative peak height accuracy and, therefore, quantitation (e.g., isotope ratios). The obvious cure for this problem, increasing the scan time from 1 ms to forever, which would yield a flat power spectrum, is clearly worse than the disease. In analogy to a method to suppress solvent signals in FTNMR, Marshall and co-workers, have recently introduced a major breakthrough on this problem that will have a far-reaching effect on how the experiment will be performed in the future. Using what they have recently termed "stored waveform inverse Fourier transform" or SWIFT excitation, virtually any excitation frequency profile can be applied to the cell [4, 239]. In the simplest terms, SWIFT or tailored excitation involves first choosing the desired excitation frequency profile and specifying it as a discrete frequency domain spectrum. Next, an inverse discrete Fourier transform is used to generate the equivalent discrete time-domain waveform, which is then converted to an analog signal, amplified, and applied to the FTMS cell. SWIFT excitation introduces a tremendous flexibility into the FTMS experiment as exemplified by Fig. 29, which shows three general types of excitation suggested by Marshall: flat power (for more accurate peak heights), flat power with windows (e.g., to eject all but one selected ion, as in MS/MS), and power at several selected m/z ratios (for simultaneous multiple-ion monitoring or ejection). The most recent application

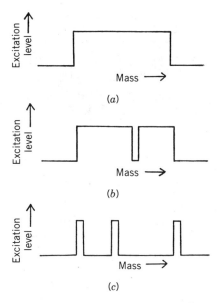

Fig. 29. Schematic excitation power spectra for producing: (a) uniform excitation, (b) excitation windows for MS/MS, and (c) excitation peaks for multiple-ion monitoring. Each effect may be produced by applying an excitation waveform generated from the inverse Fourier transform of the desired spectrum (see text). (Reprinted from Ref. [239] with permission.)

of SWIFT from Marshall's laboratory was to demonstrate its use in extending the dynamic range of the FTMS experiment [102]. The dynamic range of FTMS is determined by a detection limit of about 100 ions in the cell and an upper limit (for a cubic cell) of about 100,000 ions due to adverse effects arising from space charge and ion–ion repulsion. Thus, a dynamic range of about 10^3 is typical for FTMS compared to $> 10^6$ for other mass spectrometers. In addition, because detection of all of the ions occurs simultaneously, the presence of one or more high-intensity peaks can readily "overload" the digital dynamic range of the ADC and ultimately reduce the signal-to-noise ratio for low-abundance ions. Therefore, by ejecting abundant ions, less abundant ions can be detected at higher receiver gain and/or more of the less abundant species can be generated in the ion formation step (e.g., in EI by increasing the beam duration or current). Although this concept is not new [e.g., Kleingeld and Nibbering [103] were able to generate and isolate sufficient quantities of naturally occurring $CH_3{}^{18}OH^+$ (0.2%) for mechanistic studies by ejecting $CH_3{}^{16}OH^+$], SWIFT permits for the first time a method of ejecting a large number of ions simultaneously with good resolution or selectivity. An impressive example taken from Marshall's paper is shown in Fig. 30, where 23 different ions are ejected by applying SWIFT for 17 ms. Using a conventional sweep ejection to accomplish the same task would be impossible.

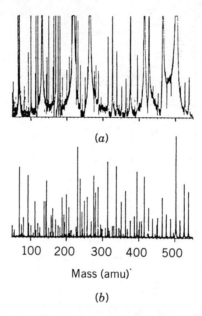

Fig. 30. (a) Vertically scale-expanded plot of the normal mass spectrum of perfluorotri-*n*-butylamine; (b) mass spectrum obtained following prior SWIFT multiple-ion ejection of all of the 23 peaks whose magnitude-mode peak heights exceed a threshold of 1.6% of the main peak. The ^{13}C isotope peak at m/z 503 ($^{13}C^{12}C_8F_{20}N$) becomes the base peak, whereas the m/z 502 ^{12}C peak is eliminated completely, even though the two peaks are only 183 Hz apart at a magnetic field strength of 3.0 T. (Reprinted from Ref. [102] with permission.)

First, to achieve the same selectivity (resolution) using a sweep would require 23×17 ms = 391 ms to accomplish, which is unacceptable since many of the desired ions would have undergone subsequent ion–molecule reactions and, second, sweeping faster would result in an unacceptable loss of selectivity. In short, since the SWIFT time-domain waveform is turned on and off only once, whereas the frequency sweep must be turned on and off N times for ejection over N distinct m/z ranges, the average SWIFT mass selectivity will be N times better than that of the frequency sweep using the same total time for ejection. The SWIFT method will also greatly improve the MS/MS technique by cutting down on the number of individual sweeps necessary for ion isolation and by allowing high MS–I or front-end resolution (i.e., high resolution of the precursor ions as demonstrated by Fig. 31). Finally, SWIFT is currently being developed at Nicolet in consultation with Marshall and will, undoubtedly, become the excitation method of choice for FTMS.

7 TOMORROW'S INSTRUMENTATION HERE TODAY

As described throughout this review, there is a basic incompatibility between many of the most important experiments that introduce a gas load to the FTMS

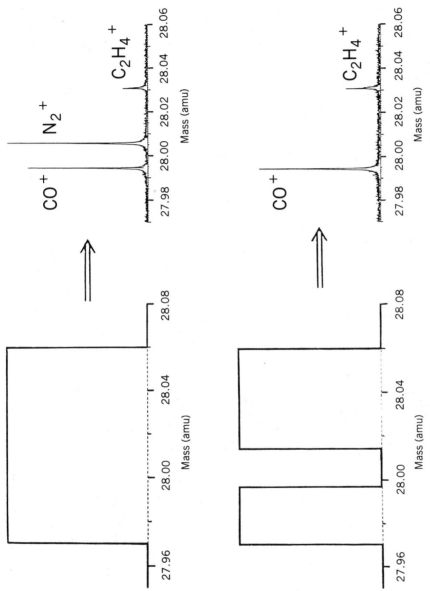

Fig. 31. Flat and windowed tailored excitation of an ionized mixture of CO, N_2, and C_2H_4. (Reprinted from Ref. [239] with permission.)

system (e.g., GC, LC, FAB, CID) and the fact that resolution in FTMS rapidly degrades with increasing pressure. Although the pulsed-valve approach has been and will continue to be very useful for many applications, it is far from optimum and is not applicable in every case (e.g., FAB). In addition, due to the high sensitivity of FTMS in CI mode where, for example, trace amounts of a substance with a high proton affinity can produce the dominant ion at long trapping times, the number of samples that can be run in a working day is limited even with a heated system.

With a general solution to this problem, McIver and co-workers introduced the tandem quadrupole-FTMS (Q-FTMS) (Fig. 32a), demonstrating the concept that ions formed externally to the trapping cell could be injected through the magnetic field into the cell for detection [240–242]. Interestingly, such a feat

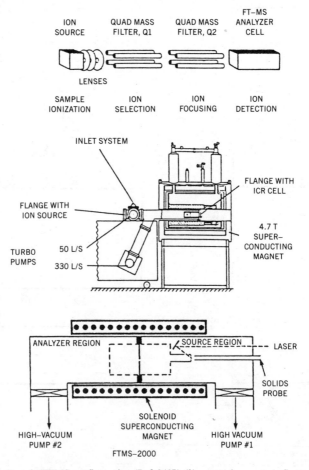

Fig. 32. (a) Quadrupole FTMS configuration (Ref. 240]); (b) external source configuration that uses electrical lenses to focus ions into the FTMS (Ref. [243]); (c) differentially pumped dual-cell configuration (Ref. [225]). (Figures reprinted with permission.)

was originally thought to be impossible due to the "magnetic mirror effect" in which ions approaching from outside the magnetic field would be deflected for the same reasons ions inside the field are contained. Trajectory calculations by the McIver group, however, suggested accelerating the ions in a "straight-line approach" would be possible, and the spectra they obtained put aside soundly any doubt. A cooperative effort involving Finnigan has resulted in a second Q-FTMS in the laboratory of Hunt at the University of Virginia [242].

The idea of having a differentially pumped external source in which only the ions are injected into a "clean," low-pressure FTMS would seem to be an ideal solution to the high-pressure/low-pressure interface problem. In this way any ionization method could be used, high gas loads could be delivered to the source, and so forth, while still maintaining the high-resolution capabilities of the FTMS. Both McIver and Hunt have published impressive results from their instruments. For example, using liquid SIMS, a resolution of 40,000 at 7.0 T was achieved for the $(M + H)^+$ ion of gramicidin S at m/z 1142. In addition a resolution of 1.7×10^6 was achieved on CH_3CO^+ (m/z 43) generated as a CID fragment ion in the second quadrupole.

Kofel and co-workers [243] have in a parallel effort developed an external source to fit the Bruker CMS-47 spectrometer, which uses electrical lenses instead of quadrupoles to focus and inject the ions (Fig. 32b). Again, analyzer pressures of 10^{-9} torr permit high-resolution spectra to be obtained readily. In addition to EI, this group has recently demonstrated laser desorption with the external source [244]. At mass 1038, a resolution of 250,000 (FWHH) was achieved using a single scan. Finally, they also demonstrated that the flight time of the ions from the external source to the FTMS cell could be used for additional mass separation [245].

Future efforts on these instruments will be aimed at improving transfer efficiency (an efficiency of less than 0.1% has been reported for the Q-FTMS experiment), interfacing GC and LC techniques, pushing up the limit of the highest mass organic ion *observed* [although $Cs(CsI)_n^+$ clusters have been observed up to m/z 16,241, thus far no organic ion has broken the 15,000 amu barrier]. Finally, Alford et al. [246] have reported successfully interfacing a supersonic metal cluster source to an FTMS to inject and study metal cluster ions. It is not unlikely that other mass spectrometry techniques such as flowing afterglow will someday be interfaced to FTMS.

At almost the same time the external source idea was being introduced, Nicolet announced at the 1984 Pittsburgh Conference the development of a differentially pumped, dual-cell instrument, the FTMS-2000 (Fig. 32c) [247]. This solution to the high-pressure problem is an elegant and ingenious one. Certainly it is simpler than the external source instruments. A conductance limit permits the two cells to be operated at different pressures (with a minimum differential of about 10^3), while permitting ions to be transferred between cells. The efficiency of transfer can be nearly 100%, and multiple transfers are readily accomplished [248]. Thus, for example, the effluent from a capillary GC can be introduced into the "source" cell maintained at

10^{-5}–10^{-6} torr, and the ions can be transferred into the "analyzer" cell maintained at 10^{-8}–10^{-9} torr for high-resolution detection. The dual-cell system has also been demonstrated [117] on a variety of other experiments, including accurate mass measurement, laser and Cs^+ desorption, high-resolution EI and CI, and, most recently, electron impact dissociation of ions [249]. The latter technique has been suggested as an alternative to CID and takes advantage of the overlap of the ion beam and electron beam that are magnetically focused through the center of the cells. In particular, electron impact dissociation requires no collision gas, permitting high resolution on the daughter ions and, unlike CID, the orbits of the ions are not increased, facilitating multiple ion transfers. Although it is too early to tell whether the external source or the dual-cell approach will be superior, an instrument having both of these features would clearly provide unusual flexibility in the types of experiments that could be performed.

8 CONCLUSION

Considering all of the exciting developments only briefly touched on in this review, you might well ask, is it time for me to buy one of these FTMS instruments? Undoubtedly, this question is raised more and more, but to date still relatively few laboratories (< 80) have an FTMS. The problem has been a general perception that while FTMS definitely has potential, it is not yet perfected and is by no means a reliable instrument for routine analysis, not to mention its high cost (> $400,000). Certainly, these are reasonable concerns, and a major thrust of the work described here has involved advances in the engineering of the instrument for analytical applications. A great deal of progress has clearly been made, for example, adapting many of the powerful new desorption methods available and in improving the GC–FTMS interface. In addition, there is a much greater fundamental understanding from these studies of the quality of the data and the factors involved, for example, in mass accuracy and quantitation. One can obtain high mass accuracy and quantitation (< 5%) MS/MS and photodissociation have well been demonstrated, and improvements involving new excitation modes and separated source and analyzer instruments look extremely promising.

The answer to whether or not to buy an FTMS now, of course, depends on the application. If an instrument is needed to run a large number of "routine" samples everyday, then there are probably less expensive instruments that would do the job. If a laboratory already has one or more mass spectrometers and would like to add new capabilities (e.g., high mass and ultrahigh resolution) or if there is some interest in doing research on, for example, developing new chemical ionization reagents or studying the properties of targeted molecules, then FTMS warrants serious consideration. It should also be noted that while the price of an FTMS is on the order of $400,000 from the major manufacturers (Nicolet and Bruker) a datasystem based on the IBM PC/AT laboratory computer is available from Ion Spec in Irvine, California, for under $100,000.

Finally, a unique feature of FTMS is that it relies heavily on computers (and software) and on magnets. Despite the addition of, for example, an external source or a GC, the basic vacuum system and cell (box) are simple. In contrast, most conventional instruments require a great deal of high-precision machining. The revolution in computers—faster speed, more memory, reduced cost—is in full gear. In addition, superconducting magnet technology is undergoing significant improvements toward higher fields and higher temperatures. There is no question that these developments suggest an even more exciting future for FTMS.

ACKNOWLEDGMENTS

Various aspects of this research were supported by the Division of Chemical Sciences in the Office of Basic Energy Sciences in the U.S. Department of Energy (DE-FG02-87ER13766) and by the National Science Foundation (CHE-8612234). The author also wishes to acknowledge his past and present students for their outstanding contributions and give special thanks to Justine Cunningham for help in preparing this manuscript. Finally, the author also wishes to thank Professors M. L. Gross, A. G. Marshall, and C. L. Wilkins for their helpful suggestions.

REFERENCES

1. "Fourier Transform Mass Spectrometry: Evolution, Innovation, and Application," M. V. Buchanon, Ed., American Chemical Society Symposium Series, V. 359, Washington, D.C., 1987.
2. D. A. McCrery, T. M. Sack, and M. L. Gross, *Spectros. Int. J.*, **3**, 57 (1984).
3. M. L. Gross and D. L. Rempel, *Science*, **266**, 261 (1984).
4. A. G. Marshall, *Acc. Chem. Res.*, **18**, 316 (1985).
5. K.-P. Wanczek, *Int. J. Mass Spectrom. Ion Proc.*, **60**, 11 (1984).
6. B. S. Freiser, *Talanta*, **32**, 697 (1985).
7. (a) D. A. Laude, Jr., C. L. Johlman, R. S. Brown, D. A. Weil, and C. L. Wilkins, *Mass Spec. Rev.*, **5**, 107 (1986); (b) C. L. Johlman, R. L. White, and C. L. Wilkins, *Mass. Spec. Rev.*, **2**, 389 (1983).
8. D. H. Russell, *Mass Spec. Rev.*, **5**, 167 (1986).
9. *Anal. Chim. Acta*, **178**, No. 1 (1985).
10. *Int. J. Mass Spectrom. Ion Proc.*, **72**, Nos. 1 and 2 (1986).
11. J. A. Hipple, H. Sommer, and H. A. Thomas, *Phys. Rev.*, **76**, 1877 (1949).
12. H. Sommer, H. A. Thomas, and J. A. Hipple, *Phys. Rev.*, **82**, 697 (1951).
13. D. Wobschall, J. R. Graham, and D. P. Malone, *Phys. Rev.*, **131**, 1565 (1963).
14. D. Wobschall, *Rev. Sci. Instrum.*, **36**, 466 (1965).
15. P. Llewellyn, U. S. Patent #3, 505, 517 (1970).
16. J. D. Baldeschwieler, *Science*, **159**, 263 (1968).
17. L. R. Anders, J. L. Beauchamp, R. C. Dunbar, and J. D. Baldeschwieler, *J. Chem. Phys.* **45**, 1062 (1966).
18. J. L. Beauchamp, *Ann. Rev. Phys. Chem.*, **22**, 527 (1971).

19. T. A. Lehman and M. M. Bursey, *Ion Cyclotron Resonance Spectrometry*, Wiley-Interscience, New York, 1976.
20. J. L. Beauchamp and J. T. Armstrong, *Rev. Sci. Instrum.*, **40**, 123 (1969).
21. J. L. Beauchamp, L. R. Anders, and J. D. Baldeschwieler, *J. Am. Chem. Soc.*, **89**, 4569 (1967).
22. J. L. Beauchamp and S. E. Buttrill, Jr., *J. Chem. Phys.*, **48**, 1783 (1968).
23. J. L. Beauchamp and R. C. Dunbar, *J. Am. Chem. Soc.*, **92**, 1477 (1970).
24. D. Holtz, J. L. Beauchamp, and J. R. Eyler, *J. Am. Chem. Soc.*, **92**, 7045 (1970).
25. J. L. Beauchamp and M. C. Caserio, *J. Am. Chem. Soc.*, **94**, 2638 (1972).
26. J. I. Brauman and L. K. Blair, *J. Am. Chem. Soc.*, **90**, 5636 (1968).
27. J. I. Brauman and L. K. Blair, *J. Am. Chem. Soc.*, **90**, 6561 (1968).
28. J. F. Wolf, R. H. Staley, I. Koppel, M. Taagepera, R. T. McIver, Jr., J. L. Beauchamp, and R. W. Taft, *J. Am. Chem. Soc.*, **99**, 5417 (1977).
29. J. E. Bartmess and R. T. McIver, Jr., in M. T. Bowers, Ed., *Gas Phase Ion Chemistry*, Academic Press, New York, 1979, Chap. 11.
30. M. T. Bowers, D. H. Aue, H. M. Webb, and R. T. McIver, Jr., *J. Am. Chem. Soc.*, **93**, 4314 (1971).
31. E. M. Arnett, F. M. Jones, III, M. Taagepera, W. G. Henderson, J. L. Beauchamp, D. Holtz, and R. W. Taft, *J. Am. Chem. Soc.*, **94**, 4727 (1972).
32. S. G. Lias, J. F. Liebman, and R. D. Levin, *J. Phys. Chem. Ref. Data*, **13**, 695 (1984).
33. D. H. Aue and M. T. Bowers, in M. T. Bowers, Ed., *Gas Phase Ion Chemistry*, Academic Press, New York, 1979, Chap. 9.
34. R. H. Staley, R. R. Corderman, M. S. Foster, and J. L. Beauchamp, *J. Am. Chem. Soc.*, **96**, 1260 (1974).
35. J. A. Jackson, S. G. Lias, and P. Ausloos, *J. Am. Chem. Soc.*, **98**, 7515 (1977).
36. D. H. Russell and M. L. Gross, *J. Am. Chem. Soc.*, **102** 6279 (1980).
37. W. D. Reents, Jr., and B. S. Freiser, *J. Am. Chem. Soc.*, **102**, 271 (1980).
38. M. S. Foster and J. L. Beauchamp, *J. Am. Chem. Soc.*, **93**, 4924 (1971).
39. R. C. Dunbar, J. F. Ennever, and J. P. Fackler, Jr., *Inorg. Chem.*, **12**, 2734 (1973).
40. R. D. Bach, J. Gauglhofer, and L. Kevan, *J. Am. Chem. Soc.*, **94**, 6860 (1972).
41. M. S. Foster and J. L. Beauchamp, *J. Am. Chem. Soc.*, **97**, 4814 (1975).
42. R. C. Dunbar, in M. T. Bowers, Ed., *Gas Phase Ion Chemistry*, Academic Press, New York, 1979, Chaps. 14 and 20.
43. L. R. Thorne and J. L. Beauchamp, in M. T. Bowers, Ed., *Gas Phase Ion Chemistry*, Academic Press, New York, 1984, vol. 3, Chap. 18.
44. P. S. Drzaic, J. Marks, and J. I. Brauman, in M. T. Bowers, Ed., *Gas Phase Ion Chemistry*, Academic Press, New York, 1984, vol. 3, Chap. 21.
45. W. N. Olmstead and J. I. Brauman, *J. Chem. Phys.*, **99**, 4219 (1977).
46. R. A. Barker and D. P. Ridge, *J. Chem. Phys.*, **64**, 4411 (1976).
47. T. Su and M. T. Bowers in M. T. Bowers, Ed., *Gas Phase Ion Chemistry*, Academic Press, New York, 1979, Chap. 3.
48. W. T. Huntress, M. M. Mosesman, and D. E. Elleman, *J. Chem. Phys.*, **54**, 843 (1971).
49. D. P. Ridge and J. L. Beauchamp, *J. Chem. Phys.*, **64**, 2735 (1976).
50. D. L. Smith and J. H. Futrell, *Int. J. Mass Spectrom. Ion Phys.*, **14**, 171 (1974).
51. P. R. Kemper and M. T. Bowers, *Int. J. Mass Spectrom. Ion Phys.*, **52**, 1 (1983).
52. R. T. McIver, Jr., *Rev. Sci. Instrum.* **41**, 555 (1970).
53. T. B. McMahon and J. L. Beauchamp, *Rev. Sci. Instrum.* **43**, 509 (1972).
54. R. T. McIver, Jr., R. L. Hunter, E. B. Ledford, Jr., M. J. Locke, and T. J. Francl, *Int. J. Mass Spectrom. Ion Phys.*, **39**, 65 (1981).

REFERENCES

55. M. B. Comisarow and A. G. Marshall, *Chem. Phys. Lett.*, **25**, 282 (1974).
56. M. B. Comisarow and A. G. Marshall, *Chem. Phys. Lett.*, **26**, 489 (1974).
57. M. B. Comisarow and A. G. Marshall, *J. Chem. Phys.*, **62**, 293 (1975).
58. M. B. Comisarow and A. G. Marshall, *J. Chem. Phys.*, **64**, 110 (1976).
59. M. B. Comisarow, *Int. J. Mass Spectrom. Ion Phys.*, **26**, 369 (1978).
60. M. B. Comisarow, *Adv. Mass Spectrom*, **7B**, 1042 (1978).
61. M. B. Comisarow, *J. Chem. Phys.*, **69**, 4097 (1978).
62. G. Parisod and M. B. Comisarow, *Chem. Phys. Lett.*, **62**, 303 (1979).
63. M. B. Comisarow, in H. Hartmann and K. Wanczek, Eds., *Lecture Notes in Chemistry*, Springer-Verlag, Berlin, 1982, vol. 31, p. 484.
64. M. B. Comisarow, *Int. J. Mass Spectrom. Ion Phys.*, **37**, 251 (1981).
65. M. B. Comisarow, *Anal. Chim. Acta*, **178**, 115 (1985).
66. A. G. Marshall, T.-C. L. Wang, and T. L. Ricca, *Chem. Phys. Lett.*, **105**, 233 (1984).
67. T.-C. L. Wang and A. G. Marshall, *Int. J. Mass Spectrom. Ion Phys.*, **68**, 287 (1986).
68. A. G. Marshall, L. Chen, A. T. Hsu, S. L. Mullen, T. L., I. Santos, R. E. Shomo, II, T.-C. L. Wang, C. R. Weisenberger, in J. F. Todd, Ed., *Advances in Mass Spectrometry*, Wiley, New York, 1986, p. 427.
69. R. L. White, E. B. Ledford, Jr., S. Ghaderi, C. L. Wilkins, and M. L. Gross, *Anal. Chem.*, **52**, 1525 (1980).
70. A. G. Marshall and D. C. Roe, *J. Chem. Phys.*, **73**, 1581 (1980).
71. E. B. Ledford, Jr., D. L. Rempel, and M. L. Gross, *Anal. Chem.*, **56**, 2744 (1984).
72. M. Allemann, Hp. Kellerhals, K.-P. Wanczek, *Int. J. Mass Spectrom. Ion Phys.*, **46**, 139 (1983).
73. R. T. McIver, *Am. Lab.*, November 1980.
74. D. L. Rempel, E. B. Ledford, Jr., S. K. Huang, and M. L. Gross, *Anal. Chem.*, **59**, 2527 (1987).
75. J. W. Cooley and J. W. Tukey, *Math. Comp.* **19**, 9 (1975).
76. J. P. Lee and M. B. Comisarow, *Appl. Spectrosc.*, **41**, 93 (1987)
77. M. B. Comisarow, *Adv. Moss Spectrom.* **8B**, 1698 (1980).
78. (a) E. B. Ledford, Jr., T. M. Sack, D. L. Rempel, and M. L. Gross, *Anal. Chem.*, in press. (b) M. Wang and A. G. Marshall, *Anal. Chem.*, **60**, 341 (1988).
79. A. Rahbee, *Chem. Phys.*, **117**, 352 (1985).
80. A. Rahbee, *Int. J. Mass Spectrom. Ion Phys.*, **72**, 3 (1986).
81. M. Allemann, Hp. Kellerhals, and K. P. Wanczek, in H. Hartmann and K.-P. Wanczek, Eds., *Lecture Notes in Chemistry*, Springer-Verlag, Berlin, vol. 31, p. 380, 1982.
82. E. Lippmaa, R. Pikver, E. Suurmaa, J. Past, J. Puskar, I. Koppel, and A. A. Tammik, *Phys. Rev. Lett.*, **54**, 285 (1985).
83. E. N. Nikolaev, Y. I. Neronov, M. V. Gorshkov, and V. L. Tal'roze, *JETP Lett.*, **39**, 534 (1984).
84. E. B. Ledford, Jr., S. Ghaderi, R. L. White, R. B. Spencer, P. S. Kulkarni, C. L. Wilkins, and M. L. Gross, *Anal. Chem.*, **52**, 463 (1980).
85. T. Francl, M. G. Sherman, R. L. Hunter, M. J. Locke, W. D. Bowers, and R. T. McIver, Jr., *Int. J. Mass Spectrom. Ion Phys.*, **54**, 189 (1983).
86. J. B. Jeffries, S. E. Barlow, and G. H. Dunn, *Int. J. Mass Spectrom. Ion Phys.*, **54**, 169 (1983).
87. R. L. White, E. C. Onyiriuka, and C. L. Wilkins, *Anal. Chem.*, **55**, 339 (1983).
88. D. A. Laude, Jr., C. L. Johlman, R. S. Brown, C. F. Ijames, and C. L. Wilkins, *Anal. Chim. Acta*, **178**, 67 (1985).
89. C. L. Johlman, D. A. Laude, Jr., and C. L. Wilkins, *Anal. Chem.*, **57**, 1040 (1985).
90. R. E. Shomo, II, A. G. Marshall, and C. R. Weisenberger, *Anal. Chem.*, **57**, 2940 (1985).

91. C. E. Brown, P. Kovacic, C. E. Wilkie, R. B. Cody, and J. A. Kinsinger, *J. Polym. Sci. Polym. Lett. Ed.*, **23**, 453 (1985).
92. M. Allemann, Hp. Kellerhals, and K.-P. Wanczek, *Chem. Phys. Lett.*, **84**, 547 (1981).
93. C. Giancaspro and M. B. Comisarow, *Appl. Spectrosc.*, **37**, 153 (1983).
94. T. J. Francl, R. L. Hunter, and R. T. McIver, Jr., *Anal. Chem.*, **55**, 2094 (1983).
95. *Collision Spectroscopy*, R. G. Cooks, Ed., Plenum Press, New York, 1978.
96. F. W. McLafferty and F. M. Bockhoff, *Anal. Chem.*, **50**, 69 (1978).
97. R. A. Yost and C. G. Enke, *J. Am. Chem. Soc.*, **100**, 2274 (1978).
98. R. B. Cody, R. C. Burnier, C. J. Cassady, and B. S. Freiser, *Anal. Chem.*, **54**, 2225 (1982).
99. M. B. Comisarow, G. Parisod, and V. Grassi, *Chem. Phys. Lett.*, **57**, 413 (1978).
100. B. S. Freiser, *Anal. Chim. Acta*, **178**, 137 (1985).
101. J. C. Kleingeld and N. M. M. Nibbering, *Tetrahedron*, **24**, 4193 (1983).
102. T.-C. L. Wang, T. L. Ricca, and A. G. Marshall, *Anal. Chem.*, **58**, 2935 (1986).
103. J. C. Kleingeld and N. M. M. Nibbering, *Org. Mass Spectrom*, **17**, 136 (1982).
104. R. B. Cody and B. S. Freiser, *Int. J. Mass Spectrom. Ion Phys.*, **41**, 199 (1982).
105. R. B. Cody, R. C. Burnier, and B. S. Freiser, *Anal. Chem.* **54**, 96 (1982).
106. F. Kaplan, *J. Am. Chem. Soc.*, **90**, 4483 (1968).
107. J. M. S. Henis, *J. Am. Chem. Soc.*, **90**, 845 (1968).
108. R. T. McIver, Jr., Workshop on FTMS, 29th Annual Conference of Mass Spectrometry and Allied Topics, Minneapolis, Minn., p. 791, 1981.
109. D. M. Fedor, R. B. Cody, D. J. Burinsky, B. S. Freiser, and R. G. Cooks, *Int. J. Mass Spectrom. Ion Phys.*, **39**, 55 (1981).
110. S. A. McLuckey, L. Sallans, R. B. Cody, R. C. Burnier, S. Verma, B. S. Freiser, and R. G. Cooks, *Int. J. Mass Spectrom. Ion Phys.*, **44**, 215 (1982).
111. J. C. Kleingeld, Ph.D. dissertation, University of Amsterdam, 1984.
112. W. D. Bowers, S.-S. Delbert, and R. T. McIver, Jr. *Anal. Chem.*, **58**, 969 (1986).
113. *Tandem Mass Spectrometry*, F. W. McLafferty, Ed., Wiley, New York, 1983.
114. F. W. McLafferty and I. J. Amster, *Int. J. Mass Spectrom Ion Proc.*, **72**, 85 (1986).
115. R. L. White and C. L. Wilkins, *Anal. Chem.*, **54**, 2211 (1982).
116. D. A. McCrery, D. A. Peake, and M. L. Gross, *Anal. Chem.*, **57**, 1181 (1985).
117. R. B. Cody, J. A. Kinsinger, S. Ghaderi, J. I. Amster, F. W. McLafferty, and C. E. Brown, *Anal. Chim. Acta*, **178**, 43 (1985).
118. J. Allison, in S. J. Lippard, Ed., *Progress in Inorganic Chemistry*, Wiley-Interscience, New York, vol. 34, 1986, p. 628.
119. D. B. Jacobson and B. S. Freiser, *J. Am. Chem. Soc.*, **105**, 736 (1983).
120. D. B. Jacobson and B. S. Freiser, *J. Am. Chem. Soc.*, **105**, 5197 (1983).
121. D. B. Jacobson and B. S. Freiser, *J. Am. Chem. Soc.*, **105**, 7484 (1983).
122. D. B. Jacobson and B. S. Freiser, *J. Am. Chem. Soc.*, **105**, 7492 (1983).
123. T. J. Carlin, L. Sallans, C. J. Cassady, D. B. Jacobson, and B. S. Freiser, *J. Am. Chem. Soc.*, **105**, 6320 (1983).
124. D. B. Jacobson and B. S. Freiser, *J. Am. Chem. Soc.*, **106**, 1159 (1984).
125. D. B. Jacobson and B. S. Freiser, *Organometallics*, **3**, 513, (1984).
126. T. C. Jackson, D. B. Jacobson, and B. S. Freiser, *J. Am. Chem. Soc.*, **106**, 1252 (1984).
127. D. B. Jacobson and B. S. Freiser, *J. Am. Chem. Soc.*, **106**, 3891 (1984).
128. D. B. Jacobson and B. S. Freiser, *J. Am. Chem. Soc.*, **106**, 3900 (1984).
129. D. B. Jacobson and B. S. Freiser, *J. Am. Chem. Soc.*, **107**, 2605 (1985).

130. D. B. Jacobson and B. S. Freiser, *J. Am. Chem. Soc.*, **107**, 4373 (1985).
131. D. B. Jacobson and B. S. Freiser, *J. Am. Chem. Soc.*, **107**, 67 (1985).
132. D. B. Jacobson and B. S. Freiser, *J. Am. Chem. Soc.*, **106**, 4623 (1984).
133. D. B. Jacobson and B. S. Freiser, *J. Am. Chem. Soc.*, **106**, 5351 (1984).
134. D. B. Jacobson and B. S. Freiser, *Organometallics*, **4**, 1048 (1985).
135. D. B. Jacobson and B. S. Freiser, *J. Am. Chem. Soc.*, **107**, 1581 (1985).
136. D. B. Jacobson and B. S. Freiser, *J. Am. Chem. Soc.*, **107**, 7399 (1985).
137. R. L. Hettich and B. S. Freiser, *J. Am. Chem. Soc.*, **107**, 6222 (1985).
138. T. C. Jackson, T. J. Carlin, and B. S. Freiser, *J. Am. Chem. Soc.*, **108**, 1120 (1986).
139. C. J. Cassady and B. S. Freiser, *J. Am. Chem. Soc.*, **108**, 5690 (1986).
140. T. C. Jackson and B. S. Freiser, *Int. J. Mass Spec. Ion Phys.*, **72**, 169 (1986).
141. R. C. Burnier, R. B. Cody, and B. S. Freiser, *J. Am. Chem. Soc.*, **104**, 7436 (1982).
142. C. J. Cassady, B. S. Freiser, and D. H. Russell, *Org. Mass Spectrom.*, **18**, 378 (1983).
143. S. W. Froelicher, R. E. Lee, R. R. Squires, and B. S. Freiser, *Org. Mass Spectrom*, **20**, 4 (1985).
144. S. W. Froelicher, B. S. Freiser, and R. R. Squires, *J. Am. Chem. Soc.*, **106**, 6863 (1984).
145. S. W. Froelicher, B. S. Freiser, and R. R. Squires, *J. Am. Chem. Soc.*, **108**, 2853 (1986).
146. (a) D. L. Bricker, T. A. Adams, Jr., and D. H. Russell, *Anal. Chem.*, **55**, 2417 (1983); (b) I. J. Amster, J. A. Loo, J. J. P. Furlong, and F. W. McLafferty, *Anal. Chem.*, **59**, 313 (1987). (c) D. H. Russell and D. L. Bricker, *Anal. Chim. Acta.*, **178**, 117 (1985).
147. J. H. Beynon and R. G. Cooks, *Int. J. Mass Spectrom. Ion Phys.*, **19**, 107 (1976).
148. A. F. Weston, K. R. Jennings, S. Evans, and R. M. Elliot, *Int. J. Mass Spectrom. Ion Phys.*, **20**, 317 (1976).
149. G. J. Louter, A. J. Boerboom, P. F. Stalmeier, H. H. Tuithof, and J. Kistemaker, *Int. J. Mass Spectrom. Ion Phys.*, **33**, 335 (1980).
150. F. W. McLafferty, *Acc. Chem. Res.*, **13**, 33 (1980).
151. F. W. McLafferty, P. J. Todd, D. C. McGilvery, and M. A. Baldwin, *J. Am. Chem. Soc.*, **102**, 3360 (1980).
152. R. B. Cody and B. S. Freiser, *Anal. Chem.*, **54**, 1431 (1980).
153. T. J. Carlin and B. S. Freiser, *Anal. Chem.*, **55**, 571 (1983).
154. T. M. Sack and M. L. Gross, *Anal. Chem.*, **55**, 2419 (1983).
155. T. M. Sack, D. A. McCrery, and M. L. Gross, *Anal. Chem.*, **57**, 1290 (1985).
156. C. L. Johlman, D. A. Laude, Jr., R. S. Brown, and C. L. Wilkins, *Anal. Chem.*, **57**, 2726 (1985).
157. C. Dass, T. M. Sack, and M. L. Gross, *J. Am. Chem. Soc.*, **106**, 5780 (1984).
158. L. Sallans, K. Lane, R. R. Squires, and B. S. Freiser, *J. Am. Chem. Soc.*, **105**, 6352 (1983).
159. L. Sallans, K. R. Lane, R. R. Squires, and B. S. Freiser, *J. Am. Chem. Soc.*, **107**, 4379 (1985).
160. D. B. Jacobson and B. S. Freiser, *J. Am. Chem. Soc.*, **108**, 27 (1986).
161. M. Bensimon and R. Houriet, *Int. J. Mass Spectrom. Ion Proc.*, **72**, 93 (1986).
162. R. A. Forbes, L. M. Lech, and B. S. Freiser, *Int. J. Mass Spectrom. Ion Proc.*, **77**, 107 (1987).
163. H. Kang and J. L. Beauchamp, *J. Am. Chem. Soc.*, **108**, 5663 (1986).
164. N. Aristov and P. B. Armentrout, *J. Am. Chem. Soc.*, **106**, 4065 (1984).
165. M. T. Kinter and M. M. Bursey, *J. Am. Chem. Soc.*, **108**, 1797 (1986).
166. R. P. Clow and J. H. Futrell, *Int. J. Mass Spectrom. Ion Phys.*, **4**, 165 (1970).
167. B. S. Freiser and J. L. Beauchamp, *J. Am. Chem. Soc.*, **98**, 3136 (1976).
168. B. S. Freiser and J. L. Beauchamp, *Chem. Phys. Lett.*, **35**, 35 (1975).
169. B. S. Freiser and J. L. Beauchamp, *J. Am. Chem. Soc.*, **96**, 6260 (1974).
170. B. S. Freiser and J. L. Beauchamp, *J. Am. Chem. Soc.*, **99**, 3214 (1977).

171. T. E. Orlowski, B. S. Freiser, and J. L. Beauchamp, *Chem. Phys.*, **16**, 439 (1976).
172. E. S. Mukhar, I. W. Griffiths, F. M. Harris, and J. H. Beynon, *Int. J. Mass Spectrom. Ion Phys.*, **37**, 159 (1981).
173. C. E. Hamilton, M. A. Duncan, T. S. Zwier, J. C. Weisshaar, G. B. Ellison, V. M. Bierbaum, and S. R. Leone, *Chem. Phys. Lett.*, **94**, 4 (1983).
174. M. L. Vestal and J. H. Futrell, *Chem. Phys. Lett.*, **28**, 559 (1974).
175. R. J. Hughes, R. E. March, and A. B. Young, *Int. J. Mass Spectrom. Ion Phys.*, **47**, 85 (1983).
176. (a) R. L. Hettich and B. S. Freiser, *J. Am. Chem. Soc.*, **108**, 2537 (1986); (b) R. L. Hettich and B. S. Freiser, *J. Am. Chem. Soc.*, **109**, 3543 (1987).
177. R. L. Hettich, T. C. Jackson, E. M. Stanko, and B. S. Freiser, *J. Am. Chem. Soc.*, **108**, 5086 (1986).
178. J. P. Honovich, R. C. Dunbar, and T. Lehman, *J. Phys. Chem.*, **89**, 2513 (1985).
179. R. C. Dunbar and E. W. Fu, *J. Phys. Chem.*, **81**, 1531 (1977).
180. M. S. Kim and R. C. Dunbar, *Chem. Phys. Lett.*, **60**, 247 (1979).
181. N. B. Lev and R. C. Dunbar, *J. Phys. Chem.*, **87**, 1924 (1982).
182. R. C. Dunbar, J. D. Hays, J. P. Honovich, and N. B. Lev, *J. Am. Chem. Soc.*, **102**, 3950 (1980).
183. J. P. Honovich and R. C. Dunbar, *J. Am. Chem. Soc.*, **104**, 6220 (1982).
184. G. Baykut, C. H. Watson, R. R. Weller, and J. R. Eyler, *J. Am. Chem. Soc.*, **107**, 8036 (1985).
185. C. H. Watson, G. Baykut, and J. R. Eyler, 34th Conference on Mass Spectrometry and Allied Topics, Cincinnati, Ohio, 1986, p. 681.
186. M. D. A. Mabud, M. J. Dekrey, and R. Graham Cooks, *Int. J. Mass Spectrom. Ion Proc.*, **67**, 285 (1985).
187. W. D. Bowers, S.-S. Delbert, R. L. Hunter, and R. T. McIver, Jr., *J. Am. Chem. Soc.*, **106**, 7288 (1984).
188. E. B. Ledford, Jr., R. L. White, S. Ghaderi, C. L. Wilkins, and M. L. Gross, *Anal. Chem.*, **52**, 2450 (1980).
189. R. L. White and C. L. Wilkins, *Anal. Chem.*, **54**, 2443 (1982).
190. C. L. Wilkins, *Science*, **222**, 291 (1983).
191. D. A. Laude, G. M. Brissey, C. F. Ijames, R. S. Brown, and C. L. Wilkins, *Anal. Chem.*, **56**, 1163 (1984).
192. T. M. Sack, D. A. McCrery, and M. L. Gross, 32nd Annual Conference on Mass Spectrometry and Allied Topics, San Antonio, 1984, p. 591.
193. S. L. Mullen and A. G. Marshall, *Anal. Chim. Acta*, **178**, 17 (1985).
194. (a) D. A. Laude, Jr., C. L. Johlman, J. R. Cooper, and C. L. Wilkins, *Anal. Chem.*, **57**, 1044 (1985); (b) R. B. Spencer, S. Loh, D. B. Stauffer, and F. W. McLafferty, in J. F. J. Todd, Ed., *Advances in Mass Spectrometry 1985*, Wiley, New York, 1986, vol. B, p. 1213.
195. S. G. Shore, D.-Y. Jan, W.-L. Hsu, S. Kennedy, J. C. Huffman, T.-C. Wang, and A. G. Marshall, *J. Chem. Soc., Chem. Commun.*, 392 (1984).
196. L.-Y. Hsu, W.-L. Hsu, D.-Y. Jan, A. G. Marshall, and S. G. Shore, *Organometallics*, **3**, 591 (1984).
197. R. L. Hunter and R. T. McIver, Jr., *Anal. Chem.*, **51**, 699 (1979).
198. S. Ghaderi, P. S. Kulkarni, E. B. Ledford, Jr., C. L. Wilkins, and M. L. Gross, *Anal. Chem.*, **53**, 428 (1981).
199. M. Allemann, Hp. Kellerhals, K.-P. Wanczek, *Chem. Phys. Lett.*, **75**, 328 (1980).
200. R. B. Cody, R. C. Burnier, W. D. Reents, Jr., T. J. Carlin, D. A. McCrery, R. K. Lengal, and B. S. Freiser, *Int. J. Mass Spectrom. Ion Phys.*, **33**, 37 (1980).
201. R. C. Burnier, G. D. Byrd, and B. S. Freiser, *Anal. Chem.*, **52**, 1641 (1980).

202. R. C. Burnier, G. D. Byrd, and B. S. Freiser, *J. Am. Chem. Soc.*, **103**, 4360 (1981).
203. G. D. Byrd, R. C. Burnier, and B. S. Freiser, *J. Am. Chem. Soc.*, **104**, 3565 (1982).
204. G. D. Byrd and B. S. Freiser, *J. Am. Chem. Soc.*, **104**, 5944 (1982).
205. M. B. Wise, D. B. Jacobson, Y. Huang, and B. S. Freiser, *Organometallics*, **6**, 346 (1987).
206. T. J. Carlin, Ph.D. Thesis, Purdue University, 1983.
207. R. A. Forbes, E. C. Tews, B. S. Freiser, M. B. Wise, and S. P. Perone, *Anal. Chem.*, **58**, 684 (1986).
208. M. Lombarski and J. Allison, *Int. J. Mass Spectrom. Ion Phys.*, **49**, 281 (1983).
209. (a) D. A. Peake and M. L. Gross, *Anal. Chem.*, **57**, 115 (1985); (b) D. A. Peake, S.-K. Huang, and M. L. Gross, *Anal. Chem.*, **59**, 1557 (1987).
210. D. W. Squire and R. B. Bernstein, *J. Phys. Chem.*, **88**, 4944 (1984).
211. J. J. Yang, J. D. Simon, and M. A. El-Sayed, *J. Phys. Chem.*, **88**, 6091 (1984).
212. M. D. Morse, G. P. Hansen, P. R. R. Langridge-Smith, L. S. Zheng, M. E. Geusic, D. L. Michalopoulos, and R. E. Smalley, *J. Chem. Phys.*, **80**, 5400 (1984).
213. M. P. Irion, W. D. Bowers, R. L. Hunter, F. S. Rowland, and R. T. McIver, Jr., *Chem. Phys. Lett.*, **93**, 375 (1982).
214. T. J. Carlin and B. S. Freiser, *Anal. Chem.*, **55**, 955 (1983).
215. C. H. Watson, G. Baykat, M. A. Battiste, and J. R. Eyler, *Anal. Chim. Acta*, **178**, 125 (1985).
216. M. A. Posthumus, P. G. Kistemaker, H. L. Meuzelaar, M. Ten Noever de Braun, *Anal. Chem.*, **50**, 985 (1978).
217. K. Takayama, N. Qureshi, K. Hyver, J. Honovich, R. J. Cotter, P. Mascagni, and H. Schneider, *J. Biol. Chem.* **261**, 10624 (1986).
218. T. A. Dang, R. J. Day, and D. M. Hercules, *Anal. Chem.*, **56**, 866 (1984).
219. D. V. Davis, R. G. Cooks, B. N. Meyer, and J. L. McLaughlin, *Anal. Chem.*, **55**, 1302 (1983).
220. (a) W. D. Reents, Jr., M. L. Mandich, and V. E. Bondybey, *Chem. Phys. Lett.*, **131**, 1 (1986); (b) M. L. Mandich, V. E. Bondybey, and W. D. Reents, Jr., *J. Chem. Phys.*, **86**, 4245 (1987).
221. D. A. McCrery, E. B. Ledford, Jr., and M. L. Gross, *Anal. Chem.*, **54**, 1435 (1982).
222. C. L. Wilkins, D. A. Weil, and C. L. Ijames, *Anal. Chem.*, **57**, 520 (1985).
223. D. A. McCrery and M. L. Gross, *Anal. Chim. Acta*, **178**, 91 (1985).
224. D. A. McCrery and M. L. Gross, *Anal. Chim. Acta*, **178**, 105 (1985).
225. R. B. Cody, J. A. Kinsinger, S. Ghaderi, I. J. Amster, F. W. McLafferty, and C. E. Brown, *Anal. Chim. Acta*, **178**, 43 (1985).
226. M. L. Coates and C. L. Wilkins, *Biomed. Mass Spectrom.*, **12**, 424 (1985).
227. R. B. Brown, D. A. Weil, and C. L. Wilkins, *Macromolecules*, **19**, 1255 (1986).
228. (a) R. E. Shomo, II, and A. G. Marshall, *Int. J. Mass Spectrom. Ion Proc.*, **72**, 209 (1986); (b) A. T. Hsu and A. G. Marshall, *Anal. Chem.*, in press.
229. C. L. Wilkins and C. L. C. Yang, *Int. J. Mass Spectrom. Ion Proc.*, **72**, 195 (1986).
230. M. G. Sherman, J. R. Kingsley, J. C. Hemminger, and R. T. McIver, Jr., *Anal. Chim. Acta*, **178**, 79 (1985).
231. M. E. Castro and D. H. Russell, *Anal. Chem.*, **56**, 578 (1984).
232. M. E. Castro and D. H. Russell, *Anal. Chem.*, **57**, 2290 (1985).
233. I. J. Amster, F. W. McLafferty, M. E. Castro, D. H. Russell, R. B. Cody, and S. Ghaderi, *Anal. Chem.*, **58**, 483, (1985).
234. D. F. Torgerson, R. P. Skowranski, and R. D. MacFarlane, *Biochem. Biophys. Res. Commun.*, **60**, 616 (1974).
235. R. D. Macfarlane and D. F. Torgerson, *Science*, **191**, 920 (1976).

236. R. D. Macfarlane, *Anal. Chim.*, **55**, 1247A (1983).
237. S. K. Viswanadham, D. M. Hercules, R. R. Weller, and C. S. Giam, *Biomed. Environ. Mass Spectrom.*, **14**, 43 (1987).
238. A. G. Marshall, *Chem. Phys. Lett.*, **63**, 515 (1979).
239. A. G. Marshall, T.-C. Wang, and T. L. Ricca, *J. Am. Chem. Soc.*, **107**, 7893 (1985).
240. R. T. McIver, Jr., R. L. Hunter, and W. D. Bowers, *Int. J. Mass Spectrom. Ion Proc.*, **64**, 67 (1985).
241. D. F. Hunt, J. Shabanowitz, R. T. McIver, Jr., R. L. Hunter, and J. E. P. Syka, *Anal. Chem.*, **57**, 765 (1985).
242. D. F. Hunt, J. Shabanowitz, J. R. Yates, R. T. McIver, Jr., R. L. Hunter, J. E. P. Syka, and J. Amy, *Anal. Chem.*, **57**, 2728 (1985).
243. P. Kofel, M. Allemann, Hp. Kellerhals, and K.-P. Wanczek, *Int. J. Mass Spectrom. Ion Proc.*, **65**, 97 (1985).
244. P. Kofel, M. Allemann, and P. Grossmann, 34th Annual Conference on Mass Spectrometry and Allied Topics, Cincinnati, Ohio, June 1986, p. 886.
245. P. Kofel, M. Allemann, Hp. Kellerhals, and K.-P. Wanczek, *Int. J. Mass Spectrom. Ion Proc.*, **72**, 53 (1986).
246. J. M. Alford, P. E. Williams, D. J. Trevor, and R. E. Smalley, *Int. J. Mass Spectrom. Ion Proc.*, **72**, 33 (1986).
247. S. Ghaderi and D. P. Littlejohn, 33rd Conference on Mass Spectrometry and Allied Topics, San Diego, Calif. May 1985. p. 727
248. C. Giancaspro., F. R. Verdun, and J. F. Muller, *Int. J. Mass Spectrom. Ion Proc.*, **72**, 63 (1986).
249. R. B. Cody and B. S. Freiser, *Anal. Chem.*, **59**, 1054 (1987).
250. A. G. Marshall, Proceedings from the Industry-University Cooperative Chemical Program Symposium on New Directions in Chemical Analysis, Texas A & M University, College Station, Texas, 1985, p. 111.

Chapter **III**

NEUTRAL PRODUCTS FROM ELECTRON BOMBARDMENT FLOW STUDIES

Thomas Hellman Morton

---

1 **Introduction**
   1.1 Comparison of Electron Impact and Photoionization
   1.2 Origins of Recovered Products
2 **Scope and Limitations of the EBFlow Technique**
   2.1 Scope
   2.2 Limitations
   2.3 Experimental Plan
       2.3.1 Cationic Rearrangements
       2.3.2 Corroboration by Mass Spectrometry
       2.3.3 Effects of Added Base or Increased Pressure
       2.3.4 Isotopic Labeling
3 **Apparatus and Operating Conditions**
   3.1 Ion Capture Radius (r_i)
       3.1.1 Magnetron Motion
       3.1.2 Cyclotron Motion
       3.1.3 Molecular Collisions
   3.2 Apparatus Design
       3.2.1 Electron Gun
       3.2.2 Apparatus A
       3.2.3 Apparatus B
       3.2.4 Apparatus C
   3.3 Operating Conditions
4 **Reaction Types**
   4.1 Unimolecular Ion Decompositions
       4.1.1 EBFlow Radiolyses of Alkyl Phenyl Ethers
       4.1.2 Comparison with Other Reactions
       4.1.3 Comparison with Free Alkyl Cations
   4.2 Bimolecular Proton Transfer Reactions
   4.3 Ion–Molecule Association Reactions
   4.4 Bimolecular Lewis Acid–Base Reactions
Acknowledgment
References

1 INTRODUCTION

The electrically uncharged products of ionic reactions in the gas phase often contain information that is otherwise inaccessible to the mass spectroscopist. Only occasionally can the structure of a gaseous ion be inferred from its mass-to-charge ratio (m/z) and the mass-to-charge ratios of its reaction products. Usually it is difficult even to ascertain whether a given m/z value corresponds to a single isomer or a mixture of several isomers. But analysis of the neutral product distribution from an ion–molecule reaction frequently reveals features of the ion's structure, its choice of reaction channels, and the dynamics of the reaction. The electron bombardment flow (EBFlow) reactor, first reported at a scientific meeting in 1974 and published in 1976 [1], represents a general design for collecting and identifying the uncharged products of homogeneous, gas-phase ion–molecule reactions.

This chapter will describe the electron bombardment flow reactor, principles of its operation, and strategy of its application to the study of gaseous positive ions at pressures $\leqslant 10^{-3}$ torr. The heart of the reactor is a cylindrical reaction vessel with an electron beam focused along its central axis. Reagent gases flow through the reaction vessel and are condensed in a liquid nitrogen-cooled trap. Three versions of the apparatus have been built. They are described in Section 3 and are designated as A, B, and C. The acronym for the technique, EBFlow (pronounced "ebb flow"), serves as a reminder that no complete theory has been developed to describe ion motion within the reaction vessel and that several control experiments are often needed to provide empirical validation for any interpretation of EBFlow results.

Several methods have been reported for identifying neutral products of ion–molecule reactions. Lieder and Brauman have reported a mass spectrometric technique based on ion cyclotron resonance (ICR) for analyzing the products of gas-phase nucleophilic displacements by halide anions [2]. Ellison and co-workers have collected neutral products from anionic reactions in flowing afterglow experiments [3]. Charge stripping of negative ions ($R^- \to R^{\cdot} \to R^+$, in which the neutral product is the intermediate radical) has been used to probe the structures of stabilized carbanions [4].

With regard to positive ions, the EBFlow reactor is designed to collect stable, neutral products under conditions that closely approximate those inside a mass spectrometer or ICR. Other methods for examining neutral products of gas-phase cationic reactions include photoionization [5], γ-radiolysis [6], and analysis of the radioactive products from β-decay of multiply tritiated molecules [7]. These techniques, in general, operate at total pressures between 1 torr and 1 atm, while the EBFlow operates at pressures below 10^{-3} torr. Other low-pressure methods complement the EBFlow technique. Neutralization of positive ions by charge transfer from metal vapors to afford unstable or metastable products [8] offers an extraordinary way to probe ion structures, although the structure of the neutral products is not examined directly. Workers at DuPont have chemically intercepted radicals from ion decomposition and neutralization in the mass spectrometer and have subsequently characterized the adducts by reionization [9]. Research groups at Ottawa and Utrecht have characterized expelled radicals by collisional ionization and decomposition [10].

1.1 Comparison of Electron Impact and Photoionization

Electron bombardment efficiently produces positive ions at low pressures. In terms of ion production rate as a function of incident power, there is no readily available method more efficient than 30–70-eV electron impact. The following comparison with photolysis provides order-of-magnitude estimates. Suppose we wish to form positive ions in a gas at 10^{-4} torr (corresponding to a number density $\rho = 3.2 \times 10^{12}$ molecules/cm^3). The cross sections for ionization of organic molecules by 70-eV electrons have been widely measured, and empirical relationships for predicting these cross sections are available [11–13]. Ionization of a molecule with molecular weight ≈ 100 has a cross section on the order of $\sigma_I = 10^{-15}$ cm^2 ($= 10$ Å^2). At the number density corresponding to 10^{-4} torr, the mean free path of a 70-eV electron is $1/\sigma_I \rho = 3$ m. For an electron current of 0.2 mA (1.2 × 10^{15} electrons/s, corresponding to an incident power of 14 mW, easily obtainable in the EBFlow reactor) traveling a distance of 1 m, positive ions are produced by primary electrons at a rate of 0.6 nmol s^{-1}. This corresponds to an ion production of 3 μmol per Ampere-second (i.e., 3 μmol/C of bombarding electrons) and neglects ions that may be produced by secondary electrons.

Compare this with the photon flux required from an argon resonance lamp (11.6–11.8 eV) to afford the same ion production rate. The extinction coefficient for absorption of photons is $\varepsilon < 10^5 M^{-1}$ cm^{-1} (usually very much less) with a quantum yield for ionization $\phi < 1$. Passing the photons through a 1-m path length of gas at 10^{-4} torr yields at most one ion per 20 photons (as opposed to six ions per twenty 70-eV electrons). In order to produce positive ions at a rate of 0.6 nmol s^{-1}, continuous irradiation with > 13 mW of ionizing photons would be needed, at least as much incident power as for 70-eV electrons. The usable output of a typical argon resonance lamp, however, is on the order of 0.2 mW (10^{14} photons/s) [14]. Other types of lamp [15] are capable of producing 10^{15} photons/s but at longer wavelengths where the quantum yield for ionization is substantially less than unity (e.g., $\phi < 0.3$ for C_4 hydrocarbons at 1236 Å [14]).

1.2 Origins of Recovered Products

What proportion of the radiolysis products from 70-eV electron bombardment result from cation formation? This has been explored for alkyl bromides, and the data in Table 1 are representative. The 70-eV mass spectra of n-alkyl bromides are sufficiently uncomplicated to permit a dissection of the neutral products into two categories: rearranged products that must have come from $[M - Br]^+$ ions ($C_4H_9^+$, which is 23% of the total ionization from 1-bromobutane, or $C_5H_{11}^+$, which is 14% of the total ionization from 1-bromopentane [18]), and everything else.

Table 1. Relative Yields of Linear Alkenes and Principal Cationic Rearrangement Products from 70-eV Electron Impact on n-Alkyl Bromides ($C_nH_{2n+1}Br$) at a Total Pressure of 4×10^{-4} torr ([Alkyl Bromide]/[Ether] = 3) in apparatus A

$C_nH_{2n+1}Br$	Ether	Linear C_nH_{2n}	Cationic Rearrangement Products
1-bromobutane [16]	None	1-butene = 1	2-butenes = 0.5; isobutene = 0.4; methylcyclopropane = 0.2
1-bromobutane [16]	Et_2O	1-butene = 1	2-butenes = 1.8; isobutene = 0.8; methylcyclopropane = 0.4
1-bromopentane [17]	Me_2O	1-pentene = 1	2-methyl-1-butene = 1.4; 2-methyl-2-butene = 1.2; ethylcyclopropane = 0.1

All of the C_4H_8 products from 1-bromobutane must come from butyl cations except for 1-butene, which might also arise via n-butyl radicals (which have never been observed to undergo isomerization) or via elimination of HBr. EBFlow radiolysis of neat 1-bromobutane afforded 1-butene as approximately half of the C_4H_8 yield. We can envisage that, in the absence of a suitable base, $C_4H_9^+$ cations are not efficiently converted to C_4H_8. Indeed, addition of diethyl ether (which is more basic than any C_4H_8 isomer) increased the proportion of the C_4H_8 products that must have arisen from $C_4H_9^+$ cations—cis and trans-2-butene, isobutene, and methylcyclopropane (no cyclobutane was observed)—to the extent that 1-butene represented only one-quarter of the C_4H_8 yield in the presence of ether. The effect of base represents one of the control experiments that can be used to assess the contributions of homogeneous gas-phase reactions to the neutral product yield. In the absence of ether, cations either must react with the parent molecule or else be neutralized via a heterogeneous reaction (e.g., collision with a wall). A likely outcome from the heterogeneous reaction is formation of radicals, which will have to encounter other radicals (represented as R·) in order to terminate. As Eq. 1 depicts, C_4H_8 is only one of several possible products from the reactions of C_4H_9·radical:

$$C_nH_{2n+1}^+ \xrightarrow{\text{wall}} C_nH_{2n+1}\cdot \xrightarrow{R\cdot} \begin{cases} C_nH_{2n+2} \\ C_nH_{2n} \\ C_nH_{2n+1}R \end{cases} \qquad (1)$$

When base is added, alkyl cations are efficiently deprotonated in the gas phase, and the products whose yields increase in the presence of base can be inferred to result from homogeneous deprotonation.

Radiolysis of 1-bromopentane in the presence of dimethyl ether produced 2-methylbutenes and ethylcyclopropane (neither methylcyclobutane nor dimethylcyclopropanes were detected) as at least half of the C_5H_{10} yield. Since 1-pentyl radicals are known to rearrange to 2-pentyl radicals [19], 2-pentenes cannot be unambiguously assigned as cation products (as could the 2-butenes from 1-bromobutane). Rearrangement of the carbon skeleton, however, is a clear-cut indication of the intermediacy of cations.

Could other cations have contributed to the C_nH_{2n} yields from EBFlow radiolysis of $C_nH_{2n+1}Br$? The mass spectra of the n-alkyl bromides are simple enough to allow this to be ruled out. The 70-eV mass spectrum of 1-bromobutane shows that $C_4H_9^+$ represents three-fourths of all primary ions with four carbons, and it is more abundant than $C_4H_8^{\cdot+}$ by a factor of 6 and more abundant than $C_4H_9Br^{\cdot+}$ by a factor of 10. Although these odd-electron ions might conceivably have yielded the observed products by surface neutralization, their contribution must have been small in comparison to the products of $C_4H_9^+$. Similarly, mass spectrometry of 1-bromopentane shows that $C_5H_{11}^+$ represents four-fifths of all ions with five carbons and that it is 16 times more abundant than $C_5H_{10}^{\cdot+}$ and 12 times more abundant than $C_5H_{11}Br^{\cdot+}$. Here, too, the odd-electron ions must have contributed negligibly to the recovered neutral products. Insofar as ion–molecule reactions of other fragments are concerned, alkyl bromides are either unreactive with their daughter ions or else react predominantly via halide transfer, as Eq. 2 represents:

$$C_nH_{2n+1}Br + R^+ \rightarrow RBr + C_nH_{2n+1}^+ \qquad (2)$$

Ions with less than four carbons are prominent in the mass spectrum of 1-bromopentane and they, too, may have contributed to the neutral product yield. Linear C_3 ions ($C_3H_5^+$, $C_3H_6^{\cdot+}$, and $C_3H_7^+$) together constitute half of the total ionization, Σ. Not surprisingly, neutral products that we would expect from homogeneous and heterogeneous reactions of the C_3 ions, propene and allene (in approximately a 3:1 ratio), were recovered (together in roughly the same yield as 2-methyl-2-butene). We can also envisage many nonionic processes that could have yielded these products. As a gauge of the extent of nonionic fragmentation of the carbon skeleton, note that the only prominent C_4 ion in the mass spectrum is $C_4H_7^+$ (8% Σ), which is unlikely to yield butenes or more highly saturated C_4 products via homogeneous processes. Nonetheless 1-butene and isobutene (in yields of 0.25 and 0.2 relative to 1-pentene) were recovered as the predominant C_4H_8 isomers, as well as butane and isobutane (yields of 0.5 and 0.2, relative to isobutene). Products with higher degrees of unsaturation (e.g., 1, 3-butadiene) were also observed as, of course, was pentane. In terms of mass balance, the cationic rearrangement products in Table 1 represented 10% of the yield of volatile hydrocarbons from 70-eV electron impact on 1-bromopentane in the EBFlow reactor. Note that dimethyl ether and $C_5H_{11}Br$ were both ionized by the electron bombardment in proportion to their partial pressures and their 70-eV ionization cross sections (calculated to be 8.7 and 20.5 Å^2, respectively [12]). The $C_5H_{11}^+$ ion therefore constituted approximately one-eighth of the total ionization within the reaction vessel, and the neutral products of its deprotonation, Eq. 3, were recovered as an appropriate proportion of the products of 70-eV electron impact:

$$CH_3CH_2CH_2CH_2CH_2Br \xrightarrow[\text{electron impact}]{70\,eV} (CH_3)_2\overset{+}{C}CH_2CH_3$$
$$\textit{tert}\text{-amyl cation}$$
$$\xrightarrow{Me_2O} (CH_3)_2C{=}CHCH_3 + H_2C{=}C(CH_3)CH_2CH_3 \quad (3)$$

Comparison with other electron impact methods is appropriate. Electrical discharge and inductively coupled plasmas have been widely explored [20]. Under most experimental conditions, products are observed that are of much higher molecular weight than the reactants (including substantial quantities of polymer), whereas EBFlow products have, in general, molecular weights comparable to those of the reactants. In cases in which plasma products are of comparable size to reactants, mechanisms have been inferred to operate via excitation of neutral molecules and not via ionic processes. In Section 4.1, a comparison will be made between EBFlow radiolysis and electrical discharge results for butyl phenyl ether.

2 SCOPE AND LIMITATIONS OF THE EBFLOW TECHNIQUE

The EBFlow reactor is used in conjunction with mass spectrometric techniques to characterize ion structures and elucidate reaction pathways. Many homogeneous, gas-phase reactions of positive ions produce electrically uncharged products in sufficient yield for them to be identified and quantitated, but careful control experiments are needed to provide verification. The design of the EBFlow reactor enables us to collect these products with a minimum of contamination from heterogeneous processes. Once collected, the reaction mixture can subsequently be analyzed by gas chromatography (GLPC), GLPC-mass spectrometry (GC–MS), nuclear magnetic resonance (NMR), or other conventional methods. The products must be readily distinguishable from starting material, since unreacted starting material is recovered in large excess. And, of course, the appropriate controls must be feasible. Interpreting the neutral products from ion–molecule reactions depends critically upon the specific chemistry under investigation. For that reason, the case histories in Section 4 of the chapter provide the best primer.

2.1 Scope

Experiments performed to date embrace cation–molecule reactions that obey both first- and second-order kinetics. The identities of the neutral products reveal the structural features of the precursor ion. Beyond this, however, additional questions can be posed.

1. How many product molecules result from a given reaction? This is represented schematically by Eq. 4:

$$A^+ + Q \begin{array}{c} \longrightarrow R^+ + XZ \\ \longrightarrow R^+ + X + Z \end{array} \qquad (4)$$

There are cases in which known reagents, ion A^+ and neutral Q, yield a known ionic product, R^+. Nevertheless, the reaction pathway may still be ambiguous if there is more than one plausible, exothermic channel, for instance, one that gives a single molecule of neutral product (XZ) and another that gives two molecules of neutral product (X and Z). A case in which this type of question was answered is presented in Section 4.4.

2. What distribution of structural isomers is produced? Often this will tell whether a reactant ion of given m/z is a mixture of isomers. On the one hand, for example, the deprotonation of $C_4H_9^+$ that yielded the products listed in Table 1 demonstrated that m/z 57 did not correspond to a single classical structure, since linear, branched, and cyclized product were all formed. On the other hand, less variety was seen among the products of butyl cations when they are deprotonated in ion–molecule reactions that obey first-order kinetics, as discussed in Section 4.1.

3. What distribution of positional isomers is produced? When the structure of an ionic reactant is well known, the distribution of positional isomers often tells whether the reaction is under thermodynamic or kinetic control. Section 4.2 describes such cases.

4. What distribution of geometrical or stereoisomers is recovered? This provides mechanistic information about ion–molecule reactions, and Section 4.3 and 4.4 describe examples.

5. Are products formed that are larger than the reactants? Ordinarily, such products are absent; e.g., there is no evidence for formation of $\geqslant C_6$ products from EBFlow radiolysis of 1-bromopentane (Table 1). However, reaction conditions can be arranged so that ion–molecule association products are observed, as Section 4.3 discusses.

6. What is the extent, distribution, and location of isotopic label? GC–MS, although useful, has been of limited use in assessing the isotopic labeling pattern of neutral products, for it cannot in general give the location of the label. Nuclear magnetic resonance has proved more useful in this regard, particularly for fluorine-substituted products. Deuterium-induced chemical shift differences in the proton NMR and the ^{19}F NMR are often large enough to resolve resonances of different deuterated analogs of the same molecule. The pattern corresponding to each extent of deuteration usually tells through spin–spin splitting the distribution and location of the isotopic label. For this reason, NMR analysis of neutral products has greatly broadened the range of the EBFlow technique, as discussed in Section 4.4.

2.2 Limitations

The ability to answer these questions illustrates the scope of the EBFlow technique. Its limitations can be similarly enumerated.

1. The yield of product is small. With a theoretical primary ion yield on the order of 1 nmol s^{-1} and a collection efficiency of approximately 75%, a 1 hour run can be expected to afford no more than 3 μmol of product (unless it comes from a chain reaction). This means that products must usually be identified by comparison with authentic samples as standards in order for high sensitivity analysis (GLPC, GC–MS, or NMR) to provide unambiguous identifications. Where hydrocarbons have been identified by GLPC, practice has been to use polar stationary phases (dimethylsulfolane, n-octane on Porasil, or β,β-oxydipropionitrile) near room temperature. The retention times of most volatile hydrocarbons have been measured under these conditions and used for calibration. Since virtually nothing else elutes on these columns as rapidly as hydrocarbons with seven or fewer carbons, retention time alone (when compared with authentic samples of virtually every other possibility) can provide a clear identification. It becomes impractical to use this approach for larger hydrocarbons and almost impossible (except under very special circumstances) for molecules with heteroatoms. Therefore, EBFlow studies have examined light hydrocarbons or molecules easily identified by NMR.

2. Conversion is low. In order to eliminate products of pyrolysis on the filament, the electron source region of the EBFlow reactor is differentially pumped. For this reason, a continuous flow of reagents is needed; otherwise, the contents of the reaction vessel would be thoroughly depleted within a few minutes. The flow rate needed to maintain steady-state pressure in the range 10^{-4}–10^{-3} torr is of the order of micromoles to millimoles per second. Ordinarily, this means that unreacted starting material is recovered in a 10^2–10^4-fold excess over reaction products. Analysis techniques must therefore not only be capable of great sensitivity and resolution, but also must have wide dynamic range. Moreover, purity of starting materials is of critical importance, and it must be possible to distinguish ion–molecule reaction products from any that might arise from the reagents simply upon standing (e.g., cyclohexene from bromocyclohexane plus ammonia).

3. Nonionic and heterogeneous reactions take place. Pyrolysis on the filament contributes negligibly to recovered product yields, as demonstrated by control experiments in which the filament was turned on without permitting ionizing electrons to enter the reaction vessel. It is more difficult to control for products of excitation (electronic or vibrational) by electron impact within the reaction vessel. The chemistry under investigation must therefore be carefully chosen so that such effects can be ruled out. Interferences can also arise from products formed when ions strike the walls of the reaction vessel and are neutralized (e.g., Eq. 1 above). As will be discussed in Section 3, the design of the reaction vessel has evolved so as to minimize heterogeneous contributions of this nature.

4. Side products are formed. In some cases ions can react by a variety of pathways to yield products of interest. Either different ions can give the same product, or a given ion can lead to similar products by different pathways. For example, ions are well known to cluster at pressures $> 10^{-3}$ torr; for this reason

the EBFlow is operated at or below this pressure range. Some molecules (alcohols, e.g.) cluster efficiently even at these low pressures, and the cluster ions may provide a separate ion–molecule route that competes with a bimolecular pathway. It is possible in some cases to intercept cluster formation and assess their contribution.

2.3 Experimental Plan

These limitations impose stringent constraints on the reactions whose neutral products can be studied. Apart from the design of the EBFlow reactor, four further aspects of the experimental plan enable us to overcome the limitations.

2.3.1 Cationic Rearrangements

One of the major concerns is to discriminate ionic products from products of free radicals. A distinguishing feature of cations is their propensity for undergoing rearrangements that free radicals do not. As noted above, n-butyl radicals have never been observed to undergo rearrangement, whereas the 1, 2-hydrogen shifts and other rearrangements of butyl cations that lead to the neutral products summarized in Table 1 are well precedented in solution as well as in the gas phase. In some cases free radicals enjoy rearrangement pathways that are not accessible to cations, for instance, the cyclization shown in Eq. 5, which has been widely used as a diagnostic for free radical mechanisms [21]:

$$H_2C=CHCH_2CH_2CH_2CH_2\cdot \;\rightarrow\; \underset{}{\bigcirc}\!\!-CH_2^{\cdot} \;\Big/\; \underset{}{\bigcirc}^{\cdot} \;\simeq 50 \qquad (5)$$

The choice of substrates that undergo characteristic rearrangements permits meaningful analysis and interpretation of the tiny quantities of product collected from the EBFlow.

2.3.2 Corroboration by Mass Spectrometry

Neutral product studies are designed to complement mass spectrometry. Without mass spectrometric data it is usually not possible to interpret EBFlow results. We must ordinarily have three types of data in hand before studying the neutral products of electron impact:

1. The fragmentation pattern of the reactants
2. The products of ion–molecule reactions between fragment ions and their neutral precursors
3. The products of ion–molecule reactions with other reagents in the reaction vessel.

Consequently, ICR, FTMS, or high-pressure mass spectrometric results usually must be available in order to interpret a reaction. In the best of cases, fragment ions that are not of interest will either produce more of the ion under study (e.g., Eq. 2) or will "dead end." An example of this latter case is the phenyl ethers discussed in Section 4.1. The stable parent ions do not yield interfering products, since the worst they can do is form free radicals upon surface neutralization. But what happens to the $PhOH^{\cdot+}$ fragments, which are produced in copious quantities? ICR studies [22] show that they simply undergo charge transfer (Eq. 6) to form the comparatively innocuous, stable parent ion, $ROPh^{\cdot+}$:

$$PhOH^{\cdot+} + ROPh \rightarrow PhOH + ROPh^{\cdot+} \tag{6}$$

In the specific examples in Section 4, frequent reference will be made to the use of mass spectrometric studies to evaluate the corroborate EBFlow results.

2.3.3 Effects of Added Base or Increased Pressure

Most of the reactions studied in the EBFlow have been Brønsted or Lewis acid–base reactions where the ion plays the role of acid. When a bimolecular reaction is under study, competition among gas phase (homogeneous) ion–molecule reactions and between homogeneous and heterogeneous (e.g., ion–surface collisions) reactions is an important variable. Variation of the identity of the base has been used to differentiate between two homogeneous pathways that form *sec*-butyl ether from 2-butanol (Section 4.3). Whereas heterogeneous processes are of slight importance in mass spectrometry (since the ion typically loses its charge and becomes undetectable), they can be a dominant concern in neutral product studies. One way to control for heterogeneous processes is to vary the homogeneous reaction rate and examine the effect on the neutral product yield. Table 1 provides an example of this in the case of butyl ions from 1-bromobutane. The addition of a Brønsted base increased the yield of cationic rearrangement products and corroborated their genesis as ion–molecule reaction products. A similar approach has been used to assess the efficiency by which unreacted ions exit from the reaction vessel via the clown cap in apparatus C (Section 4.2). Sometimes the reaction under study occurs between a fragment ion and its parent neutral. In such a case, where there is only one reactant in EBFlow, increasing pressure is equivalent to adding more base (Section 4.4).

2.3.4 Isotopic Labeling

Just as isotopic labeling is a standard feature of mass spectrometric investigations, it also plays a role in neutral product studies. The specific examples in Section 4 provide illustrations. Sometimes ionic and nonionic reactions lead to the same neutral product, but with different extents of labeling when a specifically deuterated precursor is used (Section 4.1). It has even been possible to distinguish two different ionic routes to the same product in a similar way (Section 4.2). Fluorine NMR has turned out to be a remarkably straightforward method to measure the extent and positions of deuteration of neutral product mixtures, since the resonances of differently labeled analogs can often be resolved

(Section 4.4). Labeling studies can be used to probe ion structures as well as reaction pathways.

3 APPARATUS AND OPERATING CONDITIONS

Chemical considerations dictated the design of the EBFlow reactor. The pressure regime ($\leqslant 10^{-3}$ torr) was chosen so as to maximize ion production rate and ion–molecule reaction yield without promoting clustering reactions and ion–electron recombination, processes that require third-body stabilization. Electron–ion neutralizations posed an especially serious concern, for if positive ions were lost by neutralization rapidly compared with ion–molecule reactions the side products would have presented a problem. At low pressure the third-order rate constants in polar gases are on the order of $k_3 \simeq 10^{-23}$ cm^6 molecule^{-2} s^{-1} [23]. At 3×10^{-4} torr this yields an effective second-order rate constant of $\rho k_3 = 10^{-10}$ cm^3 molecule^{-1} s^{-1}, which means that even at the highest plausible electron densities (10^{10} cm^{-3}) an ion persists for 1 s on average before neutralization. A higher pressure regime would not have been compatible with the large electron currents ($> 1\,\mu$A) needed to secure measurable quantities of products. As a consequence, the EBFlow regime turns out to be difficult to characterize in terms of the physics of ion motion; this is because the simplifying assumptions that are ordinarily used do not apply.

Ions and electrons move under the influence of magnetic and electric fields with collisional damping of their motion. Although electric fields are not large (on the order of 0.1–1 V cm^{-1}), the low pressure of gas means that electric field energies are comparable to thermal energies. Charges probably do not reach their terminal velocities, as the mean free path for ion–neutral collisions, λ, is comparable to the dimensions of the EBFlow, and the ratio of electric field to gas density is $\geqslant 100$ townsend (Td) (1 Td = 10^{-17} V cm^2). Reduced ion mobilities vary with electric field in this regime [24], and it is not yet within the scope of transport theory to portray migrations of polyatomic ions in field gradients through polyatomic gases. Similarly, other models for ion motion do not work for the EBFlow, and only a qualitative description of ion motion can be given at the present time.

A consideration of EBFlow geometry will further illustrate the limitations of a theoretical description. The EBFlow reaction vessel is enclosed in a solenoid, which confines reactant ions for times on the order of milliseconds or longer. Three versions of the apparatus have been built, and a simplified schematic of one of them (apparatus B) is shown in Fig. 1. Electrons from a directly heated cathode (immersed in a magnetic field) are focused through a small aperture down the axis of the reaction vessel. The contents of the EBFlow are near the dividing line between a collection of free charges and a plasma. The minimum average ion density required in order to produce a detectable yield of neutral products is on the order of $n = 10^8$ ions cm^{-3}. Every time a positive ion is formed by < 100-eV electron impact, two electrons are formed, one with a kinetic energy of, at most, a few electrons volts [25]. If we assume that this

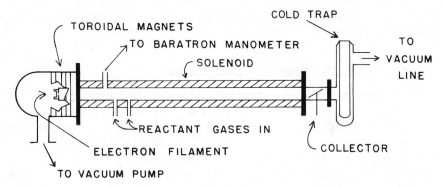

Fig. 1. Schematic of an EBFlow Reactor (apparatus B).

slow electron and the ion both acquire thermal velocities by collisions with neutral molecules, then the Debye shielding length λ_D can be defined as

$$\lambda_D = \left(\frac{kT}{4\pi q^2 n}\right)^{1/2} \qquad (7)$$

The value of λ_D for $n = 10^8$ ions cm^{-3} is 10^{-2} cm, which is considerably less than the dimensions of the reaction vessel and corresponds to the quasineutrality condition of a plasma [26]. However, λ_D is not very large compared to the average distance between ions: In fact λ_D is only five times as large as $(1/n)^{1/3}$ in the above estimate, and the two distances get closer in magnitude at higher ion densities. Therefore, ions are largely shielded from one another, but the EBFlow will not, as a whole, tend to exhibit the collective properties of a plasma. The distribution of charges is not uniform, though, and there are heterogeneities that fluctuate with time. Portions of the EBFlow therefore probably do behave as a plasma. Electron and ion motions in the reaction vessel have not been analyzed in detail, but their qualitative behavior can be described in terms of three microscopic phenomena: magnetron motion, cyclotron motion, and collisions between charged and neutral species (i.e. molecular diffusion).

3.1 Ion Capture Radius (r_i)

Figure 2 displays a qualitative picture of the motion of a positive ion off axis outside the region of quasineutrality. This picture neglects collisions. The large orbit corresponds to the magnetron motion. The small cycloidal orbits correspond to cyclotron motion. A dimension pertinent to apparatus design, the ion capture radius r_i (in meters), can be defined as follows for a singly charged positive ion of mass m (in kilograms per mole):

$$r_i = \left(\frac{2m}{\pi kT}\right)^{1/2}\left(\frac{97I}{B\sqrt{E}} + \frac{310kT}{8mv_c}\right) \qquad (8)$$

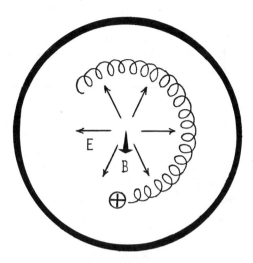

Fig. 2. Cross-sectional view of the EBFlow reaction vessel showing a schematic representation (not to scale) of the magnetron and cyclotron motion of an ion in the absence of collisions. Axial magnetic field B is perpendicular to the page. Radial electric field E is due to the axial electron beam.

where energies are in electron volts, I is electron beam current, E its energy, B the magnetic field, and v_c the cyclotron frequency. Note that r_i is independent of pressure. The discussion below describes the derivation and significance of r_i.

3.1.1 Magnetron Motion

Electrons enter the reaction vessel through an aperture (1 mm diameter) that is much smaller than the radius, r_0, of the reaction vessel. The entering beam can be portrayed as a line of charges whose linear density is proportional to $I/\sqrt{E}$, where I is the beam current (in amperes) and E is the electron energy (in electron volts). The radial electric potential (normal to the vessel axis) is $V = 3.0 \times 10^4 (I/\sqrt{E}) \ln(r/r_0)$, where V is in volts. Repulsion of the electrons by one another (the space-charge effect) will tend to defocus the electrons. The axial magnetic field lines, however, bend electron paths, and, above a critical field B_c, will prevent them from reaching the vessel wall. The critical field (in tesla) is $B_c \cong (8Vm/qr_0^2)^{1/2}$, where the mass-to-charge ratio of the electron is $m/q = 5.68 \times 10^{-12}\,\mathrm{kg\,C^{-1}}$. For apparatus A (with the smallest vessel radius, $r_0 = 12$ mm) the value of B_c is 2.4×10^{-4} T (2.4 G) for a 10^{-5} A beam of 20-eV electrons. For apparatus C (with the highest electron beam currents), B_c has the same value for a 2×10^{-4} A beam of 70-eV electrons. EBFlow experiments are run with magnetic fields that are 10–1000 times more intense than B_c, so that, in the absence of collisions with molecules, the radius of the electron beam remains small compared to that of the reaction vessel. When gases are introduced, the electron beam is scattered. Scattering probabilities are roughly proportional to molecular weight [27], so the effect of introducing gas into the EBFlow reactor is a function or the mass density. In apparatus A, with a magnetic field of 10^{-3}–10^{-2} T, at least half of a 70-eV incident beam is scattered to the wall over the length of the reaction vessel (46 cm) at gas densities corresponding to 2–3 μg/L. Apparatus B and apparatus C have larger diameters ($r_0 = 30$ and 45 mm, respectively) and more intense magnetic fields ($B = 0.02$ and 0.2 T, respectively) to cut down the extent of scattering.

For electrons with kinetic energies above the ionization potential, the scattering cross section decreases with increasing energy, while the ionization cross section increases. Although there have not been many published reports of total scattering cross sections by organic molecules, available data suggest that approximately 30% of scattering collisions at 35 eV are ionizing, as opposed to 15% at 20 eV [13, 27]. The increased proportion of nonionizing collisions as energy is lowered may cause substantial attenuation of the electron beam at energies below 30 eV.

Just as the space charge of the beam tends to drive electrons off axis, it will tend to focus positive ions toward the axis. Near the axis, where electron density is high and quasineutrality obtains, the net inward velocity becomes vanishingly small. But if a positive ion is knocked away from the axis by a collision, the ion will be accelerated back toward the axis until it returns to the region of quasineutrality. This restoring force exists because a collision displaces a charged particle a distance on the order of its cyclotron radius, which is much larger for an ion than for an electron.

3.1.2 Cyclotron Motion

The cyclotron frequency, v_c, of charges in a magnetic field is extensively discussed in Chapter I on ICR and Chapter II on FTMS in this volume. Unlike the magnetron motion, v_c does not depend on the electric field. The cyclotron radius, r_c, is a function of the velocity component, v_r, in the radial direction, $r_c = v_r/2\pi v_c$, in the absence of an electric field.

Cyclotron motion contributes to the overall path length as a charge moves parallel to the axis of the reaction vessel, since it travels a helical path and not a linear one. If the maximum value of r_c of the electrons in a beam is taken to be the radius of the aperture through which the beam enters the reaction vessel (0.5 mm), then the maximum value of v_r is 4×10^8 cm s^{-1} for the apparatus shown in Fig. 1, where the electron gun was immersed in a 0.05 T field. Since the field within the reaction vessel was 0.02 T, the maximum value of r_c inside the reaction vessel would therefore be 1.25 mm. Experimental studies on an apparatus of similar geometry [28] have shown that the value of v_r is, in fact, much less than the maximum value and corresponds to energies on the order of 0.1% of the beam voltage, or $v_r \approx 1.6 \times 10^7$ cm s^{-1} for a 70-eV beam. Although the path length of an electron spiraling down the axis of the reaction vessel is greater than the length of the vessel, the difference turns out to be negligible.

The cyclotron frequency of an ion is much lower than that of an electron. For an ion of m/z 80 in a field $B = 0.02$ T, its value is $v_c = 4$ kHz, versus 0.6 GHz for an electron. The average value of $\bar{v}_r$ is taken to be the mean thermal velocity in two dimensions, $\bar{v}_r = (\pi k T/2m)^{1/2}$. If v_r has the mean thermal velocity at

300 K, $\bar{v}_r = 2 \times 10^4$ cm s^{-1}, the ion's cyclotron radius will be $r_c = \bar{v}_r/2\pi v_c = 9$ mm. For a particle moving isotropically in the absence of any field, the total distance traveled between collisions is, on average, $\sqrt{3}$ times the distance traveled in any given direction. All that the magnetic field does to an ion is to curve its path. A thermalized ion will therefore travel, on average, an axial distance of $\lambda/\sqrt{3}$ (where λ is the mean free path) between collisions provided that the cyclotron orbit does not bring the ion into contact with the wall.

A useful approximation is describing ions in magnetic fields is the guiding center approximation, which describes the motion of the center of the cyclotron orbit. Unfortunately, this cannot be easily applied to the EBFlow, since the approximation assumes a uniform force on the ion throughout a cyclotron orbit. The electric field from the electron beam is not constant on the scale of r_c. If the center of the ion's orbit is $2r_c$ away from a 100-μA beam of 70-eV electrons, the potential energy of the ion at its orbit's furthest distance from the axis is 0.4 eV greater than at the orbit's closest approach to the axis, an order of magnitude greater than kT at 300 K.

3.1.3 Molecular Collisions

The Langevin treatment of ion–molecule collisions predicts a second-order rate constant on the order of 1×10^{-9} cm^3 molecule^{-1} s^{-1}. At 3×10^{-4} torr, the collision rate will be $\xi = 10^4$ s^{-1} and the mean free path at thermal velocities $\lambda = 3$ cm. In the absence of an electric field, collisions knock particles off axis with a mean radial excursion per collision of r_c. Clearly ions will be deflected more than electrons. For an average step size of 9 mm in apparatus B (whose reaction vessel has a radius of $r_0 = 30$ mm), a two-dimensional random walk will bring an ion to the wall in, on average, $(30/9)^2/2 = 6$ collisions. An alternative way of describing diffusion in the absence of the electric field would be to treat the problem macroscopically. The magnetic field decreases the effective radial diffusion coefficient by a factor of $1 + 2\pi v_c^2/\xi^2$. For an m/z 80 ion at 0.02 T, $2\pi v_c^2/\xi^2 \simeq 5.85$. In other words, the diffusion constant in the axial direction, D_a, will be 6.85 times as great as that in the radial direction, D_r. In the time required for diffusion from a point source on the axis to the wall [24], the ion will diffuse axially an average distance of $r_0(D_a/D_r)^{1/2} = 8$ cm.

The effect of the radial electric field can be assessed when $2\pi v_c \gg \xi$. With axial magnetic field $B = 0.2$ T this condition is met, and the inward velocity of a positive ion at distance r off axis is given by Eq. 9 [29]:

$$v_r = \frac{3.0 \times 10^4 \xi I}{2\pi v_c B r \sqrt{E}} \qquad (9)$$

For a 10^{-4}-A beam of 70-eV electrons, an m/z 80 ion drifts inward with a velocity of 7.5 m s^{-1} at $r = 0.009$ m. During a time interval of $1/\xi$ the ion moves inward, on average, 0.9 mm, which, fortuitously, is equal to r_c at 0.2 T. In other words, for $r < 9$ mm the inward drift caused by the electric field exceeds the average outward deflection per collision that would have occurred in the absence of an electric field. This ion capture radius provides an estimate of how far out the electric field can be expected to focus ions toward the axis. The kinetic energy gained by the ion from the electric field between collisions at the ion capture radius $r_i = 9$ mm is 0.03 eV, about the same as the thermal energy that the ion already possesses. Because v_r is inversely proportional to r and the electrical potential is inversely proportional to $\ln r$, the energy gained between collisions increases with decreasing r until the ion enters the region near the axis where quasineutrality obtains (quasineutrality must begin at a radius $> v_r/\xi$ in order for energy not to diverge). Since the ion temperature increases as r decreases, the Debye shielding length (Eq. 7), λ_D, also increases. The conditions of a plasma are met in a sheath region whose outer radius is presumed to be less than the ion capture radius r_i.

3.2 Apparatus Design

Three versions of the EBFlow have been constructed. The first, apparatus A, was designed to demonstrate the feasibility of the technique. Subsequent versions, B and C, have enlarged the dimensions of the reaction vessel. The same electron gun has been used in all three versions.

3.2.1 Electron Gun

The electron gun for the EBFlow is illustrated in Fig. 3. The design consists of a directly heated filament and two planar, stainless-steel electrodes (a control lens and an anode), each of which has a 1-mm-diameter hole to transmit the electron beam. The same gun has been used in all three versions of the EBFlow: The anode is mounted on a Vespel [30] base, and the control lens is set parallel to it (at a distance of 1 mm) with boron nitride supports. The filament assembly is mounted on a boron nitride crosspiece that is set on a pair of Vespel [30] posts so that the ribbon filament is held 1 mm from the control lens. The control lens not only focuses the beam but also helps to shield the filament from gas that spews in through the anode aperture from the reaction vessel. Originally, the gun also included a repeller mounted behind the filament [1], but this was found to be of little advantage and has not been used subsequent to the first published experiments. A number of different filament materials has been tried in order to improve durability under the comparatively high-pressure conditions of its operation. A solid block of lanthanum hexaboride (LaB_6) mounted on carbon rods was first tried, then LaB_6-coated rhenium ribbon prepared by cataphoretic deposition of LaB_6 followed by sintering at 1800 K was tried [31]. Both of these were rapidly poisoned by organic vapors from the reaction vessel, and all studies that have been published so far on apparatuses A and B have used a thoria-coated iridium filament. A 1.5 cm length of 0.030-in. iridium ribbon (0.001 in. thick) was tack welded to 0.035-in. tungsten rods with pieces of 0.0005-in. nickel foil between the filament and the rods. Thoria (ThO_2) was then cataphoretically deposited on the ribbon to a thickness of 0.025–0.05 mm by immersing the filament in a magnetically stirred suspension of ThO_2 in

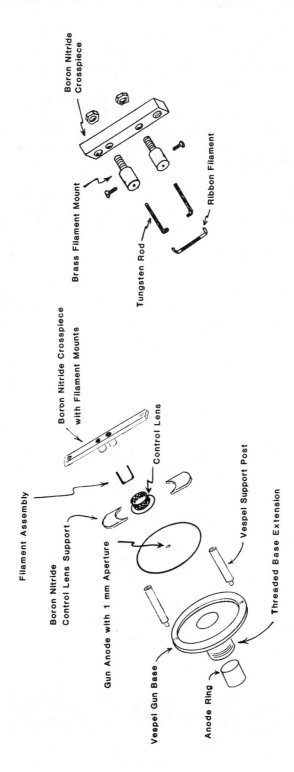

Fig. 3. Exploded views of the electron gun and the filament mounting assembly.

methanolic thorium nitrate and passing a current of 2–3 mA through this electrolysis solution for 15–30 minutes [32]. The gun was then mounted in the EBFlow reactor and degassed by slowly raising the heating current at a pressure $< 10^{-6}$ torr until its temperature reached 1300 °C (corrected optical pyrometer reading), corresponding to a current of 4–4.5 A and a voltage drop across the gun assembly of 1.5–1.6 V. With continued use, the current required to sustain electron emission from the filament (and its operating temperature) increased. For mounting in apparatuses B and C, an extension of the Vespel base of the gun assembly was threaded so that it could be screwed into the upstream end plate of the stainless-steel reaction vessel. A stainless-steel ring inside this extension is maintained at the same voltage as the gun anode. In apparatus C, in which the electron gun is immersed in the same magnetic field as the reaction vessel ($B = 0.2$–0.3 T), the filament is thoriated rhenium. Rhenium ribbon is more easily welded to the tungsten rods than iridium (nickel foil is not necessary), and cataphoretic deposition of ThO_2 appears to improve the filament's stability.

3.2.2. Apparatus A

Figure 4 shows a schematic diagram of Apparatus A. The housing was made entirely of glass so that the filament temperature could be measured with an

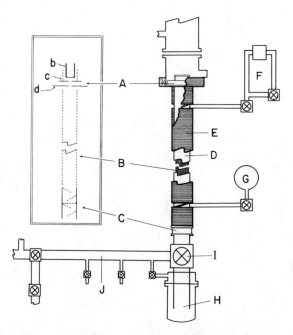

Fig. 4. Schematic diagram of apparatus A: A, electron source (insert: b, filament; c, control lens; d, anode); B, cage within reaction vessel; C, collector; D, cylindrical glass reaction vessel; E, solenoid; F, Baratron manometer; G, gas sample bulb; H, liquid nitrogen cooled trap; I, three-way vacuum stopcock; J, vacuum transfer line.

optical pyrometer. Any photoemission from the electron beam would have been visible through gaps in the solenoid, but no light could be seen from the reaction vessel (either with or without gaseous contents) above the background from the filament glow when the laboratory was darkened. The reaction vessel was lined with stainless-steel screen to prevent charge from building up on the walls, and electron energy was measured as the difference between the bias voltage of this screen and that of the filament. At pressure $< 10^{-6}$ torr a 70-eV beam of up to 1 mA could be passed the length of the reaction vessel with only a few percent of the electrons striking the walls (as measured by the current flowing from the screen through a meter). With gaseous contents $> 10^{-4}$ torr, however, the upper limit was approximately 200 μA, and at least 50% of the current (and often as much as 95%) would strike the screen. Control grid biases on the order of 150 V were typically required for operation, even when the air-cooled solenoid was operated at its highest magnetic field, 0.016 T (where the solenoid becomes hot to the touch).

The rate of flow of gas through apparatus A was found to be proportional to steady-state pressure with a rate of 24 μmol torr^{-1} s^{-1} (and not 24 mmol torr^{-1} s^{-1}, as inadvertently printed in Ref. 1). With a reaction vessel volume of 230 cm^3 this means that the average residence time of a gas molecule was 0.5 s. Because the cyclotron radius r_c of a typical ion was approximately the same as the vessel radius r_0, many control experiments were required to show that recovered neutral products resulted from homogeneous gas-phase reactions rather than wall collisions. One control was to measure the effect of magnetic field on the products [31]. Typically, product distributions depended on magnetic field below 0.01 T but ceased to vary at higher field strengths. This was taken to imply that the combined effects of the field of the electron beam and the axial magnetic field were sufficient to prevent ions from striking the wall before undergoing enough gas-phase collisions for ion–molecule reactions to go to essential completion. Where a reaction has been run in both apparatuses A and C, only small differences are seen (see Section 4.2).

3.2.3 Apparatus B

Because the electron beam current was found to depend on magnetic field, a new apparatus was constructed in 1978 with a larger air-cooled solenoid (capable of operating at 0.2 T without heating appreciably above room temperature). This apparatus is shown in Fig. 1. The reaction vessel was made of stainless steel and was substantially larger than in A, which necessitated horizontal instead of vertical mounting. Although it was feared that this geometry might impair the efficiency of the liquid-nitrogen-cooled trap, collection of neutral products was no less efficient than in apparatus A. The increased length of the reaction vessel (1.2 m) was designed to increase the fraction of the electron beam yielding ionization, and two separate inlets were mounted so that reactant gases would not have to be mixed before entering the reaction vessel. The design permitted the electron gun to be surrounded by a bank of 0.05 T toroidal permanent magnets (Indiana General), which improved the gun's efficiency. As in apparatus A, a collector grid was set at the downstream end of the reaction vessel to prevent charges from reaching the cold trap where products are collected.

3.2.4 Apparatus C

In order to minimize products of surface neutralization, a third version of the EBFlow was constructed at the University of California at Riverside in 1981. Instead of a collector at the downstream end, this reaction vessel has a conical Faraday plate that is truncated at the apex. This clown-cap design permits unreacted ions and electrons to pass out of the reaction vessel into a differentially pumped chamber, so that products of surface neutralization will not mix with the products of homogeneous reactions. Figure 5 shows a schematic of this version, and Fig. 6 shows a longitudinal section (to scale) of the reaction vessel and the differentially pumped region, with equipotentials corresponding to typical bias settings. The aperture at the apex of the clown cap is 2 cm in diameter, comparable to the ion capture radius, r_i, under experimental conditions. In apparatus C (unlike A or B) r_i at 300 K is an order of magnitude greater than r_c when $I \simeq 10^{-4}$ A, $E = 70$ eV, and $B \simeq 0.2$ T (the value of r_i is meaningless unless $r_i > r_c$). The combined effects of magnetic and electric fields not only inhibit ions from reaching the wall, but will also allow them to be extracted at the downstream end.

The solenoid electromagnet (fabricated by Stonite Corporation, Yardville, New Jersey) has seven parallel wraps of hollow copper conductor (for cooling water to pass through), which can generate a field of up to 0.3 T. This magnet

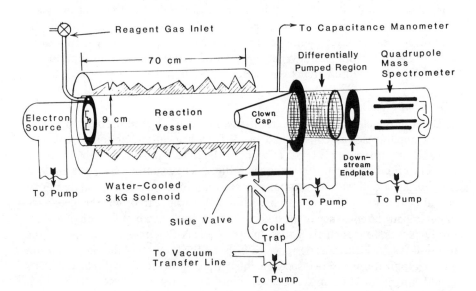

Fig. 5. Schematic diagram of EBFlow Apparatus [33]. The electron gun is the same as shown in Fig. 3, and the vacuum transfer line is the same as shown in Fig. 4.

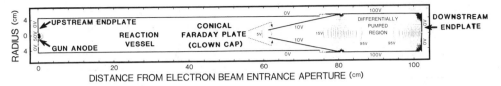

Fig. 6. Static electrostatic potentials in the EBFlow in the absence of the electron beam [33]. The 0.1 V equipotentials in the reaction vessel are 7 cm downstream of the gun anode and 5 cm upstream of the 5 V equipotential of the clown cap.

generates a field within the enclosed volume that maintains $>80\%$ of its maximum value as one goes out in the radial direction from the center and, as one goes out along the axis, to <5 cm from the end. Operating at 45 V and 250 A, it produces a maximum field of $B = 0.31$ T at the center without any appreciable heating of the reaction vessel (although the cooling water in the outer layer of the solenoid approaches the boiling point). The electron gun is mounted on axis 5 cm inside the magnet, and the exit to the cold trap (which has a diameter of 6 cm) is at a right angle to the reaction vessel immediately outside the solenoid. The clown cap extends 15 cm into the reaction vessel and, as control experiments described in Section 4.2 show, appears to transmit approximately 90% of unreacted ions into the differentially pumped region. The radius of the reaction vessel is 4.5 cm, more than 50 times greater than the cyclotron radius, r_c, of m/z 80 at 0.2 T. The diameter is large enough for three reactant inlets to affix to the upstream end plate, right next to the Pyrex envelope that houses the electron gun. The reaction vessel is stainless steel, the differentially pumped region is Pyrex lined with a stainless-steel screen, and O-ring joints connect the glass components to the stainless-steel ones. In an experiment using bromocyclohexane vapor, the gas throughput was found to be 4.5 mmol s^{-1} torr^{-1}, and 75% of $C_6H_{11}Br$ put in at the upstream end was collected in the cold trap. The mean residence time of a gas molecule is 50 ms.

The pressure inside the reaction vessel is monitored using an MKS Baratron capacitance manometer with a lower limit of 1×10^{-5} torr. The detection head is connected to the reaction vessel by a 0.8 m length of 3/8 in tubulation. The reference to the head is continuously evacuated with a 2 l/s VacIon pump. Operation of the manometer at its lowest range requires not only a temperature compensator unit, but also several centimeters of insulation (glass wool) enclosed in a polystyrene box surrounding the manometer head. Reactants are let in via variable leak valves from glass gas bulbs (whose contents are replenished by periodically opening a stopcock to a reservoir containing liquid). Because many organic compounds tend to condense near the construction of a leak valve during expansion, the liquid reservoir remains isolated from the leak valve most of the time, except for a few seconds from time to time when the gas bulb is replenished. Even so, the conductance of leak valves tends to decrease with time unless they are heated to 40–50 °C. For this reason, Varian 951–5106 variable leaks (which are bakeable to 450 °C) are now used for reactants that are not gases at room temperature. To ensure that reactants are thermalized before entering the EBFlow, the variable leaks are connected to the reaction vessel by 0.6 m lengths of stainless-steel bellows tubing.

A UTI 100C quadrupole mass spectrometer connects to the differentially pumped region via a 1-cm aperture in the downstream end plate with a homemade slide valve to allow it to be isolated from the EBFlow when not in use. This has been used primarily to analyze the neutral gases in the EBFlow (see Section 4.2). Although ions from the EBFlow have been detected directly, the presence of the ionizer in the mass spectrometer prevents the spectrometer from being used to quantitate ions emerging from the reaction vessel.

3.3 Operating Conditions

The general operating procedure has been the same for all three versions of the EBFlow. A mixture of gases in the reaction vessel is continuously bombarded by the electron beam as it flows from the inlet to the cold trap. All reactants are equally subject to ionization by the beam, and experiments have to be planned with that in mind. After a period of radiolysis (from 10 minutes to several hours) the electron beam is turned off, the cold trap is isolated from the reaction vessel and allowed to warm up, and the contents are vacuum transferred to a liquid-nitrogen-cooled sample tube over a period of many hours. Often a known quantity of an internal standard has been frozen in the sample tube before transferring the radiolysis products.

Conditions for operation of apparatus C will be described below. The electron energy is set by the filament bias (negative) relative to the reaction vessel, which (including the upstream end plate) is grounded. Separate biases are set on the control lens, the gun anode, the clown cap, the cage of the differentially pumped region, and the downstream end plate. Currents striking these circuit elements and the wall of the reaction vessel are monitored. The gun anode is biased positive (ordinarily $+50$ V) and the control lens adjusted so as to optimize beam current before sample is admitted to the reaction vessel. Often the control lens bias is the same as the anode bias. The clown cap is biased at $+10$ V, and the cage of the differentially pumped region is biased at $+100$ V. The downstream end plate is usually grounded, and the slide valve that connects it to the mass spectrometer can be opened and shut during a run.

The bias of the clown cap turns out to have an effect on the collected product. The setting of $+10$ V gives the same product distribution as 0 V, but the electron current transmitted to the differentially pumped region and the downstream end plate is much greater if the clown cap is positive. The electron current increases further if the bias is increased, but neutral product ratios begin to exhibit variations suggestive of high ion kinetic energies when the bias is as high as $+50$ V.

Although the differentially pumped region is positive relative to the reaction vessel, positive ions do make it through to the downstream end plate, as can be demonstrated by opening the slide valve to the mass spectrometer. With 3×10^{-4} torr bromocyclohexane in the reaction vessel, the principal fragment ion (m/z 83) can be seen in the mass spectrometer without turning on its ionizer. At first it may seem surprising that positive ions can pass through a region in which the potential is so much greater than their kinetic energy, but the space charge of the electron beam is not negligible. Also, if the ions are contained in a plasma, quasineutrality allows them to pour through the aperture in the clown cap as though they were an uncharged fluid.

Before a run the liquid-nitrogen-cooled baffles on the vacuum pumps are filled and allowed to equilibrate. The cold trap is charged with liquid nitrogen, the solenoid is set at 0.2–0.3 T, and the filament current is turned up slowly. Before sample is admitted (pressure in the reaction vessel $<10^{-6}$ torr), the electron beam is set at a desired current, which is usually stable in the range 0.1–0.5 mA measured at the cage of the differentially pumped region and the downstream end plate combined. Then sample is admitted, and electron beam current tends to drop. Steady-state partial pressures of each reactant are set using the variable leaks with the Baratron to monitor total pressure.

During a run most of the electron current strikes the cage of the differentially pumped region. With 70-eV electron bombardment of 3×10^{-3} torr bromocyclohexane plus 1×10^{-4} torr ammonia, a current of 0.3 mA at the cage was accompanied by 0.05 mA on the clown cap and 0.01 mA at the downstream end plate. The current striking the reaction vessel depended on the electron beam current. With $\leqslant 0.1$ mA, a positive ion current struck the wall of the reaction vessel. The ion capture radius, r_i, decreases with the electron beam current, and, accordingly, the current striking the reaction vessel wall becomes an electron current as the total beam current is raised substantially above 0.1 mA. With a total beam current of 0.3 mA, the electron current at the reaction vessel wall was approximately 0.02 mA. These currents fluctuate during a run, and the filament current and control lens bias are frequently readjusted so as to maintain stability as much as possible.

After a run is through and sample has been transferred, the receiving tube is sealed off under vacuum (if the receiving tube is a 5-mm NMR tube, it will have been previously charged with solvent before sample transfer). The sample is then kept at liquid-nitrogen temperatures until it is analyzed. GLPC analyses are performed using a flame ionization detector, whose response is proportional to the number of carbons in a hydrocarbon eluent. Therefore, the area of a peak divided by the sum of peak areas is, to a close approximation, a measure of the mass fraction that that peak represents. If a known quantity of internal standard is present, then the product yield can be gauged in moles and divided by the average electron beam current and the run time to get normalized yield. For NMR analysis, normalized conversions are expressed in units of $\%\,mA^{-1}$. The conversion for a product is measured as the ratio of its peak area to the area of a peak corresponding to the unreacted starting material and divided by the mean total measured electron beam current. Since product peak fractions are typically of the order of one part per thousand, large experimental uncertainties require further calibration to convert $\%\,mA^{-1}$ to units of μmol $A^{-1}s^{-1}$. For low pressures, where $<20\%$ of the electron beam causes ionizations, the yield can be expressed as the pressure-normalized yield, μmol A^{-1} s^{-1} torr^{-1}, based on the assumption that the yield will vary linearly with pressure until the electron beam is substantially attenuated.

4 REACTION TYPES

Use of the EBFlow reactor demands that many control experiments be performed to rule out alternative sources for the recovered neutral products. The cases below present specific examples (by no means an exhaustive list) of pertinent controls applied to reactions that have been studied with this technique.

4.1 Unimolecular Ion Decomposition

Most ion–molecule reactions can be identified as such by the fact that they exhibit second- or higher-order kinetics. It is possible, however, for an ion–molecule reaction to obey first-order kinetics, too. If, for example, ion and molecule are both generated via a unimolecular bond cleavage, as Eq. 10 depicts,

$$R-X^+ \xrightarrow{k_{sciss}} \begin{bmatrix} R^+ \\ X \end{bmatrix} \xrightarrow{k_{esc}} R^+ + X$$
$$\text{ion–molecule} \quad \xrightarrow{k_{reac}} \quad \text{ion–molecule}$$
$$\text{complex} \qquad\qquad \text{reaction} \qquad (10)$$
$$\text{products}$$

and then react before escaping from one another, the overall reaction will obey first-order kinetics. This is analogous to geminate-pair reactions in solution, where the reaction partners are born simultaneously and are held together by a solvent cage. In the gas phase, ion–dipole forces can play a role analogous to that of a solvent cage, holding the ion and molecule within 5–10 Å of one another. Identifying this process requires that the products be characterized as typical of an ion–molecule reaction, as opposed to some more conventional type of unimolecular decomposition.

EBFlow studies have played a major role in demonstrating the existence of ion–molecule reactions that obey first-order kinetics. Because application of the EBFlow technique has been so intimately connected with the chemistry under investigation, it will be worthwhile to review the logic of the experimental investigation. The objective of the EBFlow study was to exhibit the necessity for inferring a rapid bond scission, whose rate constant is represented by k_{sciss} in Eq. 10, that leads to an intermediate ion–molecule complex. The following EBFlow experiments were indicated:

1. Study the products of RX$^+$ for which the rate of escape from the ion–molecule complex, k_{esc}, could be inferred, on the basis of mass spectrometric evidence, to be much smaller than the ion–molecule reaction rate, k_{reac}. This criterion is met, for example, by the molecular ions of alkyl

phenyl ethers, ROPh$^{\cdot+}$. The putative ion–molecule reaction product, PhOH$^{\cdot+}$, is, in general, at least an order of magnitude more abundant than R$^+$, the escape product.
2. Demonstrate that conventional unimolecular ion decomposition and nonionic reaction afford products different from Eq. 10.
3. Compare the products from ion–molecule reactions of free R$^+$ with those of R$^+$ within the ion–molecule complex. Alkyl cations are well known to undergo skeletal rearrangements. Rearrangements that are faster than k_{reac} take place, whereas those that are slower than k_{reac} do not.

4.1.1 EBFlow Radiolyses of Alkyl Phenyl Ethers

The first EBFlow study was of butyl phenyl ether [1], for which Eq. 11 had been the subject of a number of mass spectrometric investigations:

$$CH_3(CH_2)_3OPh \xrightarrow[\text{impact}]{\text{electron}} PhOH^{\cdot+} + C_4H_8 \qquad (11)$$

Since the commercially available material contained too many impurities, the reactant was synthesized and purified by preparative GLPC. Figure 7 plots the normalized C_4H_8 yield as a function of electron energy. Units on the y axis are μmol A^{-1}s^{-1} (10.4 μmol A^{-1}s^{-1} corresponds to one molecule of product for every bombarding electron). The ionization cross section for butyl phenyl ether is estimated to be 27.3 Å^2 at 70 eV [12] and 11.6 Å^2 at 20 eV [13], and Eq. 11 corresponds to 45%Σ [34] and 68%Σ [35], respectively, at these energies. Predicted normalized yields (neglecting any contribution from secondary electrons and assuming 100% collection efficiency) would be 4.4 μmol A^{-1}s^{-1} and 3.1 μmol A^{-1}s^{-1} in apparatus A. Fluctuations in the observed normalized C_4H_8 yield between 30 and 100 eV are taken to be random, since an analysis of variance ($F_{5,12} = 2.61$) reveals no statistically significant differences at the $p < .05$ level.

The linearity of the yield with ionizing electron current was tested using LaB$_6$-coated rhenium filaments, which put out currents in the range 0.1–1 mA during their brief operating lifetimes (on the order of 1 hour). In a 70-eV run where the mean current was 950 μA, approximately one-third of the butyl phenyl ether was converted to products. Average normalized yields of all observed hydrocarbon products are compared in Table 2 at 10, 20, 30, 40, and 70 eV [36]. The normalized yields of C_4H_8 in the high-current experiment lie within 95% confidence limits of the values from the low-current experiments that used thoriated iridium filaments.

Below 20 eV, the C_4H_8 yield fell off precipitously, as does the PhOH$^{\cdot+}$ yield in the mass spectrum of butyl phenyl ether [36]. This energy dependence represents one way of confirming that recovered neutral products resulted from Eq. 11. The identities of the neutral products were, of course, the focus of the EBFlow study. The variety of $\leqslant C_4$ hydrocarbons is sufficiently limited so that unambiguous identifications can be based on GLPC retention times on a

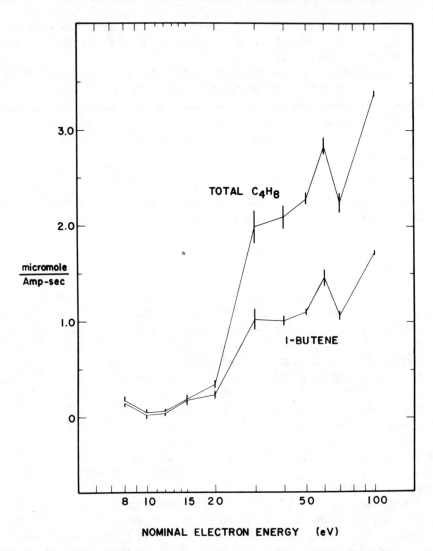

Fig. 7. Normalized yield of C_4H_8 products from EBFlow radiolysis of butyl phenyl ether in apparatus A at 2.5×10^{-4} torr [1]. Units are $\mu mol\, A^{-1} s^{-1}$ on the y axis ($10.4\, \mu mol\, A^{-1} s^{-1}$ corresponds to one molecule of recovered product per bombarding electron). Each point represents at least 3 independent determinations with total bombarding electron current in the range 2–4 μA. Error bars represent sample standard deviations. [Reprinted by permission from F. B. Burns and T. H. Morton, *J. Am. Chem. Soc.* **98**, 7308–7313 (1976). Copyright 1976 American Chemical Society.]

dimethylsulfolane column at room temperature. The product ratio, *cis*-2-butene:*trans*-2-butene: 1-butene, was $1:1.10 \pm 0.05:2.3 \pm 0.4$ for energies $\geqslant 30\, eV$ and $1:1.12 \pm 0.03:7.8 \pm 0.8$ at $20\, eV$ (errors represent sample standard deviations). Control experiments (EBFlow radiolysis of pure hydrocarbons) showed that these products were stable under the reaction conditions. The

Table 2. Net Normalized Yields of Hydrocarbons (μmol A^{-1} s^{-1}) from Butyl Phenyl Ether (BuOPh), 4.5:1 1-Bromobutane:diethyl Ether (BuBr), and 2-n-Butylcyclopentanone (BuCP) at Selected Electron Energies in apparatus A [36][a]

Reactant[b]	BuOPh (2.5)	BuOPh (2.5)	BuOPh (2.5)	BuOPh (2.5)	BuOPh (2.5)	BuOPh (3)	BuBr (5)	BuCP (2.5)
Energy[c]	10 eV(4)	20 eV(3)	30 eV(3)	40 eV(3)	70 eV(3)	70 eV(1)	70 eV(1)	70 eV(2)
Mean Current	2.9 μA	3.5 μA	2.6 μA	2.4 μA	3.1 μA	950 μA	76 μA	3.7 μA
Product								
Propane	0.006	0.082	0.06	0.04	0.03	0.001	0.08	0.22
Propene	0.02	0.04	0.15	0.20	0.23	0.18	0.59	1.2
n-Butane	0.16	0.26	0.28	0.16	0.11	0.04	0.29	0.07
Isobutane	d	d	d	d	d	0.003	0.09	d
Acetylene	0.06	0.03	0.14	0.34	0.65	0.90	1.3	0.52
1-Butene	0.15	0.35	1.1	1.1	1.1	0.75	0.48	2.1
Isobutene	d	d	0.01	0.01	0.05	0.05	0.38	0.03
Allene	d	d	0.06	0.05	0.09	0.11	0.17	0.22
trans-2-Butene	0.02	0.045	0.47	0.51	0.57	0.36	0.60	0.07
Methylcyclopropane	d	d	0.01	0.01	0.01	0.02	0.19	d
cis-2-Butene	0.01	0.04	0.43	0.49	0.50	0.33	0.17	0.06
1,3-Butadiene	d	d	0.04	0.09	0.09	0.06	0.12	0.35
Methylacetylene	d	d	0.03	0.08	0.10	0.05	0.18	0.15

[a] Values for methane, ethane, and ethylene exhibit wide variability and have therefore been omitted. Mean currents represent total electron flux striking the reaction vessel wall and the collector.
[b] Numbers in parentheses denote mean total pressure times 10^4 torr^{-1}.
[c] Nominal ionizing energy. Numbers in parentheses denote the number of independent runs that were averaged.
[d] Less than 0.005 μmol A^{-1} s^{-1}.

pressure-normalized yield of C_4H_8 isomers from, for example, 70-eV radiolysis of methylcyclopropane at 4×10^{-4} torr was 4 mmol $A^{-1} s^{-1}$ torr^{-1}, so low that rearrangement of products from butyl phenyl ether under EBFlow conditions could be neglected. The stability of reaction products in control experiments provides convincing evidence that other processes in the reaction vessel (e.g. heating by the electron beam) do not interfere.

The ionization cross section at 70 eV is estimated to be 1.2 times greater than at 35 eV [13]. The average normalized yield of 2-butenes at 70 eV was 1.2 times greater than that at 30–40 eV (although this was not statistically significant, as the sample standard deviations were on the order of 20%). The importance of this observation is its implication that ionization by secondary electrons is not substantial. When a < 100-eV electron ionizes a molecule, two electrons of lower energy are produced, one with kinetic energy of at most a few electron volts. If we guess that the centroid of its He(I) photoelectron spectrum [1], 12–13 eV, is the average ionization potential (IP) of butyl phenyl ether, then the average energy of the other scattered electrons should be on the order of the electron energy minus the IP, or ≈ 55 eV for 70-eV electrons and ≈ 15 eV for 30-eV electrons. Clearly, secondary electrons cannot be contributing to the ionization if the normalized yield does not vary significantly between 30 and 70 eV. The normalized yield of 2-butenes falls off by a factor of 11 ± 1 between 30 and 20 eV, much more precipitously than would have been predicted by the cross section for Eq. 11. This result, together with the insignificance of secondary electrons in producing ionization products, suggests that nonionizing collisions attenuate a < 30 eV electron beam as a function of the ratio of nonionizing to ionizing collisions (*vide supra*, Section 3.1.1).

EBFlow experiments with a deuterated analog, shown in Eq. 12, were analyzed using GLPC-mass spectrometry (GC–MS):

$$PhOCH_2CD_2CH_2CH_3 \xrightarrow[\substack{\text{EBFlow} \\ \text{radiolysis}}]{70 \text{eV}} \frac{C_4H_6D_2}{C_4H_7D} \geqslant 1.5 \tag{12}$$

The butene yields were the same, within experimental error, as from undeuterated butyl phenyl ether. There were two major sources of experimental uncertainty in the GC–MS analysis of the level of deuteration of the recovered butenes: (1) contributions of $M-1$ peaks to the mass spectra of C_4H_8 and deuterated analogs, which interfere with measurement of the extent of deuteration based on molecular ion intensities, and (2) isotopic fractionation on the GLPC column (which manifested itself as a systematic variation in the m/z 56:57:58 ratio over a single peak). The extent of deuteration of each of the three butene isomers from Eq. 12 was found to be nearly the same within a large uncertainty, $\pm 15\%$. The GC–MS analysis was calibrated with authentic samples of 1-butene-d_2 (6%-d_4, 19%-d_3, 54%-d_2, 18%-d_1) and *cis*-2-butene-d_2 (10%-d_3, 71%-d_2, 13%-d_1, 4%-d_0) from Raney nickel-catalyzed reduction of methylallene by D_2 and with 2-deuterio-1-butene (> 95%-d_1) [36].

Although every effort was made to fit the experimental data to the then prevailing conventional view of unimolecular ion decompositions (viz. bond-forming steps precede bond-breaking steps) in the first publication of these results [1], it became apparent shortly thereafter that the neutral products resulted from ion–molecule reactions operating in a first-order regime [16, 22].

4.1.2 Comparison with Other Reactions

As an example of a conventional mass spectrometric rearrangement, the McLafferty rearrangement of 2-n-butylcyclopentanone (BuCP) was examined. EBFlow radiolysis afforded a copious yield of 1-butene, as would be expected from the conventional γ-hydrogen rearrangement mechanism. Although this product cannot be distinguished from the products expected from excited neutral BuCP molecules, the yield data in the right-hand column of Table 2 show an important result: The yields of other C_4H_8 isomers are tiny. At the very least this confirms that, as a reaction product, 1-butene is stable under the EBFlow conditions.

EBFlow yields from butyl phenyl ether $\leqslant 20\,eV$ exhibit product distributions that differ from those at higher energies. Since, as noted above in Section 3.2.1, nonionizing collision cross sections increase markedly at this energy whereas ionization cross sections fall off, the 1-butene yield at 20 eV was used to assess an upper limit for C_4H_8 production from excited neutral butyl phenyl ether. To confirm this interpretation, EBFlow radiolysis of the deuterated butyl phenyl ether was studied at 20 eV. The yield of 2-butenes was too low to permit GC–MS analysis of isotopic content, but the 1-butene from this experiment was approximately 15%–d_2 and 85%–d_1. Clearly most of the 1-butene at 20 eV came from a different reaction than the 1-butene at 70 eV. This interpretation is consistent with results of Danon and co-workers at Orsay, who have studied neutral products of $\leqslant 20$-eV electron impact on hydrocarbons. In their study of neopentane, for instance, they report that the yield of excited neutral molecules at 13 eV is 1.2 times the yield of ions. The presence of unsaturation in the target molecule appears to increase the probability of excitation; for instance, at 11 eV the yield of excited neutrals from electron bombardment of propene is > 1.25 times as great as that for neopentane [37].

Gas-phase photolyses were compared to the EBFlow results. Photolysis of butyl phenyl ether vapor in quartz tubes at 254 m yielded butane as the principal product, $> 100:1$ relative to C_4H_8. This result obtained even when the photolysis was performed in the presence of a twofold excess of phenyl acetate (as a source of phenoxy radicals). Within the experimental error of GC–MS analysis, the butane produced from 254-nm photolysis of $CH_3CH_2CD_2CH_2OPh$ or of $CH_3CH_2CH_2CD_2OPh$ was dideuterated to the same extent as starting material. The butane produced from 20-eV radiolysis of $CH_3CH_2CD_2CH_2OPh$ was also analyzed, and it showed the same level of isotopic substitution as the photolysis products. The inference that n-butane from EBFlow radiolysis was a product of excited neutrals therefore seems justified.

Electrical discharge was also examined. A Geissler discharge through 0.2 torr of butyl phenyl ether vapor yielded a volatile product distribution very different from the EBFlow. With a discharge tube whose volume was 110 cm^3 and a voltage drop between electrodes of 10 kV, passage of 10 mA for 1000 s yielded 11 nmol of $\leqslant C_4$ hydrocarbons in the following molar proportions: ethylene (65%), ethane (8%), propane (18%), acetylene (4%), butane (3%), propene (2%), and isobutane (1%). No butenes were detected among the products. The product distribution is not easy to interpret in terms of any of the mechanisms discussed above, although isobutane was also an unexplained minor product in the EBFlow radiolysis of bromopentane (*vide supra*, discussion of Table 1).

Pyrolysis of butyl ethers also showed a C_4H_8 product different from $\geqslant$ 30-eV electron bombardment. The major C_4 products from pyrolysis of dibutyl ether or butyl phenyl ether in vacuo at 770–780 °C were 1-butene and *n*-butane, in ratios of 1:0.08 (Bu$_2$O) and 1:0.22 ± 0.02 (BuOPh). Ethylene was also recovered in a variable yield from 1.1 to 2.5 times the molar abundance of 1-butene. All other products were recovered in less than half the butane yield. In an EBFlow control experiment the filament was turned on, but electrons were not admitted into the reaction vessel. Butane and butene were collected in a $\approx$ 2:3 ratio, with an ethylene yield approximately equal ($\pm$ 50%) to the C_4 yield. These are presumed to be products of pyrolysis on the filament that found their way back into the reaction vessel, and the rate of C_4H_8 production was less than one-tenth the rate of 1-butene formation from $\geqslant$ 30-eV EBFlow radiolysis (in which ethylene was rarely as much as one-third the molar abundance of 1-butene).

Three conclusions can be drawn from these control experiments. First, there appears to be no alternative to Eq. 11 as the source of 2-butenes or to equation 12 as the source of dideuterated 1-butene. Second, sources of recovered neutral C_4H_8 products from $\geqslant$ 30 eV electron impact can be weighted as follows: positive ions $\gg$ excited neutrals $>$ filament pyrolysis. Third, reactions of excited neutrals contribute about as much as ion-molecule reactions at 20 eV.

4.1.3 Comparison with Free Alkyl Cations

The C_4H_8 isomer distribution from Eq. 10 stands in contrast to the second-order deprotonation of free butyl cations. Table 2 includes a comparison of normalized yields, and the proportions of isobutene and methylcyclopropane for BuBr there and in Table 1 are much greater than the few percent they represent in the C_4H_8 yield for butyl phenyl ether.

This result typifies the short lifetime (10^{-11}–10^{-6} s) of the ion–molecule complex [22]. Its duration is long enough for facile processes to take place, such as the interconversion of linear butyl cations in Eq. 13, but too short for structural isomerizations, such as the higher barrier processes depicted in Eq. 14:

$$[CH_3CH_2CH_2CH_2^+] \rightarrow CH_3CH_2\overset{+}{C}HCH_3 \rightleftarrows CH_3\overset{+}{C}HCH_2CH_3 \qquad (13)$$

$$\overset{+}{CH_3CHCH_2CH_3} \rightarrow CH_3-\underset{}{\overset{H\diagdown \;\; \overset{H^+}{\diagup} H}{\triangle}} \rightarrow (CH_3)_3C^+ \qquad (14)$$

The deuterium labeling experiment represented in Eq. 12 corroborates this interpretation and is also consistent with mass spectrometric data reported for deuterated analogs [16].

Comparison with mass spectrometric results provides a critical test of EBFlow results. For instance, 70-eV electron bombardment of neopentyl phenyl ether (Eq. 15) was studied:

$$(CH_3)_3CCH_2OPh \xrightarrow[\text{electron impact}]{70\,eV} [(CH_3)_2\overset{+}{C}CH_2CH_3 \quad PhO^\bullet]$$

$$\rightarrow \frac{H_2C=C(CH_3)CH_2CH_3}{(CH_3)_2C=CHCH_3} = 1.1 \qquad (15)$$

The normalized yield in apparatus B was $4.5\,\mu mol\,A^{-1}s^{-1}$ at 2×10^{-4} torr, close to the theoretical yield. EBFlow radiolysis afforded nearly the same ratio of 2-methylbutenes as the deprotonation of $C_5H_{11}^+$ in Table 1. (The yield was once again corrected for a control in which the filament was on, but no ionizing electrons entered the reaction vessel.) In terms of Eq. 10, the isomer distribution implied that deprotonation of *tert*-amyl cation by an oxygen base (an ether or a phenoxy radical) gives the same product distribution regardless of whether it takes place in an ion–molecule complex or via a free cation. Since the mass spectrum of neopentyl phenyl ether predicts $k_{reac}/k_{esc} = 13$, there was every likelihood that the EBFlow experiment accurately recorded the outcome of Eq. 15, but the following mass spectrometric experiment was used to confirm this supposition. The proportions of $PhOD^{\bullet+}$ and $PhOH^{\bullet+}$ from $(CD_3)_3CCH_2OPh$ and $(CH_3)_3CCD_2OPh$ provided a measure of the proportions of γ-hydrogen transfer (which yields 2-methyl-1-butene) and α-hydrogen transfer (which yields 2-methyl-2-butene). These proportions, 56% γ and 44% α, were in good agreement with the EBFlow results.

Another instance in which mass spectrometry has provided a useful confirmation of EBFlow results is in the case of ω-alkenyl phenyl ethers. As Eq. 16 depicts,

$$H_2C=CH(CH_2)_nOPh \xrightarrow[\text{electron impact}]{70\,eV} [H_2C=CH(CH_2)_n^+ \quad PhO^\bullet]$$
$$\downarrow$$
$$\text{side products}$$

$$\rightarrow [\overbrace{(CH_2)_{n+1}CH^+} \quad PhO^\bullet] \rightarrow \text{cycloalkene} + PhOH^{\bullet+} \qquad (16)$$

the nascent cation in the ion–molecule complex can undergo intramolecular cyclization. The expected ultimate products are cycloalkenes, and those are observed, as the GLPC traces from EBFlow studies for $n = 3$ and 4 in Fig. 8 and 9 show. These experiments were performed in apparatus B at 1×10^{-4} torr. Clearly, many other neutral products are recovered, too. How were we to assess whether they were side products of reaction 16? One way to do so was

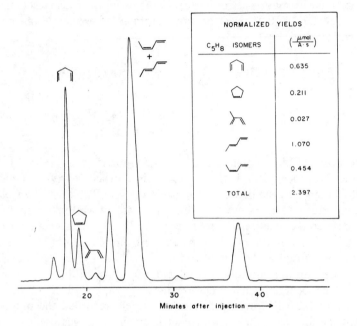

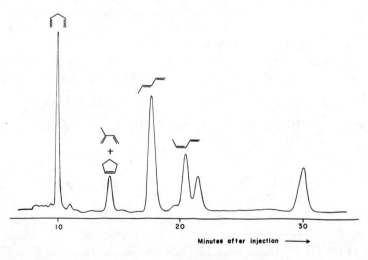

Fig. 8. GLPC traces and tabulated normalized yields of C_5H_8 products from 70 eV EBFlow radiolysis of $H_2C=CH(CH_2)_3OPh$ at 1×10^{-4} torr in apparatus B. Because of fortuitous overlaps of retention times, analyses were performed both on n-octane on Porasil and on β,β-oxydipropionitrile packed columns using a flame ionization detector.

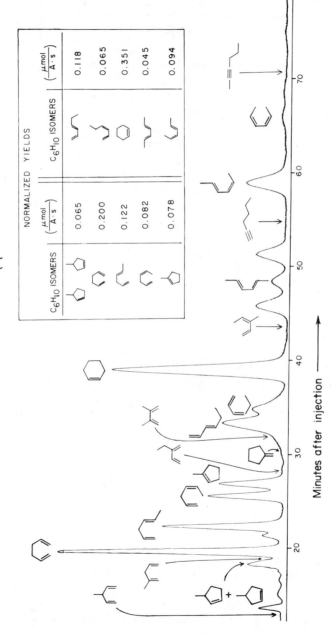

Fig. 9. GLPC trace and tabulated normalized yields of C_6H_{10} products from 70 eV EBFlow radiolysis of $H_2C=CH(CH_2)_4OPh$ at 1×10^{-4} torr in apparatus B. Analysis was performed on a 20 ft × 1/8 in. 12% β,β-oxydipropionitrile packed column at 25°C using a flame ionization detector. Calibration with selected isomers confirmed that the detector response was the same for different C_6H_{10} structures.

to examine the proportion of PhOD$^{\cdot+}$ in a separate mass spectrometric experiment, since cyclization renders the CD_2 equivalent to a CH_2. If isotope effects are neglected, the cycloalkyl cation ought to transfer a deuteron as often as a proton (unless it undergoes subsequent rearrangement).

The normalized 70 eV yield of neutral C_5H_8 from 4-pentenyl phenyl ether ($n = 3$) was 2.4 μmol A^{-1} s^{-1}, and the yield of C_6H_{10} from 5-hexenyl phenyl ether ($n = 4$) was 1.2 μmol A^{-1} s^{-1}. Cycloalkene composed 8% (cyclopentene) of the C_5H_8 and 28% (cyclohexene) of the C_6H_{10}. The mass spectrometric experiment showed that 5% of the phenol daughter ions from $D_2C=CH(CH_2)_3OPh$ contained deuterium, in agreement with the prediction that the proportion of PhOD$^{\cdot+}$ in the mass spectrum should be approximately equal to half the proportion of cycloalkene among the neutral products. In the $n = 4$ case, there is ample precedent for rapid hydride migrations around the six-member ring. This puts the CD_2 group in nonacidic positions and therefore decreases the statistical proportion of deuterons transfered from the cycloalkyl cation from one-half to as little as one-eleventh of the cycloalkene yield (if the deuteria become randomized throughout the ring). Experimentally, the proportion of PhOD$^{\cdot+}$ from $D_2C=CH(CH_2)_4OPh$ was 5%, approximately two-fifths the amount expected from an unrearranged cyclohexyl cation [50].

There is additional evidence that the cyclohexyl cation is formed in the ion–molecule complex. If free radicals were produced, they would not be expected to yield cyclohexenes. Internal cyclization would instead lead to cyclopentyl methyl radical (Eq. 5), and the expected C_6H_{10} product would be methylenecyclopentane. As Fig. 9 shows, that C_6H_{10} isomer is absent. The evidence argues that the recovered products come from ionic precursors, just as did virtually all the neutral C_nH_{2n} products from 70-eV electron bombardment of $C_nH_{2n+1}OPh$ above. The ratio k_{reac}/k_{esc} for $H_2C=CH(CH_2)_3OPh$ at 70 eV is about 50, based on the mass spectrum, so there is little question about the role of free cations. But free $C_nH_{2n-1}^+$ is more abundant in the mass spectrum of $H_2C=CH(CH_2)_4OPh$, corresponding to $k_{reac}/k_{esc} = 9$. Here, the escape product ought to yield a different cyclized isomer, since free cyclohexyl cations are well known to rearrange in solution as well as in the gas phase, as Eq. 17 depicts:

cyclohexyl cation   1-methylcyclopentyl cation   (17)

The rearrangement product, 1-methylcylopentyl cation, is deprotonated by weak bases (e.g., phenyl ethers) to yield 1-methylcyclopentene. This product is observed among the recovered neutral products. The short lifetime of the ion–molecule complex probably does not admit this rearrangement (cyclohexyl cation is estimated to have a lifetime on the order of 1 ms in the gas phase [7a]), and the free cation is probably the source of 1-methylcyclopentene, a small fraction of the cyclohexene, and possibly some of the hexadienes.

Based on correlations between EBFlow and mass spectrometric studies, the following general conclusion can be drawn about first-order ion–molecule reactions. Ion–molecule complexes live long enough for very rapid cationic rearrangements (viz., hydride shifts or internal cyclizations) to take place but not long enough for rearrangements of the carbon skeleton (e.g., conversion of unbranched secondary to branched tertiary cations). Skeletal rearrangements are, instead, characteristic of free cations. Unfortunately, as aliphatic cations get larger than five or six carbons, many rearrangement processes become accessible. The variety of products then becomes so large that the complexity of EBFlow reaction mixtures begins to exceed the capability of GLPC or GC–MS for complete product analysis.

4.2 Bimolecular Proton Transfer Reactions

For free alkyl cations there is often a single structure that is more stable than all other isomers. The tertiary cations in Eqs. 3, 14, and 17 all represent the thermodynamic minima corresponding to their molecular formulas. Given suitable precursors, these cations will predominate in the EBFlow. The *tert*-amyl and 1-methylcyclopentyl cation are of special interest because each can give two neutral products when deprotonated. To study these in the EBFlow, it has been necessary to choose neutral precursors that will not afford the same products via radical pathways.

The tertiary cations were prepared via cationic rearrangements from 70-eV electron impact on alkyl bromide precursors, *tert*-amyl from isoamyl bromide, Eq. 18,

$$(CH_3)_2CHCH_2CH_2Br \xrightarrow[\text{electron impact}]{70\,eV} [(CH_3)_2CHCH_2CH_2^+] \rightarrow (CH_3)_2\overset{+}{C}CH_2CH_3$$

isoamyl bromide   primary isoamyl   *tert*-amyl cation
                                      cation   (18)

and 1-methylcyclopentyl from bromocyclohexane via Eq. 17. The abundances of pertinent mass spectrometric fragments were available from the literature [18]. Although the EBFlow yielded neutral products corresponding to other isomeric ions, the tertiary cations were precursors to most of the C_5H_{10} from isoamyl bromide and to nearly all the rearranged C_6H_{10} from bromocyclohexane. The effect of base was especially informative, particularly in apparatus C, where unreacted ions for the most part pass via the clown cap into the differentially pumped region.

The use of isoamyl bromide to prepare *tert*-amyl cation presented a special problem, for this was a case in which another fragment ion (M-HBr) (*m/z* 70) could yield the same products as deprotonation of *tert*-amyl cation. How could the two reactions be dissected? RRKM calculations on *tert*-amyl predicted that if the primary isoamyl cation was initially formed by electron impact, the excess energy of the consequent *tert*-amyl would be sufficient for it to undergo rapid scrambling of all of its hydrogens on the time scale of collision time,

$1/\xi = 0.05$ ms at a total pressure of 6×10^{-4} torr [38]. Therefore, a monodeuterated cation would yield predominantly d_1 product. Mass spectrometry of $(CH_3)_2CDCH_2CH_2Br$ showed that the odd-electron fragment ion at m/z 70 contained no deuterium. In other words, the γ-deuterio analog expelled only DBr and not HBr, so its neutral products would be d_0. Neutral product of the troublesome m/z 70 ion could then be discriminated from the deprotonation products of *tert*-amyl cation simply by mass spectrometric analysis of the recovered 2-methyl-butenes. Unfortunately, isotopic fractionation on the GLPC introduced an unacceptably high level of uncertainty into GC–MS analysis, and it was necessary to purify the tiny quantities of 2-methyl-2-butene and 2-methyl-1-butene by preparative GLPC in order to collect entire peaks for mass spectrometric analysis of deuterium content. The $d_0:d_1$ ratios were measured for the separated 2-methyl-butene products of 70-eV EBFlow radiolysis of $(CH_3)_2CDCH_2CH_2Br$ and of $(CH_3)_2CHCH_2CHDBr$. The isomer distribution from deprotonation was taken to be the distribution of d_1-2-methylbutenes from the γ-d_1-analog. The 2-methylbutene ratio from deprotonation of *tert*-amyl cation by 2×10^{-4} torr triethylamine was inferred to be $H_2C=C(CH_3)CH_2CH_3/(CH_3)_2C=CHCH_3 = 2.6$ on this basis. An independent estimate of this ratio was based on the observation that m/z 70 became less abundant relative to *tert*-amyl cation in the mass spectrum of $(CH_3)CDCH_2CH_2Br$ when compared to undeuterated isoamyl bromide. The gross 2-methylbutene ratio was 20% larger for the γ-d_1-isomer, corresponding to a 2-methylbutene isomer ratio of 2.1 ± 0.4 from deprotonation of *tert*-amyl cation. In other words, 70–80% of the 2-methylbutene yield from 70-eV electron bombardment of isoamyl bromide came from *tert*-amyl cation, while the remainder (almost entirely 2-methyl-2-butene) came from neutralization of m/z 70 [39]. The more abundant isomer from deprotonation of *tert*-amyl cation is the thermodynamically less stable one. Obviously, gas-phase deprotonation does not take place under thermodynamic control.

The above discussion briefly summarizes the rather involved procedure needed to dissect two different pathways that form the same product in the EBFlow. Happily the situation was not nearly so complicated for bromocyclohexane. Deprotonation of the methylcyclopentyl cation by triethylamine afforded a C_6H_{10} ratio of 1-methylcyclopentene: methylenecyclopentane of 1.22 (standard deviation 0.02) in apparatus A at the highest magnetic field strengths. The ratio in apparatus C was 1.24 (standard deviation 0.045). The C_6H_{10} ratio with ammonia as base had the value 2.75 (with larger error limits) for apparatuses B and C [33, 38]. The experiments in apparatus C represent the state of the art at this writing, and Table 3 summarizes experimental results for different bases (and no base) at 3×10^{-4} torr of bromocyclohexane plus 1×10^{-4} torr of base [40].

The normalized yield of 1-methylcyclopentene and methylenecyclopentane together was $3.8 \, \mu mol \, A^{-1} s^{-1}$ when 1×10^{-4} torr of trimethylamine was present and $0.3 \, \mu mol \, A^{-1} s^{-1}$ when base was omitted. The ionization cross section of bromocyclohexane by 70-eV electrons is $22.7 \, \text{Å}^2$, whereas that of

Table 3. Proportion of 1-Methylcyclopentene in the Neutral Products of Deprotonation of 1-Methylcyclopentyl Cation by Various Bases from 70-eV EBFlow Radiolysis of 3×10^{-4} torr of Bromocyclohexane[a]

Base	Reported Proton Affinity, kJ/mol	(ratio)	σ
Triethylamine	972	.554	0.009
Trimethylamine	942	.573	0.006
Ammonia	854	.733	0.012
Dipentyl ether	859	.829	0.006
Dibutyl ether	852	.885	0.038
No Base	(795)	>.985	

[a]The sample standard deviation, σ, is not shown for the experiment where no base was present, since methylenecyclopentene was not detected. Proton affinities are taken from S. G. Lias, J. F. Liebman, and R. D. Levin, *J. Phys. Chem. Ref. Data* **13**, 695–808 (1984). The value used for the case of no base is the estimated proton affinity of bromocyclohexane.

trimethylamine is 7.4 Å^2 [12]. At 70 eV, $C_6H_{11}^+$ constitutes 44% Σ from bromocyclohexane [18]. Four-fifths of the electron beam produces ionizations as it passes through the reaction mixture, of which 40% yields $C_6H_{11}^+$ directly. But virtually all of the other fragment ions react with bromocyclohexane (unless they are intercepted by base) in a fashion analogous to Eq. 2 to produce $C_6H_{11}^+$. Therefore the theoretical normalized yield of C_6H_{10} from deprotonation is somewhere between 3.3 and 7.5 μmol A^{-1} s^{-1}, and the observed yield in the presence of trimethylamine is consistent with the 75% trapping efficiency estimated for the cold trap in apparatus C (vide supra Section 3.2.4), assuming that deprotonation occurs on every collision. The C_6H_{10} yield falls off for weaker bases and drops sharply when base is omitted. The rearranged C_6H_{10} isomers in the absence of base are presumed to result largely from surface neutralization. The efficiency with which ions are removed through the clown cap into the differentially pumped region is estimated to be $1 - (0.3/3.8)$.

The precision of EBFlow measurements of product ratios is ordinarily quite good as the values of the sample standard deviations, σ, attest. The exception is the measurement for dibutyl ether as base, for which experimental product ratios varied from 4:1 to 15:1 in nine independent runs that were performed as closely as possible under identical conditions. The large variation for the weakest base is taken to signify that ion temperature fluctuates widely in the EBFlow. For highly exothermic reactions, product ratio does not depend strongly upon ion temperature. But when one product channel becomes only slightly exothermic, its proportion may vary greatly with ion kinetic energy. Deprotonation of methylcyclopentyl cation by dibutyl ether to yield methylenecyclopentane is exothermic by 12 kJ/mol, and it seems this lies within a range of ΔH where the efficiency of the reaction channel depends greatly on ion temperature. Since ion kinetic energies vary with electron beam current (which fluctuates over the course of a run), any result that is strongly affected will exhibit a large variation from run to run.

In addition to the bases listed in Table 3, deuterated ammonia and dipentyl ether were investigated. Because the deuteria of ND_3 readily exchange with water on the walls of the apparatus, the quadrupole mass spectrometer was used to assess the isotopic purity of ND_3 in the reaction vessel. When first let into the reaction vessel, the ammonia gave a mass spectrum that showed m/z 19 more abundant than m/z 20, indicating that ND_3 was exchanging label. Several days of continuous exchange of the apparatus (during which 1 g of D_2O was slowly passed through) were required before it was possible to have confidence that the base was indeed ND_3. Deuterating the ammonia turned out to have a negligible effect on the product distribution [33]. Needless to say, prolonged exchange with H_2O was necessary to bring the apparatus back to status quo ante.

Further experiments on methylcyclopentyl cation support the conclusion that the effects of perdeuteration on neutral product distributions are not large. Deprotonation of $C_6H_{11}^+$ by $(n-C_5D_{11})_2O$ yielded a 1-methylcyclopentene: methylenecyclopentane ratio of 4.3 ($\sigma = 0.3$), within experimental error of the value for undeuterated pentyl ether [40]. With completely deuterated Brønsted acid the effects are also slight. Reaction of $C_6D_{11}^+$ with NH_3 gave a ratio of deuterated 1-methylcyclopentene: methylenecyclopentane equal to 3.1, which is statistically significantly larger than the ratio in the perprotio case (by a factor of only 1.1) [33]. With pentyl ether the isomer ratio from $C_6D_{11}^+$, 4.7 ($\sigma = 0.4$), is within experimental error of the value in the perprotio case.

4.3 Ion–Molecule Association Reactions

For the association reaction shown schematically in Eq. 19 to take place,

$$R^+ + Q \rightleftarrows [RQ^+]^{\neq} \xrightarrow{Q} RQ^+ \tag{19}$$

a third molecule is usually needed so that the reverse reaction does not destroy the product. A third-order rate constant $k_3 > 10^{-28}\,\mathrm{cm^6\,mol^{-2}\,s^{-1}}$ is needed for the association to occur to the extent of $>5\%$ in apparatus A at 10^{-3} torr. In other words, the initially formed $[RQ^+]^{\neq}$ must persist for times on the order of $0.01/\xi$ for it to live long enough to experience a stabilizing collision. Vibrationally excited aggregates do not ordinarily have lifetimes on the order of microseconds unless they lie in a very deep potential well. Lampe and co-workers [41] have reported $[RQ^+]^{\neq}$ species with lifetimes of the order of

10^{-5} s when R^+ is SiH_3^+ and Q is ethylene. Is it possible that unimolecular rearrangement might lead to kinetic stabilization of collision complexes?

EBFlow techniques offered an opportunity to test this hypothesis for $R^+ = (CH_3)_2CH^+$ and Q = ethylene. RRKM calculations [41] predicted that the initially formed collision complex (where RQ is portrayed as a corner-protonated cyclopropane) would have a lifetime < 10 ns unless it rearranged, but that rearrangement to *tert*-amyl cation could compete with this dissociation. Once the tertiary cation was formed, it could live for milliseconds without collisional stabilization.

Equations 20–22 show the sequence of reactions studied in apparatus A:

$$CH_3CH_2CH_2OCH_2CH_2CH_3 \xrightarrow{70\,eV} CH_3CH_2CH_2^+ \longrightarrow (CH_3)_2CH^+ \qquad (20)$$

$$(CH_3)_2CH^+ \quad \begin{matrix}CH_2\\ \|\\ CH_2\end{matrix} \rightleftharpoons \left[(CH_3)_2CH^+ \begin{matrix}-CH_2\\ |\\ -CH_2\end{matrix}\right]^{\ddagger} \longrightarrow \quad \underset{+}{\diagup\!\!\!\diagdown\!\!\!\diagup} \qquad (21)$$

$$\underset{+}{\diagup\!\!\!\diagdown\!\!\!\diagup} \quad \xrightarrow{Pr_2O} \quad \diagup\!\!\!\diagdown\!\!\!\diagup\!\!\!/ \qquad \diagup\!\!\!\diagdown\!\!\!=\!\!\!\diagdown \qquad (22)$$

$$ 1.2 \quad : \quad 1$$

Propyl ether (3×10^{-4} torr) was ionized by 70-eV electron impact in the presence of ethylene (1×10^{-3} torr). The initially formed $C_3H_7^+$ fragment rearranged to isopropyl cation. Addition to ethylene yielded the *tert*-amyl cation, presumably as shown in Eq. 21, and subsequent deprotonation led to 2-methylbutenes. Needless to say, many products that could be ascribed to free radical addition to ethylene were also observed: 3-methyl-1-butene [from isopropyl radicals produced by surface neutralization of $(CH_3)_2CH^+$] and 1-pentene and *cis* and *trans*-2-pentene (from *n*-propyl radicals). Remarkably, the 2-methylbutenes represented approximately three-eights of the recovered C_5H_{10}, with a normalized yield of 0.17 μmol A^{-1}s^{-1} [39]. Since there was no other conceivable route to the 2-methylbutenes, it was possible to conclude that unimolecular rearrangement could effectively compete with dissociation. (An analogous mechanism could account for Lampe's results.)

Aggregation of alcohols with their protonated parents (Eq. 23)

$$ROH_2^+ + ROH \rightleftharpoons (ROH)_2H^+ \xrightarrow{ROH} \xrightarrow{ROH} \cdots \xrightarrow{ROH} (ROH)_nH^+ \qquad (23)$$
$$\downarrow NX_3$$
$$HNX_3^+$$

is one association reaction that takes place with high efficiency even below

10^{-3} torr. This complex grows by adding more alcohol molecules and can be viewed as a very tiny droplet of solvent. At what stage does this droplet begin to take on the properties of bulk solvent? An EBFlow study in apparatus B used the competition between S_N1 and S_N2 reactions as a probe.

When R is an optically active group like *sec*-butyl, backside displacement will produce a *meso*-isomer of the protonated ether from an optically pure starting material, as depicted in Eq. 24:

$$\text{(S)} \quad \text{R-OH} + \text{R-OH}_2^+ \longrightarrow \text{meso (S,R)} \tag{24}$$

In the presence of an amine base (NX_3), deprotonation of this product led to *sec*-butyl ether, which was detected by GLPC of the 70-eV EBFlow radiolysis products on a didecyl phthalate open tubular column. Since this product was only 0.01% of the unreacted starting material, the GLPC conditions had to be carefully controlled so as to achieve the best separation of the diastereomeric *sec*-butyl ethers without their being swamped by the enormous peak due to starting material that came before.

When the amine base (at 2×10^{-4} torr) was strong enough (X = *n*-propyl or *n*-butyl) to intercept the growing cluster as depicted in Eq. 23, the ratio of *meso* to *d,l*-*sec*-butyl ether was 6:1. In other words, the second-order reaction (Eq. 24) went predominantly by the S_N2 route, with inversion of stereochemistry. But when the amine base was not strong enough to intercept the growing cluster (X = H), then the cluster, too, contributed to the yield of *sec*-butyl ether, as evidenced by the production of equal quantities of the *meso* and *d,l*-*sec*-butyl ethers [42]. At 5×10^{-4} torr of 2-butanol, the clusters cannot be growing much larger than $n = 2$ or 3 in apparatus B. But it appears that even at this size, the S_N1 type process shown in Eq. 25,

$$(ROH)_nH^+ + ROH \rightarrow (ROH)_{n+1}H^+ \rightarrow [(ROH)_nH_2O \quad R^+]$$
$$\rightarrow R_2O(ROH)_{n-1}H_3O^+ \tag{25}$$

(which randomizes stereochemistry) is taking place instead of the backside displacement. Insofar as their ability to promote solvolysis of protonated alcohols is concerned, very small clusters can behave like bulk solvent. Since backside attack ostensibly cannot take place on a *tert*-butyl group, the ICR observation that *tert*-butanol reacts with its protonated parent via water expulsion from the proton-bound dimer [43] suggests that a cluster size $n = 1$ may be enough for solvolysis to take place when R^+ is a tertiary cation.

4.4 Bimolecular Lewis Acid–Base Reactions

If GLPC and GC–MS were the only ways to analyze EBFlow reaction product mixtures, the previous sections would have covered the range of neutral product studies feasible with this technique. Recently, however, it has been shown that it is possible to use NMR to identify products in the crude reaction mixture without any prior separation. This development was necessitated by the desire to study monofluorinated hydrocarbons, many of which readily lose HF under GLPC conditions. The unexpected lability of these molecules dictated that neutral product characterization not heat the mixture far above room temperature nor expose it to high surface areas of alumina or silica. Moreover, the huge variety of isomers for even simple molecular formulas would have demanded extraordinary efforts to prepare a large number of authentic samples in order for GC–MS to provide unambiguous identifications.

As it turns out, with high-field Fourier transform NMR instruments now routinely available, proton and ^{19}F NMR that show products whose normalized yields are on the order of 0.1 μmol A^{-1}s^{-1} can be secured within a few hours. The major obstacle is not the sensitivity limit of NMR nor the chemical shift dispersion, but rather the dynamic range needed to measure signals in the presence of 10,000-fold excess of unreacted starting material. Although signal suppression techniques are available, it happens that a 1000-Hz separation between two resonances is often sufficient to permit the discrimination of peaks of such widely different intensities. Determination of relative peak areas is sometimes more easily done by cutting and weighing than by electronic integration, but otherwise the use of NMR turns out to be perfectly conventional in its approach.

The first study was performed on *tert*-butyl fluoride, for which the pertinent issue was an instance of Eq. 4. The principal fragment ion, $(CH_3)_2CF^+$, was known to react with parent neutral to yield *t*-butyl cation [44]. Did this reaction take place via Lewis (fluoride abstraction) or via Brønsted (proton transfer) acid–base chemistry? Equation 26 illustrates the two alternatives:

$$(CH_3)_2CF^+ + (CH_3)_3CF \begin{array}{c} \xrightarrow{\text{Brønsted}} (CH_3)_3C^+ + CH_3CF=CH_2 + HF \\ \xrightarrow{\text{Lewis}} (CH_3)_2CF_2 + (CH_3)_3C^+ \end{array} \quad (26)$$

The question was whether one neutral molecule, $(CH_3)_2CF_2$, or two, $CH_2{=}CFCH_3$ plus HF, are produced by the reaction. To answer this question all that was needed was to examine the NMR of the EBFlow radiolysis product of *tert*-butyl fluoride.

A pressure variation experiment was performed to assess pathways and side products [45]. At a low pressure (3×10^{-5} torr), $(CH_3)_2CF^+$ was unlikely to encounter parent neutrals frequently enough for Eq. 26 to proceed to any appreciable extent. As Fig. 10 shows, the ^{19}F NMR shows only $CH_3CF{=}CH_2$

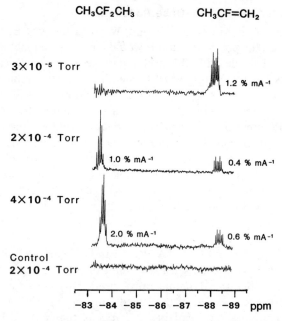

Fig. 10. Products of 70 eV EBFlow radiolysis of *tert*-butyl fluoride in apparatus C at various pressures analyzed by 282 MHz fluorine NMR [45]. Normalized conversions for 2,2-difluoropropane and 2-fluoropropene are expressed as percent of recovered starting material divided by ionizing electron current (sum of mean currents measured at the reaction vessel wall, clown cap, cage, and downstream endplate). Chemical shifts relative to $CFCl_3$. [Reprinted by permission from E. W. Redman, K. K. Johri, R. W. K. Lee, and T. H. Morton, *J. Ann. Chem. Soc.* **106**, 4639–4640 (1984). Copyright 1984 American Chemical Society.]

at this low pressure. As pressure is raised, a new product grows in with increasing yield, namely, the Lewis product $(CH_3)_2CF_2$. In a control experiment (filament on, but no ionizing electrons entering the reaction vessel) neither product was seen.

Two features of these NMR spectra are worthy of note. With electron beam current on the order of 0.1 mA, the products represented at most a few parts per thousand of the unreacted starting material. The huge peak from *tert*-butyl fluoride, which was nearly 12,000 Hz upfield from these peaks, did not interfere with their detection. Second, the yield of CH_2=$CFCH_3$ does not appear to be vanishing at the highest pressure.

Here was a case in which, it turned out, two reactions of interest were taking place. Equation 26 appears to have gone exclusively (>90%, within experimental uncertainty) via the Lewis pathway, but there is another fragment ion that also abstracts fluoride from the parent neutral. This ion is $CH_3\overset{+}{C}$=CH_2. There is a prominent m/z 41 peak in the mass spectrum of *tert*-butyl fluoride, and it had been assumed to be allyl cation (which in separate experiments can be shown to abstract fluoride from *tert*-butyl fluoride to yield allyl fluoride [51]). But no allyl fluoride was seen, and authentic sources of $CH_3\overset{+}{C}=CH_2$ also formed the product $CH_2=CFCH_3$.

To look at these authentic sources, $(CD_3)_3CF$ was used as the Lewis base [46]. The products of its fragment ions were, of course, fully deuterated. EBFlow radiolysis in the presence of $CH_3CBr=CH_2$, isobutene, or $CH_3CH=CHBr$ produced undeuterated $CH_3CF=CH_2$, whose ^{19}F resonance was 1.3 ppm away from the d_5-analog [51]. The large upfield deuterium isotope effect on ^{19}F chemical shifts is well known, and its magnitude appears to correlate with the spin–spin coupling of the fluorine to the substituted hydrogen [47]. This means that ^{19}F NMR can be used to assess the extent and positions of deuteration in a mixture of products with different degrees of isotopic substitution. In fact, clearly resolved spectra of $CH_2=CFCD_3$, $CD_2=CFCH_3$ together with the d_5 and d_0 analogs in an EBFlow product mixture have been published to demonstrate that label does not scramble in the partially deuterated 2-propenyl cation [46]. Another important control was a confirmation that F^+ or $F\cdot$ were not giving rise to the observed fluorinated products. Reactions of these atomic species with organic molecules ordinarily produces HF rather than a fluorocarbon [48], but a double check by EBFlow radiolyses of CF_4 (a copious source of F^+ and $F\cdot$) with C_3H_5Br showed a rate of production of $CH_2=CFCH_3$ that could not account for more than a small fraction of the products ascribed to $CH_3\overset{+}{C}=CH_2$ [46].

Isotope shifts of the proton NMR also turn out to be useful. Figure 11 shows that ^1H NMR of a mixture of $(CH_3)_2CF_2$ and $CD_3CF_2CH_3$ from an EBFlow experiment. There is no evidence of $CD_2HCF_2CH_2D$, which argues that hydrogen scrambling does not take place in the $(CH_3)_2CF^+$ cation.

The stability of α-fluoro and vinylic cations has been a subject of great interest, and EBFlow radiolysis of $(C_2H_5)_3CF$ at 3×10^{-4} torr yields proportions of recovered products that closely match the relative abundances of the mass spectrometric fragments. Table 4 shows a comparison of the ion abundances and the EBFlow radiolysis results. Such a close correlation of relative abundances argues that the neutral products must have resulted from fluoride abstraction by the fragment ions. One of the most interesting results is the E/Z isomer ratio in 3-fluoro-2-pentene, which is close to the thermodynamic ratio. This suggests that the steric bulk of the triethylcarbinyl group that is donating fluoride to the vinylic cation $CH_3CH=\overset{+}{C}CH_2CH_3$ does not have a major effect on determining which isomer is formed.

The ability to analyze extent and position of deuteration in mixtures of isotopically substituted products makes fluoride abstraction an excellent way to probe gaseous ion structures. The use of fluorine NMR to analyze neutral products of ion–molecule reactions gives gas-phase ion chemistry the ability to answer detailed questions about organic reaction mechanisms in the gas phase. Neutral product studies establish a common basis upon which ion–molecule reactions in the gas phase can be compared with reactions of cations in solution.

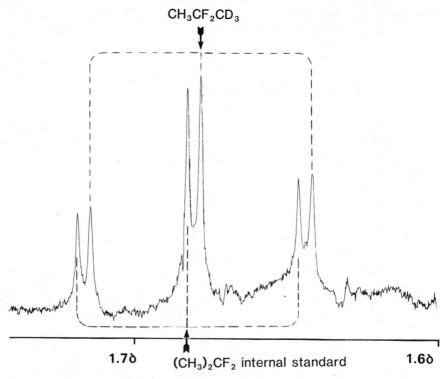

Fig. 11. Proton NMR (500 MHz) of partially deuterated 2, 2-difluoropropane from 70 eV EBFlow radiolysis of $CH_3(CD_3)_2CF$ at 2×10^{-4} torr in apparatus C [46]. Recorded at the Southern California Regional NMR Facility supported by NSF Grant CHE 79-16324. In the fluorine NMR of this same reaction mixture, $(CD_3)_2CF_2$ and $CD_3CF_2CH_3$ are seen at -84.4 and -84.0 ppm (relative to $CFCl_3$), well upfield of the resonance of $(CH_3)_2CF_2$ at -83.6 ppm. The proton NMR shows what the fluorine NMR cannot, namely that the label in the partially deuterated product are all on the same carbon. [Reprinted by permission from E. W. Redman, K. K. Johri and T. H. Monton, *J. Am. Chem. Soc.* **107**, 780–784 (1985). Copyright 1985 American Chemical Society.]

Table 4. Correlation of the Relative Intensities of the Principal Ions in the 70-eV Mass Spectrum of Triethylcarbinyl Fluoride (Recorded on a VG-ZAB Double Sector Instrument) with the Relative Abundances of Fluorinated Neutral Products from the 70-eV EBFlow Radiolysis [49]

m/z	Relative Intensities in 70 eV MS	Structure of Ion	Corresponding Neutral	Relative Integrals from ^{19}F NMR
89	1	Et_2CF	Et_2CF_2	1
69	2.7	$C_2H_5\overset{+}{C}=CHCH_3$	$C_2H_5CF=CHCH_3$	2.9 (Z/E = 2.3)
55	1.2	$CH_3CH_2\overset{+}{C}=CH_2$	$CH_3CH_2CF=CH_2$	1.3
27	0.7	$H_2C=CH$	$H_2C=CHF$	1.1

ACKNOWLEDGMENT

This work has been supported by the donors of the Petroleum Research Fund, administered by the American Chemical Society, and the National Science Foundation.

REFERENCES

1. F. B. Burns and T. H. Morton, *J. Am. Chem. Soc.*, **98**, 7308–7313 (1976).
2. C. A. Lieder and J. I. Brauman, *J. Am. Chem. Soc.*, **96**, 4028–4030 (1974).
3. (a) M. A. Smith, R. M. Barkley, and G. B. Ellison, *J. Am. Chem. Soc.*, **102**, 6851–6852 (1980); (b) M. E. Jones, S. R. Kass, J. Filley, R. M. Barkley and G. B. Ellison, *J. Am. Chem. Soc.*, **107**, 109–115 (1985).
4. M. M. Bursey, D. J. Harvan, C. E. Parker, and J. R. Hass *J. Am. Chem. Soc.*, **105**, 6801–6804 (1983) and references contained therein.
5. P. Ausloos and S. G. Lias in C. Sandorfy, P. J. Ausloos, and M. B. Robin, Eds., *Chemical Spectroscopy and Photochemistry in the Vacuum Ultraviolet*, Reidel, Dordrecht, Holland, 1974, pp. 465–482.
6. (a) P. Crotti, F. Macchia, A. Pizzabiocca, G. Renzi, and M. Speranza, *J. Chem. Soc. Chem. Commun.*, 486–487 (1986); (b) G. de Petris, P. Giacomello, T. Picotti, A. Pizzabioca, G. Renti, and M. Speranza, *J. Am. Chem. Soc.*, **108**, 7491–7495 (1986); (c) F. Cacace, G. de Petris, S. Fornarini, and P. Giacomello, *J. Am. Chem. Soc.*, **108**, 7495–7501 (1986) and references contained therein.
7. (a) M. Attina, F. Cacace, R. Cipollini, and M. Speranza, *J. Am. Chem. Soc.*, **107**, 4824–4828 (1985); (b) S. Fornarini and M. Speranza, *J. Am. Chem. Soc.*, **107**, 5358–5363 (1985); (c) G. Occhiucci, F. Cacace, and M. Speranza, *J. Am. Chem. Soc.*, **108**, 872–876 (1986); (d) F. Cacace and M. Speranza, present volume, Chap. VI.
8. (a) G. I. Gellene and R. F. Porter, *Acc. Chem. Res.*, **16**, 200–207 (1983); (b) C. Wesdemiotis, R. Feng, and F. W. McLafferty, *J. Am. Chem. Soc.*, **108**, 656–5657 (1986); (c) C. Wesdemiotis, R. Feng, P. O. Danis, E. R. Williams, and F. W. McLafferty, *J. Am. Chem. Soc.*, **108**, 5847–5853 (1986); (d) W. J. Griffiths, F. M. Harris, A. G. Brenton, and J. H. Beynon, *Int. J. Mass Spectrom. Ion Proc.*, **74**, 317–321 (1986); (e) A. B. Raksit and R. F. Porter, *Int. J. Mass Spectrom. Ion. Proc.*, **76**, 299–306 (1987); (f) A. B. Raksit and R. F. Porter, *J. Chem. Soc. Chem. Commun.*, 500–502 (1987); (g) A. B. Raksit and R. F. Porter, *Org. Mass Spectrom.* **22**, 410–417 (1987).
9. C. N. McEwen, *Mass Spectrom. Rev.*, **5**, 521–547 (1986) and references contained therein.
10. J. L. Holmes, C. E. C. A. Hop, and J. K. Terlouw, *Org. Mass Spectrom.*, **21**, 776–778 (1986) and references contained therein.
11. J. A. Beran and L. Kevan, *J. Phys. Chem.*, **73**, 3866–3871 (1969).
12. W. L. Fitch and A. D. Sauter, *Anal. Chem.*, **55**, 832–835 (1983).
13. H. Deutsch and M. Schmidt, *Beitr. Plasmaphys.*, **25**, 475–484 (1985).
14. R. Gorden, Jr., R. E. Rebbert, and P. Ausloos, *Rare Gas Resonance Lamps*, NBS Technical Note 496, U. S. Government Printing Office, Washington, D.C., 1969.
15. D. Davis and W. Braun, *Appl. Opt.*, **7**, 2071–2074 (1968).
16. T. H. Morton, *J. Am. Chem. Soc.*, **102**, 1596–1602 (1980).
17. T. H. Morton, unpublished results, 1976.
18. S. R. Heller and C. W. A. Milne, *EPA/NIH Mass Spectral Data Base*, National Bureau of Standards, Washington, D.C., 1978: vol. I, NSRDS-NBS 63.
19. (a) L. Endrenyi and D. J. LeRoy, *J. Phys. Chem.*, **70**, 4081–4084 (1966); (b) K. W. Watkins and D. R. Lawson, *J. Phys. Chem.*, **75**, 1632–1640 (1971); (c) K. J. Mintz and D. J. LeRoy, *Can. J. Chem.*, **51**, 3534–3538 (1973).
20. (a) N. Friedmann, H. H. Bovee, and S. L. Miller, *J. Org. Chem.*, **36**, 2894–2897 (1971); (b) S. L.

Miller, *Adv. Chem. Phys.* **55**, 85–107 (1984); (c) L. L. Miller, *Acct. Chem. Res.*, **16**, 194–199 (1983); (d) Y. Mizobuchi and L. L. Miller, *J. Org. Chem.*, **50**, 318–321 (1985).

21. A. L. J. Beckwith and C. H. Schiesser, *Tetrahedron*, **41**, 3925–3941 (1985).
22. T. H. Morton, *Tetrahedron*, **38**, Report 137, 3195–3243 (1982).
23. D. A. Armstrong, *Radiat. Phys. Chem.*, **20**, 75–86 (1982).
24. (a) E. W. McDaniel and E. A. Mason, *The Mobility and Diffusion of Ions in Gases*, Wiley, New York, 1973; (b) H. W. Ellis, E. W. McDaniel, D. L. Albritton, L. A. Viehland, S. L. Lin, and E. A. Mason, *Atomic Data Nucl. Data Tables*, **22**, 179–217 (1978); (c) H. W. Ellis, M. G. Thackston, E. W. McDaniel, and E. A. Mason, *Atomic Data Nucl. Data Tables*, **31**, 113–151 (1984).
25. E. W. McDaniel, *Collision Phenomena in Ionized Gases*, Wiley, New York, 1964.
26. F. F. Chen, *Introduction to Plasma Physics and Controlled Fusion*, 2nd ed. Plenum, New York, 1984, vol. 1.
27. (a) F. Schmieder, *Z. Elektrochem.* **36**, 700–704 (1930); (b) R. P. Brode, *Rev. Mod. Phys.*, **5**, 257–279 (1933).
28. R. K. Asundi, *Proc. Phys. Soc. (London)*, **82**, 372–374 (1963).
29. R. C. Dunbar, J. H. Chen, and J. D. Hays, *Int. J. Mass Spectrom. Ion Proc.*, **57**, 39–56 (1984).
30. Vespel is a registered trademark of E. I. DuPont de Nemours Company.
31. (a) L. J. Favreau, *Rev. Sci. Instrum.* **36**, 856–857 (1965); (b) L. J. Favreau and D. F. Koenig, *Rev. Sci. Instrum.*, **38**, 841 (1967).
32. T. E. Hanley, *J. Appl. Phys.*, **19**, 583–589 (1948).
33. E. W. Redman and T. H. Morton, *J. Am. Chem. Soc.*, **108**, 5701–5708 (1986).
34. S. R. Heller and C. W. A. Milne, *EPA/NIH Mass Spectral Data Base, Supplement* **1**, National Bureau of Standards, Washington, D.C., 1980, NSRDS-NBS 63-1.
35. Recorded by ICR, with peak intensities normalized by m/z. T. H. Morton, unpublished results.
36. F. B. Burns, Ph.D. Thesis, Brown University, 1976.
37. (a) R. Derai, P. Nectoux, and J. Danon, *J. Phys. Chem.*, **80**, 1664–1676 (1976); (b) R. Derai and J. Danon, *J. Phys. Chem.*, **81**, 199–206; (c) R. Derai and J. Danon, *Chem. Phys.*, **15**, 331–344 (1976); (d) J. Danon, *J. Chem. Phys.*, **76**, 1051–1057 (1979).
38. T. H. Morton, *Radiat. Phys. Chem.*, **20**, 29–40 (1982).
39. W. J. Marinelli and T. H. Morton, *J. Am. Chem. Soc.*, **100**, 3536–3539 (1978); **101**, 1908 (1979).
40. E. W. Redman and T. H. Morton, paper presented at the 32nd Annual Conference on Mass Spectrometry and Allied Topics, San Antonio, Texas, May 1984.
41. A Oppenstein and F. W. Lampe, *Rev. Chem. Intermed.*, **6**, 275–333 (1985).
42. D. G. Hall, C. V. Gupta, and T. H. Morton, *J. Am. Chem. Soc.*, **103**, 2416–2417 (1981).
43. J. L. Beauchamp, M. C. Caserio, and T. B. McMahon, *J. Am. Chem. Soc.*, **96**, 6243–6251 (1974).
44. J. Y. Park and J. L. Beauchamp, *J. Phys. Chem.*, **80**, 575–584 (1976) and references therein.
45. E. W. Redman, K. K. Johri, R. W. K. Lee, T. H. Morton, *J. Am. Chem. Soc.*, **106**, 4639–4640 (1984).
46. E. W. Redman, K. K. Johri, and T. H. Morton, *J. Am. Chem. Soc.*, **107**, 780–784 (1985)
47. (a) J. B. Lambert and L. G. Greifenstein, *J. Am. Chem. Soc.*, **95**, 6150–6152 (1973); (b) J. B. Lambert and L. G. Greifenstein, *J. Am. Chem. Soc.*, **96**, 5120–5124 (1974); (c) P. E. Hansen, F. M. Nicolaisen, and K. Schaumburg, *J. Am. Chem. Soc.*, **108**, 628–629 (1986).
48. M. Hamdan, N. W. Copp, K. Birkinshaw, J. D. C. Jones, and N. D. Twiddy, *Int. J. Mass Spectrom. Ion Phys*, **69**, 191–195 (1986).
49. K. K. Johri, E. W. Redman, and T. H. Morton, paper presented at the 33rd Annual Conference on Mass Spectrometry and Allied Topics, San Diego, Calif., May 1985.
50. D. G. Hall and T. H. Morton, *J. Am. Chem. Soc.*, **102**, 5686–5688 (1980).
51. D. A. Stams, K. K. Johri, and T. H. Morton, *J. Am. Chem. Soc.*, **110**, 699–706 (1988).

Chapter **IV**

FLOWING AFTERGLOW AND SIFT

Nigel G. Adams
and
David Smith

---

1 Introduction
2 Flowing Afterglow
   2.1 Experimental Method
   2.2 Flow Dynamics and Data Analysis
   2.3 Some Early Successes of the FA
   2.4 Some Recent FA Studies
   2.5 Other Uses of the FA
3 Selected Ion Flow Tube
   3.1 Experimental Method
   3.2 Some Studies Illustrating the Versatility of the SIFT
      3.2.1 Binary Reactions of Ground State Positive and Negative Ions
      3.2.2 Ternary Association Reactions
      3.2.3 Binary Reactions of Excited Positive Ions
      3.2.4 Reactions of the Structural Isomers of Some Positive Ions
4 Selected Ion Flow Drift Tube
   4.1 Variable-Temperature SIFDT
5 Concluding Remarks
References

---

1 INTRODUCTION

The flowing afterglow (FA) and the selected ion flow tube (SIFT) are examples of fast flow tube reactors in which rate coefficients (or rate constants) are determined for ion–neutral reactions under multiple collision conditions and hence conditions for which the kinetic (translational) temperature of the reactants, and usually the internal temperature, can be fully defined. In both the FA and SIFT, charged particles (electrons, positive ions, and negative ions) are created in or injected into the upstream region of a carrier gas and are convected along a flow tube by the carrier gas with which (ideally) they are unreactive. Multiple collisions of the relatively low-number density charged particles with the more numerous carrier gas atoms or molecules ensure that their velocity distribution rapidly become Maxwellian in the upstream region at the temperature of the carrier gas. Gases can then be introduced into the downstream region where they may react with the now kinetically thermalized charged particles.

The FA is a development of the stationary afterglow (SA), and the SIFT is a further development of the FA. Each development has represented an enormous step forward in the study of ionic reactions at thermal energies. In the SA, an ion source gas and a reactant gas are mixed at relatively low concentrations with an inert buffer gas in a glass or stainless-steel vessel and ionization is then created in the mixture, usually by a short duration dc, rf, or microwave discharge. Reactions between the ions and the reactant gas are then studied in the afterglow plasma following the cessation of the discharge, usually by monitoring the mass-analyzed wall currents of the ions. A discussion of SA methods is given in the textbook by McDaniel and Mason [1]. Whereas a good deal of the early data relating to positive ion–neutral reactions were acquired using SA methods, the disadvantages are several, mostly due to the fact that the ion source gas and the reactant gas are both present in the reaction vessel and are thus exposed to the discharge, that excited ions are produced and persist in the afterglow (especially metastable states), and that the method lacks flexibility. Also, the ambipolar electric fields in the SA are such that negative ions are inhibited from diffusing to the vessel walls [2], and thus negative ion reactions cannot be studied with any confidence except for the special circumstances of an ion–ion afterglow plasma [1].

The success of the FA is that it overcomes most of the disadvantages of the SA [3, 4]. In the FA, ionization is created in the carrier gas upstream of the position at which ion source gas is added, and hence the latter is not exposed to the dc or microwave discharge or to the electron impact ion source that is commonly used to create the FA plasma. Reactant ions can then be produced by a more controlled ion chemistry, i.e., via reactions between carrier gas ions (e.g., He^+ when the carrier gas is helium, as is usually the case) and the added source gas, the flow of which can be carefully controlled to optimize the rate of production of the secondary ion species. Sequential additions of gases at different positions along the flow tube allow a wide variety of ions to be produced and hence extend the range of reactions that can be studied well beyond that possible using the SA. The reactions of negative ions can also be studied since they are equally efficiently convected downstream to the mass spectrometer sampling orifice through which the positive or negative ions are transmitted, thence mass selected and detected. Again, since the carrier gas is only weakly ionized (typical ionization number densities are $\sim 10^7$–$10^9 \, cm^{-3}$ and typical carrier gas number densities are $\sim 10^{16} \, cm^{-3}$), the ions are rapidly kinetically relaxed in collisions with carrier gas atoms well before they have been convected to the downstream (reaction) region. Note that the flow tube is typically ~ 100 cm long, the flow velocity is $\sim 10^4$ cm s^{-1} and hence the ion residence time is $\sim 10^{-2}$ s, i.e., much longer than the collision time between ions and the carrier gas. The rate coefficients for both positive and negative ion–neutral reactions can be determined under kinetically thermalized conditions by observing the reduction in the current of a given ionic species (monitored by the downstream mass spectrometer/detection system) as controlled amounts of reactant gas are introduced into the afterglow plasma. Product ions can also be identified by the mass spectrometer. The presence of internally (vibrationally and electronically) excited reactant ions is a potential problem. Fortunately the presence of free electrons in the FA acts to minimize the concentrations of excited ions (via superelastic collisions between excited ions and electrons). Gases that "quench" the excited ions to their ground states, but that do not react chemically, can also be added to the afterglow. Metastable helium atoms can be removed from the FA by the addition of, say, argon atoms that undergo Penning ionization with the metastable atoms. So, in general, ions in the reaction zone of a FA will be in their ground states, and hence rate coefficients will generally refer to reactions between truly thermalized ions and neutrals. The reactions of a wide variety of positive and negative ion types with a wide variety of neutral atoms and molecules have been studied using variable temperature FA apparatuses in some cases over a wide range of temperature [5]. Further aspects of the FA technique and some illustrative results of the many FA studies of ion–neutral reactions will be discussed in Section 2, which also includes a brief discussion of the flowing afterglow/Langmuir probe (FALP) method [6] that has been developed and exploited to study other reaction processes at thermal energies, including positive ion/electron recombination, electron attachment, and positive ion/negative ion neutralization.

The FA has, however, several undesirable features, i.e. the presence of the ion source gas in the reaction zone that introduces complications due to reactions between the ions and their source gas, the presence of electrons and the possible occurrence of ion electron dissociative recombination that can distort ion product distributions, the presence of metastable atoms that are potential sources of ionization in the afterglow, and the presence of energetic photons originating from the ion source. These are all avoided in the SIFT since positive or negative ions are created in a remote ion source, and, after mass selection, they are injected into the carrier gas whence they are convected along the flow tube and sampled by the downstream mass spectrometer/detection system [4]. Hence a swarm of positive or negative ions, not an electron/ion plasma, is convected along the flow tube by the inert carrier gas in the absence of the ion source gas. Thus the potential complicating factors listed above for the FA are simply avoided in the SIFT, and the reactions of any ionic species that can be created in an ion source and injected into the carrier gas without being collisionally dissociated can be studied over a wide range of temperatures and with a wide range of reactant neutrals. As will be illustrated in Section 3, this includes reactions of ion types ranging from weakly bound cluster ions to energetic doubly charged ions. A further great advantage of the SIFT over the FA is that accurate ion product distributions can be determined even when multiple products are evident in a reaction. The rate coefficients are determined in a manner identical to that adopted for the FA, and they always refer to kinetically and rotationally relaxed ionic and neutral reactants. However, a fraction of the ions injected into the SIFT may well be vibrationally and/or electronically excited, and this excitation can be retained against multiple collisions with the carrier gas and thus excited ions may reach the downstream reaction zone. Careful checks must be made for such excitation (see Section 3) if the reactions of ground-state ions are to be studied, and then appropriate action needs to be taken to minimize or remove excited ions from the swarm. Conversely, the presence of excited ions can be deliberately encouraged (by, for example, adjustment of the ion source conditions), and then reactions of the excited ions can be studied. A good deal of such work has recently been carried out and is referred to in Section 3 together with a discussion of the variable-temperature SIFT (VT-SIFT) technique and of some data illustrative of the many reactions that have been studied.

The most recent development in fast flow tube techniques is the variable-temperature selected ion flow *drift* tube (VT-SIFDT) [7], which followed from the flow drift tube developed some 10 years previously [8]. Descriptions of the various drift tubes and flow drift tubes have been given in the review by Lindinger and Smith [9]. The VT-SIFDT consists of a VT-SIFT into which has been included a series of equally spaced metal rings coaxial with the flow tube axis in the downstream reaction zone. These rings are used to establish a uniform electric field along this reaction zone. The field imparts a drift velocity to the reactant ions, thus increasing the center-of-mass energy between the ions and both the buffer gas and the reactant gas atoms or molecules. Thus the energy and temperature dependences of ion–neutral reaction rate coefficients and ion product distributions can be studied in this one apparatus. Center-of-mass energies in excess of about 1 eV can be achieved, equivalent to appreciable fractions of the bond energies in some molecules. Hence the value of the VT-SIFDT for chemical kinetics studies is obvious. The rate coefficient data obtained can be compared with those derived from ion beam measurements of the cross sections for ion–neutral reactions [10]. The VT-SIFDT method is discussed more fully in Section 4, where illustrative results are also presented.

The considerable amount of data relating to ion–neutral reactions, as obtained from FA and SIFT experiments, has been used to great effect to elucidate the ion chemistry of media as diverse as the terrestrial ionosphere, interstellar gas clouds, laser plasmas, and surface etchant plasmas. Some of the data relevant to these media will be referred to in the following sections.

2 FLOWING AFTERGLOW

The FA was first developed to study ion–molecule reactions at thermal energies in the early 1960s by Fehsenfeld, Ferguson, and Schmeltekopf in the

NOAA Laboratories in Boulder, Colorado. The major review of the technique by these workers stands as the most thorough appraisal of the technique to date [3] (although other reviews have been written [4, 11, 12]). Reviews have also been published that summarize the results of studies of ion–molecule reactions carried out by exploiting the several flowing afterglows that have been built in the NOAA laboratory and in other laboratories [12–16]. A major motivation to the NOAA group for developing the FA was its interest in elucidating the ion chemistry of the ionosphere, and they were enormously successful in this [17–19]. However, the exploitation of the FA resulted in an even greater prize, viz., a rapid increase in the understanding of the fundamental nature of ion–molecule reactions at thermal energies that has implications in several other areas of research, some of which we refer to in this review. From the time of its inception to the development of the SIFT technique [20–22] the FA was used exclusively to study ion–molecule reactions, in the main positive ion–molecule reactions but with some notable studies of negative ion–molecule reactions [5]. However, since 1976 the FA has largely been replaced by the SIFT for the study of positive ion reactions while the FA has been increasingly used to study negative ion reactions and processes for which the large electron and ion number densities attainable are essential. For example, the FA has been used to study infrared (IR) emissions from the products of ion–molecule reactions [23] and also to study several other plasma reaction processes including electron–ion recombination, ion–ion neutralization, and electron attachment [24]. Flowing afterglows are also increasingly being used as sources of relaxed ions and cluster ions (both positive and negative ions) for SIFT experiments [25] and ion beam experiments [23, 26, 27].

In this section, we review briefly the essential features of the FA technique, including the elements for the flow dynamics and the relevant rate equations, and discuss the method of deriving reaction rate coefficients. Then we present some results that illustrate the range, capabilities, and successes of the technique. Some of the most recent results are presented, especially those relating to negative ion reactions and to the IR luminescence studies. At the end of this section, we describe the FALP technique and mention some of the recent interesting results obtained from its exploitation.

2.1 Experimental Method

A schematic diagram of a typical flowing afterglow apparatus is shown in Fig. 1. It consists of a flow tube (made either of glass or stainless steel) that is usually about 100 cm long and about 8 cm in diameter (this conforms to the original FA design [3]) coupled to which is a large mechanical pump (usually a Roots-type pump) having a large pumping speed (about several hundred liters per second) at pressures in the range 0.1 to a few torr, these being the typical operating pressures of FA experiments. A carrier gas that is usually pure helium (pure argon, nitrogen and hydrogen have also been used) is introduced at the upstream end of the flow tube and then flows along it due to the action of the pump at a velocity of about 10^4 cm s^{-1}. Ionization is created either in the pure

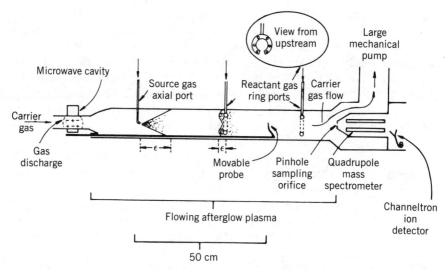

Fig. 1. A schematic of a flowing afterglow, in this case including a Langmuir probe and using a microwave cavity discharge to generate a high-density afterglow plasma for recombination studies (this is the FALP apparatus [6, 37, 74]). For studies of ion–neutral reactions only low ion number densities are necessary, and then lower power electron impact or dc discharge sources are used. Both axial and ring parts are illustrated (see text) via which gases can be introduced into the afterglow. The carrier gas is constrained to flow along the flow tube by the action of a large mechanical pump. The ions present in the downstream region of the afterglow are identified by a quadrupole mass spectrometer/detection system. The earliest flow tubes were constructed from glass, but they are now of stainless steel and can be varied over a wide temperature range.

carrier gas or in the carrier gas containing a small admixture of another gas from which particular positive or negative ions are generated by electron impact using a hot cathode electron emitter, although microwave and cold cathode discharges have also been used. Hence a gaseous plasma is produced near to the ionizer in which the charged particles, in particular the electrons, are suprathermal. A fraction of the positive ions (and hence of electrons also) are convected along the flow tube by the carrier gas whence they "cool" in collisions with carrier gas atoms or molecules. Thus a thermalized afterglow plasma is created in the downstream region of the flow tube. An axial gradient in the charged particle number density exists due to the loss of these charged particles via their ambipolar diffusion to the flow tube walls where efficient surface recombination of the positive ions with electrons (or negative ions) occurs. The charged particle number densities in the afterglows can be varied over orders of magnitude by adjusting the ion source current and/or the carrier gas pressure (which determines the ambipolar diffusive loss rate), but the maximum electron and ion number densities are of the order of 10^{11} cm^{-3}. Although such large number densities are necessary to study recombination processes (such as in the FALP experiments [24]), they are most undesirable for the study of ion–molecule reactions for which recombination is a disturbing influence. Ions that arrive at the downstream end of the flow tube can be detected after they enter a pinhole orifice (typically 0.1–1 mm diameter) located in a disk positioned at the apex of a nose cone that forms part of a differentially pumped mass spectrometer housing. The ions are then accelerated into a quadrupole mass spectrometer and detected with an electron multiplier device that is positioned off axis to reduce the background count rate due to energetic photons from the ionization source, due to metastable atoms or molecules and—in the case of negative ion sampling—due to electrons that are not efficiently filtered out by the quadrupole mass spectrometer. In the FA experiment of Grabowski, the mass analyzed ions are accelerated onto a metal plate where they eject electrons (for positive ion impact) and positive ions (for negative ion impact), which are then accelerated into an electron multiplier [28]. Further details of the ion sampling and the detection system are given in a recent review [7]. Rate coefficients are then determined by measuring the decay of the primary ion count rate as a function of the flow rate of reactant gas into the afterglow [3, 4, 7]. The analysis of the data requires an appreciation of the carrier gas and plasma flow dynamics, and this is discussed in Section 2.2.

A great variety of primary positive and negative ions can be produced by variously introducing controlled amounts of ion source gas either upstream of the ionization source or into the afterglow downstream. Unfortunately, it is often difficult to generate just one primary ion species, and although the presence of two or more primary ions does not usually prevent an accurate determination of the rate coefficient for a given reaction it can complicate, and sometimes prevent, the identification of ion products of the reaction. (The accurate determination of ion product distributions is one of the many advantages of the SIFT technique [21].) Also, the presence in the afterglow of energetic species such as helium metastable atoms can complicate the "product" ion spectra due to Penning ionization of reactant gases [29–31]. Notwithstanding these potential complications, by careful thought and good experimental technique, accurate rate coefficients and product ion identifications have been obtained for a large number of ion–molecule reactions, in some cases over the very wide temperature range 80–900 K in a variable-temperature FA [32, 33]. A compilation of the data obtained from FA (and the earliest SIFT) experiments has been given by Albritton [5].

2.2 Flow Dynamics and Data Analysis

Consider the hypothetical positive ion reaction

$$A^+ + B \rightarrow C^+ + D \tag{1}$$

in which the primary ion A^+ reacts with the neutral species B producing C^+ and D; any one or all of the reactants and products may be molecular. In practice, A^+ is generated by the addition of a suitable source gas to the afterglow plasma, and the count rate of A^+ at the downstream detector is monitored as the number density of B, i.e. [B], is varied in the plasma afterglow. [B] is determined from a measurement of Q, the absolute flow rate of B using the expression $Q = \pi a^2 v_0 [B]$, where a is the diameter of the flow tube and v_0 is the flow velocity of B that is identical to the bulk flow velocity of the carrier gas. The latter is obtained from a determination of the flow rate and the number density (pressure) of the carrier gas according to an expression identical to that given for Q above. Under conditions such that A^+ ions are not lost from the plasma at a significant rate by dissociative recombination with electrons (in practice this requires that the electron number density is less than $\sim 10^8$ cm^{-3}), then the only loss processes for A^+ are reaction with B and ambipolar diffusion. Then the appropriate continuity equation for $[A^+]$ is

$$v_i(r) \frac{\partial [A^+]}{\partial z} = D_a \nabla^2 [A^+] - k[A^+][B] \qquad (2)$$

where $v_i(r)$ is the radially dependent flow velocity of A^+ along the flow tube axis (z coordinate), k is the rate coefficient for the reaction 1, and D_a is the ambipolar diffusion coefficient that describes the loss of A^+ by diffusion with electrons to the flow tube walls. An analogous equation to Eq. 2 can be written for the reaction of negative ions but with D_a replaced by an appropriate diffusion coefficient [34]. The solution to Eq. 2 is of the form

$$[A^+]_z = [A^+]_0 \exp - \left(\frac{\Delta D_a}{a^2} + \Gamma k [B] \right) \frac{z}{v_0} \qquad (3)$$

where $[A^+]_0$ is the number density A^+ at the reactant gas inlet port, $[A^+]_z$ is $[A^+]$ at the downstream ion sampling orifice, Δ is a numerical factor that is weakly dependent on the carrier gas pressure, and Γ is the ratio of the bulk carrier gas flow velocity to the mean ion flow velocity, i.e., v_0/v_i [35, 36]. Detailed independent treatments of the flow dynamics of the FA have indicated that v_i exceeds v_0 and that v_i/v_0 ($=\Gamma^{-1}$) should be within the range 1.4–1.6 over the range of conditions appropriate to FA experiments [3]. This is in excellent agreement with measurements of v_i/v_0 in FA (and in SIFT) experiments that give a value of 1.45 at 300 K, rising to 1.65 at 80 K at pressures of ~ 0.5 torr. A more detailed discussion of the dynamics is given in a recent review [7]. The z in Eq. 3 is, by implication, the distance between the reactant gas inlet port and the ion sampling orifice. However, in practice, this should be replaced by an effective reaction distance, $(z + \varepsilon)$, where ε is an "end correction" to z, which is necessary because the reactant gas is not dispersed uniformly into the carrier gas at precisely $z = 0$. The magnitude and sense of ε depend on the type of inlet port used. For the simple axial type (see Fig. 1), ε is positive by several centimeters; for the radial type ε is negative by several centimeters. However, for the ring inlet ports shown in Fig. 1, which we have adopted in both our FA and SIFT apparatuses [7, 37], ε is very small, being only about $+1$ cm if the reactant gas is initially directed upstream (the contraflow technique). Hence for a typical z of, say, 40 cm, the end correction is insignificant using such ring ports. The determination of the rate coefficient for a given reaction is invariably carried out at a constant carrier gas pressure, whence the parameter $\Delta D_a/a^2$ in Eq. 3 is constant. Hence, relating $[A^+]_z$ to n_0 and n_i the count rates of A^+ at the mass spectrometer for zero flow and finite flows of B into the afterglow respectively, it is clear that Eq. 3 can be reduced to the simpler equation

$$\ln \frac{n_i}{n_0} = -k[B]\Gamma \frac{z}{v_0} \qquad (4)$$

By replacing z by $(z + \varepsilon)$ to account for any end correction and replacing $[B]$ by $Q/(\pi a^2 v_0)$, where Q is the flow rate of B into the afterglow, then

$$\ln \frac{n_i}{n_0} = -\frac{kQ(z+\varepsilon)}{\pi a^2 v_0 v_i} \qquad (5)$$

Clearly a plot of the natural log of the reactant ion count rate against the reactant gas flow rate should be linear, and a value of the rate coefficient for the reaction can then be readily obtained from the slope of the line if v_i can be measured directly. Numerous sample plots of data have appeared in the early literature; they are quite analogous to the example of SIFT data given in Section 3. Also, although the analysis has been developed here for a binary (two-body) reaction, it is directly applicable to ternary (three-body) reactions in which the effective binary rate coefficient, k_{eff}, is determined as described above but as a function of the carrier gas pressure (or number density $[M]$). Then the ternary rate coefficient, k_3, is deduced from the simple relation $k_{eff} = k_3[M]$.

2.3 Some Early Successes of the FA

Several reviews have been published that outline much of the early work carried out using flowing afterglows [12–16]. The great beauty of the technique lies in its flexibility for the study of a wide range of reactions. A variety of carrier gases can be used, all manner of gases and vapors (including metal vapors [38]) can be used to generate a great variety of positive and negative ions, and sequential ionic reactions along the flow tube can be used to extend the range of ion types. The accuracy of the rate coefficients determined with the FA (which originally centered around uncertainties in the flow dynamics but these are now well understood) has been tested against other established (but less versatile) techniques such as stationary afterglows, mass spectrometer ion sources, and drift tubes. Rate coefficients can now be readily measured in the FA to $\pm 20\%$ and in favorable cases to $\pm 10\%$. The earliest measurements [39, 40] concentrated on quite elementary reactions that were considered to be of importance in the terrestrial ionosphere, such as the reactions of He^+ with N_2 and O_2; e.g.,

$$He^+ + N_2 \rightarrow N_2^+ + He \qquad (6a)$$

$$\rightarrow N^+ + N + He \qquad (6b)$$

The branching into the two product channels is observed in the FA experiment, but determination of the branching ratio is difficult and is best achieved using a SIFT (see Section 3.1). Reaction 6 and the $He^+ + O_2$ reaction both proceed on every collision between the reactants, that is, the rate coefficients, k, are equal to their respective Langevin or collisional rate coefficients, k_c [41, 42]. The measured value of $k(6)$ at 300 K is 1.2×10^{-9} cm^3 s^{-1} compared to the calculated k_c of 1.7×10^{-9} cm^3 s^{-1}. Other elementary reactions were shown to be much less efficient, including the reactions of N_2^+ with O_2 and of O^+ with both N_2 and O_2; e.g.,

$$O^+ + O_2 \rightarrow O_2^+ + O \tag{7}$$

$k(7)$ at 300 K was measured to be 1.9×10^{-11} cm^3 s^{-1} compared to the $k_c(7)$ of 9.1×10^{-10} cm^3 s^{-1}, indicating that the reaction occurred on only about 1 collision in 50 [33]. These measured rate coefficients agreed precisely with those determined previously in stationary afterglow experiments [39, 43]. Furthermore, the temperature variation of $k(7)$ had been determined quite precisely in a stationary afterglow study [44] and shown to vary as $T^{-1/2}$, and precisely this dependence was reproduced in the first variable-temperature FA study [33]. FA studies of the variation with temperature of the k for a number of binary positive ion–molecule reactions have shown that little variation of k is observed for fast reactions (i.e., reactions for which $k \sim k_c$), whereas the k for slower reactions invariably change with temperature, commonly exhibiting an increasing k with decreasing temperature [32, 33]. This phenomenon has been described in more detail in a recent review [45]. A further elementary reaction worthy of note is

$$O^+ + N_2 \rightarrow NO^+ + N \tag{8}$$

FA studies [32] showed that $k(8)$ increases by about a factor of 3 as the temperature reduces from about 600 to 80 K, and, in a classic FA experiment [46], $k(8)$ was shown to increase dramatically as the vibrational state of the N_2 was increased by discharge excitation. Reaction 8 is an extremely important reaction in the terrestrial ionosphere, and as such it has received considerable attention in thermal energy experiments and also in drift tubes [47].

Many other ionospheric reactions have been studied in flowing afterglows [17–19]. Notable among these studies is the elucidation of the ion chemistry that leads to the production of hydrated hydronium ions $H_3O^+(H_2O)_n$ in the lower ionosphere. It has been shown that an important first stage is the ternary association reaction:

$$NO^+ + H_2O + M \rightleftharpoons NO^+ \cdot H_2O + M \tag{9}$$

In this reaction M is an abundant atmospheric molecule, i.e. N_2 or O_2. Further, similar reactions lead to $NO^+(H_2O)_3$, which can then react with another H_2O

molecule eliminating HNO_2:

$$NO^+(H_2O)_3 + H_2O \rightarrow H_3O^+(H_2O)_2 + HNO_2 \tag{10}$$

Another route to $NO^+ \cdot H_2O$ in the lower ionosphere is via the ternary reaction

$$NO^+ + N_2 + N_2 \rightleftharpoons NO^+ \cdot N_2 + N_2 \tag{11}$$

which is followed by the rapid ligand switching reaction

$$NO^+ \cdot N_2 + H_2O \rightarrow NO^+ \cdot H_2O + N_2 \tag{12}$$

Ligand switching reactions involving both positive and negative ions were first studied in the FA. These reactions were shown to occur with nearly unit collisional efficiency when they are exothermic, and this provided information on relative bond strengths in cluster ions [48]. Much similar FA work has been carried out on the negative ion chemistry of the lower ionosphere. Although this chemistry is somewhat less clear than the positive ion chemistry, the basic processes that occur in the region have been identified and are discussed in several reviews [17–19].

Also notable among the early FA studies were those of Bohme and his colleagues who have made comprehensive studies of proton transfer reactions involving both positive ions and negative ions [15]. Such reactions are exemplified by

$$H_3^+ + N_2 \rightarrow N_2H^+ + H_2 \tag{13}$$

and

$$CH_3O^- + C_2H_2 \rightarrow C_2H^- + CH_3OH \tag{14}$$

It was observed that when proton transfer is exothermic, the reactions are invariably fast. The measured rate coefficients at 300 K for a large number of such reactions have been compared with various theoretical predictions for their respective collisional rate coeffficients, k_c. Indeed, proton transfer has now been adopted as a test of the accuracy of theoretical predictions of k_c [42, 49].

The early FA studies of associative detachment reactions of negative ions are worthy of note. Such reactions are exemplified by

$$F^- + H \rightarrow HF + e \tag{15}$$

$$S^- + O_2 \rightarrow SO_2 + e \tag{16}$$

The results of this work have been reviewed and discussed in an interesting paper by Fehsenfeld [14]. Studies of reactions such as 15 obviously required the development of techniques for flowing adequate quantities of atomic radicals into the FA and for correctly measuring the atom flow rate (H, N, and O atom reaction have been successfully studied). An interesting finding of these studies was that when the reaction mechanism is relatively simple, such as in 15 in which the atomic species exothermically unite to form a diatomic molecule and the electron is released, then the reaction proceeds rapidly (i.e., $k \sim k_c$). However, when the insertion of an atom is necessary to produce the most stable configuration of the product molecule, such as in reaction 16, then the reaction is relatively slow at 300 K, presumably due to activation energy barriers. In many examples of this latter class of associative detachment reactions, no reaction occurred even when very exothermic product channels were available. Studies of associative detachment also embraced reactions of cluster ions with H atoms [50]; e.g.,

$$OH^-(H_2O)_3 + H \rightarrow 4H_2O + e \qquad (17)$$

This reaction and the similar reactions of other members of the $OH^-(H_2O)_n$ series are quite rapid and are implicated in ionospheric chemistry.

2.4 Some Recent FA Studies

The above summary represents only a tiny fraction of the many FA studies of ion–molecule reactions that have been carried out. A few of these FA programs are continuing, but now most flow tube studies of ion–molecule reactions are exploiting SIFT apparatuses (see Section 3.2). However, some recent outstanding FA studies must be mentioned here since they further emphasize the value and versatility of the FA for studies of ionic reactions.

A most exciting development in FA work has been the observation of infrared emissions from nascent product molecules formed in various types of ion–molecule reactions occurring in the afterglows. The first to be studied was the associative detachment reaction:

$$O^- + CO \rightarrow CO_2 + e \qquad (18)$$

The IR radiation emitted by the product CO_2 indicated that the reaction energy is quite efficiently channeled into the vibrational modes of the CO_2 [51]. Other associative detachment reactions have been studied, including the reactions of F^-, Cl^-, and CN^- with H atoms producing vibrationally excited HF, HCl, and HCN, respectively, and these studies indicated that associative detachment strongly favors the population of the highest accessible vibrational states of the product molecules. A similar interesting study of the reaction of SF_6^- with H and D atoms in which F atom abstraction occurs generating HF and DF has also been carried out [52]. These beautiful experiments are directed toward gaining an understanding of the energy partition in the products of ion–molecule reactions and of the reaction dynamics. For these reasons, other types of reaction have been similarly studied, including the proton transfer reactions of F^-, Cl^-, O^-, and CN^- with various hydrogen halides, e.g.,

$$F^- + HBr \rightarrow HF + Br^- \qquad (19)$$

In these reactions the product molecules are also vibrationally excited. The general observation is that all energetically accessible vibrational states of the product molecules are populated. Product molecules in vibrationally excited states were detected by their IR emissions, and those in the ground vibrational state were detected using laser-induced fluorescence (LIF). This work has been extended to study the states of the product molecular ions formed in the reactions of atomic positive ions with diatomic molecules, such as the reaction

$$N^+ + O_2 \rightarrow NO^+ + O \qquad (20)$$

The NO^+ was observed to be excited up to $v = 14$. In a later experiment, it was also established that the majority of the product O atoms were excited to the ^1D state, and the remainder were in the ^3P ground state [53]. In other experiments by the same group the rovibrational states of the CO^+ and N_2^+ product ions of the thermal energy charge transfer reactions of N^+ with CO and Ar^+ with N_2 were studied using LIF [54, 55]. Further details can be obtained from the review paper by Bierbaum et al. [23].

A good deal of the recent FA work has centered around the chemistry of organic negative ions. Much of this work has been summarized in several reviews, which also include summaries of related SIFT studies [56, 57]. The technique adopted in these FA studies has been to create simple inorganic negative ions such as F^-, OH^-, and NH_2^- by flowing small concentrations of ion source gases (NF_3, NH_3, etc.) through the upstream ion source (where the negative ions are formed in dissociative attachment reactions) and then to add organic molecules to the afterglow to form organic negative ions via proton transfer reactions such as

$$OH^- + CH_3CN \rightarrow CH_2CN^- + H_2O \qquad (21)$$

Ions like OH^- and NH_2^- are among the strongest bases known, and they can abstract a proton from most organic molecules. A wide range of organic negative ions has been formed in this way, and their reactions studied at thermal energies [56, 57]. Thus a good deal of effort has been put into understanding the mechanisms of isotope exchange in the reactions of negative ions with deuterated polar molecules such as D_2O, ND_3, CH_3OD, and CF_3CH_2OD. It was observed that isotope exchange is facile between negative ions and exchange reagents that are as much as 20 kcal mol^{-1} less acidic. This elegant FA work on H/D exchange follows and extends comprehensive isotope exchange studies carried out using the SIFT technique (see Section 3.2.1). Other studies have included reactions of NH_2^- and PH_2^- with many species in which elimination of stable molecules occurs with the production of more exotic negative ions that are difficult or impossible to generate by dissociative attachment reactions. Such reactions include

$$NH_2^- + CO_2 \rightarrow NCO^- + H_2O \qquad (22)$$

and
$$NH_2^- + N_2O \rightarrow N_3^- + H_2O \tag{23}$$

This provides the opportunity to study the reactions of species such as the azide ion N_3^-. The FA has also been used extensively to study, and thus to explain, the fundamental ion chemistry of organic compounds [58–61].

Ion chemistry occurring in the FA, in which the energies and states of the reactants are in general quite precisely defined, can be used to determine fundamental parameters such as electron affinities. A case in point is the bracketing of the electron affinity of SF_4 [62]. The observation that SH^- transfers an electron to SF_4 whereas NO_2^- does not, brackets the electron affinity (EA) of SF_4 between that of SH (EA = 2.32 eV) and NO_2 (EA = 2.38 eV). The EA of SO_2 has also recently been established in this way in a concerted series of FA and SIFT measurements [63].

Little has so far been mentioned about ternary association reactions. Many such studies have been carried out in recent years using SIFT apparatuses, and these are discussed in Section 3.2.2. However, some notable FA studies of the association reactions of F^- and Cl^- with BF_3 and BCl_3 have been carried out at 300 K [64]. As mentioned previously, the effective binary rate coefficients, k_{eff}, for association reactions are determined as a function of the number density (pressure) of the stabilizing third body. Normally, a plot of k_{eff} against pressure is linear (in the "low-pressure regime" [65, 66]) and extrapolates through the origin of coordinates. However, for the reactions mentioned above, e.g., the association of F^- with BF_3 to form BF_4^-, a nonzero intercept is indicated that has the same magnitude for three different carrier gases (i.e., He, Ar, and N_2). The slopes of the plots were also finite, but these depended on the nature of the carrier gas. These interesting observations have been interpreted in terms of the simultaneous occurrence, under the pressure conditions of the FA, of both ternary association and binary (radiative) association. Thus, for example, in the reaction

$$F^- + BF_3 + He \rightarrow BF_4^- + He \tag{24a}$$

$$F^- + BF_3 \rightarrow BF_4^- + h\nu \tag{24b}$$

both the collisionally stabilized reaction 24a and the radiatively stabilized reaction 24b occur in parallel. This work (which is of fundamental importance) has been repeated and extended in a recent variable-temperature SIFT study from which additional evidence for the occurrence of radiative association in these reactions has been obtained [67].

2.5 Other Uses of the FA

Although the FA since its inception has been exploited to study ion–neutral reactions, it has also been developed to study different reaction types using the FALP technique and as a source of relaxed ions for spectroscopic and photofragmentation studies.

In the FALP technique [6, 68], the chemical versatility of the FA is exploited to create afterglow plasmas in which processes such as electron–ion recombination, ion–ion neutralization, and electron attachment individually dominate the loss of ionization from the plasma. In the FALP, in addition to the downstream mass spectrometer, a Langmuir probe is included that can be located at any position along the axis of the afterglow column where it can be used to determine the electron number density, n_e, or the positive or negative ion number density, n_+ or n_-. Hence the axial gradient of n_e, n_+, n_- as appropriate can be measured.

To determine positive ion–electron recombination coefficients α_e, the chemical versatility of the FA is used to create positive ion/electron afterglow plasmas in which one species of molecular positive ion exists at sufficiently high n_+ and n_e (typically $10^{10}\,\mathrm{cm}^{-3}$) and sufficiently high carrier gas pressure (typically 1 torr) that recombination loss of ions and electrons (rather than ambipolar diffusion) is the dominant loss process along the afterglow column. As an example, consider the determination of α_e for O_2^+ ions in a helium carrier gas. Initially, He^+ (and metastable helium atoms) are created upstream via a microwave discharge. A few centimeters downstream, and therefore in the afterglow, sufficient Ar is added to destroy the metastables (and any He_2^+ ions formed in reactions of He^+ with the He atoms). Hence an afterglow plasma consisting of He^+ and Ar^+ and electrons is created in which the only loss process for ionization is the relatively slow ambipolar diffusion (since atomic ions do not recombine at a significant rate with electrons). Note that the downstream mass spectrometer is routinely used to identify the ion types present in the afterglow. Thus the axial gradient of n_e (and n_+) determined using the Langmuir probe is very small. O_2 is then added to the plasma in sufficient quantities to convert all the He^+ and Ar^+ ions to O_2^+ ions, and this immediately initiates dissociative recombination; i.e.,

$$O_2^+ + e \to O + O \qquad (25)$$

This is manifest by a marked increase in the axial gradient of n_e, i.e., $\partial n_e/\partial z$, which is determined by the Langmuir probe and is described by the simple relationship

$$v_p \frac{\partial n_e}{\partial z} = -\alpha_e n_e n_+ = -\alpha_e n_e^2 \quad \text{since } n_+ = n_e \qquad (26)$$

where v_p is the plasma flow velocity ($=v_e$ and v_+). Equation 26, which is only valid when diffusive loss is insignificant (but diffusive loss could be accounted for if necessary), can be readily integrated to yield

$$\frac{1}{(n_e)_{z_1}} - \frac{1}{(n_e)_{z_2}} = \alpha_e \frac{(z_2 - z_1)}{v_p} \qquad (27)$$

and so a plot of $1/n_e$ vs. z should be linear with slope α_e/v_p. Since v_p can be readily calculated (from v_0) or directly measured, then α_e can be determined.

Experiments can be carried out with the FALP over the temperature range 95–600 K, and α_e for many ions have been determined over this temperature range, including the α_e for ions involved in the chemistry of the terrestrial ionosphere [69] and interstellar gas clouds [70]. An outstanding discovery [71] has been that H_3^+ ions in their ground vibrational state do not dissociatively recombine with electrons at a measureable rate, contrary to previous experimental findings [72], but in accordance with recent theoretical work [73]. This has had a profound influence on thinking in interstellar chemistry. Afterglow plasmas consisting of H_3^+ as the dominant species are readily produced, and since H_3^+ readily transfers a proton to almost any molecule, e.g.,

$$H_3^+ + N_2, CO, CH_4, H_2O \rightarrow N_2H^+, HCO^+, CH_5^+, H_3O^+ + H_2 \qquad (28)$$

then, by the addition of these molecules (separately) to the H_3^+ afterglow plasma, a wide range of plasmas can be created in which specific positive ion species are dominant and for which the respective α_e can be readily determined.

The FALP technique can equally successfully be used to determine electron attachment coefficients, β, at thermal energies, also over a range of temperatures. For example, to determine β for the reaction

$$CCl_4 + e \rightarrow Cl^- + CCl_3 \qquad (29)$$

a measured flow of CCl_4 is added to the thermalized He^+/Ar^+/electron afterglow and the z gradient of n_e determined with the Langmuir probe in the usual way. For these studies, however, n_e (and n_+) must be sufficiently small so that loss of electrons due to the recombination of any positive ions formed in reactions of He^+ or Ar^+ with the attaching gas is negligibly small compared to loss via the attachment reaction 29. (In this example the reactions of He^+ and Ar^+ with CCl_4 can lead to CCl_2^+ and CCl_3^+.) Again, if ambipolar diffusion is also unimportant, then it is a straightforward procedure to derive the β from $\partial n_e/\partial z$, v_p, and the number density of attaching molecules introduced into the afterglow (see [74] for the analytical procedure). The β's for a number of attachment reactions have been determined at 300 K and for some reactions over various ranges of temperature. For example, dissociative attachments to freons such as CCl_3F, CCl_2F_2, CF_3Br, and direct attachment to form parent negative ions, e.g., to SF_6, C_6F_6, and C_7F_{14}, have been studied (see [74–76]). Very recently dissociative attachment reactions involving the superacids H_2SO_4, FSO_3H, and CF_3SO_3H have been studied [77]; e.g.,

$$FSO_3H + e \rightarrow FSO_3^- + H \qquad (30)$$

Reactions such as 30, which are extremely facile gas-phase reactions, can be viewed as being analogous to Brønsted acid reactions in the liquid phase in that they donate a proton to the electron (which can be considered as a weaker acid or a stronger base!).

The introduction of sufficient quantities of electron-attaching (negative ion-forming) gases into the FA can convert the positive ion/electron plasma to a positive ion/negative ion plasma that is completely devoid of electrons. This provides an opportunity to study neutralization reactions of positive ions with negative ions, reactions that are potentially important in ionized gas containing electronegative gases such as the lower terrestrial atmosphere [78] and laboratory surface etchant plasmas [79]. The FALP technique has been successfully applied to such studies, and the most reliable measurements to date of ion–ion recombination coefficients, α_i, have been made for reactions such as [34]:

$$NO^+ + NO_2^- \rightarrow NO^* + NO_2 \tag{31}$$

The star on the product NO in reaction 31 indicates that in this case the reaction energy appears as electronic excitation of the NO. This has been shown by spectroscopic observations of the recombining ion–ion plasma [80]. In addition to reactions such as 31, some ion–ion reactions involving only atomic species (e.g., $Ar^+ + Cl^-$) have been studied, reactions that proceed only very slowly for reasons well understood [81], and some reactions involving cluster ions such as $H_3O^+(H_2O)_3 + NO_3^- \cdot H_2O$ have been studied [82], reactions that are important in the lower terrestrial atmosphere.

Very few studies of the neutral products of reactions occurring in FA experiments have been attempted. The study of the NO product of reaction 31 represents the beginning of what will be a concerted spectroscopic study of the products of ion–molecule reactions and recombination reactions in the coming years. One unusual experiment in which a cryogenic trapping column has been configured with a FA to trap and then identify by chromatography the neutral products of a negative ion–molecule reaction is worthy of note [83]. However, this intrinsically difficult technique is not expected to have wide application.

Finally in this section, it is worth noting that flowing afterglows are finding useful applications as versatile sources of a wide variety of relaxed ions. Some years ago an FA was coupled to a SIFT apparatus and used as a source of hydrated hydronium ions (including their partially deuterated analogs) and other ion types [25, 84], and recently an FA source has been used as a valuable source of ions containing rare isotopes [85]. Again, the value of the FA as an ion source is in its chemical versatility. Flowing afterglows are also being used as sources of cold ions for high-resolution spectroscopic studies of ions [86]. They are also being succcessfully used in studies of the structure and of electron detachment energies of all manner of simple and polyatomic negative ions using photoelectron spectroscopy [27]. So, although FA are diminishing in importance for the study of ion–molecule reactions as the SIFT technique takes over this area, they are by no means obsolete for the study of other reaction processes and remain a vital aid to other molecular physics and chemistry experiments.

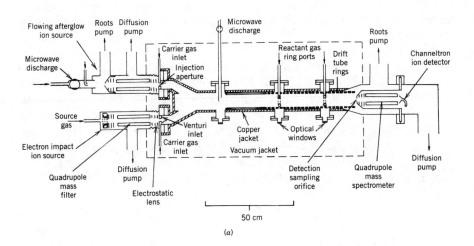

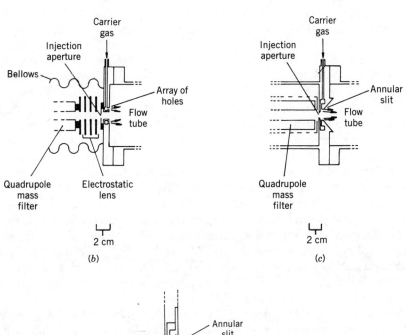

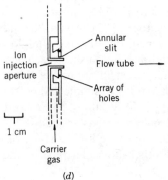

3 SELECTED ION FLOW TUBE

The SIFT technique was initially conceived, developed, and exploited by Adams and Smith [20, 21] in response to a need for laboratory data relating to the ion chemistry of interstellar clouds [87–89]. A major requirement is for the rate coefficients and, importantly, the product ions for the reactions of positive ions (including organic ions) with polyatomic molecules (again including organic molecules). These are the kind of reactions that cannot be studied easily using FA methods, largely because the introduction of polyatomic molecules into the FA as source gases usually results in the production of several primary ions that react rapidly with their parent molecules. The difficulty of unambiguously identifying product ions of particular reactions is then obvious, especially if the ion species under study is a minority ion. To be able to introduce a single ion type into a fast flowing gas and exclude the parent gas is clearly desirable, and this has been achieved in the SIFT, the basic principles of which are presented in Section 1. In this section, we briefly describe the general features of the SIFT including the variable temperature SIFT (the VT–SIFT), describe the methods for determining rate of coefficients and ion product distributions, refer to the wide range of reaction types that have been studied, present some data that illustrate the enormous versatility of the technique, and mention some of the new insights gained from SIFT experiments about the fundamentals of ion–neutral reactions and on the chemistry of various media such as interstellar gas clouds.

3.1 Experimental Method

A schematic diagram of a SIFT apparatus is shown in Fig. 2 (in fact the Birmingham double SIFDT is represented). The dimensions of the flow tube are similar to those typical of most FA apparatuses; indeed some SIFT apparatuses have been developed by modifying existing FA apparatuses. The essential difference between the SIFT and the FA is the ion injection system. Positive

---

Fig. 2. (*a*) A schematic of the Birmingham VT–SIFDT indicating the essential features of most SIFT and SIFDT apparatuses, but with some additional features including two SIFT ion injection systems together with different ion sources [4, 7]. Quadrupole mass filters are used in the upstream (injection) system and the downstream (detection) system. Ring ports, as indicated (and see Fig. 1), are generally used as reactant gas inlets because of the associated small end corrections. The wider bore axial port shown upstream is used to introduce into the flow tube radicals produced in a microwave discharge through a reactant gas/carrier gas mixture. Suitable potentials applied to the drift tube rings establish a uniform *E*-field in the downstream region, thus converting the SIFT to a SIFDT. The copper jacket, which is clamped to the stainless-steel flow tube, minimizes temperature gradients during heating or cooling. The dashed line is the outline of the large vacuum jacket that surrounds the complete flow tube and acts as a thermal insulator. (*b*) The original Birmingham SIFT injector in which the carrier gas enters the flow tube via an array of small holes [4]. (*c*) The NOAA/York injector in which the carrier gas flows through an annular slit [92]. (*d*) The central venturi element of the latest design used in Birmingham in which approximately half of the carrier gas flows through an annular slit and the remainder flows through the concentric array of holes. This combines the best features of (*b*) and (*c*) and is an effective venturi inlet.

or negative ions are created in an ion source (of which various forms have been used, see below) from which they are extracted and focused into a quadrupole mass filter. Ions of a selected mass-to-charge ratio pass through the mass filter, and a fraction of them then pass through a small aperture (~ 1 mm diameter) and into the flow tube. These ions are then convected down the flow tube as an ion swarm by a fast flowing carrier gas (at a pressure of typically 1 torr) in which they kinetically thermalize to the temperature of the carrier gas. They are sampled downstream via a small orifice located in the center of a thin molybdenum disk that is coupled to, but electrically insulated from, the nose cone part of the housing containing the mass spectrometer detection system. Reactant gases can be added to the carrier gas, and the rate coefficient for reaction with the ions in the swarm can be determined in a manner identical to that used for FA studies (but see the discussion below centered around Eq. 3). The crucial difference, however, between the FA and the SIFT is that in the SIFT only a single primary ion type exists in the carrier gas at a relatively low number density (typically 10^2–10^4 cm^{-3}) compared to the much higher number densities of electrons and ions in the FA plasma. Also, metastable atoms or energetic photons are not present in the SIFT since the ion source (or electrical discharge) is not located in the carrier gas. Thus these interfering species are not present to confuse data interpretation and also therefore cannot produce troublesome "dark current" at the detection system electron multiplier. Thus, in a SIFT, the electron multiplier can be positioned "on axis" to improve the detection efficiency for ions (not possible in a FA), especially when the SIFT injection systems are placed off axis whence there is clearly no line of sight between the ion source and the electron multiplier (see Fig. 2). This very "clinical" situation allows accurate ion product distributions (branching ratios) to be determined for ion–neutral reactions since a simple method is now available to account for the mass discrimination that inevitably occurs in quadrupole mass filters and that can (if not accounted for) introduce serious errors in ion product distribution determinations [21].

This method is as follows. The current of injected primary ions arriving at the insulated orifice disk mounted at the entrance to the detection system (see Fig. 2) is measured with a picoammeter, simultaneously with a measurement of the count rate of the ions arriving at the electron multiplier after they have passed through the quadrupole spectrometer. Thus an "efficiency factor" for the collection of an ion species of a given mass-to-charge ratio is obtained for a particular arrangement of lens-focusing potentials, resolution setting of the quadrupole, and gain of the particle multiplier and amplifiers. This procedure is then repeated for different ion species, and an "efficiency factor" is derived for each. Hence a mass discrimination factor curve can be constructed that can be used to correct apparent ion product distributions to true ion product distributions. This procedure is not possible in the FA because the presence of electrons, photons, and metastable atoms confuses attempts to measure orifice disk currents, and it is this, coupled with the problems associated with the occurrence of secondary reactions between primary ions and their parent (ion source) gas, that combined to frustrate attempts to determine accurate ion product distributions in FA experiments.

The crucial feature of the SIFT is that ions from the source have to be injected from the low-pressure quadrupole mass filter (SIFT) chamber through a small aperture into a carrier gas at relatively high pressure (~ 1 torr) with a sufficiently low energy to avoid collisonal dissociation on the carrier gas. This requires that the backflow of carrier gas through the aperture is insufficient to raise the pressure in the SIFT chamber to a level that prevents the mass filter from operating (in practice this pressure must be less than about 10^{-4} torr). This is made possible by introducing the carrier gas into the flow tube via a venturi-type inlet (see Fig. 2). In the original Birmingham SIFT injector [4], the carrier gas enters the flow tube via a series of small holes that are concentric with the ion entrance aperture. As the gas enters the flow tube, it is flowing near supersonically and is directed down the flow tube and away from the ion entrance aperture. Hence the backflow of gas into the SIFT chamber is minimized, and the pressure in the SIFT chamber can be readily held below the critical pressure by a diffusion pump (typically a 4-in. pump is adequate for the purpose). The amount of gas that does backstream does not seriously inhibit the passage of ions into the flow tube even when they are injected at low energies (see below). Several centimeters downstream of the gas injection point the carrier gas flow is subsonic and becomes laminar (after a short "settling-down distance") well before reaching the reaction zone of the SIFT. The settling-down distance depends on the type of injector and on the gas flow rate through it. For the Birmingham design the settling down to laminar flow requires about 20 cm for a carrier gas flow rate of 10^2 torr l s^{-1}. For the injector designed by the NOAA group at Boulder, Colorado [90], and the York University (Toronto) group [91], through which the carrier gas enters the flow tube via an annular slit concentric with the ion entrance aperture, the settling-down distance is significantly greater than that of the Birmingham injector for the same gas flow rate. Indeed, shock cell propagation along the flow tube is much more evident for the NOAA/York injector than for the Birmingham injector, although the former design has the merit that it is a more effective venturi (and so less backflow of carrier gas occurs [92]). To avoid extensive shock cell propagation when the NOAA/York injector is used, it is now the practice to introduce only a fraction (about 30%) of the carrier gas into the flow tube via the venturi inlet and the remainder subsonically via a series of holes in the injector plate, that is, a combination of the NOAA/York and the Birmingham designs (see Fig. 2d). A more detailed discussion of the merits of the various injector designs is given in a very recent review [7].

As mentioned above, two SIFT injection systems that necessarily have to be coupled to the flow tube by angled tubes are shown in Fig. 2. Rather than being a disadvantage, these angled pipes act to dissipate any shock cells that originate at the injector and thus reduce the settling-down distance. Most SIFT apparatuses that have only a single injector usually have the injector coaxial with the flow tube (except the original Birmingham single SIFT for which the injector was off axis [20, 21]). The double SIFT facility is useful in that two different varieties of ion source can be used simultaneously, generating different types of ions, including positive ions and negative ions (this was used to great effect in studying the reactions of hydrogen atoms; see Section 3.2.1). Several types of ion source have been used, including microwave discharge sources [20], low- and high-pressure electron impact sources [93, 94], and flowing afterglow sources [25, 85], the choice depending on the type of ions required. For example, low-pressure sources have been used to generate double-charged ions [95–98] and high-pressure and flowing afterglow sources to generate cluster ions [84, 99]. The wide range of ion types used in SIFT studies is indicated by the illustrative results presented in Section 3.2.

The flow dynamics of the SIFT are quite analogous to the FA dynamics discussed in Section 2.2. A point of difference worthy of note is that diffusive loss of ion in the SIFT is via free diffusion of the single ionic species through the carrier gas (usually helium) to the flow tube walls rather than ambipolar diffusion that occurs in the FA plasma. Free diffusion is a slower process [1], and thus proportionately fewer ions are lost to the walls of the SIFT as they are convected along the flow tube than are lost by ambipolar diffusion in the FA under otherwise identical conditions. Thus, the continuity equation describing the loss rate of an injected ion species in a SIFT ion swarm under settled-down conditions is given, as for the FA, by Eqs. 2 and 3 but with D_a replaced by D, the free diffusion coefficient for the particular ion type. A detailed discussion of the flow dynamics of the SIFT is given in a recent review [7].

The experimental approach for determining the rate coefficient, k, for an ion–neutral reaction is straightforward. An appropriate type of ion source containing a suitable source gas is interfaced to the SIFT. The particular ion of interest is selected using the mass filter and injected into the flow tube at an appropriately low energy. Typical laboratory energies used are $\sim 20\,\text{eV}$ for atomic ions and robust molecular ions, but energies as low as a few electron volts are necessary for weakly bound cluster ions in order to prevent fragmentation on the carrier gas, especially when heavier gases such as Ar or N_2 are used as carrier gases. These laboratory energies, E_{lab}, correspond to center-of-mass energies $E_{\text{cm}} = E_{\text{lab}} \times M_c/(M_i + M_c)$, where M_c and M_i are the masses of the carrier gas and the injected ion respectively.

Reactant gas is added through a gas inlet port (see Fig. 2) into the reaction zone at a controlled flow rate while the count rate of the injected ion at the downstream detection system is monitored. A plot of $\ln n_i/n_0$ vs. Q is plotted in accordance with Eq. 5, and k is derived from the slope of the line. Samples of SIFT data for both a negative ion reaction and a positive ion reaction are illustrated in Figs. 3 and 4, respectively, which show the quality of the data that is obtained routinely. These data also illustrate another point, which is that fewer products are generally observed for negative ion reactions than for positive ion reactions (largely due to energetic constraints). The data in Fig. 3

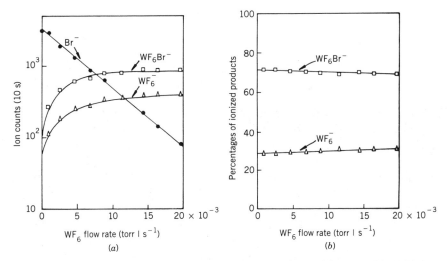

Fig. 3. (a) A typical SIFT data set, in this case for reaction 32 of Br$^-$ with WF$_6$ at 300 K. The rate coefficient for the reaction is derived from the slope of the linear semilogarithmic decay plot of the Br$^-$ ion count rate (10-s integration) vs. the WF$_6$ flow rate. (b) This illustrates the standard approach to the determination of ion product distributions (branching ratios) [4, 21]. The percentages of each ion arising from the primary reaction and/or secondary reactions are plotted as a function of the reactant gas flow rate, and the true product distribution is that indicated at zero reactant gas flow. This procedure accounts for any differential loss of the ion products due to secondary reactions. In this reaction both WF$_6^-$ (28%, formed by charge transfer) and WF$_6$Br$^-$ (72%, formed by association of the reactants) are primary products (see Sections 3.1 and 4.1).

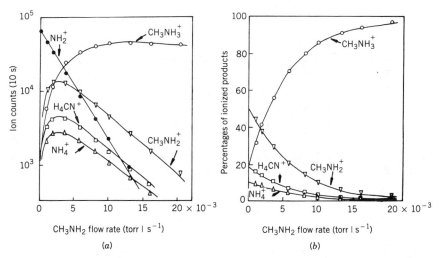

Fig. 4. (a) SIFT data for the reaction of NH$_2^+$ with CH$_3$NH$_2$ at 300 K indicating the formation of four possible products of the reaction (taken from Adams et al. [100]), three of which are clearly undergoing reactions with CH$_3$NH$_2$. (b) The conventional plot from which the ion product distribution is determined for the primary reaction (see Fig. 3b), indicating a product distribution of CH$_3$NH$_2^+$ (50%), CH$_3$NH$_3^+$ (20%), H$_4$CN$^+$ (20%), NH$_4^+$ (10%). It is also evident that CH$_3$NH$_2^+$, H$_4$CN$^+$, and NH$_4^+$ react with CH$_3$NH$_2$ to produce CH$_3$NH$_3^+$.

are for the reaction of Br^- with WF_6 at 300 K:

$$Br^- + WF_6 \rightarrow WF_6Br^- \qquad (32a)$$

$$\rightarrow WF_6^- + Br \qquad (32b)$$

Only two product are observed, that leading to the adduct ion WF_6Br^- and to the charge transfer product WF_6^-. The ion product distribution (branching ratio) is determined by plotting the count rate of each ion, as a percentage of the total product ion count rate, versus the flow rate of the reactant gas as shown in Fig. 3b. Then the true percentage product distribution is obtained by extrapolating each line to zero reactant gas flow rate. The near parallelism of the two lines indicates that secondary reactions of the WF_6Br^- (72%) and WF_6^- (28%) with the WF_6 reactant gas are not occurring at a significant rate. This reaction is mentioned again in Section 4. However, the form of the data illustrated in Fig. 4 for the reaction of NH_2^+ with CH_3NH_2 is quite different. In this positive ion reaction, four primary products are evident, which are NH_4^+ (10%), H_4CN^+ (20%), $CH_3NH_2^+$ (50%), and $CH_3NH_3^+$ (20%). As can be seen from Fig. 4a, the NH_4^+, H_4CN^+, and $CH_3NH_2^+$ undergo secondary reactions with CH_3NH_2, and this is confirmed by Fig. 4b, which shows the decrease in the percentages of these three ions as the CH_3NH_2 flow rate increases. These data forcibly illustrate the value of this approach for determining ion product distributions, i.e., the necessity to extrapolate plots such as in Figs. 3b and 4b to zero reactant gas flow rate to account for secondary reactions. Clearly, it would be quite inaccurate to determine the ion product distributions at a finite flow of the CH_3NH_2 (but nothing like so erroneous for a finite flow of WF_6 in reaction 32). It should be noted again that if significant mass discrimination is evident in the downstream detection system, then corrections to the ion product distributions are required in accordance with the procedure outlined previously in this section.

The rate coefficients and ion product distributions for a large number and a wide variety of reactions, including a good fraction over appreciable ranges of temperature have been determined using VT–SIFT apparatuses in recent years [7]. The Birmingham VT–SIFT illustrated in Fig. 2 can be operated over the temperature range 80–600 K, which adds greatly to the already enormous versatility of the SIFT. In the next section, some illustrative results are presented that indicate the range and capabilities of the technique.

3.2 Some Studies Illustrating the Versatility of the SIFT

The facility to create both positive and negative ions in a remote ion source under a variety of temperature and pressure conditions very different from those that pertain to the carrier gas into which they are to be injected, and the exclusion of the source gas from the reaction zone of the SIFT, allow the reactions of a much wider variety of ions to be studied than are possible using flowing afterglows. In this section, we present the results of some SIFT studies that illustrate the enormous versatility of the technique. Thus, for example, the reactions of ions in recognizable series (e.g., the ions derived from methane, CH_n^+, $n = 0-5$) can be studied with a great variety of reactant atomic and molecular gases and vapors to investigate, for instance, the influence of the degree of hydrogenation of the reactant ion on reactivity. The forward and the corresponding reverse rate coefficients of nearly thermoneutral reactions can be measured as a function of temperature to determine enthalpy and entropy changes and hence relative bond strengths, proton and electron affinities, etc. Isotopic labeling of the reactant ions or reactant neutrals can be used to study reaction mechanisms and to elucidate the factors that govern isotope exchange rates. Ternary association reactions of a particular ionic species can be studied with a range of ligand molecules and with several carrier gases as third bodies without interference from binary reactions of the ion with its source gas, which are often quite rapid relative to the slow ternary reactions. The reactions of some vibrationally and electronically excited ions can be studied since often such excitation is not quenched in the carrier gas but often is efficiently quenched in collisions with the source gas. Rate coefficients (and ion product distributions) can be determined sufficiently accurately to allow comparison with theory, and this has been particularly profitable in the cases of the reactions of ions with polar molecules and for some ternary association reactions that we discuss in Sections 3.2.1 and 3.2.2 together with other topics. Excited ion reactions are discussed in Section 3.2.3, and some interesting recent work relating to the reactions of the isomeric forms of some ions is described in Section 3.2.4.

3.2.1 Binary Reactions of Ground State Positive and Negative Ions

The earliest SIFT studies were directed toward reactions that were considered to be involved in the synthesis of interstellar molecules [88, 89]. This synthesis occurs via many parallel and sequential positive ion–molecule reactions in which atomic and diatomic ions (e.g., C^+, CH^+, N^+, O^+) are converted to the polyatomic ions that are the precursors of the observed polyatomic interstellar molecules. Thus, studies of the reactions of ions in recognizable series, including ions derived from methane (CH_n^+, $n = 0-5$) [101–105], ammonia (NH_n^+, $n = 0-4$) [100], formaldehyde (H_nCO^+, $n = 0-3$) [106], and hydrogen sulphide (H_nS^+, $n = 0-3$), have been carried out [107]. Also, reactions of hydrocarbon ions containing two, three, and four carbon atoms have been studied (i.e., $C_2H_n^+$, $C_3H_n^+$, and $C_4H_n^+$) [91, 103, 108, 109]. The reactions of many of these ions have been studied with a large number of stable molecular gases, vapors, and some radical atomic species (see below). These studies illustrate one of the valuable features of the SIFT, i.e., the rapidity of data acquisition. For example, to study the reactions of the CH_n^+ series of ions, methane is introduced into the ion source, and each of the ions C^+, CH^+, CH_2^+, CH_3^+, CH_4^+, and CH_5^+ (and also C_2- and C_3-bearing ions that are produced via secondary reactions at sufficiently high pressures of methane) can be separately injected into the flow tube, and their reactions can be quickly studied with a particular neutral species under identical conditions of carrier gas flow, pressure, temperature, and so on. Hence relative rate coefficients can be obtained very accurately (to within a few percent) since any systematic errors in flow velocities, pressures, etc., are effectively eliminated. It is then a simple process to study the reactions of the same series of ions with several different reactant gases. Such studies are valuable not only to interstellar chemistry, for which they are revealing the reactions involved in building up interstellar molecules, but also in helping to discover how factors such as the degree of hydrogenation and recombination energies of ions influence ionic reactivity. The variety and complexity of reaction mechanisms can also be revealed by these studies. For example, in the CH_n^+ series, the C^+, CH^+, and CH_2^+ ions are extremely reactive with most species, whereas CH_3^+ is less reactive. Also the propensity to form products containing the strong C—O and C—N bond is clear; e.g.,

$$C^+ + O_2 \rightarrow O^+ + CO \qquad (62\%) \qquad (33a)$$
$$\rightarrow CO^+ + O \qquad (38\%) \qquad (33b)$$

and

$$CH^+ + NH_3 \rightarrow H_2CN^+ + H_2 \qquad (68\%) \qquad (34a)$$
$$\rightarrow NH_3^+ + CH \qquad (17\%) \qquad (34b)$$
$$\rightarrow NH_4^+ + C \qquad (15\%) \qquad (34c)$$

(The ion product distributions given in parentheses are those determined at 300 K.) Both reactions 33 and 34 occur on every collision (i.e., $k = k_c$). Note that for reaction 34 both a charge transfer channel 34b and a proton transfer channel 34c are evident and that the channel 34a in which new bonding arrangements are forged is by far the majority channel. This implies that the CH^+/NH_3 interaction is intimate and that the exothermic channels 34b and 34c, which in principle could occur by direct charge transfer and proton transfer, most probably occur within a $(CH^+NH_3)^*$ complex. These SIFT studies also revealed the propensity of CH_3^+ to undergo ternary association reactions; e.g.,

$$CH_3^+ + CO + He \rightarrow CH_3^+ \cdot CO + He \qquad (35)$$

these being especially rapid at low temperatures [104, 110]. These studies led directly to the conclusion that binary (radiative) association of CH_3^+ ions with a range of molecules is an important mechanism for molecular synthesis in interstellar clouds [89, 111]. Ternary association reactions are discussed further in Section 3.2.2.

There is a wealth of information contained in the studies of the reactions of ions in series, some of which has been summarized in review papers [45, 112]. The reactions of ions in the NH_n^+ ($n = 0-4$) series are particularly interesting in that there are obvious differences in the recombination energies of, and the proton binding energies in, the various ions, and these differences are clearly reflected in their reactivities [100]. Similar trends are also evident for the ions in the H_nS^+ ($n = 0-3$) series [107].

The accuracy of the rate coefficients, k, derived from SIFT studies ensures that they are valuable checks of the various theories that predict collisional rate coefficients, k_c [42]. It has been known for some years that exothermic proton transfer reactions invariably proceed at the collisional rate ($k = k_c$), and so they have become the reactions of choice for checking predictions of k_c. Such studies have indicated that, at a temperature of 300 K, the average dipole orientation (ADO) theory closely predicts k_c for proton transfer reactions involving polar molecules [15, 113]. Recently, however, other theories including the adiabatic capture centrifugal sudden approximation (ACCSA) theory have predicted a steep increase in k_c with decreasing temperature for reactions involving polar molecules [114]. This has been checked using a VT–SIFT for the reactions of H_3^+ with HCN and HCl [50]; e.g.,

$$H_3^+ + HCN \rightarrow H_2CN^+ + H_2 \tag{36}$$

The very accurate data obtained down to 200 K has confirmed the ACCSA predictions, and this has had a profound influence on the development of our understanding of interstellar chemistry [115]. VT–SIFT experiments at very low temperatures are not feasible in the SIFT for polar reactants because of condensation and dimerization problems. (An expanding jet technique has recently become available and is solving some of the problems [116].) However, SIFT experiments at 200 K have resulted in some interesting data on the reactions of $(HCN)_2$ dimers.

Growing numbers of SIFT experiments are being carried out with radical atomic species (H, D, N, O) as reactants [117–121]. These are generated in sufficient quantities by flowing a mixture of the helium carrier gas and the appropriate stable molecular gas (H_2, D_2, N_2, etc.) through a microwave discharge and then into the reaction zone of the SIFT. Oxygen atoms have been produced by titrating NO into a stream that contains N atoms, and these react with the NO forming O atoms and N_2. In this way, the reactions of O and N have been studied with several hydrocarbon ions [117]; e.g.,

$$CH^+ + N \rightarrow CN^+ + H \tag{37a}$$
$$\rightarrow H^+ + CN \tag{37b}$$

and similarly

$$CH^+ + O \rightarrow CO^+ + H \tag{38a}$$
$$\rightarrow H^+ + CO \tag{38b}$$

These reactions proceed at appreciable fractions of the collisional rate and illustrate again the propensity for the formation of the strong C—N and C—O bonds. Studies of these ion–atom reactions are of great value in understanding the chemistry of diffuse interstellar clouds [89, 115].

The reactions of some negative ion species with some atoms and molecules have been studied recently in VT–SIFT experiments [121]. A common reaction mechanism is associative detachment (see Section 2.3, reactions 15 and 16 and reaction 39), and a study has been made of the temperature dependence of the rate coefficients for several such reactions including those of O^- with NO and S^- with CO and O_2; e.g.,

$$S^- + CO \rightarrow OCS + e \qquad (39)$$

These reactions also proceed at appreciable fractions of the collisional rate, and the k is found to vary approximately as $T^{-0.75}$ [122]. Such studies are much more complicated using a FA because of the destruction of the negative ions by competing reactions. Data on these and other negative ion reactions have been summarized in a recent review [123]. A similar study of the associative detachment reactions of F^-, Cl^-, Br^-, and I^- with H atoms has recently been carried out, and, in the same laboratory, the reverse dissociative attachment reactions have been studied using a FALP apparatus [121]. All four associative detachment reactions are fast and for the reactions

$$I^- + H \underset{k_r}{\overset{k_f}{\rightleftharpoons}} HI + e \qquad (40)$$

both the forward (k_f) and the reverse (k_r) rate coefficients are close to the respective k_c values, indicating that the reactions are closely thermoneutral; i.e., that the enthalpy change is essentially zero.

The VT–SIFT technique is admirably suited to the determination of the forward and reverse rate coefficients as a function of temperature for ion–molecule reactions that are close to thermoneutral, i.e., reactions for which both k_f and k_r can be measured accurately. Then from a van't Hoff plot of $\ln k_f/k_r$ vs. $1/T$, the enthalpy change, ΔH, and the entropy change, ΔS, can be determined. This has been achieved for the reactions

$$O_2H^+ + H_2 \underset{k_r}{\overset{k_f}{\rightleftharpoons}} H_3^+ + O_2 \qquad (41)$$

The O_2H^+ ions are created in a high-pressure ion source containing a mixture of H_2 and O_2. The very small ΔH determined for the reaction (-0.33 kcal mol^{-1}) is equivalent to the difference between the proton affinities (PA) of O_2 and H_2; i.e., PA(H_2)–PA(O_2) = 0.33 kcal mol^{-1} [124]. Before these VT–SIFT measurements, it had been in doubt as to which PA was the greater. The ΔS determined from these measurements also established that O_2H^+ has a triplet ground-state configuration. Similar studies of k_f and k_r for the reactions

$$H_2Cl^+ + CO \underset{k_r}{\overset{k_f}{\rightleftharpoons}} HCO^+ + HCl \qquad (42)$$

established that PA(CO) – PA(HCl) is equal to 2.7 kcal mol^{-1} [125] although very recent SIFT studies of Birmingham have shown this difference to be greater at $\sim 7\,\text{kcal mol}^{-1}$. This type of temperature-dependence SIFT study could be used to great effect to establish accurate PA scales in support of the considerable amount of such work that has already been carried out using room temperature ion cyclotron resonance techniques [126]. Not so precise, but nevertheless worthwhile, is the bracketing of proton affinities. The guiding principle is that when a proton transfer reaction occurs at the collisional rate, then the PA of the receptor species exceeds that of the donor species. Thus, for example, HCS^+ rapidly transfers a proton to acetone but is totally unreactive with HCN and so $PA(CH_3COCH_3) > PA(CS) > PA(HCN)$. Such SIFT studies have closely bracketed the PA of the CS radical [127]. In this way the PA of C_2N_2 and HC_3N have been bracketed [109, 128]. In a similar manner, in FA experiments the relative electron affinities of some stable and unstable (radical) species have been determined, including H_2SO_4 and ClO [129, 130], and more recently in a SIFT that of WF_6 has been determined (principally via the observation that electron transfer between Br^- and WF_6 is nearly thermoneutral [131], see Eq. 32 and Fig. 3).

Among the most elegant SIFT studies carried out to date are those of isotope exchange in reactions involving both positive and negative ions [45, 57, 132]. These are exemplified by the reactions

$$^{13}C^+ + {}^{12}CO \underset{k_r}{\overset{k_f}{\rightleftharpoons}} {}^{12}C^+ + {}^{13}CO \tag{43}$$

The k_f and k_r for these reactions can be measured using an appropriately labeled gas in the ion source (^{12}CO or ^{13}CO for reaction 43), injecting in turn the $^{12}C^+$ and $^{13}C^+$ ions and then reacting them with ^{13}CO and ^{12}CO, respectively, in the usual way [133]. At room temperature the k_f and k_r differ by about 10% but this increases to about 50% at 80 K. This is because the reverse reaction is endothermic due mainly to the difference in the zero-point-energies (zpe) of the ^{12}CO and the ^{13}CO. A van't Hoff plot of the data (i.e., a plot of $\ln k_f/k_r$ vs. T^{-1}) provides a value for ΔH of $-0.08\,\text{kcal mol}^{-1}$, in agreement with that calculated from the zpe difference. An interesting feature of this and the many other isotope exchange reactions studied is that k_f invariably approaches k_c at low temperatures [132] (when, of course, the k_r must decrease). This observation has been crucial in understanding the phenomenon of isotope fractionation in interstellar molecules by which the heavy isotopes of some elements, notably deuterium, are clearly enriched in some interstellar molecules.

D/H exchange has been studied in many ion–molecule reactions, including those in the $H^+ + H_2$, $H_3^+ + H_2$, and $CH_3^+ + H_2$ systems [134–137]. These comprehensive studies involved the reactions between all of the partially and totally deuterated ions and neutral molecules. When D replaces H in a molecule, a relatively large change in the zpe results and hence the ΔH for the reactions are large (e.g., relative to that for reaction 43). For the reaction

$$D^+ + H_2 \underset{k_r}{\overset{k_f}{\rightleftharpoons}} H^+ + HD \tag{44}$$

the ΔH is $-0.92\,\text{kcal mol}^{-1}$. Unlike reaction 43, there is also an entropy change in reaction 44. The ΔS has been calculated from the relevant partition functions, and the experimentally derived value is in good agreement with the calculated value [134]. The very thorough VT–SIFT study of D/H exchange in the $CH_3^+ + H_2$ system [136], i.e., involving the reactions of CH_3^+, CH_2D^+, CHD_2^+, and CD_3^+ with H_2, HD, and D_2, including the reactions

$$CH_3^+ + HD \underset{k_r}{\overset{k_f}{\rightleftharpoons}} CH_2D^+ + H_2 \qquad (45)$$

has provided an accurate value for the difference in the C—H and C—D bond strengths in CH_3^+-like ions. This again demonstrates the value of VT–SIFT measurements in the provision of thermochemical data.

Studies have also been made of D/H exchange in the reactions of some polyatomic molecules, M, with their protonated species, MH^+. Of particular interest is the $H_3O^+ + H_2O$ system, since the $H_3O^+ \cdot H_2O$ cluster is quite stable. Both the reaction of H_3O^+ with D_2O and the reaction

$$D_3O^+ + H_2O \rightarrow D_2HO^+ + HDO \qquad (46a)$$

$$\rightarrow DH_2O^+ + D_2O \qquad (46b)$$

have been studied [138, 139]. Previous similar work carried out using an ion cyclotron resonance spectrometer had apparently shown that only the deuteron transfer (channel 46b) occurred and that the measured k was approximately $k_c/2$, consistent with the formation of a deuteron-bound dimer $D_2O \cdot D^+ \cdot H_2O$, which could separate into D_3O^+ or DH_2O^+ with equal probability [140]. This is contrary to the SIFT study at 300 K for reaction 46, the data for which are shown in Fig. 5. Clearly, D_2HO^+ is the major product, and, in fact, it is favored in the ratio of 2:1 with respect to DH_2O^+. This is precisely to be expected if a total scrambling of the H and D atoms occurs within the $(D_3O^+ \cdot H_2O)^*$ complex before its decomposition. The same result was obtained for the mirror reaction $H_3O^+ + D_2O$. Thus these $(H_3O^+ \cdot H_2O)^*$ ion–molecule complexes live long enough to allow total scrambling to occur. Also, the measured k in the SIFT experiment was hardly distinguishable from k_c, and this again is quite consistent with the scrambling model (see [138, 139] for further discussion). The analogous reactions of the hydrated hydronium ions $H_3O^+(H_2O)_{1,2,3}$ with D_2O and their mirror reactions were also studied with similar results. These cluster ions were generated in a flowing afterglow ion source in which it was also possible to generate the partially deuterated clusters and to inject adequate currents of them into the SIFT. Total scrambling is allowed in these intermediate complexes by virtue of their long lifetime, which is a manifestation of the strong bonding of the H_2O ligands and the large number of vibrational modes in the complexes. Similar studies have been carried out for the $NH_4^+ + ND_3$ and the $CH_5^+ + CD_4$ reactions and their mirror reactions [139]. The weaker bonding relative to that

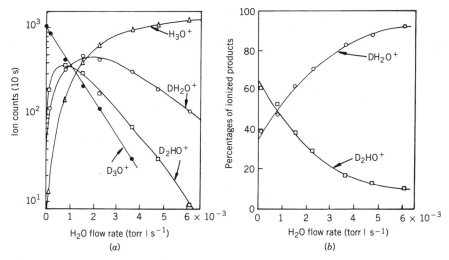

Fig. 5. SIFT data for the near-thermoneutral D/H exchange reaction 46 between D_3O^+ and H_2O at 300 K (taken from Smith et al. [84]). The slope of the reactant ion decay plot in (a) indicates that the reaction proceeds at a rate close to the collisional rate, and the product distribution derived from (b) indicates that complete mixing of the D and H atoms occurs in the $(D_3H_2O_2^+)^*$ complex before its dissociation to the observed products (for further discussion see Section 3.2.1).

in the $H_3O^+ \cdot H_2O$ system, in the $NH_4^+ \cdot NH_3$ and $CH_5^+ \cdot CH_4$ systems, especially for the last, and the resulting shorter complex lifetimes do not allow for complete scrambling of the H and D atoms among the products of the reactions, and this is clearly shown by the SIFT experiments. The several interesting aspects of these reactions are discussed in detail in the relevant papers [132, 139, 140].

H/D exchange has also been comprehensively studied in negative ion–molecule reactions using both SIFT and FA apparatuses. This work has been reviewed by DePuy [57]. Many reactions of the kind

$$OD^- + MH \to OH^- + MD \tag{47}$$

have been studied, where MH is H_2, NH_3, C_6H_6, and C_2H_4 to mention only a few [141]. The kinetic data obtained from the SIFT measurements revealed a correlation between the efficiency of isotope exchange, the relative acidities of the reactants, and the polarizabilities of the reactant molecules. Clearly, in negative ion reactions also, the lifetimes of the ion–molecule complexes will to some extent determine whether isotopic scrambling occurs in the reactions. In the reaction

$$D_2N^- + NH_3 \to H_2N^- + NHD_2 \quad (73\%) \tag{48a}$$

$$\to HDN^- + NH_2D \quad (27\%) \tag{48b}$$

even though thermoneutral proton transfer can occur (analogous to the

$D_3O^+ + H_2O$ reaction), the H and D are certainly not scrambled among the products, and this is indicative of a very short lifetime of the intermediate complex [57, 141].

Isotopic labeling of the ion or molecule can of course be used to investigate the mechanisms of *exothermic* ion–molecules reactions. This has not yet been exploited to any great extent, although the SIFT technique is admirably suited to such studies. One simple example of the method will suffice here. Does the rapid exothermic reaction

$$NH_3^+ + NH_3 \rightarrow NH_4^+ + NH_2 \tag{49}$$

proceed via proton transfer from the NH_3^+ or by atom abstraction from the NH_3? By injecting ND_3^+ into the SIFT and observing its reaction with NH_3,

$$ND_3^+ + NH_3 \rightarrow NH_3D^+ + ND_2 \quad (85\%) \tag{50a}$$

$$\rightarrow ND_3H^+ + NH_2 \quad (15\%) \tag{50b}$$

it is seen that (at 300 K) deuteron transfer (channel 50a) occurs in the majority of the collisions but not to the exclusion of atom abstraction [100]. It would be interesting to determine the ion product distribution of reaction 50 at lower temperatures to investigate the role of the complex lifetime in determining the product distribution. A further interesting application of isotope labeling is in the study of the more complex reaction of NH_2^+ with C_2H_4 in which the product ions are not unambiguously identified by their mass (e.g., $C_2H_4^+$ and H_2CN^+ are both 28 amu). The ion product distribution for this reaction has been elucidated using isotopic labeling [142].

3.2.2 Ternary Association Reactions

We have previously referred to some FA studies of ternary association reactions involving positive and negative ions (e.g., reactions 9 and 24). As mentioned before, the presence of the ion source gas and also metastable atoms in the FA can often seriously disturb the determination of relatively small association rate coefficients because of faster competing binary reactions of the primary ion with the source gas and the production of the primary ions by Penning ionization. The SIFT technique avoids these complications, and quite accurate association rate coefficients, k_3, can be determined over appreciable temperature ranges, albeit with only a few third bodies (carrier gases), and these k_3 can be compared with theoretical predictions. As mentioned at the end of Section 2.2, the k_3 for an association reaction is determined by measuring the effective binary rate coefficient for the process, k_{eff}, as a function of the carrier gas number density. A careful study of the association reactions of CH_3^+ ions with H_2, N_2, CO, and O_2 has been made in a helium carrier gas over the approximate temperature range 80–550 K [110]. The overall reaction, e.g.,

$$CH_3^+ + H_2 + He \rightarrow CH_5^+ + He \tag{51}$$

proceeds via the formation of an excited intermediate complex $(CH_5^+)^*$ of lifetime τ_d, which can decompose to reactants or be stabilized to CH_5^+ in a collision or collisions with He atoms. These VT–SIFT studies have shown that the k_3 for these reactions vary over the temperature range stated above as $k_3 \sim T^{-2.5}$, and this is in good agreement with theory that predicts that, for these simple reactants, the temperature exponent should be approximately equal to $l/2$, where l is the total number of rotational degrees of freedom in the reactants (i.e., five for polyatomic ion/diatomic molecule reactants) [143, 144]. Small deviations from $l/2$ arise if the efficiency of the collisional stabilization step is temperature dependent. Thus, VT–SIFT studies of the reaction

$$N_2^+ + N_2 + He \rightarrow N_4^+ + He \tag{52}$$

indicate that the k_3 varies at $T^{-2.3}$ (rather than the predicted T^{-2}) [145]. However, studies of the analogous reaction

$$N_2^+ + N_2 + N_2 \rightarrow N_4^+ + N_2 \tag{53}$$

indicate k_3 to vary as $T^{-1.7}$ [146]. This difference has been attributed to a temperature dependence in the efficiency of the stabilization of $(N_4^+)^*$ complex in the superelastic collisions, and it has been shown that stabilization in reaction 53 occurs via the switching of N_2 molecules between the $(N_4^+)^*$ in collisions with N_2 i.e., quite different stabilization mechanisms are involved in reactions 52 and 53 [147] (see Section 4.1).

It is clear that the magnitude of k_3 for a given reaction is dependent on τ_d, which can itself be estimated from a measurement of k_3. (Note that the τ_d so obtained is necessary in estimating the rate coefficients for vibrational predissociation in excited ion–molecule collisions [148] that we refer to in Section 3.2.3). The k_3 for the $CH_3^+ + CO$ association reaction exceeds that for the $CH_3^+ + O_2$ association reaction by some two orders of magnitude, which is a manifestation of the much stronger bonding in the CH_3CO^+ complex and its larger τ_d compared to that in the $CH_3O_2^+$ complex. The τ_d, and hence the k_3, for association reactions can also be enhanced due to the phenomenon of "isotopic refrigeration" and "endothermic trapping" in which endothermic bond rearrangements occur within the intermediate complexes (driven by the ion–dipole energy). This, at least in part, explains why the k_3 for the reaction

$$CD_3^+ + H_2 + He \rightarrow CD_3H_2^+ + He \tag{54}$$

exceeds the k_3 for reaction 51 by about a factor of 8 [136] and why the k_3 for the reaction

$$C_2H_2^+ + H_2 + He \rightarrow C_2H_4^+ + He \tag{55}$$

is very rapid [149].

If the reactant ions or neutrals are significantly vibrationally excited, it is entirely to be expected that the k_3 for association reactions will be small, since the additional energy in the ion–molecule complex will surely diminish τ_d [150, 151]. Low-frequency vibrational modes in some molecules will be excited even in room temperature carrier gas, and this has been invoked to explain the VT–SIFT observations of very rapid temperature variations of k_3 for the association reactions of cluster negative ions [152, 153]; e.g.,

$$NO_3^-(HNO_3) + HCl + He \rightarrow NO_3^-(HNO_3) \cdot HCl + He \qquad (56)$$

For such reactions, the k_3 varies rapidly with temperature, much faster than the $T^{-2.5}$ predicted in the absence of vibrational excitation but as fast as T^{-6}, although the exponent itself is also temperature dependent in these reactions for which more of the "floppy" bonds in the cluster ions become excited with increasing temperature. The reactions of such complex species could not be studied with any certainty in a FA since mixtures of polar (reactive) molecules coexisted (e.g., HNO_3, HCl), which result in the presence of mixed cluster ions, the occurrence of switching reactions, etc. The SIFT technique makes such studies possible. These types of studies have added greatly to the understanding of the ion chemistry of the lower atmosphere and of interstellar gas clouds. Although ternary association reactions cannot occur at the very low pressures pertaining to interstellar clouds, VT–SIFT measurements of k_3 for ternary association at low temperatures, and the values of τ_d that they provide, allow estimates to be made of radiative association rate coefficients [111]. For example, the measured values of k_3 for reaction 51 have provided an estimate for the analogous radiative association rate coefficient for the production of CH_5^+ in interstellar clouds, which is an important step in the production of interstellar methane [89, 111].

3.2.3 Binary Reactions of Excited Positive Ions

Electron impact ionization of atoms and molecules at energies above the threshold invariably generates electronically excited positive ions some of which will be long-lived (metastable). These metastable ions are usually efficiently quenched or destroyed in collisions with their parent (source) atoms or molecules. Since this cannot happen in low-pressure ion sources (in which collisions are infrequent), appreciable fractions of the ions emerging from the ion source may be excited. Helium is a very inefficient quencher of low-lying electronic (and vibrational) states of many ions, and thus such ions can survive in the helium carrier gas of a SIFT. These excited ions can usually be distinguished from their ground-state ions by their different reactivity. When both the ground state and an excited state of a given ion species are present in the reaction zone of a SIFT, then any difference in reactivity with an added gas is manifest by curvature in the ion decay plot. When only two different states that react at very different rates are present, the decay curve will generally have two distinct linear portions (see Fig. 6) from which the rate coefficients, k, for the reactions of both ionic

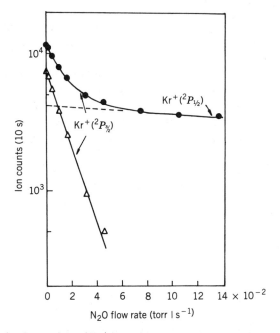

Fig. 6. SIFT data for the reactions of Kr^+ ions with N_2O at 300 K (taken from Adams et al. [158]). The Kr^+ ions were derived from a low-pressure electron impact ion source containing Kr, and so both the $Kr^+(^2P_{1/2})$ and $Kr^+(^2P_{3/2})$ spin-orbit states are generated and injected into the flow tube. The curvature on the plot of total Kr^+ count rate vs. N_2O flow rate (filled circles) is a manifestation of the different reactivity of the two states with N_2O. The rate coefficient for the $Kr^+(^2P_{1/2})$ reaction is derived from the linear slope of the plot at high N_2O flow, and the extrapolation of this line to zero N_2O flow indicates the fraction of the injected Kr^+ that is in the $^2P_{1/2}$ state. The rate coefficient for the $Kr^+(^2P_{3/2})$ reaction is determined from the slope of the line (open triangles), which is derived by subtracting the count rates indicated by the dotted line from the total Kr^+ count rate. This great difference in reactivity of the two states of Kr^+ with N_2O means that it can be used as a filter gas for $Kr^+(^2P_{3/2})$ (see Section 3.2.3).

species can be obtained. When the k for the reactions of the two species are similar, a curve-fitting procedure can be used to extract the two k values from a curved plot. Alternatively, the "monitor gas"/"monitor ion" technique or the "filter gas" technique can be used.

The monitor gas technique involves the addition of a gas, via an inlet port that is only just upstream of the detection system sampling orifice in the SIFT. This gas must react with only one of the ion species (usually the excited state) to produce a particular ion product that is, of course, detected by the mass spectrometer/detection system. This ion is then used as a monitor ion of the particular state of the reactant ion from which it is produced. For example, a monitor gas for metastable $O_2^+(a^4\pi_u)$ ions is Ar with which it rapidly charge transfers while the ground-state $O_2^+(X^2\pi_g)$ state does not, due to energetic constraints in this case [154, 155]:

$$O_2^+(a^4\pi_u) + Ar \to Ar^+ + O_2 \tag{57}$$

So the downstream count rate of the monitor ion Ar^+ is monitored as a reactant gas (which reacts with the $O_2^+(a^4\pi_u)$) is added to the upstream reaction zone of the SIFT, and this results in a decrease of the Ar^+ count rate. The k for the metastable ion reaction is then derived in the usual way. A complication arises when both a chemical reaction (by which we also mean charge transfer) and collisional quenching occur between an excited ion and a reactant gas, e.g.,

$$NO^+(a^3\Sigma) + N_2 \to N_2^+ + NO \tag{58a}$$

$$\to NO^+(X^1\Sigma^+) + N_2^* \tag{58b}$$

In this reaction, charge transfer 58a and energy transfer/quenching 58b occur, but both channels destroy the metastable $NO^+(a^3\Sigma)$ ions. When quenching does occur, ground-state (unreactive) NO^+ is produced. Thus, use of the monitor gas method allows the separate contributions of reaction and quenching to be obtained.

In the "filter gas" technique, a gas is added upstream of the SIFT reaction zone to remove completely one of the states of the injected ion either by reaction or quenching or both, thus leaving only the other state, the reactions of which can then be studied directly. Thus, for example, in the case of the reaction 58 the addition of sufficient N_2 upstream will totally remove the $NO^+(a^3\Sigma)$ and leave the NO^+ ground state in the reaction zone. Using these methods, the rate coefficients, including in some cases the quenching rate coefficients, and ion product distributions for the ground and metastable states of O_2^+, NO^+, and $O^+(^4S, {}^2D, {}^2P)$ with several gases have been determined [155]. In a similar SIFT study, the reactions of the ground and metastable states of C^+, N^+, S^+, and N_2^+ were studied [156]. These and other studies of excited ion reactions are summarized in a recent review [9].

The reactions of the ground and metastable states of some doubly charged ions have also been studied in SIFT experiments. Actually, the first significant study at thermal energies of the reactions of doubly charged ions was made in a FA experiment in which the rate coefficients were measured for the single charge transfer reactions of Ca^{2+} and Mg^{2+} with several neutral species [157]. Such studies are only possible when the doubly charged ions do not react at a significant rate with the carrier gas. Hence a detailed study has been carried out of the reactions of doubly charged xenon in the ground state $Xe^{2+}(^3P)$ and the metastable states $Xe^{2+}(^1D_2)$ and $Xe^{2+}(^1S_0)$, none of which react with helium [95]. Extensive use was made of the filter gas technique to remove selectively one or two of the components from the swarm of injected Xe^{2+} ions created in a low-pressure ion source. It was found that the ground-state ions $Xe^{2+}(^3P)$

reacted with Ar

$$Xe^{2+}(^3P) + Ar \rightarrow Xe^+(^2P_{3/2}) + Ar^+(^2P_{3/2}, {}^2P_{1/2}) \tag{59}$$

whereas the higher energy metastable states $Xe^{2+}(^1D_2, {}^1S_0)$ did not react at a significant rate and so Ar was used to filter out the $Xe^{2+}(^3P)$. The reactions of the metastable ions could then be measured within the remaining two component mixture of ions. Also, it was discovered that the addition of N_2 to the ion swarm removed the $Xe^{2+}(^3P)$ and $Xe^{2+}(^1D_2)$ ions leaving the $Xe^{2+}(^1S_0)$ in the helium carrier gas, and so the reactions of $Xe^{2+}(^1S_0)$ could be studied separately. Also, $Xe^{2+}(^3P)$ could be generated separately by reducing the electron energy in the ion source below that required to generate the metastable ions. Hence, the reactions of the $Xe^{2+}(^3P)$ ions could also be studied directly. Thus, in this way the reactions of the excited and ground states of these doubly charged ions with several molecular gases and the rare gases have been studied. Single charge transfer was the most commonly observed process (exemplified by reaction 59), and quenching of the $Xe^{2+}(^1D_2)$ was also observed in the reactions with molecular gases (i.e., H_2, N_2, O_2, and CO_2). It is the great flexibility of the SIFT technique coupled with the filter gas approach that allows such detailed studied to be made. A good illustration is the demonstration that, at thermal energies, the Xe^+ product ions of reaction 59 were produced selectively in the lower energy $^2P_{3/2}$ state of the spin-orbit doublet. This was achieved using a reaction that distinguished between the $Xe^+(^2P_{1/2})$ and $Xe^+(^2P_{3/2})$ ions (see below and [158]). In a combined SIFT and drift tube study, it was further shown that the higher energy $Xe^+(^2P_{1/2})$ became increasingly favored as the $Xe^{2+}(^3P)$/Ar interaction energy increased [97].

Similar studies have been made of the reactions of the ground and metastable states of Kr^{2+}, Ar^{2+}, and Ne^{2+} [96, 98]. As the recombination energy of these species increases, the possibility arises that they might react with helium carrier gas. This happened for $Ar^{2+}(^3P)$:

$$Ar^{2+}(^3P) + He \rightarrow Ar^+ + He^+ \tag{60}$$

However, the $Ar^{2+}(^1S_0)$ did not react with helium, and its reactions could be studied directly since, in effect, the helium carrier gas acted as a filter gas for the ^3P states. No evidence was obtained in these SIFT studies for the presence of $Ar^{2+}(^1D_2)$ ions. It was also discovered that $Ar^{2+}(^3P)$ ions did not react with Ar, and so their reactions were studied using Ar as the carrier gas. A summary of the results of these SIFT studies of doubly charged ion reactions has been given in [9] and reference to some reactions of the ionospherically important O^{2+} ions in [159] and to the doubly charged CO_2^{2+} molecular ion in [160].

The filter gas technique has been used to great effect to study the separate reactions of the $^2P_{1/2}$ and $^2P_{3/2}$ spin-orbit states of the singly charged Kr^+ and Xe^+ ions [158]. When Kr^+ and Xe^+ were generated in a low-pressure ion source and injected into the SIFT, it was found that the ion swarm contained both of the spin states in the population ratio 2:1 in accordance with the statistical weights of the 3/2 and 1/2 states. It was also found that N_2O and O_2 very effectively filtered out $Kr^+(^2P_{3/2})$ ions and that N_2O and CH_4 very effectively filtered out $Xe^+(^2P_{1/2})$ ions. Thus, a detailed study of the separate reactivities of these two states of both Kr^+ and Xe^+ has been carried out. The most common reaction mechanism observed in the reactions of these ions with several molecular gases, including CH_4, N_2O, O_2, COS, H_2S, and NH_3, was charge transfer as dictated by the energetics of the reactions. The most interesting result in all the reactions studied was that the rate coefficients for the reactions of the lower energy $^2P_{3/2}$ ions exceeded those for the higher energy $^2P_{1/2}$ ions (except where both spin states reacted at the collisional rate or where the $^2P_{3/2}$ ions could not charge transfer because of their lower energy). This interesting result has been rationalized on the basis of symmetry effects using correlation diagrams, although this explanation is still not entirely convincing [161].

Quenching of the $^2P_{1/2}$ state of Kr^+ to the $^2P_{3/2}$ state occurs in collisions with some molecular gases. A very recent SIFDT study of the reaction

$$Kr^+(^2P_{1/2}) + N_2 \rightarrow Kr^+(^2P_{3/2}) + N_2^* \tag{61}$$

has been carried out using O_2 as a monitor gas for $Kr^+(^2P_{3/2})$ [162] from which conclusions are drawn concerning the nature of the ion–molecule interaction (which, it is argued, is largely repulsive rather than the more usual attractive potential for ion–molecule interactions; see below). The guiding principles for this work have been established by detailed studies of the quenching of vibrationally excited O_2^+ and NO^+ ions [148]. In order to study reactions such as

$$O_2^+(v > 0) + CO_2 \rightarrow O_2^+(v = 0) + CO_2^* \tag{62}$$

in which no chemistry (or charge transfer) but only vibrational relaxation of the ion occurs, a monitor gas for the $O_2^+(v > 0)$ is required; i.e., a gas that distinguishes between $O_2^+(v = 0)$ and vibrationally excited O_2^+. Such a gas is Xe, which has an ionization energy that lies between the recombination energies of $O_2^+(v = 0)$ and $O_2^+(v = 1)$ [163]. Hence the charge transfer reaction rates of Xe with these ions are very different. In the same way, SO_2 is a valuable monitor gas for $O_2^+(v > 1)$ as is H_2O for $O_2^+(v > 2)$. Also CH_3I has been used as a monitor gas for $NO^+(v > 1)$ by virtue of its ionization energy [164]. Using these techniques, a systematic SIFT study (together with some SIFDT studies) has been carried out of the collisional quenching rate coefficients for the reactions of $O_2^+(v)$ and $NO^+(v)$ with several atomic and molecular gases [148]. A correlation has been observed between the efficiency of quenching and the strength of the bond between the ion and the quenching neutral. Also the quenching rate coefficients generally are observed to decrease with increasing relative collision energy. This indicates that the vibrational quenching process is dominated by long-range attractive forces (compare the conclusions with those regarding reaction 61). A model has been proposed that envisages quenching to occur within a transient ion–molecule or ion–atom complex, which then vibrationally predissociates resulting in a vibrationally relaxed ion. Using predominantly SIFT measurements of ternary association, rate coefficients [165], values are being provided of the complex lifetimes, τ_d (see Section 3.2.2), and estimates of vibrational predissociation rate coefficients have been obtained. A most interesting result is that NO^+ $(v > 0)$ is rapidly quenched by electronically excited $O_2(^1\Delta_g)$ molecules but only very slowly by ground state $O_2(^3\Sigma)$ molecules [166]. This observation is considered within a wider discussion of vibrational relaxation in recent reviews by Ferguson [148, 167] in which reference is also made to the first measurements of vibrational relaxation of neutral molecules by ions; e.g.,

$$O_2^+(v=0) + N_2(v=1) \to O_2^+(v=1) + N_2(v=0) \tag{63}$$

In these experiments, the $N_2(v=1)$ was created by discharging a mixture of He and N_2 before its entry into the SIFT flow tube and Xe was used as the monitor gas for the $O_2^+(v=1)$ product ion [168].

3.2.4 Reactions of the Structural Isomers of Some Positive Ions

There is a growing interest among users of the SIFT technique in the reactions of the structural isomers of ions. As the complexity (atomicity) of the ions increases, a greater number of structural isomers is possible. For relatively simple species, theoretical calculations are sometimes available that indicate which isomers are likely to be stable and also provide heats of formation. Experimentally, the presence of two or more stable ionic species of the same molecular weight is often manifest by curvature on decay plots due to their different reactivities (similar to that which occurs for excited and ground-state ions; see Section 3.2.3). Then the major challenge is to determine if the different reactivity is due to the presence of excited states of a given structural isomer or the presence of other structural isomers or even some mixture of excited ions and structural isomers. However, there is much optimism that the SIFT together with monitor gas and filter gas techniques will make considerable advances in this area, and indeed some recent experiments support this optimistic view.

The existence of the isomers HCO^+ and COH^+ has been suspected for many years, and interest in them heightened when it was suggested that they might both be present in interstellar clouds [169]. An "excited state" of the ion of mass 29 amu derived from a variety of precursor molecules (including H_2CO and $HCOOH$) was detected in the earliest Birmingham SIFT experiments by virtue of the fact that only a fraction of the ions in the SIFT swarm transferred a proton to both CH_4 and N_2. This process is endothermic for ground-state HCO^+ ions because the PA of CH_4 and N_2 are less than the PA of CO. However, the possibility could not be excluded that the "excited ion" was a vibrationally excited state of HCO^+ rather than the higher energy structural isomer COH$^+$ [170]. Recent ion cyclotron resonance (ICR) [171] and SIFT [172] studies coupled with theoretical calculations of the PA of CO at both C (forming HCO$^+$) and O (forming COH$^+$) have conclusively demonstrated the existence of the isomer COH$^+$, principally by the observation of the occurrence of the near thermoneutral proton transfer reaction 64a:

$$COH^+ + H_2 \rightarrow H_3^+ + CO \qquad (64a)$$

$$\rightarrow HCO^+ + H_2 \qquad (64b)$$

This established the PA of the "excited ion" to be closely equal to that of H_2 (101.3 kcal mol^{-1}) in agreement with the theoretical predictions [170]. Using this diagnostic, it can now be shown that COH$^+$ is produced in some degree by electron impact on a variety of molecules (including CD$_3$OH [171], H$_2$CO, and HCOOH, which now explains the observation in the earliest Birmingham SIFT studies mentioned above). COH$^+$ is also produced almost selectively at 300 K in the ion–molecule reaction:

$$C^+ + H_2O \rightarrow COH^+ + H \quad (\sim 80\%) \qquad (65a)$$

$$\rightarrow HCO^+ + H \quad (\sim 20\%) \qquad (65b)$$

Isomeric forms of the triatomic ions are now being recognized. Two differently reacting forms of the ion C$_2$N$^+$ are generated by electron impact on CH$_3$CN, HC$_3$N, and C$_2$N$_2$ as determined by recent SIFT experiments using CH$_4$ as the reactant gas [173]. The question arises, as always, as to what are the natures of the two species? For example, it has to be questioned whether they are the singlet and triplet states of the carbene cation [174]. The heats of formation of the isomers CNC$^+$ (lower energy) and CCN$^+$ (higher energy) are known from calculations [175] and experiments [176]. It can therefore be shown that only the lowest energy CNC$^+$ isomer can be produced in the reaction of ground state C$^+$(^2P) ions with HCN and C$_2$N$_2$; e.g.,

$$C^+(^2P) + HCN \rightarrow CNC^+ + H \qquad (66)$$

The rate coefficients for the reactions of this product CNC$^+$ with some neutral species can therefore be used as a diagnostic for CNC$^+$. However, it was observed in the SIFT experiment that when the C$^+$ was derived by electron impact on certain molecules (e.g., CO$_2$, CCl$_4$) and reacted with HCN, two forms of C$_2$N$^+$ were produced, as indicated by their different reactivities with CH$_4$. This has been shown to be due to the presence of metastable C$^+$(^4P) ions in the injected ion beam that have sufficient energy to generate CCN$^+$ ions in reaction with HCN (and also with C$_2$N$_2$). Interestingly, this can be used as a diagnostic for the presence of C$^+$(^4P) in an injected C$^+$ ion swarm. Similar studies have been carried out on the ions HC$_2$N$^+$ (possible isomers are HCCN$^+$ and HCNC$^+$) and CH$_3$NCH$^+$ and CH$_3$CNH$^+$ [174].

A further demonstration of the power of the SIFT technique in this type of study is the determination of the structure of the product $CH_3O_2^+$ ion of the much-studied reaction of O_2^+ with CH_4 [177]. Initially, it had been argued (from studies of isotope exchange with D_2O and from collisional breakup studies) that the $CH_3O_2^+$ product was protonated formic acid, $HC(OH)_2^+$. Clearly, the reactions of $HC(OH)_2^+$ and $CH_3O_2^+$ can be studied individually in the SIFT, the $HC(OH)_2^+$ being formed directly by electron impact on formic acid and injected into the flow tube, whereas the $CH_3O_2^+$ is formed by injecting O_2^+ ions and adding CH_4 to the flow tube. The reactions of these two isomeric ions have been studied with a large number of molecular gases and vapors, studies that have clearly indicated that these ions have different structures. The reactivity pattern of the ion derived from the $O_2^+ + CH_4$ reaction demonstrates convincingly that it has the structure CH_2OOH^+ and is thus the methylene hydroperoxy cation. Again, supporting evidence is obtained from the calculation of the heat of formation of the ions [177].

Clearly, the scope for such studies is enormous. One final example will suffice: Some recent SIFT studies at Birmingham of the reactions of the ions C_3H^+, $C_3H_2^+$, and $C_3H_3^+$ (being carried out as part of an on-going study of the chemistry of interstellar clouds) have revealed the presence in the flow tube of two different forms of both the $C_3H_2^+$ and the $C_3H_3^+$ ions derived by electron impact on methylacetylene (CH_3CCH) [178]. Figure 7 shows the raw SIFT data for the association reaction of $C_3H_3^+$ with CO. Similar plots showing two linear portions were obtained for the $C_3H_2^+ + CO$ reaction. Again the important question is whether the two reacting species of $C_3H_2^+$ and $C_3H_3^+$ are excited states of a ground-state ion or are structural isomers. The addition of H_2 to the ion swarm in relatively large quantities does not interconvert one of the ionic species to the other, an interconversion that might be expected if the two species were a vibrationally or electronically excited state and the ground state of a given isomer. This evidence and further evidence obtained from other detailed studies demonstrated that the two ionic species of both $C_3H_2^+$ and $C_3H_3^+$ are linear (or open chain) and cyclic structural isomers [178]. Which of the two isomers was the cyclic (c-$C_3H_3^+$) and the linear (l-$C_3H_3^+$) was established using the fact that only c-$C_3H_3^+$ can be produced in the reaction of C_3H^+ with CH_4 (from energetic constraints [179]). Hence, C_3H^+ was injected into the flow tube and reacted with CH_4, and the reaction of the product c-$C_3H_3^+$ with CO was studied. The rate coefficient obtained for this reaction was identical to one of the isomers produced by electron impact on CH_3C_2H (mentioned above), this isomer then being identified as c-$C_3H_3^+$. Thus, the reactions of CO with c-$C_3H_3^+$ (slowly reacting) and l-$C_3H_3^+$ (rapidly reacting) have been used to differentiate between these two isomers in the SIFT and to determine the fraction of each isomer in the injected beam. Both linear and cyclic isomers of $C_3H_3^+$ have also been detected in an ICR cell during studies of the reactions of $C_3H_3^+$ ions with several gases [180]. The CO reaction has also been used to establish the fractions of linear and cyclic isomers of $C_3H_3^+$, which are produced in some ion–molecule reactions. For

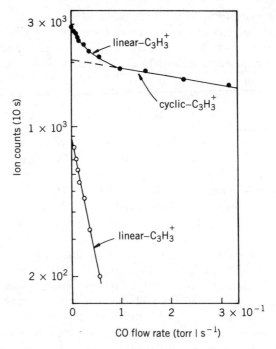

Fig. 7. SIFT data for the association reaction of $C_3H_3^+$ ions (generated in an electron impact ion source containing CH_3CCH) with CO at 300 K. The curvature on the plot of the total $C_3H_3^+$ count rate vs. CO flow rate (filled circles) indicates the presence of two forms of $C_3H_3^+$ that react at different rates with CO. The rate coefficient for the slowly reacting forms is derived from the shallow slope of the line at large CO flows and is attributed to cyclic-$C_3H_3^+$ (the cyclopropenium ion) [178]. The steeper line (open circles) is derived from the total $C_3H_3^+$ count rate in the same way as described for $Kr^+(^2P_{3/2})$ ions in the caption to Fig. 6, and the slope of the line yields the rate coefficient for the reaction with CO of the linear-$C_3H_3^+$ ion (more properly described as the open-chain isomer, the propargyl ion, $HCCCH_2^+$) [178].

example, it has been shown that only c-$C_3H_3^+$ is produced in the reaction of CH_3^+ with C_2H_2. The very different reaction rates of l-$C_3H_2^+$ and c-$C_3H_2^+$ with CO have also been used to identify these isomers. Thus, it has been shown that at 300 K in the reaction

$$C_3H^+ + H_2 \xrightarrow{He} C_3H_3^+ \quad (67a)$$

$$\rightarrow C_3H_2^+ + H \quad (67b)$$

both c-$C_3H_3^+$ (55%) and l-$C_3H_3^+$ (45%) are produced and that the $C_3H_2^+$ product ions are 100% in the c-$C_3H_2^+$ form. It has thus been argued that reaction 67a is the route to the production of the cyclic C_3H_2 molecules recently observed in interstellar clouds [181].

4 SELECTED ION FLOW DRIFT TUBE

Static drift tube techniques have been used for decades to determine the mobilities of ions in gases (as a function of the parameter E/N, where E is the electrostatic field established in a gas of number density N) and to study the reactions between ions and neutrals within the center-of-mass energy range from about 0.1 to a few eV. Much of this early work is reviewed in some well-known texts [1, 182]. With the development of the flow drift tube (FDT) in 1973 [183–185], a leap forward was made in the measurement of ionic mobilities and rate coefficients for ion–neutral reactions at the suprathermal energies. The FDT basically consists of a FA in which the downstream section of the flow tube consists of a series of identical stainless-steel rings coupled together with O-rings to form the vacuum seal but electrically decoupled from each other to allow an electrostatic field to be established along the flow tube direction. Ions are created in the carrier gas (as in the case of the FA) and are convected downstream, where ions of a given sign of charge (positive or negative ions) are constrained to drift through the flowing carrier gas by the E field with a drift velocity v_d in the field direction (characterized by their mobility in the particular carrier gas, $\mu = v_d/E$). The mean kinetic energy (in the laboratory frame) of the ions, E_i, is given by [186]

$$E_i = \tfrac{1}{2}(M_i + M_c)v_d^2 + \tfrac{3}{2}k_b T \tag{68}$$

where M_i and M_c are the masses of the drifting ion and the carrier gas atoms or molecules, respectively, k_b is the Boltzmann constant, and T is the carrier gas temperature. In reaction kinetics, the important energy is the center-of-mass energy between the reactant ion and reactant neutral of mass M_r, which we designate E_{ir}. It can readily be shown [184] that E_{ir} is given by

$$E_{ir} = \left(\frac{M_r}{(M_i + M_r)}\right)(E_i - \tfrac{3}{2}k_b T) + \tfrac{3}{2}k_b T \tag{69}$$

The center-of-mass energy between the drifting ion and the carrier gas particles, E_{ic}, is also an important parameter, and this can be calculated by replacing M_r by M_c in Eq. 69. E_{ic} is important when the drifting ions are molecular since its magnitude will determine whether internal excitation of the ion can occur. We discuss this interesting phenomenon in Section 4.1. It is a straightforward matter to determine the v_d for a positive or negative ion species as a function of E/N using the FDT, and so the E_{ir} (and the E_{ic}) for an ion–neutral interaction can readily be obtained. Hence the rate coefficients as a function of E_{ir} (or E_{ic}) can be studied for a wide range of reactions using the well-known versatility of the FA method to generate many different reactant ion species. Many interesting reactions have been studied in this way [9, 47], and a combined experimental and theoretical study has been made of the speed distribution of atomic ions in both helium and argon carrier gases [187].

Following the development of the SIFT technique, a SIFT injector was coupled to the FDT to create a selected ion flow drift tube SIFDT [188]. This natural development extended the versatility of the FDT to allow the study of a wider range of ion–neutral reactions and of other reaction processes such as collisional dissociation of complex ions, and vibrational excitation and relaxation of molecular ions. Several SIFDT apparatuses are currently being exploited to study many of these processes. Much of the data obtained before 1983 have been summarized in a recent review [9].

Notable among the most recent room temperature SIFDT studies (i.e., at a carrier gas temperature of 300 K) are the measurements of the quenching rate coefficients of vibrationally excited O_2^+ ions and NO^+ ions as a function of E_{ir} [148] which we referred to at the end of Section 3.2.3 (and in relation to reactions 61 and 62). These studies are contributing greatly to an understanding of the quenching phenomenon. Recently, the reverse process of vibrational excitation of O_2^+ and N_2^+ ions in helium carrier gas in a SIFDT has been observed [189]. This process was previously considered to be too inefficient to influence the reactions of these diatomic ions drifting through helium, although it is known to occur efficiently when ions drift through Ar carrier gas [167]. Clearly vibrational excitation can become significant when E_{ic} becomes comparable to the energy required to excite the lowest vibrational state of the molecule. This process may therefore be facile at only modest values of E/N for the low-energy bending modes in polyatomic molecules (evidence for this, in relation to CH_3^+ ions, is referred to in Section 4.1).

SIFDT experiments are being used increasingly to study the onset of endothermic reactions as E_{ir} is increased. The most straightforward reactions to study are those involving atomic reactant ions (i.e., no internal excitation). The reaction

$$C^+ + H_2 \rightarrow CH^+ + H \tag{70}$$

is endothermic by 0.4 eV (the threshold energy), and so the k is negligible at 300 K. However, for an E_{ir} equal to the threshold energy, the rate coefficient has been measured to be $\sim 5 \times 10^{-11}$ cm^3 s^{-1}, increasing to larger values with increasing E_{ir} [190, 191]. This reaction is considered to be important in generating the abundant CH^+ ions observed in the shocked regions of interstellar gas [190, 192]. Similarly, the endothermic reaction

$$S^+ + H_2 \rightarrow SH^+ + H \tag{71}$$

may be the first stage in the synthesis of H_2S in shocked interstellar clouds. Reaction 71 has been studied as part of a wider program of selected ion drift tube measurements of the reactions of SH_n^+ ($n = 0, 1, 2$) with H_2 and of SIFDT measurements of the reactions of SH_n^+ ($n = 1, 2, 3$) with H atoms [193]. The reaction

$$N^+ + H_2 \rightarrow NH^+ + H \tag{72}$$

is only slightly endothermic, and this reaction is probably important in the synthesis of NH_3 in interstellar clouds [190]. Reaction 72 and the analogous reactions of N^+ with HD and D_2 have been studied in SIFDT experiments, and these have established that the endothermicity of reaction 72 is only (11 ± 3) meV [194] in acceptable agreement with other work [195, 196]. This then establishes precisely the absolute proton affinity of N atoms to be 3.531 eV.

A room temperature SIFDT has also been used to study negative ion–molecule reactions [57]. These studies have shown that negative ion reactions can be initiated by increasing E_{ir} (and E_{ic}). For example, the H/D exchange reaction

$$DO^- + CH_2CH_2 \rightarrow HO^- + CH_2CHD \tag{73}$$

can readily be initiated at modest values of E/N. Also the reaction of D_2N^- ions with CH_2CH_2 results only in H/D exchange at zero field, but at increased translational energies the reaction

$$D_2N^- + CH_2CH_2 \rightarrow CH_2CH^- + NHD_2 \tag{74}$$

is dominant, producing CH_2CH^-. Thus an opportunity is provided to study the reactions of CH_2CH^-, which is otherwise difficult to create [57]

4.1 Variable-Temperature SIFDT

All of the SIFDT experiments referred to above have been performed using SIFDT apparatuses capable of operating only at room temperature. The first VT–SIFDT has now been developed in Birmingham, and this allows reactions to be studied as a function of E_{ir} and E_{ic} at any carrier gas temperature within the available range 80–600 K [4, 7]. This has been achieved by including a drift tube section in the downstream region of a VT–SIFT as is illustrate in Fig. 2. The drift section consists of some 50 stainless-steel rings, insulated from each other and from the (grounded) flow tube with machineable ceramic spacers. Details of the structure of the apparatus have been given in a recent review [7]. This development adds yet another dimension to fast flow tube/drift tube techniques. Clearly, with the VT–SIFDT, the separate influences, on reaction rate coefficients and on ion product distributions, of true temperature and of center-of-mass energy (E_{ir} and E_{ic}) can be studied in the same apparatus. Also the formation and reaction rates of weakly bound ions, which are thermally decomposed in warm carrier gas, can be studied in low-temperature gas as a function of E_{ir} and E_{ic}. Since the technique is very new, it is only beginning to be exploited; but several more VT–SIFT apparatuses are being converted to VT–SIFDT apparatuses [197, 198] and the data flow from these exciting systems will inevitably increase. The following will suffice to illustrate the potential of the technique.

In Section 3.2.4 we referred to the much-studied reaction

$$O_2^+ + CH_4 \rightarrow CH_2OOH^+ + H \tag{75}$$

To cast further light on the mechanism of this reaction, a study has been carried out of the temperature and energy dependence of its rate coefficient, k, using the VT-SIFDT [199]. It was found that, for a fixed carrier gas (helium) temperature of 200 K, an *increase* in E_{ir} initially resulted in a *decrease* in the k, towards a minimum value and then increased with a further increase in E_{ir}. This behavior mirrors the variation of k with the carrier gas temperature T. The experiment was repeated for carrier gas temperatures of 80, 300, 420 K, and at each fixed temperature, the sense of the change in k with increasing E_{ir} mirrored the change with increasing temperature. However, the most interesting results of these experiments was that, independent of T, the change of E_{ir}, i.e., ΔE_{ir}, and the change in T, i.e., ΔT, required to produce a given change in k were in a constant ratio such that $\Delta E_{ir}/k_b \Delta T \approx 8$. It has been shown that if all the 15 available vibrational modes in the $(O_2CH_4)^{+*}$ intermediate complex were active in energy storing, then the above ratio should equate to 10. The lower value of 8 implies that 2 modes are not active in energy storage, these inactive modes presumably being the high-frequency C—H and O—H stretching modes. A similar VT–SIFDT study has been made of the $Br^- + WF_6$ reaction 32, but in this case the experimentally observed ratio of $\Delta E_{ir}/k_b \Delta T \approx 34$ [200] is in good agreement with the expected value of 36 [7]. Much more work along these lines could be carried out to probe the nature of energy storage in excited ion–molecule intermediate complexes.

Among the first experiments performed using the VT–SIFDT were the measurements of the ternary association rate coefficients, k_3, as a function of E_{ic} and E_{ir} for a number of reactions of CH_3^+ [201], including reaction 51 and also reaction 52 involving the association of N_2^+ with N_2 to form N_4^+ [147]. As was indicated in Section 3.2.2, the mechanism of stabilization of the $(N_4^+)^*$ excited intermediate complex depends on whether the reaction takes place in He or N_2 carrier gas. It was deduced from the SIFDT observation that, when N_4^+ ions are vibrationally excited (to the point of dissociation to N_2^+ and N_2), the switching reaction

$$^{14}N_2\,^{14}N_2^+ + {}^{15}N_2 \rightarrow {}^{14}N_2\,^{15}N_2^+ + {}^{14}N_2 \tag{76}$$

becomes facile, whereas for vibrationally relaxed N_4^+, reaction 76 does not occur. Hence it has been argued that stabilization of $(N_4^+)^*$ in N_2 occurs via the efficient switching of N_2 molecules, whereas stabilization in He is via the less efficient superelastic collisions between $(N_4^+)^*$ and He atoms [147].

In the determination of ternary association rate coefficients, k_3, in FA and SIFT experiments, it is usually necessary to introduce relatively large quantities of the reactant gas into the carrier gas to obtain a sufficient decrease in the reactant ion count rate to make the measurement acceptably accurate. This is because the k_3 are often small and then the $k_{eff}(=k_3[M])$ at the available carrier gas number densities [M] are also small. Problems can arise if too much reactant gas is added since the diffusive loss rate of the reactant ions is reduced. This phenomenon is especially problematical in FDT and SIFDT experiments in which the presence of the E field also reduces the residence (reaction) time of the ions thus requiring an even greater addition of reactant gas to achieve an acceptable decrease in the reactant ion count rate. If a large amount of reactant gas is added, it not only influences the diffusive loss of the ions but also changes their mobility (and hence their residence time in the drift field and their energy). If such large additions are necessary, then account must be taken of the changes in diffusion loss and in the residence time. One of the great advantages of the VT–SIFDT for such studies can now be appreciated in that the carrier gas temperature can be reduced to enhance the truly thermal (zero E field) k_3, thus allowing measurements to be made over appreciable ranges of E_{ic} and E_{ir}. Of course, when the thermal k_3 is large at room temperature (and above), then k_3 can be determined as a function of E_{ic} and E_{ir} for any temperature within the available range of the VT–SIFDT (see below). Since the association reaction 52 of N_2^+ with N_2 in helium at 300 K is relatively slow, it was studied in the VT–SIFDT at a helium carrier gas temperature of 80 K [147]. The results obtained for the k_3 vs. E_{ir} were at first quite perplexing. It was observed that a plot of $\ln k_3$ vs. $\ln E_{ir}$ was nonlinear, unlike the corresponding plot of $\ln k_3$ vs. $\ln T$ obtained using the VT–SIFT (see Section 3.2.2 and [145]), which was quite linear with a slope of -2.3 indicating that $k_3 \sim T^{-2.3}$. However, a plot of $\ln k_3$ vs. $\ln E_{ic}$ was quite linear with a slope of -1.5, indicating that $k_3 \sim E_{ic}^{-1.5}$. A subsequent study of the association reactions of CH_3^+ with N_2 and CO also revealed that the plots of $\ln k_3$ vs. $\ln E_{ir}$ for these reactions were nonlinear and that the corresponding E_{ic} plots were quite linear with a slope close to -1.5 (see Fig. 8 compared with the slopes of close to -2.5 for the $\ln k_3$ vs. $\ln T$ plots) [201]. The power law relationship involving E_{ic} rather than E_{ir} can be understood if the following assumptions are made: (1) that the rotational temperature of the drifting ions, T_{rot}, is equilibrated with E_{ic}; i.e. $\frac{3}{2}k_b T_{rot} = E_{ic}$, and (2) that the temperature of the excited intermediate ion–molecule complexes is controlled by E_{ir}. Assumption (1) is surely justified because of the many collisions the ions undergo with the helium carrier gas atoms before they interact with the reactant gas (N_2 or CO in these examples). This allows equilibrium to be reached among the rotational states of the CH_3^+ before they associate with N_2, CO, etc. Assumption (2) is justified because E_{ir} is much greater than E_{ic} by virtue of the fact that the mass of the reactant gas is greater than that of He (see Eq. 69 for confirmation). With these assumptions and from a consideration of the partition functions for the reactants and for the intermediate complex along the lines indicated for association reactions under truly thermal conditions [143, 144], it has been shown that k_3 should indeed vary as $\sim E_{ic}^{-1.5}$ for these CH_3^+ association reactions as the VT–SIFDT data indicate [201]. For the association reaction of CH_3^+ with H_2, i.e., for a reaction for which the reactant molecule is of smaller mass than the He carrier gas atoms, it was found that the plot of $\ln k_3$ vs. $\ln E_{ir}$ (rather than $\ln E_{ic}$!) was linear with a slope close to -1.5, and this is also quite consistent with expectations based on the above approach using the partition functions.

For fast association reactions, such as the $CH_3^+ + CO$ reaction 35, the k_3

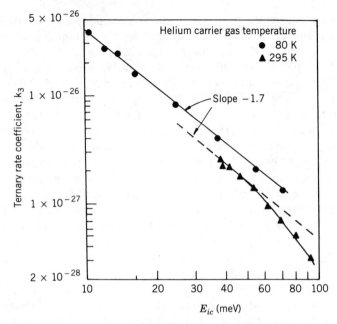

Fig. 8. SIFDT data obtained for the ternary association reaction of CH_3^+ with CO in helium carrier gas at 80 and 295 K. The ternary association rate coefficient, k_3, is plotted as a function of the center-of-mass energy between the CH_3^+ ions and He carrier gas atoms (a procedure justified by both theory and experiment [201]). The slope of the line (-1.7) is consistent with theoretical expectations based on a consideration of translational and rotational partition functions. The steepening slope of the plot obtained at a carrier gas temperature of 295 K at higher E_{ic} is attributed to vibrational excitation of the CH_3^+ ions. The displacement of the two lines is due to the additional rotational energy possessed by the CO at 295 K (compared to 80 K) [201].

can also be studied as a function of E_{ic} in the VT–SIFDT carrier gas at temperatures of 300 K and above; then larger values of E_{ic} and E_{ir} are acccessible (principally due to the lower carrier gas number densities at the higher temperatures), and vibrational modes in the reactant ions can be excited in the collisions with the carrier gas atoms. This has been observed for CH_3^+ ions by studying reaction 35 at 300 K and is manifest by a departure from linearity of the $\ln k_3$ vs. $\ln E_{ic}$ plot to a steeper slope (see Fig. 8) [201]. This more rapid reduction in k_3 with increasing E_{ic} is due to the additional energy brought into the interaction in the form of vibrational energy of the CH_3^+. This effect has been dramatically illustrated by studying the $CH_3^+ + CO$ reaction using Ar carrier gas in the VT–SIFDT. It is well known that collisions with a heavy carrier gas efficiently induce vibrational excitation in drifting ions, and this was manifest by a very rapid decrease in k_3 for the reaction with only small increases in E_{ic} (for which no such effect was observed in He carrier gas). Such studies, combined with developments in the theory, offer the opportunity for quantitative studies of the influence on ion–molecule reactions of vibrational excitation in the reactant ions.

It is interesting to note that the potentials along the drift tube in the VT–SIFDT can be arranged to maintain a finite E field in the upstream region and a zero E field region downstream. This can then be used to preexcite vibrational modes in ions, the reactions of which can then be studied under zero E field conditions (i.e., for translationally and rotationally thermalized ions) in the downstream region in the usual manner [189]. When this was done for CH_3^+ ions it was observed that the induced vibrational excitation was very rapidly quenched in collisions with He atoms in the zero field region. A similar result has been obtained for N_4^+ vibrationally excited ions [202].

At a helium carrier gas temperature of 80 K, reaction 35 is somewhat "pressure saturated" even at the lowest He pressure at which measurements could be made, and the log–log plot of the *apparent* k_3 vs. E_{ic} was nonlinear at low E_{ic}. However, at higher E_{ic} (equivalent to higher ion–molecule interaction temperatures), the plot became linear with the expected slope. Then an extrapolation of the linear part of the plot to low E_{ic} provides a value for the thermal k_3, thus circumventing the problem of "pressure saturation". This approach has been used to determine the k_3 for the very fast association reaction of CH_3^+ with HCN (which at 300 K in the SIFT is severely "pressure saturated"), and the value obtained was in close agreement with the value obtained in the most recent low-pressure ICR experiment [203].

The potential of the VT–SIFDT for the study of a wide variety of ion–neutral reactions is clear from the above brief discussion. It is also worth noting that it has been used in other studies. For example, it has been used to estimate the fragmentation energy of the $CH_3O_2^+$ ion and thus helped in the elucidation of its structure [199]. Its potential for the study of collisional dissociation of weakly bound cluster ions under controlled conditions is clear, and such studies will surely be pursued in the near future.

5   CONCLUDING REMARKS

This review has shown the massive strides that have been made in the study of ion–neutral reactions at thermal energies during the two decades since the advent of the FA. For 10 years the FA was unrivaled in these studies; then the much more versatile SIFT was conceived of and developed in the mid-1970s, and this has now largely replaced the FA for the study of ion–neutral reactions. More than 20 SIFT apparatuses, including a few VT–SIFT apparatuses, are being exploited to study a wide range of positive ion and negative ion reactions in several laboratories around the world. An enormous amount of kinetic data has thus been obtained from FA and SIFT experiments, from which much thermochemical data have been derived. These kinetic and thermochemical data are furthering the understanding of the ion chemistry of gaseous plasmas as diverse as laser and surface etchant plasmas, planetary atmospheres, and intersteller gas clouds. The FA (specifically the FALP techinque) is finding complementary applications in the study of other plasma reaction processes, such as recombination and electron attachment, and also as a versatile source of exotic ions for SIFT apparatuses.

Further developments of the SIFT technique are inevitable as the need arises to study different types of ionic reaction processes. Recently, after several years of use of the Birmingham and NOAA/York SIFT injectors, further attention has been given to new injector designs with very satisfying results. The features of the Birmingham and NOAA/York designs have been combined with spectacular results, for example, in the Birmingham VT–SIFDT (see Fig. 2 and its caption), and the use in the Boulder Chemistry Department SIFDT of large diffusion pumps on the injection quadrupole chamber coupled with a FA source has resulted in a massive increase in the injected ion currents. These advances are paving the way for spectroscopic studies on sifted ions and for the detection of radiation from ion–neutral reactions in SIFT experiments. Already these much more efficient ion injectors are allowing studies of the reactions of ions that are only minor components of a cracking pattern (such as C_3^+, C_4^+, C_5^+ from hydrocarbon molecules) to be carried out.

The development of the first VT–SIFDT is very exciting in that it is now possible to study in one experiment the influence of both temperature change and ion–neutral center-of-mass energy change on the course of ion–neutral reactions. The first results obtained from the VT–SIFDT, as presented in this review, illustrate the potential of the technique. It is obvious that both the VT–SIFT and the VT–SIFDT will both be developed and exploited for years to come for the study of ion–neutral reactions at thermal and suprathermal energies, thus providing greater insights into the kinetics and mechanisms of ionic reactions.

REFERENCES

1. E. W. McDaniel and E. A. Mason, *The Mobility and Diffusion of Ions in Gases*, Wiley, New York, 1973.
2. D. Smith, A. G. Dean, and N. G. Adams, *J. Phys. D*, **7**, 1944 (1974).
3. E. E. Ferguson, F. C. Fehsenfeld, and A. L. Schmeltekopf, *Adv. At. Mol. Phys.*, **5**, 1 (1969).
4. D. Smith and N. G. Adams, in M. T. Bowers, Ed., *Gas Phase Ion Chemistry*, Academic Press, New York, 1979, vol. 1, pp. 1–44.
5. D. L. Albritton, *Atom. Data Nucl. Data Tables*, **22**, 1 (1978).
6. D. Smith and N. G. Adams, in F. Brouillard and J. W. McGowan, Eds., *Physics of Ion-Ion and Electron-Ion Collisions*, Plenum, New York, 1983, pp. 501–531.
7. D. Smith and N. G. Adams. *Adv. At. Mol. Phys.*, **24**, 1988.
8. M. McFarland, D. L. Albritton, F. C. Fehsenfeld, E. E. Ferguson, and A. L. Schmeltekopf, *J. Chem. Phys.*, **59**, 6610 (1973).
9. W. Lindinger and D. Smith, in A. Fontijn and M. A. A. Clyne, Eds., *Reactions of Small Transient Species*, Academic Press, London, 1983, pp. 387–455.
10. K. M. Ervin and P. B. Armentrout, *J. Chem. Phys.*, **84**, 6738, 6750 (1986).
11. E. W. McDaniel, V. Cermak, A. Dalgarno, E. E. Ferguson, and L. Friedman, Eds., *Ion-Molecule Reactions*, Wiley, New York, 1970, pp. 37–53.

12. F. C. Fehsenfeld, *Int. J. Mass Spectrom. Ion Phys.*, **16**, 151 (1975).
13. E. E. Ferguson, in J. L. Franklin, Ed., *Ion-Molecule Reactions*, Butterworths, London, 1972, vol. 2, pp. 363–393.
14. F. C. Fehsenfeld, in P. Ausloos, Ed., *Interactions between Ions and Molecules*, Plenum, New York, 1975, pp. 387–412.
15. D. K. Bohme, in P. Ausloos, Ed., *Interactions between Ions and Molecules*, Plenum, New York, 1975, pp. 489–504; D. K. Bohme, *Trans. Roy. Soc. Canada*, Series IV, **XIX**, 265 (1981).
16. E. E. Ferguson, *Ann. Rev. Phys. Chem.*, **26**, 17 (1975).
17. E. E. Ferguson, in P. Ausloos, Ed., *Interactions between Ions and Molecules*, Plenum, New York, 1975, pp. 313–339.
18. E. E. Ferguson, F. C. Fehsenfeld, and D. L. Albritton, in M. T. Bowers, Ed., *Gas Phase Ion Chemistry*, Academic Press, New York, 1979, vol. 1, pp. 45–82.
19. D. Smith and N. G. Adams, *Topics in Current Chemistry*, Springer Verlag, Berlin, 1980, vol. 89, pp. 1–43.
20. N. G. Adams and D. Smith, *Int. J. Mass Spectrom. Ion Phys.*, **21**, 349 (1976).
21. N. G. Adams and D. Smith, *J. Phys. B.*, **9**, 1439 (1976).
22. D. Smith and N. G. Adams, *Ap. J.*, **217**, 741 (1977).
23. V. M. Bierbaum, G. B. Ellison, and S. R. Leone, in M. T. Bowers, Ed., *Gas Phase Ion Chemistry*, Academic Press, Orlando, Florida, 1984, vol. 3, pp. 1–39.
24. D. Smith and N. G. Adams, in W. Lindinger, T. D. Märk, and F. Howorka, Eds., *Swarms of Ions and Electrons in Gases*, Springer-Verlag, Wien, 1984, pp. 284–306.
25. D. Smith and N. G. Adams, *J. Phys. D.*, **13**, 1267 (1980).
26. D. R. Guyer, L. Hüwel and S. R. Leone, *J. Chem. Phys.*, **79**, 1259 (1983).
27. D. G. Leopold, K. K. Murray, A. E. Stevens-Miller, and W. C. Lineberger, *J. Chem. Phys.*, **83**, 4849 (1985).
28. J. J. Grabowski, private communication (1986). A. G. Harrison, *Chemical Ionization Mass Spectrometry*, CRC Press, Boca Raton, Florida, 1983, pp. 49–50.
29. A. L. Schmeltekopf and F. C. Fehsenfeld, *J. Chem. Phys.*, **53**, 3173 (1970).
30. R. C. Bolden, R. S. Hemsworth, M. J. Shaw, and N. D. Twiddy, *J. Phys. B.* **3**, 61 (1970).
31. W. Lindinger, A. L. Schmeltekopf, and F. C. Fehsenfeld, *J. Chem. Phys.*, **61** 2890 (1974).
32. W. Lindinger, F. C. Fehsenfeld, A. L. Schmeltekopf, and E. E. Ferguson, *J. Geophys. Res.*, **79**, 4753 (1974).
33. D. B. Dunkin, F. C. Fehsenfeld, A. L. Schmeltekopf, and E. E. Ferguson, *J. Chem. Phys.*, **49**, 1365 (1968).
34. D. Smith and M. J. Church, *Int. J. Mass Spectrom. Ion Phys.*, **19**, 185 (1976).
35. R. C. Bolden, R. S. Hemsworth, M. J. Shaw, and N. D. Twiddy, *J. Phys. B.*, **3**, 45 (1970).
36. N. G. Adams, M. J. Church, and D. Smith, *J. Phys. D.*, **8**, 1409 (1975).
37. E. Alge, N. G. Adams, and D. Smith, *J. Phys. B.*, **16**, 1433 (1983).
38. K. G. Spears, F. C. Fehsenfeld, M. McFarland, and E. E. Ferguson, *J. Chem. Phys.*, **56**, 2562 (1972).
39. J. Sayers and D. Smith, *Discuss. Faraday Soc.*, **37**, 167 (1964).
40. E. E. Ferguson, F. C. Fehsenfeld, D. B. Dunkin, A. L. Schmeltekopf, and H. I. Schiff, *Planet. Space Sci.*, **12**, 1169 (1964).
41. G. Gioumousis and D. P. Stevenson, *J. Chem. Phys.*, **29**, 294 (1958).
42. T. Su and M. T. Bowers, in M. T. Bowers, Ed., *Gas Phase Ion Chemistry*, Academic Press, New York, 1979, vol. 1, pp. 83–118.
43. M. J. Copsey, D. Smith, and J. Sayers, *Planet. Space Sci.*, **14** 1047 (1966).
44. D. Smith and R. A. Fouracre, *Planet. Space Sci.*, **16**, 243 (1968).

45. N. G. Adams and D. Smith, in A. Fontijn and M. A. A. Clyne, Eds., *Reactions of Small Transient Species*, Academic Press, London, 1983, pp. 311–385.
46. A. L. Schmeltekopf, E. E. Ferguson, and F. C. Fehsenfeld, *J. Chem. Phys.*, **48**, 2966 (1968).
47. D. L. Albritton in P. Ausloos, Ed., *Kinetics of Ion–Molecule Reactions*, Plenum, New York, 1979, pp. 119–142.
48. N. G. Adams, D. K. Bohme, D. B. Dunkin, F. C. Fehsenfeld, and E. E. Ferguson, *J. Chem. Phys.*, **52** 3133 (1970).
49. D. C. Clary, D. Smith, and N. G. Adams, *Chem. Phys. Lett.* **119**, 320 (1985).
50. C. J. Howard, F. C. Fehsenfeld, and M. McFarland, *J. Chem. Phys.*, **60**, 5086 (1974).
51. V. M. Bierbaum, G. B. Ellison, J. H. Futrell, and S. R. Leone, *J. Chem. Phys.* **67**, 2375 (1977).
52. C. E. Hamilton, V. M. Bierbaum, and S. R. Leone, *J. Chem. Phys.*, **80**, 1831 (1984).
53. A. O. Langford, V. M. Bierbaum, and S. R. Leone, *J. Chem. Phys.*, **84**, 2158 (1986).
54. C. E. Hamilton, V. M. Bierbaum, and S. R. Leone, *J. Chem. Phys.*, **83**, 2284 (1985).
55. C. E. Hamilton, V. M. Bierbaum, and S. R. Leone, *J. Chem. Phys.*, **84**, 2180 (1986).
56. C. H. DePuy and V. M. Bierbaum, *Acc. Chem. Res.*, **14**, 146 (1981).
57. C. H. DePuy in M. A. Almoster Ferreira, Ed., *Ionic Processes in the Gas Phase*, Reidel, Dordrecht, Holland, 1984, pp. 227–242.
58. D. K. Bohme, G. I. Mackay, and S. D. Tanner, *J. Am. Chem. Soc.*, **101**, 3742 (1980); **102**, 407 (1980).
59. R. N. McDonald and A. K. Chowhury, *J. Am. Chem. Soc.*, **102**, 6146 (1980).
60. R. N. McDonald, A. K. Chowhury, and D. W. Setser, *J. Am. Chem. Soc.*, **102**, 6491 (1980).
61. S. R. Kass, J. Filley, J. M. Van Doren, and C. H. DePuy, *J. Am. Chem. Soc.*, **108**, 2849 (1986).
62. L. M. Babcock and G. E. Streit, *J. Chem. Phys.*, **75**, 3864 (1981).
63. J. J. Grabowksi, J. M. Van Doren, C. H. DePuy, and V. M. Bierbaum, *J. Chem. Phys.*, **80**, 575 (1984).
64. L. M. Babcock and G. E. Streit, *J. Phys. Chem.*, **88**, 5025 (1984).
65. M. Meot-Ner in M. T. Bowers Ed., *Gas Phase Ion Chemistry*, Academic Press, New York, 1979, vol. 1, pp. 197–271.
66. D. Smith and N. G. Adams, *Chem. Phys. Lett.*, **54**, 535 (1978).
67. C. R. Herd and L. M. Babcock, *J. Phys. Chem.*, **91**, 2372 (1987).
68. D. Smith, N. G. Adams, A. G. Dean, and M. J. Church, *J. Phys. D.*, **8**, 141, (1975).
69. E. Alge, N. G. Adams, and D. Smith, *J. Phys. B.*, **16**, 1433 (1983).
70. D. Smith and N. G. Adams, *Ap. J. (Lett.)*, **284**, L13 (1984).
71. N. G. Adams, D. Smith, and E. Alge, *J. Chem. Phys.*, **81**, 1778 (1984).
72. M. T. Leu, M. A. Biondi, and R. Johnsen, *Phys. Rev.*, **A8**, 413 (1973).
73. H. H. Michels and R. H. Hobbs, *Ap. J. (Lett.)*, **286**, L27 (1984).
74. D. Smith, N. G. Adams, and E. Alge, *J. Phys. B.*, **17**, 461 (1984).
75. E. Alge, N. G. Adams, and D. Smith, *J. Phys. B.*, **17**, 3827 (1984).
76. N. G. Adams, D. Smith, E. Alge, and J. Burdon, *Chem. Phys. Lett.*, **116**, 460 (1985).
77. N. G. Adams, D. Smith, A. A. Viggiano, J. F. Paulson, and M. J. Henchman, *J. Chem. Phys.*, **84**, 6728 (1986).
78. D. Smith and N. G. Adams, *Geophys. Res. Lett.*, **9**, 1085 (1982).
79. D. Smith and N. G. Adams, *Pure Appl. Chem.*, **56**, 175 (1984).
80. D. Smith, N. G. Adams, and M. J. Church, *J. Phys. B.*, **11**, 4041 (1978).
81. M. J. Church and D. Smith, *J. Phys. D.*, **11**, 2199 (1978).
82. D. Smith, N. G. Adams, and E. Alge, *Planet. Space. Sci.*, **29**, 449 (1981).

83. M. E. Jones, S. R. Kass, J. Filley, R. M. Barkley, and G. B. Ellison, *J. Am. Chem. Soc.*, **107**, 109 (1985).
84. D. Smith, N. G. Adams, and M. J. Henchman, *J. Chem. Phys.*, **72**, 4951, (1980).
85. J. M. Van Doren, S. E. Barlow, C. H. DePuy and V. M. Bierbaum, *Int. J. Mass. Spectrom. Ion Proc.*, **81**, 85 (1987).
86. J. C. Hansen, C. H. Kuo, F. J. Grieman, and J. T. Moseley, *J. Chem. Phys.*, **79**, 1111 (1983).
87. E. Herbst and W. Klemperer, *Ap. J.* **185**, 505 (1973).
88. A. Dalgarno and J. H. Black, *Rep. Prog. Phys.*, **39**, 573 (1976).
89. D. Smith and N. G. Adams, *Int. Rev. Phys. Chem.*, **1**, 271 (1981).
90. F. Howorka, F. C. Fehsenfeld, and D. L. Albritton, *J. Phys. B.*, **12**, 4189 (1979).
91. G. I. Mackay, G. D. Vlachos, D. K. Bohme, and H. I. Schiff, *Int. J. Mass Spectrom. Ion Phys.*, **36**, 259 (1980).
92. G. Dupeyrat, B. R. Rowe, D. W. Fahey, and D. L. Albritton, *Int. J. Mass Spectrom. Ion Phys.*, **44**, 1 (1982).
93. D. Smith, N. G. Adams, and D. Grief, *J. Atmos. Terr. Phys.*, **39**, 513 (1977).
94. J. F. Paulson and F. Dale, *J. Chem. Phys.*, **77**, 4006 (1982).
95. N. G. Adams, D. Smith, and D. Grief, *J. Phys. B.*, **12**, 791 (1979).
96. D. Smith, N. G. Adams, and D. Grief, *Int. J. Mass Spectrom. Ion Phys.*, **30**, 271 (1979).
97. D. Smith, N. G. Adams, E. Alge, H. Villinger, and W. Lindinger, *J. Phys. B.*, **13**, 2787 (1980).
98. N. G. Adams and D. Smith, *Int. J. Mass Spectrom. Ion Phys.*, **35**, 335 (1980).
99. M. J. Henchman, J. F. Paulson, and P. M. Hierl, *J. Am. Chem. Soc.*, **105**, 5509 (1983).
100. N. G. Adams, D. Smith, and J. F. Paulson, *J. Chem. Phys.*, **72**, 288 (1980).
101. D. Smith and N. G. Adams, *Int. J. Mass Spectrom. Ion Phys.*, **23**, 123 (1977).
102. D. Smith and N. G. Adams, *Chem. Phys. Lett.*, **47**, 145 (1977).
103. N. G. Adams and D. Smith, *Chem. Phys. Lett.*, **47**, 383 (1977).
104. D. Smith and N. G. Adams, *Chem. Phys. Lett.*, **54**, 535 (1978).
105. N. G. Adams and D. Smith, *Chem. Phys. Lett.*, **54** 530 (1978).
106. N. G. Adams, D. Smith, and D. Grief, *Int. J. Mass Spectrom. Ion Phys.*, **26**, 405 (1978).
107. D. Smith, N. G. Adams, and W. Lindinger, *J. Chem. Phys.*, **75**, 3365 (1981).
108. A. B. Raksit and D. K. Bohme, *Int. J. Mass Spectrom. Ion Proc.*, **55**, 69 (1983/4).
109. J. S. Knight, C. G. Freeman, M. J. McEwan, N. G. Adams, and D. Smith, *Int. J. Mass Spectrom. Ion Proc.*, **67**, 317 (1985); J. S. Knight, C. G. Freeman, M. J. McEwan, S. C. Smith, N. G. Adams, and D. Smith, *Mon. Not. R. Astron. Soc.* **219**, 89 (1985).
110. N. G. Adams and D. Smith, *Chem. Phys. Lett.*, **79**, 563 (1981).
111. D. Smith and N. G. Adams, *Ap. J. (Lett.)*, **220**, L87 (1978).
112. D. Smith and N. G. Adams, in P. Ausloos, Ed., *Kinetics of Ion–Molecule Reactions*, Plenum, New York, 1979. pp. 345–376.
113. S. D. Tanner, G. I. Mackay, A. C. Hopkinson, and D. K. Bohme, *Int. J. Mass Spectrom. Ion Phys.*, **29** 153 (1979).
114. D. C. Clary, *Mol. Phys.*, **54**, 605 (1985); K. Sakimoto and K. Takayanagi, *J. Phys. Soc. Jpn.*, **48**, 2076 (1980).
115. N. G. Adams, D. Smith, and D. C. Clary, *Ap. J. (Lett.)*, **296**, L31 (1985); T. J. Millar, N. G. Adams, D. Smith, and D. C. Clary, *Mon. Not. R. Astron. Soc.*, **216**, 1025 (1985); D. Smith, *Phil. Trans. R. Soc. London*, **A323**, 269 (1987).
116. J. B. Marquette, B. R. Rowe, G. Dupeyrat, G. Poissant, and C. Rebrion, *Chem. Phys. Lett.*, **122**, 431 (1985).

117. A. A. Viggiano, F. Howorka, D. L. Albritton, F. C. Fehsenfeld, N. G. Adams, and D. Smith, *Ap. J.*, **236**, 492 (1980).
118. W. Federer, H. Villinger, F. Howorka, W. Lindinger, P. Tosi, D. Bassi, and E. E. Ferguson, *Phys. Rev. Lett.*, **52**, 2084 (1984).
119. N. G. Adams and D. Smith, *Ap. J. (Lett.)*, **294**, L63 (1985).
120. W. Federer, H. Villinger, W. Lindinger, and E. E. Ferguson, *Chem. Phys. Lett.*, **123**, 12 (1986).
121. D. Smith and N. G. Adams, *J. Phys. B.*, **20**, 4903 (1987).
122. A. A. Viggiano and J. F. Paulson, *J. Chem. Phys.*, **79**, 2241 (1983).
123. A. A. Viggiano and J. F. Paulson, in W. Lindinger, T. D. Märk, and F. Howorka, Eds., *Swarms of Ions and Electrons in Gases*, Springer Verlag, Wien, 1984, pp. 218–240.
124. N. G. Adams and D. Smith, *Chem. Phys. Lett.*, **105**, 604 (1984).
125. D. Smith and N. G. Adams, *Ap. J.*, **298**, 827 (1985).
126. S. G. Lias, J. F. Liebman, and R. D. Levin, *J. Phys. Chem. Ref. Data*, **13**, 695 (1984).
127. D. Smith and N. G. Adams, *J. Phys. Chem.*, **89**, 3964 (1985).
128. A. B. Raksit and D. K. Bohme, *Int. J. Mass Spectrom. Ion Proc.*, **57**, 211 (1984).
129. A. A. Viggiano, R. A. Perry, D. L. Albritton, E. E. Ferguson, and F. C. Fehsenfeld, *J. Geophys. Res.*, **85**, 4551 (1980).
130. I. Dotan, D. L. Albritton. F. C. Fehsenfeld, G. E. Streit, and E. E. Ferguson, *J. Chem. Phys.*, **68**, 5414 (1978).
131. A. A. Viggiano, J. F. Paulson, F. Dale, M. J. Henchman, N. G. Adams, and D. Smith, *J. Phys. Chem.*, **89**, 2264 (1985).
132. D. Smith and N. G. Adams, in M.A. Almoster Ferreira, Ed., *Ionic Processes in the Gas Phase*, Reidel, Dordrecht, Holland, 1984, pp. 41–66.
133. D. Smith and N. G. Adams, *Ap. J.*, **242**, 424 (1980).
134. M. J. Henchman, N. G. Adams, and D. Smith, *J. Chem. Phys.*, **75**, 1201 (1981).
135. N. G. Adams and D. Smith, *Ap. J.*, **248**, 373 (1981).
136. D. Smith, N. G. Adams, and E. Alge, *J. Chem. Phys.*, **77**, 1261 (1982).
137. D. Smith, N. G. Adams, and E. Alge, *Ap. J.*, **263**, 123 (1982).
138. D. Smith, N. G. Adams, and M. J. Henchman, *J. Chem. Phys.*, **72**, 4951 (1980).
139. N. G. Adams, D. Smith and M. J. Henchman, *Int. J. Mass Spectrom. Ion Phys.*, **42**, 11 (1982).
140. T. B. McMahon, P. G. Miasek, and J. L. Beauchamp, *Int. J. Mass Spectrom. Ion Phys.*, **21**, 63 (1976).
141. J. J. Grabowski, C. H. DePuy, and V. M. Bierbaum, *J. Am. Chem. Soc.*, **105**, 2565 (1983).
142. D. Smith and N. G. Adams, *Chem. Phys. Lett.*, **76**, 418 (1980).
143. D. R. Bates, *J. Phys. B.*, **12**, 4135 (1979).
144. E. Herbst. *J. Chem. Phys.*, **70**, 2201 (1979).
145. H. Böhringer, F. Arnold, D. Smith, and N. G. Adams, *Int. J. Mass Spectrom. Ion Phys.*, **52**, 25 (1983).
146. H. Böhringer and F. Arnold, *J. Chem. Phys.*, **77**, 5534 (1982).
147. D. Smith, N. G. Adams, and E. Alge, *Chem. Phys. Lett.*, **105**, 317 (1984).
148. E. E. Ferguson, *J. Phys. Chem.*, **90**, 731 (1986).
149. E. E. Ferguson, D. Smith, and N. G. Adams, *J. Chem. Phys.*, **81**, 742 (1984).
150. L. M. Bass and K. R. Jennings, *Int. J. Mass Spectrom. Ion Proc.*, **58**, 307 (1984).
151. A. A. Viggiano, *J. Chem. Phys.*, **84**, 244 (1986).
152. A. A. Viggiano, *J. Chem. Phys.*, **81**, 2639 (1984).
153. A. A. Viggiano, F. Dale, and J. F. Paulson, *J. Geophys. Res.*, **90**, 7977 (1985).

REFERENCES 219

154. W. Lindinger, D. L. Albritton, M. McFarland, F. C. Fehsenfeld, A. L. Schmeltekopf, and E. E. Ferguson, *J. Chem. Phys.*, **62**, 4105 (1975).
155. J. Glosik, A. B. Rakshit, N. D. Twiddy, N. G. Adams, and D. Smith, *J. Phys. B.*, **11**, 3365 (1978).
156. M. Tichy, A. B. Rakshit, D. G. Lister, N. D. Twiddy, N. G. Adams, and D. Smith, *Int. J. Mass Spectrom. Ion Phys.*, **29**, 231 (1979).
157. K. G. Spears, F. C. Fehsenfeld, M. McFarland, and E. E. Ferguson, *J. Chem. Phys.*, **56**, 2562 (1972).
158. N. G. Adams, D. Smith, and E. Alge, *J. Phys. B.*, **13**, 3235 (1980).
159. F. Howorka, A. A. Viggiano, D. L. Albritton, E. E. Ferguson, and F. C. Fehsenfeld, *J. Geophys. Res.*, **84**, 5941 (1979).
160. J. D. C. Jones, A. S. M. Raouf, D. G. Lister, K. Birkinshaw, and N. D. Twiddy, *Chem. Phys. Lett.*, **78**, 75 (1981).
161. T. T. C. Jones, K. Birkinshaw, J. D. C. Jones, and N. D. Twiddy, *J. Phys. B.*, **15**, 2439 (1982).
162. W. Lindinger, *Int. J. Mass Spectrom. Ion Proc.*, **70**, 213 (1986).
163. H. Böhringer, M. Durup-Ferguson, D. W. Fahey, F. C. Fehsenfeld, and E. E. Ferguson, *J. Chem. Phys.*, **79**, 4201 (1983).
164. W. Dobler, W. Federer, F. Howorka, W. Lindinger, M. Durup-Ferguson, and E. E. Ferguson, *J. Chem. Phys.*, **79**, 1543 (1983).
165. E. E. Ferguson, D. Smith, and N. G. Adams, *Int. J. Mass Spectrom. Ion Proc.*, **57**, 243 (1984).
166. I. Dotan, S. E. Barlow, and E. E. Ferguson, *Chem. Phys. Lett*, in press.
167. E. E. Ferguson, in W. Lindinger, T. D. Märk, and F. Howorka, Eds., *Swarms of Ions and Electrons in Gases*, Springer-Verlag, Wien, 1984, pp. 126–145.
168. E. E. Ferguson, N. G. Adams, D. Smith, and E. Alge, *J. Chem. Phys.*, **80**, 6095 (1984).
169. R. C. Woods, C. S. Gudeman, R. L. Dickman, P. F. Goldsmith, G. R. Huguenin, W. M. Irvin, A. Hjalmarson, L. A. Nyman and H. Olofsson, *Ap. J.*, **270**, 583 (1983).
170. D. J. DeFrees and A. D. McLean, *J. Computer Chem.*, **7**, 321 (1986); D. A. Dixon, A. Komornicki, and W. D. Kraemer, *J. Chem. Phys.*, **81**, 3603 (1984).
171. W. Wagner-Redeker, P. R. Kemper, M. F. Jarrold, and M. T. Bowers, *J. Chem. Phys.*, **83**, 1121 (1985).
172. C. G. Freeman, J. S. Knight, J. G. Love, and M. J. McEwan, *Int. J. Mass Spectrom. Ion Proc.*, **80**, 255 (1987).
173. M. J. McEwan, private communication (1987); A. B. Raksit and D. K. Bohme, *Can. J. Chem.*, **63**, 854 (1985).
174. J. S. Knight, C. G. Freeman, and M. J. McEwan, *J. Am. Chem. Soc.*, **108**, 1404 (1986).
175. N. H. Haese and R. C. Woods, *Ap. J. (Lett.)*, **246**, L51 (1981).
176. P. W. Harland and B. J. McIntosh, *Int. J. Mass Spectrom. Ion Proc.*, **67**, 29 (1985).
177. J. M. Van Doren, S. E. Barlow, C. H. DePuy, V. M. Bierbaum, I. Dotan, and E. E. Ferguson, *J. Phys. Chem.*, **90**, 2772 (1986).
178. D. Smith and N. G. Adams, *Int. J. Mass Spectrom. Ion Proc.*, **76**, 307 (1987).
179. D. K. Bohme, *Nature*, **319**, 473 (1986).
180. K. C. Smyth, S. G. Lias, and P. Ausloos, *Combustion Sci. Technol.*, **28**, 147, (1982).
181. N. G. Adams and D. Smith, *Ap. J. (Lett.)*, **317**, L25 (1987).
182. I. R. Gatland, in W. Lindinger, T. D. Märk, and F. Howorka, Eds., *Swarms of Ions and Electrons in Gases*, Springer-Verlag, Wien, 1984, pp. 44–59.
183. M. McFarland, D. L. Albritton, F. C. Fehsenfeld, E. E. Ferguson, and A. L. Schmeltekopf, *J. Chem. Phys.*, **59**, 6610 (1973).
184. M. McFarland, D. L. Albritton, F. C. Fehsenfeld, E. E. Ferguson and A. L. Schmeltekopf, *J. Chem. Phys.*, **59**, 6620 (1973).

185. M. McFarland, D. L. Albritton, F. C. Fehsenfeld, E. E. Ferguson, and A. L. Schmeltekopf, *J. Chem. Phys.*, **59**, 6629 (1973).
186. G. H. Wannier, *Bell Syst. Tech. J.*, **32**, 170 (1953).
187. D. L. Albritton, I. Dotan, W. Lindinger, M. McFarland, J. Tellinghuisen, and F. C. Fehsenfeld, *J. Chem. Phys.*, **66**, 410 (1977).
188. F. Howorka, I. Dotan, F. C. Fehsenfeld, and D. L. Albritton, *J. Chem. Phys.*, **73**, 758 (1980).
189. W. Federer, H. Ramler, H. Villinger, and W. Lindinger, *Phys. Rev. Lett.*, **54**, 540 (1985).
190. N. G. Adams, D. Smith, and T. J. Millar, *Mon. Not. R. Astron. Soc.*, **211**, 857 (1984).
191. N. D. Twiddy, A. Mohebati, and M. Tichy, *Int. J. Mass Spectrom. Ion Proc.*, **74**, 251 (1986).
192. B. T. Draine and N. S. Katz, *Ap. J.*, **306**, 655 (1986).
193. T. J. Millar, N. G. Adams, D. Smith, W. Lindinger, and H. Villinger, *Mon. Not. R. Astron. Soc.*, **221**, 673 (1986).
194. N. G. Adams and D. Smith, *Chem. Phys. Lett.*, **117**, 67 (1985).
195. J. A. Luine and G. H. Dunn, *Ap. J. (Lett.)*, **299**, L67 (1985).
196. J. B. Marquette, B. R. Rowe, G. Duperat, and E. Roueff, *Astron. Astrophys.*, **147**, 115 (1985).
197. J. F. Paulson, private communication (1986).
198. M. J. McEwan, private communication (1986).
199. N. G. Adams, D. Smith, E. E. Ferguson, *Int. J. Mass Spectrom. Ion Proc.*, **67**, 67 (1985).
200. A. A. Viggiano, J. F. Paulson, F. Dale, M. J. Henchman, N. G. Adams, and D. Smith, *J. Phys. Chem.*, **89**, 2264 (1985).
201. N. G. Adams and D. Smith, *Int. J. Mass Spectrom. Ion Proc.*, **81**, 273 (1987).
202. M. Tichý, N. D. Twiddy, D. P. Wareing, N. G. Adams, and D. Smith, *Int. J. Mass Spectrom.*
203. P. R. Kemper, L. M. Bass, and M. T. Bowers, *J. Phys. Chem.*, **89**, 1105 (1985).

Chapter **V**

PULSED ELECTRON HIGH PRESSURE MASS SPECTROMETER

Paul Kebarle

---

1 Introduction
2 PHPMS Apparatus and Measurements
   2.1 Typical Measurements and Brief Description of PHPMS Apparatus
   2.2 Typical Layout of PHPMS Components
   2.3 Pumping Systems in PHPMS
   2.4 Electron Guns in PHPMS
   2.5 PHPMS Ion Source
       2.5.1 Ion Source Construction and Ion and Electron Slits
       2.5.2 Temperature Control of PHPMS Ion Source
       2.5.3 Electrical Discharge from Ion Source and Other Insulation Problems
   2.6 Ion Accleration and Focusing
   2.7 Mass Analysis
       2.7.1 Calibration for Mass Dependent Transmission
       2.7.2 Magnetic Sector Versus Quadrupole for PHPMS
   2.8 Ion Detection and Data Storage and Handling
   2.9 Gas-Handling Plant
   2.10 Pulsing Circuitry Mode
3 Determination of Rate and Equilibrium Constants
   3.1 Electrical Plasma Conditions in PHPMS Ion Source
   3.2 Diffusion Coefficients of Ions and Ion–molecule Rate and Equilibria Measurements
   3.3 Some Typical PHPMS Ion–Molecule Kinetics Measurements
       3.3.1 Treatment of Reactions with Inclusion of Ion Diffusion to Wall
       3.3.2 Normalization to the Total Ion Intensity: Method
4 Ion–Molecule Equilibria Measurements with the PHPMS
   4.1 Transfer (Exchange) Ion–Molecule Equilibria
       4.1.1 Kinetic Aspects
       4.1.2 Evaluation of Thermochemical Data
   4.2 Ion–Molecule Association and Clustering Equilibria
       4.2.1 Kinetic Aspects and Some Typical Measurements
       4.2.2 Conditions for Association Equilibria Measurements
       4.2.3 Thermochemical Data from Association and Clustering Equilibria
References

1 INTRODUCTION

The pulsed electron high ion source pressure mass spectrometry technique (PHPMS), also often referred to in the literature as HPMS, has proven to be one of the most successful if not the most successful method for measurement of ion–molecule equilibria. The resulting thermochemical information is having a significant impact on the development of chemistry.

The measurement of equilibria is intimately connected with measurements of the rates of approach to equilibrium. The PHPMS technique is basically a kinetic technique in which the approach and achievement of equilibrium are observed. Thus, it provides information on ion–molecule reaction kinetics in its own right, particularly on the temperature dependence of ion–molecule rates. Since conditions are achieved in which all reactions involving the ions are (pseudo) first order, even complex consecutive and parallel reaction systems also involving reversible reactions can be subjected to simple mathematical analysis that yields all the rate constants of the individual elementary reactions. Thus, the PHPMS technique can, when needed, also provide kinetic data for rather complex reaction systems.

The most important types of ion equilibria to which the technique has been applied are listed below, together with typical examples and the type of thermochemical data that are obtained:

1. Ion–ligand and ion–solvent molecule equilibria:

$$Na^+(NH_3)_{n-1} + NH_3 = Na^+(NH_3)_n \tag{1}$$

$$K^+(H_2O) + CH_3OH = K^+(CH_3OH) + H_2O \tag{2}$$

2. Proton transfer, gas phase acidities and basicities

$$NH_4^+ + CH_3NH_2 = NH_3 + CH_3NH_3^+ \tag{3}$$

$$CH_3COO^- + CH_2ClCOOH = CH_3COOH + CH_2ClCOO^- \tag{4}$$

3. Electron transfer, electron affinities, and ionization energies

$$C_6H_5NO_2^- + SF_6 = C_6H_5NO_2 + SF_6^- \tag{5}$$

$$\text{Naphthalene}^+ + \text{anthracene} = \text{naphthalene} + \text{anthracene}^+ \tag{6}$$

4. Hydride and chloride transfer: hydride and chloride ion affinities, stabilities of carbocations.

$$\textit{tert-}C_4H_9^+ + C_6H_5CH_2X = \textit{tert-}C_4H_9X + C_6H_5CH_2^+ \tag{7}$$

A more detailed discussion of the thermochemical measurements and data is given in Sections 3 and 4.

The earliest systematic measurements of ion–molecule reactions, Talroze and Lyubimova [1], Stevenson and Schissler [2], and Field et al. [3], were performed with conventional mass spectrometers operated at somewhat elevated ion source pressures. Typical ion source pressures used were 10^{-5}–10^{-2} torr. The presence of a weak electric field applied via the repeller electrode while helping in the extraction of the ions led to ion–molecule collisions occurring at different center of mass energies and thus to variable and nonthermal conditions. There followed a gradual development toward higher ion-source pressures while maintaining tolerably low mass analyzer pressures. This was achieved by closing down the area of the ion exit and the electron entrance slits to reduce gas outflow from the ion soure and increasing the pumping speed at the vacuum housing containing the ion source. By 1961 Field [4] had achieved the "ultrahigh" ion source pressure of 0.35 torr, and in 1969 Wexler and Pobo [5] were working at pressures up to 1 torr and using 2-MeV protons as a penetrating ionizing medium. Some of the products observed under high-pressure conditions required the occurrence of as many as six consecutive reactive collisions.

Much of the impetus for the early ion–molecule work was due to the interest in the role that ion–molecule reactions played in the radiolysis of gases. Research in radiation chemistry was stimulated in its own turn by the then recent development of nuclear fission. Conventional radiolysis studies generally involved irradiation of gases at near atmospheric pressures. Kebarle and Godbole [6], having become involved with some mass spectrometric research aimed at solving problems in radiation chemistry and realizing that the ion–molecule reaction phenomena may be quite different at near atmospheric pressures, decided to develop apparatus for ion–molecule reaction studies at near atmospheric pressures. The first version, Kebarle [7], used an α-particle source for the irradiation of gases and operated at up to 350 torr pressures. Apart from the penetrating ionizing medium, the apparatus used a very high capacity (relative to the usage at the time) 6 in. pumping system and a very small ion exit leak. The origin of ions observed when hydrocarbon gases like ethylene were irradiated at 300 torr pressure was very difficult to interpret, (Kebarle and Godbole [7] and Kebarle and Hogg [8, 9]) since the pressure gap between known and unknown high-pressure ion–molecule chemistry was too large and the reaction systems too complicated. However, gases such as water or ammonia led to mass spectra that consisted of the simple ion series $H_3O^+(H_2O)_n$ and $NH_4^+(NH_3)_n$, respectively, and it was soon realized that the ion clusters observed may be in thermodynamic equilibrium (Kebarle [8, 10, 11]). The enormous potential of such equilibrium measurements was recognized, and tests for equilibria as well as instrumental modifications were undertaken (Kebarle et al. [10–12]) in order to develop apparatus specially suited to ion–molecule equilibria measurements. It was realized that conditions for equilibrium measurements are better at pressures in the 5–10 torr range since the problems connected with unadulterated ion sample transfer to the mass analyzer are less serious in this range (Kebarle and Hogg [10–11]). An important modification of the apparatus occurred with the introduction of a pulsed electron beam (Kebarle et al. [13, 14]). The return to electrons, but with 2 keV energy for penetrating power, was a question of convenience, i.e., electrons are readily produced with a heated filament, and the electron beam is easily pulsed. Pulsed electron ion sources had been used earlier by Talroze and Frantevitch [15] and Shannon et al. [16] but at much lower ion source pressures ($\sim 10^{-3}$ torr). Pulsing in the torr range, although technically slightly more complicated since the filament and electron gun had to be located further away from the ion source, had a great advantage. At low pressures the ion lifetimes are very short since the ions can reach the walls of the ion source in free flight, and this takes only microseconds. At high pressures, the ions must diffuse through the gas to reach the walls. Thus, the ions are trapped by the gas and their lifetime, before discharge on the wall, is increased to hundreds of microseconds. Therefore, pulsing in the PHPMS allowed one to observe reactive ion-intensity changes over times as long as 10 ms after the short ($\sim 10\,\mu s$) electron pulse and this with a time resolution of $\sim 10\,\mu s$. This ability combined with the higher neutral reactant concentrations accessible at high pressure expanded the scale of rate measurements enormously.

The present version of the PHPMS, which has changed little since 1972 (see Cunningham et al. and Payzant [17]) will be described in Sections 2. A preliminary description and some typical rate and equilibria measurements will be given first in order to illustrate the instrumental requirements. This will be followed by some instrumental details intended to facilitate construction of similar apparatus in other laboratories. Section 3 describes typical kinetic measurements and techniques, and Section 4 deals with ion–molecule equilibria measurement techniques.

Even though the PHPMS apparatus has proven to be a veritable ion-thermochemistry factory, at present fewer research groups (Meot-Ner and Field [18]; Jennings et al. [19]; Stone et al. [20]; Meot-Ner and Sieck [21–23]; Hiraoka et al. [24]; McMahon [25]; Grimsrud [26], and Kebarle et al. [17]) are using this method compared to groups using the flowing afterglow (FA) apparatus selected ion flow tube (SIFT), or the ion cyclotron resonance ICR and Fourier transform ICR apparatus; see articles on these methods in the present volume. One reason may be the fact that the PHPMS is not commercially available. Another reason could be that the apparatus was never described in detail. We hope that the present article will remove this latter problem. The PHPMS apparatus is easily assembled and of relatively low cost. About five instruments were constructed in Kebarle's laboratory over the years, and four of these are still in use. In two of these instruments, commercial mass spectrometers discarded by analytical mass spectrometrists were used. The conversion of existing mass spectrometers for PHPMS work is not difficult provided there is some instrumental know how and a machine shop. The cost of the completed conversion need not exceed $30,000.

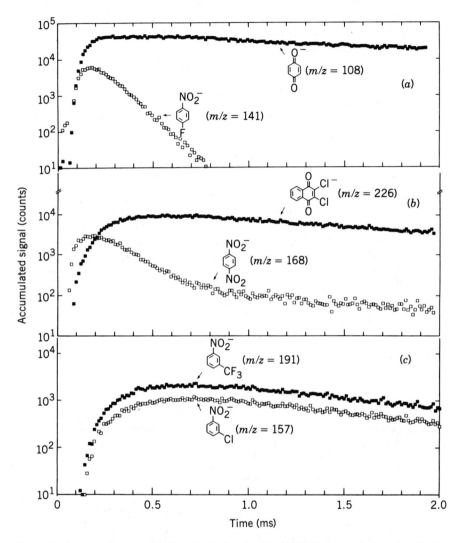

Fig. 1. Ion intensity changes with time after a short ($\sim 10\,\mu s$) 2000-V electron beam pulse. The ion source (see Fig. 2) contains 3.5 torr of bath gas (CH_4) and is at 150°C. Secondary electrons formed by the ionization of the bath gas are slowed down by collisions with the bath gas and then captured by compounds A and B, which have positive electron affinities (EA). A and B engage in electron transfer: $A^- + B = A + B^-$. In Fig. 1a, A = fluoronitrobenzene (2 mtorr) and B = benzoquinone (0.3 mtorr). Since B has a much higher electron affinity, electron transfer to B is irreversible. Slope of A^- decay leads to rate constant for electron transfer reaction. In Fig. 1b, A = dinitrobenzene 0.5 mtorr and B = dichloronaphthoquinone. B has somewhat higher EA, and electron transfer equilibrium is reached after some 0.9 ms. In Fig. 1c, A = trifluoronitrobenzene (0.53 mtorr) B = chloronitrobenzene (7.6 mtorr). Electron transfer reaction reaches equilibrium rapidly. (From Grimsrud [27]).

2 PHPMS APPARATUS AND MEASUREMENTS

2.1 Typical Measurements and Brief Description of PHPMS Apparatus

The time dependence of the observed ion intensities after the short electron pulse for three typical experiments is shown in Fig. 1. All three cases involve the electron transfer reaction 8:

$$A^- + B = A + B^- \tag{8}$$

The apparatus is shown in Figs. 2 and 3. The ion source-reaction chamber shown schematically in Fig. 2 contains about 5 torr of the major (bath) gas, which is methane in the present case, and a few to tens of millitorr of compounds A and B that have positive electron affinities. Secondary electrons are created by the ionization of the bath gas by the primary electron pulse. The secondary electrons are slowed down to near thermal velocities by collisions with the bath gas. Some of the slow electrons are captured by A and B. The resulting A^- and B^- are initially internally excited by the exothermicity of the electron capture reaction. The excited A^- and B^- are gradually thermalized by collisions with the bath gas. As the ions diffuse toward the walls, they also engage in the electron transfer reaction. In Fig. 1a, the compound B is of much higher electron affinity, and the electron transfer reaction proceeds to completion from left to right. The linear logarithmic decay of A^- permits the determination of the rate constant. In Fig. 1b, the two compounds have closer electron affinities. Initially A^- decreases due to electron transfer, and then the reaction reaches equilibrium. This type of run allows the determination of the forward rate constant and the equilibrium constant. The run in Fig. 1c represents a case in which the equilibrium establishes rapidly. The rate constant cannot be obtained, but the

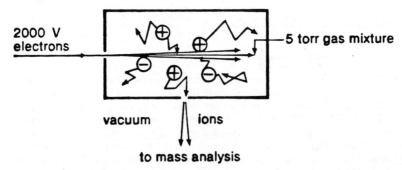

Fig. 2. Schematic diagram of pulsed electron high-pressure ion source. Ions produced by short ($\sim 10\,\mu s$) 2000-V electron pulse diffuse to the walls of the source where they become discharged. Due to diffusion at relatively high pressure of 5 torr, ions are "trapped" by the gas; i.e., diffusion to wall for some ions may take hundreds of microseconds. Ions coming to the vicinity of very small ion exit slit escape into vacuum and are mass analyzed.

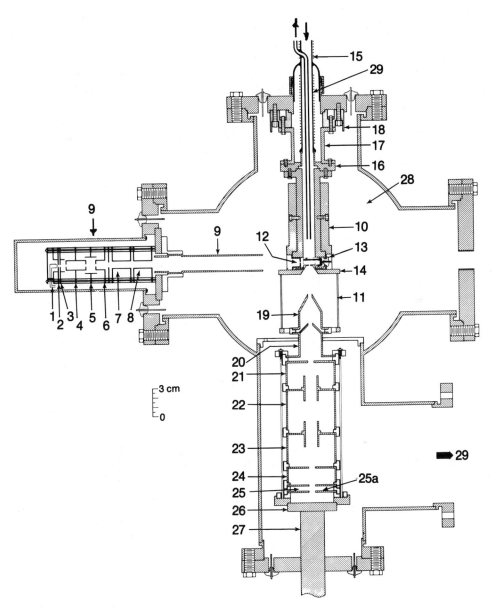

Fig. 3. PHPMS apparatus (Cunningham et al. [17]) used with 90° magnetic sector. Pulsed electron gun: 1, filament; 2, draw out; 3, extractor; 4, 5, 6, einzel lens; 7, 8, electron beam deflection plates in X and Y direction; 9, electric and magnetic shield. Thermostatted ion source with in- and outflow of reactant gas mixture: 10, heating and cooling jacket; 11, wire mesh electric shield; 12, electron entrance slit; 13, electron trap and ion repeller; 14, ion exit slit flange; 15, glass gas in and out circulation tube; 16, ion source lid flange with kovar seal; 17, support of ion source; 18, insulator of support. Ion acceleration tower: 19–24, cylindrically symmetrical electrodes; 25 and 25a, collimating and ion deflecting slits; 26, flange carrying entrance into mass analyzer slit; 27, mass analyzer tube. Vacuum system: 28, to 6-in. pump; 29, to 4-in. pump.

conditions are ideally suited for the determination of the equilibrium constant, K_1.

The slow decrease of ion intensity in Fig. 1 after the kinetic stage is completed (see B^- in Figs. 1a and 1b and A^- and B^- in Fig. 1c at $t > 1$ ms) is due to the disappearance of ions due to diffusion to the wall followed by ion discharge on the wall.

The ion intensities shown in Fig. 1 are not observed in situ in the ion source. The ion sampling is achieved by allowing gas to escape in near molecular flow

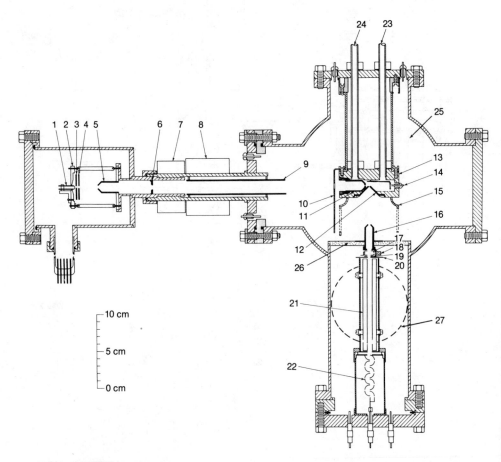

Fig. 4. PHPMS apparatus (Durden et al. [13, 29], Lau [30]) with quadrupole MS. Pulsed electron gun: 1, filament; 2, filament support and electron reflector plate; 3, 4, draw out and pulsing electrodes; 5, acceleration; 6, collimation electrode; 7, solenoid providing axial magnetic field for electron collimation; 8, TV yoke for X and Y deflection of beam; 9, deflection plates, not in use; 10, fluorescent beam focusing screen. Ion source: 11, electron entrance slit carrying flange; 12, ion exit slit carrying cone flange; 13, ion source block with heaters; 14, electron trap; 15, wire mesh electrical shield. Ion acceleration and focusing: electrodes, 16–20. Mass analysis and detection: 21, quadrupole; 22, electron multiplier; 23, 24, reactant gas circulation glass tubes for in- and outflow. Pumping system: 25, to 6-in. pump; 26, differential pumping lid; 27, to 4-in. pump.

from a small ion exit slit in the wall of the ion source. Ions diffusing to the vicinity of the slit escape also. The gas exiting through the slit is pumped out by the high capacity differential pumping system. The ions are captured by electric fields, accelerated, and subjected to mass analysis obtained either with a magnetic sector or with a quadrupole filter. The arrangement is illustrated in Fig. 3. After mass analysis the ions are detected with an electron multiplier. The m/z ion intensities shown in Fig. 1 were obtained by collecting one given m/z sorted in function of time by a given group of channels of a multichannel analyzer. The trigger of the electron beam pulse triggers also the multichannel collection sweep. The accumulated counts with $10\,\mu s$ per channel shown in Fig. 1 were obtained typically after the collection of a given m/z for about 10,000 electron pulses. The time between electron pulses is typically 5 ms. Thus the collection time of the ions with a given m/z is $\sim 50\,s$. Generally, about four to five different ions amount to over 90% of the total ion current. Thus, the collection of all these ions in the multiscaler requires typically a total of 5–20 minutes. The conditions in the source must be kept as constant as possible for this duration. This is achieved by passing the reaction mixture in slow flow in and out of the ion source; see Fig. 3. The flow constantly replenishes the ion source gas mixture and removes electron impact induced reaction products. A flow that exceeds the outflow through the electron and ion exit slit by a factor of about 5 is generally used.

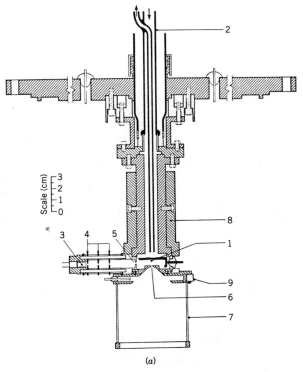

(a)

Fig. 5. (*Contd.*)

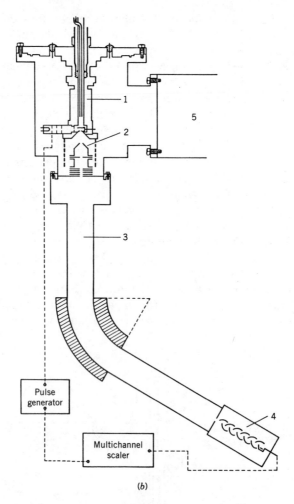

Fig. 5. PHPMS used with commercial magnetic sector mass spectrometers CH_4 and CH_5. From French and Kebarle [31, 32], Ikonomou [33]. Fig.a: ion source and flange: 1, ion source; 2, gas in and out glass tubes; 3–5, pulsed electron gun; 6, ion exit slit; 7, wire mesh shield; 8, heating block; 9, auxiliary low-pressure ionization electron filament. Fig. b:layout with commercial CH_4 mass spectrometer: 1, ion source as in A; 2, ion acceleration and focusing electrodes; 3, mass analyzer tube; 4, electron multiplier; 5, to 6-in. pumping.

The reaction mixture used is prepared in a 4-L glass bulb that is part of a gas-handling plant, and the gas flow through the ion source is controlled by means of a valve.

2.2 Typical Layout of PHPMS Components

The diagram in Fig. 3 (Cunningham et al. [17]) shows one possible layout of the major components of the PHPMS. This arrangement was used with a

15-cm radius 90° magnetic sector whose geometry goes back to the Canadian mass spectrometry pioneers Lossing [28] and Graham et al. [28]. A layout used with a commercial quadrupole mass filter is shown in Fig. 4 (Durden [13, 29], Lau [30]), and an ion source arrangement used with two commercial magnetic sector (MS) instruments, Atlas Mat CH4 and CH5, is shown in Fig. 5 (French and Kebarle [31], French [32], and Ikonomou [33]). The figures are to a large extent self-explanatory. In Figs. 3 and 4 the electron guns are mounted separately and can be removed for cleaning and filament change. The vacuum housing with a crosslike manifold is particularly well suited when the other components are light, i.e., mass analysis is obtained with a quadrupole. In this case the crosslike manifold is mounted directly on top of the 6-in. pump, see Fig. 4. The arrangement in Fig. 5 is suitable when a commercial magnetic mass spectrometer with the usual heavy magnet is used. In such cases the mass analysis tube cannot be moved. A cylindrical vacuum housing is fitted to the mass spectrometer tube flange; see Fig. 5b. The lid of this housing carries all the new components, i.e., electron gun and ion source; see Fig. 5a.

2.3 Pumping Systems in PHPMS

The ion exit slit and electron entrance slit of the high-pressure ion sources have a combined conductance $F_i \approx 4 \times 10^{-3}$ L/s (for air at 300 K); see Section 2.5. To handle the gas load resulting from gas outflow from the ion source, all versions of the instruments in Figs. 3–5 have a 6-in. pumping system providing a pumping speed $S_h \approx 400$ L/s at the housing containing the ion source. Thus, for a given ion source pressure P_i the housing pressure P_h can be calculated with Eq. 9 (see Dushman [34]). For a typical ion source pressure,

$$P_h = P_i \frac{F_i}{S_h} \approx P_i \times 10^{-5} \qquad (9)$$

$P_i = 5$ torr, the resulting $P_h = 5 \times 10^{-5}$ torr, and pressures in that range were actually measured in the vacuum housing. The ion acceleration tower, Fig. 3, and the quadrupole housing, Fig. 4, are differentially pumped by a 4-in. system providing a pumping speed of $S_m \approx 100$ L/s. The pressure P_m in that volume is $P_m \approx 0.1 P_h$ or 5×10^{-6} torr for $P_i = 5$ torr. The electron multiplier volume of the apparatus of Fig. 3 was differentially pumped by a 2-in. system such that the electron multiplier is at a pressure $P_{em} \approx 10^{-6}$ torr for $P_i = 5$ torr. Low pressures $P_m \sim 10^{-5}$ torr in the ion acceleration and mass analysis region are very desirable, particularly for ion-cluster equilibria measurements; see Section 4.2.2. Low pressure in the electron multiplier housing prevents electrical discharges and also seems to increase the life of the multiplier. The instruments (Fig. 5) did not have a 4-in. differential pumping system for the mass analyzer but relied on the pumping systems provided by the manufacturer for that region of the apparatus (~ 20 L/s at mass analyzer).

All pumping was obtained with oil diffusion pumps charged with polyphenyl ether (Monsanto-Santovac-5). These gave essentially service-free performance for periods as long as 5 years. The ion acceleration electrodes and the mass analysis tube had to be cleaned only once every 5–10 years, even though bakeout provisions for these components were not available. Thus, the cleaner cryo or turbo pumps are a possible alternative, but are not essential.

2.4 Electron Guns in PHPMS

Three different electron guns are shown in Figs. 3–5. The guns of Figs. 3 and 4 are designed for placement of the filament at some considerable distance from the ion source and are mounted on a separate flange. Locating the filament away from the electron entrance slit of the high-pressure ion source reduces exposure of the filament to gases escaping from the ion source. Exposure to gases decreases the lifetime of the filament, and some gases reduce the electron emission, presumably due to reaction with the filament surface. Since the pressure of escaping gas decreases rapidly with distance from the slit, much shorter distances, s in Fig. 5, are almost equally effective. The electron gun in Fig. 3 uses electrostatic focusing. The design of the electrodes follows conventional cathode ray gun geometry and optics. The electrodes were from nonmagnetic stainless steel mounted on a tower and insulated with glass spacers. Typical voltages used are given in Table 1.

The electron gun in Fig. 4 is a "home" design. The potentials used at the various electrodes, 1–5, are given in Table 2. Electron beam focusing was obtained by an axial magnetic field from an electric coil, 7 (1000 turns of 22 AWG copper wire with 0.2–1.5 A required for focusing). X and Y deflection of the beam was obtained from a television (TV) tube yoke, 8, which was outside

Table 1. Electron Gun and Ion Acceleration Potentials used with PHPMS (Fig. 3)

	Electron Gun[a]			Ion Source and Acceleration[b]	
1	Filament	0	10–14	Ion source	2000
2	Drawout	5	19	First cone	1920
3	Extractor	275	20	Second cone	1720
4		2000	21	Aperture	2080
5	Einzel Lens	260	22	First cylinder	1400
6		2000	23	Second cylinder	1100
7	X_a Deflection	2000	24	Exit hole	1000
	X_b	2000 ± 50			
			25	Collimating slit	0
8	Y Deflection	2000	25a	Deflection plates	0 and 0 ± 100
	Y_b Deflection	2000 ± 50	26	Collimating slit	0
			27	Mass analyzer	0

[a]Potentials in volts given for positive ions operation. To obtain approximate potentials used with negative ions, add −4000 V to each electrode potential. For additional details, see Payzant [17].
[b]Potentials in volts used for positive ions. To obtain approximate potentials for negative ions, reverse sign.

Table 2. Electron Gun and Ion Acceleration Potentials used with PHPMS (Fig. 4)

	Electron Gun[a]			Ion Source and Acceleration[a]	
1	Filament	−2000	10–15	Ion source	+8
2	Reflector	−2000	16	Cone	−250
3	Draw out	−1955	17–18	Low p source[b]	0
4	Focus	−1700	19	Aperture	−45
5	Cone	0	20	Quadrupole entrance	0
6	Collimator	0	21	Quadrupole	0

[a]All potentials in volts relative to ground for positive ion mode. For negative ion mode electron gun essentially the same and ion source and acceleration with reversed sign. Additional information can be found in Durden [13, 29] and Lau [30].
[b]Low-pressure ion source, axially open, not in operation when high-pressure ion source is used.

the vacuum envelope. The yoke produces magnetic fields for the deflection. The two solenoids of the yoke were powered by a low-voltage (7 V) power supply, providing up to 20 mA for each solenoid.

The metal shield (10, Fig. 4), which was lightly painted with a phosphor and had a small ~4-mm-diameter hole in the center, was used for focusing of the electron beam after a new filament had been installed. The shield, 10, could be observed through a window. The deflected beam striking the shield was focused with choke, 7, to bright spot of 3–6 mm diameter, and then the spot was guided by means of the X, Y deflection, 8, into the hole in the shield. A similar arrangement was used also with the apparatus shown in Fig. 3. This focusing mode was useful but not absolutely necessary. A focusing based entirely on electron trap, 14 (Fig. 4) current readings was also possible, but more laborious.

The simplest electron gun used was that in Fig. 5a, shown in more detail in Fig. 6. The gun is mounted directly on the small flange that carries the electron entrance slit to the ion source and bolts airtight to the ion source. The typical voltages used are given in Fig. 6. Due to proximity of the electron gun, no special beam focusing was required.

The electron currents obtained with the three guns are shown in Table 3. As can be seen, all three guns deliver approximately the same current inside the ion source. This current is very low, i.e., in the 3×10^{-8} A range. This is a consequence of the very small electron entrance slits used $(0.01 \times 1 \, \text{mm})$. However, low ionizing currents are not a drawback since *low* ion densities are required for the achievement of space charge free conditions that are needed for the *thermal* ion–molecule reaction measurements.

The electron guns used with the magnetic mass analyzers (Figs. 3 and 5) had to be extensively magnetically shielded. The shielding is required because even very weak stray fields from the magnet deflected the electron beam and reduced the ionizing electron current. In the absence of magnetic shielding, the magnetic mass scan produced higher stray fields and thus lower ionizing currents at higher mass. This could be demonstrated by the fact that the XY focusing of the electron beam for maximum ion intensity was mass dependent and also

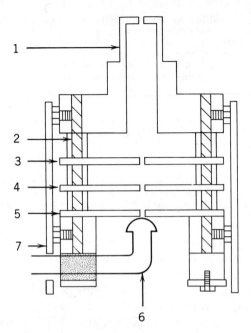

Fig. 6. Detail of electron gun (Ikonomou [33]) used with apparatus of Fig. 5. 1, Stainless flange carrying electron entrance into ion source slits and bolted to ion source (3000 V); 2, vespel rods and glass spacers; 3, two half-plates for electron beam deflection (1000 and 1000 ± 100 V), 4, acceleration slit (850 V); 5, drawout (30 V); 6, filament (0 V); 7, magnetic shielding cylinder.

by direct measurements of the stray fields with a Hall probe. In the absence of shielding, fields as high as 100 G were measured in the electron gun area (Fig. 5) at highest mass. Extensive shielding with multiple layers of high μ metal sheets (Netic-conetic) removed this effect, and no X, Y refocusing of the electron beam at high mass was required Cunningham et al. [17], Ikonomou [33].

Conventional filament material was used. For oxygen as bath gas, thoriated

Table 3. Electron Currents[a] from Electron Guns (Figs. 3–5)

Apparatus	Fig. 3 (Payzant [17])	Fig. 4 (Lau [30])	Fig. 5a (Ikonomou [33])
Total emission[b]	100	100	100
Ion source[c]	~5	50	5
Electron trap[d]	~0.03	0.04	0.04–0.08

[a]All currents in microamperes for continuous (nonpulsed) operation.
[b]Electron current leaving filament equals total emission. Total emission could be increased up to 1000 μA. Other currents increased roughly in proportion.
[c]Electron current hitting ion source.
[d]Electron current entering ion source, collected on electron trap inside ion source, in absence of gas in ion source. (Electron entrance slit, 1 mm × 10^{-2} mm.)

iridium ribbon was found to last longest; for methane and other gases either thoriated iridium or 75/25 tungsten/rhenium was used.

2.5 PHPMS Ion Source

2.5.1 Ion Source Construction and Ion and Electron Slits

Only the modern version of the ion source shown in Fig. 3 (Payzant [17]) will be described since this design is used in all instruments in the author's laboratory. The massive nonmagnetic stainless-steel cylindrical block ion source has a long circular 0.5-in. diameter channel that preheats the flowing gas. The ion source block is longer than usual. This is partly a consequence of the need to fill the available space of the ~ 10-in. wide vacuum chamber, 28, leading to the 6-in. pump. At the bottom end of the source a 0.5-in. diameter channel, which is perpendicular to the main axis channel, forms the ion source proper. This channel is closed off on both sides by two hatlike flanges, 12 and 13, that carry the electron entrance slit, 12, and the electron trap cum repeller, 13. At the bottom of the source a third flange carries the ion exit slit. All three flanges are made vacuum tight by means of thin gold wire gaskets and held by four (entrance and trap) or six (ion exit) small bolts.

The electron entrance and ion exit slit were made by spot welding two razor blades pieces, roughly trimmed to half-moon shape, across the 1 mm hole in the flange. The slit width used was 0.01 mm so that a slit of 1×0.01 mm resulted. A special jig was used such that the blade aligning and spot welding could be done on the object table of a microscope, i.e., under microscopic observation. After the two razor blade pieces were spot welded, their protruding edges could be easily trimmed with ordinary scissors. Various commercial stainless razor blades were used with equal success. The jig used was simple: a cylindrical brass plate ~ 0.5 in. high and 2 in. diameter with a central 0.25 in. hole. This plate was the one pole of the spot welder. The slit-holding flange was bolted to the brass plate. The razor blade half-moons were then maneuvered over the slit plane and spot welded with the other spot welder electrode, which was a handheld copper pencil.

The ion exit slits, manufactured as explained above, fulfilled the special purpose of letting a considerable volume of gas, and thus also ions, out (large slit area) without leading to dynamic flow in which adiabatic expansion and cooling of the gas occurs. See Section 4.2.1. Furthermore, a robust slit is obtained that can hold 1 atm pressure differential without breaking. The razor blade edges lead to effectively very thin slit walls and thus minimal ion loss by discharge on passage of the ions through the slit "channel."

An earlier version (Hogg et al. [12]) fulfilling the same flow and minimum discharge criteria consisted of a square array of 30 holes, each of 10 μm diameter, spaced out over 1 mm^2. The holes were produced by laser beam fusion of a 2.5-μm-thick stainless foil.

The hole array had the drawback of plugging rather easily. Therefore, this type of leak was abandoned after it was established that the same clustering equilibrium constants were obtained also with the razor blade leak. Razor blade slits, particularly for the ion exit slit, have been used for about 15 years in this laboratory, and we believe that they are well worth the trouble required to make them.

Since no external fields are applied inside the ion source and the ion exit slit is at the same nominal potential as the ion source, the ions exiting the ion source have near-thermal energies. We expect that even very minor surface charge (i.e., up to millivolts) could prevent the ions from exiting the source.

From time to time ion signal loss occurred. Such signal loss appeared to be associated with the presence of particular gases and higher ion source temperatures. The signal was recovered after installation of new slits. The surface appearance of the slits was not a reliable guide to slit induced ion loss. In the average, slit replacement was necessary about two or three times a year.

2.5.2 Temperature Control of PHPMS Ion Source

The ion source block could be heated or cooled for kinetic or equilibria measurements at temperatures other than room temperature. Heating was obtained by inserting six cartridge heaters. (Hotwatt Inc., Danvers, Mass.) into circular channels in the ion source block. The version in Fig. 3 has a separate heating mantle block, 10, which carries the heaters. Later ion source versions did not have a separate mantle, but the "mantle" was part of a single stainless block containing the ion source. The advantage of the single block is a more rapid establishment of the final temperature, as pointed out by Hiraoka et al. [24]. Measurements at temperatures up to 600°C (Cunningham et al. [17]), where the ion source block glowed cherry red, were performed with this arrangement.

The temperature of the gas in the ion source was measured by measuring the ion source wall temperature with iron constantan thermocouples. A small diameter blind hole was drilled radially into the cylinder block roughly in position 1, Fig. 5a. The thermocouple was inserted from the outside into the hole, and its tip was firmly pressed against the end of the hole. As the hole terminated near the inner wall, i.e., near the ion source cavity, it was assumed that this arrangement led to a correct reading of the ion source temperature.

Since the gas and ions whose temperature is to be accurately known originate from a region that is within a few millimeters of the ion exit slit, we might be concerned that the ion source block heaters may not be providing a uniform temperature and, particularly, that the slit region may be cooler. The ion source (Fig. 5a) also had a set of four cartridge heaters embedded in the bottom plate, 9, which pressed firmly against the ion slit flange. Measurements of clustering equilibria with this ion source gave similar results (French and Kebarle [31] and French [32]) to those observed earlier in this laboratory, and it was assumed that heaters in plate 9 are not essential. Unfortunately, these experiments did not include equilibria measurements at the highest temperatures, i.e., where the expected error in the temperature measurement would be the highest.

In general, it was observed that whenever two thermocouples were embedded near the wall of the ion source cavity, but in somewhat different positions, they would read somewhat different temperatures when the ion source did not contain gas. However, admission of gas at normal experimental pressures (~ 4 torr) led to near or complete equalization of the temperature readings. Thus, the thermal conductivity of the gas helps equalize the temperatures.

Since the temperature is such an important parameter in the PHPMS measurements, it would be worthwhile embedding heaters also in the first electrode, i.e., 19, in Fig. 3 and maintaining this electrode at ion source temperatures in high-temperature experiments. Unfortunately, so far, this experiment has not been done.

Low temperatures were obtained by circulating cooled fluids through channels in the ion source block. Temperatures down to $-170°C$ could be obtained (Hiraoka and Kebarle [35]) by passing N_2 gas from a compressed gas container through a copper coil immersed into liquid nitrogen, into the ion source cooling channels. Good temperature control could be obtained with this simple arrangement. To achieve lowest temperatures, the outflow of the cooled copper coil had to be very close to the cooling inflow lead to the ion source flange. A good mechanical drawing of the cooling leads arrangement can be found in Hiraoka [24b].

2.5.3 Electrical Discharge from Ion Source and Other Insulation Problems

Since the ion source is at high positive or negative voltage when the magnetic sector is used for mass analysis, electrical gas discharges can occur between the ion source and the gas-handling plant. The glass tubing of ~ 10 mm diameter connecting the ion source to the gas-handling plant is deliberately not too short (~ 60 cm) in order to reduce the chance for discharge at a given voltage and pressure; see the Paschen breakdown curve [36]. Electrical discharges were observed at ion source pressures in the mtorr range, i.e., considerably below normal PHPMS operating pressure. To prevent discharges, the high voltage was turned off when the ion source was pumped out or brought up to operating pressure. As an additional precaution, the metal valves in the gas-handling plant box (see Section 2.8) were at a floating potential, i.e., had no connection to ground, and the valves were turned via Lucite (insulating) handles.

Since the thermocouples measuring the ion source temperature were in contact with the metal, the thermocouple readout circuits were at a floating potential, and the temperature was read with the high voltage off. The cartridge heaters of the ion source were also floating. They were supplied with ac power via isolation transformers.

2.6 Ion Acceleration and Focusing

The ions escaping from the ion exit slit of the high-pressure ion source have thermal energies. Since also the locus of their origin, i.e., the ion exit slit, is very small, they are easy to focus. The focusing and acceleration systems used with a magnetic and a quadrupole mass spectrometer are shown in Figs. 3 and 4, and the typical voltages applied to the electrodes are given in Tables 1 and 2.

An important design consideration was the provision of sufficient space for the rapid pumpout of the gas escaping from the ion source and particularly the ion source slit. Thus, the first electrode after the ion source is conically shaped, see electrode 19, Fig. 3, and 16, Fig. 4. The opening of this electrode is the limiting gas conductance for the second stage (differential) pumping. The series of electrodes 19–26 used with the magnetic instrument are spaced out to provide the necessary distance accommodating the 6-in. lead of the 4-in. pumping system. These electrodes provide the necessary acceleration and focusing. The potentials shown in Table 1 indicate that electrodes 20–22 act like an einzel lens. The last electrodes before the mass analyzer, i.e., 25, 25a, 26, were slits whose long dimension was in the z direction, i.e., the direction of the magnetic field. Electrode 25 consists of two half-plates that functioned as deflection plates. By offsetting the voltage of the one plate relative to the other, the beam could be swept across the collimating slit 25a. The slit 26 was the ion entrance slit into the magnetic analyzer, which together with the mass analyzer exit slit, not shown in Fig. 3, fixes the mass resolution. The slit width of 25a and 25 was 0.5 mm.

The ion focusing for the quadrupole mass analyzer, see Fig. 4, is even simpler since no small orifice collimating electrodes are required. Typical potentials used are shown in Table 2. The electrodes 17–20 were present in the commercial quadrupole used, which had a low pressure, axial (to the quadrupole), molecular beam ion source. This ion source was not removed and fulfilled a very useful function for the calibration of the mass dependent transmission of the quadrupole mass analyzer (see Section 2.7).

2.7 Mass Analysis

2.7.1 Calibration for Mass Dependent Transmission

The mass analysis in PHPMS is in its nature no different than the mass analysis for other purposes. Since very adequate descriptions of mass analysis can be found in the literature, we will not consider this part of the apparatus here. However, two points are of interest; mass dependent discrimination and choice of magnetic sector vs. quadrupole mass analysis (see Section 2.7.2).

The ion transmission of magnetic sector instruments using constant acceleration voltage and magnetic scanning changes (decreases) only very slowly with mass. Therefore, corrections for the mass dependent transmission in PHPMS measurements were not made when magnetic sector instruments were used.

The ion transmission of quadrupole mass analyzers is strongly dependent on the parameters used, and mass scans at constant U/V, which are often used, lead to a rapid fall off of transmission with mass (Dawson [37]). Since the PHPMS applications often require the determination of ion intensity ratios of ions with different mass, viz. ion–molecule equilibria, these ratios are to be equal to the fluxes of ions exiting the ion source. Therefore, a correction for mass dependent discrimination in the mass analysis is required. The most convenient calibration was obtained with use of the low-pressure ion source, 18, Fig. 4. A gas, like perfluorokerosene, was bled via the high-pressure source, with the high-pressure filament off, into the low-pressure vacuum chamber such that constant pressures $\sim 10^{-6}$ torr due to the gas was obtained in that region. The mass spectrum observed with the low-pressure filament was recorded. The mass spectrum of the same compound had been obtained with a magnetic sector instrument. Plots of the ion intensity ratios in the normalized mass spectra (base peak = 100) obtained with the two instruments resulted in curves that gave the mass dependent transmission of the quadrupole, assuming the magnetic sector to be with a flat response.

In general several compounds were used, and although some scatter of points obtained with different compounds was observed, smooth interpolated transmission curves could be obtained. The quadrupole ion intensities observed with the PHPMS were corrected by multiplying each mass with the mass dependent factors determined from the transmission curves.

2.7.2 Magnetic Sector Versus Quadrupole for PHPMS

Here we examine the relative merits of the magnetic sector and quadrupole mass analyzer in PHPMS. The quadrupole has a number of advantages:

1. Due to the low weight of the quadrupole mass analyzer and its simple mounting on a flange, it is much easier to incorporate a quadrupole into a 6- and 4-in. differential pumping system; see Fig. 4. With the magnetic sector, the heavy vacuum system has to be mated to the heavy mass analysis system, and this requires special arrangements for gradually bringing the two systems together for bolting.
2. Since the ion source in magnetic instruments is at high positive or negative voltage, electrical gas discharge and insulation problems are encountered (see Section 2.5.3); such discharge problems do not exist for the quadrupole.
3. The radially symmetric ion focusing system and the larger ion entrance into the mass analysis region aperture that is used with the quadrupole is simpler to set up.
4. In actual PHPMS measurements, at times we encounter ions at a given m/z whose identity cannot be deduced from the gas mixture that was prepared and from expected ion–molecule reactions. Either high resolution allowing element identification or MS–MS with collisional activation would be very useful in such instances. These methods have not yet been applied to PHPMS. Due to high ion energies and high accelerating voltages, the incorporation of an MS–MS facility is somewhat more complicated when a magnetic sector is used. On the other hand, the replacement of a simple quadrupole with a triple quadrupole in an existing PHPMS system would be only a question of adapting flanges.

The drawback of the quadrupole mass analyzer relative to the magnetic sector for PHPMS is the need for ion-transmission calibration (see Section 2.7.1) and the more rapid deterioration of performance with use experienced with these analyzers, i.e., the more frequent cleaning of the mass filter required.

Magnetic sectors were much used in Kebarle's laboratory because of previous experience with them. Also, there are many magnetic sector instruments discarded by analytical mass spectrometrists that can be obtained at very low or no cost. The use of a high-resolution double-focusing instrument for PHPMS would be very desirable because of the access to high mass and elemental analysis that such systems provide.

2.8 Ion Detection and Data Storage and Handling

Since the detection and recording of ions in PHPMS is the same as that with other MS techniques, only a brief description will be given. The ions were counted after detection with a secondary electron multiplier (Spiraltron 4219 EIC from Gallileo Electric Optics). The pulses were amplified and counted with conventional devices. The pulses were collected in a multiscaler and dedicated computer [Le Croy 3500 computer with Le Croy multiscaler interface MCS 3521A; an alternate is the IBM personal computer (PC) with the ORTEC-MCS 913 on an IBM PC plug-in card].

2.9 Gas-Handling Plant

As described in Section 2.1, gas mixtures of a major carrier gas at 1–5 torr containing generally at least two (reactant) gases at partial pressures in the 10^{-2} mtorr range are supplied to the ion source in slow flow. A convenient device for the preparation of the gas mixtures is the gas-handling plant (GHP) shown in Fig. 7. The gas mixture is prepared in a 5-L glass bulb. The glass bulb has a glass tube extension that carries a rubber septum. The bulb is evacuated via valves 1 and 2 (see Fig. 7) and filled up to 1 atm with carrier gas from a high-pressure gas storage cylinder attached to gas inlet valve 4 or 5. The pressure in the bulb is read with an absolute manometer (variable capacitance Baratron or variable reluctance Validyne) attached with a glass line to valve 6. The minor components, most often liquids or solutions, are injected via the septum into the bulb with valve 1 closed off. After 20 minutes mixing time, valve 1 is opened to the previously evacuated manifold, and the flow through the ion source is controlled through ion source valve 7. The gas exiting the ion source passes through a 10-mm-diameter glass lead (see Fig. 3), then through the wide open valve 8, and then through a flow controlling glass capillary that is followed by the pump. The ion source pressure is measured with a 0–10-torr variable capacitance absolute manometer connected with a glass line to the glass tube connecting the ion source and valve 8.

The GHP is contained in a box that is temperature controlled and generally operated at $\sim 140°C$.

The PHPMS measurements may be made very difficult by presence of minor impurities in the gas mixture used. For example, previously used compounds with high electron affinities adsorbed in the walls of the GHP can take over the negative charge and prevent the measurement of the electron transfer

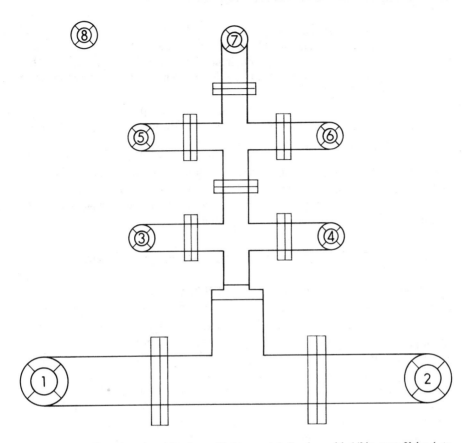

Fig. 7. Gas-handling plant. 1 and 2 valves with 1 in. port, 3–7 valves with 1/2 in. port. Valve 1, to 5 L storage bulb; valve 2, to liquid N_2 trap and diffusion pump; valve 3, to rough pump; valves 4, 5 gas inlets; valve 6, to manometer; valve 7, to ion source. Materials stainless steel, glass, Cu gaskets, no elastomer gaskets.

equilibrium $A^- + B = A + B^-$ involving A and B compounds of lower EA. The GHP and valve system are designed to reduce this problem. Only all metal valves are used. The valves 1 and 2 exhausting the bulb are wide bore, 1 in. whereas the rest are ~ 0.5 in. A liquid N_2 trap is used in the 2-in. diffusion pumped system, and the rubber septum is changed when a new series of measurements with new compounds is started.

2.10 Pulsing Circuitry Mode

A conventional square wave pulse generator at near ground potential produced the master pulse that was used to trigger the multichannel analyzer sweep and the electron gun on–off pulse generator. The electron gun pulse generator was floated to the filament potential and triggered via a laser diode that provided insulation from the ground pulser. In the pulsed mode, the electron gun was turned off by applying a permanent -50 V relative to filament potential to the first (draw out) electrode. The gun was turned on by a short, generally 10 μs, positive pulse to the draw out from the floating pulse generator that temporarily restored the draw out electrode to the normal on voltage given in Table 1.

A second floating pulse generator connected to the repeller-electron trap was used to destroy the remaining ions before repetition of the cycle. A short, $\sim 100\ \mu s + 50$ V, pulse relative to the ion source was applied to the repeller-electron trap about 500 μs before repetition of the electron on pulse. Ion destruction with the second pulser is not absolutely necessary and was not used in every instrument. Lengthening the time between electron on pulses so that the ion intensity has decayed to insignificant values can be used also. This mode requires somewhat longer ion collection times.

3 DETERMINATION OF RATE AND EQUILIBRIUM CONSTANTS

3.1 Electrical Plasma Conditions in PHPMS Ion Source

Plots of the intensity of ions as a function of time after the electron pulse obtained with PHPMS apparatus and used for ion–molecule rate constants and equilibrium constants determinations were shown in Fig. 1. The use of such PHPMS results for rate constants and equilibrium constants determination is straightforward. The assumption is made that the intensity of a given ion A^- at time t, $(A^-)_t$, is proportional to the concentration $[A^-]_t$ in the ion source; see Eq. 10. A second assumption is that the intensity ratio $(A^-)_t/(B^-)_t$ of two ions observed in the same experiment is equal to the ion concentration ratio $[A^-]_t/[B^-]_t$ in the ion source; see Eq. 11:

$$(A^-)_t = C_A [A^-]_t \tag{10}$$

$$\frac{(A^-)_t}{(B^-)_t} = \frac{[A^-]_t}{[B^-]_t} \tag{11}$$

$$C_A = C_B \tag{12}$$

Different levels of assumptions are required for the different determinations. For example, for the simple irreversible reaction in which the pseudo-first-order

$$A^- + B = A + B^- \tag{8}$$

rate constant $v_8 = k_8[B]$ is determined from a plot of $\ln(A^-)$ vs. t (see Fig. 1), the required condition is that the proportionality constant C_A in Eq. (1) is time independent. On the other hand, for the equilibrium constant determinations $K_8 = [A](B^-)/[B](A^-)$, the condition $C_A = C_B$ is required but the proportionality constants may be time dependent. In the general treatment of the

PHPMS results, both time independence of the proportionality constants and (approximate) equality $C_A = C_B$ have been assumed.

With these assumptions, the ion intensities were treated as if they represented in situ concentrations in the ion source and as if these concentrations were independent of the spatial coordinates in the ion source, as would be the case had stable ions in solution been present. A justification of assumptions inclusive Eqs. 10–12 can be made a posteriori; i.e., the rate constants and equilibrium constants obtained are sensible since the data obtained fulfill expected conditions. Thus, in the rate constant determinations the $\ln(A^-)$ vs. t plots are linear, the slope v_8 obey the relationship $v_8 = k_8[B]$ when v_8 are measured for different B concentrations and finally the rate constant k_8 values agree with determinations obtained with other methods. Similar tests can be applied to the equilibria determinations. Thus, K_8 must be independent of t after equilibrium is reached, K_8 must be independent of [A] and [B], the van't Hoff plots must be linear, the resulting ΔH^0, ΔG^0, and ΔS^0 values must be consistent when each is combined in thermodynamic cycles (see Figs. 25–27, Section 4.2), and finally the values must agree with determinations from other methods, or if such determinations are not available, as was the case for the first cluster binding energies (Hogg et al. [12]), then the numbers obtained should make physical sense. This type of approach was taken in the PHPMS method development. In general, experimental conditions were sought where the PHPMS data that were obtained conformed to the above kinetic and equilibrium tests. The empiricism of the approach was largely a question of necessity since the conditions in the ion source and the sampling process are quite complex. Some understanding of the detailed processes is required, however, and the present section is devoted to a description of the actual conditions and their relationship to the PHPMS measurements.

Experimental and theoretical studies of systems in which ions and electrons are created in a gas by an ionizing pulse (photons, electrons, or microwave radiation) and then the history of the system followed after the pulse is shut off are known as stationary afterglow studies in physics, where stationary is used as a distinction from the flowing afterglow (FA) (see Chapter IV). The PHPMS system is thus a stationary afterglow system. A comparison of PHPMS (Kebarle [38]) with stationary afterglow results of Puckett and Lineberger [39] is given in Figs. 8 and 9. There is a considerable similarity in the ion-time profiles of the two figures, although Lineberger's plots also show reactive changes due to ion–molecule reactions, whereas the PHPMS plot gives only the total ion intensities. The time scale in the Lineberger experiment is approximately 100 times longer. This is due to the much larger vessel used (cylinder of 45 cm diameter and 90 cm length vs. ~ 1 by 3 cm for PHPMS).

Since the ion sampling orifice replaces a given wall area, the detected ion intensity in Figs. 8 and 9 is due to the ion wall current to the orifice. The observed changes are (1) linear decrease in $\log(A^+)$ in the time interval 20–170 ms (0.01–0.4 ms), PHPMS times from Fig. 9 in brackets, where (A^+) is the total positive ion current measured, assumed proportional to the orifice wall current;

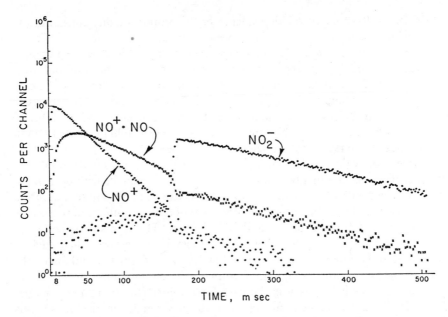

Fig. 8. Temporal profiles of ion intensities of $NO^+ \cdot NO$ and NO_2^- observed in large vessel stationary afterglow experiment by Puckett Lineberger [39]. Pressure of NO = 50 mtorr = total pressure. [Reprinted by permission from L. J. Puckett and W. C. Lineberger, *Phys. Rev.*, **A7**, 1635 (1970).]

(2) a sharp increase of negative ion current starting at 170 ms (0.4 ms) and a sharp decrease of positive ion current by a factor of about 2 at 170 ms (0.4 ms); (3) Linear decay of logarithm of ion current with same slope for positive and negative ions from 190 ms (1 ms) onward to the longest times observed.

Puckett and Lineberger's [39] analysis of the observed changes based on earlier work by Oskam [40] and Kregel [41] serves as a good introduction to the ion plasma phenomena involved in stationary afterglow experiments. The diffusion current density Γ (ions/area·time) may be expressed by Eqs. 13–15, where D is the diffusion coefficient for free diffusion, μ is the low-field mobility of the respective species, E is the electric field produced by the noncharge neutrality of the plasma, the term $D\nabla N$ describes the free diffusion due to a number density gradient, and $N\mu E$ gives the field induced charged particle drift in the gas:

$$\Gamma_+ = -D_+ \nabla N_+ + N_+ \mu_+ E \tag{13}$$

$$\Gamma_- = -D_- \nabla N_- - N_- \mu_- E \tag{14}$$

$$\Gamma_e = -D_e \nabla N_e - N_e \mu_e E \tag{15}$$

Although there are no applied electric fields in the "afterglow," a "self-field"

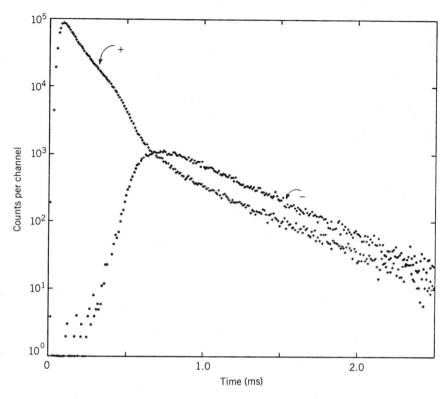

Fig. 9. Temporal profile of ion intensities observed with PHPMS (Kebarle [38]). The shapes of the positive and negative ion curves display the same features as those observed in the stationary afterglow (Fig. 8). However, the time scale is much shorter here. This is due to the much smaller dimensions of the PHPMS in source. $p(CH_4) = 4$ torr, 40 mtorr H_2O, traces of $C_2H_5ONO_2$, temperature 127°C. +, Positive ions $(H_3O^+(H_2O)_n)$; −, negative ions $(NO_2^-(H_2O)_n)$.

develops due to the initial rapid diffusion of the secondary electrons produced by the ionizing pulse. This diffusion produces a charge separation between positive ions and electrons that decreases the electron and enhances the positive ion diffusion current density. At sufficiently high charge densities, determined by the value of the Debye length (McDaniel [42]), this effect leads to the diffusion currents becoming the same, Eq. 16:

$$\Gamma_e = \Gamma_+ = \Gamma \qquad (16)$$

and since the initial number density of positive ions and electrons is the same, the number densities of positive ions and electrons are also the same, Eq. 17:

$$N_+ = N_e = N \qquad (17)$$

Using Eqs. 13, 14, 16, and 17, we can eliminate E thus obtaining Eq. 18:

$$\Gamma = -\left(\frac{D_+\mu_e + D_e\mu_+}{\mu_+ + \mu_e}\right)\nabla N \tag{18}$$

$$\Gamma = D_{+,e}\nabla N \tag{19}$$

The expression in parentheses in Eq. 18 is called the positive ion–electron ambipolar diffusion coefficient $D_{+,e}$, see Eq. 19. Substituting Einstein's Eq. 20, see McDaniel [42], into Eq. 18

$$\frac{\mu}{D} = \frac{e}{kT} \tag{20}$$

and taking into account that $D_e \gg D_+$, we obtain Eqs. 21 and 22:

$$\Gamma_+ = -2D_+\nabla N_+ \tag{21}$$

$$\Gamma_e = -2D_+\nabla N_e \tag{22}$$

which means that the ions and electrons diffuse twice as fast as the free ions would (at very low charge densities).

The free diffusion coefficient of the electrons D_e and the mobility of the electrons μ_e are a factor of 10^5 greater than those of the positive ions, and the electron diffusion and mobility terms are in opposition to each other due to the restraining direction of the self-field. Because of the large magnitude of D_e and μ_e, a departure of only ~ 1 part in 10^5 in the magnitude of the diffusion and mobility terms is sufficient to maintain $\Gamma_e = \Gamma_+$ (Puckett and Lineberger [39]).

The initial linear $\log(A^+)$ decay in Figs. 8 and 9, stage a, is due to positive ion–electron ambipolar diffusion.

In addition to being lost through positive ion–electron diffusion to the wall, electrons may be lost by electron attachment reactions that form negative ions. The electric field opposes the diffusion of negative ions, and therefore in the positive ion–electron ambipolar stage Eq. 23 holds. This is evidenced by

$$\Gamma_- \simeq 0 \tag{23}$$

the very low negative ion currents in stage a, seen in Figs. 8 and 9, even though a considerable number of negative ions is present in the ion source.

A collapse of the electric field in the plasma is evidenced at the termination of stage a; see Figs. 8 and 9. In the absence of an electric field, the remaining electrons are lost very rapidly due to their very high free diffusion coefficient. At the same time, the positive diffusion current Γ_+ decreases from $2D_+N_+$ to $D_+\nabla N_+$, which means a drop of positive ion intensity by a factor of 2 at this time; see Fig. 8. Furthermore, there is a continuous reduction of the diffusion current decay by a factor of 2, i.e., the slope of $\log(A^+)$ in the free diffusion stage c is only about half that of stage a; see Figs. 8 and 9. The collapse of the field also terminates the containment of the negative ions, and $\Gamma_- \simeq 0$ changes to $\Gamma_- \simeq \Gamma_+$ in stage c.

The relatively rapid ion intensity changes observed at the collapse of the electron diffusion caused self-field in stage b in Figs. 8 and 9 is not a desirable feature for the study of ion–molecule reaction kinetics. Furthermore, the two different diffusion coefficients in stages a and c are also not desirable. It should be noted that in Figs. 8 and 9 only very small concentrations of electron capturing compounds were present such that the reduction of the electron number density due to electron capture is slow. Addition of larger concentrations of electron capturing compounds leads to a very rapid capture of the free electrons. The large majority of the electrons can be captured very early, and then we observe positive ion–negative ion ambipolar diffusion as illustrated by the PHPMS results in Fig. 10. The positive–negative ion ambipolar diffusion coefficient (McDaniel [42]) is given by Eq. 24:

$$D_{+-} = \frac{D_+ \mu_- + D_- \mu_+}{\mu_+ + \mu_-} \quad (24)$$

It will be shown in Section 3.2 that the free diffusion coefficients of positive and negative ions are nearly equal (in practice), and the same is true for the mobilities, so that Eq. 24 reduces to Eq. 25:

$$D_{+-} \approx D_+ \approx D_- \quad (25)$$

The approximate validity of Eq. 25 is supported by the nearly equal slopes in the logarithmic plots of the positive and negative ions over the complete time interval observed in Fig. 10. Since positive–negative ion ambipolar diffusion must change at long t to free $+$ and $-$ diffusion, a change of slopes would have been observed at that point. The absence of such a change attests to the approximate equality in Eq. 25.

It is clear that the conditions depicted in Fig. 10 are better suited for the study of ion–molecule reactions involving either positive or negative ions, and such conditions were generally used in the kinetic studies; see Section 3.3. Since the electron capture rate constants can be as large as $k_{ec} = 10^{-7} \, \text{cm}^3$ molecule^{-1}s^{-1} (Christophorou [43]), we need to add only 10^{-4} torr (3×10^{12} molecules/cm^3) in order to capture the electrons within some 10 μs. Dissociative electron capture agents like CCl_4 or $CHCl_3$, Eq. 26,

$$e + CCl_4 = Cl^- + CCl_3 \quad (26)$$

were found the most suitable because of their high k_{ec}. In some rare cases, the electron capture agents may interfere with the reaction studied even when

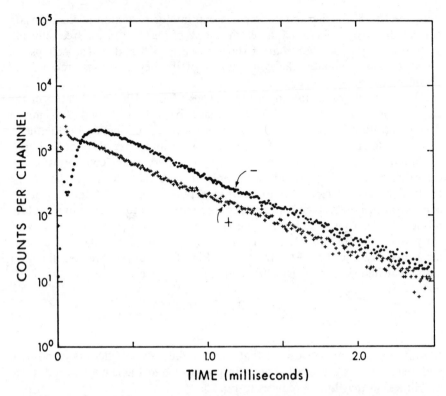

Fig. 10. Temporal profile of ion intensities observed with PHPMS. Due to larger concentrations of electron capture agent, $C_2H_5ONO_2$, secondary electrons are captured much faster than is the case in Fig. 9 so that no ambipolar positive ion–electron diffusion develops. $p(CH_4) = 4$ torr, 40 mtorr H_2O, few mtorr $C_2H_5ONO_2$, temperature 127°C. +, Positive ions ($H_3O^+ \cdot H_2O$); −, negative ions ($NO_2^- \cdot H_2O_n$).

present at very low concentrations. This is the case in association equilibria studies $A^+ + B = AB^+$, where B is much more weakly binding than the electron capture molecule.

The concentration change due to diffusion of a given ion, whose concentration (number density) is n, is given by Fick's diffusion law, Eq. 27,

$$\frac{\partial n}{\partial t} = D\nabla^2 n(x, y, z, t) \tag{27}$$

where D is the diffusion coefficient; see McDaniel [44]. Assuming that an ion distribution $n_0(x, y, z)$ was established at $t = 0$ when the ionizing radiation is turned off, we can separate the variables if D is independent of position; see Eq. 28. Substituting Eq. 28 into Eq. 27, we obtain Eqs. 29 and 30:

$$n(x, y, z, t) = n_0(n_x, n_y, n_z)T(t) \tag{28}$$

$$T(t) = e^{-t/\tau} \tag{29}$$

$$\nabla^2 n_0 + \frac{n_0}{D\tau} = 0 \tag{30}$$

Equation 29 gives the integrated time dependent function $T(t)$, where τ is the separation constant. The differential equation for the time independent positional dependence of the number density is given by Eq. 30. The solution of Eq. 30 depends on the shape of the container, which determines the boundary conditions. It is assumed that the ion density is zero at the walls of the container. Solutions of Eq. 30 are available for containers of various shapes (McDaniel and Mason [44]), including a cylindrical container of radius $r = r_0$ and length l, i.e., a shape similar to that used for the PHPMS ion source, where the electron beam goes axially through the cylinder at $r = 0$. The integration of the second order partial differential equation, eq. 30, and the imposition of the boundary conditions that fix the values of the two integration constants leads to a condition also on the separation constant τ (eigenvalue problem). τ is allowed to have only certain eigenvalues, and there is a solution (Eigenfunction) $n_0(n_x, n_y, z)_{k,l,m}$ for each eigenvalue $\tau_{k,l,m}$. However, the effect of the exponential form of the time-dependent Eq. 29 removes the contribution of the eigenfunctions for higher $\tau_{k,l,m}$ (higher diffusional modes) after some initial time interval since these decay much faster with time than the fundamental mode for which $\tau_{k,l,m}$ has the lowest value.

For the simplified case of a cylindrical cavity where the axial length l is much larger than r_0, Eq. 30 depends only on r. The solution for the fundamental diffusional mode is given by Eq. 31, where J_0 is the zero-order Bessel function, $n(0, 0)$ is the number density at $t = 0$ and $r = 0$, and the fundamental time constant τ_1 is given by Eqs. 32 and 33, where λ, the fundamental diffusion length of the cavity, depends on the dimensions of the cavity (McDaniel and Mason [44]):

$$R(r) = n(0, 0) J_0\left(\frac{2.405 r}{r_0}\right) \tag{31}$$

$$\tau_1 = \frac{\lambda^2}{D} \tag{32}$$

$$\frac{1}{\lambda^2}\left(\frac{2.405}{r_0}\right)^2 \tag{33}$$

The zero-order Bessel function J_0 in function of r is shown in Fig. 11.

The time-dependent solution for the fundamental diffusional mode of a cylindrical cavity of radius r_0 and length l in the z direction is obtained in a similar manner (McDaniel and Mason [44]) and is given by Eq. 34, while Eq. 35 gives the dependence of the fundamental time constant and the fundamental

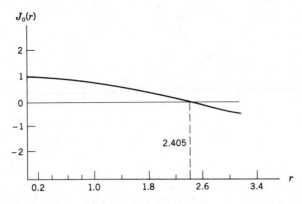

Fig. 11. Bessel function: $J_0(r)$.

diffusion length λ:

$$n(r, z, t) = GJ_0\left(\frac{2.405r}{r_0}\right)\left(\cos\frac{\pi z}{l}\right)e^{-t/\tau} \quad (34)$$

$$\tau = \frac{\lambda^2}{D} \quad \frac{1}{\lambda^2} = \left(\frac{2.405}{r_0}\right)^2 + \left(\frac{\pi}{l}\right)^2 \quad (35)$$

The origin of z is at $l/2$. Thus, $\cos(\pi z/l)$ equals zero at $+l/2$ and $-l/2$ and equal to 1 at $z = 0$. G is the number density at $t = 0$, $r = 0$, $z = 0$.

In the presence of ion–molecule reactive change (see, e.g., the reaction in Eq. 8), the differential Eq. 27 is extended with a reactive term as shown in Eq. 36:

$$\frac{\partial[A^-]}{\partial t} = D\nabla^2[A^-] - v_{\text{im}}[A^-] \quad (36)$$

where $v_{\text{im}} = k_{\text{im}}[B]$ is the ion molecule reaction frequency, and the number density n of the ion has been shown as $[A^-]$. The solution of Eq. 36 is similar to Eq. 27 and is shown in Eq. 37:

$$[A^-]_{r,zt} = [A^-]_0 J_0\left(\frac{2.405r}{r_0}\right)\left(\cos\frac{\pi z}{l}\right)e^{-vt} \quad (37)$$

$$\frac{1}{\tau} = v = \frac{D}{\lambda^2} + v_{\text{im}} \quad v_{\text{im}} = k_{\text{im}}[B] \quad (38)$$

Equation 37 is of identical form as Eq. 34, but the time constant τ given by Eq. 38 now includes the effect of the removal of A^- by the ion molecule reaction: $A^- + B = A + B^-$ or any other first-order reactive change. λ is the same as before (see Eq. 35), and $[A^-]_0$ is the concentration of A^- at $t = 0, r = 0, z = 0$.

As mentioned earlier, the direct observable is not the ion source concentration $[A^-]$ but the mass analyzed ion count rate (intensity) (A^-). Assuming that (A^-) is proportional to the wall current to the ion exit slit and the wall current is a diffusion current driven by the density gradient, we can evaluate the density gradient by differentiating Eq. 37 with respect to r at constant t. For a slit at $z = 0$, the z dependence may be neglected, and we obtain Eq. 39. Substituting some appropriate slit distance r_s into the differential with respect to r, we obtain Eq. 40:

$$(A^-)_t \propto \left(\frac{\partial [A^-]}{\partial r}\right)_t = \left(\frac{\partial [J_0(2.405r/r_0)]}{\partial r}\right)[A^-]_0 e^{-vt} \qquad (39)$$

$$(A^-)_t \propto R_0 [A^-]_0 e^{-vt} \quad \text{where } R_0 \text{ is a constant} \qquad (40)$$

$$v = \frac{D}{\lambda^2} + v_{\text{im}} = v_D + v_{\text{im}} \qquad (41)$$

Thus the observed intensity $(A^-)_t$ is proportional to $[A^-]_0 e^{-vt}$, and the observed changes of $(A^-)_t$ with time can be used to evaluate v. For example, a plot of $\ln(A^-)_t$ vs. t will have $-v$ as slope. Experiments in which v is determined using different known neutral reactant concentrations $[B]$ permit the evaluation of the diffusional term v_D and of the reactive term $v_{\text{im}} = k[B]$; see Section 3.31. Thus the experimental use of the ion intensities $(A^-)_t$ is justified by the results of the above treatment and particularly Eq. 40. Obviously, only rates of reactions that are first order in the ion concentration can be measured with the technique, since the separation of the time dependence as in Eq. 36 will not be possible with higher orders in the ion concentration. Furthermore, measurements of $(A^-)_t$ should be used only after the system has reached the fundamental diffusional mode. This stage will be recognized by the appearance of a straight line plot of $\ln(A^-)_t$ vs. t in the absence of ion–molecule reactive changes.

The discussion above was largely directed toward ion-source plasma conditions most suited for the measurement of ion–molecule reactions; i.e., conditions in which the electrons were rapidly converted to negative ions. A more general and very useful appraisal of the mass-spectrometer-stationary afterglow combination can be found in Smith and Plumb [45] and Adams et al. [46]. Hiraoka [24b] also provides a useful discussion of the plasma conditions in PHPMS in the absence of electron capturing agents.

3.2 Diffusion Coefficients of Ions and Ion–Molecule Rate and Equilibria Measurement

Equation 40 derived in Section 3.1 predicts an exponential time dependence of the ion current $(A^\pm)_t$ through the exit slit. The exponential coefficient v depends on the diffusional loss of the ions to the wall and on the ion–molecule reactive changes. The v for the simplest case of ion–molecule reactive change via the unidirectional reaction $A^{\pm} + B = A + B^{\pm}$ is given by Eq. 41.

Any additional ion–molecule reactions to which $A^{\pm}$ is the precursor will also be first order in the ion concentrations, due to the preponderance of the neutral reactants over the ion concentrations. Consecutive reactions $A^{\pm} \to B^{\pm} \to C^{\pm}$, consecutive and parallel reactions, and consecutive and parallel reversible reactions all lead to integrated rate expressions consisting of algebraic sums of exponential terms; see Rodiguin and Rodiguina [47]. It is clear that these terms containing subsequent changes can be added to Eq. 40 such that Eq. 40 in this extended form will account for these additional ion–molecule changes. The resulting equations predicting the ion intensity change with time $(A^{\pm})_t$, $(B^{\pm})_t$, $(C^{\pm})_t$, etc., will be the same as the conventional equations of chemical kinetics (Rodiguin [47]), but for each ion there will be one additional "reaction" corresponding to the diffusional loss to the wall. The reaction frequency for this reaction will be v_D; see Eq. 42, which follows from Eq. 41:

$$v_D = \frac{D}{\lambda^2} \tag{42}$$

An essential simplification, if this approach is to be taken, would occur for the case in which the diffusion coefficients of the different ions A^-, B^-, C^- are very similar so that they can be assumed to be the same. Thus, as far as diffusion is concerned and diffusional mode, the ions only change label, from A^- to B^- to C^-, but not their diffusional property. This is indicated by the scheme shown:

$$A^- \underset{v_{-1}}{\overset{v_1}{\rightleftarrows}} B^- \underset{v_{-2}}{\overset{v_2}{\rightleftarrows}} C^-$$
$$\downarrow v_D \quad \downarrow v_D \quad \downarrow v_D$$

Lindinger and Albritton [48] have reported reduced zero field mobilities μ_0 for positive and negative ions extending over a mass range from 2 to 300 amu observed in He and Ar. They found fairly good agreement with the Langevin expression, Eq. 43, see McDaniel and Mason [44], particularly for the heavier gas, Ar. From

$$\mu_0(\text{Langevin}) = \frac{13.876 \text{ cm}^2}{(\alpha m_r)^{1/2} \text{ V} \cdot \text{s}} \tag{43}$$

From Einstein's relationship, Eq. 20, and Eq. 43, we obtain Eq. 44 for the diffusion coefficients:

$$D = \frac{\mu_0 kT}{e} \propto \frac{kT}{e(\alpha m_r)^{1/2}} \tag{44}$$

where m_r is the reduced mass, α the polarizability of the gas, and e the charge of the electron. Equation 44 can be used to demonstrate the very small change of D with increase of ion mass. Most of the PHPMS experiments are performed with ions in the mass range between 30 and 600 amu. For the most commonly used bath gas, CH_4, the reduced mass change is small such that the predicted diffusion coefficient decreases by only 14% over the above mass range. Since the PHPMS ion–molecule experiments are generally conducted under conditions in which the reaction frequency v_{im} is much larger than the diffusive loss frequency, it follows that the assumption of equal diffusion coefficients for the ions is generally justified.

3.3 Some Typical PHPMS Ion–Molecule Kinetics Measurements

3.3.1 Treatment of Reactions with Inclusion of Ion Diffusion to Wall

A typical measurement of a rate constant of ion–molecule reactions proceeding at near unit collision efficiency is shown in Figs. 12 and 13 (Grimsrud

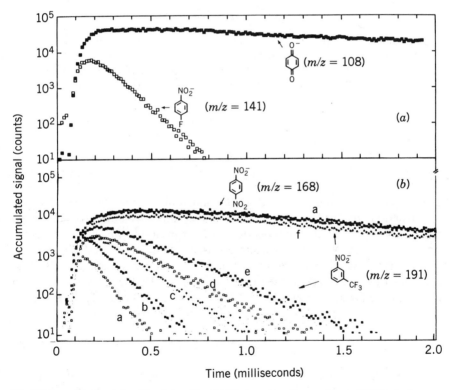

Fig. 12. Logarithmic plots of ion intensities: (a) Electron transfer reaction: p-FNB$^-$ + quinone = p-FNB + quinone$^-$, removes p-FNB$^-$ ion. (b) Electron transfer reaction: m-CF$_3$NB$^-$ + p-NO$_2$NB = m-CF$_3$NB + p-NO$_2$NB$^-$ removes m-CF$_3$NB$^-$ ion. Slope of linear ln plots leads to $v = v_D + k[p$-NO$_2$NB]. $p(p$-NO$_2$NB) in different runs: a, 0.5; b, 0.33; c, 0.25; d, 0.17; e, 0.12; f, 0.0 mtorr. CH_4, 4 torr; m-CF$_3$NB, 15 mtorr in all runs, temperature 150°C. [Reprinted by permission from E. P. Grimsrud, G. Caldwell, S. Chowdhury, and P. Kebarle, *J. Am. Chem. Soc.*, **107**, 4627 (1985). Copyright 1985 American Chemical Society.]

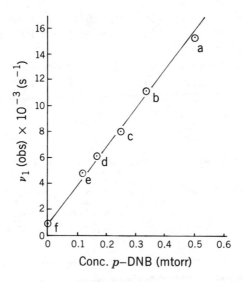

Fig. 13. Plot of $v = v_0 + k[p\text{-}NO_2NB]$ for v obtained at different $[p\text{-}NO_2NB]$ in Fig. 12b. Slope leads to rate constant for electron transfer Eq. 45, and intercept leads to ion diffusion loss frequency v_D. [Reprinted by permission from E. P. Grimsrud, G. Caldwell, S. Chowdhury, and P. Kebarle, *J. Am. Chem. Soc.*, **107**, 4627 (1985). Copyright 1985 American Chemical Society.]

[27]). The ion–molecule reaction is the electron transfer, Eq. 45, where NB stands for nitrobenzene:

$$m\text{-}CF_3NB^- + p\text{-}NO_2NB = m\text{-}CF_3NB + p\text{-}NO_2NB^- \qquad (45)$$

Since the electron affinity of $p\text{-}NO_2NB$ is much higher than that of $m\text{-}CF_3NB$, the reaction proceeds to completion. The intensity-time dependence of $m\text{-}CF_3NB^-$ observed in a number of separate runs a–f in which the concentration of $p\text{-}NO_2NB$ was stepwise decreased is shown in Fig. 12. The slopes of the linear logarithmic plots correspond to the decay frequency v of the $m\text{-}CF_3NB^-$ ion. This frequency is composed of the reactive and diffusive decay as shown in Eq. 46; see Section 3.1 and Eq. 40. A plot of v vs. the known $[p\text{-}NO_2NB]$ for runs a–f is shown in Fig. 13. As expected from Eq. 46,

$$v = k_{45}[p\text{-}NO_2NB] + v_D \qquad (46)$$

a straight line is obtained. The slope of the line is equal to the bimolecular rate constant k_{45}, and the intercept is equal to v_D. The results shown in Figs. 12 and 13 represent a more careful and somewhat more laborious determination of the rate constant k. Almost equally good results are obtained by obtaining v from one or two runs at two known, high concentrations of the neutral reactant, for example, runs a and b in Fig. 12, and evaluating k_{45} directly from Eq. 46. The $v_{D'}$, which is very small, is approximated on basis of v_D values determined from ion decays of ions of similar mass in the absence of reactive changes. Grimsrud et al. [49] gives Eq. 47, which provides an approximate v_D for the mid-mass range and a typical PHPMS ion source:

$$v_D = 1.84 \times \frac{10^{-2}T}{P} \quad (\text{s}^{-1}) \tag{47}$$

The numerical factor is for use with degrees Kelvin and torr and for CH_4 gas.

The rate constants of about 90 electron transfer reactions, $A^- + B = A + B^-$, were determined recently with the above method; Kebarle et al. [27, 50–54]. The majority of these were fast reactions proceeding at near collision rates. However, very slow reactions can also be measured without any difficulty. We need only increase the concentration of B. Since the PHPMS operates at relatively high pressures, the pressure of B can be increased into the torr range. This permits the determination of rate constants as low as $k = 10^{-14}$ cm$^3 \cdot$molecules^{-1}s^{-1}. For an example, see the slow electron transfer reactions involving SF_6 and perfluorocycloalkanes; Grimsrud et al. [50].

The temperature dependence of the rate constants can be also routinely determined by making measurements at different ion source temperatures. The temperature dependence of the rate constants is of particular interest when slow bimolecular ion molecule reactions are involved, since the data provide information on the internal energy barrier for the reaction. For slow electron transfer reactions, see Grimsrud et al. [50], and for S_N2 reactions involving negative ions, see Caldwell et al. [55].

3.3.2 Normalization to the Total Ion Intensity:

It was shown in Section 3.2 that the diffusion coefficients of ions with masses above 30 amu change very slowly with increase of mass in CH_4 or other light carrier gases and that, therefore, it is a good approximation to assume that the diffusion coefficients are the same. A convenient approach, using this assumption, is the normalization to the total ion current technique, where we work with the individual ion intensities expressed as a fraction or percentage of the total ion intensity. Since, as assumed, all ions decay with the same v_D due to diffusion to the wall, the normalization amounts to eliminating diffusion to the wall as a competitive reaction, and the treatment of the reaction system reduces to the conventional kinetic treatments as used for systems in which all reactants are neutral; Rodiguin and Rodiguina [47]. A simple proof can be given for the special case of the first-order bimolecular reaction:

$$A^+ + B \xrightarrow{v_{im}} C^+ + D$$

$$A^+ \xrightarrow{v_D} \text{wall}$$

$$C^+ \xrightarrow{v_D} \text{wall}$$

The integrated rate equations are given in Eqs. (48):

$$[A^+] = [A_0^+] \exp[-(\nu_{im} + \nu_D)t] \quad (48a)$$

$$[C^+] = [A_0^+]\{\exp(-\nu_D t) - \exp[-(\nu_{im} + \nu_D)t]\} \quad (48b)$$

Using Eq. 39, which showed that the ion intensities (A^+) and (C^+), respectively, are proportional to the expressions on the right side of Eq. 48, we can form the total intensity fraction of A^+ shown on the left of Eq. 49:

$$\frac{(A^+)}{(A^+)+(C^+)} = \exp(-\nu_{im} t) \quad (49)$$

Due to cancellations, see Eq. 48, the resulting Eq. 49 reduces to the simple form shown. Thus, plotting the natural logarithm of the fraction of the total

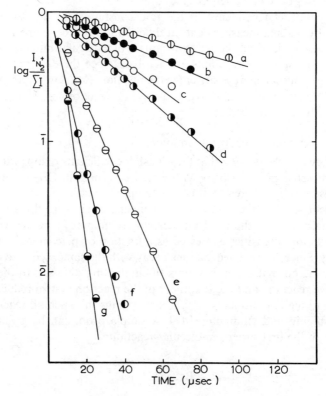

Fig. 14. Logarithmic plot of normalized N_2^+ ion intensity vs. t observed in pure N_2 at different $p(N_2)$: a, 0.65; b, 0.84; c, 1.04; d, 1.26; e, 2.05; f, 2.65; and g, 3.42 torr. Ion source temperature 380 K. Slopes provide ion–molecule reaction frequency ν_{im}. [Reprinted by permission from A. Good, D. A. Durden, and P. Kebarle, *J. Chem. Phys.*, **52**, 212, 222 (1970)].

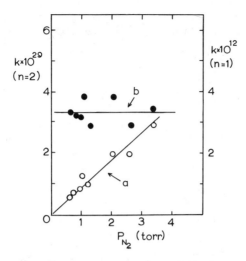

Fig. 15. Plot of data from Fig. 14, established order of reaction: $N_2^+ + nN_2 = N_4^+ + (n-1)N_2$. (a) Data treated taking $n = 1$ predict a rate constant increasing linearly with N_2. (b) Data treated taking $n = 2$ lead to an invariant (third) order rate constant. [Reprinted by permission from A. Good, D. A. Durden, and P. Kebarle, *J. Chem. Phys.*, **52**, 212, 222 (1970).]

ion intensity for A^+ vs. t, we obtain a linear plot whose slope equals $-v_{im}$. The fraction of the total ion intensity of A^+ has been often called "the normalized intensity of A^+."

The normalization to the total ion current has been extensively used in rate studies of simple bimolecular, trimolecular, and complex ion–molecule reaction systems. An illustration of an early measurement, Good et al. [56], that of the third body dependent reaction, Eq. 50,

$$N_2^+ + 2N_2 = N_4^+ + N_2 \tag{50}$$

are the results shown in Figs. 14 and 15. The logarithmic decay of the normalized ion intensity of N_2^+ at different N_2 pressures is shown in Fig. 14. The slopes of the linear plots lead to v_{im}. The results in Fig. 15 examine the order of the reaction and establish that the reaction is third order, i.e., second order in the N_2 concentration ($v_{im} = k[N_2]^2$). The temperature dependence of k was determined also; Good et al. [56].

The results displayed in Figs. 16 and 17 illustrate the rate constant determination for the reversible reaction Eq. 51:

$$O_2^+ + 2O_2 \rightleftharpoons O_4^+ + O_2 \tag{51}$$

Since the reaction reaches equilibrium quite rapidly, the forward reaction is not easily isolated. However, the standard kinetic plot for first-order reversible reactions shown in Fig. 17 permits the determination of the forward rate constant

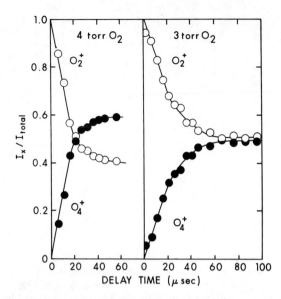

Fig. 16. Plot of normalized ion intensity of O_2^+ and O_4^+ observed at 3 and 4 torr of pure O_2 gas at 298 K. Formation of O_4^+ due to reaction $O_2^+ + 2O_2 = O_4^+ + O_2$. Reaction reaches equilibrium at long t. [Reprinted by permission from D. A. Durden, P. Kebarle, and A. Good, *J. Chem. Phys.*, **50**, 302 (1969).]

(from Durden [13]). The temperature dependence of the forward rate constants based on measurements like those in Figs. 16 and 17 undertaken over a range of temperatures was obtained also (Durden [13]) and is shown in Fig. 18. For rate constants leading to O_6^+ see Payzant et al. [57].

Results for a complex but important reaction system are shown in Fig. 19. Given are the observed normalized ion intensities as a function of time of the ions in moist oxygen gas, Good et al. [56]. The results in Figs. 19a and b directly demonstrate the long reaction sequences by which O_2^+ is converted to $H^+(H_2O)_n$. The complete scheme, which involves 15 consecutive and parallel and reversible reactions, is given in Table 1; Good et al. [56]. The rate constants for 12 out of the 15 reactions could be determined from an analysis of intensity–time plots like those in Fig. 19. The detailed approach by which this complex reaction system was unraveled is given in Good et al. [56]. Here we wish to bring out some of the general features of the approach. Starting with the primary ion O_2^+, plots of the natural log of the normalized intensity of O_2^+ are found to be linear and lead to v_a. Examination of the concentration dependence of v_a by determining v_a at different concentrations of O_2 and H_2O shows that $v_a = k_a[O_2]^2$. This means that O_2^+ disappears by the three-body

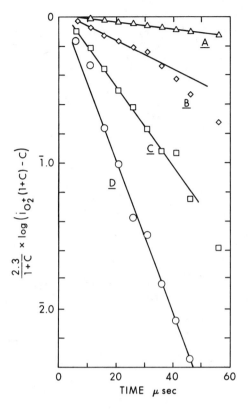

Fig. 17. Determination of the forward rate constant k_f for the reaction $O_2^+ + 2O_2 = O_4^+ + O_2$ from data like those in Fig. 16. C is the equilibrium ratio: $(O_2^+)/(O_4^+)$. Slope of plots leads to $v = k_f[O_2]^2$. [Reprinted by permission from D. A. Durden, P. Kebarle, and A. Good, *J. Chem. Phys.*, **50**, 302 (1969).]

reaction a:

$$O_2^+ \xrightarrow[k_a[O_2]^2]{v_a} O_4^+ \xrightarrow[k_b[H_2O]]{v_b} O_2^+ \cdot H_2O \xrightarrow[k_c[H_2O][O_2]]{v_c} O_2^+(H_2O)_2$$

$$O_2^+ + 2O_2 = O_4^+ + O_2 \tag{a}$$

$$O_4^+ + H_2O = O_2^+ \cdot H_2O + O_2 \tag{b}$$

$$O_2^+ \cdot H_2O + H_2O + O_2 = O_2^+(H_2O)_2 + O_2 \tag{c}$$

Subjecting the normalized intensity of O_4^+ to the integral plot, vide infra, we find v_b; and by examining the concentration dependence of v_b, we establish

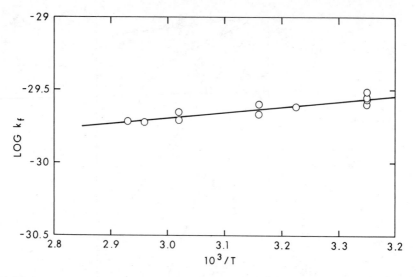

Fig. 18. Temperature dependence of third-order rate constant k_f for reaction $O_2^+ + 2O_2 = O_4^+ + O_2$. [Reprinted by permission from D. A. Durden, P. Kebarle, and A. Good, *J. Chem. Phys.*, **50**, 302 (1969).]

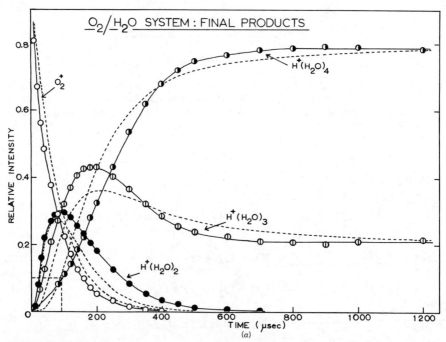

Fig. 19. Normalized ion intensities observed in moist oxygen. (*a*) 2 torr O_2, 4 mtorr H_2O, 307 K. Dashed lines represent theoretical fits based on mechanism. (*b*) Initial products of reaction system corresponding to changes occurring in square shown at lower left corner in Fig. 19*a*. [Reprinted by permission from A. Good, D. A. Durden and P. Kebarle, *J. Chem. Phys.*, **52**, 212, 222 (1970).]

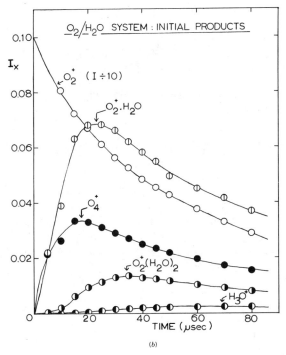

Fig. 19. (*Contd.*)

that the reaction removing O_4^+ is the ligand exchange reaction b. Furthermore, the known concentration dependence of $v_b = k_b[H_2O]$ leads to a determination of k_b. Proceeding in this manner from short to long times in the ion intensity plots we can obtain rate constants for all the important reactions that participate in the ion–molecule mechanism.

The "integral plot" is a very useful method for obtaining rate constants from ion intensities time plots like that shown in Fig. 19. Consider the general reaction sequence

$$A^+ \xrightarrow{v_1} B^+ \xrightarrow{v_2} C^+ \xrightarrow{v_3} D^+$$

Assuming that there is visible intensity of B^+, we can obtain v_2 by plotting the normalized intensities of products of reaction 2; i.e., $(C^+)_t$ and $(D^+)_t$ vs. the time integral of $(B^+)_t$. According to Eq. 52,

$$(C^+) + (D^+) = v_2 \int_0^t (B^+) \, dt \tag{52}$$

which is simply the integrated rate Eq. 53,

$$\frac{d[(C^+)+(D^+)]}{dt} = v_2(B^+) \tag{53}$$

the integral plot should lead to a straight line whose slope is v_2. An example of an integral plot for the reaction 8

$$H^+(H_2O)_2 \xrightarrow{v_8} H^+(H_2O)_3 \rightarrow H^+(H_2O)_4 \rightarrow H^+(H_2O)_n \tag{54}$$

from the study of Good et al. [56] is given in Fig. 20, where the integral of $H^+(H_2O)$ is plotted versus the sum of the intensities of all the products resulting from reaction 8; i.e., the sum of all the hydrates $H^+(H_2O)_n$ with $n > 2$. The different slopes give v_8 obtained at different concentrations of O_2 and H_2O. In this manner the dependence $v_8 = k_8[H_2O][O_2]$ was established and the value of k_8 obtained.

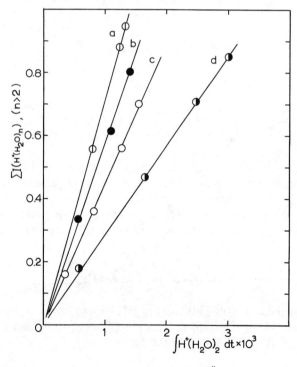

Fig. 20. Integral plot used to determine v for reaction $H^+(H_2O)_2 \xrightarrow{v} H^+(H_2O)_n$, where $n > 2$. Runs a–d obtained with different H_2O and O_2 concentrations. Slopes obtained correspond to v. [Reprinted by permission from A. Good, D. A. Durden, and P. Kebarle, *J. Chem. Phys.*, **52**, 212, 222 (1970).]

More efficient determinations of mechanism and rate constant for complex reaction sequences can be obtained by "wiring" the reaction mechanism into an analog computer. Preliminary inspection of the ion intensities time plots is required to establish the most likely mechanism. The visual display of the output of the analog computer permits rapid choice of the reaction frequencies v (resistor settings), which lead to ion intensity plots most closely matching the experimental results. Such an analog computer fit is shown in Fig. 21 for the ion intensities observed in a reaction system where NO^+ is converted to $H^+(H_2O)_n$; see French et al. [58]. Analog computer fitting or graphical methods were also used for rate constant determinations in several other complex systems. (Kebarle et al. [59–62]).

Attempts were made to develop a general routine program for fitting normalized ion intensities with a digital computer; Magnera [63]. Approaches using the available analytical forms of the integrated rate equations (Rodiguin and Rodiguina [47]) and treating the rate constants as the unknowns were not successful. The successful digital computer fitting programs presently used in this laboratory are equivalent to the graphical methods described above and particularly the integration method illustrated by Eqs. 52 and 53 and Fig. 20. The integral of (B^+) is obtained by summing the collected counts in the channels of the multiscalar that were used for the B^+ ion. Since the time width per

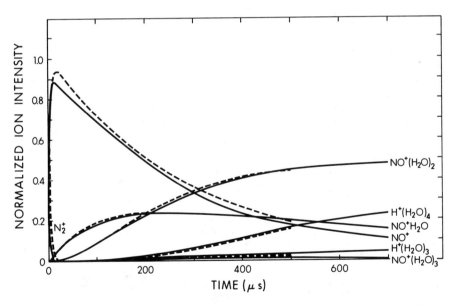

Fig. 21. Normalized intensities observed in moist nitrogen containing some NO. Pressures: N_2, 2; NO, 0.1; H_2O, 0.004 torr, temperature 306 K. Reactions sequence: $N_2^+ \rightarrow NO^+ \rightarrow NO^+(H_2O) \rightarrow NO^+(H_2O)_2 \rightarrow NO^+(H_2O)_3 \rightarrow H^+(H_2O)_3 \rightarrow H^+(H_2O)_4$. Dashed curves obtained from analog computer fit of above mechanism. [Reprinted by permission from M. A. French, L. P. Hills, and P. Kebarle, *Can. J. Chem.*, **51**, 4561 (1973).]

channel is only 10 μs, this summation leads to a satisfactory integration (Magnera [63]).

Meot-Ner (Mautner) and Field have reported extensive studies of ion–molecule reaction rates with use of the PHPMS. The significance of that work to the development of ion–molecule reaction theory has been well summarized in a review chapter, Meot-Ner [64]. Here we will consider some aspects of technique. The apparatus and experimental conditions have been only briefly described in Meot-Ner et al. [18, 65, 66]. These authors worked at pressures (0.8–2.2 torr), which are lower by a factor of about 3 than those used in the PHPMS experiments in the laboratory of the present author. The lower pressure should lead to a faster decrease of the total ionization. Correspondingly, Meot-Ner et al. adjusted the neutral reactant partial pressures to be higher such that the studied reaction half-lives are 20–50 μs, i.e., considerably shorter than those generally used in the present laboratory. A large amount of very useful data concerning the temperature dependence of hydride transfer reactions,

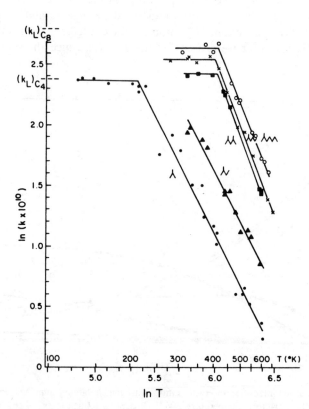

Fig. 22. Temperature dependence of rate constants for hydride transfer reactions: $s\text{-}C_3H_7^+ + RH = C_3H_8 + R^+$. RH is the hydrocarbon indicated in the figure. [Reprinted by permission from M. Meot-Ner and F. H. Field, *J. Chem. Phys.*, **64**, 277 (1976).]

Eq. 55, were obtained by working with a suitable reaction mixture of constant composition:

$$R_0^+ + RH = R_0H^+ + R^+ \tag{55}$$

The total pressure of the mixture was adjusted to maintain a constant reactant concentration [RH] in the runs undertaken at different constant temperatures. The rate constants were then evaluated from logarithmic plots of the normalized ion intensities (R_0^+). Linear slopes gave v, which led to k via $v = k[RH]$. Occasional checks with different concentrations [RH] were made in order to check that v is proportional to [RH] (Meot-Ner [65]). Some of the results for the hydride transfer reactions, Eq. 55, are summarized in Fig. 22. The reactant ion R_0^+ in all the determinations is s-$C_3H_7^+$, whereas the different RHs used are shown in Fig. 22. The results in Fig. 22 and earlier similar results by Meot-Ner [64] were particularly interesting since they showed for the first time that many slow bimolecular ion–molecule reactions have rate constants that increase substantially with temperature decrease and reach unit collision efficiency rates at low temperatures.

4 ION–MOLECULE EQUILIBRIA MEASUREMENTS WITH THE PHPMS

4.1 Transfer (Exchange) Ion–Molecule Equilibria

4.1.1 Kinetic Aspects

Equilibria of the form shown in Eq. 56,

$$A^\pm + B \underset{k_r}{\overset{k_f}{\rightleftharpoons}} C^\pm + D \tag{56a}$$

$$K = \frac{[C^+]_e[D]}{[A^+]_e[B]} \tag{56b}$$

$$K = \frac{k_f}{k_r} \tag{56c}$$

require measurement conditions that are somewhat different than the association equilibria $A^\pm + B = AB^\pm$, and therefore these two classes of equilibria will be discussed in separate sections. In general, chemical reactions of the Eq. 56 type that are reversible are transfer or exchange reactions, i.e., reactions in which a given, generally simple, group is exchanged between the reactants. As already mentioned in the introduction, important classes of such reactions are ligand exchange reactions; see Eq. 2, proton transfer reaction, Eqs. 3 and 4, electron transfer reactions, Eqs. 5 and 6, hydride and halide transfer reactions; Eq. 7.

In practice, the measurement of equilibria, Eq. 56, is dependent not only on the presence of reversibility but also on the absence of dominant competing side reactions and on rates in the forward and reverse direction that are faster than the ion diffusion rates v_D to the wall. Since the neutral concentrations [A] and [B] are many orders of magnitude larger than the ion concentrations, the forward and reverse reactions are first order and have pseudo-first-order rate constants: $v_f = k_f[B]$ and $v_r = k_r[D]$. The relaxation time τ_{eq} of the equilibrium is thus given by Eq. 57a:

$$\tau_{eq} = \frac{1}{k_f[B] + k_r[D]} \tag{57a}$$

$$\tau_D = \frac{1}{v_D} \tag{57b}$$

$$\tau_D \approx 10^{-3} \, \text{s} > \tau_r \tag{57c}$$

It was shown in Section 3.3, see Eq. 47, that v_D is in the neighborhood of $10^3 \, \text{s}^{-1}$, such that $\tau_D \approx 10^{-3}$ s. Thus, conditions must be selected where $\tau_{eq} < \tau_D$ in order to minimize the effect of the somewhat different diffusion coefficients of the ions.

For reactions Eq. 56, which are only mildly exoergic such that the equilibrium constant K is not larger than say $K = 50$ ($\Delta G^0 \approx 3$ kcal/mol for $T \approx 400\,K$), the equilibrium conditions are easily met when the exothermic direction has a rate constant of the Langevin-collision rate magnitudes $k_f \approx 10^{-9}\,\text{cm}^3$ molecule^{-1}s^{-1}. Such rate constants occur for the vast majority of the proton transfer reactions between oxygen and nitrogen bases, Eqs. 3 and 14, for the vast majority of electron transfer reactions, Eqs. 5 and 6, for halide transfer reactions, Eq. 7, and for ligand exchange reactions where the number of ligands is small. For example, a choice of the typical conditions: $p_B = 2$ mtorr, $[B] \approx 6 \times 10^{13}$ molecules/cm^3 and $p_D \approx 10$ mtorr, $[D] \approx 30 \times 10^{13}$ molecules/cm^3, for $K = 50$, leads at equilibrium to $[C^+]/[A^+] = 10$, which is an easily measured ratio, and $\tau_{eq} = 1.6 \times 10^{-5}$ s, evaluated from Eq. 57a. The relaxation time is much shorter than τ_D as required. Thus, it is advantageous to select equilibria where K is not too large. The majority of the measured exchange equilibria are determined under such conditions. Examples of the ion intensity time dependence for electron transfer equilibria were given in Figs. 1b and c.

Although the ΔG^0 values that are measured in this manner are each only a few kilocalories per mole, series of interconnected equilibria can be obtained, and thus this permits the construction of an ΔG^0 scale (ladder) that may extend over as much as 100 kcal/mol. Such a scale for electron transfer reactions involving negative ions is shown in Section 4.1.2.

Occasionally it is desirable to measure much larger ΔG^0 steps. This need may arise when intermediate compounds permitting a small steps ladder do not exist, as in the case for proton affinity determinations in the region between water and methane (McMahon and Kebarle [67]) or in situations in which suitable intermediate compounds may exist but are not available to the experimentalist.

Compared to the other techniques, ICR and flowing afterglow, the PHPMS permits the measurement of the largest ΔG^0 steps, i.e., the largest K. The results in Fig. 23 give an example. The equilibrium:

$$tert\text{-butyl}^+ + 2\text{-Me-norbornane} = \text{isobutane} + 2\text{-Me-2-norbornyl}^+$$

between the *tert*-butyl cation and 2-Me-norbornane was measured (Sharma et al. [68]) and led to a $K = 7.5 \times 10^5$.

The ability of the PHPMS to measure such large K values derives from the possibility to measure large equilibrium $R_i = (C^+)_{eq}/(A^+)_{eq}$ intensity ratios and the possibility to use large neutral $R_n = [D]/[B]$ ratios. For example, in Fig. 23 $R_i = 1420$ and $R_n = 527$. Large R_i ratios can be measured only when the reaction system is very clean. For example, in the hydride transfer

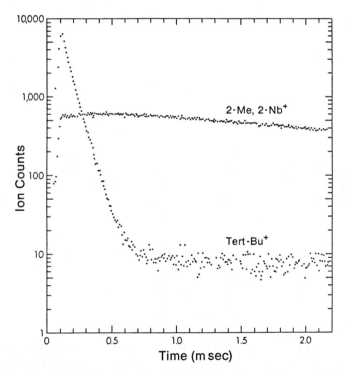

Fig. 23. Plot of logarithm of ion intensities for hydride transfer reaction: *tert* Bu$^+$ + 2-Me-2-norbornane = iso-butane + 2-Me-2-norbornyl$^+$. Equilibrium is achieved were intensities become equidistant (~ 1 ms). MeNB$^+$ was collected for 10 s and *tert*-Bu$^+$ for 200 s; iso-butane, 4.2 torr; norbornane, 8 mtorr. Resulting equilibrium constant: $K = 7.5 \times 10^5$. [Reprinted by permission from R. B. Sharma, D. K. Sen Sharma, K. Hiraoka, and P. Kebarle, *J. Am. Chem. Soc.*, **107**, 3747 (1985). Copyright 1985 American Chemical Society.]

after an initial $\sim 200\,\mu s$ only the two ions participating in the equilibrium were observed over the whole mass range. Thus, the chances that a low-intensity ion of the same mass as $tert\text{-Bu}^+$ may be contributing to the ion intensity is extremely small. Instrumentally the measurement of large R_i ratios presents no special problem. When the intensity of the less abundant ion is very low, we collect this ion for a longer time in the multiscalar. Subtraction of the noise from the accumulated counts permits measurement of R_i that are much larger than 10^3. Large neutrals ratios up to $R_n = 10^4$ are possible by using partial pressures for compound D that are in the torr range. It is often possible to use D as the carrier gas.

Although the partial pressure of D can be increased up to the torr range, i.e., the maximum pressure that the instrument will tolerate, the pressure of B cannot be decreased to any desired value since the reaction should reach equilibrium within no longer than a millisecond.

Using the well-known expression for reversible first-order reactions, Eq. 58a,

$$(k_f[B] + k_r[C])t = \ln\left(\frac{[A^+]_0 - [A^+]_{eq}}{[A^+]_t - [A^+]_{eq}}\right) \tag{58a}$$

$$R_n = \frac{[D]}{[B]} \tag{58b}$$

$$R_i = \frac{[C^+]_{eq}}{[A^+]_{eq}} \tag{58c}$$

we can, for cases like that in Fig. 23 where R_i is very large, make the good approximations shown in Eq. 59. Assuming we wish to calculate the time t_2, where $[A^+]_{t_2}$ has decreased to a value where it is twice as large as $[A^+]_{eq}$, we can easily show that Eq. 58a and 59 lead to the simple Eq. 60 for t_2:

$$[C^+]_{eq} = [A^+]_0 \quad \text{and} \quad R_i = \frac{[A^+]_0}{[A^+]_{eq}} \tag{59}$$

$$t_2 = \frac{(\ln R_i)}{k_f[B]} \quad \text{where } [A^+]_{t_2} = 2[A^+]_{eq} \tag{60}$$

It should be noted that $[C^+]_{t_2} \approx [C^+]_{eq}$. The absence of k_r from Eq. 60 may appear surprising. The elimination of k_r is obtained through the equilibrium expression Eqs. 56b and c. At t_2 the equilibrium quotient Q will be equal to only half of the value of the equilibrium constant, which may be considered as a maximum acceptable error in K. Even for forward reactions proceeding at collision rates, $k_f \approx 10^{-9}$ molecule$^{-1} \cdot$cm$^3 \cdot$s^{-1}, a large R_i ratio, say $R_i \approx 1000$, leads to a requirement that $[B]$ should not be much lower than 3×10^{13} molecule/cm^3, i.e. $p_B \approx 1$ mtorr if t_2 predicted by Eq. 60 is to be less than 1 ms.

The ability to use relatively high reactant gas pressures is of decisive advantage also in the measurement of exchange equilibria, Eq. 56, where the exothermic direction proceeds at lower than collision rates. For example, the hydride transfer reactions, Eq. 55, generally proceed at low rates at ordinary temperatures; see Fig. 22. In spite of this, Solomon et al. [69–71] using PHPMS were able to measure a large number of hydride transfer equilibria and provide important data on the thermochemistry of carbocations.

Similarly, proton transfer reactions involving carbon acids are slow even in the exothermic direction (Brauman [72]), and some of the proton transfer equilibria involving carbon acids are difficult to measure with the ICR technique. Use of the PHPMS apparatus (McMahon and Kebarle, Cumming and Kebarle [72]) at higher reactant gas pressures permitted the measurement of such equilibria.

Exothermic methyl cation transfer reactions, which probably proceed by an $S_N 2$ type of mechanism, often have large internal energy barriers and are slow (Sen-Sharma and Kebarle [61]). In spite of this, PHPMS measurements of methyl cation transfer equilibria at suitably high reactant pressures were

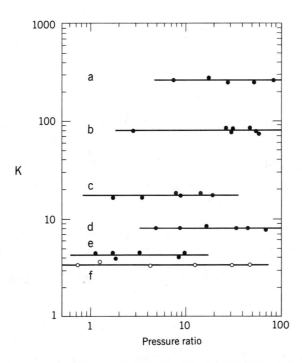

Fig. 24. Equilibrium constants K for equilibria $B_1 H^+ + B_2 = B_1 + B_2 H^+$ involving proton transfer between various nitrogen bases. Plot demonstrates that K remains invariant for different neutral reactant partial pressure ratios. [Reprinted by permission from R. Yamdagni and P. Kebarle, *J. Am. Chem. Soc.*, **95**, 3504 (1973). Copyright 1973 American Chemical Society.]

successful and have provided a methyl cation affinity scale involving some 30 Lewis bases (McMahon et al. [73]).

4.1.2 Evaluation of Thermochemical Data

Equilibrium constants K obtained from determinations of the ion intensity ratios after the equilibrium has been achieved (see Section 4.10 and Fig. 1 and 23) can be determined at the same temperature for different neutral reactant ratios. Independence of K from the neutral reactant concentrations represents one more check of equilibrium conditions. The results shown in Fig. 24 show data representing such a check. The measurements (Yamdagni and Kebarle [74]) involved proton transfer between different nitrogen bases. The data in

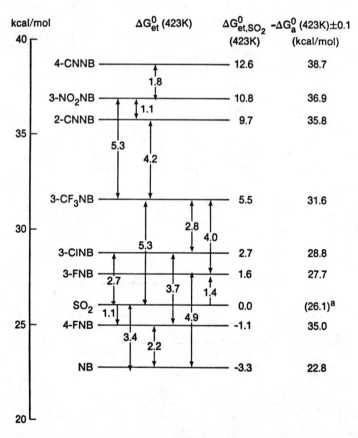

Fig. 25. Free energy ladder (scale) from ΔG_{et}^0 obtained from individual measurements of electron transfer equilibria: $A^- + B = A + B^-$. NB stands for nitrobenzene; $\Delta G_{etSO_2}^0$ corresponds to $A^- + SO_2 = A + SO_2^-$, and ΔG_a^0 corresponds to $e + A = A^-$. Absolute ΔG_a^0 obtained by calibration to literature value for SO_2. [Reprinted permission from S. Chowdhury, T. Heinis, E. P. Grimsrud and P. Kebarle, J. Phys. Chem., **90**, 2747 (1986). Copyright 1986 American Chemical Society].

Fig. 24 demonstrate that K remains independent for changes of neutral concentration ratio by factors as large as 500.

The equilibrium constants K can be used to evaluate the free change ΔG^0; see Eq. 61:

$$\Delta G^0 = -RT \ln K \qquad (61)$$

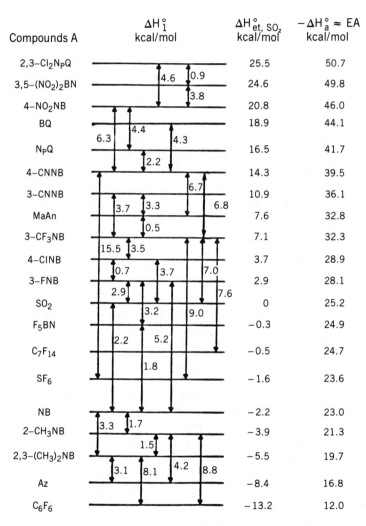

Fig. 26. Enthalpy changes for electron transfer reactions $A^- + B = A + B^-$ obtained from van't Hoff plots shown in Fig. 28. ΔH_a values correspond to electron attachment reaction $A + e = A^-$. These absolute values based on literature value for $e + SO_2 = SO_2^-$. [Reprinted by permission from S. Chowdhury, T. Heinis, E. P. Grimsrud, and P. Kebarle, *J. Phys. Chem.*, **90**, 2747 (1986). Copyright 1986 American Chemical Society.]

The determination of a number of such equilibria involving a series of compounds is then used to construct a ΔG^0 ladder or scale. Such a ladder is shown in Fig. 25 for electron transfer equilibria (Grimsrud et al. [27]). The ladder shown includes only nine compounds. The full ladder of electron transfer ΔG^0 values contains about 120 compounds and extends over 50 kcal/mol (Kebarle and Chowdhury [54]). The electron transfer equilibria were determined at 150°C, a temperature at which the measurements were most conveniently made, since at high temperatures compounds with low electron affinities thermally detach electrons (Grimsrud [49]) and at lower temperatures adduct ions A_2^-, AB^-, B_2^-, form removing thus the reactants A^- and B^- and (Kebarle and Chowdhury [54]).

For many of the compounds the equilibria can be measured over a range of temperatures, and this allows the construction of van't Hoff plots and the experimental determination of the enthalpy and entropy change ΔH^0 and ΔS^0 for the reactions. The availability of the ΔH^0 and ΔS^0 values in turn can be used for the construction of ΔH^0 and ΔS^0 ladders. Such ladders are shown in Figs. 26 and 27 for the electron transfer equilibria. Some of the van't Hoff plots are shown in Fig. 28 (from Chowdhury et al. [75]).

The determinations shown in Figs. 25–27 contain multiple thermodynamic cycles that permit checks of the thermodynamic consistency of the results. For example, in Fig. 25 there are three measurements providing the ΔG^0 difference between SO_2 and 3-CF_3NB—a direct one of 5.3 kcal/mol and two others, one via the intermediate compound 3-ClNB leading to 5.5 kcal/mol and the other via 3-FNB leading to 5.4 kcal/mol. The consistency of the ΔG^0 measurements appears to be within 0.2–0.3 kcal/mol, and inconsistencies of this magnitude have also been generally found for other transfer or exchange equilibria.

The ΔH^0 and ΔS^0 values obtained from the van't Hoff plots have somewhat larger errors. The results in Fig. 28 for reaction 6 illustrate part of the problem. Reaction 6 was measured on two different instruments of similar design by two different investigators. Although the points of the two determinations are very close, we find that the ΔG^0 differs, on the average by ~ 0.3 kcal/mol whereas the ΔH^0 and the ΔS^0 differ by 1.6 kcal/mol and 2 cal/degree (Chowdhury et al. [75]). The reasons for these small discrepancies are not clear. Obviously, the available temperature range over which reliable measurements can be obtained is often rather narrow, a common problem for data obtained from reaction equilibria and van't Hoff plots. Determination of van't Hoff plots with the same instrument also leads to reproducibility that is seldom better than 1 kcal/mol for the ΔH and 1–2 cal/degree·mol in ΔS^0; see, for example, Fig. 2 in Chowdhury et al. [75]. Correspondingly, the consistency in ΔH^0 scales obtained from van't Hoff plots is generally in the 1 kcal/mol and of ΔS^0 scales in the 2 cal/deg·mol range. The absolute errors cannot be estimated but are probably roughly twice as large.

The equilibria measurements and the resulting ΔG^0, ΔH^0, and ΔS^0 provide relative values for quantities such as electron affinities, proton affinities, and hydride affinities. Absolute values are obtained by anchoring the scales to known

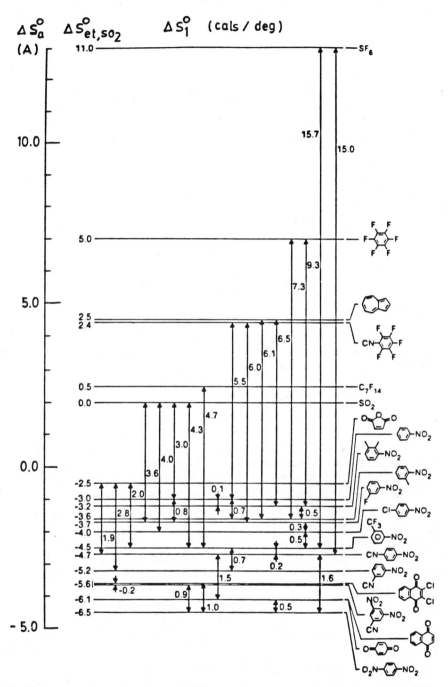

Fig. 27. Entropy changes for electron transfer reactions $A^- + B = A + B^-$. ΔS_a corresponds to entropy change for reaction $e + A = A^-$. [Reprinted by permission from S. Chowdhury, T. Heinis, E. P. Grimsrud, and P. Kebarle, *J. Phys. Chem.*, **90**, 2747 (1986). Copyright 1986 American Chemical Society.]

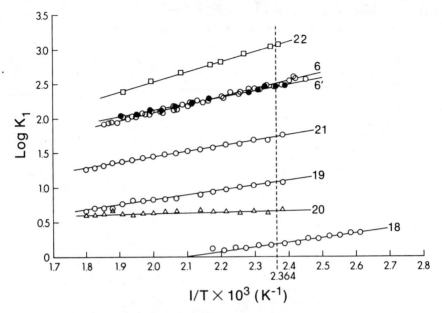

Fig. 28. Van't Hoff plots of equilibrium constants for electron transfer reactions $A^- + B = A + B^-$. Plots 6 and 6' are based on measurements for the same reactions by two different investigators in the author's laboratory on two different PHPMS instruments. [Reprinted by permission from S. Chowdhury, T. Heinis, E. P. Grimsrud, and P. Kebarle, *J. Phys. Chem.*, **90**, 2747 (1986). Copyright 1986 American Chemical Society.]

literature absolute values for one or more compounds that are part of the scale. For example, the electron affinity scales were anchored (Chowdhury et al. [75]) to the known electron affinity of SO_2, which had been determined by electron photodetachment (Celotta et al. [76]), whereas the entropy scale was calibrated to the entropy for electron attachment $e + SO_2 = SO_2^-$, which can be calculated from available vibrational frequencies and moments of inertia for SO_2 and SO_2^-; see Table III in Chowdhury et al. [75].

The availability of more than one good external standard within the scale permits us to check the reliability of the data obtained from the equilibria measurements.

Scales have been obtained for proton transfer reactions involving positive ions, see Eq. 3 (gas phase basicities and proton affinities of neutral bases) and proton transfer involving negative ions, see Eq. 4 (gas acidities of neutral acids and proton affinities of anions). As a result of proton transfer equilibria (Eq. 3) measurements over more than a decade by PHPMS [67, 74, 77–80], ICR [81, 82], and flowing afterglow [83], a continuous scale of proton affinities starting from the lowest proton affinity H_2 to the highest 1,8-*bis* (dimethylamino) naphthalene (proton sponge), containing some 700 compounds and covering a proton affinity difference of 150 kcal/mol, has been obtained. The agreement between different methods generally has been good [81, 84]. For a critical review of the data, see Lias et al. [84].

A continuous scale of gas phase acidities obtained from equilibria (Eq. 4), PHPMS [72b, 85], and ICR [86, 87] extending from the weak acid methanol to the strongest acid $C_2F_5CO_2H$ and containing some 300 acids has also been obtained.

Electron transfer equilibria involving positive ions (Eq. 6) have been measured by PHPMS (Meot-Ner [88, 89]). These measurements are particularly useful for systems in which the lowest energy structure of the positive ion is very different from that of the molecule, since in such cases the adiabatic ionization energies cannot be obtained by conventional spectroscopic techniques due to Franck-Condon restrictions.

Examples of electron transfer equilibria leading to electron affinities were considered earlier in this section. The hydride and chloride transfer equilibria were also discussed.

Ligand transfer reactions, see, for example, Eq. 62,

$$X^-A + B = X^-B + A \tag{62}$$

$$F^-HOH + HF = F^-HF + H_2O$$

have been extensively studied by ICR (Larson and McMahon [90]). PHPMS determinations have also been made; Caldwell and Kebarle [91], Hiraoka et al. [92].

4.2 Ion–Molecule Association and Clustering Equilibria

4.2.1 Kinetic Aspects and Some Typical Measurements

The ion–molecule association reactions of the type Eqs. 63 and 64,

$$A^\pm + B + M = (AB)^\pm + M \tag{63}$$
$$(AB_{n-1})^\pm + B + M = (AB_n)^\pm + M \tag{64}$$

require third-body collisions for the removal of the exothermicity of the association reaction. The energy transfer mechanism using the strong collision assumption represents the kinetics; see Eq. 65:

$$A^\pm + B \underset{k_d}{\overset{k_c}{\rightleftarrows}} (AB^\pm)^* \underset{k_a[M]}{\overset{k_s[M]}{\rightleftarrows}} AB^\pm \tag{65}$$

A steady-state treatment leads to Eq. 66a for the rate constant of the forward reaction:

$$k_f = \frac{k_c k_s[M]}{k_d + k_s[M]} \tag{66a}$$

$$k_f = \frac{k_c k_s [M]}{k_d} \quad \text{for } k_d \gg k_s[M] \tag{66b}$$

$$k_f = k_c \quad \text{for } k_d \ll k_s[M] \tag{66c}$$

For a review and PHPMS work on association kinetics, see Meot-Ner [65]. Three pressure regimes are distinguished. At very low pressure, the lifetime of the excited $(AB^\pm)^*$ is short relative to the time between stabilizing collisions, and Eq. 66a reduces to Eq. 66b, which predicts a first-order dependence on the pressure of M and overall third order. The intermediate range where k_d and $k_s[M]$ are of similar magnitude is given by Eq. 66a and the high-pressure range where decomposition is much slower than stabilization, such that the reaction is independent of M and thus second order is given by Eq. 66c. For most cases of interest, the association reactions are in the third-order regime (66b) or the intermediate region (66a) in the PHPMS pressure range of 1–10 torr. The rates are generally sufficiently rapid and the association–dissociation equilibria (Eqs. 63 and 64) can be measured. In general, the pressures used for the ICR and FTICR technique are too low and the association reactions, which are in the third-order regime at these pressures, are too slow so that association products $AB^\pm$ are not observed and the association equilibria cannot be measured. The higher pressures used in the FA technique permit the measurements of many association equilibria.

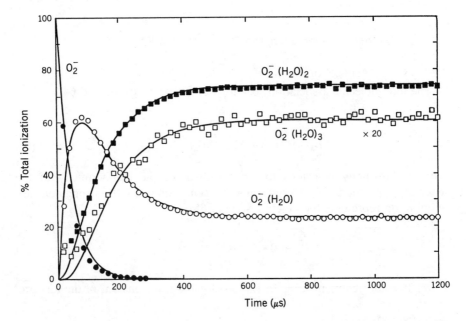

Fig. 29. Normalized, ion intensities observed in moist O_2. Results show hydration sequence $O_2^- \rightarrow O_2^- \cdot H_2O \leftrightarrows O_2^- \cdot (H_2O)_2 \leftrightarrows O_2^- (H_2O)_3$ where the mono, di, and trihydrate reach equilibrium. [Reprinted by permission from J. Payzant and P. Kebarle, *J. Chem. Phys.*, **56**, 3083 (1972).]

Conditions in which association equilibria establish are easily found with the PHPMS technique. An example of the establishment of such an equilibrium was given in Fig. 16. Shown in Fig. 29 are the normalized intensity-time plots observed for the hydration of the superoxide anion O_2^- (Payzant and Kebarle [93]). The figure shows clearly the achievement of the equilibria $(n-1, n)$:

$$O_2^-(H_2O) + H_2O = O_2^-(H_2O)_2 \qquad (1,2)$$

$$O_2^-(H_2O)_2 + H_2O = O_2^-(H_2O)_3 \qquad (2,3)$$

Increase of the water partial pressure would lead to the equilibria shifting to higher $(n-1, n)$. Similarly, increase of temperature at constant water pressure leads to shift of equilibria to low $(n-1, n)$, whereas decrease of temperature leads to high $(n-1, n)$.

The results in Fig. 30 illustrate tests of the invariance of the equilibrium constant $K_{2,3}$ with a change of water pressure by a factor of 100. Determinations of the equilibrium constants K from plots like that in Fig. 30 obtained at different temperatures lead to van't Hoff plots and these to the corresponding ΔH^0 and ΔS^0 values. Examples of such plots are given in Figs. 31 and 32. The plots in Fig. 31 are for the equilibria equation $(n-1, n)$ of the general reaction

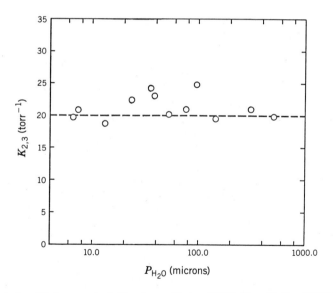

Fig. 30. Plot of equilibrium constants $K_{2,3}$ for equilibrium $O_2^-(H_2O)_2 + H_2O = O_2^-(H_2O)_3$ demonstrating invariance of $K_{2,3}$ with water pressure change by a factor of ~ 200. Dashed line gives results from earlier measurement. [Reprinted by permission from J. Payzant and P. Kebarle, *J. Chem. Phys.*, **56**, 3083 (1972).]

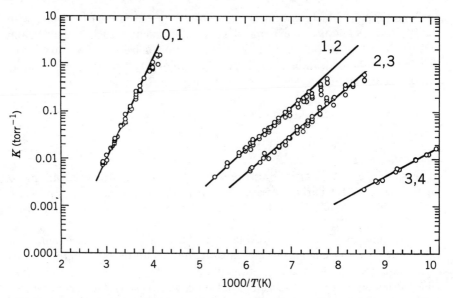

Fig. 31. Van't Hoff plots for equilibria $H_3^+(H_2)_{n-1} + H_2 = H_3^+(H_2)_n$. [Reprinted by permission from K. Hiraoka and P. Kebarle, *J. Chem. Phys.* **62**, 2267 (1975).]

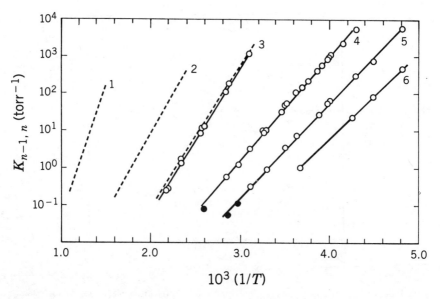

Fig. 32. Van't Hoff plots for equilibria $H_3O^+(H_2O)_{n-1} + H_2O = H_3O^+(H_2O)_n$. [Reprinted by permission from Y. K. Lau, S. Ikuta, and P. Kebarle, *J. Am. Chem. Soc.*, **104**, 1462 (1982). Copyright 1982 American Chemical Society.]

involving clustering of H_2 around H_3^+, from

$$H_3^+(H_2)_{n-1} + H_2 = H_3^+(H_2)_n$$

$$H_3O^+(H_2O)_{n-1} + H_2O = H_3O^+ \cdot (H_2O)_n$$

(Hiraoka and Kebarle [35]), whereas those in Fig. 32 are for the clustering of H_2O around H_3O^+ (from Lau et al. [69]). The plots also illustrate the wide temperature range available. Thus, the lowest temperature measurement in Fig. 31 was performed at $-173°C$, whereas the highest temperature measurement in Fig. 32 was performed at $600°C$. It is interesting to note that the pattern of the van't Hoff plots of these two very different systems is quite similar. A big gap occurs between $n = 1$ and 2 and a second big gap between the $n = 3$ and 4 plots. This is a result that reveals the similar bonding pattern in the hydrogen bonded systems $H_3^+(H_2)_n$ and $H_3O^+(H_2O)_n$. In each of these systems there are abrupt stability decreases after $n = 1$ and after $n = 3$; see Hiraoka and Kebarle [35] and Lau [60].

4.2.2 Conditions for Association Equilibria Measurements

Association equilibria measurements are more prone to error than those of transfer-exchange equilibria. The transfer of the ions into the vacuum and the mass analysis can lead to more severe problems for the association equilibria case. The sampling problems were already identified in the first measurements, and (Hogg and Kebarle [11]) as:

1. Cooling of the gas due to adiabatic expansion through ion exit leak into vacuum
2. Collision-induced dissociation in the region of high-density gas immediately outside the ion exit leak
3. Unimolecular dissociation of the ions during mass analysis
4. Collision-induced dissociation during mass analysis

The cooling of the gas was essentially suppressed by using ion exit leaks through which the flow is near molecular. The effect is illustrated in Fig. 33 (from Hogg et al. [12]), which shows that replacement of the molecular flow leak (square array of 30 small pinholes, see Section 2.5.1) with a single circular hole of the same total area, from which hydrodynamic flow is expected, leads to a shift of cluster populations to higher n. This shift is caused by the adiabatic cooling. For experiments in which adiabatic cooling is expressly used to produce cluster growth, see Searcy and Fenn [94].

Adiabatic cooling, if present, will also affect measurements of exchange equilibria; however, in general the error will be less serious, since the exchange equilibria generally measured have a much smaller temperature coefficient, smaller $|\Delta H^0|$, than the association equilibria. Furthermore, in clustering

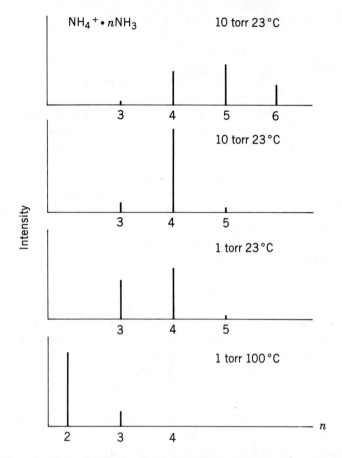

Fig. 33. Schematic representation of ion intensities for clustered $NH_4^+ \cdot nNH_3$. Intensities are expressed as fractions of total ion current. Condition of 1 torr and 100° shows cluster of largest concentration to be $NH_4^+ \cdot 2NH_3$. Reduction of temperature at constant pressure (23°, 1 torr) causes clusters to grow. Increase of pressure at constant temperature (23°, 10 torr) produces further cluster growth. Spectrum at top (23°, 10 torr) shows effect of using large sampling pinhole (70-μ diameter) producing dynamic flow. The resultant adiabatic cooling causes further nonequilibrium growth of clusters. [Reprinted by permission from A. M. Hogg, R. M. Haynes, and P. Kebarle, *J. Am. Chem. Soc.*, **88**, 28 (1966). Copyright 1966 American Chemical Society.]

equilibria the very nature of the reactants is changed by the cooling, as illustrated in Fig. 33.

The danger for collision-induced dissociation of the adducts or clusters in the higher pressure region immediately outside the leak is also reduced by using molecular flow leaks. A further reduction is achieved by using weak electric fields between the ion source and the first ion accelerating electrode; see Fig. 3 and Table 2. Furthermore, due to the requirement of momentum conservation, the energy E_{cm} made available by the collision as internal energy (center of mass energy) is only a fraction of the kinetic energy E_{lab} of the accelerated ion as shown in Eq. 67:

$$E_{cm} = E_{lab} \frac{m}{(M+m)} \qquad (67)$$

where M is the mass of the ion and m the mass of the carrier gas molecules (generally CH_4). Since, in general, M is much bigger than m, little energy is available in the region near the ion exit slit. For recent research work on accelerated ion–molecule collisions, see Ervin and Armentrout [95].

The possibility of unimolecular dissociation of the adduct or cluster ions during their flight in the vacuum of the mass analysis region was investigated by Sunner and Kebarle [96]. Since the ion populations exiting the ion source are believed to be very close to thermal, only the fraction f_{ex} of the Maxwell–Boltzmann distribution of ions that has internal energy higher than the cluster dissociation energy (critical energy) E_0 can dissociate. Generally, f_{ex} of adducts and clusters is very low such that only a very small fraction can dissociate. However, when E_0 is not large and when the cluster has a large internal energy, due to the presence of a large number of low-frequency (soft) vibrations, f_{ex} can become large, and a substantial fraction of the clusters will be able to dissociate. Such conditions are present for clusters of high n that distinguish themselves by weak bonding, i.e., low E_0 and large number of soft vibrations. Estimates of the rate constants k_d for ion molecular dissociation for the potassium hydrates $K^+(OH_2)_n$ on the basis of RRKM theory showed that due to the relatively long ion flight time (tens of microseconds) most of the clusters whose energy is higher than E_0, i.e., nearly the whole fraction f_{ex}, will dissociate. Evaluation of the Maxwell–Boltzmann energy distribution $P(E)$ of the clusters and evaluation of $k_d(E)$ on the basis of estimated vibrational frequencies showed that as much as 80% of the $K^+(H_2O)_6$ cluster may dissociate under unfavourably chosen ion source reaction equilibria conditions (Sunner and Kebarle [96]). Ligands with many internal degrees of freedom like dimethylformamide relative to H_2O aggravate the problem. For potassium clusters with such ligands, problems can arise already with $n = 3$ clusters (Sunner and Kebarle [96, 97]).

"Observed" equilibrium constants $K_{n-1,n}$ were calculated, which included the effect of the predicted unimolecular decomposition of the clusters, and the van't Hoff plots of these were compared with the van't Hoff plots of equilibrium constants that were not affected by decomposition. The "observed" van't Hoff plots showed some curvature, which may remain unnoticed in plots over a narrow range of temperatures. The following conclusions were made (Sunner and Kebarle [96]).

Unimolecular decomposition of ion clusters can occur in the mass analysis vacuum region. The effect of the decomposition under improperly chosen (low values of $K_{n-1,n}$) conditions is to steepen the slope, i.e., increase $-\Delta H^0_{n-1,n}$ and $-\Delta S^0_{n-1,n}$. The problem occurs for $n > 4$ and is aggravated by ligands with a large number of internal degrees of freedom. To minimize the problem, we must measure equilibrium constants $K_{n-1,n}$ at the lowest experimentally accessible temperatures. This amounts to measurements where $K_{n-1,n}$ are large, i.e., in the $10-10^4$ (torr^{-1}) range. Since, as will be shown below, it is desirable to keep the ratio $[A^{\pm}B_n]/[A^{\pm}B_{n-1}]$ fairly close to unity, large $K_{n-1,n}$ are achieved with low clustering gas B pressures. To obtain the required equilibrium relaxation time of less than 1 ms, we need high third gas M pressures (see Eqs. 65 and 66). In practice, third gas pressures in the torr range and B in the 1–10 mtorr range lead to the desired conditions. Fortunately, gases B like H_2 that are very weakly clustering and with which a third gas cannot be used since the third gas replaces B as ligand have very low internal heat capacities and are thus less prone to unimolecular dissociation.

Tests for the occurrence of collision-induced dissociation by residual gas in the mass analysis region were performed by deliberately increasing the gas pressure in the magnetic sector instruments and measuring changes of ion intensities and metastable ion intensities arising from ion dissociations: $A^{\pm}(B)_n = A^{\pm}(B)_{n-1} + B$ (Sunner and Kebarle [96], Lau et al. [60]). It was found [60] that the collision-induced metastable transitions had cross sections in the 10 Å^2 range. The cross sections increased with the size of the cluster and with the ion source temperature. Thus, the expected collisional attenuation of the clusters was both temperature and n dependent. In order to eliminate errors due to collisional dissociation, mass analyzer residual gas pressures that are less than 10^{-6} torr are desirable. Such pressures are easily achieved with differential pumping; see Sections 2 and 3.

Transfer or exchange equilibria (see Section 4.1) are generally less affected by unimolecular dissociation or collision-induced dissociation in the mass analysis region because the ions engaged in these equilibria are generally more stable. When the exchange equilibria are not ligand exchange, dissociation will generally require energies in the 30 kcal/mol or higher region. Even when ligand exchange equilibria are involved, it will be possible to measure these at low temperatures where the internal energy and thus f_{ex} of the clusters are extremely small. This would suppress both unimolecular and collisional dissociation.

Since ion association and clustering equilibria are more prone to sampling error, measurements in which the observed ion ratio $A^{\pm}B_n/A^{\pm}B_{n-1}$ is very large should be avoided. For example, in the hydronium hydrate $H_3O^+(H_2O)_n$ measurements, low intensities of H_3O^+ are observed at low temperature where H_3O^+ should not be present at all, i.e., where only the higher hydrates are stable. These H_3O^+ ions are probably produced in the sampling process due to collisional breakup of higher clusters. The observed ratio $H_3O^+ \cdot H_2O$ to H_3O^+ is large and changes very slowly with temperature under these conditions and if used in an equilibrium constant leads to completely erroneous results.

4.2.3 Thermochemical Data from Association and Clustering Equilibria

A very large amount of thermochemical data, ΔG^0, ΔH^0, and ΔS^0, have been obtained for association and clustering equilibria. Early reviews have been written by Kebarle [98–102]. Recent reviews have been prepared by Castleman

[103, 104]. Castleman [104] gives the most up-to-date compilation of data. These include PHPMS data from this laboratory, HPMS and PHPMS data due to Field, Meot-Ner, Sieck, Hiraoka, and Stone's laboratory, as well as the determinations obtained with Castleman's drift ion source. Data obtained with the flowing afterglow FA and SIFT measurements and other techniques are also included. On the whole, the agreement between the different methods is good. ΔH^0 values are generally within ± 2 kcal/mol, whereas ΔG^0 determination when undertaken at the same temperature are within ± 1 kcal/mol. The ΔS^0 determinations are within ± 4 cal/degree mol. The absolute errors in some of the determinations may be somewhat larger since all of the major techniques rely on extraction of the ions into vacuum and are thus subject to similar, i.e., parallel sources of error.

REFERENCES

1. V. L. Talroze and A. K. Lyubimova, *Dokl. Akad. Nauk SSSR*, **86**, 909 (1952).
2. D. T. Stevenson and D. O. Schissler, *J. Chem. Phys.*, **23**, 1353 (1955); **29**, 282 (1958).
3. F. H. Field, J. L. Franklin, and F. W. Lampe, *J. Am. Chem. Soc.*, **79**, 2419 (1957).
4. F. H. Field, *J. Am. Chem. Soc.*, **91**, 7233 (1969).
5. S. Wexler and L. G. Pobo, *J. Am. Chem. Soc.*, **91**, 7233 (1969).
6. P. Kebarle and E. W. Godbole, *J. Chem. Phys.*, **36**, 302 (1962).
7. P. Kebarle and E. W. Godbole, *J. Chem. Phys.*, **39**, 1131 (1963).
8. P. Kebarle and A. M. Hogg, *Adv. Mass Spectrom.*, **3**, 401 (1964).
9. P. Kebarle and A. M. Hogg, *J. Chem. Phys.*, **42**, 668 (1965).
10. P. Kebarle and A. M. Hogg, *J. Chem. Phys.*, **42**, 798 (1965).
11. A. M. Hogg and P. Kebarle, *J. Chem. Phys.*, **43**, 449 (1965).
12. A. M. Hogg, R. M. Haynes, and P. Kebarle, *J. Am. Chem. Soc.*, **88**, 28 (1966).
13. D. A. Durden, P. Kebarle, and A. Good, *J. Chem. Phys.*, **50**, 302 (1969).
14. A. Good, D. A. Durden, and P. Kebarle, *J. Chem. Phys.*, **52**, 212 (1970); **52**, 222 (1970).
15. V. L. Talroze and E. L. Frankevitch, *Russ. J. Phys. Chem.*, **34**, 1275 (1960).
16. T. W. Shannon, F. Meyer, and A. G. Harrison, *Can. J. Chem.*, **43**, 159 (1965).
17. A. J. Cunningham, J. D. Payzant, and P. Kebarle, *J. Am. Chem. Soc.*, **94**, 7627 (1971); J. D. Payzant Kinetics and Equilibria of Ion Molecule Reactions, Ph.D. Thesis, Chemistry Department, University of Alberta (1973).
18. M. Meot-Ner and F. H. Field, *J. Am. Chem. Soc.*, **99**, 998 (1977); J. J. Solomon, M. Meot-Ner, and F. H. Field, *J. Am. Chem. Soc.*, **96**, 3727 (1974).
19. D. K. Bohme, J. A. Stone, R. S. Mason, R. S. Stradling, and K. R. Jennings, *Int. J. Mass Spectrom. Ion Phys.*, **37**, 283 (1981).
20. J. A. Stone, D. A. Splinter, and S. Y. Kong, *Can. J. Chem.*, **60**, 910 (1982).
21. M. Meot-Ner and L. W. Sieck, *J. Am. Chem. Soc.*, **105**, 2956 (1983).
22. L. W. Sieck and M. Meot-Ner, *J. Am. Chem. Soc.*, **105**, 2956 (1983).
23. L. W. Sieck, *J. Phys. Chem.*, **89**, 5552 (1985).
24. (a) K. Hiraoka, K. Morise, and T. Shoda, *Int. J. Mass Spectrom. Ion Proc.* **67**, 11 (1985); (b) K. Hiraoka, K. Morise, T. Nishijima, S. Nakamura, M. Nakazato, K. Ohkuma, *Int. J. Mass Spectrom. Ion Proc.*, **68**, 99 (1986).

25. T. B. McMahon, Chemistry Department, University of Waterloo, private communication (1986).
26. E. P. Grimsrud, Chemistry Department, University of Montana, Bozeman, private communication (1986).
27. E. P. Grimsrud, G. Caldwell, S. Chowdhury, and P. Kebarle, *J. Am. Chem. Soc.*, **107**, 4627 (1985).
28. F. P. Lossing, *Can. J. Chem.*, **35**, 305 (1957); F. P. Lossing, A. W. Tickner, and W. A. Bryce, *J. Chem. Phys.* **19**, 1254 (1951); R. L. Graham, A. L. Harkness, and H. G. Thode, *J. Sci. Instrum.*, **24**, 119 (1947).
29. D. A. Durden, Thermal Ion–Molecule Reactions at Pressures up to 10 torr with a Pulsed Mass Spectrometer, Ph.D. Thesis, Chemistry Department, University of Alberta (1969).
30. Y. K. Lau, Gas Phase Basicities and Proton Affinities, Ph.D. Thesis, Chemistry Department, University of Alberta (1979).
31. M. French and P. Kebarle, *Can. J. Chem.*, **53**, 2268 (1975).
32. M. A. French, Thermal Ion Molecule Reactions and Equilibria, Ph.D. Thesis, Chemistry Department, University of Alberta (1977).
33. M. G. Ikonomou, Direct Insertion Probe PHPMS, M.Sc. Thesis, Chemistry Department, University of Alberta (1987).
34. S. Dushman, *Scientific Foundations of Vacuum Technique*, Wiley, New York, 1958.
35. K. Hiraoka and P. Kebarle, *J. Chem. Phys.*, **62**, 2267 (1975).
36. D. T. A. Blair, in J. M. Meaks and J. D. Craggs, Eds. *Electrical Breakdown in Gases*, Wiley Interscience, 1978, p. 533.
37. P. H. Dawson, Ed., *Quadrupole Mass Spectrometry and Its Applications*, Elsevier, Amsterdam, 1976.
38. P. Kebarle, Thermochemical Information from Ion Equilibria, in P. Ausloos, Ed., *Interactions between Ion and Molecules*, Plenum, New York, 1975.
39. L. J. Puckett and W. C. Lineberger, *Phys. Rev. A*, **7**, 1635 (1970); W. C. Lineberger and L. J. Puckett, *Phys. Rev.*, **186**, 116 (1969).
40. H. J. Oskam, *Philips. Res. Rept.*, **13**, 401 (1958).
41. M. D. Kregel, *J. Appl. Phys.*, **41**, 55 (1978).
42. E. W. McDaniel, *Collision Phenomena in Ionized Gases*, Wiley, New York, 1964.
43. L. G. Christophorou, *Adv. Electronics Electron. Phys.*, **46**, 55 (1978).
44. E. W. McDaniel and E. A. Mason, *The Mobility and Diffusion of Ions in Gases*, Wiley, New York, 1973, pp. 13–28.
45. D. Smith and I. C. Plumb, *J. Phys. D.*, **6**, 3517 (1975).
46. N. G. Adams, A. G. Dean, and D. Smith, *Int. J. Mass. Spectrom. Ion Phys.*, **10**, 63 (1972).
47. N. M. Rodiguin and E. N. Rodiguina, *Consecutive Chemical Reactions*, D. Van Nostrand, New York, 1964.
48. W. Lindinger and D. L. Albritton, *J. Chem. Phys.*, **62**, 3517 (1975).
49. E. P. Grimsrud, S. Chowdhury, and P. Kebarle, *J. Chem. Phys.*, **83**, 3983 (1985).
50. E. P. Grimsrud, S. Chowdhury, and P. Kebarle, *Int. J. Mass Spectrom. and Ion Proc.*, **83**, 1059 (1985).
51. E. P. Grimsrud, S. Chowdhury, and P. Kebarle, *Int. J. Mass Spectrom. and Ion Proc.*, **68**, 57 (1986).
52. S. Chowdhury, E. P. Grimsrud, T. Heinis, and P. Kebarle, *J. Am. Chem. Soc.*, **108**, 3630 (1986).
53. S. Chowdhury and P. Kebarle, *J. Am. Chem. Soc.*, **108**, 5453 (1986).
54. P. Kebarle and S. Chowdhury, *Chem. Rev.*, **87**, 513 (1987).
55. G. Caldwell, T. F. Magnera, P. Kebarle, *J. Am. Chem. Soc.*, **106**, 959 (1984).

56. A. Good, D. A. Durden, and P. Kebarle, *J. Chem. Phys.*, **52**, 212, 222 (1970).
57. J. D. Payzant, A. J. Cunningham and P. Kebarle, *J. Chem. Phys.*, **59**, 5615 (1973).
58. M. A. French, L. P. Hills, and P. Kebarle, *Can. J. Chem.*, **51**, 4561 (1973).
59. D. K. Sen Sharma and P. Kebarle, *J. Am. Chem. Soc.*, **100**, 5826 (1978).
60. Y. K. Lau, S. Ikuta, and P. Kebarle, *J. Am. Chem. Soc.* **104**, 1462 (1982).
61. D. K. Sen Sharma and P. Kebarle, *J. Am. Chem. Soc.*, **104**, 19 (1982).
62. A. T. Blades and P. Kebarle, *J. Chem. Phys.*, **78**, 783 (1983).
63. T. F. Magnera, Studies of Ion–Molecule Reactions and Equilibria, Ph.D. thesis, University of Alberta (in preparation).
64. M. Meot-Ner (Mautner), Temperature and Pressure Effects in the Kinetics of Ion Molecule Reactions, in M. T. Bowers, Ed., *Gas Phase Ion Chemistry*, Academic Press, New York, 1979, vol. 1.
65. M. Meot-Ner and F. H. Field, *J. Am. Chem. Soc.*, **97**, 2014 (1975).
66. M. Meot-Ner and F. H. Field, *J. Chem. Phys.*, **64**, 277 (1976).
67. T. B. McMahon and P. Kebarle, *J. Am. Chem. Soc.*, **107**, 2612 (1985).
68. R. B. Sharma, D. K. Sen Sharma, K. Hiraoka, and P. Kebarle, *J. Am. Chem. Soc.*, **107**, 3747 (1985).
69. J. J. Solomon and F. H. Field, *J. Am. Chem. Soc.*, **97**, 2624 (1975).
70. J. J. Solomon and F. H. Field, *J. Am. Chem. Soc.*, **98**, 1025 (1976).
71. M. Meot-Ner, J. J. Solomon, and F. H. Field, *J. Am. Chem. Soc.*, **98**, 7891 (1976).
72. (a) T. B. McMahon and P. Kebarle, *J. Am. Chem. Soc.*, **98**, 3399 (1976); (b) J. B. Cumming and P. Kebarle, *Can. J. Chem.*, **76**, 1 (1978). (c) W. E. Farneth, and J. I. Brauman, *J. Am. Chem. Soc.* **98**, 7891 (1976).
73. T. B. McMahon, G. Nicol, T. Heinis, and P. Kebarle (to be published).
74. R. Yamdagni and P. Kebarle, *J. Am. Chem. Soc.*, **95**, 3504 (1973).
75. S. Chowdhury, T. Heinis, E. P. Grimsrud, and P. Kebarle, *J. Phys. Chem.*, **90**, 2747 (1986).
76. R. J. Celotta, R. A. Bennett, and J. L. Hall, *J. Chem. Phys.*, **59**, 1740 (1974).
77. R. Yamdagni and P. Kebarle, *J. Am. Chem. Soc.*, **98**, 1320 (1976).
78. Y. K. Lau, P. P. S. Saluja, P. Kebarle, and R. W. Alder, *J. Am. Chem. Soc.*, **100**, 7328 (1978).
79. R. B. Sharma, A. T. Blades, and P. Kebarle, *J. Am. Chem. Soc.*, **106**, 510 (1984).
80. M. Meot-Ner (Mautner), *J. Am. Chem. Soc.*, **105**, 4906 (1983).
81. J. F. Wolf, R. H. Staley, I. Koppel, M. Taagepera, R. T. McIver, J. L. Beauchamp, and R. W. Taft, *J. Am. Chem. Soc.*, **99**, 5417 (1977).
82. D. H. Aue, M. T. Bowers, M. T. Bowers, Ed., in *Gas Phase Ion Chemistry*, Academic Press, New York, 1979, vol. 2, p. 1.
83. H. I. Schiff and D. K. Bohme, *Int. J. Mass Spectrom. Ion Phys.*, **16**, 167 (1975).
84. S. G. Lias, J. F. Liebman, and R. J. Levin, *J. Phys. Chem. Ref. Data*, **13**, 699 (1985).
85. G. Caldwell and P. Kebarle, *J. Am. Chem. Soc.* (to be published).
86. J. E. Bartmess, J. A. Scott, and R. T. McIver, *J. Am. Chem. Soc.*, **101**, 6076 (1979).
87. J. E. Bartmess and R. T. McIver, in M. T. Bowers, Ed., *Gas Phase Ion Chemistry*, Academic Press, New York, 1979, vol. 2, p. 88.
88. M. Meot-Ner (Mautner), *J. Phys. Chem.*, **84**, 2724 (1980).
89. M. Meot-Ner (Mautner), S. F. Nelsen, M. F. Willi, and T. B. Frigo, *J. Am. Chem. Soc.*, **106**, 7384 (1984).
90. J. W. Larson and T. B. McMahon, *J. Am. Chem. Soc.*, **105**, 2944 (1985).
91. G. Caldwell and P. Kebarle, *Can. J. Chem.*, **63**, 1399 (1985).
92. K. Hiraoka, E. P. Grimsrud, and P. Kebarle, *J. Am. Chem. Soc.*, **96**, 3359 (1974).

93. J. Payzant and P. Kebarle, *J. Chem. Phys.*, **56**, 3083 (1972).
94. J. Q. Searcy and J. B. Fenn, *J. Chem. Phys.*, **61**, 5282 (1974); J. Q. Searcy, *J. Chem. Phys.*, **63**, 4114 (1975).
95. K. M. Ervin and P. B. Armentrout, *J. Chem. Phys.*, **84**, 6750 (1986).
96. J. Sunner and P. Kebarle, *J. Phys. Chem.*, **85**, 327 (1981).
97. J. Sunner and P. Kebarle, *J. Am. Chem. Soc.*, **106**, 6135 (1984).
98. P. Kebarle, *Adv. Chem. Ser.*, **72**, 24, (1968); *Am. Chem. Soc.*
99. P. Kebarle, in J. L. Franklin, Ed., *Ion–Molecule Reactions*, Plenum, New York, 1972, vol. 2.
100. P. Kebarle, in M. Szwarc, Ed., *Ions and Ion-Pairs in Organic Reactions*, Wiley Interscience, New York, 1972.
101. P. Kebarle, Gas Phase Ion Equilibria and Ion Solvation, in B. E. Conway and J. O. M. Bockris Eds., *Modern Aspects of Electrochemistry*, Plenum, New York, 1974, vol. 9.
102. P. Kebarle, *Ann. Rev. Phys. Chem.*, **28**, 445 (1977).
103. A. W. Castleman, Jr., and R. G. Keese, *Chem. Rev.*, **86**, 589 (1986).
104. R. G. Keese and A. W. Castleman, Jr., *J. Phys. Chem. Ref. Data*, **5**, 1011 (1982).

Chapter **VI**

NUCLEAR-DECAY TECHNIQUES

Fulvio Cacace
and
Maurizio Speranza

---

1 Introduction
2 Principles of the Decay Technique: The Ionogenic Process
   2.1 Decay of Isolated Tritium Atoms
   2.2 Decay of Chemically Bound Tritium Atoms
        2.2.1 Theoretical Results
        2.2.2 Experimental Results
        2.2.3 Summary of the Chemical Effects of the Decay in Isolated Molecules
3 Outline of the Decay Technique
   3.1 Detection and Characterization of the Decay Products by Multiple Labeling
   3.2 Prevention of Radiolytic Artifacts
4 Experimental Aspects of the Technique
   4.1 Synthesis of Multitritiated Compounds
   4.2 Dilution of Multitritiated Compounds
   4.3 Purification and Characterization of Multitritiated Compounds
   4.4 Preparation of the Reaction Mixture
        4.4.1 Choice of the Specific Activity of the Mixture
        4.4.2 Choice of the Neutral Substrate(s)
        4.4.3 Choice of the Experimental Conditions
   4.5 Analysis of the Reaction Products
5 Applications
   5.1 Proof of Existence of Gaseous Cyclobutyl Cation
   5.2 Proof of Existence and Isomerization of Cyclohexyl Cation
   5.3 Automerization of Phenylium Ion
   5.4 Gas-Phase Protonation of Five-Membered Heteroaromatic Rings
   5.5 Addition of Vinyl Cation to Methane
6 Concluding Remarks
References

1 INTRODUCTION

The purpose of this chapter is to illustrate the role of radiochemical methods, in particular of those based on nuclear decay, in the study of gaseous ions. In discussing this topic, we will focus attention largely on the β decay of tritiated molecules, in part because of the authors' own interest and in part because no other methods have been the subject of such a wide range of theoretical and experimental studies. The approach known today as the "decay technique," first introduced in 1963 [1], has since been the subject of a number of reviews covering its expanding applications [2–12]. The sustained and currently growing interest in this technique, in spite of its demanding experimental requirements, is justified by unique features that set it apart from other approaches to gas-phase ion chemistry. In short, the decay technique allows cations of known structure and initial charge location to be generated in any environment of interest, from low-pressure gases to dense gases, liquids and solids, and their reactions to be studied by standard procedures, including pressure and temperature-dependence studies and competition experiments. Of particular interest is the actual isolation of the neutral end products, whose identification and characterization provide precious information on the structure, the isomeric composition, and the stereochemical features of the gaseous ions investigated.

As it happens, the most attractive features of the decay approach, namely, the unrestricted applicability and the high structural discrimination, pertain to aspects of gas-phase ion chemistry whose study has severely been hampered by recognized limitations of the mass spectrometric methods, confined to a pressure range that seldom exceeds 1 torr, and characterized by an inherently low structural resolution. Such a state of affairs makes mass spectrometric and nuclear-decay techniques eminently complementary, and in fact their concerted application has proved an extremely powerful tool in the structural, mechanistic, and kinetic study of gaseous cations.

As to the organization of the chapter, Section 2 presents a concise survey of the principles of the decay technique, Section 3 gives a general outline of the method, and Section 4 illustrates in some detail typical laboratory techniques used in the decay experiments. Finally, Section 5 presents selected applications of the decay technique to structural and reactivity problems, and Section 6 gives concluding remarks.

2 PRINCIPLES OF THE DECAY TECHNIQUE: THE IONOGENIC PROCESS

In contrast with other approaches, relying on external agents such as electron beams, high energy photons, and fast atoms for the ionization of neutral molecules, in the decay technique the ions are formed by an internal event, i.e., the spontaneous decay of a radioactive atom contained in the molecule. The nuclear nature of the ionogenic process makes the latter entirely insensitive to environmental influences, with the important consequence that the technique can be used in a large variety of media, from gases to solids, as mentioned in the introductory section. The nucleogenic ions, as, for that matter, those from more conventional ionization techniques, are generally formed with excess internal energy whose specific source, nature, and extent depend on the particular nuclear transition involved and on chemical factors, e.g., the nature of the molecule affected and the intramolecular position of the decayed atom.

The following sections are devoted to a discussion of the atomic and molecular consequences of the decay of tritium without attempting to cover the more complex ionization and excitation phenomena triggered by the decay of heavier nuclei, including the ones (^{76}Kr, ^{77}Kr, ^{80m}Br, ^{125}I, etc.) used in mechanistically oriented decay experiments. For a more comprehensive treatment, see previous specialized reviews [1,13].

2.1 Decay of Isolated Tritium Atoms

Tritium, ^{3}H, hereafter indicated with the symbol T, has a half-life of 12.26 years, decaying into ^{3}He following the emission of an antineutrino and of a β^- particle (maximum energy, 18.6 KeV; mean energy, 5.6 KeV) according to the equation

$$T \xrightarrow[\text{decay}]{\text{beta}} {}^3\text{He}^+ + \beta^- + \bar{\nu} \qquad (1)$$

The decay entails the change of the nuclear charge and, therefore, of the chemical identity of the radioactive atom, an event that has profound chemical consequences. In addition, the emission of the negatively charged β^- particle requires that the overall charge of the system must increase by one unit in order to comply with the principle of charge conservation. In the frequent case in which neutral atoms or molecules are involved, the β decay yields a daughter cation. The sudden increase of the nuclear charge and of the associated electric field affects the electronic cloud of the atom, which can lead to excited electronic states of the daughter ion according to a mechanism known as electron "shaking." The total energy released in a decay event is shared among the daughter ion, the antineutrino, and the β^- particle, which traverses the atom in a time $< 10^{-16}$ s, far too short to allow the adiabatic readjustment of the electron cloud to the increased nuclear field (Table 1). As a consequence, a fraction of the energy that in an ideal, adiabatic transition would be carried away by the outgoing β^- particle can be converted into electronic excitation of the ^{3}He$^+$ ion. Another way of looking at the shaking effect, perhaps more familiar to the chemist, proceeds from the realization that the 1s orbitals of the T atom and of the ^{3}He$^+$ ion do not overlap exactly. As a consequence, following the sudden nuclear change undergone by the atom, its electron has a finite probability to occupy some ^{3}He$^+$ orbital other than the ground-state (1s) one; i.e. an electronically excited daughter ion is formed.

The very simple species involved and the applicability of the sudden perturbation approach have allowed an accurate theoretical analysis of the excitation mechanism as early as 1941. In that year, Migdal [14] showed that

Table 1. Time Scale of Events in Ionogenic Processes

Time(s)	Event
10^{-18}	1 MeV electron traverses molecule
10^{-17}	Lifetime of an electron vacancy
10^{-16}	1 MeV α-particle traverses molecule 5-eV electron traverses molecule Excitation or ionization by charged particle
10^{-15}	Auger effect Charging time of vacancy cascades
10^{-14}	Autoionization of superexcited states C-^{3}He$^+$ bond breaking (β decay) Transfer of energy to vibrational modes
10^{-13}	Molecular vibration Ion–molecule collisions in condensed phase
10^{-12}	Molecular collisions in condensed phase Internal conversion to lowest excited state in polyatomic molecule
10^{-11}	Electron is solvated in polar media
10^{-9}	Ion–molecule collisions in gases at 1 atm Molecular collisions in gases at 1 atm
10^{-8}	Fluorescence
10^{-6}	Lifetime of metastable ions
10^{-5}	Scavenging of radicals
10^{-3}	Phosphorescence

the probability of the decay-promoted transition from the 1s electronic ground state of tritium to a final electronic state of the ^{3}He$^+$ daughter ion, characterized by the n and l quantum numbers, can be computed from the square of the overlap integral:

$$P\{1s(T) \rightarrow n, l(^3He^+)\} = \left| \int \psi_{n,l}^* \psi_{1s} d\tau \right|^2$$

where $d\tau$ is the volume element. The sum of the probabilities of the transitions to all bound states of the electron converges to a definite limit that, subtracted from the unity, gives the probability of ejection of the electron into the continuum, namely, of formation of the doubly charged ^{3}He^{2+} ion. In the decay of *isolated* tritium atoms, ca. 70% of the transitions lead to ground-state ^{3}He$^+$ ions, ca. 25% to the first excited level (40.5 eV), ca. 2.5% to higher excited states, and ca. 2.5% to the formation of ^{3}He^{2+} ions.

Subsequent studies have shown that the results of Migdal's calculations were essentially correct, independent support being provided by theoretical and experimental investigations on the decay of chemically bound T atoms (vide infra), which excluded other mechanisms of electronic excitation operative in heavier nuclei, e.g., direct collision of the outgoing β^- particle with orbital electrons, vacancy cascades, and conversion of internal bremsstrahlung [13, 15].

Passing from electronic to translational excitation, we must consider the momentum transfer to the nucleus from the β^- particle and the antineutrino, which imparts excess kinetic energy ("recoil" energy) to the daughter ion. Under the assumption that both the β^- particle and the antineutrino are emitted in exactly the same direction, simple calculations show that the distribution of "recoil" energies follows that of the emitted β^- particle, ranging from zero to a maximum value of ca. 3.6 eV, most ($>80\%$) of the ^{3}He$^+$ ions acquiring kinetic energies not exceeding 0.08 eV [16].

In conclusion, only two excitation mechanisms are operative in the decay of isolated T atoms:

1. The shaking effect, yielding electronically excited, or doubly charged, daughter ions in some 25% of the transitions
2. The momentum transfer, imparting only a small excess of translational energy to the ^{3}He$^+$ ions from most transitions.

2.2 Decay of Chemically Bound Tritium Atoms

The decay of a covalently bound T atom has, in general, significant chemical consequences. In many cases the very change of identity undergone by the radioactive hydrogen isotope, suddenly converted into a noble-gas ion, is sufficient to cause bond disruption. Furthermore, the excitation mechanisms effective in the decay of isolated T atoms (the shaking and recoil effects) are operative as well in the decay of chemically bound atoms.

The problem has been the subject of extensive theoretical and experimental studies whose major conclusions are briefly outlined in the following sections.

2.2.1 Theoretical Results

All theoretical approaches to the problem concur in the conclusion that the molecular consequences of tritium decay largely depend on the nature of the partner atom. Thus, ^{3}HeX$^+$ (X = H, T), ^{3}HeLi$^+$, and B$_2$H$_5^3$He$^+$ ions from the decay of the corresponding tritiated parents are predicted to be stable, whereas the C—He, N–He, O—He interactions are purely repulsive, the corresponding daughter ions undergoing loss of ^{3}He in a time comparable to the bond-vibration period.

The consequences of the decay of hydrogenlike molecules XT are best understood as a result of a sustained theoretical effort. It has long been recognized that the ground-state ^{3}HeX$^+$ ion is quite stable, being the isotopic counterpart of the well-known ^{4}HeH$^+$ cation whose eigenfunctions and eigenvalues have accurately been calculated. In other words, the molecular ions from the decay process 2,

$$XT \xrightarrow[\text{decay}]{\text{beta}} {}^3\text{HeX}^+ + \beta + \bar{\nu} \qquad (2)$$

can survive indefinitely in the collision-free space *if formed in their ground state*. As a consequence, most of the theoretical studies have been aimed at assessing the fraction of the decay events yielding ground-state ^{3}HeX$^+$ ions. Various techniques of different sophistication have been used, including time-dependent perturbational methods [17], the sudden-perturbation approximation [18,19], molecular orbital (MO) calculations based on Whitten's type orbitals [20], STO-3G and STO-6G methods [21,22], and other approaches [23,24]. In view of the widely different level of the calculations and their extended time span (over 40 years), the results are surprisingly consistent, showing that ^{3}HeX$^+$ ions in their electronic ground state are formed from the decay of TX molecules with a probability ranging from 60 to 90%, which compares with experimental values clustering around 90% (vide infra).

As to the effects of the momentum transfer, not all the recoil energy from the decay of a chemically bound atom is available for bond fission, a fraction necessarily being converted into translational energy of the entire molecule. This factor, and the generally low amount of the recoil energy, make the latter quite insufficient for bond rupture in the large majority of the decay events. Nevertheless, the momentum-transfer mechanism can cause *vibrational* excitation of the daughter molecule. As an example, early calculations suggest that some 20% of the ^{3}HeT$^+$ ions from the decay of T$_2$ are formed in the $v = 1$ vibrational level of the electronic ground state [17]. The measurement of the infrared (IR) emission spectrum of T$_2$ gas decaying at 20 K is grossly consistent with the theoretical estimate, assigning a 0.4 ± 0.2 probability to the formation of ^{3}HeT$^+$ ions in the $v = 1$ vibrational level [25].

Of course, the theoretical treatment of the decay effects becomes much more difficult in passing from hydrogenlike species to more complex molecules. In the first place, we must assess the stability of the newly formed bond between ^{3}He and atoms other than hydrogen. In the second place, we must estimate the extent of the electron shaking and of the recoil effects in the specific molecule of interest and must model the distribution of the excess internal energy from the decay to derive the fragmentation pattern of the daughter molecular ion. Finally, the decay-formed ions contain excess internal energy of a peculiar type ("deformation" energy), arising from the fact that the decay event occurs in a time far too short to allow relaxation of the structure typical of the neutral parent to the (generally different) ground-state geometry of the daughter ion (vide infra).

Owing to the complex nature of the problem, its theoretical studies have followed step-by-step approaches. The preliminary question concerning the inherent stability of He bonding has been tackled by computing the relevant potential energy surfaces with ab initio and semiempirical methods. The results show that the surfaces of the He—C, He—N, He—O, He—F, and He—Be pairs are repulsive in contrast with those of He—H and He—Li bonds [19,20]. A consequence of particular interest to the present discussion is that the decay of tritiated hydrocarbons leads, in a very short time ($\sim 10^{-14}$ s), to the loss of ^{3}He as a neutral atom on account of its higher ionization potential, thus affording a route to free carbocations (Table 1).

$$-\overset{|}{\underset{|}{C}}-T \xrightarrow[\text{decay}]{\text{beta}} \left[-\overset{|}{\underset{|}{C}}-{}^3He\right]^+ \xrightarrow[100\%]{\text{very fast}} -\overset{|}{\underset{|}{C}}{}^+ + {}^3He \qquad (3)$$

The next step, i.e., the calculation of the decay-induced fragmentation pattern on purely theoretical grounds, is an exceedingly difficult problem, even for simple molecules. In fact, most published studies are semiempirical, aimed at rationalizing experimentally measured fragmentation patterns. Apart from simple molecules, e.g., HTO, NH_2T, TF, and BeHT [19, 20], special attention has been devoted to tritiated alkanes [26, 27], including CH_3T [20], the two isomeric monotritiated propanes, and the four isomeric monotritiated toluenes [28, 29]. Recently, the decay of dimers of tritiated amides, acids, amines, and heterocycles has been the subject of theoretical studies aimed at investigating the H^+-transfer processes in the proton-bound adducts formed from the decay [30–32].

In view of the structural and mechanistic applications of the decay technique, it is of interest to evaluate the excess internal energy of polyatomic daughter ions, in particular carbenium ions, formed in their electronic ground state. In analogy with the conclusions concerning diatomic ${}^3HeX^+$ species, the amount of vibrational excitation from the recoil effect can safely be neglected in most of the decay events. The deformation energy associated with the "wrong" geometry of the decay ions, a source of vibrational and rotational excitation first noted in 1978 [33], is much more significant in most cases, as shown by theoretical calculations on pertinent examples. Consider, for instance, the decay of a tritiated alkane:

$$\underset{R_1 \ R_2}{\overset{T}{\underset{|}{C}}}\!\!\diagdown\! R_3 \xrightarrow[\text{decay}]{\text{beta}} \underset{R_1 \ R_2}{\overset{{}^3He}{\underset{|}{C^+}}}\!\!\diagdown\! R_3 \xrightarrow[\text{fast, 100\%}]{-{}^3He} \underset{R_1 \ R_2}{\overset{}{C^+}}\!\!\diagdown\! R_3 \longrightarrow \underset{R_2}{\overset{R_1}{\underset{|}{C^+}}}\!\!\diagdown\! R_3 \qquad (4)$$

$$\qquad\qquad\qquad\qquad\qquad\qquad\qquad\qquad\quad I \qquad\qquad II$$

Relaxation from the pyramidal structure I, reminiscent of that of the alkane, to the ground-state, planar geometry II releases an amount of energy corresponding to the stability difference between I and II. Depending on the nature and the mass of the R groups, a fraction of such "deformation" energy can be carried away by the outgoing ${}^3He$ atom, yet a large fraction will appear as internal energy of the ion, causing vibrational and rotational excitation of II. In the case of CH_3T, the energy difference between pyramidal and planar methyl cations is calculated around 30 Kcal mol^{-1} [34, 35]. Theoretical calculations set the upper limit of the excess internal energy of $C_2H_3^+$ ions from the decay of C_2H_3T near 50 Kcal mol^{-1} [36, 37] and of Ph^+ ions from the decay of PhT between 25 and 32 Kcal mol^{-1} [38, 39].

2.2.2 Experimental Results

The decay-induced fragmentation pattern of gaseous tritiated molecules has been investigated with a specialized technique, the so-called "charge" mass spectrometry. It is convenient to think of the instruments used as conventional mass spectrometers fitted to a large-volume tank, which replaces the ion source. The tritiated gas investigated enters from a capillary leak into the tank, maintained at low pressure by an adequate pumping system. No ionization devices are required since the ions are produced by the spontaneous decay of the tritiated molecules. An array of electrodes, maintained at suitable potentials, drives and focuses the ions formed anywhere within the tank into the entrance slit of the mass spectrometer, which analyzes them in the usual way (Fig. 1). A large-volume tank is required to ensure an adequate rate of formation of the decay ions, since the pressure of the tritiated gas cannot be raised much above 10^{-6} torr to prevent ionization by the β^- particles and secondary ion–molecule reactions [40,41]. The design of the "charge" mass spectrometers has undergone significant changes over the years. Quadrupole mass filters have replaced magnetic-sector analyzers [42], and, perhaps more significantly, commercial ion cyclotron resonance (ICR) spectrometers, fitted with time-averaging computers, have been found perfectly adequate to measure decay-induced fragmentation patterns without requiring special modification [43]. Irrespective of the spectrometer used, careful consideration must be given to the health-physics problems associated with the contamination of the equipment and with the release of radioactive gases (up to several millicurie per run) from the pumping system.

The mass spectrometric results are, in general, fairly consistent with those of theoretical calculations, showing, however, that the latter tend to overestimate the excitation arising from the beta decay. As a rule, the fraction of the molecular daughter ions that survive undissociated after the relatively long time (10^{-5} s)

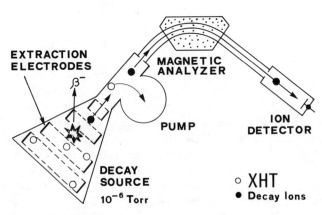

Fig. 1. A "charge" mass spectrometer for the study of molecular fragmentation following the decay of tritiated molecules (Ref. [41]).

Table 2. Abundances of Major Daughter Ions from the Decay of Isolated Tritiated Molecules

Molecule	Daughter Ion	Abundance (%)	Reference
CH_3T	CH_3^+	82	46
C_2H_5T	$C_2H_5^+$	80	47
$CH_3CH_2CH_2T$	$C_3H_7^+$	41	28
CH_3CHTCH_3	$C_3H_7^+$	56	28
C_6H_5T	$C_6H_5^+$	72	48
$o\text{-}CH_3C_6H_4T$	$C_7H_7^+$	78	28
$m\text{-}CH_3C_6H_4T$	$C_7H_7^+$	79	28
$p\text{-}CH_3C_6H_4T$	$C_7H_7^+$	76	28
$C_6H_5CH_2T$	$C_7H_7^+$	79	28
$c\text{-}C_4H_7T$	$C_4H_7^+$	80	43
$c\text{-}C_5H_9T$	$C_5H_9^+$	75	43
CH_2IT	CH_2I^+	56	49

required for detection in the charge mass spectrometers, is invariably higher than predicted by theoretical estimates. Thus the decay of TX molecules (X = H, T) gives abundances of undissociated daughter ions ranging from 89.5 to 94.5%, minor fragments being $^3He^+$, X^+, and $^3He^{2+}$ [44,45]. A similar situation prevails in the case of tritiated molecules, e.g., alkanes, whose decay gives daughter ions unstable with respect to ^3He loss. A typical example is provided by CH_3T, whose decay yields undissociated CH_3^+ ions as the major species (82%), together with CH_2^+ (4.9%), CH^+ (4.0%), C^+ (4.9%), with much smaller abundances of other fragment ions [46]. Analogous fragmentation patterns characterize the decay of other tritiated hydrocarbons RT, the organic cation R^+ representing by far the most abundant daughter ion, as shown by the mass spectrometric results listed in Table 2.

2.2.3 Summary of the Chemical Effects of the Decay in Isolated Molecules

The overall conclusions from the available theoretical and experimental evidence can be outlined as follows:

1. The decay of isolated, hydrogenlike molecules TX gives high yields (over 90%) of the $^3HeX^+$ daughter ions, in their electronic ground state, containing a moderate excess of vibrational energy. Such ions, in particular $^3HeT^+$, are used as extremely powerful protonating agents in gas-phase ion chemistry.
2. Tritiated hydrocarbon molecules RT, decaying in isolation, give high yields (up to 80%) of the corresponding carbenium ions R^+ in their electronic ground state. The major source of excitation is the deformation energy of

the daughter ions, whose extent depends on the specific molecule concerned. In the lack of confirmatory experimental evidence, the available theoretical results [28] suggest that the above conclusions can be extended to the decay of T_2O, yielding the hydroxyl cation OT^+, the only tritiated species other than $^3HeT^+$ and carbenium ions actually used in gas-phase ion chemistry studies [50].

3   OUTLINE OF THE DECAY TECHNIQUE

In a typical decay experiment, labeled cations generated in a gaseous or liquid system by the decay of a multiply tritiated precursor are allowed to react with suitable gaseous nucleophiles. The neutral end products from the ion–molecule reactions triggered by the decay event, hereafter referred to as decay products, are analyzed by radio chromatography and, if necessary, isolated by preparative techniques in order to carry out their chemical degradation.

The general design of a mechanistically oriented study in the field of gas-phase ion chemistry based on the decay technique and the salient experimental features of the latter are outlined in the following sections.

3.1   Detection and Characterization of the Decay Products by Multiple Labeling

To minimize the extent of unwanted radiolytic processes, the activity of the tritiated precursors used in the decay experiments must be kept as low as possible, typically below 10 mCi per mol. As a consequence, the number of the decay ions, hence the amount of their neutral end products formed in any reasonable period of time, is comparatively quite small. Even in a storage period as long as 1 year, the decay of 10 mCi of a monotritiated precursor in 1 mol of an inactive gas ("bulk gas") yields only about 10^{16} daughter ions. Under the favorable assumption that all ions give a single end product, its concentration would be of the order of only a few parts per billion (ppb), posing formidable analytical problems. Even more seriously, the decay products would be swamped by those arising from the radiolysis of the bulk gas, promoted by the β^- particles of tritium. In fact, given the mean energy of the latter, 5.6 KeV, each decay event can be expected to generate some 200 radiolytic ions together with *a single* daughter ion.

The above difficulties can be circumvented, resorting to multiply labeled precursors (e.g., CT_4) whose decay yields daughter ion (e.g., CT_3^+) still containing radioactive atoms that act as a convenient label. Multiply tritiated precursors serve three useful purposes. First, the analysis of the labeled decay products can easily be accomplished by extremely sensitive radiometric techniques, e.g., radio gas–liquid chromatography (GLC) or HPLC, capable of detecting an extremely small number of tritiated molecules. Referring to the previous example, 10^{16} molecules of decay products, each containing one T atom, have the respectable activity of about 0.5 mCi, exceeding the detection limit by a factor of at least 10^7.

Second, the decay ions from a multiply tritiated precursor are necessarily labeled, which is not the case of the ions, and eventually of the products, formed from the radiolysis of the unlabeled bulk gas. Thus, multiple labeling provides a way of distinguishing the decay products from those formed via radiolytic processes (vide infra).

Finally, in those cases in which the position of the tritium atoms in the molecule of the precursor is precisely known, measuring the intramolecular tritium distribution in the products can provide mechanistic and structural insight, as illustrated in the example reported on page 311.

An obvious drawback of the multiple labeling approach is the need of precursors containing at least two radioactive atoms *in the same molecule*, whose synthesis, purification, isotopic analysis, and storage pose severe experimental problems. In principle, the precursor could contain as a label any suitable radioactive nuclide, in addition to tritium. Thus, $^{14}CH_3T$ could conceivably be used instead of CT_4 as a source of ^{14}C-labeled, rather than T-labeled, methyl cations. However, the different half-lives of the two nuclides reduce significantly the sensitivity of the radiometric analysis. For instance, the ratio of ^{14}C atoms to T atoms necessary to obtain a given activity exceeds 500. This consideration and the difficulties encountered in the synthesis of compounds containing two different radionuclides in the same molecule have in practice restricted the choice of the precursors to doubly, or multiply tritiated compounds.

The random nature of the radioactive decay of T imposes an additional experimental constraint. In order to obtain daughter ions having a single structure and charge location, the radioactive atoms must be contained in equivalent positions within the molecule of the multitritiated precursor.

3.2 Prevention of Radiolytic Artifacts

As previously mentioned, we have to be wary of the role of the β^- radiation of tritium as a source of radiolytic products that could superimpose on those of the decay ions. An effectively way of dealing with the problem is based on the fact that all daughter ions from the decay of a multitritiated precursor are necessarily labeled, whereas radiolytic decomposition affects both tritiated and unlabeled molecules. Consider, as a simplified example, the decay of a multi-tritiated precursor in a large excess of unlabeled, but otherwise identical, molecules. According to the decay law, the formation rate of the daughter ions, and eventually of their tritiated products, is proportional to the tritium activity in the system, i.e., to the precursor concentration [P]. On the other hand, the rate of formation of *tritiated* radiolytic products depends in two ways on [P], which determines, in the first place, the intensity of the β^- radiation and therefore the overall rate of the radiolysis. Furthermore, the probability that radiolytic processes affect, in particular, tritiated molecules is proportional, in the example chosen, to their mole fraction, i.e., again to [P]. The two combined effects make the formation rate of *tritiated* radiolytic products proportional to a power of [P] *higher than unity*. Consequently, decreasing [P], namely, the tritium activity in the system, depresses the formation rate of decay products to a smaller extent than of *tritiated* radiolytic products whose yields become comparatively negligible at sufficiently low activity levels. In actual decay experiments in which the tritiated precursor is, in general, chemically different from the bulk gas, the assumption that radiolytic processes are entirely indiscriminate is questionable owing to the possible occurrence of selective charge and energy transfer, selective ionic and radical reactions, etc. Nevertheless, the experimental evidence concerning a large variety of systems shows that decreasing the tritium activity in the gas is indeed an effective way of reducing to negligible levels the radiolytic route to tritiated products. In fact, it has been shown [1, 12] that the occurrence of radiolytic artifacts can easily be detected carrying out a "blank" run under exactly the same set of conditions prevailing in the actual decay experiment, except for the fact that the multitritiated precursor is replaced by the same activity of the corresponding monotritiated compound. Since the decay of the latter gives unlabeled daughter ions, any tritiated products formed must necessarily arise from radiolytic pathways, whose role can thus exactly be assessed. The evidence from such "blank" runs shows that, except in a few systems prone to radiation-induced chain reactions, the yields of tritiated radiolytic products are negligible in comparison with those of decay products when the tritium activity does not exceed about 100 mCi per mol of bulk gas. Unfortunately, the possibility of radiolytic artifacts has not been perceived by some workers, who have used large activities of tritiated precursors, sometimes as high as several hundred curies per mole [51], for the sake of reducing the duration of the experiments and of easing the analytical procedures. The results of these early studies are to be used with considerable caution owing to possible artifacts from the incursion of radiolytic processes.

4 EXPERIMENTAL ASPECTS OF THE TECHNIQUE

Having dealt with the principles and salient features of the decay technique, we can now proceed to examine the general outline of a typical application to structural and mechanistic problems in the realm of gas-phase ion chemistry. The major steps involved can be itemized as follows:

1. Synthesis of a monotritiated precursor of the cation of interest and study of its decay-induced fragmentation pattern under isolated-molecule conditions by mass spectrometric techniques.
2. Synthesis and purification of a suitable multitritiated precursor, followed by its characterization by mass spectrometric, nuclear magnetic resonance (NMR), or chemical degradation techniques, in order to establish the isotopic composition and the intramolecular tritium distribution.
3. "Blank" runs carried out under the same conditions as the planned decay experiments, except for the use of a monotritiated precursor, to verify the absence of radiolytic artifacts.
4. Actual decay experiments, involving storage of the multitritiated precursor in gaseous systems of carefully controlled composition, at appropriate pressures and temperatures, for periods of several months. In certain cases,

extension of the study to the liquid phase can be in order for comparative purposes.

5. Identification and quantitative analysis of the labeled decay products by radio chromatographic techniques, in particular by radio GLC and HPLC.
6. Isolation of the products by preparative chromatography, following addition of the corresponding inactive carriers, and application of chemical derivatization and/or degradation techniques aimed at measuring the intramolecular distribution of tritium.

The above experimental steps can be complemented by theoretical calculations, which are particularly useful to estimate the excess internal energy of the daughter ions arising from their "deformed" structure.

4.1 Synthesis of Multitritiated Compounds

The simplest and most readily available starting reagent for the synthesis of multitritiated molecules is molecular tritium itself, T_2, which finds a number of industrial applications, and is commercially available in a state of high purity ($>99\%$), corresponding to a specific activity of 58,250 Ci mol^{-1}. Other multitritiated compounds are not listed in the radiochemicals catalogs (even though a few can be supplied on special order by commercial labeling services) and must be synthesized in the laboratory.

Several types of synthetic approaches to multitritiated compounds have now been described, all of them using molecular tritium as the starting reagent. In this way, no-carrier added (NCA) T_2O, specific activity, ca. 58,000 Ci mol^{-1}, can readily be obtained by oxidizing T_2 molecules with CuO at 450°C [52–55]. NCA T_2O is a key reagent required by a number of hydrolytic processes leading to multitritiated organic compounds.

Thus, NCA CT_4, specific activity 116,000 Ci mol^{-1}, was prepared by treatment of pure aluminium carbide, Al_4C_3, with NCA T_2O at 150°C [52–57]:

$$T_2 \xrightarrow[450°C]{CuO} T_2O \xrightarrow[150°C]{Al_4C_3} CT_4 \tag{5}$$

If, instead of Al_4C_3, calcium carbide, CaC_2, or lithium carbide, Li_2C_2, are treated with T_2O at 150°C, we obtain bitritiated acetylene of very high specific activity, ca. 58,000 Ci mol^{-1} [58]:

$$T_2O \xrightarrow[150°C]{CaC_2 \text{ or } Li_2C_2} C_2T_2 \tag{6}$$

Room-temperature trimerization of C_2T_2 over a silica-alumina catalyst, activated with 0.2% w/w K_2CrO_4, gives almost quantitative yields of hexatritiated benzene, C_6T_6, whose specific activity exceeds 170,000 C mol^{-1} [58]. When propyne (10 mol%) is added to C_2T_2, ring-multitritiated toluene is formed together with C_6T_6 [58].

Multitritiated arenes, labeled at specific positions, have been obtained by an alternative procedure, namely, by decomposition of the corresponding Grignard compounds by T_2O. Thus, for instance, 1,4-T_2-benzene was successfully prepared by hydrolysis of 1,4-phenylenebis-[bromomagnesium] with NCA T_2O [59]:

$$T_2 \xrightarrow[-Cu]{+CuO} T_2O$$

(7)

By the same approach, 1,4-T_2-butane was synthesized from the corresponding 1,4-butylenebis-[bromomagnesium] precursor [60].

An adaptation of the method described by Leblanc et al. [61] was used to prepare side-chain multitritiated toluene by interaction of benzotrichloride with a Zn mirror at 200°C, followed by reaction with T_2O [58]:

$$C_6H_5CCl_3 + 3T_2O + 3Zn \rightarrow C_6H_5CT_3 + 3TCl + 3ZnO \qquad (8)$$

A number of multitritiated aliphatic and alicyclic compounds has been prepared adopting an entirely different strategy based on the catalytic addition of a T_2 molecule to double and triple bonds. 1,2-T_2-ethane [62], -propane [63], and -cyclopentane [64] were obtained from the corresponding olefins by hydrogenation with T_2 over a special Cr_2O_3 catalyst at -10 to $-7°C$. Under such conditions, H/T scrambling processes, which affect reactions carried out over metallic catalysts, leading to mixtures of products containing a different number of T atoms are largely suppressed. Randomly multitritiated propane [65], cyclobutane [43], and cyclohexane [66] were instead prepared from the corresponding alkenes and cycloalkenes by reaction with pure T_2 over a platinum black catalyst. Finally, the preparation of NCA-multitritiated ethene was performed by addition of T_2 to acetylene in the presence of a Lindlar catalyst [67]. The catalyst had been previously poisoned by quinoline vapor in order to increase its selectivity toward the triple bond and to minimize further reduction of ethene as it was formed.

In principle, one of the most convenient ways to overcome the formidable problems associated with the NCA synthesis of more complex multitritiated compounds required for chemical and biological applications is the nuclear-decay method itself. For instance, the decay of CT_4 is known to give, irrespective of the environment, nearly quantitative ($>82\%$) yields of methyl ions, whose subsequent attack on practically all nucleophiles leads to the formation of multitritiated products, which generally retain the radioactive atoms exclusively in the CT_3 group. In spite of the low decay rate (ca. 0.46% per month), this synthetic approach can provide within reasonable periods of time quite respectable activities of multiply tritiated products, largely sufficient for most applications, provided that sufficiently high activities of the starting CT_4 precursor are used. As an example, 2.8 mCi (86.1%) of methanol, labeled exclusively in the methyl group, were obtained by allowing 0.6 Ci of multitritiated methane to decay for 14.5 days in water [56]. Another example is afforded by the simultaneous preparation of multitritiated methyl (4.4 mCi, 47%) and ethyl bromide (2.3 mCi, 24%) from the decay of multitritiated methane in a $HBr:CH_4:O_2 = 600$ torr:137 torr:10 torr mixture for 43 days at room temperature [56]. One-step simultaneous preparation of multitritiated 3-methyl- and 2-methyl-furan was attained by allowing CT_4 to decay in gaseous furan (150 torr) for 30 days at 100°C [68].

Remarkable advantages of the decay synthesis over the conventional procedures are the following: first, multitritiated products from the decay are obtained in an essentially carrier-free state. Their continuous build up makes the procedure a convenient and long-lasting source of the desired multilabeled material. Finally, after the synthesis, the bulk of the undecayed, multitritiated precursor (e.g., CT_4) can be recovered virtually unchanged from the reaction mixture.

4.2 Dilution of Multitritiated Compounds

The preparation of multitritiated compounds represents an inherently difficult task owing to the high specific activity of tritium (ca. 60,000 Ci mol^{-1}), which causes fast self-radiolytic decomposition of the synthesized tritiated product and formation of radiolytic by-products, including those containing a lower number of T atoms per molecule. In some instances, self-radiolysis of the tritiated species may lead to its extensive polymerization, as in the case of multitritiated ethene and acetylene, which causes rapid loss of the tritiated monomer.

As mentioned earlier, in order to reduce such undesired radiolytic processes, it is necessary to decrease the specific activity of the sample to sufficiently low levels by diluting the multitritiated product with an inactive compound.

Generally, the diluent used is the unlabeled analog of the multitritiated product, which can facilitate the subsequent purification and characterization of the radioactive sample. The necessary dilution level can be estimated with a calculation based on the $G_{(-M)}$ value of the compound involved and the radiation dose received by the system. As an example, it has been found that the rate of formation of tritiated radiolytic products in alkanes is reasonably small at specific activity levels below 0.1 Ci mol^{-1} [69].

In some instances, simple dilution of the multitritiated compound with the inactive carrier is not sufficient to inhibit secondary processes, since the inactive compound itself may concur actively to maintaining secondary chain processes, such as isotope scrambling and polymerization, induced by the nuclear decay of the labeled molecule. For instance, the radiolysis of methane by the energetic β particles from the nuclear transition leads to formation of CH_4^+ and, eventually, of CH_5^+ ions [70], which promote a fast quasiresonant proton transfer process,

$$CH_5^+ \xrightarrow[-CH_4]{+CT_4} CT_4H^+ \xrightarrow[-CH_4T^+]{+CH_4} CT_3H \xrightarrow[-CH_4]{+CH_5^+} CT_3H_2^+ \to \to CTH_3 \qquad (9)$$

resulting in a complete randomization of the T atoms among the methane molecules of the mixture [71].

The extent of multiple isotope scrambling 9 can be minimized by addition of an efficient proton acceptor (e.g., NH_3) to the gaseous mixture. Alternatively, the CT_4 can be diluted by an inert additive (e.g., Ar) unable to maintain a chain isotope scrambling similar to sequence 9 [72].

The same strategy can also be adopted in the systems, such as multitritiated ethene at low specific activity, undergoing fast self-radiolytic polymerization. Addition of both NH_3 and O_2 to the gaseous mixture inhibits the radiation-induced polymerization of the alkene, initiated, respectively, by cationic and radical species [67].

In the case of bitritiated acetylene, addition of the same additives does not appreciably reduce the rate of polymerization and the NCA-labeled molecule must therefore be diluted with an inert gas [58].

4.3 Purification and Characterization of Multitritiated Compounds

After dilution with a large excess of inactive carrier, the crude multitritiated compound is subjected to a rigorous purification by preparative chromatographic procedures. Once chemically pure, the multitritiated product is assayed for isotopic purity by conventional analytical techniques. Both mass spectrometry and ^{3}H-NMR spectrometry have frequently been used for this purpose. However, whereas mass spectrometry only allows measurement of the overall T atom content in a labeled molecule, ^{3}H-NMR provides otherwise inaccessible information about the relative position of the labels in the multitritiated molecule. For instance, isotopic analysis of the purified tritiated benzene from reaction 7 by ^{3}H-NMR spectroscopy showed the major component (56%) of the product to be 1, 4-T_2-benzene, the remainder being mainly T-benzene (Fig. 2). No evidence for the presence of significant amounts of 1, 3-T_2- and 1, 2-T_2-benzene was obtained [59].

Dealing with relatively simple molecules, their purification and the characterization can be achieved in a single step. This happens, for instance, in the case of CT_4, generated via reaction 5 together with variable proportions of CT_3H, CT_2H_2, and CTH_3 [52]. Once diluted with a large excess of CH_4, the radioactive mixture was completely resolved by a preparative gas chromatographic (GC) procedure over a special capillary column [73]. Figure 3 shows the separation achieved in the analysis of the crude multitritiated methanes mixture, and Fig. 4 illustrates the analysis of the CT_4 fraction recovered from a preparative GC step.

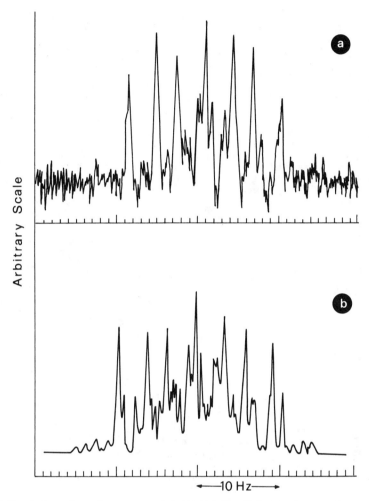

Fig. 2 (*a*) Resolution-enhanced proton-coupled 213.47-MHz FTT (Fourier Transform Tritium) NMR spectrum of the T_x-benzene ($x = 1, 2$) mixture from the reaction of the appropriate Grignard compound with T_2O (Ref. [59]). (*b*) Calculated 213.47-MHz FTT NMR spectrum of a 44:56 $C_6H_5T/p\text{-}C_6H_4T_2$ mixture (Ref. [59]).

4.4 Preparation of the Reaction Mixture

By virtue of the nuclear nature of its ionogenic process, the decay method allows the study of the reactions of the ion of interest with a specific compound, alone or in competition with a reference molecule, in any state of aggregation. Three experimental parameters play an important role in the design of the experiment: (1) the specific activity level of the mixture, (2) the neutral substrate(s) involved, and (3) environmental conditions, e.g., the range of temperature and pressure.

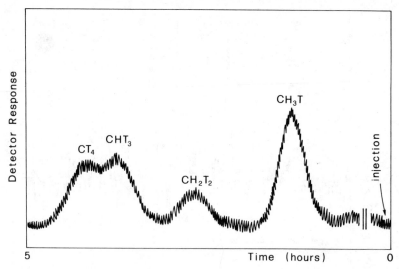

Fig. 3. Isotopic gas–solid chromatographic analysis of a mixture of tritiated methanes from the reaction of T_2O with Al_3C_4 (Ref. [53]).

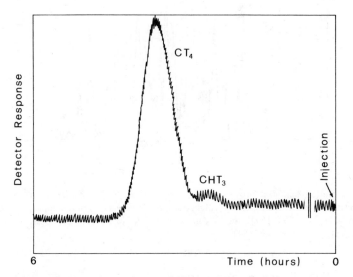

Fig. 4. Isotopic composition of the CT_4 sample following purification by preparative gas–solid chromatography (Ref. [53]).

4.4.1 Choice of the Specific Activity of the Mixture

As discussed earlier in the chapter, an excessive specific activity of the mixture favors the occurrence of undesired radiolytic processes, which may lead to formation of labeled products indistinguishable from those of the ion–molecule reactions under investigation. In order to minimize the effects of self-radiolysis, the labeled specimen must be diluted with either the neutral substrate(s) or some suitable inert additive (e.g., a noble gas) to a specific activity of the order of 0.1 Ci mol^{-1}.

4.4.2 Choice of the Neutral Substrate(s)

The main component of the decay mixture is normally represented by a neutral reactant, which serves as an interceptor of the ion. If the aim of the study is to generate and characterize a specific nucleogenic ion, and/or to evaluate its unimolecular isomerization, the neutral interceptor chosen must satisfy several requisites. It should present only a single nucleophilic center in order to yield the simplest possible pattern of products. It must produce simple ionic derivatives, wherein any conceivable secondary isomerization or fragmentation or loss of the label is excluded. Finally, it must be chemically and radiochemically inert toward the labeled precursor of the nucleogenic ion. If, instead, the purpose of the study is to evaluate the reactivity pattern of the ion in a given substrate, it could be necessary to resort to competition kinetics by adding to the substrate another suitable reactant. The presence of an inert moderator in the reaction mixture is needed if the interest is focused on the effects of the excess internal energy of the nucleogenic ions on their reactivity. In some cases, a suitable neutral compound can be introduced into the reaction system in order to convert the decay ions into some other charged species of mechanistic interest. This is illustrated by the decay of CT_4 in a large excess of carbon monoxide, where the multitritiated methyl cations are converted into acetylium cations,

$$CT_4 \xrightarrow{\beta\,decay} CT_3^+ \xrightarrow{CO} CT_3CO^+ \xrightarrow{S} products \qquad (10)$$

which subsequently interact with the neutral substrate [74].

4.4.3 Choice of the Experimental Conditions

Once the neutral interceptor(s) of a given nucleogenic ion is defined, the experimental conditions most suitable to obtain the maximum amount of mechanistic information remain to be established. As shown in Table 1, the position occupied by the process of interest in the time scale of events following the nuclear decay of the labeled molecule determines the experimental conditions to be adopted in the decay experiment to obtain information on that process. Thus, for instance, if we are interested in defining the initial structure of a nucleogenic ion, efficient trapping of the ion soon after its formation is needed in order to prevent any conceivable secondary isomerization processes. Accordingly, the ion must be generated in the liquid phase in the presence of a very efficient interceptor in order to make its lifetime as short as possible (ca. 10^{-13} s). Of course, kinetic information upon structural rearrangement of the ion occurring at longer lifetimes may be achieved by trapping the ion in the gas phase with the same interceptor in any convenient range of pressure

(10^{-10}–10^{-8} s). A hint into the thermodynamically most stable structure of the same ionic species can be provided by the results of experiments carried out at the lowest pressure (ca. 1 torr), i.e., at the highest ion lifetime (up to 10^{-7} s) attainable in a decay experiment.

The same considerations apply to the ionic intermediates formed from an exothermic ion–molecule reaction. If we are interested in determining the site selectivity of the ion toward a multidentate substrate, care must be taken in eliminating secondary isomerization and selective fragmentation of the corresponding ionic adducts excited by the exothermicity of their formation process.

In this case, the reaction has to be carried out in the liquid phase or, if we wish to deal with unsolvated species, in the gas phase at the highest attainable pressure, i.e., under conditions in which extensive secondary isomerization and fragmentation of the ionic adducts are prevented by efficient collisional quenching by the bath gas. An additional way to prevent secondary processes in the ionic adducts from an ion–molecule reaction is to favor their neutralization via deprotonation processes, which is achieved by adding significant concentrations of a suitable base into the reaction mixture. If, instead, we wish to look at the inter- or intramolecular isomerization of the excited ionic adducts from exothermic ion–molecule reactions, experimental conditions are required that increase the lifetime of the adducts; i.e., the experiment should be carried out in the gas phase at low pressures and in the absence of bases.

In some instances, interest can be focused on the effects of solvation upon the reactivity of an ionic species. In these cases, *the same* ion–molecule reaction must be studied both in the liquid phase and in the gas phase at different

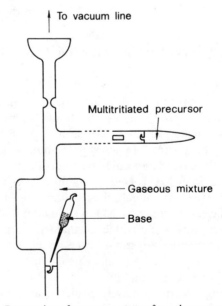

Fig. 5. Preparation of a gaseous system for a decay experiment.

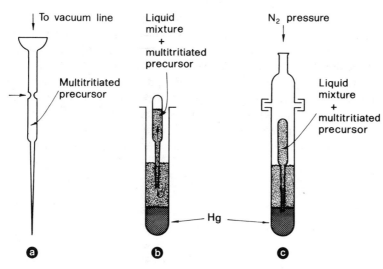

Fig. 6. Preparation of a liquid system for a decay experiment: (*a*) filling the vial with the multitritiated precursor, (*b*) filling the vial with the liquid substrate, (*c*) closing the vial with a mercury seal and eliminating gaseous bubbles.

pressures and temperatures, namely, under conditions in which ion solvation is known to be largely different.

Gas-phase nuclear-decay experiments are normally carried out by diluting the multitritiated precursor of the ion of interest with the substrate(s) until a conveniently low specific activity is attained. The mixture is then sealed off in break-seal tipped Pyrex vessels containing the necessary additives, e.g., a base (Fig. 5). In the decay experiments carried out in the liquid phase, care must be taken to avoid that the ion–molecule reaction of interest occurs as well in the residual volume normally occupied by the vapor in equilibrium with the liquid. This represents a particularly compelling problem when the multitritiated precursor of the ion is highly volatile (e.g., in the case of CT_4) and therefore tends to reside mostly in the vapor phase over the liquid sample. To overcome this problem, a new setup of the liquid-phase decay experiment has been worked out, which eliminates the formation of vapor from the liquid samples by subjecting them to the hydrostatic pressure of a Hg column (Fig. 6).

The sealed vessels are normally stored in the dark, at the selected temperature, for a suitable growth time of the products, which depends upon the specific activity of the mixture and the sensitivity of the analytical method adopted for product analysis. Typical growth periods range from several months to 1 year.

4.5 Analysis of the Reaction Products

The relatively low activity of the decay ions formed during the experiment and consequently the low activity of their neutral derivatives, coupled with the fact that the latter are formed in the presence of a large amount of the undecayed multitritiated precursor, call for an analytical methodology that must combine sensitivity and resolving power. Radio gas chromatography and radio-HPLC represent to date the most convenient methods for the qualitative and quantitative determination of the labeled products from the decay experiments. However, many investigations require that the determination of the nature and the yields of the products be complemented by an accurate determination of the tritium atoms' distribution within their molecule. This information is particularly relevant in mechanistic studies involving the possibility of structural rearrangements. Determination of the distribution of tritium within a labeled molecule is usually accomplished by measuring the decrease of the molar activity when T atoms bound to a given position (or to equivalent positions) are substituted by a suitable inactive group. Such analysis has been carried out, for example, in the study of the tritiated products from the tritonation of toluene by $^3HeT^+$ ions [75, 76] using the substitution reactions shown in Scheme 1.

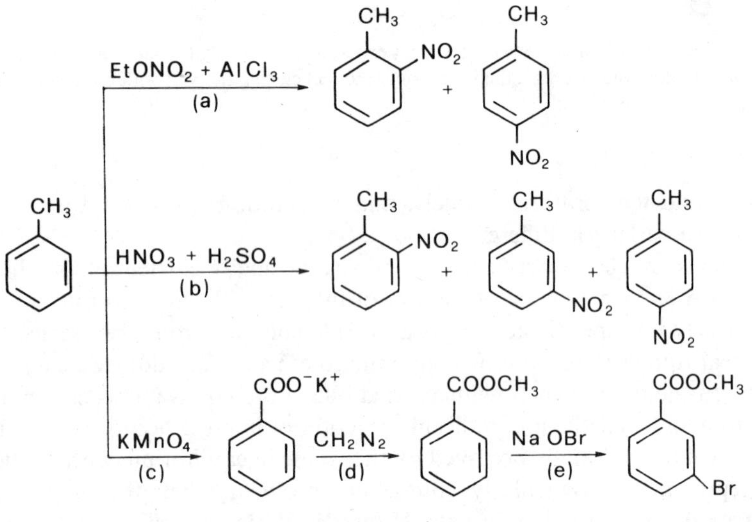

Scheme 1

Cross-degradations of the same compound [e.g., b vs. c; a + c vs. b] must provide the same results within the experimental error. In general, the reactions adopted must exhibit no significant isotope effect [77] nor induce any appreciable T scrambling within the labeled product nor loss of T from the product to the reaction medium.

5 APPLICATIONS

In this section, several examples of application of the decay technique to gas-phase ion chemistry problems are illustrated. The first three examples deal with the structural characterization of a specific ion and the study of its isomerization processes. The subsequent examples concern the study of the

APPLICATIONS 309

reactivity of a given ionic species toward neutral gaseous substrates unaffected by the complicating environmental factors, such as solvation and ion pairing, encountered in solution chemistry.

5.1 Proof of Existence of Gaseous Cyclobutyl Cation

Gaseous $C_4H_7^+$ ions have long been detected by mass spectrometry, which, however, could not provide any evidence about their structure (cyclic or open) owing to its limited structural discrimination. Theoretical calculations on the existence and the relative stability of cyclic $C_4H_7^+$ ions in the gas phase led to somewhat conflicting conclusions.

Earlier STO-3G calculations identified the cyclopropylcarbinyl cation in the bisected configuration 1 as the most stable cyclic $C_4H_7^+$ structure into which others, such as the planar 3 and puckered 2 cyclobutyl cations, were predicted to collapse immediately without activation energy [78, 79].

$$\text{(2)} \quad \longrightarrow \quad \text{(1)} \quad \longleftarrow \quad \text{(3)} \qquad (11)$$

A conclusive answer to this structural problem was provided by the decay of multitritiated cyclobutane, $c\text{-}C_4X_8$ (X = H, T). Preliminary ICR studies [43] demonstrated that $C_4H_7^+$ ions, of unknown structure(s), are by far the most abundant (85%) daughter ions formed from the nuclear decay of monotritiated cyclobutane and survive undissociated for at least 10^{-6} s.

Generation of $C_4X_7^+$ ions from the decay of $c\text{-}C_4X_8$ in a large excess of suitable gaseous nucleophiles, such as H_2O (6–25 torr) or NH_3 (400 torr), yielded a complex mixture of labeled products [80]. Nevertheless, formation of significant amounts of cyclobutyl (4–35%) and cyclopropylcarbinyl derivatives (3–36%) from all systems investigated provides compelling evidence for the existence of their cyclic ionic precursors, namely, cyclobutyl and cyclopropylcarbinyl cation. The reactions of the gaseous ions with the nucleophile NuH (NuH=H_2O or NH_3), followed by proton transfer from the intermediate cyclic onium ions 4 to a gaseous base B, contained in the system, accounts for the formation of the observed products:

$$c\text{-}C_4X_8 \xrightarrow{\beta\,\text{decay}} c\text{-}C_4X_7^+ \xrightarrow[(H_2O\,\text{or}\,NH_3)]{NuH} c\text{-}C_4X_7NuH^+ \xrightarrow[-BH^+]{+B} c\text{-}C_4X_7Nu$$

$$\text{(4)} \qquad\qquad (12)$$

It was concluded that both cyclobutyl and cyclopropylcarbinyl cations do indeed exist in the dilute gas state at least for the time ($> 10^{-9}$ s) necessary to be trapped by the nucleophile and, therefore, must be regarded as fully legitimate ionic intermediates characterized by significant local minima on the $C_4H_7^+$ potential surface.

Following these unequivocal results, theoretical reinvestigation of the $C_4H_7^+$ potential surface using the more flexible split-valence 4–31 G and 6–31 G* basis sets led to the discovery of a pair of equivalent asymmetrical minimum energy forms in the region of the bisected cyclopropylcarbinyl cation **1** corresponding to a $C_4H_7^+$ structure, which is consistent with the conclusions of the nuclear-decay experiments [81].

5.2 Proof of Existence and Isomerization of Cyclohexyl Cation

The cyclohexyl cation is recognized as a legitimate ionic intermediate in a variety of reactions occurring in solution [82]. The situation is entirely different for the *free* cyclohexyl cation in media of very low nucleophilicity or in the gas phase. Any attempt to detect $c\text{-}C_6H_{11}^+$ ions in superacidic solutions invariably failed, the 1-methylcyclopentyl ion being the only observable species. Careful mass spectrometric studies supported this conclusion, showing that the cyclohexyl cation from various neutral precursors apparently rearranges to the 1-methylcyclopentyl structure already in its "incipient" state, thus excluding the existence of the cyclohexyl cation in the gas phase [83]. This rationalization of the mass spectrometric results, although undoubtedly correct, refers, however, to the composition of the ionic population after the relatively long delay ($\geqslant 10^{-5}$ s) between generation of ions in the source and their detection. Clearly, such evidence does not exclude that the gaseous cyclohexyl cation is actually formed and, in the absence of collisional stabilization, undergoes isomerization to the more stable 1-methylcyclopentyl cation before the late structural assay allowed by mass spectrometric techniques.

A conclusive answer to this question was provided by the decay technique [66]. Multitritiated cyclohexane, with an average T content of 2.2 ± 0.2 atoms per molecule, was allowed to decay at room temperature in mixtures containing a large excess of an inert moderator (CMe_4, $SiMe_4$, $c\text{-}C_6H_{12}$) in the presence of smaller amounts (0–30 mol%) of a suitable nucleophilic interceptor (MeOH, 1,4-$C_4H_8Br_2$), either in the gas phase at pressures ranging from 5 to 735 torr or in solution. Radio gas chromatographic analysis of the decay products from liquid systems revealed the almost quantitative (96–100%) retention of the cyclohexyl structure, irrespective of the specific nucleophile used to trap the decay ions.

In gaseous hydrocarbons, isomerization of the decay ions to the 1-methylcyclopentyl structure occurs to an extent depending on the total pressure of the system (e.g., 68–74% at 720 torr, 97–100% at 50 torr) rather than on the nature of the specific nucleophile used.

In both gaseous and liquid systems, no evidence for the formation of isomerization derivatives other than those of 1-methylcyclopentyl cation was obtained, in spite of a specific search.

The results provide compelling evidence for the existence of free cyclohexyl cation in the dilute gas state, as well as in the liquid phase, with a lifetime of at least 10^{-8}–10^{-7} s. It follows that the cyclohexyl structure must correspond to a local minimum on the $C_6H_{11}^+$ energy surface. However, the extensive isomerization observed in the gas phase, even at pressures as high as 720 torr, indicates that such a local minimum must be relatively shallow. Classical RRKM (Rice-Ramsperger-Kassel-Marcus) calculations on a nucleogenic cyclohexyl cation with an excess vibrational energy estimated around 30 kcal mol^{-1} led to an approximate value of the energy barrier for isomerization to 1-methylcyclopentyl cation around 10 kcal mol^{-1}. These findings provide a satisfactory explanation of the persistent failure to detect and characterize gaseous c-$C_6H_{11}^+$ with other experimental techniques, in particular with "structurally diagnostic" mass spectrometric methods [83]. In fact, taking into account the remarkably low activation barrier involved, rearrangement of c-$C_6H_{11}^+$ to 1-methylcyclopentyl ion is expected to reach completion within the time lag between the generation of the ion in the source of the mass spectrometer and its structural assay ($> 10^{-6}$ s).

Concerning the c-$C_6H_{11}^+$ isomerization mechanism, no neutral compounds suggestive of the intervention of cyclopentylmethylium or bicyclohexylium ions could be detected among the radioactive products from the gaseous and liquid systems, including those consisting of neat liquid MeOH. Such negative evidence suggests that if the above species are actually involved in the ring contraction, they are best characterized as ionic transition states (lifetime $< 10^{-13}$ s), and therefore the isomerization of cyclohexyl cation to the 1-methylcyclopentyl structure is likely to proceed via concerted mechanisms.

5.3 Automerization of Phenylium Ion

The existence, the reactivity, and the electronic configuration of phenylium ion **5** (singlet **5a** vs. triplet **5b**), have been a matter of debate since Waters'

(5a)   (5b)

proposal [84] of its intermediacy in the decomposition of benzenediazonium salts [85]. A contribution to the characterization of the phenylium ion, both in the dilute gas state and in solution, has been provided by the decay technique [86].

1,4-T$_2$-benzene [59] was allowed to decay in liquid methanol or in gaseous methanol at pressures ranging from 5 to 65 torr. The relative distribution of the tritiated products recovered from the decay experiments is reported in Fig. 7. Under all conditions, anisole is the major product, arising from the attack of a singlet nucleogenic phenylium ion **5a** on the n-type center of methanol.

The tritium distribution within the recovered anisole was determined by chemical degradation procedures, *via* systematic substitution of the T atoms of the ring by suitable inactive groups, as shown in Scheme 2:

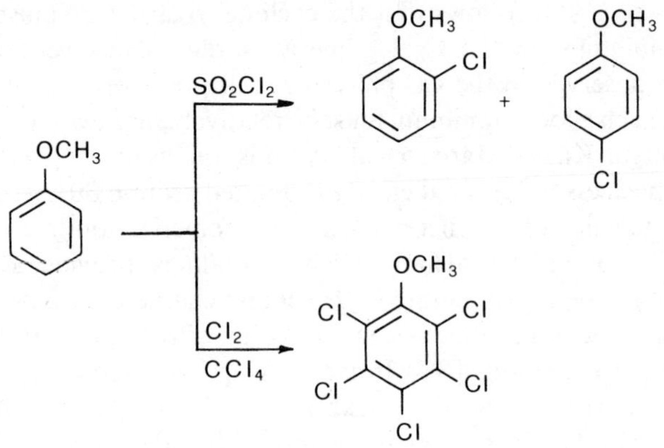

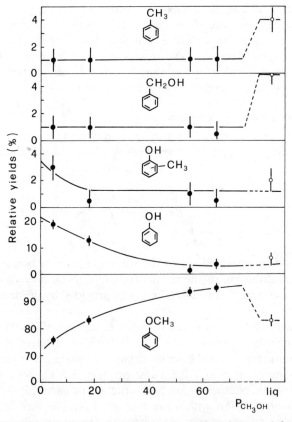

Scheme 2

Fig. 7. Relative distribution of the tritiated aromatic products recovered from the decay of benzene-1,4-T_2 in liquid (open circles) or gaseous (solid circles) MeOH (Ref. [86]).

The relevant data are reported in Fig. 8. The label migration, observed in the anisole from the gas-phase decay experiments, was ascribed to a degenerate rearrangement of the nucleogenic phenylium ion (path a in Scheme 3) *before* addition to MeOH, since other conceivable isomerization processes taking place in the oxygen-protonated anisole intermediate from addition of **5a** to MeOH (path b in Scheme 3) were excluded by the results of ancillary radiolytic experiments [86]:

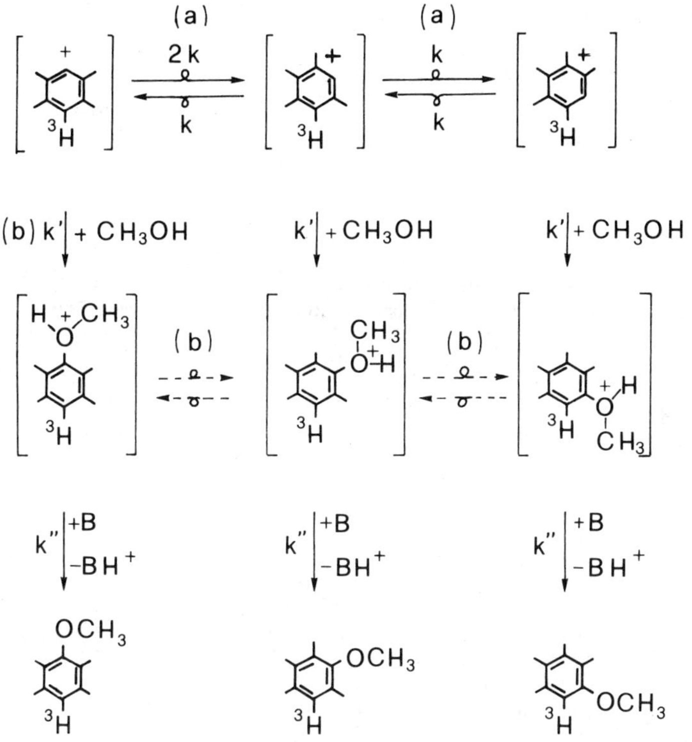

Scheme 3

The complete retention of the original T distribution observed in the anisole from the liquid-phase reaction indicated that automerization is slow in comparison with the collision frequency in the condensed phase. An estimated value of $k = 10^7\text{–}10^8 \, \text{s}^{-1}$ for the phenylium ion automerization rate constant (Scheme 3) was obtained from the best fitting of the experimental points of Fig. 8, with integrated rate equations calculated for the automerization reaction network of Scheme 3.

Observation of a relatively fast phenylium ion automerization process in the gas phase was in contradiction with theoretical predictions of high energy barriers ($\Delta E^{\neq} = 44\text{–}77 \, \text{kcal mol}^{-1}$) for the same process [38], irrespective of the nature of the intermediate species or transition states involved. On these grounds, occurrence of the phenylium ion automerization sequence (path a of Scheme 3) was questioned by Dewar and Reynolds [87], who attributed the

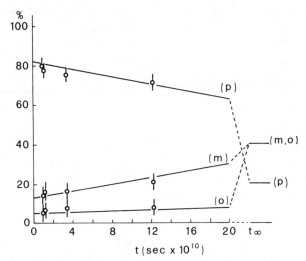

Fig. 8. Phenylium ion automerization extent as a function of the partial pressure of CH_3OH. The curves represent the integrated rate equations from path a of Scheme 3 (relative yield of T-phenylium ions: p, p-T-$C_6H_4^+$; m, m-T-$C_6H_4^+$; o, o-T-$C_6H_4^+$). The open circle denote the relative distribution of the corresponding isomeric T-anisoles from the gas-phase decay runs.

scrambling of the label in tritiated anisole to consecutive hydrogen shifts within the protonated anisole formed by addition of a *para*-T-phenylium ion on MeOH:

(13)

However, further compelling evidence against sequence 13 and in favour of path a of Scheme 3 was obtained [88] by using different nucleophilic interceptors of phenylium ion, such as methyl halides CH_3X (X = F, Cl, Br). In fact, the corresponding methylphenylhalonium intermediates **6** from the gas-phase attack

(14)

of phenylium ion on the *n*-center of CH_3X, can hardly undergo tritium shifts within the aromatic ring, owing to the lack of mobile hydrogens.

A close correspondence between the T distribution within the neutral halobenzenes from reaction 14 and that within labeled anisole from the previous MeOH decay experiments was obtained. These findings indicate that the same mechanism is operative in both CH_3X and CH_3OH decay systems, which requires that phenylium ions undergo automerization *before* being trapped by the nucleophile (path a of Scheme 3).

The criticism leveled in early theoretical studies at the results of the decay experiments failed to consider that the automerization of *nucleogenic* phenylium ions can be allowed by their vibrational excitation arising from the relaxation of the regular hexagonal structure of tritiated benzene into the highly distorted geometry of the ground-state, singlet Ph^+ ion. In essence, the problem boils down to an appraisal of the difference between the energy barrier of the automerization and the deformation energy of the nucleogenic Ph^+ cation. Recent theoretical calculations have gone a long way toward removing the discrepancy with the results of decay experiments. In fast, a refined study at the MP2/6 31G** level has set a limit of over $32 \, kcal \, mol^{-1}$ to the deformation energy of the nucleogenic phenylium ions, whereas the latest estimate of the barrier for their automerization is down to about $40 \, kcal \, mol^{-1}$ [39].

5.4 Gas-Phase Protonation of Five-Membered Heteroaromatic Rings

Extensive applications of the decay technique in the field of electrophilic aromatic substitutions in the gas phase proved of value in estimating the *intrinsic* reactivity and selectivity of a variety of free carbocations and in comparing the results with those from the study of the corresponding reactions in solution.

The same approach has recently been used to investigate the intrinsic reactivity of heteroaromatic substrates toward a free ionic electrophile.

In fact, it is well known that in solution electrophilic substitution in pyrrole and furan occurs predominantly at the α-carbon [89,90], although most theoretical calculations [91–95] at the semiempirical and *ab initio* levels assign a larger electronic density to the β than to the α positions. Moreover the corresponding molecular electrostatic potentials [96,97] established within the encounter pair from an isolated heteroaromatic molecule and a positive point charge indicate favorable pathways of electrophilic attack at the β ring carbons, as well as at the heteroatom.

Such discrepancies between theory and experimental data can, in principle, be traced either to a failure of the calculation methods or to profound effects of the reaction environment, which may mask the *intrinsic* directive properties of the heteroaromatic compounds.

To discriminate between these possibilities, the $^3HeT^+$ ion, conveniently generated by the β decay of T_2, was allowed to react in the gas phase with pyrrole, N-methylpyrrole, furan, and thiophene [98]. The ability of the nuclear-decay method of generating free, unsolvated ions in the gas phase, coupled with the recognition that $^3HeT^+$ ion is a reasonable approximation of the positive point charge considered in theoretical investigation of heteroaromatic reactivity, make the decay approach best suited to provide experimental results against which theoretical predictions can correctly be compared.

The tritiated heteroaromatic substrate was the major product from all systems investigated, and its intramolecular T distribution was determined by the following degradation procedure (Scheme 4):

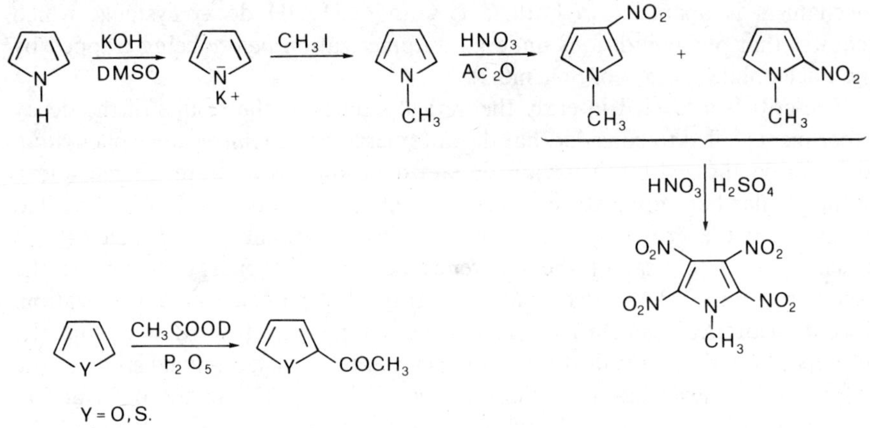

Scheme 4

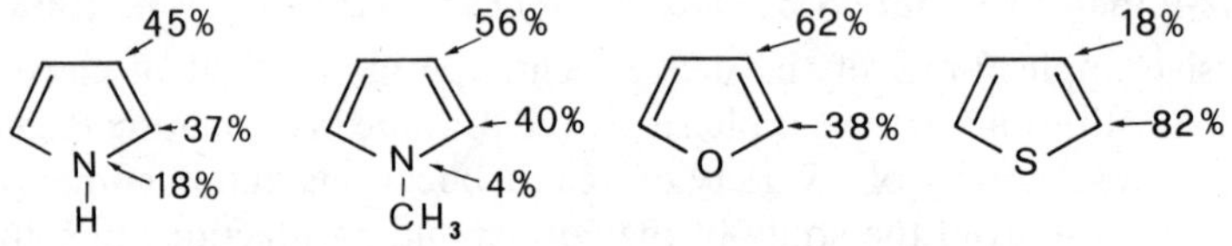

Fig. 9. Positional selectivity of gaseous ^{3}HeT$^+$ ions toward simple heteroaromatic molecules (Ref. [98]).

From T atom distribution reported in Fig. 9, it is concluded that the *intrinsic* directive properties of the selected substrates toward a free protonating agent, such as ^{3}HeT$^+$, correlate well with theoretical predictions based upon the molecular electrostatic potential established in the encounter pair [96, 97].

Clearly, no correspondence exists between the above gas-phase results and those of more conventional reactions in solution, where predominant α substitution is invariably observed [89, 90]. The discrepancy points to the influence of the reaction medium (solvation, ion pairing, catalyst, etc.) as the main factor responsible for the "classical" reactivity and selectivity features of heteroaromatics toward electrophilic substitution in solution.

5.5 Addition of Vinyl Cation to Methane

The direct study of the insertion reaction of vinyl cation into the C—H bonds of methane is inaccessible to conventional mass spectrometric methods. In fact, unless effectively stabilized by collisional deactivation, the primary $C_3H_7^+$ adduct 7, excited by the exothermicity (59 kcal mol^{-1}) of the reaction, undergoes complete fragmentation into $C_3H_5^+$ ions of unknown structure:

$$C_2H_3^+ + CH_4 \rightarrow [C_3H_7^+]_{exc} \begin{array}{l} \rightarrow C_3H_7^+ \quad (15a) \\ \rightarrow C_3H_5^+ + H_2 \quad (15b) \end{array}$$
(7)

In summary, sequence 15b, whose overall exothermicity is 23 kcal mol^{-1}, is the only process detectable under mass spectrometric conditions. Its specific rate, deduced from mass spectrometric kinetic measurements [99–101], ranges from $0.87 \pm 0.12 \times 10^{-10}$ to $2.2 \pm 0.2 \times 10^{-10}$ cm^3 molecule^{-1} s^{-1}, depending upon the kinetic approach used.

Application of the decay technique to this problem required the synthesis of multitritiated ethene (C_2X_4; $X = H, T$) as the source of labeled vinyl cation. Preparation of the precursor by partial reduction of acetylene with T_2 over a quinoline-treated Lindlar catalyst gave a 56:44 mixture of $C_2H_2T_2$ and C_2H_3T [67].

The purified mixture was allowed to decay in CH_4 (60–720 torr), containing a radical scavenger (O_2, 4 torr), a gaseous base (NMe_3, 0.3 torr), and a suitable trapping nucleophile, e.g. 1,4-$C_4H_8Br_2$, benzene or MeOH [102]. The latter were used to sample the population of gaseous cations by fast reactions, yielding products whose structure unequivocally reflects that of the intercepted cation, e.g.,

$$C_2X_3^+ + 1,4\text{-}C_4H_8Br_2 \rightarrow C_2X_3Br + C_4H_8Br^+ \qquad \Delta H^0 = -40 \text{ kcal mol}^{-1} \tag{16a}$$

$$C_2X_5^+ + 1,4\text{-}C_4H_8Br_2 \rightarrow C_2X_5Br + C_4H_8Br^+ \qquad \Delta H^0 = -7 \text{ kcal mol}^{-1} \tag{16b}$$

$$i-C_3X_7^+ + 1,4\text{-}C_4H_8Br_2 \rightarrow i\text{-}C_3X_7Br + C_4H_8Br^+ \qquad \Delta H^0 = -6 \text{ kcal mol}^{-1} \tag{16c}$$

$$i-C_3X_7^+ + 1,4\text{-}C_4H_8Br_2 \rightarrow C_3X_6 + C_4H_8Br_2H^+ \qquad \Delta H^0 = \sim 0 \text{ kcal mol}^{-1} \tag{16d}$$

where $C_4H_8Br^+$ denotes a tetramethylenebromonium ion. The other nucleophiles used yield as well structurally diagnostic products, e.g., styrene, allylbenzene, cumene, and propene from C_6H_6, methyl vinyl ether, allyl methyl ether, isopropyl methyl ether, and propene from MeOH, etc. Sampling the same ionic population with different nucleophiles is clearly useful to overcome problems arising from inefficient or "blind" channels affecting the reactivity of a specific nucleophile.

The decay experiments have allowed to detect vinyl, allyl, and, for the first time, isopropyl ions in the systems investigated. In fact, the high efficiency of collisional stabilization in the range 60–720 torr makes branch 15a not only detectable, but largely predominant, as illustrated in Fig. 10, which refers to systems containing C_6H_6 as the trapping nucleophile. Analysis of the tritiated products demonstrates that most (>90%) of the $C_3X_7^+$ complexes formed by addition of nucleogenic $C_2X_3^+$ ions to CH_4 and stabilized by collisional deactivation have the isopropyl ion structure when trapped by the nucleophile, ca. 10^{-9} s after their formation. Since the insertion of a vinyl ion into a C—H

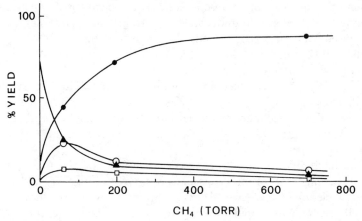

Fig. 10. Pressure dependence of the relative yields of *i*-propyl (solid circles), *n*-propyl (open circles), allyl (open squares), and vinyl (solid triangles) products from vinyl cation attack on gaseous methane (Ref. [102]).

bond of methane cannot yield i-$C_3X_7^+$ without some kind of rearrangement, we can speculate as to whether such structural change involves the intermediacy of a protonated cyclopropane moiety, as suggested by the results of an ion-beam scattering study at energies in the 0.5–3.2-eV range [103]. The remarkably low yields of n-propyl derivatives and the lack of c-C_3X_6 among the products fail to support this view, at least as long as the formation of a persistent c-$C_3H_7^+$ intermediate (lifetime $> 10^{-9}$s) is concerned. The product's pattern shows that the limited fraction of the $C_3X_7^+$ ions that retain sufficient energy to fragment yields allyl, rather than 2-propenyl cations, consistent with the higher activation energy for the dissociation into the 2-propenyl structure established by metastable transition studies [104].

Finally, the linear dependence of the

$$\{[C_3X_5^+] + [C_3X_7^+]\}:[C_2X_3^+]$$

ratio on the [CH$_4$]:[nucleophile] ratio, illustrated in Fig. 11, allows a direct estimate of the specific rate of the addition of vinyl ion to methane, based on the reasonable assumption that trapping of vinyl, isopropyl, and allyl ions by the nucleophiles is a relatively fast process. The k value obtained, $5.5 \pm 1.0 \times 10^{-10}$ cm^3 molecule^{-1} s^{-1}, is comparable to, if slightly higher than, those deduced from mass spectrometric methods, a truly remarkable agreement if we consider the assumptions made in both sets of experiments and especially the widely different reaction environment. The calculated value corresponds to about one-half of the Langevin collision limit, 1.2×10^{-9} cm^3 molecule^{-1} s^{-1}, pointing to the high efficiency of the reaction.

In conclusion, the decay technique has allowed the first *direct* study of the insertion of vinyl ion into the bonds of methane, unaffected by secondary

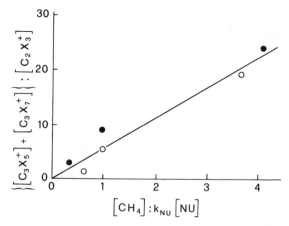

Fig. 11. Dependence of the relative yields of labeled products from vinyl cation attack on gaseous CH_4/Nu mixtures [Nu:benzene, solid circles; methanol, open circles) as a function of the system composition (Ref. [102]).

fragmentation, and the characterization of the charged adducts formed, whose structure could only be speculated upon based on circumstantial mass spectrometric evidence.

6 CONCLUDING REMARKS

The examples reported in the previous section and a survey of published studies bear witness to the powers of the decay technique as a tool for the positive structural characterization of gaseous ions. Two unique features concur to this effect, namely, the actual isolation of structurally diagnostic derivatives of the ions studied and the unrestricted pressure range accessible to the experiments.

Deducing the structure of a charged intermediate from that of a suitable neutral derivative is common practice in solution, and the decay technique allows its extension to the gas phase. Access to pressures exceeding by many orders of magnitude those of mass spectrometric methods is equally essential to structural analysis. In fact, the high frequency of ion–molecule collisions in dense gases, typically of the order of $10^{10}\,s^{-1}$ at 760 torr (Table 1), allows effective stabilization and/or trapping of short-lived ions whose inherent instability or excess internal energy would lead to complete fragmentation or isomerization.

The mechanistic study of ion–molecule reactions, another major application of the decay technique, benefits as well from its high structural resolution, which allows, for instance, determination of the isomeric composition of products and, consequently, the evaluation of crucial kinetic parameters, such as partial rate factors and free energy correlations.

Furthermore, much scope exists for extending the study of ion–molecule reactions to virtually unexplored "high-pressure" domains far beyond the limits of mass spectrometric approaches. For one thing, under conditions ensuring effective collisional stabilization of excited ions, it becomes feasible to study ion–molecule reactions under conditions of *kinetic*, rather than thermodynamic, control of products. Of intrinsic kinetic interest is also the transition from the electrostatic activation mechanism, peculiar to interactions under low-pressure, isolated-pair conditions, to ordinary thermal kinetics prevailing at the high-pressure limit. Only at sufficiently high pressures can the study of gas-phase ionic processes serve one of its major purposes, i.e., to provide generalized and simplified models of the ionic reactions in condensed media, which are driven by thermal rather than electrostatic activation mechanisms.

Finally, owing to the insensitivity of its nuclear ionogenic process to environmental influences, which allows introduction of *the same* free cation in any system of interest, the decay technique shows considerable promise as a means of establishing kinetic and mechanistic correlations between gas-phase and condensed-phase ion chemistry.

ACKNOWLEDGMENTS

The authors are indebted to the Italian National Research Council (CNR) and to the Ministry of Pubblica Istruzione for financial support. They wish also to express their appreciation to G. Naruli and M. Viola for their invaluable help in the preparation of the manuscript.

REFERENCES

1. F. Cacace, in J. Sirchis, Ed., *Proceedings of the Conference on the Methods for Preparing and Storing Marked Molecules, Bruxelles 1963*, Euratom, Bruxelles, 1964, p. 719.
2. G. Stöcklin, *Chemie Heisser Atome*, Verlag Chemie, (1969) and references therein.
3. F. Cacace, in V. Gold, Ed., *Advances in Physical Organic Chemistry*, Academic Press, New York-London, vol. 8, 1970, p. 79.
4. F. Cacace, in F. S. Rowland, Ed., *Hot Atom Chemistry Status Report*, IAEA, Vienna, 1975, p. 229.
5. F. Cacace, in P. Ausloos, Ed., *Interactions Between Ions and Molecules*, Plenum, New York-London, 1975, p. 527.
6. G. P. Akulov, *Usp. Khim.*, **45**, 1970 (1976).
7. G. P. Akulov, *Russ. Chem. Rev.*, **45**, 1008 (1976).
8. F. Cacace, in P. Ausloos, Ed., *Kinetics of Ion-Molecule Reactions*, Plenum, New York-London, 1979, p. 199.
9. F. Cacace, *Adv. Chem. Ser.*, **197**, 33 (1981).
10. M. Speranza, *Gazzetta*, **113**, 37 (1983).
11. F. Cacace, in T. Matsuura, Ed., *Hot Atom Chemistry*, Kodansha Ltd., Tokyo, 1984, p. 161.
12. F. Cacace, in D. A. Bromley, Ed., *Treatise on Heavy Ion Science*, Plenum, New York-London, 1985, vol. 6, p. 63.
13. S. Wexler, in M. Haissinsky, Ed., *Actions Chimiques et Biochimiques des Radiations*, 8ième série, Masson et Cie., Paris, 1965, p. 110, and references therein.
14. A. Migdal, *J. Phys. (USSR)*, **4**, 449 (1941).

15. S. Ikuta, K. Yoshihara, and T. Shiokawa, *Radiochem. Radioanal. Lett.*, **28**, 435 (1977).
16. R. R. Edwards and T. H. Davies, *Nucleonics*, **2** (June), 44 (1948).
17. M. Cantwell, *Phys. Rev.*, **101**, 1747 (1956).
18. M. Wolfsberg, *J. Chem. Phys.*, **24**, 24 (1956).
19. L. Wolniewicz, *J. Chem. Phys.*, **43**, 1087 (1965).
20. S. Ikuta, K. Okuno, K. Yoshihara, and T. Shiokawa, *J. Nucl. Sci. Technol.* (*Tokyo*), **14**, 720 (1977).
21. S. Ikuta, S. Iwata, and M. Imamura, *J. Chem. Phys.*, **66**, 4671 (1977).
22. S. Ikuta, K. Okuno, K. Yoshihara, and T. Shiokawa, *J. Nucl. Sci. Technol.* (*Tokyo*), **14**, 131 (1977).
23. H. H. Michels, *J. Chem. Phys.*, **44**, 3834 (1966).
24. H. M. Schwartz, *J. Chem. Phys.*, **23**, 400 (1955).
25. R. Raitz, K. Luchner, H. Micklitz, and V. Wittwer, *Phys. Lett.*, **A47**, 301 (1974).
26. S. Ikuta, K. Okuno, K. Yoshihara, and T. Shiokawa, *Radiochem. Radioanal. Lett.*, **23**, 213 (1975).
27. S. Ikuta, K. Yoshihara, and T. Shiokawa, *J. Nucl. Sci. Technol.* (*Tokyo*), **14**, 661 (1977).
28. S. Wexler, G. R. Anderson, and L. A. Singer, *J. Chem. Phys.*, **32**, 417 (1960).
29. K. Okuno, K. Yoshihara, and T. Shiokawa, *Radiochimica Acta*, **25**, 21 (1978).
30. S. Ikuta, S. Hashimoto, and M. Imamura, *Chem. Phys.*, **42**, 269 (1979).
31. S. Ikuta, S. Hashimoto, and M. Imamura, *Int. J. Quantum Chem.*, **18**, 515 (1980).
32. S. Ikuta and M. Imamura, *Radiochimica Acta*, **28**, 123 (1981).
33. F. Cacace and P. Giacomello, *J. Chem. Soc. Perkin Trans.* 2, 652 (1978).
34. J. E. Williams, Jr., V. Buss, L. C. Allen, P. v. R. Schleyer, W. A. Lathan, W. J. Hehre, and J. A. Pople, *J. Am. Chem. Soc.*, **93**, 6867 (1971).
35. J. Burdon, D. W. Davies, and G. del Conde, *J. Chem. Soc. Perkin Trans.* 2, 1193 (1976).
36. R. Sustmann, J. E. Williams, Jr., M. J. S. Dewar, L. C. Allen, and P. v. R. Schleyer, *J. Am. Chem. Soc.*, **91**, 5350 (1969).
37. A. C. Hopkins, K. Yates, and I. G. Csizmadia, *J. Chem. Phys.*, **55**, 3835 (1971).
38. J. D. Dill, P. v. R. Schleyer, J. S. Binkley, R. Seeger, J. A. Pople, and E. Haselbach, *J. Am. Chem. Soc.*, **98**, 5428 (1976).
39. P. v. R. Schleyer, A. J. Kos, and K. Ragavachari, *J. Chem. Soc. Chem. Commun.*, 1296 (1983).
40. T. A. Carlson, *Phys. Rev.*, **130**, 2361 (1963).
41. S. Wexler, *Proceedings of the Symposium on the Chemical Effects of Nuclear Transformations*, Prague, 1960, IAEA, Vienna 1961, V. I., p. 115 and references therein.
42. K. Okuno, M. Hiraga, K. Yoshihara, and T. Shiokawa, *Mass Spectroscopy*, **24**, 245 (1976).
43. L. G. Pobo, S. Wexler, and S. Caronna, *Radiochimica Acta*, **19**, 5 (1973).
44. A. H. Snell, F. Pleasonton, and H. E. Leming, *J. Inorg. Nucl. Chem.*, **5**, 112 (1957).
45. S. Wexler, *J. Inorg. Nucl. Chem.*, **10**, 8 (1959).
46. A. H. Snell and F. Pleasonton, *J. Phys. Chem.*, **62**, 1377 (1958).
47. S. Wexler and D. C. Hess, *J. Phys. Chem.*, **62**, 1382 (1958).
48. T. A. Carlson, *J. Chem. Phys.*, **32**, 1234 (1960).
49. K. Nishizawa, K. Narisada, H. Teramatsu, H. Iwami, and M. Shinagawa, *Mass Spectroscopy*, **21**, 199 (1973).
50. V. D. Nefedov, G. P. Akulov, and S. V. Volkovich, *Zh. Obshch. Khim.*, **46**, 904 (1976).
51. V. D. Nefedov, E. N. Sinotova, G. P. Akulov, and M. V. Korsakov, *Zh. Org. Khim.*, **6**, 1214 (1970).

52. G. Ciranni and A. Guarino, *J. Labelled Compds.*, **2**, 198 (1966).
53. F. Cacace and M. Schüller, *J. Labelled Compds.*, **11**, 313 (1975).
54. V. V. Leonov, E. N. Sinotova, and M. V. Korsakov, *Radiokhimiya*, **16**, 564 (1974).
55. E. N. Sinotova, M. V. Korsakov, and B. A. Shishkunov, *Radiokhimiya*, **22**, 466 (1980).
56. F. Cacace, G. Ciranni, and M. Schüller, *J. Am. Chem. Soc.*, **97**, 4747 (1975).
57. J. H. Jones and M. Goldblatt, *J. Mol. Spectrosc.*, **2**, 103 (1958).
58. F. Cacace, M. Speranza, A. P. Wolf, and R. Ehrenkaufer, *J. Labelled Compds. Radiopharm.*, **19**, 905 (1982).
59. G. Angelini, M. Speranza, A. L. Segre, and L. J. Altman, *J. Org. Chem.*, **45**, 3291 (1980).
60. V. D. Nefedov, E. N. Sinotova, Yu. M. Arkhipov, and E. O. Kalinin, *Radiokhimiya*, **25**, 561 (1983).
61. M. L. Leblanc, A. T. Morse, and L. C. Leitch, *Can. J. Chem.*, **34**, 354 (1956).
62. B. Aliprandi, F. Cacace, and A. Guarino, *J. Chem. Soc.* (*B*), 519 (1967).
63. F. Cacace, M. Caroselli, and A. Guarino, *J. Am. Chem. Soc.*, **89**, 4584 (1967).
64. L. Babernics and F. Cacace, *J. Chem. Soc.* (*B*), 2313 (1971).
65. R. Cipollini and M. Schüller, *J. Labelled Compds.*, **15**, 703 (1978).
66. M. Attinà, F. Cacace, R. Cipollini, and M. Speranza, *J. Am. Chem. Soc.*, **107**, 4824 (1985).
67. S. Fornarini and M. Speranza, *Tetrahedron Lett.*, **25**, 869 (1984).
68. M. Speranza, G. Angelini, and C. Sparapani, *J. Labelled Compds. Radiopharm.*, **19**, 39 (1982).
69. F. Cacace, G. Ciranni, and A. Guarino, *J. Am. Chem. Soc.*, **88**, 2903 (1966).
70. P. Ausloos, S. G. Lias, and R. Gorden, Jr., *J. Chem. Phys.*, **39**, 3341 (1963).
71. R. H. Lawrence, Jr., and R. F. Firestone, *J. Am. Chem. Soc.*, **87**, 2288 (1965).
72. V. D. Nefedov, E. N. Sinotova, G. P. Akulov, and V. A. Syreishchikov, *Radiokhimiya*, **10**, 602 (1968).
73. F. Bruner and G. P. Cartoni, *J. Chromatog.*, **18**, 390 (1965).
74. P. Giacomello and M. Speranza, *J. Am. Chem. Soc.*, **99**, 7918 (1977).
75. F. Cacace and S. Caronna, *J. Am. Chem. Soc.*, **89**, 6848 (1967).
76. F. Cacace and S. Caronna, *J. Chem. Soc. Perkin Trans.* 2, 1604 (1972).
77. J. F. Eastham, J. L. Bloomer, and F. M. Hudson, *Tetrahedron*, **18**, 653 (1962).
78. W. J. Hehre and P. C. Hiberty, *J. Am. Chem. Soc.*, **94**, 5917 (1972).
79. W. J. Hehre and P. C. Hiberty, *J. Am. Chem. Soc.*, **96**, 302 (1974).
80. F. Cacace and M. Speranza, *J. Am. Chem. Soc.*, **101**, 1587 (1979).
81. B. A. Levi, E. S. Blurock, and W. J. Hehre, *J. Am. Chem. Soc.*, **101**, 5537 (1979); M. L. Mc Kee, *J. Phys. Chem.*, **90**, 4908 (1986).
82. G. A. Olah, P. v. R. Schleyer, Eds., *Carbonium Ions*, Wiley, New York, 1970.
83. C. Wesdemiotis, R. Wolfschutz, and H. Schwarz, *Tetrahedron*, **36**, 275 (1979).
84. W. A. Waters, *J. Chem. Soc.*, 266 (1942).
85. H. B. Ambroz and J. T. Kemp, *Chem. Soc. Rev.*, **8**, 353 (1979).
86. G. Angelini, S. Fornarini, and M. Speranza, *J. Am. Chem. Soc.*, **104**, 4773 (1982).
87. M. J. S. Dewar and C. H. Reynolds, *J. Am. Chem. Soc.*, **104**, 3244 (1982).
88. M. Speranza, Y. Keheyan, and G. Angelini, *J. Am. Chem. Soc.*, **105**, 6377 (1983).
89. A. R. Katritzky and J. M. Lagowski, *The Principles of Heterocyclic Chemistry*, Academic Press, New York, 1968.
90. R. A. Jones, *Adv. Heterocycl. Chem.*, **11**, 383 (1970).
91. R. B. Hermann, *Int. J. Quantum Chem.*, **2**, 165 (1968).

92. H. J. T. Preston and J. J. Kaufmann, *Int. J. Quantum Chem. Symp.*, **7**, 207 (1973).
93. J. J. Kaufmann, H. J. T. Preston, E. Kerman, and L. C. Cusach, *Int. J. Quantum Chem. Symp.*, **7**, 249 (1973).
94. J. Kao, A. L. Hinde, and L. Radom, *Nouv. J. Chim.*, **3**, 473 (1979).
95. F. R. Cordell and J. E. Boggs, *J. Mol. Struct.*, **85**, 163 (1981).
96. P. Politzer and H. Weinstein, *Tetrahedron*, **31**, 915 (1975).
97. D. Chou and H. Weinstein, *Tetrahedron*, **34**, 275 (1978).
98. G. Angelini, G. Laguzzi, C. Sparapani, and M. Speranza, *J. Am. Chem. Soc.*, **106**, 37 (1984).
99. F. H. Field, J. L. Franklin, and M. S. B. Munson, *J. Am. Chem. Soc.*, **85**, 3575 (1963).
100. N. G. Adams and D. Smith, *Chem. Phys. Lett.*, **47**, 383 (1977).
101. S. Fornarini, R. Gabrielli, and M. Speranza, *Int. J. Mass Spectrom. Ion. Proc.*, **72**, 137 (1986).
102. S. Fornarini and M. Speranza, *J. Phys. Chem.*, **91**, 2154 (1987).
103. S. N. Senzer, K. P. Lim, and F. W. Lampe, *J. Phys. Chem.*, **88**, 5314 (1984).
104. R. D. Bowen and D. H. Williams, *J. Chem. Soc. Perkin Trans. 2*, 1479 (1976).

Chapter **VII**

ION-BEAM METHODS

James M. Farrar

It's all a matter of intensity.
—S. Millman [1]

1 Introduction
2 Cross Sections, Beam Intensities, and Signal Levels
3 Experimental Configurations
   3.1   Ion-Beam–Collision Cell Geometries
   3.2   Crossed-Beam Instruments
   3.3   Merged Beams
   3.4   Guided Beams
4 Survey of Experimental Techniques
   4.1   Ion-Beam Production
        4.1.1   Electron Impact
        4.1.2   Chemical Ionization
        4.1.3   Thermal and Surface Ionization
        4.1.4   Discharge Ion Sources
        4.1.5   Photoionization
        4.1.6   Sputtering and Laser Vaporization
   4.2   Ion Transport and Mass Selection
   4.3   Neutral Sources
   4.4   Product Detection
        4.4.1   Energy Analyzers
        4.4.2   Particle Detectors
        4.4.3   Time-of-Flight Methods
5 Kinematic Analysis
6 Ion-Beam Studies of Reaction Dynamics: A Sampling
   6.1   Beam–Gas Cell Measurements
   6.2   Crossed-Beam Experiments
   6.3   Merged-Beam Studies
   6.4   Guided-Beam Experiments
7 Prognosis
References

1 INTRODUCTION

The study of gas-phase chemical reaction dynamics has undergone a revolution in the past 20 years [2-9]. When we use the term "reaction dynamics," we refer to studies that address, beyond the issue of rate, the more detailed questions of preferred modes of reactant energy in promoting reactions, energy partitioning in elementary reactions, preferred collision geometries, and a number of additional issues that relate to the forces that the approaching reactants and separating products exert on one another during the course of chemical reactions. All experiments performed to address these issues take place at gas densities sufficiently low, and with observation times sufficiently short, that the products observed in the reaction result from single collisions. One of the key methods developed in the past few decades to meet these goals is the molecular beam technique [3, 4, 10-12], and studies of ions have figured prominently in such research [13-18]. The purpose of this chapter is to present to the nonspecialist an overview of the use of beam techniques in the study of ion-molecule reactions in the gas phase. Because of the didactic nature of the review, we will not cover every advance in the state of the art but will choose examples to illustrate important principles required to perform and interpret experiments. We will cover the nuts and bolts of beam techniques, with emphasis on anticipated signal levels, experimental geometries, and kinematics of collisions. In order to illustrate what information and insight a given technique yields, we will present a number of recent and representative examples of experimental results from the literature. These examples will not represent an exhaustive survey of recent accomplishments but will be selected to illustrate the capabilities and limitations of a given technique. We will emphasize rearrangement collisions, but occasionally we will find that experimental results in charge transfer or inelastic scattering will provide a particularly good example of a technique's power. A number of reviews and treatises [19] that emphasize new results and techniques have appeared in the literature in the past several years; many of these articles, although written for the specialist, amplify considerably upon the material in this chapter and will be noted where appropriate. A particularly thorough article by Gentry [13] surveys a number of techniques and includes an exhaustive review of results through 1979.

We will emphasize nonoptical methods in this chapter; the broad area of optical methods for product detection in ion-molecule reactions has been reviewed in detail in Chapter IX by Tsuji. Photoionization methods have also found great use in beam studies of ion-neutral interactions, and that subject has been discussed extensively by Ng in Chapter VIII.

Because the goal of studies in chemical dynamics in general, and beam techniques specifically, is to understand collision phenomena in terms of the potential energy surfaces that govern the nuclear motion, we must develop probes of such surfaces over a range of energies and internuclear separations. Ion-molecule interactions offer advantages over neutral systems in that the experimental control of collision energy is quite simple; wide variation in this parameter allows studies at very low collision energies where long-range electrostatic forces dominate interactions, whereas studies at higher collision energies allow probes of potential wells and short-range repulsive interactions. We will focus attention on the regime from thermal energies up to a few electron volts, comparable to bond energies in covalent systems.

The fact that anions and cations may be focused with electrostatic and magnetic fields into directed rays of particles led to early development of beam techniques for the study of ion–molecule reactions [20]. Many of the earliest experiments on ion–molecule reactions involved studies of reactions occurring in the ion source of a magnetic sector mass spectrometer. Although such studies rigorously involve the use of beam techniques since the products of the reaction are analyzed with a mass spectrometer, we will not include a discussion of such phenomena in the present review. Our review will cover studies in which at least one of the reactants is formed into a collimated beam and interacts with a collision partner, thereby defining a direction for the initial relative velocity vector of the collision.

In the sections that follow, we will discuss cross sections and signal levels, considerations that must be the first step in the design of any molecular beam experiment. We will then discuss experimental geometries for beam studies, methods for ion production and beam formation, and detection techniques. Following a brief discussion of collision kinematics, we will review some of the recent accomplishments in beam studies of gas-phase ion–molecule reactions, indicating the diversity of systems that are now amenable to study with these powerful techniques.

2 CROSS SECTIONS, BEAM INTENSITIES, AND SIGNAL LEVELS

Beam experiments were developed to define more precisely than in a bulb the initial and final conditions in a collision. By collimating reactants with known speeds and allowing these beams to intersect at a precisely defined angle, we can define the magnitude and direction of the initial relative velocity vector for a collision. The measured quantity in any beam experiment is a cross section rather than a rate constant, and the attributes of the cross section that a given experiment measures depend upon the nature of the averaging over initial conditions and summing over final states inherent in the technique. We can understand the simplest kind of cross section, the total cross section, in terms of the attenuation of one of the reactant beams as it passes through the volume in which interaction with the second reactant occurs or in terms of the total rate of product formation per unit volume. In the first case, the primary beam intensity I is attenuated according to a Beer-Lambert law:

$$I = I_0 \exp(-\sigma n L) \tag{1}$$

where

I_0 = incident beam intensity
I = final beam intensity
n = target number density (cm^{-3})
L = attenuation length (cm)
σ = cross section (cm^2)

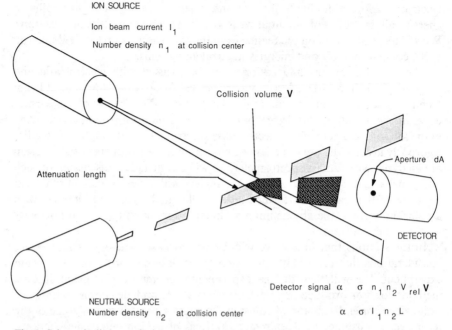

Fig. 1. Schematic diagram of a beam experiment, relating number densities, fluxes of reagents, and cross sections to rate of product formation. The neutral beam, and thus the products, are chopped in order to allow phase sensitive detection.

In this expression, the scattering cross section is analogous to the absorption coefficient in optical spectroscopy.

The use of beam conditions for one or both of the reactants in a collision experiment implies that chemical reactions will occur only in the volume V in which the reactants overlap. Figure 1 shows an experimental arrangement in which two beams intersect at an angle, defining a collision volume V. The total rate of product formation within this volume is given by the following equation:

$$\frac{dN}{dt} = \sigma v_{\text{rel}} n_1 n_2 V \qquad (2)$$

where

σ = total reaction cross section
v_{rel} = relative speed of the collision partners
n_1 = number density of the primary (ion) beam
n_2 = number density of the collision gas
V = collision volume

Since ion-beam currents are readily measured with an electrometer during the course of an experiment, Eq. 2 is more useful if expressed in terms of primary ion-beam current rather than number density. In addition, the collision volume is conveniently expressed as the product of the cross-sectional area A of the ion beam and the length L over which the ion beam intersects the gas from the neutral target. With these substitutions, the rate of product formation may be rewritten as

$$\frac{dN}{dt} = 6.25 \times 10^{-7} \sigma I_1 n_2 L \tag{3}$$

with σ expressed in square angstroms, I_1 in nanoamperes, n_2 in cm^{-3}, and L, the attenuation length, in centimeters. An ion beam of current 1 nA, intersecting a target of length 1 cm at a pressure of 10^{-3} torr, corresponding to a number density of 3.5×10^{13} cm^{-3} at standard temperature and pressure (STP), and reacting with a cross section of 1 Å^2 yields a total rate of product formation of 2×10^7 s^{-1}. This relatively small product yield may be scattered over a wide range of solid angles and into many product quantum states, thus making efficient product detection a critical design criterion for beam experiments.

The total cross section is only one of many cross sections that we may wish to measure in a beam experiment [21, 22], and in order to understand the relations among these quantities, we first describe an "ideal" experiment in which we collide reactants with well-specified quantum numbers [23] collectively denoted **i** at a precisely defined relative velocity and resolve products in quantum states **f** scattered through center-of-mass angle θ. Such an experiment is impossible with present technology, but as we shall see, great progress has been made in recent years toward this lofty goal. We must generally settle for a less precisely defined experiment in which we average over initial conditions or sums over final states. For example, integration of the "ideal" differential cross section over product scattering angles yields state-to-state total cross sections that may be averaged over a Maxwell-Boltzmann distribution of relative velocities to yield detailed rate constants. Other averages yield less detailed quantities until we reach the conventional thermal rate constant $k(T)$. The relationships among the cross section and rate constants are shown in Fig. 2. As we move from more highly averaged quantities to the more detailed cross sections, the signal levels must decrease accordingly. Consequently, we find that the trade-offs of sensitivity and detailed information content dictate that a range of complementary techniques rather than a single method is required to extract maximum dynamical insight from a given chemical system.

3 EXPERIMENTAL CONFIGURATIONS

Ion beam techniques generally divide into four different classifications based on experimental configuration. Each configuration may be categorized according to sensitivity, the accessible range of relative collision energies, angular and energy dispersion of reaction products, and the potential for reactant quantum state selection or product state resolution. The choice of a particular configuration must be dictated by the strong coupling of increased resolution

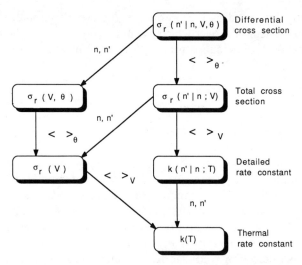

Fig. 2. Relationship between cross sections for state-to-state processes and the averaging over initial and final states n and n', respectively, that results in cross sections and rate constants with less information content. Angle brackets ($\langle \; \rangle$) indicate averaging over relative velocity V and scattering angle θ.

with decreased sensitivity. In the next several paragraphs we will review each of the four major beam configurations briefly, and in later sections discuss in significantly greater detail the detailed operation of key components.

3.1 Ion-Beam–Collision Cell Geometries

This experimental arrangement, one of the first to be used in beam studies of gas-phase ionic processes, consists of an ion beam impinging on a static gas cell maintained at a pressure low enough to attenuate the primary ion beam by no more than a few percent. Under such circumstances, the probability for multiple collisions is small enough to ignore, making the observed products the result of single collisions with the target. Although the primary ions will undergo only a single collision with a target molecule, the target molecules may undergo numerous collisions among themselves. Obviously, the collision cell method can only be used to study reactions of stable gases. In a typical experiment, the ion beam is prepared from a sector magnetic analyzer that selects the momentum of the primary ionic projectile. When the intrinsic energy spread of the ion source is small, such an analyzer behaves as a mass selector. In some recent examples, the combination of a mass filter and electrostatic energy analyzer allows us to select both mass and energy of the incident ions. Detector configurations have been developed that integrate over a large range of scattering angles for total cross-section determinations, whereas analyzers that rotate about the geometric center of the collision cell and view only a small angular range have been developed for angular distribution measurements. The term "tandem mass spectrometer" has been used to describe the general experimental configuration in which a mass spectrometer prepares a primary ion beam that impinges on a target and products are detected by a second mass analyzer.

In evaluating Eq. 2 for signal levels for this experimental configuration, we must recognize that the products may be scattered through an angular range, which can be as large as 4π steradians, although the dynamics of a given reaction may limit this solid angle range. The fraction of the signal received by the detector depends on the solid angle subtended by the entrance slit to the detector. Because the dimensions of the entrance slit may determine both the energy and angular resolution of the detection system, it is clear that the price we pay for increased resolution is decreased sensitivity. Equation 2 must be multiplied by a detector efficiency factor that expresses the fraction of collision events subtended by the entrance slit of the detector. The signal reduction that arises from this finite solid angle may be calculated easily. A square detector aperture 3 mm on a side, located 3 cm from the interaction volume subtends a solid angle $d\Omega = dA/r^2 = (0.3/3)^2 = 0.01$ steradians (sr). In the most pessimistic case of reaction products distributed over 4π sr, this detector aperture will allow a fraction $d\Omega/4\pi = 8 \times 10^{-4}$ of the total signal to pass through the detector entrance slit. Since number densities and beam currents for reactants are limited by the sources available for a given ion or molecule, much of the sensitivity in a given experiment depends on the distance L over which the reactants interact as well as the acceptance angle of the detector. The key advantage of the beam–gas cell configuration is the increase in signal level that accompanies a long interaction path length L with fairly high number densities for the neutral reactants.

The relative kinetic energy of the ionic and neutral collision partners having masses m_I and m_N and speeds v_I and v_N, respectively, is given by the following expression:

$$E_{\rm rel} = \frac{1}{2} \frac{m_I m_N}{m_I + m_N} (v_I^2 + v_N^2) \tag{4}$$

In most collision cell instruments, the speed of the primary ion beam is much greater than that of the neutral; furthermore, the isotropic speed distribution of the neutral makes its velocity zero. With the inequality $v_I \gg v_N$, we may relate the relative collision energy to the lab energy of the primary ion beam through the equation

$$E_{\rm rel} \approx \frac{m_N}{m_I + m_N} E_{\rm lab} \tag{5}$$

Thus, we see that the relative collision energy is simply related to the laboratory energy of the primary ion beam, and that the lowest relative collision energies may be achieved when a heavy projectile ion impinges upon a light target. One of the principle characteristics of the collision cell geometry is the isotropic velocity distribution of the target gas. The finite speed distribution of the neutral gas particles creates significant dispersion in the initial relative kinetic energy at low ion beam energies and also can create a very broad distribution of reactant relative velocity vectors formed as the vector sum of a well-defined ion beam vector and an isotropic distribution of neutral target vectors.

Chantry [24] has analyzed the effect of random thermal motion on the energy dispersion in a beam–gas cell experiment and has shown that the distribution of relative collision energies has a full width at half-maximum (FWHM) as given by the following expression:

$$\Delta E = (11.1 \gamma k T E_{c.m.})^{1/2} \tag{6}$$

where kT characterizes the thermal energy in the gas cell, $E_{c.m.}$ is the relative collision energy (c.m. is center of mass), and γ is the ratio of the reactant ion mass to the total mass of the system. When a heavy ion impinges on a light target, γ approaches its maximum value of unity, and at low energies, the distribution of relative energies approaches a Maxwell-Boltzmann distribution characterized by an effective temperature of γT. This dominance of the distribution by the thermal motion of the target creates serious problems for low energy studies. For example, an F^+ cation at a laboratory energy of 1 eV has a speed of $3.2 \times 10^5 \, \mathrm{cm \, s^{-1}}$, whereas a hydrogen molecule at 300 K has a most probable speed of $1.58 \times 10^5 \, \mathrm{cm \, s^{-1}}$. Because of the isotropic velocity distribution of neutral molecules in a collision cell, the vector sum of the incident F^+ velocity with the distribution of hydrogen velocity vectors creates a very broad distribution of relative velocity vectors and relative collision energies, which effectively defeats the purpose of doing beam experiments. An obvious approach to this problem is to cool the collision cell to reduce the thermal motion of the target, but this solution is clearly of limited scope. Under favorable conditions, we may be able to deconvolute the effect of the thermal energy of the target gas, but at low collision energies, crossed-beam techniques offer important improvements in kinematic resolution.

The situation is more serious in experiments in which we wish to measure energy and angular distributions of reaction products. Under such conditions, the extraction of meaningful center-of-mass energy and angular distributions requires a precise definition of the relative velocity vector and centroid vector of the approaching reactants. As we shall discuss in Section 3.2, the isotropic velocity distribution of neutrals in the collision cell broadens the distribution of centroid vectors significantly. This problem is less severe at high collision energies but becomes quite serious and limits the information content of experiments at lower collision energies at which chemical effects become important.

The primary advantages of the beam–collision cell arrangement are simplicity and reasonably high sensitivity. Collision cells are fairly simple devices relative to crossed neutral beam sources; moreover, the interaction path length that can be achieved in a collision cell is normally at least one order of magnitude larger than for a crossed beam, and the number densities used in collision cells (partial pressures of 10^{-3} torr) are typically factors of $10-10^2$ larger than for a crossed-beam geometry. The beam–gas cell configuration therefore yields signal levels two to three orders of magnitude larger than a crossed-beam arrangement. The beam–gas cell method has also been used by several investigators [25–27] in a novel configuration in which the reaction products created in excited electronic states are detected through their characteristic chemiluminescence. In this variation of the technique, photons rather than ions are detected. This technique has been applied fruitfully most notably by Ottinger and co-workers [26, 27], whose studies of atomic ions with H_2 and D_2 have led to the discovery of new excited electronic states of XH^+ for $X = B, C, N, Al$, and P. The method is capable of optical resolution at the rotationally resolved level but can only be used to detect reaction products formed in emitting electronically excited states.

3.2 Crossed-Beam Instruments

The kinematic dispersion that arises from the isotropic distribution of neutral speeds in a collision cell geometry can be removed from beam experiments by replacing the static gas cell with a crossed neutral beam. Such a replacement comes at a significant cost, both in terms of the mechanical complexity of a neutral beam source and the loss in signal arising from the lower number densities and smaller interaction lengths achievable with neutral beams. Such losses may amount to factors between 10^2 and 10^3. Despite such losses in sensitivity, crossed-beam techniques afford several significant advantages over collision cell techniques. In the environment of a molecular beam, the reactants do not suffer collisions among themselves [28], thereby preserving any internal state distribution imparted to them. With the development of reliable supersonic molecular beam sources, the inherent cooling of internal degrees of freedom of molecular species during the expansion provides a means for delivering internally "cold" reactants to the collision region [29]. We may wish to use this cold distribution directly or exploit the collisionless environment of the beam to prepare specific quantum states of the neutral reactant by laser photoexcitation, for example. This ability to prepare quantum states of the neutral, although used only to a limited extent in extant studies, makes the crossed-beam geometry a good candidate in moving toward the "ideal" experiment in which we select reactant states and resolve individual quantum states of the products. The collisionless environment of the molecular beam also allows the technique to be used for the production of transient species such as free radicals and open shell atoms. Although the low number densities associated with ion beams suggest that low intensity neutral beams of transient species would yield prohibitively low signal levels, Ding and collaborators [30, 31] have used microwave discharge techniques to prepare beams of transient atomic species for ion–neutral scattering measurements. In principle, such methods could be extended to reactive scattering studies.

If the ion and neutral beams with speeds v_I and v_N intersect at an angle γ, the relative collision energy may be computed from

$$E_{\text{rel}} = \tfrac{1}{2}\mu(v_I^2 + v_N^2 - v_I v_N \cos \gamma) \tag{7}$$

where μ is the reduced mass of the reactants. The relative energy is also approximately related to the lab energy of the ion beam and a mass factor as in Eq. 5, but the neutral beam may contribute as well. In general, attaining low collision energies requires low ion–beam energies.

The principal advantage to crossed-beam techniques is the precise kinematic definition that they afford. The directed nature of a neutral beam produced by supersonic expansion eliminates the isotropic distribution of neutral velocity vectors and results in a precisely defined distribution of reactant relative velocity vectors, leading to more precisely defined relative kinetic energy distributions. Because all neutral molecules have a well-defined direction, the vector sum of ion-beam and neutral-beam velocities results in a well-defined direction for the center-of-mass vector of the approaching reactants. Such definition is critical for accurate determinations of product scattering angles and speeds in the center-of-mass collision system as required for meaningful differential cross-section measurements. Consider, for example, a 1-eV beam of F^+ colliding with H_2 in a collision cell held at 77 K. If we assume that the ion beam is energy selected and has an energy distribution characterized by an FWHM of 0.10 eV and the neutral molecules have a Maxwellian distribution, the most probable collision energy is 0.15 eV, with a FWHM relative energy spread of ± 0.08 eV. Only 0.009 eV comes from the energy spread of the ion beam when transformed to the center of mass. The FWHM spread in centroid angles is 5.6°. In contrast, if we replace this Maxwellian scattering cell with a collimated effusive beam with the same speed distribution, but intersecting the ion beam at 90°, the relative energy spread reduces tenfold to ± 0.009 eV, and the FWHM spread in centroid angles drops to 1.8°.

We have already noted that such kinematic sharpening exacts a great price through reduced signal levels. Determination of the precise scattering volume defined by the overlap of the beams, coupled with difficulties in measuring precise number densities at the collision center, make this method poorly suited for determination of absolute differential cross sections or angular integration to yield total cross sections. Furthermore, the crossed-beam geometry strongly couples the lower energy limit to the primary ion-beam energy. These shortcomings in absolute cross-section determination and accessibility of low relative energies have led to the development of merged-beam and guided-beam methods as described in the next sections of this chapter.

3.3 Merged Beams

The beam methods described above are limited by the energy range that they can cover as well as the chemical diversity of the reactants amenable to study. The merged-beam method [32] offers a powerful solution to the problem of accessing very low collision energies and extends the range of chemical species that can be studied within certain constraints. The merged-beam method achieves very low relative velocities by allowing the reactants beams to merge in a common volume with parallel velocity vectors. Under these conditions, the relative velocity of the approaching reactants is the difference in their lab velocities. With very careful control of the energies of the reactants, this relative velocity, and therefore the relative collision energy, can be made quite low. Gentry and co-workers [33] have reported achieving relative kinetic energies of 0.002 eV, 10 times lower than thermal energy at room temperature! At such low collision energies, the precise form of the long-range attractive potential that the reactants experience may be probed with high accuracy. Because kinetic energy is a quadratic function of speed, a constant spread in the kinetic energies of the ions at low collision energy results in a large fractional energy spread, but that same ΔE at high relative energies or speeds translates into a small spread in relative velocities. Thus, the usual configuration for a merged-beam experiment involves the interaction of two beams with laboratory energies of a few keV, leading to very low relative energies with small spread in relative collision energy.

Because both beams in a merged-beam experiment must have laboratory energies of several keV, the study of ion–neutral collisions in merged beams requires that the neutral beam be created by charge exchange neutralization of an ion after it has been accelerated to its full energy. This method allows us to produce labile reactants such as open shell atoms and free radicals by neutralization of the appropriate ionic precursor, but frequently the process that generates these precursors yields a distribution of internal states that is further perturbed by the charge exchange neutralization process. Such methods allow some chemical diversity since the precursor ion can be mass selected, but this diversity is at least partially negated by the potential for a very poorly characterized reactant state distribution.

The large laboratory velocity of both of the reactant beams, coupled with relatively small center-of-mass recoil speeds, effectively removes the possibility of measuring differential cross sections in energy and angle. However, the projection of all center-of-mass vectors onto the relative velocity vector can be extracted from careful measurements of the product kinetic energy spectrum. Such a measurement contains center-of-mass energy and angular details, and the dynamical information content of such a measurement can be considerable.

Unlike the beam–gas cell and crossed-beam configurations described previously, the merged-beam method lends itself well to measurements of absolute cross sections, since high-energy ions and neutrals can be detected with high efficiency and the volume over which the reagents interact can be mapped out precisely. Despite this advantage, as well as the ability to achieve subthermal collision energies with "exotic" reagents, the merged-beam technique remains one of the more difficult and challenging methods for the study of ion–

molecule reactions. The principle reasons for this are that merged-beam experiments are quite complex, using two charged particle beams with their attendant difficulties, and the fact that the attenuation of the beams is very small, of order 10^{-10}–10^{-14}, making merged-beam experiments exceedingly challenging and difficult. Neynaber [34] has reviewed the critical aspects of such experiments in two clear review articles.

3.4 Guided Beams

The guided-beam technique, pioneered by Teloy and Gerlich [35] at the University of Freiburg, uses electrode configurations to which radio-frequency (rf) potentials are applied to transport ions from one region of space to another. It has been known for some time from classical mechanics [36] that ions in a potential field that changes direction rapidly compared to ion transit times between electrodes are held in a "pseudopotential" that is flat near the centerline of the device and rises steeply near the electrodes. The frequencies and amplitudes of applied fields can be chosen to create pseudopotential wells whose depth is greater than the kinetic energies of the ions contained in the electrode structure, thereby creating an ion trap analogous to a collision cell for neutrals. Such guided ion-beam traps have been used for two primary purposes in the study of ion–molecule reactions: as "containment vessels" in which collisions quench nascent internal excitation in electron impact created ions and to provide true 4π-sr collection of the products of ion–neutral interactions for accurate total cross-section studies. In the latter case, guided-beam systems, primarily in octupole configurations, have been configured to capture all products of chemical reactions by creating well depths greater than the maximum kinetic energy release in a reactive collision. Since detection sensitivities reach 100% in such geometries, only a careful measurement of pressure in the collision cell, located within the ion guide, is required for an accurate absolute total cross-section determination. Thus, the primary advantage of this technique is very high sensitivity and accuracy for total cross-section measurements. In conjunction with ion trap methods for production of primary ions in a beam experiment, the internal excitation of the ion beam can be eliminated, and the translational energy spread of the ions can be reduced to figures characteristic of thermal distributions at temperatures near 300 K. Such a reduction in the energy spread of the primary ions with energy selection leads to better defined kinematic relations for scattering studies.

Gentry [13] has prepared a semiquantitative summary of the relative merits of each of these experimental methods for studying gas-phase ionic reactions. It is clear that each method has benefits and weaknesses and that more than one method will be required to achieve a complete description of a reactive process.

In the next sections, we will review individual components in ion–neutral scattering instruments, emphasizing the modular nature of ion and neutral sources and detectors and discussing the merits of individual techniques for preparing and detecting ions and neutrals.

4 SURVEY OF EXPERIMENTAL TECHNIQUES

The experimental methods required for ion-beam studies of collision processes divide into the areas of ion formation and transport, neutral reactant production, and product analysis and detection. A large body of literature exists in each of these areas, and our purpose is to provide an overview of the methods in use.

4.1 Ion-Beam Production

Ion production is a central concern in mass spectrometry, and many methods for generating ions have been developed over the last few decades. Beam experiments require high-intensity sources of ions with well-defined kinetic energy spreads and well-characterized internal state distributions. Because these requirements are often conflicting, all techniques involve compromises, and in order to achieve chemical diversity, the ion beam experimentalist must be familiar with many different ion sources.

Before examining the various ion sources in use today, we must consider the fundamental issue of space charge arising from the mutual coulombic repulsion of ions [37, 38]. Space charge provides a limit on the divergence of a beam of specified current density at a given energy. This limitation can be realized in ion sources as the maximum current that the source can produce and inject into an ion-beam transport system, e.g., ion focusing optics and mass or energy selectors. The space-charge limit also applies to the minimum divergence that an ion beam at a given energy and current density will experience in a beam transport system of a specified length. The space-charge limit is fundamental, and no source or optical system can be designed to overcome its constraints. In simplest terms, the space-charge limit can be expressed as the maximum current at a given energy E for ions of mass M that can be transported through a tube of diameter d and length L [38]; the maximum current is computed from the following equation:

$$I_{max} = 9.0 \times 10^{-7} E^{3/2} M^{-1/2} \left(\frac{d}{L}\right)^2 \text{ A} \qquad (8)$$

For example, we may wish to transport a beam of diameter d of 3 mm through a distance of 10 cm at an energy of 1 eV. For ions of mass 28, corresponding to N_2^+, the maximum current allowed by the space-charge limit is 1.53×10^{-10} A. Transporting larger currents at the same energy requires that the beam diverge during its travel. An alternative statement of the space-charge limit can be formulated in terms of the fractional spread W that an ion beam of energy E, mass M, and current density j experiences as it travels a length L [39]:

$$\frac{W}{L} = (4.1 \times 10^6 M^{1/2} E^{-3/2} j)^{1/2} \text{ cm}^{-1} \qquad (9)$$

This equation is valid for $2 < W < 5$. If we wish to transport a 1-eV beam of

H_2^+ with a beam diameter of 1 mm a distance of 30 cm with a maximum spread of a factor of 2, the maximum current consistent with these boundary conditions is 7.6×10^{-12} A. Clearly then, the space-charge limit provides an important constraint in the design of ion sources and optical systems. Faced with these limitations, the design of an optical system for low-energy beams should minimize the distances over which the beam is at the lowest energies. In a system with mass analysis, for example, the beam may be analyzed at a higher energy where space charge is not a serious problem, decelerating the beam to the desired low energy only as the last step in a multistage system.

4.1.1 Electron Impact

A venerable technique in mass spectrometry, electron impact ionization produces atomic and molecular ions by colliding electrons emitted by a hot cathode with neutral precursor atoms or molecules. When the kinetic energy of the electron is greater than the ionization potential of the precursor, ionization may occur. This method is capable of producing intense ion beams but often creates atomic and molecular ions in excited electronic states and molecular ions with broad distributions of vibrational energies. The rate of ion production depends on the cross section for electron impact ionization, and such cross sections increase rapidly from the ionization threshold as the electron kinetic energy increases [40]. In order to achieve adequate ion yields, electron energies may be increased to levels high enough above threshold to create significant populations of excited vibrational and electronic states.

Figure 3 illustrates the essential nature of electron impact ionization. The incident electron excites the molecular precursor through an energy transfer

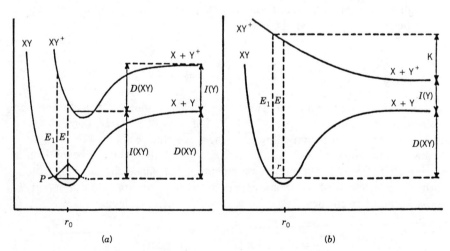

Fig. 3. Diagrams showing Franck-Condon transitions for electron impact leading to (a) formation of stable XY^+ molecular ions from XY precursors and (b) formation of Y^+ ions by dissociative ionization through a repulsive potential energy surface with kinetic energy release K.

collision in a Franck-Condon transition to excited electronic states. For small molecules, ionization or fragmentation can occur depending on the nature of the potential energy surface accessed in the initial transition. Figure 3 suggests that parent ion formation occurs in a distribution of vibrational states; at sufficiently high excitation, this distribution leads to dissociative ionization forming fragments. The potential curves for H_2 and H_2^+ indicate, for example, that Franck-Condon factors for ionization produce a broad distribution of vibrational levels from $v = 0$–18 [41]. For larger molecules, a statistical theory of mass spectra, the Quasiequilibrium Theory (QET) [42], assumes that internal conversion from electronically excited states to the ground electronic state of the parent cation occurs fast relative to dissociation, and fragmentation occurs through vibrational predissociation. These ionization mechanisms indicate that internal energy distributions of parent ions created by electron impact ionization are often very broad. Excited states that decay by allowed electronic transitions will not survive the flight time to the interaction region, but vibrationally excited states and metastable electronic states will not be quenched in this manner. Some effort has been made in recent years to use chemical quenching to reduce the initial distribution of internal energy states produced by electron impact. For example, Cotter and Koski [43] have demonstrated that in the production of F^+, the metastable ^1D state can be quenched out by adding paramagnetic NO to the ion source, leaving only ^3P ground-state ions. Herman and Pacak [44] have demonstrated that the broad distribution of H_2^+ vibrational states created by electron impact ionization of H_2 can be quenched partially by adding neon gas to the source. Under such conditions, the Ne reacts with H_2^+ vibrational states with $v \geqslant 2$ in an exothermic reaction, whereas the reactions of the $v = 0$ and 1 states to form NeH^+ are endothermic.

Such chemical quenching has also been used in the production of ground-state C^+ for reactive scattering studies [45, 46]. By mixing CO in a large excess of He, the probability of electron impact on He is much higher than on CO. He^+ thereby produced has enough energy to undergo dissociative ionization with CO to make ground state (^2P) C^+, but the reaction

$$He^+ + CO \rightarrow He + C^+(^4P) + O \qquad (10)$$

to form the first excited metastable state is endothermic. These examples indicate that although electron impact does produce a broad distribution of initial vibrational and electronic energies in the ions, careful source and quenching conditions can reduce these spreads to acceptable levels under appropriate conditions. Because metastable and ground-state ions often have very different scattering cross sections with suitably chosen targets, differential attenuation measurements [259, 260] can be used in favorable situations to assess the excited-state composition of an ion beam produced with a given method. In such experiments, the attenuation of an ion beam through a collision cell is measured as a function of the cell pressure. If one or more excited states in the beam have scattering cross sections very much different than for the ground state, the transmitted beam intensity will show nonexponential attenuation as a function of pressure. The attenuation curve can then be decomposed into contributions from ground and excited states in the ion beam.

More recently, van Koppen et al. [47] have used a drift cell filled with an appropriate gas following the ion source to relax the distribution of electronic states created by electron impact to the ground state by collisional quenching. A series of guard rings maintains a uniform electric field within the cell, allowing the nascent ions to enter the cell, drift through, and exit in the lowest electronic state. We will present a number of examples of experiments performed with this cell in a later section, demonstrating its effectiveness in preparing ground electronic state ions.

Figure 4 shows a sketch of an electron impact source first used by Udseth

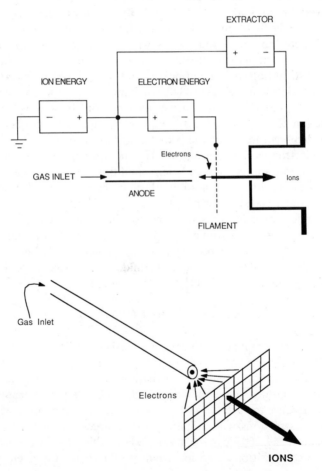

Fig. 4. In-line electron impact source. Gas entering the source through the anode tube is ionized within the tube by electrons emitted from the negatively biased tungsten mesh. This bias potential also extracts the ions from the source.

et al. [48] to study proton scattering. The source has also found use in several other beam studies of ion–molecule reactions [49, 50]. Source gas enters the anode tube held at the desired ion energy. Electrons from a heated tungsten mesh are accelerated into the anode tube in which ionization takes place in the equipotential volume inside the tube. The bias voltage of the planar cathode extracts the ions through the mesh and into a mass spectrometer. This in-line source produces very intense beams of many ions with narrow energy distributions, but because the filament heats the anode tube, thermally labile precursor molecules may not be used for ion production. Crossed-beam ionizers such as those used in commercial mass spectrometers avoid this difficulty. Conventional electron impact sources can also be used for negative ion production under favorable conditions. The process of dissociative attachment

$$e^- + AB \rightarrow A^- + B. \tag{11}$$

by low-energy electron impact [51, 52] is efficient for several species, with one of the most common examples being O^- production from N_2O gas [52].

4.1.2 Chemical Ionization

In analytical mass spectrometry, one of the most serious problems with electron impact ion production is the extensive fragmentation that accompanies the ionization of large molecules. Many so-called "soft" ionization techniques have been developed to circumvent this serious problem, with chemical ionization among the most popular [53]. The chemical ionization (CI) method uses a chemical reaction between ions generated from a reactant gas and the species to be ionized to create primary ions. The reactant gas in CI is usually chosen to be a species such as CH_4 or H_2, which may be ionized by electron impact, following which reactions with the neutral gas lead to protonated species such as CH_5^+ and H_3^+. These species may then undergo proton transfer reactions with the species M to be ionized. Such reactions generally lead to MH^+ ions in thermodynamic equilibrium at the ion source temperature with very little fragmentation [54–56]. The CI method was developed to accentuate parent ion intensities in mass spectra, and its full power has not been applied to ion sources for beam-scattering experiment, but the method may be used for the production of ions with low internal temperatures. For example, Futrell and co-workers have used a CI source with hydrogen gas to make beams of internally relaxed H_3^+ [57] and with traces of H_2O and Ar in hydrogen to make H_3O^+ [58] and ArH^+ [59], respectively. The CI method holds promise for the production of reasonably internally relaxed ions that are the most thermodynamically stable species produced by ion–molecule reactions in the ion source.

4.1.3 Thermal and Surface Ionization

The ionization of metals of low ionization potential at the surface of a low work function metal such as Pt is a venerable technique for the production of alkali ions [60]. This method of surface ionization holds an important place in the history of molecular beam kinetics [61]. For a species of ionization potential I impinging on the surface of a metal of work function ϕ, the Langmuir-Kingdon equation describes the ionization probability [62]

$$\frac{N_+}{N_0} = \frac{w_+}{w_0} \exp\left(-\frac{(I-\phi)}{kT}\right) \qquad (12)$$

where N_+/N_0 is the ratio of positive ions to neutrals, and w_+/w_0 is the degeneracy ratio for the ground states of the ion and the neutral.

Equation 12 indicates that ion yields are highest when $I - \phi$ is negative, a condition satisfied for low ionization potential metals such as alkali metals, on surfaces of tungsten ($\phi = 4.52\,\text{eV}$) and rhenium ($\phi = 4.80\,\text{eV}$), although operation of filaments at higher temperatures makes higher ionization potential materials amenable to this technique.

In its simplest form, such a source consists of a heated, positively biased filament near a source of metal vapor. Atoms adsorbed on the filament surface are evaporated as ions. Commercially available alkali emitters [63] use porous tungsten plugs impregnated with the desired metal; internal heaters may then be activated to produce ions by surface ionization (SI). A closely related method involves the preparation of mixtures of Al_2O_3, SiO_2, and the oxide of the metal to be ionized [64, 65]. After the fusion of a glasslike bead of this oxide mixture onto a platinum or rhenium filament, resistive heating of the bead leads to thermionic emission of positive ions. This method was suggested by Blewitt and Jones [66] half a century ago for production of many ions, including alkali ions, some alkaline earths, and Al^+.

More recent applications of the SI method have been found in studies of metal ion chemistry. Armentrout and Beauchamp [67, 68] in particular have used variations of the surface ionization method to study reaction dynamics of ground-state metal cations. In their method, radiative heat transfer from a rhenium ribbon filament to a small crucible containing a halide salt of the desired metal vaporizes the salt, which then decomposes on the heated filament. The metal atoms produced then ionize and are focused into an beam. Equation 12 suggests that Boltzmann factors for low-lying excited metastable states may be large enough to yield impurities of such states of a few percent in beams that are primarily ground-state cations. The intensities from such sources are low, with beam currents typically 10^{-12} A, and their use has been restricted to beam–gas cell and guided-beam experiments with high collection efficiency and sensitivity. Metal ions from thermionic emitters have also been used recently in chemical ionization studies [261].

Although less widely used, negative surface ionization is a technique for producing anions of species having high electron affinities such as halogen atoms. The "negative electron affinity" material LaB_6, when in contact with an alkali halide, for example, will evaporate halide anions from its surface [69]. A surface based on this design has been used for studies of Br^- with CH_4 [70].

4.1.4 Discharge Ion Sources

An electrical discharge, either direct current (dc) or alternating current (ac), can serve as an efficient medium for ion production. In such sources, electrical breakdown of the gas in the source creates a plasma that can couple to an external dc or ac field [71]. In an ion source operating in the microwave region, for example, electrons in the low-pressure gas will oscillate in quadrature with the applied field, gaining energy from collisions with the gas molecules. The mean kinetic energy of electrons in such a discharge can be as low as a few electron volts, and ions are produced by collisions of precursor neutrals with these electrons. Because their energy is low in comparison with excitation energies in atomic and molecular ions, such electrons generally produce ground electronic state ions [72]. The production of atomic ions from molecular precursors is thought to proceed through a sequential process of electron-induced dissociation of the molecule to atoms, followed by electron impact ionization. In contrast, a dc-discharge source, an example of which is shown in Fig. 5, operates with electron energies between 50 and 200 eV, resulting in significant excited-state populations [73]. A particular advantage of this source is the possibility of vaporizing by radiative heat transfer a solid charge held in a crucible coaxial with the source cathode. Dissociative ionization of the vaporized solid extends the range of ions that can be prepared with such sources; Kasdan and Lineberger [74], for example, have used such a discharge source with an alkali halide charge to produce beams of alkali anions.

Because the kinetic energy distribution of ions is a reflection of the local environment in which they are created, the potential distributions in plasmas

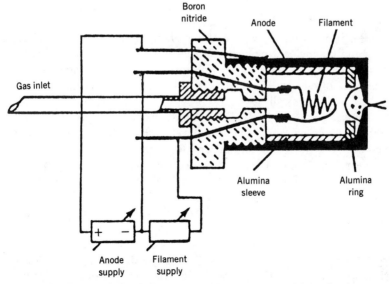

Fig. 5. Schematic diagram of DC-discharge source. Ions are extracted from the plasma confined by insulating inner surfaces of the source.

can play a crucial role in determining ion-beam energy spreads. In the dc-discharge source in Fig. 5, positive ions are thought to be created in the so-called "positive column" of the discharge [73] where the potential is reasonably uniform and the energy spread of the source is small (< 0.5 eV). Hasted [75] has pointed out that the acceleration of ions from the plasma through the plasma sheath may lead to broadened energy distributions. The characteristics of individual sources are therefore critical in determining the initial conditions of a beam-scattering experiment.

4.1.5 Photoionization

Perhaps the most elegant and selective method for ion production is the photoionization technique. The precise control of wavelength for ionization of molecules allows us to produce molecular ions in vibrationally selected states, and this phenomenon has been used with ultraviolet (UV) sources for single photon ionization, as well as with laser sources for multiphoton ionization, to study ion–molecule collisions. The subject of photoionization methods in conjunction with beam techniques for studying gas-phase ion chemistry is an area that has experienced impressive growth in the past few years, and Chapter VIII by Cheuk-Yiu Ng provides a detailed and comprehensive review of this exciting area of research.

4.1.6 Sputtering and Laser Vaporization

The ejection of atoms and ions from a substrate bombarded by a directed energy source, either high-energy ions or intense laser beams, has been developed into a valuable tool for the production of a number of ions ranging from monatomic metal ions [76–78] to ionic clusters of metals [79, 80] and nonmetals, including silicon [81] and carbon [82, 83].

Sputtering ion sources operate with a high-energy ion beam that impacts on a surface, and the incident particles transfer momentum to the substrate atoms and eject them from the surface [84]. Sputtering processes are capable of ejecting large amounts of material from the surface along with molecules adsorbed on the surface, and chemically complex processes occurring in this high-density plasma can lead to the production of a variety of chemical species. For example, Corderman and co-workers [85] have developed a negative ion sputter source in which Cs^+ ions from a thermionic emitter impact on a conical metal surface. Negative ions are formed in such a source when ejected atoms from the bulk metal undergo charge transfer with a layer of cesium atoms on the surface. These workers have indicated that the source may operate in two modes: In the first, at high Cs^+ current density, the metal surface remains clean and anions of the bulk metal predominate in the ion yield. In the second operating mode, molecules adsorbed on the sputtering substrate may fragment and ionize, or undergo reactions with substrate atoms, to yield ions characteristic of the adsorbed species rather than the bulk. Feigerle and co-workers have used this method to produce CH_2^- from cycloheptatriene adsorbed on iron, as well as numerous transition metal anion clusters [86].

Although this specific variation of sputtering technology has not been applied to beam studies of ion–molecule reactions, Anderson and co-workers [87, 88] have used an argon ion gun with selected metal surfaces to sputter monatomic and cluster ions for use as reactants in studies of reaction dynamics. The earliest work to come from this group [87] indicated that the kinetic energy distribution of sputtered clusters is very broad but that it can be reduced significantly by collisional cooling in a gas cell used in conjunction with a guided ion geometry [89]. Clearly, the sputtering technique, in conjunction with methods to reduce the spread in internal and translational energies, holds great promise in extending the range of species that can be studied in beam experiments.

The laser vaporization method is closely related to sputtering by ion impact on a surface and has been developed in several laboratories as a source for monatomic and cluster ions. Intense laser pulses depositing large amounts of energy on a surface concentrated in a small area and for a very short time result in the ejection of large numbers of anions, cations, and neutrals, both monatomic and cluster species. First reported by Friichtenicht et al. [76] for the production of pulsed ions beams of materials such as Al, Fe, and Ti, the method was then adapted by numerous workers, including Byrd and collaborators to pulsed ion cyclotron resonance (ICR) experiments [77], and combined with a helium flowing afterglow apparatus by Tonkyn and Weisshaar [78]. Smalley and co-workers [80, 90] developed a similar method that combined pulsed laser vaporization of a metal rod with entrainment of the pulsed vapor in a supersonic helium expansion to cool the internal degrees of freedom of the metal atoms and enhance cluster formation in the flow. When this flow is intercepted by an excimer laser pulse, significant ionization of the clusters takes place, and the resultant cluster ions are formed into a beam for further studies. Recent development of this technique by this group [80] has led to the ability to inject such a beam into the cell of a Fourier transform mass spectrometer (FTMS) where mass selection takes place and chemical reactions of the selected cluster can be studied by adding reactant gases to the resonance region of the FTMS. Freiser's review of FTMS in Chapter II discusses the capabilities of this method, and we leave the subject to that chapter.

Very recently, two applications of the laser vaporization technique have been used directly as ion sources in beam studies of ion–molecule reactions. Jarrold and Bower [91] have used pulsed laser vaporization of aluminum, followed by electron impact ionization, supersonic expansion, and mass selection by a quadrupole mass spectrometer to produce Al_n^+ clusters, examining their reactions with molecular oxygen. Because a pulsed laser is used in these studies, the duty cycle for such experiments is quite low, making sensitivity an important constraint. In order to overcome limitations imposed by pulsed ion sources, Armentrout and co-workers have developed a "quasi-continuous-wave (quasi-cw)" laser vaporization source that uses a high repetition rate Cu vapor laser operating at 8 kHz [92]. These investigators report that cluster ions are observed with this laser source with minimum average laser powers of 11 W and that an external ionizing source is not required. The ions produced are entrained in helium and expanded into vacuum, where they are mass analyzed. The high repetition rate of the copper vapor laser combined with the width of the helium gas pulse leads to a nearly cw beam of cold clusters. It is clear that this important innovation will lead to many new studies of gas-phase ion-cluster chemistry.

4.2 Ion Transport and Mass Selection

Once ions are produced, they must be transported as a beam to the interaction region of the instrument. The topic of charged particle optics is a vast one, and many excellent monographs have been written on the subject of beam transport [93–96]. Because ions of many masses may be produced by the sources described in the previous section, one of the most important beam transport applications is that of mass analysis, and momentum analysis with a magnetic sector is a key technique. Balancing the forces that an ion of mass m experiences when moving with velocity v in a magnetic field of strength B perpendicular to the trajectory shows that the radius of curvature R_0 of the trajectory depends on the particle's momentum:

$$R_0 = \frac{mv}{eB} \tag{13}$$

In addition to imposing a momentum-dependent radius on the trajectory, sector magnetic fields also have focal properties that allow them to be used as mass selective prisms in ion-beam experiments. The optical properties of such prisms can be summarized with a Newtonian lens equation of the following form [97]:

$$(l_0 - g)(l_i - g) = f^2 \tag{14}$$

where

$$g = R_0 \cot \phi \quad \text{and} \quad f = R_0 \csc \phi \tag{15}$$

The relationships of sector angle ϕ, radius R_0, object distance l_0, and image distance l_i are shown in Fig. 6, and for the case of $\phi = 60°$, the object and image distances are equal to $\sqrt{3}$ times the sector radius. This geometry, first introduced by Nier [98], moves the object and image points well away from the magnetic field, simplifying the coupling of such a magnet with ion sources and detectors. This magnetic field configuration has enjoyed great popularity in analytical instruments as well as in ion beam scattering machines. Because a sector magnet is a momentum analyzer, the resolution of such an instrument depends on the ratio of slit width S to the trajectory radius, and the ratio of ion source energy spread ΔE to the beam analysis energy:

$$\frac{S}{R_0} = \frac{\Delta M}{M} + \frac{\Delta E}{2E} \tag{16}$$

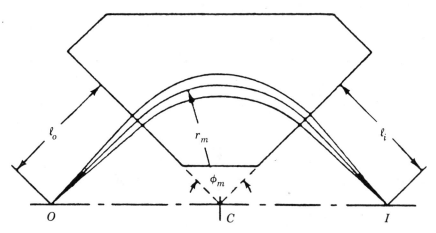

Fig. 6. Sketch of magnetic sector mass spectrometer focal conditions, indicating image distance l_i, object distance l_o, sector angle ϕ_m, and radius of curvature of ion trajectories.

Most magnetic sectors operate at acceleration energies of a few hundred to a few thousand electron volts, minimizing the second term's contribution to the mass resolution, but because ion energies of only a few electron volts are desirable for most collision experiments in the chemically interesting regime, such analyzers must be used with ion-beam deceleration systems.

Magnetic prisms only focus trajectories in the plane perpendicular to the direction of the magnetic field. In order to compensate for the absence of focusing in the plane of the field, astigmatic lenses must be used. Quadrupole doublets placed before the magnetic field convert the cylindrically symmetric beam into a rectangular ribbon in which ion velocity components along the field are zero. After momentum analysis, the emerging ribbon-shaped beam may be converted back to cylindrical symmetry with a second quadrupole doublet that effectively compensates for the astigmatism of the first doublet. The use of astigmatic doublets has been discussed in the literature [99, 100], and some of the instruments we review in Section 6 use such lenses.

Magnetic analyzers operate by deflection of ionic trajectories and therefore place geometric constraints on the construction of the instrument with which they are used. In many applications, a mass analyzer that is collinear with the path from the ion-beam source to the interaction region leads to simplifications in instrument design, and the Wien filter, a geometry with crossed electric and magnetic fields, affords such an in-line geometry. Such a device, an example of which is shown in Fig. 7, has a long history, going back to Thomson's electric deflection experiments for the determination of the charge-to-mass ratio for the electron. By balancing the electrostatic force with the Lorentz force that an ion experiences, a Wien filter serves as a velocity selector for charged particles. For ions that are produced with a reasonably monochromatic energy distribution, velocity selection may also allow mass selection, and the Wien filter has become

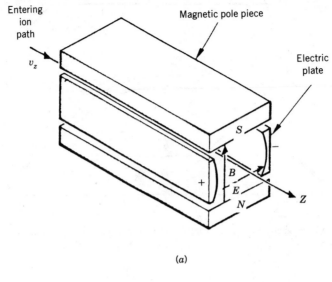

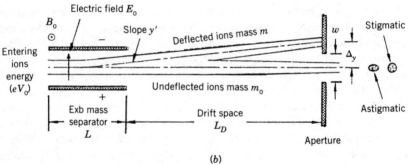

Fig. 7. Curved electrode Wien filter. In this version of the Wien filter, ion trajectories are focused stigmatically to a nearly cylindrically symmetric beam, appropriate for beam studies of reaction dynamics.

a popular mass analyzer operating at kinetic energies of several hundred electron volts. Uniform electric and magnetic fields yield astigmatic focusing conditions in which a cylindrically symmetric object is focused to a line. Such focusing conditions are poorly suited for collision work in which a cylindrically symmetric image is desired, but curved electrodes may be used to effect reduced astigmatism in the focusing conditions [101, 102]. Legler [101] has demonstrated a geometry with a uniform magnetic field and an inhomogeneous electric field leads to partial stigmatic operation. Designs in which the magnetic field is inhomogeneous have also been reported [102] and provide the basis for the commercial Colutron mass analyzer. Mass resolution $M/\Delta M$ for such a device is typically of the order 50.

During the past 20 years, quadrupole mass filters have become important

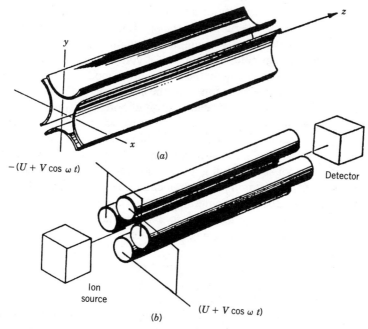

Fig. 8. Schematic of electrode structure and potentials required for operation of a quadrupole mass filter. (a) The upper panel shows a geometry with hyperbolic electrodes; (b) circular rods are used in the lower panel.

tools for mass selection and detection in ion–neutral experiments. As mass selectors, quadrupoles have the advantage of being in-line instruments similar to the Wien filter. The appropriate potential Φ consisting of dc and rf fields applied to a quadrupole structure such as that shown in Fig. 8 behaves as a mass filter [103]. Referring to Fig. 8 for a definition of potentials, the ratio of U/V determines the resolution $M/\Delta M$ for the quadrupole, whereas the absolute values of the U and V components of the applied potential determine the mass transmitted by the filter. Ions transmitted through the filter execute sinusoidal trajectories, whereas ion masses that are rejected experience exponential trajectories leading to collisions with the electrode structure. The principle advantages of the quadrupole mass filter are its compact size, easily variable resolution by control of the ratio U/V, and reasonably energy independent transmission. Although the device was initially considered a low-resolution mass spectrometer, recent developments in the injection optics for mass filters [104] have increased the resolution $M/\Delta M$ to nearly 10,000. In its first applications to ion–molecule collision dynamics, the mass filter was used as a compact particle detector, but in recent years it has also been used for mass selection of primary ion beams for collision studies. The details of quadrupole mass filter operation, stability, and resolution are topics well studied in the literature, and we refer you to those articles for greater detail [103, 105, 106].

The laboratory kinetic energy of an ion at the point at which it interacts with a collision partner is determined by the potential energy difference between the point of its formation and the interaction volume. In many applications, particularly in low-energy collision studies, it is convenient to hold the collision volume at ground potential, with the ion source anode at the desired low potential. Since these low-energy ions must be accelerated to a higher potential for mass analysis and then decelerated to ground potential, some care must be exercised in the beam transport optics in order to maximize ion transmission. Low-energy ion beams are particularly subject to defocusing from stray electrostatic fields arising from surface charge. Cleanliness is obviously an important criterion for success, but good optical design is crucial. A particularly effective design for deceleration is the exponential retarder developed by Futrell and co-workers [107, 108]. Ions entering this lens system consisting of a stack of large aperture electrodes attached to an internal resistor network that maintains an exponentially decreasing axial potential of the form

$$V(z) = V_0 \exp(-az) \tag{17}$$

are focused and decelerated by this field. This potential distribution results in angular and lateral magnifications of -6; the ions are injected through a small aperture at high energy and are decelerated and collimated electrostatically. The absence of physical slits at the low-energy end of the decelerator eliminates the possibility of surface-charge buildup, a major source of instrument degradation for low-energy collision experiments. Because all collimation of the beam is electrostatic and because the apertures in the decelerator are very large (1.27 cm) in order to avoid spherical aberration, the device functions very efficiently with very high transmission and has been adopted by several laboratories.

One of the most interesting innovations in ion-beam transport in the past several years is the guided ion-beam technique, pioneered by Teloy and co-workers [35, 109, 110]. Conceptually similar to the quadrupole mass filter, in which a combination of dc and rf fields allow passage of a narrow range of masses, the guided-beam technique exploits the effective potential that an ion experiences in an inhomogeneous multipole field to trap and transport ions with high efficiency. The effective potential for an ideal electric multipole field with $2n$ poles is given by the following expression [35]:

$$U_{\text{eff}}(r) = n^2 K \left(\frac{r}{r_0}\right)^{2n-2} \tag{18}$$

where

$$K = \frac{q^2 V_0^2}{m\omega^2 r_0^2} \tag{19}$$

for a particle of charge q and mass m inside a multipole structure of inscribed radius r_0. The rf amplitude is V_0 at frequency ω. The equations above apply within certain limits discussed in detail by Teloy and Gerlich [35]. The case in which n is 4, corresponding to an electrostatic octupole, has become a particularly interesting and widely used example of inhomogeneous fields for transporting ions. According to Eq. 18, the effective potential varies as $(r/r_0)^6$, and thus for small r, corresponding to ions near the axis of the octupole, the effective potential is fairly flat. In comparison with a quadrupole, the effective potential for an octupole is quite flat in the center of the electrode structure, rising steeply near the electrodes. The well depth K can be adjusted by choice of frequency and rf amplitude to be several volts, large enough to contain products of ion–molecule reactions created with large kinetic energies. Thus, by allowing an octupole to pass through the volume in which reactant ions and neutrals interact, all reaction products can be trapped in a sufficiently deep potential well to allow complete product collection and accurate measurement of reaction cross sections.

One of the most common applications of the rf octupole guided-beam method is in total cross-section measurements. The original application of this method by Teloy and co-workers [35, 109, 110] involved an instrument in which the reactant beam was transported with an octupole through a collision cell, allowing accurate total cross-section determination. Ervin and Armentrout at Berkeley [49] and Hanley and Anderson at Stony Brook [87] have exploited this technique to study numerous systems including monatomic and cluster metal ions, examples of which we will discuss in the next section. In addition, the guided-beam technique has been used extensively in experiments in which vibrationally state selected reactants have been prepared by UV photoionization for total cross-section measurements as a function of relative collision energy [111–114]. The details of these elegant experiments are discussed extensively by Cheuk-Yiu Ng in Chapter VIII on photoionization methods.

In addition to ensuring that the products of reactive collisions are collected with 100% efficiency, inhomogeneous electric fields can be used to trap ions in the presence of a quenching gas for times long enough to ensure that the ions have lost their internal excitation. Since the usual methods for producing ions are rather unselective with respect to internal state distributions, the possibility of quenching the nascent distribution after ionization is very important. Hanley and Anderson [87] have exploited this phenomenon by guiding ions created by sputtering, with their very broad internal and translational energy distribution, through a collision cell filled with a buffer gas used for collisional quenching. In Teloy and co-worker's original work [35, 109, 110], in addition to demonstrating the guided-beam characteristics of an electrostatic octupole, these workers also showed that inhomogeneous fields with unusual geometries can be used to quench internal excitation in nascent ions. The original design of Teloy [35], shown in Fig. 9, uses a stack of parallel electrodes to which the appropriate dc and rf potential are applied. Ions created in the center of the trap must diffuse through the U-shaped channel, undergoing many collisions, before exiting the source and entering the guiding octupole. More recently, Hanley et al. [115] have exploited this labyrinthine ion source idea to prepare

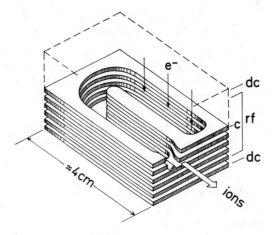

Fig. 9. Electrode arrangement for storage ion source. The upper half of the ion source is symmetric with respect to plate c and has been omitted from the diagram for clarity. The electron beam enters the right arm of the source, and ions diffuse around the U-shaped channel under the action of an rf potential and exit through holes in the central U plates. Reprinted from Ref. [35] by permission of North-Holland Physics Publishing.

collisionally relaxed metal cluster ions for reactive scattering studies. We will discuss guided-beam methods in more detail in the next section by presenting a number of recent examples of their capabilities.

4.3 Neutral Sources

Ion-beam experimentalists are accustomed to devoting the majority of their efforts to the "hard" part of the experiment, the production and transport of low-energy ion beams. The preparation of neutral reactants is just as important, however, and deserves careful attention in a well-devised experiment. For many experiments, a collision cell provides an adequate source of reactants to collide with the ion beam. As we have discussed previously, the thermal velocity spread of the neutral reactants can be prohibitively large, especially for low-energy ion beams [24]. Despite cooling the collision cell to cryogenic temperatures, the spread in initial relative velocities can be appreciable. For determination of absolute cross sections away from threshold and with suitable corrections near the threshold of an endothermic process, a gas cell provides a simple solution for introduction of neutral reactants. The high-number density that a collision cell can provide, often as high as 10^{12}–10^{13} cm^{-3}, and the possibility of long attenuation path lengths of several centimeters, make a cell an attractive option for ion–molecule reaction studies. However, for the best kinematic resolution, a crossed beam is a necessity.

In the early days of molecular beam spectroscopy and scattering, effusion of gases at low pressures through slits whose dimensions were small in comparison with the mean free path of the gas was a technique widely used for neutral-beam production by workers in these areas [116]. The speed distribution for such an effusive source is Maxwellian, and the width of the distribution may be comparable to the most probable speed. The low-pressure conditions required to satisfy the collisionless effusion criterion led to very low intensities, with number densities three to four orders of magnitude smaller than for a collision cell can provide, often as high as 10^{12}–10^{13} cm^{-3}, and the possibility of long effusive source led to unacceptably low neutral reactant concentrations, particularly for ion-beam experiments in which the ionic reactant number density is already very low.

An important innovation in neutral-beam production was first suggested by Kantrowitz and Grey [117], with pioneering experiments by Kistiakowsky and Slichter [118]. By using a high-pressure expansion from a gas stagnation region in which the mean free path is small in comparison with orifices in the source into a vacuum, a directed gas expansion can be created under conditions of hydrodynamic flow. Gas at a high pressure and stagnation temperature T_0 expand through a small pinhole into a vacuum. At some distance L downstream from the nozzle, a conical skimmer interrupts the flow, beyond which the molecules in the beam are assumed to be in a collisionless state. Numerous workers have analyzed the flow characteristics of such an expansion [119–122], with the main result that the enthalpy of the expanding gas is converted into directed translational motion of the gas. Under such conditions, the internal energy of the gas in the high-pressure stagnation region is removed collisionally and transferred to translation along flow streamlines. Such collisions cool the internal degrees of freedom of molecules in the expansion, and internal temperatures of only a few Kelvin can be achieved routinely. In addition, the collisions taking place in the expansion remove energy from translation in directions transverse to the flow. Because the local speed of sound in a gas is a measure of the random thermal motion, in this case the motion transverse to the flow, molecules moving along the flow have speeds several times greater than the local speed of sound. Therefore, such gas expansions taking place in the hydrodynamic flow regime are called supersonic expansions and may be characterized by the Mach number, the ratio of the stream velocity to the local speed of sound in the flow. Thus, the gas stream at the skimmer entrance may be characterized by a stream velocity v_s given as follows [120]:

$$v_s = \left(\frac{kT_0}{m}\right)^{1/2} \gamma M \left(1 + \frac{\gamma - 1}{2} M^2\right)^{-1/2} \tag{20}$$

In this expression, M is the Mach number, T_0 is the temperature of the stagnation chamber, and γ is the heat capacity ratio C_p/C_v.

The speed distribution for a supersonic expansion may be expressed in terms of the stream velocity v_s and the Mach number as follows [120]:

$$P(v) = N \left(\frac{v}{v_s}\right)^3 \exp\left[-\frac{\gamma M^2}{2}\left(\frac{v}{v_s} - 1\right)^2\right] \tag{21}$$

Figure 10 compares the speed distributions for supersonic expansions with selected Mach numbers with that for a Maxwell-Boltzmann distribution. Supersonic expansions show a speed distribution with characteristic narrowing and shifting to higher speeds relative to oven beams. By reducing the dispersion in initial relative velocities, this narrowing leads to vastly improved collision kinematics for crossed-beam experiments, and FWHM speed distributions corresponding to only a few percent of the most probable speed may be achieved with care. The Mach number is determined by the Knudsen (Kn) number for the flow Kn_0, the ratio of the mean free path to the nozzle diameter [119]:

$$M = 1.17 \, Kn_0^{-0.4} \qquad (22)$$

The Mach number also determines the extent of cooling in the expansion, a measure of the transfer of energy from internal degrees of freedom and translation transverse to the flow streamlines into directed motion. This cooling, often exploited in spectroscopy because of the spectral simplification that it affords [123, 124], is also advantageous in scattering studies because it allows us to prepare neutral reactants in their ground rotational states. In addition to narrower speed distributions, supersonic beams are more directional in comparison with the $\cos \theta$ distribution of effusive beams, yielding a greater centerline intensity relative to the effusive case [119]. In addition to higher directionality, the higher number densities of gas in the stagnation region of the expansion lead to higher fluxes than for effusive beams, with effective reactant concentrations at least two orders of magnitude larger. It can be shown [120] that the intensity gain G of a supersonic source over an effusive source operating at the same temperature and assuming that the effusive orifice size equals the

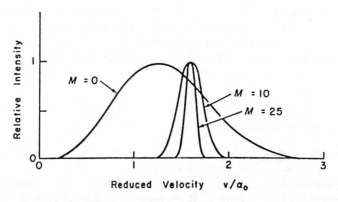

Fig. 10. A comparison of molecular speed distributions for a Maxwell-Boltzmann distribution ($M = 0$) with the axial velocity distribution of a supersonic jet of a monoatomic gas ($\gamma = 5/3$) at stagnation temperature T_0 for Mach numbers of 10 and 25. The abscissa is given in units of $\alpha_0 = (2kT_0/m)^{1/2}$.

skimmer area is given as

$$G \approx \left(\pi \frac{\gamma^3}{\gamma-1}\right)^{1/2} M^2 \approx 4.6 M^2 \qquad (23)$$

for monoatomic and diatomic gases with values $\gamma = 5/3$ and $7/5$, respectively. Thus, for a Mach number of 10, readily achieved with current technology, the gain is nearly 500.

The most probable speed in a supersonic expansion is given by the expression

$$v_{\text{peak}} = \left[\left(\frac{\gamma}{\gamma-1}\right)\frac{2kT_0}{m}\right]^{1/2} \qquad (24)$$

For a neat expansion of a single pure component, the mass m is obviously the atomic or molecular mass of the gas; but when a mixture of gases is expanded, the hydrodynamic flow of the expansion causes every species in the flow to achieve the same terminal speed. This speed can be calculated from Eq. 24 by using the average mass number of the gas mixture. By seeding a small percentage of a heavy species in a large excess of a light carrier gas such as H_2 or He, the speed of the heavy particle can be increased by many factors over its value for a neat expansion [125, 126]. As we shall note in our discussion of kinematics to follow, very different speeds for the ionic and neutral reactants in a crossed-beam experiment can severly limit kinematic resolution. The seeded-beam method provides an important way to overcome such kinematic compression in favorable cases. The seeded-beam method may also be used for producing beams of liquids with small vapor pressures at readily accessible temperatures. By sweeping a carrier gas at pressures near an atmosphere through the liquid, the vapor pressure of the liquid determines the amount of seeding that occurs in the expansion. Collisions with the diluent gas relax internal degrees of freedom in the molecule.

Although the supersonic nozzle beam technique appears best suited for permanent gases and species condensible at cryogenic temperatures, technical developments in the last several years have led to the production of beams of species such as alkali metals [127], halogens [128], and alkali halides [129] and with specialized techniques such as laser vaporization, transition metals as well [80, 90]. Beams of more exotic species such as free radicals [130] and open shell atoms [131] have been created, and it is certain that the chemical variety of neutral species available for study in supersonic expansions will continue to grow.

The widespread development of supersonic nozzles for spectroscopic and collision studies has occurred in large part because the formidable vacuum pumping requirements associated with large gas throughputs can be solved with modern vacuum technology. The ease of operating supersonic sources is sufficiently high that such sources have become the standard for neutral beam production in most molecular beam scattering laboratories. The development of pulsed valves [132–135], with low duty cycles and high peak intensities, places significantly less stringent pumping requirements on the source. At this time, however, the low duty cycle of pulsed neutral sources is poorly matched with the continuous nature of most ion production and detection techniques. As specialized methods for producing reactants by pulsed laser evaporation and photoionization become more widespread, the pulsed valve source for neutral reactants will gain in popularity.

4.4 Product Detection

The detection of low-energy ions presents challenges beyond those of preparing collimated beams of reactant ions. The specific problems to be attacked are those of energy measurement over a wide range of laboratory energies, product mass determination, and sensitive particle detection. In nonoptical methods of product state analysis, an experimenter measures the laboratory kinetic energy distribution of collision products as a function of laboratory scattering angle using a number of techniques, including electrostatic energy analysis and time-of-flight (TOF) velocity analysis. In the former case, an electrostatic analyzer can be followed by a mass spectrometer to give an unambiguous determination of the product, whereas TOF determinations of ionic speeds normally do not include a separate mass analysis step.

4.4.1 Energy Analyzers

One of the simplest energy analyzers we can construct is a plane field retarding energy analyzer [136, 137]. By applying a stopping potential to a high-transparency grid perpendicular to the flight axis of the products, the transmitted current as a function of this potential represents the integral of the kinetic energy distribution. Such an analyzer is simple to construct and is capable of fairly high resolution, determined primarily by the angular collimation of the products relative to the plane of the analyzer electrode. Because the detector signal represents the number of ions with energies greater than the stopping potential, the energy distribution must be calculated by differentiating the detector signal. Furthermore, weak features in the energy distribution may be masked by being superimposed on the strong signal of the transmitted beam. Since differentiation of the transmitted signal amplifies noise in the data and because weak features in the energy distribution may be difficult to extract if they are superimposed on the accompanying large transmitted signal, the retarding field method has serious limitations. Despite these problems, some important embellishments of this simple idea have appeared in the literature. Simpson [136] has described an efficient retarder that decelerates and focuses incoming ions to the retarding plane, reaccelerates them, and detects them; this device, called an intermediate image filter lens, has been used in ion–neutral scattering measurements by Futrell and co-workers [108]. The retarding field method has also been used in photoelectron spectroscopy, where retarding

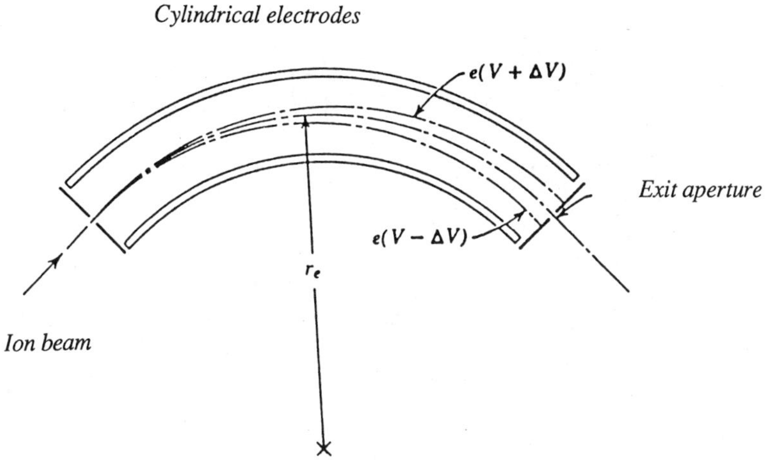

Fig. 11. Schematic diagram of an electrostatic condenser providing energy filtering of an ion beam.

potentials applied to high-transparency hemispherical grids surrounding the photoionization volume allow large solid angle collection efficiency for photoelectrons [138].

The trajectory of a charged particle in the electric field between the electrodes of a cylindrical condenser depends on the energy of the particle [139], and therefore such devices have found widespread applications as electrostatic deflection energy analyzers, the electrostatic analog of a magnetic prism. Figure 11 indicates schematically the way in which such a device disperses trajectories according to energy. The resolution of the device is a function of the slit width and the radius of curvature, the precise form of the resolution depending on the geometry of the condenser. A number of different condenser geometries exists, including plane parallel plates [140], parallel cylinders [141, 142], and concentric toroidal sections [143]. In addition, the cylindrical mirror analyzer, consisting of coaxial cylinders [144], has enjoyed popularity in surface science. Although these devices exist and have been used for specialized applications, we will restrict ourselves to a discussion of two of the most popular designs, the cylindrical and spherical analyzers.

The most general electrostatic deflector geometry is shown in Fig. 12 for an arrangement of concentric toroids. The toroids are defined by two pairs of radii: r_a and r_b define the curvature of the electrodes along the arc of the ion trajectory, and R_a and R_b define the electrode curvature orthogonal to the trajectory. The general case in which all four radii are different corresponds to an astigmatic toroid, rarely used in energy analysis. The popular cylindrical analyzer corresponds to the case in which R_a and R_b are infinite, resulting in parallel cylindrical

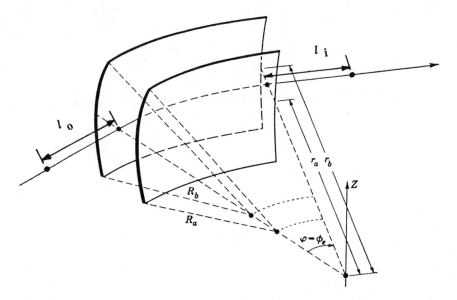

Fig. 12. Schematic diagram of focusing properties of a toroidal condenser, indicating sector angle ϕ_e, mean radius of the ion trajectory, image distance l_i, and object distance l_o.

electrodes. When $R_a = r_a$ and $R_b = r_b$, the concentric spherical analyzer results. As with a magnetic mass spectrometer, the focal properties of this configuration satisfy a Newton lens equation that relates deflection angle ϕ_e, image and object distances, and the mean trajectory radius, given as $(r_a + r_b)/2$. A particularly widely used configuration is the total immersion mode in which the image and object distances are zero, placing the image and object points at the edge of the electric field. For a cylindrical analyzer, this geometry leads to a deflection angle of 127°17′, and the focal properties of this analyzer are particularly widely studied [141, 142]. Focusing in this geometry occurs only in the plane of the electric field, leading to astigmatic operation. Although planar electrodes placed at the top and bottom of the cylinder can compensate for this defocusing, exact two-dimensional focusing can be effected with a spherical condenser [143, 145–147]. The spherical condenser also satisfies a Newton lens equation, and the total immersion mode for this analyzer, in which image and object points are at the boundary of the electrodes, occurs for a deflection angle of 180°. This particular geometry is very widely used in electron monochromators and analyzers for electron energy loss spectroscopy (HREELS, EELS) methods in surface analysis [148]. Both cylindrical and spherical analyzers have stringent angular resolution requirements for the divergence of the particles entering the gap between electrodes. This resolution, of order 1°, must be matched with the laboratory angular resolution of any defining slits in the detector.

An electrostatic analyzer functions like a prism, requiring slits at the object and image points. To first order in the angle with which ions are injected into the gap between the electrodes, the resolution of a spherical analyzer is given by [139]

$$\frac{\Delta E}{E} = \frac{w}{2R_e} \qquad (25)$$

In this expression, w is the analyzer slit width and R_e is the mean radius of the condenser.

The value of $\Delta E/E$ for a cylindrical analyzer is twice as large as that for a spherical analyzer for the same values of w and R_e, giving the hemispherical prism twice the resolution for a given size. Resolution can be increased by decreasing slit widths or increasing the physical size of the analyzer, but we reach obvious limits with such procedures. By making the pass energy E of the analyzer low, on the order of 1 eV, the ΔE for the system can be made as low as a few meV. Such operation requires that the detected ions be accelerated or decelerated to the pass energy E, a procedure requiring special optical design [149]. The focal planes for energy analyzers operating in the total immersion mode are not equipotential surfaces, and therefore simply placing slits at the field boundaries is not a satisfactory optical design. In addition, the collision center in a crossed-beam experiment cannot be placed at the object point for an energy analyzer operating in the total immersion mode. In order to solve these problems, very sophisticated optical systems [149–151] have been developed to image the collision center onto the object plane of electrostatic analyzers. The conventional way to use an energy analyzer is to operate it at constant pass energy such that ions coming from the interaction region must be accelerated or decelerated to the fixed pass energy in order to be detected. By making the energy at which the ions are analyzed a fixed value, usually only a few electron volts, with resolutions $\Delta E/E$ of a percent, for example, the bandpass ΔE convoluted with the experimental data is fixed and energy independent. a situation making data reduction much simpler. An analyzer operating at a pass energy of 1 eV with $\Delta E/E$ of 0.01 yields a resolution ΔE of only 10 meV or 80 cm^{-1}. Such resolution should allow us to resolve vibrational quanta under favorable kinematic conditions.

Ions with a distribution of kinetic energies leaving the collision center must be transported to the image plane of the energy analyzer at the pass energy in order to be detected. The optical problem to be solved is that of a variable object being focused to a fixed energy image. The lens system that performs such an operation is a "zoom" lens, and a number of designs for such lenses, using cylindrical lenses [151] and aperture lenses [150], have appeared in the literature. All such lenses are characterized by the use of three potentials from which two voltage ratios can be constructed. In a typical design, apertures at the entrance of the lens system define an object size that is then transported to the energy analyzer with no additional slits. Such a "virtual slit" system is advantageous because it avoids having slit surfaces at low energy, avoiding the production of surface charge that leads to poor transmission and loss of signal.

In addition, a virtual slit system avoids the unfavorable electrostatic condition of placing a slit in a plane that is not an equipotential surface. Zoom lens systems must be constructed carefully and be kept scrupulously clean in order to avoid the buildup of surface charge; the use of colloidal graphite reduces surface charge [152], and keeping the entire lens system heated at all times is a very effective way to optimize performance [108, 153]. The preanalyzer optical system is the single most important determinant of proper operation of the detector with high resolution and transmission.

Following the energy analyzer, most detector systems include a mass spectrometer system. Both magnetic and quadrupole systems, which we have discussed previously, have been adapted for this purpose, and the compactness of the latter, in conjunction with resolution that is readily variable by controlling the ratio of the dc to rf potentials, has made it the method of choice for detectors that rotate about the interaction volume.

4.4.2 Particle Detectors

Following the mass spectrometer, we must have a means of detection of individual charged particles. An excellent overview of many detection methods for charged particles is found in the monograph by White and Wood [154], and we review some of the most common methods here. Discrete dynode electron multipliers may be used for detecting charged particles, but the compactness of the continuous dynode electron multiplier has led to its widespread use in many particle counting systems [154]. The continuous dynode multiplier, shown schematically in Fig. 13a, relies on the process of electron amplification by secondary electron emission, and proper biasing of the multiplier makes it a detector for positively or negatively charged particles. One of the most serious limitations of this detector is the rate-dependent pulse height that dramatically erodes gain at high counting rates. This phenomenon limits the dynamic range to no more than 100 kHz. The detector is in vacuum and may be exposed to corrosive gases during operation, leading to deterioration in gain. A detector that eliminates these shortcomings, at the expense of a significant increase in complexity, is the Daly detector [155], which uses conversion dynodes to increase the pulse height for ion events. This detector, shown schematically in Fig. 13b, collimates and accelerates ions exiting from a mass selector to several keV, where they strike a highly polished cathode covered with a low atomic number metal such as aluminum. This cathode is held typically at $-30\,\text{kV}$, and each ion striking the cathode ejects several secondary electrons. This pulse of electrons is accelerated away from the cathode where it strikes a plastic scintillator. The front surface of the scintillator is coated with a thin (500 Å) coating of aluminum to block stray light and to serve as a conductor to drain the electron charge from the scintillator face. The electrons penetrate the aluminum coating and strike the scintillator surface where several photons are emitted per incident electron. The photons are then amplified by a photomultiplier tube and counted as individual events. The gain from this conversion scheme is such that each ion entering the detector results in a photon pulse

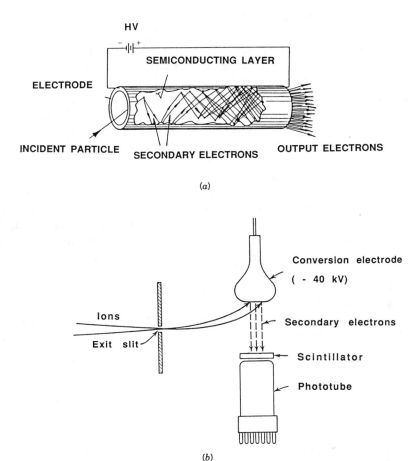

Fig. 13. (a) Continuous dynode electron multiplier. (b) Daly scintillation detector. Ions enter the detector through a series of lenses and strike the high voltage cathode. Emitted secondary electrons strike scintillator surface and emit a light pulse that is amplified by a photomultiplier tube.

containing as many as 20 photons. The height of this pulse may be distinguished readily from dark pulses from the photomultiplier by using a simple threshold discriminator. Under optimal conditions, background counting rates for the Daly detector can be no more than $0.1 \, \text{s}^{-1}$. Because the phototube is not in the vaccum envelope, it is impervious to corrosive gases, and the maximum counting rate of the detector is only limited by the pulse width of the phototube. Counting rates in excess of 10 MHz are readily achievable. Although the detector has been used primarily for positive ions, Shao and Ng have reported a modification of the Daly design that may also be used for negative ion detection [156].

4.4.3 Time-of-Flight Methods

Despite the widespread application of electrostatic energy analyzers for measuring kinetic energy distributions of charged particles, measuring speed distributions by determining the time of arrival of scattered ions traveling through a fixed distance is a technique with great potential as well [157]. Toennies and collaborators at the Max Planck Institute in Göttingen have pioneered this approach to the study of inelastic and charge transfer processes [158, 159]. An ion source followed by a mass selector and a cylindrical energy analyzer with resolution of 40 meV and electrostatic deflection plates to chop the ion beam allows a short pulse of energy-selected ions to enter the interaction region. This ion pulse intersects an intense neutral beam, and the scattered products leave the collision region, entering a 3-m flight tube where their distribution of arrival times is measured with fast pulse counting electronics. The overall resolution of the experiment may be as low as 60 meV, allowing product vibrational states to be resolved. This kind of instrumentation has been used to study state-resolved energy transfer collisions of Li^+ with a number of neutrals [158], to study vibrational energy transfer collisions of protons with a number of polyatomic molecules, including CF_4 and SF_6 [160] and, more recently, differential cross sections for proton–O_2 charge transfer collisions [159]. We will present results of this technique in the next section.

5 KINEMATIC ANALYSIS

Experimental quantities measured in the laboratory system of coordinates are dependent on the specific instrument from which they are obtained. Ready interpretation of data and comparison of results with theoretical studies generally requires that experimental data in beam experiments be transformed to coordinates of relative motion in which the motion of the center of mass of the collision system has been removed. Total cross-section measurements are the simplest to transform, since the signal in laboratory coordinates is linearly related to the cross section in the center-of-mass system, as indicated in Eq. 2. The principal experimental difficulty in measuring absolute total cross sections is the determination of absolute number density in the collision region.

Any experiment in which we measure a differential cross section, either at the quantum state level of resolution or by measuring kinetic energy distributions with finite energy and angular resolution, requires careful examination of the transformation between the lab and center-of-mass coordinate systems. Such considerations are vital since the Jacobian transformations between coordinate systems that are moving with respect to one another frequently lead to distortions in laboratory intensities that can mislead us concerning their corresponding center-of-mass values [161]. The relationship of measured laboratory cross sections to corresponding center-of-mass quantities has been discussed for crossed-beam experiments in several literature articles in the past two decades [162–168], and Friedrich and Herman [168] have published a particularly clear article whose results and conclusions have been formulated for ion–molecule scattering systems. Refer to these detailed literature references for a complete treatment of the kinematic problem. Our emphasis here is on the relationship of lab and center-of-mass cross sections.

The geometric relationship between center-of-mass and lab speeds and scattering angles can be understood by considering a kinematic diagram, or Newton diagram, for the collision process A + BC, as shown in Fig. 14. The diagram is constructed for the special case in which the reactant beams intersect at 90°. The lab scattering angle and velocity are defined by Θ and v, respectively, whereas the corresponding center-of-mass quantities are θ and u. The beam velocity vectors $\mathbf{v}_A$ and $\mathbf{v}_{BC}$ define the initial conditions, and the relative velocity vector is defined by their vector difference. The direction and velocity of the center-of-mass, or centroid, of the collision system are determined by conservation of linear momentum; the vector $\mathbf{C}$ divides $\mathbf{v}_{rel}$ in inverse proportion to the masses of A and BC:

$$\mathbf{C} = \frac{m_A \mathbf{v}_A + m_{BC} \mathbf{v}_{BC}}{M} \qquad (26)$$

$$\mathbf{u}_A = -\frac{m_{BC}}{M} \mathbf{v}_{rel} \qquad (27)$$

$$\mathbf{u}_{BC} = \frac{m_A}{M} \mathbf{v}_{rel} \qquad (28)$$

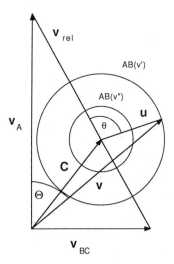

Fig. 14. Newton kinematic diagram indicating velocity vectors of reactants A and BC. The direction of the center-of-mass is indicated by the vector **C**. The relationship between lab coordinates (v, Θ) and center-of-mass coordinates (u, θ) is indicated by the vector equation $\mathbf{v} = \mathbf{u} + \mathbf{C}$. Circles indicate loci of points of constant u in the plane defined by the reagent beams for specific vibrational states v' and v'' of the AB product. In the case shown, $v'' > v'$.

M is the total mass of the reactants. An observer moving away from the laboratory origin in the direction $\Theta_{c.m.}$ with speed C would see the reactants A and BC approach along $\mathbf{v}_{rel}$ and the products retreat along the final relative velocity vector $\mathbf{v}'_{rel}$. The angle between these vectors defines the center-of-mass scattering angle θ. Because of azimuthal symmetry about $\mathbf{v}_{rel}$, the locus of points corresponding to a constant center-of-mass recoil speed $\mathbf{u}$ for a reaction product is the surface of a sphere of radius $\mathbf{u}$ with origin at the center of mass. Detection of products in a plane, often defined by the velocity vectors of the reactants, reduces this three-dimensional sphere to a circle defined by the intersection of the sphere surface with the equatorial plane containing $\mathbf{v}_A$ and $\mathbf{v}_{BC}$. Although we are not obligated to detect products in the plane of the beams, Lee [169] has outlined circumstances in reactive and inelastic collisions where detection in a plane perpendicular to the beams will prevent the observation of all products of the reaction. For a single Newton diagram, detection of products at recoil speed u and scattering angle θ requires that measurements be made at laboratory coordinates (v, Θ). If the incident beams are monochromatic, then the lab–c.m. coordinate transformation is unique when both lab scattering angle and speed are measured. However, if the reagent beams have speed distributions of finite width, then products appearing at a single lab angle and velocity correspond to a distribution of center-of-mass angles and speeds. The finite angular and energy acceptance of the detector further complicates the transformation of data and underscores the importance of performing experiments with precisely defined initial conditions. We note also that mechanical limitations on the accessibility of ranges of laboratory angle and speed may restrict our access to corresponding regions of velocity space in center-of-mass coordinates.

Linear momentum conservation also relates the center-of-mass speeds of products AB and C to the relative velocity of the separating products:

$$\mathbf{u}'_{AB} = \frac{m_C}{M} \mathbf{v}'_{rel} \tag{29}$$

$$\mathbf{u}'_C = -\frac{m_{AB}}{M} \mathbf{v}'_{rel} \tag{30}$$

The final relative kinetic energy of the separating products may be calculated from

$$E'_T = \tfrac{1}{2} \mu' v'^2_{rel} \tag{31}$$

where μ' is the reduced mass of the products.

The total energy of a collision system may be computed from the energies of the incident reagents of the energy accessible to the products:

$$E_{total} = E_T + E_{int} - \Delta D_0^\circ = E'_T + E'_{int} \tag{32}$$

E_T and E_{int} refer to the incident translational and internal energies of the reagents and the primed quantities correspond to the products. ΔD_0° represents the energy liberated in the reaction, the difference in zero point energies of the reactants and products. The maximum center-of-mass speed for the product BC arises when the available energy all appears in translation, and energy appearing in internal excitation of any product results in reduced values of v'_{rel}. The magnitude of the final relative velocity and, therefore, the center-of-mass speed, of the separating products may be computed from Eq. 33:

$$v'_{rel} = \left[\left(\frac{2}{\mu'} \right) (E_T + E_{int} - \Delta D_0^\circ - E'_{int}) \right]^{1/2} \qquad (33)$$

The quantization of the vibrational energy of AB leads to a series of discrete concentric circles about the centroid that describe the loci of final translational speeds for AB produced in specific vibrational states, as shown in Fig. 14. Rotational excitation in the products may partially fill in the region of velocity space between the circles, but under favorable conditions, some examples of which will be discussed in Sections 6.1 and 6.2, the discrete structure in velocity will persist, allowing state-resolved identification of reaction products.

In addition to the transformation between laboratory and center-of-mass angles and velocities, we must also have relationships for the transformation of intensities and cross sections between these coordinate systems. As Entemann

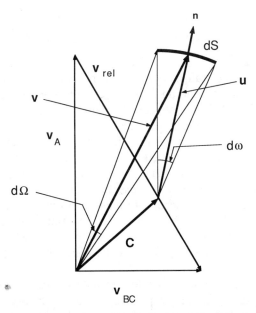

Fig. 15. Newton kinematic diagram indicating solid angles $d\omega$ and $d\Omega$ subtended by detector surface dS in center-of-mass and laboratory coordinates, respectively. Intensity transformations are discussed in the text.

[163] has pointed out, the quantity that is conserved in a transformation between two coordinate systems moving with respect to one another is the flux across a reference surface. Figure 15 shows the nature of this conservation: in laboratory coordinates, the flux into solid angle $d\Omega$ is given by $I_{\text{lab}}(\Omega)\,d\Omega$, where the lab intensity I_{lab} is the flux per unit solid angle, whereas the corresponding flux in the center of mass is $I_{\text{c.m.}}(\omega)\,d\omega$, where $I_{\text{c.m.}}$ is the flux per unit solid angle in the center of mass. The solid angles subtended by the detector in the lab and center-of-mass coordinates may be computed from the surface element dS subtended by the detector:

$$d\omega = \frac{dS}{u^2} \tag{34}$$

$$d\Omega = dS\,\frac{\cos(\mathbf{u},\mathbf{v})}{v^2} \tag{35}$$

The $\cos(\mathbf{u},\mathbf{v})$ term arises from the fact that the outward normal to the reference surface dS lies along $\mathbf{u}$ but not along $\mathbf{v}$. Using these solid angles with the flux equality

$$I_{\text{lab}}(\Omega)\,d\Omega = I_{\text{c.m.}}(\omega)\,d\omega \tag{36}$$

the intensity transformation may be expressed as follows:

$$I_{\text{lab}}(v,\Theta) = \left(\frac{v^2}{u^2\cos(\mathbf{u},\mathbf{v})}\right) I_{\text{c.m.}}(u,\theta) \tag{37}$$

The Jacobian of the transformation depends on $1/u^2$, illustrating the fact that lab intensities corresponding to points close to the centroid of the system, where u approaches zero, will appear anomalously large. The classic example of this problem occurred in the early studies of alkali atom–alkali halide exchange reactions in which preliminary observations of product angular distributions suggested that the reaction products were extremely highly internally excited, with nearly 90% of the reaction energy in product internal excitation. These initial conclusions were erroneous because they failed to account for the Jacobian singularity between lab and center-of-mass coordinates that enhances lab intensities near the collision system center of mass. Many of the early conclusions of crossed-beam experiments were based upon lab observations that indicated that products were preferentially formed with low center-of-mass recoil velocities [161].

The argument we have developed for the intensity transformation is correct for a single quantum state of the products, but if a distribution of quantum states can appear at the same center-of-mass speed, then each state has its own $\cos(\mathbf{u},\mathbf{v})$ term, and the sum over these terms becomes unity. This point has been discussed in detail in several references [163, 165, 167].

The extraction of center-of-mass fluxes from laboratory data in which the reagent beams have finite velocity spreads is a topic that has been discussed thoroughly in the literature, and direct deconvolution procedures as well as integration fitting methods have been developed to provide meaningful data reduction. We refer the reader to numerous articles on this subject [162, 168]. Once a deconvoluted center-of-mass distribution has been obtained, we can express it as a contour map or three-dimensional axonometric plot in the polar coordinates u and θ. Contour maps and their three-dimensional representations provide quick visual assessments of reaction mechanisms. By looking for the presence or absence of angular symmetry with respect to 90° in the center of mass, we can determine quickly whether a given process occurs as a direct collision or through the participation of a complex that lives at least a rotational period. To gain more quantitative insight into reaction energy partitioning, however, we need to take center-of-mass cross sections that are doubly differential in velocity and angle and integrate them appropriately to determine angle and kinetic energy distributions. The angle-averaged energy distribution and velocity-averaged angular distribution are given by the following integrals:

$$g(\theta) = \int_0^\infty I(u, \theta)\, du \qquad (38)$$

$$P(u) = \int_0^\pi I(u, \theta) \sin\theta\, d\theta \qquad (39)$$

Since $P(E'_T)dE'_T = P(u)\,du$, we may calculate the kinetic energy distribution as follows:

$$P(E_{T'}) = \frac{1}{u} \int_0^\pi I(u, \theta) \sin\theta\, d\theta \qquad (40)$$

These expressions are particularly useful in interpreting and analyzing reaction dynamics, and we shall provide examples of their use as well as the qualitative use of flux maps in the next section.

In many ion–neutral reactions, the ion-beam speed is much greater than that of the neutral. Under some circumstances, this disparity can be advantageous, but in others, it can work to the experimenter's detriment. Figure 16 illustrates some limiting cases. When the ion is light and the neutral is heavy, the centroid follows the neutral velocity vector, as in Fig. 16a. If the reaction products appear near the centroid, they will be constrained to a very narrow range of lab angles and will have low energies in the lab, making their detection difficult. The use of a seeded beam to increase the neutral speed opens up the Newton diagram, improving center-of-mass angular resolution and moving the product energy to higher values where reliable detection can be achieved, as in Fig. 16b. If the ion is heavy and the neutral is light, the centroid follows the

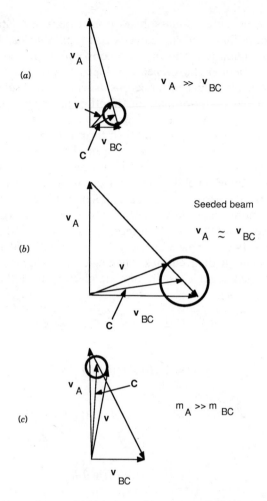

Fig. 16. Limiting kinematic cases: (*a*) Light ion and heavy neutral: The centroid follows the neutral velocity vector, and products appearing near the centroid will have low energies. (*b*) The use of a seeded beam to increase the speed of the heavy neutral opens up the Newton diagram and moves the product energies to higher values. (*c*) If the ion is heavy and the neutral is light, the centroid follows the ion, and a broad distribution of neutral speeds, all of which are much less than the ion speed, smears out the distributions of the Newton diagrams by only a small amount. The heavy circles correspond to the locus of points for a single product recoil speed.

ion, and a broad distribution of neutral speeds, all of which are much less than the ion speed, smears out the location of the centroid by only a small amount. Under such conditions, as shown in Fig. 16*c*, the kinematic sharpening that a narrower neutral speed distribution affords yields very little benefit to the experiment.

The interpretation of center-of-mass cross sections requires detailed dynamical models, but often a cursory examination of a contour map allows us to elucidate simple dynamical features by inspection. Figure 17 illustrates how contour maps may appear for direct vs. collision complex mechanisms. In a direct process, such as spectator stripping [170] or rebound mechanisms [171], in which the reactions take place in a time period comparable for the reagents to pass by one another, the flux distributions should be highly asymmetric with respect to the center of mass of the system, as illustrated in Fig. 17a. In contrast, a reaction that proceeds through a collision complex will lose all memory of the direction of approach of the reagents, and therefore its flux distribution will be symmetric with respect to the $-90°/+90°$ axis in center-of-mass coordinates [172]. The observation of a flux distribution with such symmetry, as shown in Fig. 17, is normally a signature of a reaction proceeding through a complex living several rotational periods. Both of the contour maps shown in the figure represent limiting cases, and we can often see transitions between these limiting cases in certain ranges of collision energy. For example, the observation of a collision complex at low energy may be followed by the development of asymmetry in the cross section with increasing energy as the lifetime of the complex decreases and becomes comparable to or less than a rotational period. This behavior is called "osculation" [173–175] and represents one of many dynamical variations that can be observed in beam studies of reaction dynamics. Many such dynamical variations will be found in the examples that we will discuss in Section 6.

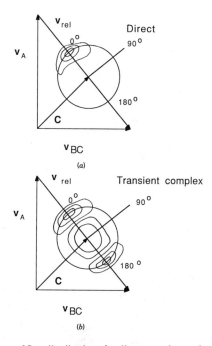

Fig. 17. Schematic diagram of flux distributions for direct reaction and complex dynamics in which a transient collision intermediate lives several rotational periods.

6 ION-BEAM STUDIES OF REACTION DYNAMICS: A SAMPLING

In the preceding sections, we outlined general experimental configurations for ion-beam experiments, as well as specific techniques for production and transport of reagents, and methods for analyzing and detecting products. We will now discuss several sets of experimental results acquired with many different instruments used to study ion–neutral interactions. In the early days of beam studies of ion–neutral interactions, very simple systems, principally involving proton and hydrogen transfer reactions with rare gases, dominated experimental results. A relatively small number of mechanisms, particularly the impulsive spectator stripping model [170], appeared to explain the majority of experimental results and ignored to a large extent the underlying potential energy surfaces. The applicability of these models to the rather high-energy processes that available technology made accessible at that time suggested that the dynamics of most ion–molecule reactions were very simple, depending only weakly on the underlying potential energy surface. The range of chemical phenomena that has been studied with beam methods since those early days has increased in scope tremendously, including increases in the chemical complexity of reactants, as well as improved selectivity in reagent preparation and product detection. The examples we have selected to discuss illustrate a number of technical capabilities, including state-resolved product analysis, reactions of highly vibrationally excited reagents, and reactions taking place at subthermal energies, as well as the chemical diversity of reactions that have been studied under beam conditions.

6.1 Beam–Gas Cell Measurements

The precise manner in which we combine various detection elements, including energy analyzers, mass spectrometers, and detectors, affords us many options for configuring an experiment. Many of the components we have discussed previously are embodied in an instrument constructed by Wendell and co-workers at Johns Hopkins University [176], shown in Fig. 18. Ions from an electron impact source are energy analyzed by a 180° spherical deflector with zoom lenses, which produces a primary ion beam with a FWHM energy distribution of 0.1 eV, followed by mass selection with a quadrupole mass spectrometer (QMS). This instrument uses a collision cell followed by a detector that is essentially the mirror image of the ion gun. Zoom lenses image the collision center onto the virtual entrance slit of a second 180° spherical electrostatic deflector, followed by a quadrupole mass filter. The entire detector rotates about an axis perpendicular to the primary ion beam, allowing complete angular and kinetic energy scans to be performed.

A particularly nice example of state-resolved product detection has been demonstrated by Jones and co-workers using this instrument [177]. Because vibrational energy levels are coarsely quantized, the creation of reaction products with a structured vibrational state distribution may result in center-of-mass product translational energy distributions that are also structured. When

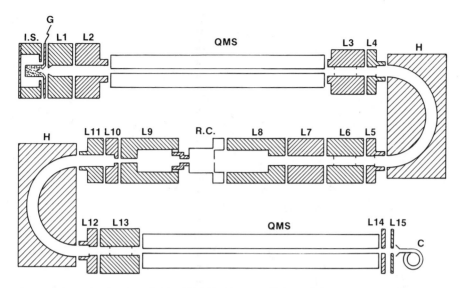

Fig. 18. Schematic diagram of Johns Hopkins beam-collision cell instrument. I.S. denotes ion source, L_1–L_{15} are lens elements, QMS denotes quadrupole mass spectrometers, R.C. denotes collision cell, and C is Channeltron particle multiplier. Reprinted from Ref. [176] by permission of American Institute of Physics.

kinematics are favorable, the laboratory kinetic energy distributions may also show such structure. Figure 19 shows kinetic energy distributions for the reaction of C^+ with H_2 and D_2 to form CH^+ and CD^+, respectively, at collision energies between 0.3 and 1.2 eV, showing clear structure arising from the population of discrete vibrational states of the reaction products. The widely spaced CH^+ vibrational levels (2700 cm^{-1}) in conjunction with very low rotational excitation in the products makes this observation of individual vibrational levels possible. Despite the kinematic imprecision of the beam–gas cell geometry, product state resolution is possible for this system. Such observations of product energy states in kinetic energy spectra, termed "translational spectroscopy," are very rare, particularly so in beam–gas cell geometries. Koski and collaborators have used their instrument to study the reactions of a number of atomic and molecular ions with isotopes of hydrogen. These studies include reactions of B^+, C^+, F^+, Cl^+, and H_2O^+ with H_2, D_2, and HD [176–182] and, more recently, reactions of CH^+, CH_2^+, and CH_3^+ and their isotopic analogs with C_2H_2 and C_2H_4 [183].

As we develop more sophisticated methods for state selection of reactants, with concomitant reductions in beam intensities, signal levels dictate that high-sensitivity methods for product detection, including gas cell techniques, be used. Zare and co-workers at Stanford University have used multiphoton ionization as a source of vibrationally state-selected NH_3^+ cations to study their abstraction and exchange reactions with D_2 [184, 185]. These reactions,

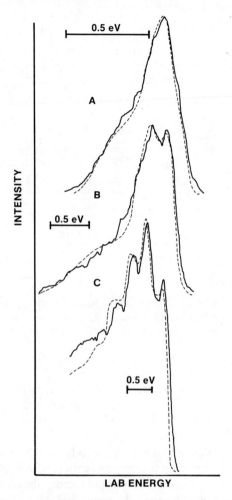

Fig. 19. Product ion laboratory energy spectra at 0° laboratory scattering angle. Structure corresponds to individual CH^+ and CD^+ vibrational states. Reprinted from Ref. [177] by permission of American Institute of Physics. (a) $C^+ + D_2 \rightarrow CD^+ + D$, $E_{rel} = 0.30$ eV; (b) $C^+ + D_2 \rightarrow CD^+ + D$, $E_{rel} = 0.85$ eV; (c) $C^+ + H_2 \rightarrow CH^+ + H$, $E_{rel} = 1.15$ eV. Dashed curves are kinematic predictions.

shown below, are interesting because they are examples of systems that are exothermic, but slow:

$$NH_3^+ + D_2 \rightarrow NH_3D^+ + D \quad \Delta H = -1.04 \text{ eV} \quad (41)$$

$$\rightarrow NH_2D^+ + HD \quad \Delta H = -0.02 \text{ eV} \quad (42)$$

This reactive system has had an interesting history, beginning with flow tube [186] and ICR studies [187] and, more recently, with ion trap [188, 189],

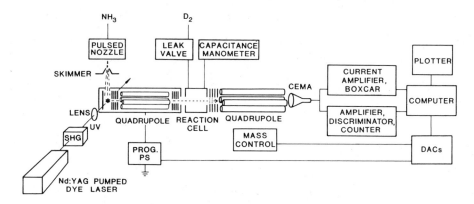

Fig. 20. Schematic diagram of tandem quadrupole mass spectrometer with laser multiphoton ionization source used by Zare and co-workers at Stanford. Reprinted from Ref. [185] by permission of American Institute of Physics.

tandem ICR [190], and multiphoton ionization studies [184, 185]. Zare and his group at Stanford [184, 185] have performed some key experiments on this system by preparing ammonia cations by laser multiphoton ionization, exploiting the fact that n-photon excitation of the neutral to a specific vibrational level in a Rydberg state, followed by single photon excitation from this Rydberg level to the same vibrational level in the ion, leads to vibrationally state-selected ions. Photoelectron energy spectra confirm that such state selection does indeed occur [191]. Franck-Condon factors suggest that the ion should be prepared in the v_2 umbrella bending mode, since the geometry of the neutral is pryamidal, whereas the ion is planar. Using this technique with the instrument shown in Fig. 20, Zare and co-workers have studied the formation of NH_3D^+ through the abstraction reaction 41 as well as the NH_2D^+ product at vibrational energies from 0 to 1 eV, corresponding to 0–10 v_2 quanta, and at translational energies from 0.5 to 10 eV. Their energy-dependent total cross-section measurements, shown in Fig. 21, suggest that at these collision energies, the onset of NH_2D^+ production correlates with a decrease in NH_3D^+ production that these workers ascribe to unimolecular decay of vibrationally excited parent ions through reaction 43:

$$NH_3D^+ \to NH_2D^+ + H \tag{43}$$

The relative yields of NH_3D^+ and NH_2D^+ in this collision energy range were shown to be consistent with production of the parent through a spectator stripping process [170] that imparts a specific fraction of the available energy in the nascent parent which then decays to the apparent exchange product, although the neutral products in the overall stoichiometry are H and D atoms rather than an HD molecule. In this collision energy range, therefore, the

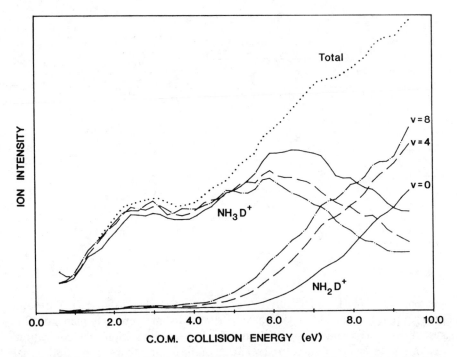

Fig. 21. Kinetic energy dependence of the product ion signals for three different NH_3^+ (v) vibrational levels. Dotted line is the total product ion yield, including NH_3^+ arising from dissociation of the nascent NH_3D^+. Reprinted from Ref. [185] by permission of American Institute of Physics.

experimental data indicate that NH_2D^+ is formed through the dissociation reaction 43 rather than through exchange process 42. Although this dynamical picture is clear at high collision energies, we shall see from tandem ICR studies discussed in Bowers' Chapter I as well as through work discussed in the next section that at lower collision energies, the dynamics of abstraction and exchange are not as simply related as the unimolecular decay mechanism discussed above suggests.

One of the most important and interesting new topics to appear in gas-phase ion chemistry in the past several years is the study of organometallic chemistry. Ion-beam studies, along with ion cyclotron resonance experiments, have played a significant role in the development of this topic [192]. Some of the early developments have taken place with ion-beam–gas-cell instruments, particularly by Armentrout and Beauchamp [67, 68, 193]. These studies have concentrated on total cross-section measurements as a function of collision energy as well as determination of bond energies through the observation of threshold energies. A particularly nice example of the mechanistic information that these experiments provide can be seen in the reaction of Co^+ with 2-methylpropane [194]:

(44)

The reaction is hypothesized to proceed by oxidative addition of a C—H or C—C bond to the metal center, followed by β-H or β-CH_3 transfer and reductive elimination of H_2 or CH_4, leading to Co^+-olefin complexes. If the initial intermediates are created with sufficiently high energy and therefore a lifetime too short for these transfers to take place, then the intermediates will decay and the identity of the fragments may be used to infer the structure of the initial intermediate. In the reaction above, the appearance of CoH^+ and $CoCH_3^+$ at higher kinetic energies as shown in Fig. 22 provides support for the oxidative addition of C—H and C—C bonds as the initial step in the reaction. Furthermore, the threshold energies for these products may be used to extract valuable thermochemical information such as Co^+—H and Co^+—CH_3 bond energies [195].

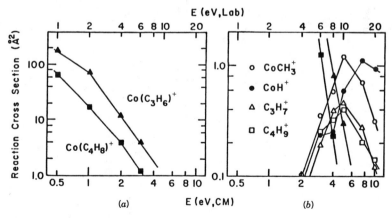

Fig. 22. Energy dependence of the cross section for interaction of Co^+ with 2-methylpropane in center-of-mass frame (lower scale) and lab frame (upper scale). (a) exothermic processes; (b) endothermic channels. Reprinted with permission from Ref. [194], copyright 1980, American Chemical Society.

Much of the work that has been done in gas-phase organometallic chemistry has involved such total cross-section measurements. We have already noted that the guided-beam method is particularly well suited for such measurements, and many of the newest developments in gas-phase ion chemistry have come from such studies; we will discuss recent topics in transition metal ion chemistry in Section 6.4.

Although the majority of studies conducted with beam–gas cell instruments have been with positive ions, a few negative ion systems have been attacked with this method. A recent example [195] is the study of the $O^- + H_2O$ system, yielding $OH^- + OH$, over the collision energy range from 0.17 to 3.5 eV. These measurements only involved integral cross sections for product formation. A number of observations, including evidence for the isotopic scrambling reaction $^{16}O^- + H_2^{18}O \rightarrow {}^{18}O^- + H_2^{16}O$ as well as energy-dependent branching ratios for the products $^{18}O^-$, $^{16}OH^-$, and $^{18}OH^-$ suggest that a long-lived collision complex mediates the reaction. The reaction has been hypothesized to occur on a double minimum potential energy surface, a topic that we discuss in detail in Section 6.2.

6.2 Crossed-Beam Experiments

Practitioners of the crossed-beam trade have made important strides in reagent specificity, product resolution, chemical complexity, and accessibility

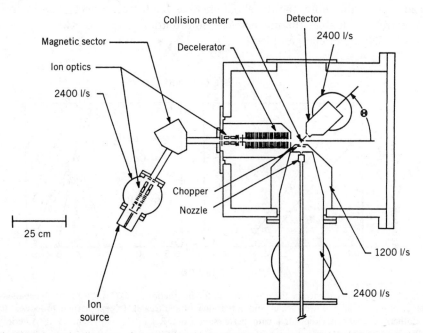

Fig. 23. Schematic diagram of crossed-beam apparatus used at the University of Rochester. Reprinted from Ref. [50] by permission of American Institute of Physics.

of low collision energies in the past several years. Figure 23 shows an instrument developed in the author's laboratory at Rochester [50] that embodies many features discussed in preceding sections. Ions are produced in an in-line electron impact source [48] that operates at high pressures, allowing the production of reagents in the presence of quenching gases. Following extraction, acceleration, focusing, and mass analysis, the ions are decelerated to ground potential by a multielectrode exponential retarder [107, 108]. Quadrupole doublets [100] before and after the momentum analyzer correct for the astigmatic focusing of the magnetic prism. The space-charge limited beam of primary ions then intersects a supersonic beam of reagent molecules that may be seeded in a carrier gas, usually hydrogen or helium, to improve kinematics. The detection system rotates about an axis perpendicular to the plane of the beams. After passing through a defining slit that establishes the detector laboratory angular resolution at 2°, the product ions pass through a three-element zoom lens [150] that accelerates or decelerates the ions to the pass energy of a 90° spherical energy analyzer operating with a resolution of 100 meV. Ions within the bandpass of the analyzer then enter a quadrupole mass filter followed by a Daly

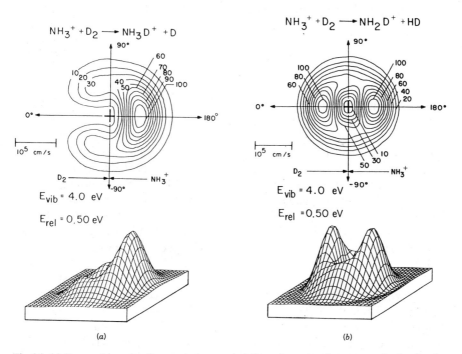

Fig. 24. (a) Barycentric polar flux contour map and three-dimensional axonometric plot for the NH_3D^+ product of the $NH_3^+ + D_2$ reaction at a collision energy of 0.50 eV and a vibrational energy of 4.0 eV. The 0°–180° line denotes the relative velocity vector. NH_3^+ comes in along this vector from the right, D_2 from the left. The contours denote the locus of constant center of mass relative flux. (b) Polar flux map for the exchange product NH_2D^+ under same reagent energy conditions. Reprinted from Ref. [196] by permission of American Institute of Physics.

scintillation detector [155]. Pulses from the particle detector are counted in synchronization with a voltage ramp applied to the energy analyzer that defines a range of lab energies over which data are collected. In addition, the neutral beam is modulated at 30 Hz with a tuning fork, and the energy analysis ramp and product detection are synchronized with the beam modulation; the entire operation of the detector is controlled by a microcomputer.

A number of experiments have been performed with this instrument over the past 5 years, and a recent study of the exchange and abstraction reactions of vibrationally energy selected NH_3^+ ions with D_2 [196] is particularly interesting in light of other studies on this system, among them, Zare's multiphoton ionization (MPI) work on the reaction [184, 185]. By diluting NH_3 in a large excess of N_2, Ar, or Kr, electron impact of the diluent gas, followed by energy resonant charge transfer to NH_3 [197] results in vibrationally energy selected ions with vibrational energy contents from 3.3 to 4.9 eV. Crossed-beam studies with these reagents at a collision energy of 0.50 eV reveal distinctly different dynamics for the exchange and abstraction reactions. The polar flux contour maps and their three-dimensional axonometric representations shown in Fig. 24 at a vibrational energy of 4.0 eV reveal that the NH_3D^+ abstraction products are backward scattered, suggesting a collision geometry with a collinear N—D—D arrangement created by approach of the D_2 bond along the C_3 axis of ammonia. In contrast, the NH_2D^+ flux distributions show near symmetry about 90° in the center of mass, suggestive of a transient intermediate living approximately a rotational period. The translational energy distributions for these products indicate that a significant fraction of the reagent vibration participates in both reactions, but the appearance of the complex in the exchange process, with a clear indication of direct dynamics for the abstraction reaction, suggests that these channels sample different regions of phase space at collision energies lower than those reported by Zare. This conclusion is also consistent with recent tandem ICR studies of these reactions [190] that demonstrate a strong vibrational energy dependence for the exchange reaction and a strong translational energy dependence for the abstraction reaction. These crossed-beam experiments involved total count rates of $0.5-3\,s^{-1}$, corresponding to total cross sections for chemical reaction of 10^{-2} Å^2, nearly three orders of magnitude below the predictions of the Langevin orbiting complex model [198].

Another set of experiments conducted with this instrument involves low-energy reactions that have been hypothesized to occur on potential surfaces with two minima. Figure 25 shows such a surface for the Li^+-induced dehydration of *tert*-butanol, to be discussed later. Double minimum potential models have been proposed by a number of workers, particularly Brauman and collaborators [199, 200], to rationalize the wide variability in the rates of ion–neutral reactions. In such a model, the reagents are drawn into an encounter complex well by their electrostatic attraction. Once in this well, the collision complex must undergo atom or group transfers before decay to products; such rearrangements occur over an isomerization barrier whose height may be less than

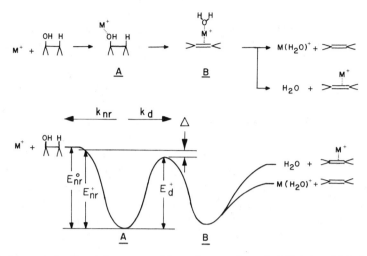

Fig. 25. Schematic double minimum potential reaction coordinate for M^+-induced dehydration of alcohols. **A** is the initial encounter complex, and E_{nr}^+ denotes the transition state energy for separation of this complex back to reagents. E_d^+ is the barrier height energy for rearrangement of this complex to **B**, the precursor to products. Reprinted from Ref. [203] by permission of American Institute of Physics.

the energy of the approaching reagents [201]. Such a barrier may not present an energetic bottleneck to isomerization and subsequent formation of products but may limit product formation through entropic constraints. Because the transition state at the top of the isomerization barrier is tight relative to that for the reagents approaching to form the initial encounter complex, i.e., the free rotations of the approaching reagents have been converted to bending vibrations as the complex isomerizes, such complexes may prefer to revert to reagents rather than convert to products. This mechanism therefore provides a route for reaction efficiencies of significantly less than unity.

Brauman and co-workers have used this model with statistical modeling of the decay rates of the encounter complex to estimate isomerization barrier heights by comparing measured efficiencies with theoretical efficiencies computed as a parametric function of the barrier height [199, 200]. However, direct tests of the assumptions of this analysis, namely, the validity of the statistical hypothesis, require measurements other than absolute rate constants. The double minimum model requires that a significant flux of encounter complexes must decay back to reagents. In a beam experiment, the relative kinetic energy of the approaching reagents can be defined precisely, but the intramolecular dynamics in the encounter well determine the distribution of relative translational energies of complexes that revert to reagents. We expect that redistribution of the incident translational energy in the complex should result in a "thermalization" of this energy, shifting the kinetic energy of reformed reagents to much lower energies. Beam experiments have the potential to verify this consequence of the double minimum potential and, by examining the precise form of the kinetic energy release distribution, to assess the validity of statistical models in the intramolecular dynamics in the transient collision complexes.

Some recent experiments [202, 203] on the dynamics of the gas-phase dehydration of *tert*-butanol by Li^+, whose reaction coordinate [204] is shown in Fig. 25, illustrate the use of crossed-beam techniques in probing double minimum potentials. The principal probe of the encounter well comes from nonreactive Li^+ flux measurements, some of which are shown in Fig. 26 at a relative energy of 0.85 eV. At small laboratory angles Θ, the flux is dominated by direct nonreactive scattering with a peak near the primary ion beam energy, whereas at larger angles, scattering near the system's centroid results in a bimodal energy distribution. The higher energy peak corresponds to direct elastic and inelastic scattering, whereas the lower energy peak corresponds to Li^+ "thermalized" in the collision complex. A detailed analysis of these data [203] yields nonreactive Li^+ recoil energy distributions that compare favorably with statistical phase space calculations. These measurements provide direct support for the validity of statistical models in analyzing the dynamics of this double minimum system. In addition, the clear identification of Li^+ scattered from the complex allows us to determine the ratio of its flux relative to the flux of reactive dehydration products. The comparison of this measured branching ratio with statistical calculations performed as a parametric function of the intermediate isomerization barrier height demonstrates that such barriers can

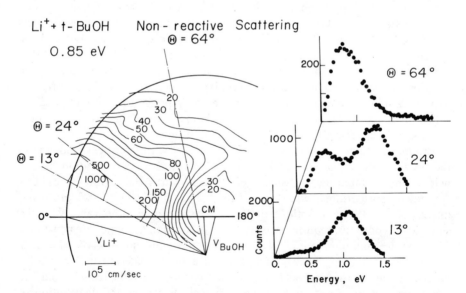

Fig. 26. Polar flux contour map and selected experimental data for Li^+ + *tert*-butanol nonreactive scattering at a collision energy of 0.85 eV. At $\Theta = 13°$, the dominant peak in the data corresponds to direct scattering of the incoming reagents from one another. At wider angles, the lower energy peak corresponds to Li^+ ejected from the complex, and at wider angles, only this peak appears in the data. Reprinted from Ref. [203] by permission of American Institute of Physics.

be determined to a precision of 0.05 eV from branching ratio measurements precise within only a factor of 2. A number of chemical systems have been hypothesized to occur on double minimum surfaces, including proton transfer [205] and nucleophilic displacement [199, 200], and beam experiments to probe barrier heights and collision complex dynamics will help elucidate the course of these reactions.

A significant increase in the chemical complexity of systems examined under crossed-beam conditions occurs in a study of the decarbonylation of acetaldehyde by the transition metal ions Fe^+ and Cr^+ [206]. The differing symmetries for the reactive fluxes for Fe^+ vs. Cr^+, as shown in the contour maps of Fig. 27, indicate that the transient intermediates in the Fe^+ case survive several rotational periods, whereas the Cr^+-induced decarbonylation distribution is

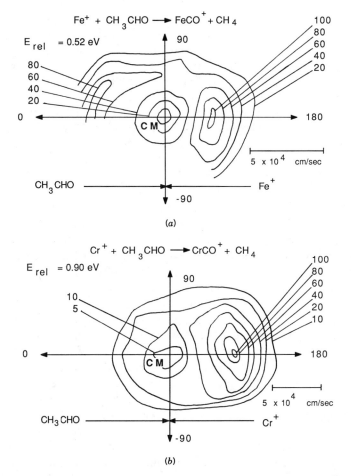

Fig. 27. (a) Polar flux contour map for decarbonylation of CH_3CHO by Fe^+ at $E_{rel} = 0.52$ eV. (b) Polar flux contour map for decarbonylation of CH_3CHO by Cr^+ at $E_{rel} = 0.90$ eV.

highly asymmetric, suggesting that the transient complex lives only a small fraction of a rotational period. Cr^+ has a d^5 closed shell electronic configuration that we would expect to be nonreactive, suggesting that excited states of the cation must be responsible for the decarbonylation reaction. The metal cations are produced by electron impact on the metal carbonyl, and in the Cr^+ case, there is clear evidence for producing excited states in this manner [207]. The fact that Cr^+—H and Cr^+—CH_3 bond energies are smaller than their Fe^+ analogs indicates that the wells corresponding to reactive intermediates [208] for decarbonylation on the potential surface for the Cr^+ system are shallower than for the corresponding Fe^+ system, and the likely participation of Cr^+ excited states indicates that the energy accessible to the transient complexes is larger in the Cr^+ case. Both of these factors should result in decreased transient complex lifetimes for Cr^+-induced decarbonylation, evidenced by the highly asymmetric contour maps in Fig. 27.

The crossed-beam method is capable of "translational spectroscopy" in which discrete internal states may be observed in kinetic energy distributions, as we discussed in Section 5, and several particularly nice examples of product state resolved scattering maps have come from the laboratory of Futrell and collaborators at the University of Delaware using an instrument very similar to that of Fig. 23 [107, 108]. The particular examples of such product state resolved studies have come from a number of charge-transfer and inelastic energy transfer collisional studies, including Ar^+ with N_2 [209, 210], N_2^+ with N_2 [211], and CO^+ with CO [211]. These studies use a high-pressure electron impact ion source that produces Ar^+ in the ground $^2P_{3/2}$ spin orbit state and creates molecular ions in their ground vibrational states. The early studies of charge transfer [209] from ground state Ar^+ to N_2 demonstrated that at collision energies near 1.7 eV, the process yields products in the $v = 1$ level, a result in agreement with LIF studies done by Hüwel and co-workers [212] at near thermal collision energies, and that the dynamics appear to be those of direct electron jump in which the momenta of the heavy particles are not changed appreciably. At lower collision energies near 0.60 eV, this picture of direct electron jump dynamics appears to dominate the scattering, although the differential cross section shows some backward scattered charge exchange products in $v = 1$, which suggests that a long-lived complex also participates in the reaction. What is particularly noteworthy and unusual about these experiments is that in a very narrow range of collision energies near 1.1 eV, a broad distribution of product states is formed with strong correlations between product quantum state and scattering angle. Contour maps for these experiments are shown in Fig. 28. Figure 28a shows a flux map for charge exchange at 0.6 eV, indicating an intense forward peak and a weaker backward peak arising from a collision complex. Figure 28b shows a remarkably different map for scattering at 1.1 eV, in which the forward scattered product appears to correspond to $v = 1$, a range of angles near 40° corresponds to $v = 2$, the backward scattered intensity appears even closer to the center of mass and has been assigned to $v = 3$, while there is some evidence for a weak backward scattered peak for

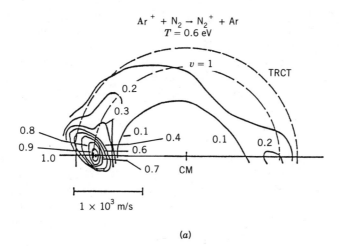

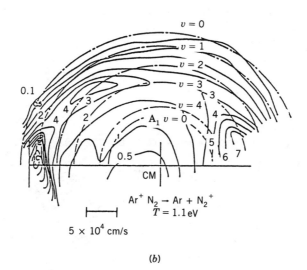

Fig. 28. (a) Contour map for charge transfer reaction $Ar^+ + N_2 \rightarrow Ar + N_2^+$ at 0.60-eV collision energy. Local maxima in the maps at 0° and 180° correspond to energy loss of 0.1 eV in the collision and production of N_2^+ in $v = 1$. The partial symmetry of the map about 90° indicates that approximately 40% of the collisions are orbiting collisions. (b) Contour map for charge transfer reaction at 1.1 eV. The flux scattered over the entire range of velocity space corresponds to production of N_2^+ in various product vibrational states as indicated. Reprinted from Ref. [210] by permission of North-Holland Physics Publishing.

$v = 4$ products. Futrell has suggested that such quantum state-specific behavior is reminiscent of similar behavior in the $F + H_2$ system [213], which has been identified with scattering resonances. Whether such an interpretation is correct or not awaits additional experimental and theoretical study, but the abrupt change in dynamics over a narrow energy range in this system, beautifully revealed by the excellent sensitivity and resolution of this instrument, is clear evidence of very interesting dynamics in what were considered well-understood systems.

A number of other examples of translational spectroscopy in charge exchange systems have appeared from this group in the past few years. Particularly interesting examples have concerned diatomic ion–diatomic molecule scattering. The $CO^+ + CO$ system [211] illustrates the capabilities of current crossed-beam technology quite well; in such symmetric systems, we cannot distinguish between energy transfer and charge-exchange collisions. The high-pressure ion source used for these experiments generates vibrationally relaxed reagents, defining the reagent quantum states precisely. At low collision energies near 0.7 eV, the collision process appears to proceed through a collision complex whose lifetime is long in comparison with a rotational period and a statistical distribution of product states is populated by the interaction. At high-collision energies, the appearance of forward scattered products is consistent with an electron jump mechanism with negligible momentum transfer between the heavy particles. At 1.08 eV, however, this system exhibits a highly structured contour map as shown in Fig. 29. The bands of intensity that surround the center of mass of the system, which arise from the production of specific vibrational states of the products, provide state-resolved information on the collision process but also reveal a

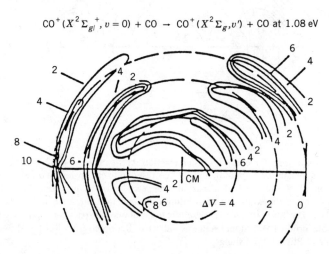

Fig. 29. Contour map for the charge transfer/inelastic scattering of CO^+ from CO at 1.08 eV relative energy. Dashed circles correspond to inelastic transitions with $\Delta v = 0$, 2, and 4. A propensity rule for transfer of even numbers of quanta via an osculating complex is suggested from these data.

propensity for even Δv changes in the vibrational quantum number. Such observations have been reported by McAfee [214] at high collision energies in the $N_2^+ + N_2$ collision system. A number of additional experiments, using isotopically labeled $^{13}CO^+$ reagents to break the nuclear symmetry, and hence the allowed rotational levels, of the linear $O^{12}C^{12}CO^+$ complex are planned to sort out the origin of the vibrational state specificity in this system.

Zdenek Herman and collaborators at the Heyrovsky Institute at Prague have performed a number of beautiful crossed-beam experiments in the past several years. Some early studies involved the reactive scattering of partially vibrationally state-selected H_2^+ with He [215], and more recent chemical reaction studies have included $H_2O^+ + H_2$ [216], $B^+ + H_2$ [217], $H_2^+ + D_2$ [218], and $CH_4^+ + CH_4$ [5]. This group has also performed a number of studies of charge-transfer reactions, including $Ar^+ + H_2$ [219], and more recently the first low-energy differential cross-section studies of doubly charged ions, particularly the $Ar^{+2} + He \rightarrow Ar^+ + He^+$ [220] and $Hg^{+2} + Kr \rightarrow Hg^+ + Kr^+$ systems [221].

An outstanding example of vibrational state resolution in a crossed-beam experiment comes from the laboratory of Toennies and collaborators at the Max Planck Institute für Strömungsforschung at Göttingen using the instrument shown in Fig. 30. This instrument has been used for a number of studies of proton energy loss collisions with polyatomic molecules, including SF_6, CF_4

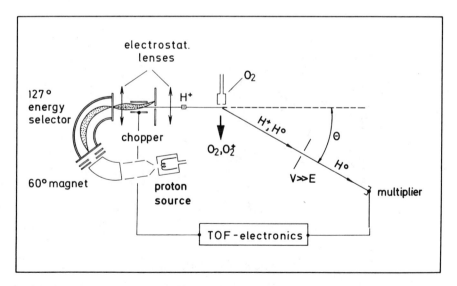

Fig. 30. Schematic diagram of the TOF instrument for proton–molecule inelastic scattering at Göttingen. Short pulses of a mass and energy selected proton beam collide with an uncollimated O_2 beam. In addition to inelastically scattered protons, the instrument can detect H atoms produced by charge exchange by blocking out scattered protons with a repeller field ($V \gg E$). Reprinted from Ref. [159] by permission of American Institute of Physics.

[160], and H_2O, N_2O, and CO_2 [222], but the capabilities of this instrument are particularly well illustrated by a recent study of charge-transfer and energy transfer collisions of protons with O_2 [159]. Protons from a dc discharge source are energy and mass selected, and a short pulse of ions then enters a collision chamber where it intersects an intense, unskimmed beam of O_2. Inelastically scattered protons are energy resolved by determining their distribution of arrival times, and charge-transfer collisions are detected by looking for the hydrogen atoms produced in the collision. In the latter case, a stopping potential much larger in magnitude than the ion-beam energy removes all protons, only allowing the hydrogen atoms to pass. Product hydrogen atoms with lab energies less than 100 eV strike the first dynode of a Be—Cu electron multiplier. Slow hydrogen atoms have a detection efficiency of only 1%, whereas ions can be accelerated before striking the first dynode, making the detection efficiency very nearly 100%. In the case of atoms, the detection mechanism is uncertain, either involving secondary electron ejection or conversion to H^- on the first dynode. Good time resolution required to resolve individual product states arises from the long (3.04 m) flight path of the products. The signal levels for this arrangement are reasonably high, and signal integration times for hydrogen atom detection range from 15 minutes at small scattering angles where signal levels are intense up to 12 hours at the widest scattering angles. Because of the more efficient product detection, data acquisition times are a factor of 20 smaller for protons.

Figure 31 shows a TOF spectrum at a scattering angle of 5° at a lab collision

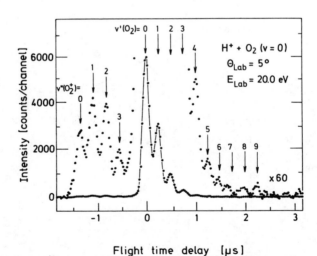

Fig. 31. Time-of-flight spectrum for $H^+ + O_2$ collisions at $\Theta = 5°$ and $E_{lab} = 20$ eV. Low-intensity regions on either side of the elastic peak at $\Delta t = 0$ are magnified by a factor of 60. The structured distribution at positive delay times corresponds to inelastically scattered O_2, whereas the structure at negative times corresponds to charge exchange to produce O_2^+ in the vibrational states indicated. Reprinted from Ref. [159] by permission of American Institute of Physics.

energy of 20 eV for the $H^+ + O_2$ system. The most intense peak at zero delay time corresponds to elastic scattering of protons from O_2, whereas the structure at positive delay times can be identified with protons having lost energy by exciting vibrational quanta in O_2. Because proton–O_2 charge transfer is exothermic by 1.5 eV, neutral hydrogen atoms produced in this process may have speeds greater than the elastically scattered protons, and the structured, low amplitude signal at negative delay times corresponds to that process. TOF spectra taken over a range of scattering angles yield a complete set of state-to-state differential cross sections for charge-exchange and vibrationally inelastic scattering. For the inelastic scattering channel, the cross sections yield transition probabilities $P(0 \to v'')$ for $v'' = 0\text{--}6$. The experimentally determined probabilities are in reasonable agreement with a Poisson distribution for the low vibrational levels, but the experimental probabilities are a factor of 2 larger than Poisson for the higher overtones $v'' = 2$, 3, and 4. The average energy transfer increases strongly with scattering angle, rising from 0.08 to 0.22 eV in the range from 4° to 11°. Over the same range, average energy transfers in the $H^+ + N_2$ system, where charge transfer is highly endothermic, are only one-fourth as large [223]. The large energy transfer observed in the O_2 system coupled with significant deviation from the Poisson distribution predictions of a simple forced oscillator model [224] led Moll and Toennies to speculate that the participation of a charge-transfer intermediate is responsible for such behavior.

Charge-transfer reaction cross sections are customarily thought of to a first approximation in terms of energy resonance and Franck-Condon factors [225]. In the former case, those product states of O_2^+ whose recombination energies are closest to the ionization potential of H are expected to be most highly populated. This argument suggests that highly vibrationally excited O_2^+ with v'' near 5, 6, and 7 should be formed preferentially. The Franck-Condon factors for production of O_2^+ from O_2 suggest that the off-resonant states 0–4 should be populated. The experimental data at small scattering angles between 0° and 2° indicate that the more resonant product states are populated preferentially, whereas the Franck-Condon predictions work well for the transition probabilities near 4°. At wider angles, a distorted Franck-Condon picture appears to work well. This model, originally proposed by Lipeles [226], allows the incoming ion to distort the neutral from its equilibrium geometry, and Franck-Condon factors for this distorted intermediate then determine the transition probability. By considering all possible curve crossings that lead to charge transfer and computing the classical displacement of the O_2 bond along straight-line trajectories for a range of impact parameters and correlating impact parameter with deflection angle, the Franck-Condon factors for the ionization of the distorted molecule can be computed. The results of such a model compare well with the experimental data in the angular range from 5° to 11°. This analysis, which relies on the motion of the system along the entire potential surface, especially near the crossings, suggests that data of the high quality that the TOF method affords will allow a more complete understanding of atom–

molecule charge transfer, beyond the simple models of resonance and Franck-Condon factors for undistorted geometries.

A few studies of negative ion reaction dynamics have been performed in crossed beams, and several of these studies have been reviewed in Gentry's article [13]. A recent differential cross-section study of the hydrogen atom transfer reaction of O^- with allene ($H_2C{=}C{=}CH_2$) by Karnett and Cross [227] over the relative collision energy range from 4.6 to 10.8 eV demonstrated direct spectator stripping dynamics, a process typical of reaction dynamics at high collision energies. A recent study of the reactions of Cl^-, Br^-, and I^- with methyl halides in the energy range from 3 to 150 eV by White et al. [228] observed significant cross sections for electron detachment, as well as elastic and inelastic scattering. The cross sections for highly inelastic scattering processes suggest that collision-induced fragmentation is a major channel, but the neutral products were not detected in these experiments. The S_N2 reaction of the form $X^- + CH_3Y \rightarrow Y^- + CH_3X$ was not observed in these studies. The dearth of studies of anion reaction dynamics, particularly at low collision energies, reflects on the difficulty of producing anion beams at low kinetic energies.

A very recent experiment performed at Orsay [262] on the reactions of $Cl^- + H_2$ in the collision energy range from 5.6 to 12 eV illustrates a number of technical innovations. By using a microchannel plate as a position sensitive detector, the angular distributions of neutral reaction products can be measured accurately, with an angular resolution of 0.3 degrees in the scattering plane. Signals from a second microchannel plate configured to detect charged particles in delayed coincidence with the neutral products allow product flight times, and therefore kinetic energy releases to be determined. Using this instrument, proton transfer to form $HCl + H^-$ and $HCl + H + e^-$, as well as electron detachment channels forming $Cl + H_2 + e^-$, $Cl + H + H^-$, and $Cl + H + H + e^-$ have been examined. The use of such imaging techniques will have an important impact in the study of ionic reactions, as they are already having on the study of laser-induced photodissociation processes [263].

6.3 Merged-Beam Studies

The merged-beam method offers a number of advantages for the study of low-energy ion–neutral reactions, among them high kinetic energy resolution and the ability to prepare exotic reagents by charge-transfer neutralization from mass selected ionic precursors. During the past 20 years several instruments have been constructed to study rearrangement collisions, charge transfer, and chemiionization reactions [32, 33, 229]. Figure 32 shows a merged-beam instrument developed by Gentry and co-workers at the University of Minnesota [33] that has a number of noteworthy features. Ion source A produces the primary ion beam, which is mass selected by a Wien filter in an in-line geometry. Ion source B prepares ionic precursors for the neutral beam that are mass selected and made coaxial with ions from source A in a 90° merging magnet. Charge-exchange gas cells allow the ions from source B (and also source A for

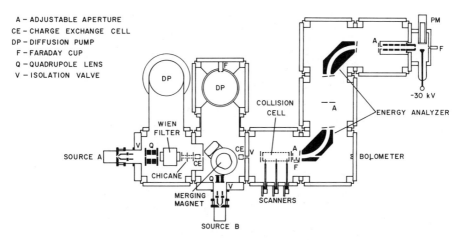

Fig. 32. Schematic diagram of merged-beam apparatus at University of Minnesota. A is an adjustable aperture, CE a charge exchange cell, DP a diffusion pump, F a Faraday cup detector, Q a quadrupole doublet lens, V an isolation valve. The edge length of each cubical chamber module is 30.5 cm. Reprinted from Ref. [33] by permission of American Institute of Physics.

chemiionization reactions) to be neutralized before reactions. The collision cell is held at a potential that determines the relative collision energy. Although the reagents overlap for a significant length before the cell, their relative energy outside the cell can be made so large that the products of any reactions outside the cell are unstable with respect to dissociation. Beyond the collision cell, products are detected by electrostatic energy analysis followed by a Daly scintillation detector.

The high kinetic energy resolution that can be achieved in a merged-beam experiment arises in large part from the fact that the reagent beams in the interaction region have very small components of velocity along the axis of the cell that defines the relative velocity vector. Three scanners located at different positions within the collision cell map out the beam shape with a spatial resolution of 0.05 mm, allowing precise determination of the three-dimensional overlap of the beams. By minimizing transverse components of the beam velocities in the interaction region, Gentry and co-workers have been able to achieve relative collision energies as low as 0.002 eV.

A comprehensive review of merged-beam results through 1979 has been prepared by Gentry [13], and we mention a few additional results here to indicate recent accomplishments. A number of merged-beam studies of the reactions of D_2^+ with O [230], N [231] and, more recently, C [232] and F [233] have been reported by the Minnesota group. In the latter two cases collision energies as low as 0.002 eV have been achieved, and absolute total cross-section measurements for deuteron transfer have been measured from 0.002 to over 10 eV; Fig. 33 shows cross-section data for the F atom reaction. The accuracy of the data allows the long-range form of the potential to be determined quanti-

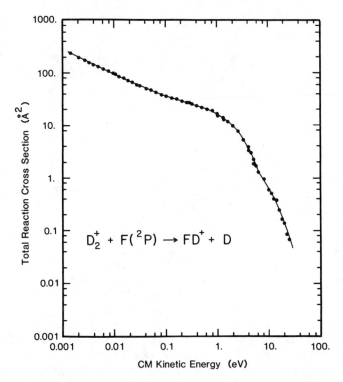

Fig. 33. Energy dependence of the absolute reaction cross section for $D_2^+ + F(^2P) \to DF^+ + D$. The smooth curve is a theoretical fit to the cross section. Reprinted from Ref. [233] by permission of American Institute of Physics.

tatively. At large separations R the potential is governed by the interaction between the charge on the ion and the permanent quadrupole moment on F, and the charge-induced dipole becomes important at smaller R. Both of these terms depend on the projection quantum number for the electronic angular momentum of ground-state fluorine atoms [234]. As R decreases, the anisotropy of the potential becomes comparable to the spin-orbit coupling energy in F, and at even shorter distances, where the anisotropy is larger, L and M_L become good quantum numbers for the F atom. A fit of the total cross-section data to potentials calculated for different values of L and M_L demonstrates that the capture collision takes place on potential energy surfaces correlating with Π state intermediates, a conclusion supported by a correlation diagram analysis. Such an analysis suggests that the reactants correlate with ground-state products $DF^+(^2\Pi)$ and $D(^2S)$ along collinear surfaces of $^1\Pi$ and $^3\Pi$ symmetry. The decrease in cross section near 2 eV correlates with onset for dissociation of the DF^+ product. The cross-section data also indicate an inflection near 8 eV, which suggests the onset of an additional endothermic channel. An electronic correlation diagram analysis suggests that at higher kinetic energies, the reactive system can access an excited $^3\Sigma$ surface that correlates with DF^+ products in the excited $^2\Sigma$ state.

A more complex chemical system that has also been studied recently in merged beams [235] is the proton transfer reaction 45:

$$H_3^+ + D_2 \rightarrow HD_2^+ + H_2 \tag{45}$$

Absolute cross sections for this reaction have been determined over the kinetic energy range from 0.002 to 11 eV. The complicated form for the anisotropy of

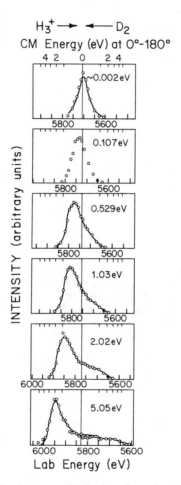

Fig. 34. HD_2^+ product lab kinetic energy distributions from the reactions of H_3^+ with D_2 for various initial c.m. energies. The central vertical line is the center energy, the lab energy of HD_2^+ product ion having the speed of the system's center of mass. The c.m. energy scale at the top of the figure corresponds to the energy a product has for scattering at 0° (on the right) and 180° (on the left). The solid lines are kinematic fits to the data. Reprinted from Ref. [235] by permission of American Institute of Physics.

the interaction potential and its dependence on the internal states of the reagents, both highly vibrationally excited in these studies, precludes a determination of its precise form, but a comparison of the measured cross section with the predictions of the Langevin-Gioumousis-Stevenson (LGS) orbiting collision model [198] suggests that at least 75% of all close collisions lead to chemical reaction. In this simple model, reaction is hypothesized to occur for every collision impact parameter for which the relative kinetic energy surmounts the combined centrifugal and long-range electrostatic potentials of the approaching reagents.

In addition to determining the energy-dependent cross section, the merged-beam method also allows the measurement of high resolution kinetic energy distributions for the products at selected collision energies. Such energy distributions represent the projection of the center-of-mass ion velocity distribution onto the direction of the relative velocity vector, the merging axis in the laboratory, corresponding to the direction of the center-of-mass 0°–180° axis. Careful kinematic analysis allows the product ion energy distributions to be

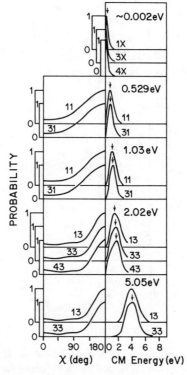

Fig. 35. HD_2^+ product c.m. kinetic energy and scattering angle distributions from numerical deconvolution of the experimental lab data. Different curves at the same energy illustrate functions that give equally good fits to the data. An isotropic angular distribution is assumed for the 0.002 eV experiment. Reprinted from Ref. [235] by permission of American Institute of Physics.

deconvoluted to yield center-of-mass energy and angular distributions. Figure 34 shows such energy scans at several collision energies. The product kinetic energy in the center-of-mass coordinate system is shown along the top of each panel, with zero denoting the "center energy," that is, the laboratory energy of a product ion having the speed of the initial system's center of mass. H_3^+ is moving faster than D_2 in these experiments, so HD_2^+ products with energies less than the center energy correspond to collisions in which the D_2 has stripped a proton from the H_3^+ and the HD_2^+ has continued in the direction of the incident D_2. The product energy distribution that is symmetric about the center energy becomes increasingly shifted to higher laboratory energies with increasing collision energy, consistent with the importance of the stripping mechanism. From these data, the energy and angular distributions shown in Fig. 35 may be extracted, confirming the qualitative importance of this direct mechanism. However, a consideration of the mean kinetic energy of the products from these distributions shows that the simple, impulsive "spectator stripping" model [170] in which D_2 removes the proton from H_3^+ with no momentum transfer to the H_2 "spectator" is quantitatively incorrect. Such observations suggest that the dynamics of this process do involve subtle features of the potential energy surface that are ignored in the simple impulsive models of direct reactions. Studies such as this illustrate the power of the merged-beam technique in examining the dynamics of elementary reactions over a broad energy range.

6.4 Guided-Beam Experiments

One of the most impressive new techniques for studying ion–neutral interactions is the guided-beam method. We have already noted that this method uses inhomogeneous electric fields to confine and transport ions, creating the possibility of quenching nascent ionic reagent internal excitation as well as resulting in very efficient product ion collection over broad collision energy ranges. The technique is best suited to quantitative measurements of total cross sections as a function of collision energy as well as accurate branching ratio measurements in multichannel reactive systems. The collection of reaction products over 4π sr gives the method very high sensitivity, making it the technique of choice for studying exotic or state-selected reagents that can only be prepared by special techniques. Originally proposed by Teloy and Gerlich over 10 years ago [35, 109, 110], the method has gained rapidly in popularity over the past several years, and a wide range of studies has been reported by several groups during this period.

Following the pioneering experiments of Teloy and Gerlich on energy-dependent total cross sections for deuterium transfer in the $Ar^+ + D_2$ system [35], dissociative charge transfer between Ne^+ and CO [35], and charge exchange and particle transfer reactions between H^+ and D_2 [109], Lee and collaborators [111–114] developed a guided-beam experiment to study vibrationally state-selected reactions of various cations, particularly H_2^+ with selected neutrals. The H_2^+ reagent was prepared by UV photoionization of H_2 from a nozzle beam, and an octupole structure guided the primary ions through a collision cell and on to a detector. Although the photoionization method ensures that the mass of the primary ion can be identified and therefore precludes the need for mass analysis before the collision cell, intensities are limited ($\sim 10^5 \, \text{s}^{-1}$), and the range of ions that can be prepared in this manner is extremely limited. We leave the important contributions of guided-beam methods to photoionization studies of ion–molecule reactions to Chapter VIII by Ng.

The addition of mass selection of the primary ion beam by Ervin and Armentrout at Utah [49] has been a key extension of the guided-beam method, opening up the possibility of studying a much wider range of chemical phenomena than had been possible heretofore. Figure 36 shows a schematic of the Utah instrument. The instrument may use a number of interchangeable sources, including in-line and crossed-beam ionizers, an electron impact source followed by a drift region through a high-pressure gas cell to thermalize excited states in the beam, and a surface ionization source. The ion focusing optics, mass selector, and deceleration optics following the source are similar to those in the instrument shown in Fig. 23. Following mass selection and deceleration, the ion beam enters an electrostatic octupole around which is wrapped a gas cell for the collision partner held at a partial pressure of up to 1 mtorr. Reaction products are trapped by the inhomogeneous electric field where they are transported to a quadrupole mass filter followed by a Daly scintillation detector [155]. A computer-controlled data acquisition system allows this instrument to measure total cross sections and branching ratio measurements as a function of collision energy. The collision energy range from 0 to 500 eV may be probed with this instrument, and Armentrout has demonstrated with retarding potential and TOF methods that the energy calibration is accurate to within 0.05 eV.

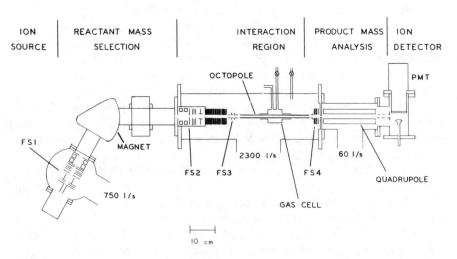

Fig. 36. Schematic diagram of the guided-ion-beam tandem mass spectrometer built by Armentrout and collaborators. Reprinted from Ref. [49] by permission of American Institute of Physics.

A number of interesting studies have been performed with this instrument, spanning the range from elementary reaction dynamics of selected first- and second-row atomic cations and rare gases with isotopic hydrogen molecules to reaction dynamics of transition metal ion clusters. The experiments illustrate how the kinetic energy dependences of such reactive processes allow us to interpret "simple" reactions through electronic orbital and state correlation diagrams as well as to gain mechanistic insight into more complex interactions relevant to understanding organometallic chemistry carried out in solution. The work of the Armentrout group illustrates a clearly defined intellectual development that moves from orbital occupancy discussions of atomic ion interactions with molecular hydrogen to adiabatic and diabatic interactions of transition metal cations with alkanes. The concepts that describe the simple systems carry through to a more complete understanding of the complex interactions associated with gas-phase collisions between atomic and metal cluster ion beams and alkanes [236].

Although studies of the rare gas (Rg) ions with isotopic hydrogen molecules are numerous, the guided-beam studies performed by the Armentrout group have led to significant new insights into their dynamics. The p^5 electron configuration of the rare gas cations leads to a ^2P ground-state term. Orbital correlation arguments [236–238] suggest that electrons in the p_z orbital of the ion, which lies along the perpendicular bisector of the H_2 bond in a C_{2v} collision geometry, lead to occupation of an antibonding orbital of the intermediate and thus result in a repulsive interaction of the approaching reagents. Therefore, we do not expect these reactions to proceed by insertion into the H—H bond to form a long-lived intermediate. Instead, the repulsion in C_{2v} geometry is avoided through a collinear collision in $C_{\infty v}$ symmetry. The reaction dynamics should be direct, and extant differential cross-section data for these systems bear out that prediction [239, 240]. Furthermore, Mahan [238] has used orbital correlation diagrams to show that in many of the rare gas systems the reactants do not correlate with ground-state products, but the charge-exchanged reactants $Rg + H_2^+$ do lead to products. Thus, the entrance channel crossing between the $Rg^+(H_2)$ and $Rg(H_2^+)$ surfaces, which depends strongly on the identity of the rare gas, controls reactivity quite closely. Figure 37 shows energy-dependent cross-section measurements for all rare gas ions except He^+ with H_2 over nearly four orders of magnitude in collision energy [236]. The Ar^+ system appears to agree most closely with the predictions of the LGS orbiting collision model [198], with a low-energy reaction cross section two-thirds of the LGS value, rising to 90% of LGS near 1 eV. The $1/E$ decline in cross section above 1 eV reflects the direct dynamics of the reaction in this region, and the more precipitous decrease in cross section above 4 eV arises from dissociation of the nascent ArH^+ product. The near-Langevin behavior at low kinetic energies reflects the efficient coupling of the $Ar^+(H_2)$ surface with the $Ar(H_2^+)$ surface correlating with the ArH^+ products. The crossing of these surfaces creates a "seam" that occurs very near the equilibrium H—H separation, allowing a strong mixing of these diabatic surfaces. The deviation from LGS behavior at

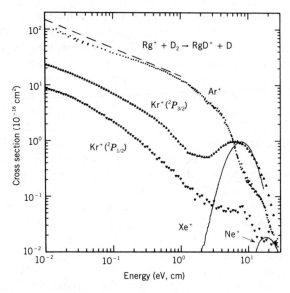

Fig. 37. Cross sections for the reaction of $Ne^+(^2P)$, $Ar^+(^2P)$, $Kr^+(^2P_{3/2})$, $Kr^+(^2P_{1/2})$, and $Xe^+(^2P)$ with D_2 as a function of relative kinetic energy. Reprinted from Ref. [236] by permission. Copyright 1987 by D. Reidel Publishing Company, Dordrecht, Holland.

the lowest attainable collision energies has been attributed to spin-orbit effects [236]. A group theoretical analysis indicates that the products of the reaction $ArH^+(^1\Sigma) + H(^2S)$ correspond to $M_J = \pm 1/2$ and have $^2A'$ symmetry in the C_s point group. Consequently, the $M_J = \pm 3/2$ components of the $Ar^+(^2P_{3/2})$ state do not correlate with the products and are of $^2A''$ symmetry in C_s, suggesting that they will not be reactive in an adiabatic collision. Only two-thirds of the reactants in a statistically prepared mixture of Ar^+ can thus react. At higher collision energies, the $^2A'$ and $^2A''$ states can mix, leading to an increase in reactivity as all components of $Ar^+(^2P_{3/2})$ react.

An even more dramatic example of spin-orbit effects appears in the Kr^+ system [241]. In this case, the fine structure states have sufficiently different reactivities with selected charge-transfer reagents that the drift cell ion source method [47] may be used to prepare state-selected Kr^+ beams for these studies. Ground-state $Kr^+(^2P_{3/2})$ is prepared in the drift cell source by resonant charge transfer with CO^+, while N_2O preferentially charge exchanges with ground-state Kr^+ in ion beams produced in both fine structure states by electron impact, leaving only the excited $^2P_{1/2}$ state [242]. The cross sections for these state-selected reagents are also shown in Fig. 37 and indicate a strong preference for reaction with the ground-state ion. As in the case with Ar^+ states, only the $M_J = \pm 1/2$ components of the ground-state ion correlate with the charge-transfer species $Kr(H_2^+)$ and asymptotic products, and the large spin-orbit splitting in Kr^+ limits the extent to which the $^2A'$ and $^2A''$ surfaces states may mix during the collision. The crossing from the $Kr^+ + H_2$ surface to the

Kr + H_2^+ surface now occurs at distances beyond the well of the latter surface, suggesting that vibrational excitation in the H_2 facilitates the crossing. Such a situation reduces the cross section for charge transfer and therefore for the hydrogen atom transfer reaction as well. This phenomenon leads to the data shown in Fig. 37, indicating that the reaction cross section is approximately one order of magnitude lower than the prediction of LGS. The data for the $^2P_{3/2}$ reaction also show a feature at higher collision energy with an onset near 1 eV that peaks near 5–6 eV. The appearance of this feature only in the ground-state ions quite likely arises from the $M_J = \pm 3/2$ components of the $^2P_{3/2}$ term. A correlation diagram analysis suggests that this feature appears to have an activation energy and may arise from a Coriolis coupling mechanism [243] between the $^2A''$ surface that evolves from these $M_J = \pm 3/2$ components and the reactive $^2A'$ surface evolving from the $M_J = \pm 1/2$ components of ground-state ions.

Data for the Ne^+ and Xe^+ reactions display behavior characteristic of processes occurring with activation energies. In the Ne^+ case, Mahan [238] showed several years ago through an orbital correlation analysis that although the formation of ground-state NeH^+ and H products is quite exothermic, the reactants correlate with an excited H product. The threshold for NeH^+ in Fig. 37 is consistent with this production of excited H, a process endothermic by 4.6 eV. The Xe^+ reaction, endothermic by 0.6 eV for the ground state, shows an apparent activation barrier in the data of Fig. 37. In contrast to the other rare gases, the $Xe^+ + H_2$ reactants correlate directly with ground-state products. In both the Ne and Xe cases, the cross section rises from a threshold and reaches a maximum, beyond which it decreases. This decrease correlates with the dissociation threshold for the RgH^+ product in its ground electronic state.

In addition to the data of Fig. 37, isotope effects have been used in these studies to gain additional insight into the dynamics of these reactions. Intermolecular isotope effects that result from comparisons of H_2 vs. D_2 reactivity, as well as intramolecular isotope effects in reactions with HD, have been very important in elucidating the collision dynamics of these elementary systems. The intramolecular isotope effect with HD is particularly revealing. In cases in which reaction dynamics are direct, as in the rare gas systems in which orbital correlation arguments indicate that collinear collisions will experience entrance channel repulsions dramatically reduced relative to C_{2v} insertions, intramolecular isotope effects can very clearly confirm the direct mechanism. If we compare the total cross sections for the reaction products, the threshold and cross-section maxima for the RgD^+ product occur at lower collision energies than for the RgH^+ product. Such behavior, noted consistently in these systems, supports the idea of a pairwise interaction between the ion and one of the atoms in the HD molecule [49, 236, 244]. The difference in reduced mass for the Rg^+–H interaction compared to the Rg^+–D interaction leads to a corresponding difference in the effective relative kinetic energy of the approaching reagents, manifesting itself in different energetic thresholds for endothermic reaction and dissociation. This pairwise interaction is a signature for a direct process, indicating that total cross-section measurements can provide mechanistic details in simple chemical systems.

The dynamical richness of results for the rare gases is quite limited relative to the varied possibilities of electronic state specific reactions of metal cations with hydrogen. Atomic metal cations are characterized by the accessibility of numerous low-lying electronic energy levels that may evolve into many electronic states as the reagents approach. In order to understand reaction dynamics of such reagents at the level of the potential energy surface, experimental methods have been developed that allow some control over the distribution of electronic states in the incident ion beam. Electron impact on a suitable precursor has the potential to create many excited metastable states of metal cations, but the use of drift cell methods [47] allows many of these states to be eliminated collisionally. The surface ionization method produces a more limited number of states, but low-lying metastable states may still be populated thermally by this method. A careful combination of these methods, however, allows a sorting out of reaction dynamics in simple systems. Of particular interest is the observation that the orbital correlation arguments that were quite successful in understanding the reactivities of first- and second-row atomic ions and rare gas ions also apply to these transition metal systems.

The effect of the method of ion preparation shows very clearly in the cross

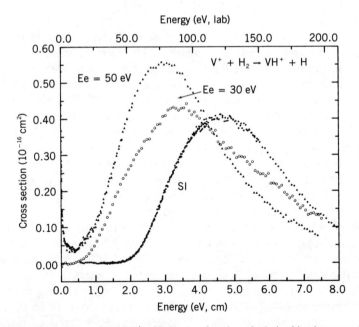

Fig. 38. Cross section for reaction of V^+ with H_2 as a function of relative kinetic energy. Data are shown for ions produced by surface ionization (SI) and by electron impact on $VOCl_3$ at electron energies (Ee) of 30 and 50 eV. Reprinted from Ref. [236] by permission. Copyright 1987 by D. Reidel Publishing Company, Dordrecht, Holland.

sections for the reaction of V^+ with H_2, shown in Fig. 38 [236, 245]. The surface ionization data denoted SI on the graph show a threshold near 2 eV, suggesting that the reacting ion is the ^5D ground state with a high spin $3d^4$ configuration. Reaction with HD yields nearly equal amounts of VH^+ and VD^+, suggesting an insertion mechanism for product formation. When V^+ is produced by electron impact on $VOCl_3$ at various electron energies as indicated in Fig. 38, the threshold for reaction lowers dramatically, indicating that electronically excited V^+ participates. Detailed threshold analysis for these data suggests that the reacting ion from V^+ produced with 50-eV electrons is the metastable ^3F state arising from a $4s3d^3$ configuration, 1.10 eV above the ground state. In contrast with the reactivity of the ^5D ground state, the triplet appears to react through a direct mechanism, evidenced by a VH^+/VD^+ ratio of 4, which Armentrout and co-workers rationalize through angular momentum disposal constraints in a near-collinear collision [245, 246].

The reactivity of Fe^+ with hydrogen isotopes provides another look at the ability of careful ion source selection to sort out the reactivities of different reagent electronic states [247]. When Fe^+ is prepared by surface ionization, the beam contains 79% ground state (^6D) ions having a $4s3d^6$ configuration and nearly 20% of the metastable ^4F state arising from a $3d^7$ configuration and lying only 0.25 eV above the ground state. Previous studies [248] suggested that the high spin coupled configuration of the ground state should be fairly unreactive, and by using a drift cell to quench out all excited states in a beam of Fe^+ produced by electron impact, the reactivities of the ground ^6D and excited ^4F states can be separated. Figure 39 illustrates both the effects of different electronic states and the intramolecular isotope effect in the HD reaction. The ground state reacts both with a smaller cross section and with a higher threshold than the excited ^4F state, confirming the claim that the high spin $4s3d^n$ configuration is less reactive than the high spin $3d^{n+1}$ configuration. The fact that the ground state produces mostly FeD^+ in the reaction with HD is evidence of a direct reaction mechanism in which the ion interacts only with the D atom on HD.

These results and those for many other transition metal reactions with isotopic hydrogen molecules suggest a number of unifying features in their reaction dynamics [236]. Transition metal ions with fewer than $5d$ electrons can avoid placing an electron in the $3d_z^2$ orbital, which correlates with an antibonding a_1^* orbital in C_{2v} approach, and therefore react via a transient complex created by insertion of the metal into the H_2 bond. The $V^+ + H_2$ system exhibits this behavior. In contrast, ions with more than $5d$ electrons must populate this $3d_z^2$ orbital, and the resultant antibonding orbital leads to repulsion of the approaching reagents. This repulsion can be avoided by going to a collinear $C_{\infty v}$ collision geometry. These concepts are precisely those that explained the trends in reactivity in reactions of first- and second-row atoms and the rare gases with molecular hydrogen [236], thus confirming the power of symmetry arguments in predicting reactivity and underscoring the elegance of these experiments in which electronic state-specific reactivities have been sorted out

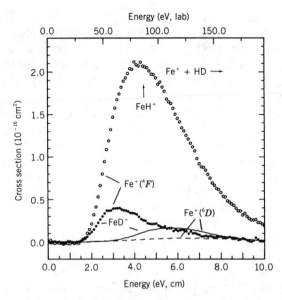

Fig. 39. Cross sections for reaction of $Fe^+(^4F)$ and $Fe^+(^6D)$ with HD to form FeH^+ (open circles and broken line) and FeD^+ (points and full line) as a function of relative kinetic energy. The arrow indicates the bond energy of HD. Reprinted from Ref. [236] by permission. Copyright 1987 by D. Reidel Publishing Company, Dordrecht, Holland,

with very careful control over methods of ion production. The extensive guided-beam work to come from Armentrout's laboratory has been summarized in an excellent review article [236].

The gas-phase activation of C—H and C—C bonds in alkanes by transition metals is a topic of widespread interest [249], and guided-beam studies with precisely known distributions of electronic states allow tests of the concepts of reactivity extracted from studies of simpler systems. The total cross sections for the reactions of Sc^+ with CH_4 are shown in Fig. 40 [236]. The reactions taking place are the following:

$$Sc^+ + CH_4 \rightarrow ScCH_2^+ + H_2 \qquad (46)$$

$$\rightarrow ScCH_3^+ + H \qquad (47)$$

$$\rightarrow ScH^+ + CH_3 \qquad (48)$$

The product $ScCH_2^+$ appears first and is the most energetically favored product when created in concert with a molecule of hydrogen. The fact that the cross section rises from a threshold suggests an activation barrier for formation of an initial adduct. The appearance of $ScCH_3^+$ and ScH^+ products with endothermic thresholds supports their identification as high energy fragmentation products, analogous to the "sampling" of intermediates that was

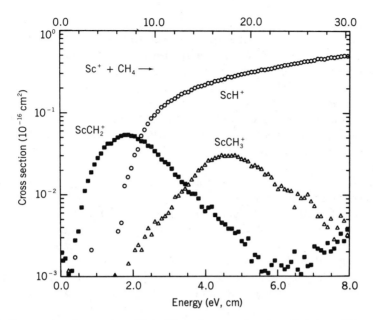

Fig. 40. Cross sections for reaction of Sc^+ with methane as a function of relative collision energy. Reprinted from Ref. [236] by permission. Copyright 1987 by D. Reidel Publishing Company, Dordrecht, Holland.

observed in the reactions of Co^+ with C_4H_{10} [194]. The scheme that Armentrout and co-workers have developed to account for these reactions is as follows:

$$M^+ + CH_4 \longrightarrow M-CH_2^+ \overset{H\ H}{|\ |} \longrightarrow M=CH_2^+ + H_2$$

$$\downarrow \text{high energy}$$

$$\longrightarrow MH^+ + CH_3$$

$$\longrightarrow MCH_3^+ + H$$

$$\longrightarrow M^+ + CH_3 + H$$

$$\longrightarrow M=CH_2^+ + H + H$$

The formation of an initial adduct by insertion of Sc^+ into the C—H bond of methane may be followed by vicinal elimination of H_2 at low energies to form the metal carbene cation and molecular hydrogen. The elimination of H_2 must occur through a tight transition state, and at higher collision energies, where the lifetime of this adduct is significantly reduced, the simple Sc—H and Sc—C bond cleavages proceed through much looser transition states and dominate the dynamics. The $ScCH_3^+$ product may also lose a hydrogen atom at higher collision energies, and the overall process is equivalent to reaction of Sc^+ with CH_4 to form the $ScCH_2^+$ carbene product and two hydrogen atoms. The bond dissociation energy of H_2 is 4.4 eV, so this secondary fragmentation process has a threshold above 5 eV. The increase in $ScCH_2^+$ near 6 eV supports this interpretation of the data. This bond insertion scheme and subsequent decomposition appear to be quite general as a mechanism for reaction of transition metal ions with CH_4.

The data of Fig. 40 also illustrate another important capability of the guided-beam method in determining thermochemical quantities. A precise measurement of the threshold for the endothermic processes 47 and 48 in conjunction with the C—H bond energy in methane allows determinations of dissocation energies for $Sc—CH_3^+$ and $Sc—H^+$ bonds, respectively. The thresholds in Fig. 40 for ScH^+ and $ScCH_3^+$ yield corresponding bond dissociation energies of 2.44 ± 0.10 eV and 2.56 ± 0.13 eV, respectively. This method of measuring an endothermic threshold is quite general and has been used to determine bond energies in numerous systems, providing important thermochemical input estimating stabilities of intermediates and products in many gas-phase organometallic reactions [194, 250–253].

An example of a much more complex reaction scheme is the $Fe^+ + C_3H_8$

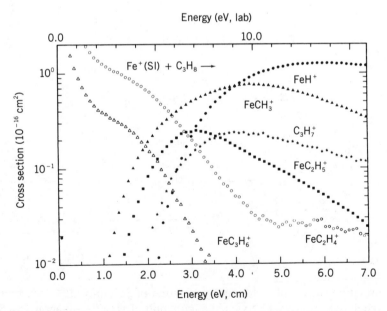

Fig. 41. Cross sections for reaction of Fe^+ (produced by surface ionization, $80\%\ ^6D + 20\%\ ^4F$) with propane as a function of relative kinetic energy. Reprinted from Ref. [236] by permission. Copyright 1987 by D. Reidel Publishing Company, Dordrecht, Holland.

collision system [236], the data for which are shown in Fig. 41. The Fe^+ reagents are produced by surface ionization, leading to 80% ^6D and 20% ^4F states in the beam. At low collision energies, the dominant reaction products are $FeC_2H_4^+$ and $FeC_3H_6^+$, arising from loss of methane and molecular hydrogen, respectively, from the initial collision complex. At higher collision energies, products arising from cleavage of C—C bonds, including $FeCH_3^+$ and $FeC_2H_5^+$, begin to appear. Interestingly, experiments performed with ground-state Fe^+ reagents prepared in a drift cell show low energy behavior virtually identical to the SI results of Fig. 41. Near 0.5 eV, the SI data show a slight increase in cross section relative to the drift cell data that indicates that the excited state only begins to react at that point. Exothermic reaction of ground-state reagents and an apparent threshold for reaction of the excited state is a very unusual observation, and Armentrout has rationalized this behavior in terms of a simple diabatic correlation diagram. From the behavior of Fe^+ (^6D) toward H_2, we would expect a repulsive interaction with the alkane, but an attractive interaction of the excited ^4F state with the C—C bond, leading to formation of an insertion complex as shown in Fig. 42. Thus, the excited state correlates diabatically with the precursor to product formation. Spin-orbit coupling can make this crossing an avoided one, leading to the adiabatic state correlations shown. At low collision energies, the ground-state system follows the adiabatic path to the insertion complex, whereas the excited-state reagents follow an adiabatic path that becomes repulsive. The excited state can only access the insertion complex when the collision energy is high enough to allow a significant fraction of the collisions to follow the diabatic pathway.

The study of metal ion clusters has become one of the most active areas of research in ionic interactions, and guided-beam techniques have made very important contributions toward an understanding of the wide variety of chemical reactions that can occur in such systems. In the late 1980s, much emphasis is being placed on the development of new techniques for production of the clusters

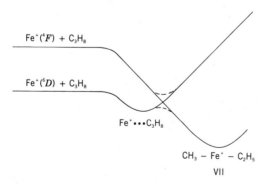

Fig. 42. Potential energy surface for C—C insertion of Fe^+ in propane. Full lines represent diabatic surfaces, and dashed lines are adiabatic surfaces. Reprinted from Ref. [236] by permission. Copyright 1987 by D. Reidel Publishing Company, Dordrecht, Holland.

and mapping out general patterns of reactivity as a function of cluster size. Some early guided-beam studies by the Armentrout group on Mn_2^+ [254] and Co_2^+ [236] produced by electron impact dissociative ionization on $Mn_2(CO)_{10}$ and $Co_2(CO)_8$, respectively, centered on collision induced dissociation (CID) characterizations of their bond energies. These studies indicated that even when the ionizing electron energy is kept near the appearance potential for the dimer cation, the internal excitation is difficult to control. Because the reactivities of these clusters depend quite sensitively on internal excitation, such control is critical to an understanding of the chemistry that they catalyze. Figure 43 shows some cross-section data for the reaction of Co_2^+ with O_2 [236]. Several processes

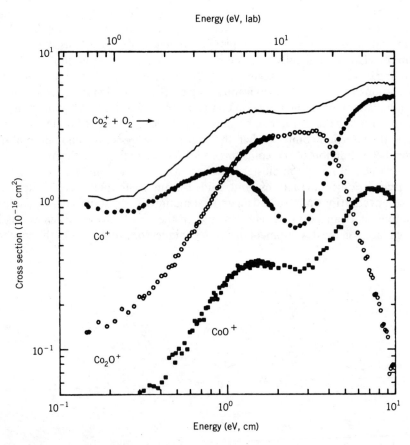

Fig. 43. Cross sections for reaction of Co_2^+ (produced by electron impact at 20 eV) with O_2 as a function of relative kinetic energy. The arrow indicates the approximate onset for collision-induced dissociation of Co_2^+ at 2.8 eV. Reprinted from Ref. [236] by permission. Copyright 1987 by D. Reidel Publishing Company, Dordrecht, Holland.

can occur:

$$Co_2^+ + O_2 \rightarrow Co^+ + CoO_2 \quad (49)$$

$$\rightarrow Co^+ + Co + O_2 \quad (50)$$

$$\rightarrow Co_2O^+ + O \quad (51)$$

$$\rightarrow CoO^+ + CoO \quad (52)$$

$$\rightarrow CoO^+ + Co + O \quad (53)$$

The Co^+ channel at low energies must arise from reaction 49, whereas the higher energy peak comes from CID, reaction 50, with a threshold at 2.3 eV, indicated by an arrow on Fig. 43. The Co_2O^+ channel, reaction 51, is an oxygen atom transfer reaction that rises from threshold and diminishes in intensity beyond 3 eV. This decrease correlates with an increase in CoO^+ intensity from process 53 and allows us to identify the lower energy component of the CoO^+ signal as arising from process 52. Thus, a complete energy-dependent cross-section measurement allows us to sort out all five reactive channels.

Laser vaporization has already been mentioned as a possible technique in conjunction with a supersonic expansion for the preparation of internally cold clusters, and the Armentrout group is currently exploiting this method [92]. The closely related sputtering technique, which we have also discussed in Section 4.1.6, has been developed recently in guided-beam instruments constructed by Anderson and co-workers at Stony Brook. In this group's initial experiments [87], ions created by sputtering are guided by an octupole through a collision cell where their broad distribution of internal energies is partially relaxed in collisions with a bath gas such as Ar. The octupole guided-beam geometry allows efficient transport of these reagents past the first collision cell and on to the interaction volume, following which the products are detected mass spectrometrically. Early experiments from this group, in which the clusters were not efficiently internally relaxed, yielded some interesting results for the reactions with ethylene of clusters of the form M_n^+ and M_nO^+ for the metals Co, Mn, V, and Cr, for $n = 1, 2, 3$, and 4. Co^+ was observed to undergo an association reaction with dehydration, to yield $CoC_2H_x^+$, with insufficient mass resolution to determine how many hydrogen atoms were lost. This reaction was not observed in the beam–gas cell measurements of Armentrout and Beauchamp [255], and the product was hypothesized to arise from fragmentation of a larger cluster. No reactions were observed for higher cobalt clusters. Similar results were obtained for Mn clusters. Both vanadium clusters V_n^+ and V_nO^+ were observed to undergo similar reactions with ethylene, yielding $V_nC_2H_x^+$ and $V_nOC_2H_x^+$, where the mass resolution of the experiment was insufficient to determine the amount of hydrogen incorporated into the products. One of the most interesting results from these initial experiments involved the reaction of the Cr_2^+ cluster, in which the appearance of the product $Cr_2CH_x^+$ indicated that C—C bond cleavage had occurred.

More recent experiment from Anderson's group [88, 89, 115, 256, 257] have attacked the problem of producing internally relaxed clusters, and the instrument shown in Fig. 44 illustrates their approach to this problem [115, 257]. As in the earlier instrument, ions are produced by sputtering; but here they pass through an ion trap with a labyrinthine geometry that allows the ions to experience many collisions before reaching the reaction zone. In a very recent set of experiments with aluminum clusters [89, 115, 256], a series of reactions with O_2, D_2O, and C_2H_4 yielded some interesting trends in the cluster size dependence of various reactive processes. Figure 45 shows some of these data for reactions of dimer, trimer, tetramer, and pentamer cluster ions with O_2 and D_2O as a function of relative collision energy. The only oxygen-containing product in the O_2 systems is the Al_2O^+ species, in agreement with the data of Jarrold and Bower [91]. Production of bare aluminum cluster fragment ions is a major reaction channel for these reagents. In the reactions with D_2O, the atomic ion reactivity is at least an order of magnitude smaller than for any of the cluster ions. Al_2OD^+ is the major product, and its decreasing cross section with increasing collision energy suggests that it is formed in an exothermic process. As with the O_2 reactions, bare aluminum cluster ions are major reaction channels, particularly at higher collision energies. Figure 45 also shows collision-induced dissociation data with Ar to gain some understanding of the relative stabilities of such clusters [115, 257]. These initial experiments suggest that the principle decomposition channel involves loss of Al^+ or a neutral Al

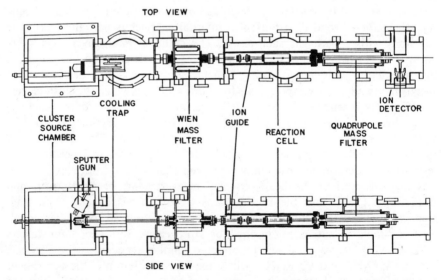

Fig. 44. Schematic diagram of guided-ion-beam mass spectrometer built by Anderson and co-workers.

Major Aluminum Cluster Ion Reaction Channels

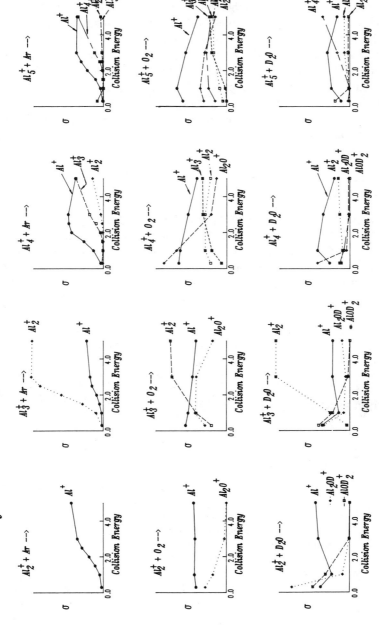

Fig. 45. Cross sections as a function of collision energy and cluster size for reactions of Al_n^+ with O_2, D_2O, and collision-induced dissociation with Ar for $n = 2, 3, 4,$ and 5.

atom, with all of the dissociation thresholds near 1 eV. The collision energy dependence for fragmentation of the clusters with Ar resembles the data for bare aluminum cluster production with O_2 and D_2O, suggesting a direct dissociation mechanism at higher collision energies in the reactive systems, although the mechanism for their production at lower collision energies undoubtedly involves more complicated chemical interactions.

Recent ab initio calculations on neutral aluminum clusters [258] provide some qualitative support for the cluster stabilities inferred from these preliminary experiments. Although even the most basic patterns of reactivity in metal clusters are still to be determined, it is clear that guided-beam methods will make an increasingly important contribution to the elucidation of their chemistry, particularly as the internal excitation of these unique reagents becomes better characterized.

7 PROGNOSIS

The experimental results we have presented here represent just a small fraction of what ion-beam techniques are capable of telling us about elementary reaction dynamics. The last 10 years have been filled with new developments, ranging from studies of much more complex chemical interactions to simple systems with increasingly microscopic probes of the reaction dynamics. Many of these developments are discussed in other chapters in this volume, particularly Tsuji's discussion of optical methods and Ng's chapter on photoionization methods. There is no question that the ingenuity of practitioners of ion-beam methods will lead to even more sophisticated techniques, expanding the chemical variety and microscopic detail that we now know is achievable.

ACKNOWLEDGMENTS

I am grateful to the many colleagues who sent me reprints, preprints, and copies of figures for this work. I also thank members of my research group for their careful reading of the manuscript.

REFERENCES

1. This often quoted aphorism is found in Ramsey's famous text, N. F. Ramsey, *Molecular Beams*, Clarendon, Oxford, 1956, Chap. XIII, p. 346.
2. R. D. Levine and R. B. Bernstein, *Molecular Reaction Dynamics and Chemical Reactivity*, Oxford, New York, 1987.
3. R. B. Bernstein, *Chemical Dynamics via Molecular Beam and Laser Techniques*, Oxford, New York, 1982.
4. K. P. Lawley, Ed., *Molecular Beam Scattering, Adv. Chem. Phys.*, **XXX** (1975); M. A. D. Fluendy and K. P. Lawley, *Chemical Applications of Molecular Beam Scattering*, Chapman and Hall, London, 1973.
5. Z. Herman and I. Koyano, *J. Chem. Soc. Faraday Trans. 2* **83**, 127 (1987).

6. I. W. M. Smith, *Kinetics and Dynamics of Elementary Gas Reactions*, Butterworths, London, 1980.
7. P. R. Brooks and E. F. Hayes, Eds., *State to State Chemistry* (Am. Chem. Soc. Symp. Ser. 56), American Chemical Society, Washington, D.C., 1977.
8. D. G. Truhlar, Ed., *Potential Energy Surfaces and Dynamics Calculations for Chemical Reactions and Molecular Energy Transfer*, Plenum, New York, 1981.
9. K. P. Lawley, Ed., *Potential Energy Surfaces, Adv. Chem. Phys.* **L**. (1982).
10. U. Buck and G. Scoles, Eds., *Atomic and Molecular Beam Methods*, Oxford, New York, 1988.
11. J. Ross, Ed., *Molecular Beams, Adv. Chem. Phys.* **X** (1966).
12. Ch. Schlier, Ed., *Molecular Beams and Reaction Kinetics*, Academic Press, New York, 1970.
13. W. R. Gentry, in M. T. Bowers, Ed., *Gas Phase Ion Chemistry*, Academic Press, New York, 1979, vol. 2, p. 221.
14. W. R. Gentry, in P. Ausloos, Ed., *Kinetics of Ion-Molecules Reactions*, Plenum, New York, 1979, p. 81.
15. J. H. Futrell, Ed., *Gaseous Ion Chemistry and Mass Spectrometry*, Wiley, New York, 1986.
16. An excellent collection of seminal papers in ion–molecule kinetics and reaction dynamics can be found in J. L. Franklin, Ed., *Ion–Molecule Reactions, Parts I and II*, Dowden, Hutchison & Ross, Stroudsburg, 1979.
17. J. L. Franklin, Ed., *Ion–Molecule Reactions*, Plenum, New York, 1972, vols. 1, 2.
18. E. W. McDaniel, V. Cermak, A. Dalgarno, E. E. Ferguson, and L. Friedman, *Ion Molecule Reactions*, Wiley-Interscience, New York, 1970.
19. An excellent collection of reviews of various facets of gaseous ion chemistry can be found in the three volume set of books edited by Bowers: M. T. Bowers, Ed., *Gas Phase Ion Chemistry*, Academic Press, New York, 1979, vols. 1, 2; 1984, vol. 3.
20. The earliest study of ion–molecule reactions was performed inadvertently in the ion source of Thomson's original mass spectrometer where he observed an anomalous line at $m/e = 3$ in a sample of hydrogen gas: J. J. Thomson, *Rays of Positive Electricity*, Longmans Green, New York, 1913.
21. E. F. Greene and A. Kuppermann, *J. Chem. Ed.*, **45**, 361 (1968).
22. M. Menzinger and R. Wolfgang, *Angew. Chem.*, **8**, 438 (1969).
23. E. E. A. Bromberg, A. E. Proctor, and R. B. Bernstein, *J. Chem. Phys.*, **63**, 3287 (1975).
24. P. J. Chantry, *J. Chem. Phys.*, **55**, 2746 (1971).
25. J. J. Leventhal, in M. T. Bowers, Ed., *Gas Phase Ion Chemistry*, Academic Press, New York, 1984, vol. 3, p. 309.
26. Ch. Ottinger, in M. T. Bowers, Ed., *Gas Phase Ion Chemistry*, Academic Press, New York, 1984, vol. 3, p. 249.
27. Ch. Ottinger, in A. Fontijn, Ed., *Gas-Phase Chemiluminescence and Chemiionization*, Elsevier, Amsterdam, 1985, p. 117.
28. D. M. Lubman, C. T. Rettner, and R. N. Zare, *J. Phys. Chem.*, **86**, 1129 (1982).
29. An article on supersonic beams written for the nonspecialist can be found in D. H. Levy, *Sci. Am.*, **250**(2), 96 (1983).
30. A. Ding, J. Karlau, and J. Weise, *J. Chem. Phys.*, **65**, 2544 (1976).
31. A. Ding, J. Karlau, and J. Weise, *Rev. Sci. Instrum.*, **48**, 1002 (1977).
32. S. M. Trujillo, R. H. Neynaber, and E. W. Rothe, *Rev. Sci. Instrum.*, **37**, 1655 (1966).
33. W. R. Gentry, D. J. McClure, and C. H. Douglass, *Rev. Sci. Instrum.*, **46**, 367 (1975).
34. R. H. Neynaber, in B. Bederson and W. L. Fite, Eds., *Methods of Experimental Physics*, Academic Press, New York, 1968, vol. 7A, p. 467; R. H. Neynaber, *Advances in Atomic and Molecular Physics*, Academic Press, New York, 1969, vol. 5, p. 57.

35. E. Teloy and D. Gerlich, *Chem. Phys.*, **4**, 417 (1974).
36. L. D. Landau and E. M. Lifshitz, *Mechanics*, Oxford, New York, 1976, 3rd ed, p. 93.
37. J. R. Pierce, *Theory and Design of Electron Beams*, Van Nostrand, Princeton, N. J. 1954, 2nd ed., p. 145.
38. R. Hutter, in A. Septier, Ed., *Focusing of Charged Particles*, Academic Press, New York, 1967, vol. II, p. 3.
39. M. von Ardenne, *Tabellen zur angewandten Physik*, Ver. Deutscher Verlag der Wissenschaften, Berlin, 1962, vol. 1, p. 619.
40. H. S. W. Massey, *Electronic and Ionic Impact Phenomena*, Clarendon, Oxford, 1969, vol. 2, pp. 910ff.
41. G. H. Dunn, *J. Chem. Phys.*, **44**, 2592 (1966); D. Villarejo, *J. Chem. Phys.*, **49**, 2523 (1968).
42. H. M. Rosenstock, M. B. Wallenstein, A. L. Wahrhaftig, and H. Eyring, *Proc. Natl. Acad. Sci. U.S.A.*, **38**, 667 (1952).
43. R. J. Cotter and W. S. Koski, *J. Chem. Phys.*, **59**, 784 (1973).
44. Z. Herman and V. Pacak, *Int. J. Mass Spectrom. Ion Phys.*, **24**, 355 (1977).
45. R. C. C. Lao, R. W. Rozett, and W. S. Koski, *J. Chem. Phys.*, **49**, 4202 (1968).
46. R. A. Curtis and J. M. Farrar, *J. Chem. Phys.*, **83**, 2224 (1985).
47. P. A. M. Van Koppen, P. R. Kemper, A. J. Illies, and M. T. Bowers, *Int. J. Mass Spectrom. Ion Phys.*, **54**, 263 (1980).
48. H. Udseth, W. R. Gentry, and C. F. Giese, *Phys. Rev. A*, **8**, 2483 (1973).
49. K. M. Ervin and P. B. Armentrout, *J. Chem. Phys.*, **83**, 166 (1985).
50. R. M. Bilotta, F. N. Preuninger, and J. M. Farrar, *J. Chem. Phys.*, **73**, 1637 (1980).
51. H. S. W. Massey, *Negative Ions*, Cambridge Press, Cambridge, Mass. 1976.
52. M. McFarland, D. L. Albritton, F. C. Fehsenfeld, E. E. Ferguson, and A. L. Schmeltekopf, *J. Chem. Phys.*, **59**, 6610, 6620, 6629 (1973).
53. A. G. Harrison, *Chemical Ionization Mass Spectrometry*, CRC Press, Boca Raton, Fla., 1983.
54. F. H. Field, *J. Am. Chem. Soc.*, **91**, 2827 (1969).
55. P. Kebarle, S. K. Searles, A. Zolla, J. Scarborough, and M. Arshadi, *J. Am. Chem. Soc.*, **89**, 6393 (1967).
56. M. Arshadi, R. Yamdagni, and P. Kebarle, *J. Phys. Chem.*, **74**, 1475 (1970).
57. M. L. Vestal, C. R. Blakley, P. W. Ryan, and J. H. Futrell, *J. Chem. Phys.*, **64**, 2094 (1976).
58. P. W. Ryan, C. R. Blakley, M. L. Vestal, and J. H. Futrell, *J. Phys. Chem.*, **84**, 561 (1980).
59. C. R. Blakley, M. L. Vestal, and J. H. Futrell, *J. Chem. Phys.*, **66**, 2392 (1977).
60. K. H. Kingdon and I. Langmuir, *Phys. Rev.*, **21**, 380 (1923).
61. D. R. Herschbach, in J. Ross, Ed., *Molecular Beams*, Wiley, New York, 1966, p. 319.
62. I. Langmuir and K. H. Kingdon, *Proc. R. Soc. London.*, **A107**, 61 (1925).
63. O. Heinz and R. T. Reaves, *Rev. Sci. Instrum.*, **39**, 1229 (1968); these sources are available from Spectra-Mat, Watsonville, Calif. 95076.
64. For example, see I. N. Tang and A. W. Castleman, Jr., *J. Chem. Phys.*, **57**, 3638 (1972).
65. H. B. Haskell, O. Heinz, and D. C. Lorents, *Rev. Sci. Instrum.*, **37**, 607 (1966).
66. J. P. Blewitt and E. J. Jones, *Phys. Rev.*, **50**, 464 (1936).
67. P. B. Armentrout and J. L. Beauchamp, *J. Am. Chem. Soc.*, **103**, 784 (1981).
68. P. B. Armentrout and J. L. Beauchamp, *Chem. Phys.*, **50**, 37 (1980).
69. N. Kashishira, E. Vietzke, and G. Zellermann, *Rev. Sci. Instrum.*, **48**, 171 (1977).
70. N. Kashishira, E. Vietzke, and G. Zellermann, *Chem. Phys. Lett.*, **39**, 316 (1976).
71. M. von Ardenne, *Tabellen der Elektronenphysik, Ionenphysik, und Übermikroskopie*, Deut. Verlag. Wiss., Berlin, 1954.

72. This has been demonstrated in, for example, K. T. Gillen, B. H. Mahan, and J. S. Winn, *J. Chem. Phys.*, **58**, 5373 (1973).
73. M. Menzinger and L. Wåhlin, *Rev. Sci. Instrum.*, **40**, 102 (1969).
74. A. Kasdan and W. C. Lineberger, *Phys. Rev. A*, **10**, 1658 (1974).
75. J. B. Hasted, *The Physics of Atomic Collisions*, Butterworths, Washington, D.C., 1964.
76. J. F. Friichtenicht, N. G. Utterback, and J. R. Valles, *Rev. Sci. Instrum.*, **47**, 1489 (1976).
77. G. D. Byrd, R. C. Burnier, and B. S. Freiser, *J. Am. Chem. Soc.*, **104**, 3565 (1982).
78. R. Tonkyn and J. C. Weisshaar, *J. Phys. Chem.*, **90**, 2305 (1986).
79. P. J. Brucat, L. S. Zheng, C. L. Pettiette, S. Yang, and R. E. Smalley, *J. Chem. Phys.*, **84**, 3078 (1986).
80. L. S. Zheng, P. J. Brucat, C. L. Pettiette, S. Yang, and R. E. Smalley, *J. Chem. Phys.*, **83**, 4273 (1986).
81. M. L. Mandich, W. D. Reents, Jr., and V. Bondybey, *J. Phys. Chem.*, **90**, 2315 (1986).
82. L. A. Bloomfield, M. E. Geusic, R. R. Freeman, and W. L. Brown, *Chem. Phys. Lett.*, **121**, 33 (1985).
83. S. W. McElvany, W. R. Creasy, and A. O'Keefe, *J. Chem. Phys.*, **85**, 632 (1986).
84. G. K. Wehner, *Phys. Rev.*, **108**, 35 (1957).
85. R. R. Corderman, P. C. Engelking, and W. C. Lineberger, *Appl. Phys. Lett.*, **36**, 533 (1980).
86. C. S. Feigerle, R. R. Corderman, S. V. Bobashev, and W. C. Lineberger, *J. Chem. Phys.*, **74**, 1580 (1981).
87. L. Hanley and S. L. Anderson, *Chem. Phys. Lett.*, **122**, 410 (1985).
88. S. L. Anderson and L. Hanley, *Proc. SPIE*, **669**, 133 (1986).
89. L. Hanley and S. L. Anderson, *Chem. Phys. Lett.*, **129**, 429 (1986).
90. T. G. Dietz, M. A. Duncan, D. E. Powers and R. E. Smalley, *J. Chem. Phys.*, **74**, 6511 (1981).
91. M. F. Jarrold and J. E. Bower, *J. Chem. Phys.*, **85**, 5373 (1986).
92. S. K. Loh, D. A. Hales, and P. B. Armentrout, *Chem. Phys. Lett.*, **129**, 527 (1986).
93. P. Grivet, *Electron Optics*, 1972, Pergamon, Oxford, 2nd ed.
94. E. Harting and F. H. Read, *Electrostatic Lenses*, Elsevier, New York, 1976.
95. A. P. Banford, *The Transport of Charged Particle Beams*, Spon, London, 1966.
96. B. Paczkowski, *Electron Optics*, Elsevier, New York, 1968.
97. L. Kerwin, in C. A. McDowell, Ed., *Mass Spectrometry*, McGraw-Hill, New York, 1963, p. 104.
98. A. O. Nier, *Rev. Sci. Instrum.*, **11**, 212 (1940).
99. C. F. Giese, *Rev. Sci. Instrum.*, **30**, 260 (1959).
100. C. Lu and H. E. Carr, *Rev. Sci. Instrum.*, **23**, 823 (1962).
101. W. Legler, *Z. Phys.*, **171**, 424 (1963).
102. R. L. Seliger, *J. Appl. Phys.*, **43**, 2352 (1972).
103. W. Paul and H. Steinwedel, *Z. Naturforsch.*, **8a**, 448 (1953).
104. W. L. Fite, *Rev. Sci. Instrum.*, **47**, 326 (1976).
105. P. H. Dawson, *Int. J. Mass Spectrom. Ion Phys.*, **14**, 317 (1974); **21**, 317 (1976).
106. P. H. Dawson, Ed., *Quadrupole Mass Spectrometry and its Applications*, Elsevier, Amsterdam, 1976.
107. M. Vestal, C. Blakley, P. Ryan, and J. H. Futrell, *Adv. Mass. Spectrom.*, **6**, 781 (1974).
108. C. R. Blakley, P. W. Ryan, M. L. Vestal, and J. H. Futrell, *Rev. Sci. Instrum.*, **47**, 15 (1976).
109. G. Ochs and E. Teloy, *J. Chem. Phys.*, **61**, 4930 (1974).
110. W. Frobin, Ch. Schlier, K. Strein, and E. Teloy, *J. Chem. Phys.*, **67**, 5505 (1977).
111. S. L. Anderson, F. A. Houle, D. Gerlich, and Y. T. Lee, *J. Chem. Phys.*, **75**, 2153 (1981).

112. F. A. Houle, S. L. Anderson, D. Gerlich, T. Turner, and Y. T. Lee, *J. Chem. Phys.*, **77**, 748 (1982).
113. S. L. Anderson, T. Turner, B. H. Mahan, and Y. T. Lee, *J. Chem. Phys.*, **77**, 1842 (1982).
114. T. Turner and Y. T. Lee, *J. Chem. Phys.*, **81**, 5638 (1984).
115. L. Hanley, S. A. Ruatta, and S. L. Anderson, in P. Jena, Ed., *The Physics and Chemistry of Small Clusters*, Plenum, New York, 1987, p. 781.
116. T. C. English and J. C. Zorn, in D. Williams, Ed., *Methods of Experimental Physics*, Academic Press, New York, 1974, vol. 3B, p. 678.
117. A. Kantrowitz and J. Grey, *Rev. Sci. Instrum.*, **22**, 328 (1951).
118. G. B. Kistiakowsky and W. P. Slichter, *Rev. Sci. Instrum.*, **22**, 333 (1951).
119. J. B. Anderson, R. P. Andres, and J. B. Fenn, in J. Ross, Ed., *Molecular Beams*, 1966, p. 275.
120. J. B. Anderson and J. B. Fenn, *Phys. Fluids*, **8**, 780 (1965).
121. J. B. Anderson, in P. P. Wegener, Ed., *Molecular Beams and Low Density Gas Dynamics*, Marcel Dekker, New York, 1974, p. 1.
122. H. Ashkenas and F. S. Sherman, in J. H. DeLeeuw, Ed., *Rarefied Gas Dynamics, 4th Symposium*, Academic Press, New York, 1969, vol. II, p. 84.
123. R. E. Smalley, L. Wharton, and D. H. Levy, *Acc. Chem. Res.*, **10**, 139 (1977).
124. D. H. Levy, L. Wharton, and R. E. Smalley, in C. B. Moore, Ed., *Chemical and Biochemical Applications of Lasers*, Academic Press, New York, 1977.
125. N. Abuaf, J. B. Anderson, R. P. Andres, J. B. Fenn, and D. G. H. Marsden, *Science*, **155**, 997 (1967).
126. T. M. Mayer, B. E. Wilcomb, and R. B. Bernstein, *J. Chem. Phys.*, **67**, 3507, (1977).
127. R. A. Larsen, S. K. Neoh, and D. R. Herschbach, *Rev. Sci. Instrum.*, **45**, 1511 (1974).
128. J. J. Valentini, M. J. Coggiola, and Y. T. Lee, *Rev. Sci. Instrum.*, **48**, 58 (1977).
129. R. P. Mariella, S. K. Neoh, D. R. Herschbach, and W. Klemperer, *J. Chem. Phys.*, **67**, 2981 (1981).
130. J. W. Farthing, I. W. Fletcher, and J. C. Whitehead, *J. Phys. Chem.*, **87**, 1663 (1983).
131. J. A. Silver, A. Freedman, C. E. Kolb, A. Rahbee, and C. P. Dolan, *Rev. Sci. Instrum.*, **53**, 1714 (1982).
132. W. R. Gentry and C. F. Giese, *Rev. Sci. Instrum.*, **49**, 595 (1978).
133. J. B. Cross and J. J. Valentini, *Rev. Sci. Instrum.*, **53**, 38 (1982).
134. M. G. Liverman, S. M. Beck, D. L. Monts, and R. E. Smalley, *J. Chem. Phys.*, **70**, 192 (1979).
135. T. E. Adams, B. H. Rockney, R. J. S. Morrison, and E. R. Grant, *Rev. Sci. Instrum.*, **52**, 1469 (1981).
136. J. A. Simpson, *Rev. Sci. Instrum.*, **32**, 1283 (1961).
137. W. E. Spicer and C. N. Berglund, *Rev. Sci. Instrum.*, **35**, 1665 (1964).
138. D. C. Frost, C. A. McDowell, and D. A. Vroom, *Proc. R. Soc. (London) A296*, 566 (1967).
139. An excellent review of energy analyzers can be found in W. Steckelmacher, *J. Phys. E.*, **6**, 1061 (1973).
140. H. D. Hagstrum, *Rev. Sci. Instrum.*, **24**, 1122 (1953).
141. A. Hughes and V. Rojanski, *Phys. Rev.*, **34**, 289 (1929); A. Hughes and J. H. McMillen, *Phys. Rev.*, **34**, 291 (1929).
142. P. Marmet and L. Kerwin, *Can. J. Phys.*, **38**, 787 (1960).
143. P. Decreau, R. Prange, and J. J. Berthelier, *Rev. Sci. Instrum.*, **46**, 995 (1975).
144. J. S. Risley, *Rev. Sci. Instrum.*, **43**, 95 (1972).
145. E. M. Purcell, *Phys. Rev.*, **54**, 818 (1939).
146. J. A. Simpson and C. E. Kuyatt, *Rev. Sci. Instrum.*, **34**, 265 (1963).

147. C. E. Kuyatt and J. A. Simpson, *Rev. Sci. Instrum.*, **38**, 103 (1967).
148. G. E. Thomas and W. H. Weinberg, *Rev. Sci. Instrum.*, **50**, 497 (1979); H. Ibach and D. L. Mills, *Electron Energy Loss Spectroscopy and Surface Vibrations*, Academic Press, New York, 1982.
149. C. L. Allyn, T. Gustaffson, and E. W. Plummer, *Rev. Sci. Instrum.*, **49**, 1198 (1978).
150. R. E. Imhof and F. H. Read, *J. Phys. E.*, **1**, 859 (1968).
151. D. W. O. Heddle, *J. Phys. E.*, **2**, 1046 (1969).
152. "Aquadag" colloidal graphite preparations are available from Acheson Colloids, Port Huron, MI 48060.
153. V. Herrmann, H. Schimidt, and F. Linder, *J. Phys. B.*, **11**, 493 (1978).
154. An excellent review of particle detection methods is found in F. A. White and G. M. Wood, *Mass Spectrometry: Applications in Science and Engineering*, Wiley, New York, 1986.
155. N. R. Daly, *Rev. Sci. Instrum.*, **31**, 264 (1960).
156. J. D. Shao and C. Y. Ng, *J. Chem. Phys.*, **84**, 4317 (1986).
157. W. C. Wiley and I. H. McLaren, *Rev. Sci. Instrum.*, **26**, 1150 (1955).
158. W. Eastes, U. Ross, and J. P. Toennies, *J. Chem. Phys.*, **66**, 1919 (1977).
159. M. Noll and J. P. Toennies, *J. Chem Phys.*, **85**, 3313 (1986).
160. U. Gierz, M. Noll, and J. P. Toennies, *J. Chem. Phys.*, **83**, 2259 (1985).
161. E. A. Entemann and D. R. Herschbach, *Faraday Discuss. Chem. Soc.*, **44**, 289 (1967).
162. F. A. Morse and R. B. Bernstein, *J. Chem. Phys.*, **37**, 2019 (1962).
163. E. A. Entemann, Ph.D. Dissertation, Harvard University, 1967.
164. P. E. Siska, *J. Chem. Phys.*, **59**, 6052 (1973).
165. K. T. Gillen, A. M. Rulis, and R. B. Bernstein, *J. Chem. Phys.*, **54**, 2831 (1971).
166. T. T. Warnock and R. B. Bernstein, *J. Chem. Phys.*, **49**, 1878 (1968).
167. G. L. Catchen, J. Husain, and R. N. Zare, *J. Chem. Phys.*, **69**, 1737 (1978).
168. B. Friedrich and Z. Herman, *Collect. Czech. Chem. Commun.*, **49**, 570 (1984).
169. Y. T. Lee, in U. Buck and G. Scoles, Eds., *Atomic and Molecular Beam Methods*, Oxford, New York, 1988.
170. A. Henglein, K. Lacmann, and G. Jacobs, *Ber. Bunsenges. Phys. Chem.*, **69**, 279 (1965).
171. J. L. Kinsey, G. H. Kwei, and D. R. Herschbach, *J. Chem. Phys.*, **64**, 1914 (1976).
172. Reference 2, pp. 412–417.
173. S. Stolte, A. E. Proctor, and R. B. Bernstein, *J. Chem. Phys.*, **65**, 4990 (1976).
174. M. K. Bullitt, C. H. Fisher, and J. L. Kinsey, *J. Chem. Phys.*, **60**, 478 (1974).
175. A recent example of osculation in an ion–molecule reaction can be found in the following study of the $C^+ + H_2O$ reaction: D. M. Sonnenfroh and J. M. Farrar, *J. Chem. Phys.*, **83**, 3958 (1985).
176. K. L. Wendell, C. A. Jones, J. J. Kaufman, and W. S. Koski, *J. Chem. Phys.*, **63**, 750 (1975).
177. C. A. Jones, K. L. Wendell, and W. S. Koski, *J. Chem. Phys.*, **66**, 5325 (1977).
178. C. A. Jones, K. L. Wendell, and W. S. Koski, *J. Chem. Phys.*, **63**, 2254 (1975).
179. C. A. Jones, I. Sauers, J. J. Kaufman, and W. S. Koski, *J. Chem. Phys.*, **67**, 3599 (1977).
180. C. A. Jones, K. L. Wendell, and W. S. Koski, *J. Chem. Phys.*, **67**, 4917 (1977).
181. K. Lin, R. J. Cotter, and W. S. Koski, *J. Chem. Phys.*, **60**, 3412 (1974).
182. K. Lin, H. P. Watkins, R. J. Cotter, and W. S. Koski, *J. Chem. Phys.*, **60**, 5134 (1974).
183. R. B. Sharma, N. M. Semo, and W. S. Koski, *J. Phys. Chem.*, **91**, 4127 (1987).
184. R. J. S. Morrison, W. E. Conaway, and R. N. Zare, *Chem. Phys. Lett.*, **113**, 435 (1985).

185. R. J. S. Morrison, W. E. Conaway, T. Ebata, and R. N. Zare, *J. Chem. Phys.*, **84**, 5527 (1986).
186. F. C. Fehsenfeld, W. Lindinger, A. L. Schmeltekopf, D. L. Albritton, and E. E. Ferguson, *J. Chem. Phys.*, **62**, 2001 (1975).
187. J. K. Kim, L. P. Theard, and W. T. Huntress, Jr, *J. Chem. Phys.*, **62**, 45 (1975).
188. J. A. Luine, Ph.D. thesis, University of Colorado, 1981.
189. S. E. Barlow, Ph.D. thesis, University of Colorado, 1984.
190. P. R. Kemper and M. T. Bowers, *J. Phys. Chem.*, **90**, 477 (1986).
191. W. E. Conaway, R. J. S. Morrison, and R. N. Zare, *Chem. Phys. Lett.*, **113**, 429 (1985).
192. J. L. Beauchamp, A. E. Stevens, and R. R. Corderman, *Pure Appl. Chem.*, **51**, 967 (1979); J. Allison, in S. J. Lippard, Ed., *Progress in Inorganic Chemistry*, Wiley, New York, 1986, p. 628.
193. P. B. Armentrout and J. L. Beauchamp, *J. Chem. Phys.*, **74**, 2819 (1981).
194. P. B. Armentrout and J. L. Beauchamp, *J. Am. Chem. Soc.*, **102**, 1736 (1980).
195. C. Lifshitz, *J. Phys. Chem.*, **86**, 3634 (1982).
196. J. W. Winniczek, A. L. Braveman, M. H. Shen, S. G. Kelley, and J. M. Farrar, *J. Chem. Phys.*, **86**, 2818 (1987).
197. P. R. Kemper, M. T. Bowers, D. C. Parent, G. Mauclaire, R. Derai, and R. Marx, *J. Chem. Phys.*, **79**, 160 (1983).
198. G. Gioumousis and D. P. Stevenson, *J. Chem. Phys.*, **29**, 294 (1958).
199. W. N. Olmstead and J. I. Brauman, *J. Am. Chem. Soc.*, **99**, 4219 (1977).
200. M. J. Pellerite and J. I. Brauman, *J. Am. Chem. Soc.*, **102**, 5883 (1980).
201. T. M. Magnera and P. Kebarle, in M. A. Almoster-Ferreira, Ed., *Ionic Processes in the Gas Phase*, Reidel, Boston, 1984, p. 135.
202. W. R. Creasy and J. M. Farrar, *J. Phys. Chem.*, **89**, 3952 (1985).
203. W. R. Creasy and J. M. Farrar, *J. Chem. Phys.*, **85**, 162 (1986).
204. R. D. Wieting, R. H. Staley, and J. L. Beauchamp, *J. Am. Chem. Soc.*, **97**, 924 (1975).
205. C. H. DePuy, in M. A. Almoster-Ferriera, Ed., *Ionic Processes in the Gas Phase*, Reidel, Boston, 1984, p. 227.
206. D. M. Sonnenfroh and J. M. Farrar, *J. Am. Chem. Soc.*, **108**, 3521 (1986).
207. L. F. Halle, P. B. Armentrout, and J. L. Beauchamp, *J. Am. Chem. Soc.*, **103**, 962 (1981).
208. L. H. Halle, W. E. Crowe, P. B. Armentrout, and J. L. Beauchamp, *Organometallics*, **3**, 1694 (1984).
209. B. Friedrich, W. Trafton, A. Rockwood, S. Howard, and J. H. Futrell, *J. Chem. Phys.*, **80**, 2537 (1984).
210. A. L. Rockwood, S. L. Howard, W-H. Du, P. Tosi, W. Lindinger, and J. H. Futrell, *Chem. Phys. Lett.*, **114**, 486 (1985).
211. J. H. Futrell, in P. Austoos, S. G. Lias, and D. Dixon, Eds., *Structure Reactivity and Thermochemistry of Ions*, Reidel, Dordrecht, Holland, 1987, p. 57, and private communication.
212. L. Hüwel, D. R. Guyer, G. H. Lin, and S. R. Leone, *J. Chem. Phys.*, **81**, 3520 (1984).
213. D. M. Neumark, A. M. Wodtke, G. N. Robinson, C. C. Hayden, and Y. T. Lee, *J. Chem. Phys.*, **82**, 3045 (1985).
214. K. B. McAfee, C. R. Szmanda, and R. S. Hosack, *J. Phys. B.*, **14**, L243 (1981); K. B. McAfee and R. S. Hosack, *J. Chem. Phys.*, **83**, 5690 (1985).
215. V. Pacák, U. Havemann, Z. Herman, F. Schneider and L. Zülicke, *Chem. Phys. Lett.*, **49**, 273 (1977);
216. A. J. Yencha, V. Pacák, and Z. Herman, *Int. J. Mass Spectrom. Ion Phys.*, **26**, 205 (1978).
217. B. Friedrich and Z. Herman, *Chem. Phys.*, **69**, 433 (1982).
218. P. M. Hierl and Z. Herman, *Chem. Phys.*, **50**, 249 (1980).

219. P. M. Hierl, V. Pacák, and Z. Herman, *J. Chem. Phys.*, **67**, 2678 (1977).
220. B. Friedrich, S. Pick, L. Hladek, Z. Herman, E. E. Nikitin, A. I. Reznikov, and S. Ya. Umanskii, *J. Chem. Phys.*, **84**, 807 (1986); B. Friedrich and Z. Herman, *Chem. Phys. Lett.*, **107**, 375 (1984).
221. B. Friedrich, J. Vancura, M. Sadilek, and Z. Herman, *Chem. Phys. Lett.*, **120**, 243 (1985).
222. B. Friedrich, G. Nieder, M. Noll, and J. P. Toennies, *J. Chem. Phys.*, **87**, 1447 (1987); G. Nieder, M. Noll, and J. P. Toennies, *J. Chem. Phys.*, **87**, 2067 (1987).
223. M. Noll, Ph.D. dissertation, Göttingen: work cited in ref. 159.
224. J. W. Gadzuk and J. K. Norskov, *J. Chem. Phys.*, **81**, 2828 (1984).
225. J. B. Laudenslager, W. T. Huntress, and M. T. Bowers, *J. Chem. Phys.*, **61**, 4600 (1974).
226. M. Lipeles, *J. Chem. Phys.*, **51**, 1252 (1969).
227. M. P. Karnett and R. J. Cross, *Chem. Phys. Lett.*, **84**, 501 (1981).
228. N. R. White, D. Scott, M. S. Huq, L. D. Doverspike, and R. L. Champion, *J. Chem. Phys.*, **80**, 1108 (1984).
229. M. Burniaux, F. Brouillard, A. Jognaux, T. R. Govers, and S. Szucs, *J. Phys. B.*, **10**, 2421 (1977).
230. D. J. McClure, C. H. Douglass, and W. R. Gentry, *J. Chem. Phys.*, **67**, 2362 (1977).
231. D. J. McClure, C. H. Douglass, and W. R. Gentry, *J. Chem. Phys.*, **66**, 2079 (1977).
232. G. F. Schuette and W. R. Gentry, *J. Chem. Phys.*, **78**, 1777 (1986).
233. G. F. Schuette and W. R. Gentry, *J. Chem. Phys.*, **78**, 1786 (1986).
234. W. R. Gentry and C. F. Giese, *J. Chem. Phys.*, **67**, 2355 (1977).
235. C. H. Douglass, G. Ringer, and W. R. Gentry, *J. Chem. Phys.*, **76**, 2423 (1982).
236. P. B. Armentrout, in P. Ausloos, S. G. Lias, and D. Dixon, Eds., *Structure/Reactivity and Thermochemistry of Ions*, Reidel, Dordrecht, Holland, 1987, p. 97.
237. K. M. Ervin and P. B. Armentrout, *J. Chem. Phys.*, **80**, 2978 (1984).
238. B. H. Mahan, *J. Chem. Phys.*, **55**, 1436 (1971); B. H. Mahan, *Acc. Chem. Res.*, **8**, 55 (1975).
239. M. Chiang, E. A. Gislason, B. H. Mahan, C. W. Tsao, and A. S. Werner, *J. Chem. Phys.*, **52**, 2698 (1970); P. M. Hierl, Z. Herman, and R. Wolfgang, *J. Chem. Phys.*, **53**, 660 (1970).
240. J. R. Wyatt, L. W. Strattan, and P. M. Hierl, *J. Chem. Phys.*, **65**, 1593 (1977).
241. K. M. Ervin and P. B. Armentrout, *J. Chem. Phys.*, **85**, 6380 (1986).
242. N. G. Adams, D. Smith, and E. Alge, *J. Phys. B*, **13**, 3235 (1980).
243. H. Laue, *J. Chem. Phys.*, **46**, 3034 (1967).
244. K. T. Gillen, B. H. Mahan, and J. S. Winn, *Chem. Phys. Lett.*, **22**, 244 (1973).
245. J. L. Elkind and P. B. Armentrout, *J. Phys. Chem.*, **89**, 5626 (1985).
246. N. Aristov and P. B. Armentrout, *J. Am. Chem. Soc.*, **108**, 1806 (1986).
247. J. L. Elkind and P. B. Armentrout, *J. Am. Chem. Soc.*, **108**, 2765 (1986); *J. Phys. Chem.*, **90**, 5736 (1986).
248. J. L. Elkind and P. B. Armentrout, *J. Chem. Phys.*, **84**, 4862 (1986).
249. The first example of gas phase C—H and C—C activation is found in J. Allison, R. B. Freas, and D. P. Ridge, *J. Am. Chem. Soc.*, **101**, 1332 (1979).
250. R. Georgiadis and P. B. Armentrout, *J. Am. Chem. Soc.*, **108**, 2119 (1986).
251. J. L. Elkind and P. B. Armentrout, *Inorg. Chem.*, **25**, 1078 (1986).
252. P. B. Armentrout, L. F. Halle, and J. L. Beauchamp, *J. Am. Chem. Soc.*, **103**, 6501 (1981); L. F. Halle, P. B. Armentrout, and J. L. Beauchamp, *Organometallics*, **1**, 962 (1982).
253. N. Aristov and P. B. Armentrout, *J. Am. Chem. Soc.*, **106**, 4065 (1984); S. K. Loh, K. M. Ervin, and P. B. Armentrout, *J. Am. Chem. Soc.*, **106**, 1161 (1984).
254. K. Ervin, S. K. Loh, N. Aristov, and P. B. Armentrout, *J. Phys. Chem.*, **87**, 3593 (1983).
255. P. B. Armentrout and J. L. Beauchamp, *J. Chem. Phys.*, **74**, 2819 (1981).

256. S. A. Ruatta, L. Hanley, and S. L. Anderson, *Chem. Phys. Lett.*, **137**, 5 (1987).
257. L. Hanley, S. A. Ruatta, and S. L. Anderson, *J. Chem. Phys.*, **87**, 260 (1987).
258. T. H. Upton, *Phys. Rev. Lett.*, **56**, 2168 (1986).
259. B. R. Turner, J. A. Rutherford, and D. M. J. Compton, *J. Chem. Phys.*, **48**, 1602 (1968).
260. B. M. Hughes and T. O. Tiernan, *J. Chem. Phys.*, **55**, 3419 (1971).
261. D. Bombick, J. D. Pinkston, and J. Allison, *Anal. Chem.*, **56**, 396 (1984).
262. M. Barat, J. C. Brenot, J. A. Fayeton, J. C. Houver, J. B. Ozenne, R. S. Berry, and M. Durup-Ferguson, *Chem. Phys.*, **97**, 165 (1985).
263. D. W. Chandler and P. L. Houston, *J. Chem. Phys.*, **87**, 1445 (1987).

Chapter **VIII**

STATE-SELECTED AND STATE-TO-STATE ION–MOLECULE REACTION DYNAMICS BY PHOTOIONIZATION METHODS

Cheuk-Yiu Ng[*,†]

---

1 Introduction
2 Experimental Techniques for Photoionization
3 General Theory for Photoionization
   3.1 Direct Ionization
   3.2 Autoionization
4 Principles for State Selection by Photoionization Methods
   4.1 Simple Photoionization Method
   4.2 Photoion–Photoelectron Coincidence Method
   4.3 Resonant-Enhanced Multiphoton Ionization Method
5 Measurements of State-Selected Ion–Molecule Reaction Cross Sections
   5.1 Threshold Electron Secondary Ion Coincidence Apparatus
      5.1.1 Okazaki Apparatus
      5.1.2 Orsay Apparatus
   5.2 Tandem Photoionization Mass Spectrometer
      5.2.1 Experimental Arrangement and Procedures
      5.2.2 Selected Experimental Results
   5.3 Crossed Ion–Neutral Beam Photoionization Apparatus
      5.3.1 Experimental Arrangement and Procedures
      5.3.2 Selected Experimental Results
6 Measurements of State-to-State Ion–Molecule Reactions Cross Sections
   6.1 Improved Crossed Ion–Neutral Beam Photoionization Apparatus
      6.1.1 Experimental Arrangement and Procedures
      6.1.2 Selected Experimental Results
   6.2 Triple-Quadrupole Double-Octopole Photoionization Apparatus
      6.2.1 Experimental Arrangement and Procedures
      6.2.2 Experimental Results

*Camille and Henry Dreyfus Teacher-Scholar
†Operated for the U.S. Department of Energy of Iowa State University under Contract No. W-7405-Eng-82. This work was supported by the Director of Energy Research, Office of Basic Energy Sciences.

7 **Conclusions and Future Developments**
References

1 INTRODUCTION

State-to-state chemistry has been the main theme of molecular reaction dynamics for more than two decades [1–3]. Microscopic reaction cross sections measured to the detail of state-to-state levels provide the most direct test of theoretical calculations. As a result of detailed spectroscopic knowledge available for neutral molecules and advances in laser technology, optical methods [4–8] have played a central role in reactant state preparations as well as product state identifications in the study of neutral–neutral interactions.

The study of state-selected ion–molecule reaction dynamics requires a comprehensive understanding of the ionization process in which reactant ions are initially prepared. Because of the threshold law and the higher energy resolution achieved in photoionization, it is preferred over electron impact ionization for the preparation of state-specific reactant ions. The ability to prepare state-selected reactant ions by photoionization directly relies on our spectroscopic knowledge of molecules in the vacuum ultraviolet (VUV) region [9], especially in the wavelength region above the ionization energy (IE) of a molecule. The lack of intense VUV light sources has slowed progress in VUV spectroscopy as well as in state-to-state ion chemistry.

Photoionization mass spectrometry (PIMS) [9] and photoelectron spectroscopy (PES) [10–14] are VUV spectroscopic techniques that are concerned with the study of ionization channels after the absorption of a VUV photon by an atom (A) or a molecule (AB):

Atomic photoionization	$A + h\nu \to A^+ + e^-$	(1)
Atomic autoionization	$A + h\nu \to A^* \to A^+ + e^-$	(2)
Molecular photoionization	$AB + h\nu \to AB^+ + e^-$	(3)
Molecular autoionization	$AB + h\nu \to AB^* \to AB^+ + e^-$	(4)
Dissociative photoionization	$AB + h\nu \to AB^+ + e^- \to A^+ + B + e^-$	(5)
Ion-pair formation	$AB + h\nu \to A^+ + B^-$	(6)

The measurements of the IEs and appearance energies (AEs) of these processes by PIMS have provided a great deal of accurate thermochemical data for both ionic and neutral species [15, 16]. Since the electrons ejected in these processes carry nearly all the kinetic energies of the ionization reactions, the analyses of the electron kinetic energy distributions at a sufficiently high resolution by PES give quantitative information about the internal state distributions of the photoions. This information is of utmost importance for using photoionization as an ionic state selection technique. Previous photoionization studies, especially the precise work of Chupka and Berkowitz and their co-workers [9, 17, 18] have laid a solid foundation for the application of photoionization methods in state-selected ion–molecule reaction studies. Indeed, the observation of strong vibrational energy dependences in the reactivities of the reactions $H_2^+(v_0'') + H_2(v_0' = 0)$ [19], $H_2^+(v_0'') + He$ (Ne or Ar) [20], and $NH_3^+(v_0'') + NH_3(v_0' = 0)$ [21] by Chupka and co-workers in 1968 marked the beginning of this field. The previous reviews [22, 23] of this subject by Chupka remain excellent references for researchers entering the field.

In recent years, the establishment of many synchrotron radiation facilities [24] all over the world has accelerated the study of the VUV chemistry [25–27] of gaseous molecules. The vibrational state distributions of many simple molecular ions resulting from autoionizing states near their ionization thresholds have been measured by PES using tunable synchrotron radiation as the ionization source [28–32]. These measurements have strengthened the database needed for vibrational-state-selected studies of ion–molecule reaction dynamics.

Since the ion–molecule reaction work of Chupka and co-workers, the most significant development in state-selected ion–molecule reaction studies has been the introduction of the photoelectron secondary ion coincidence method [33–51]. The coincidence technique has been shown to be a general method for ion internal energy selection.

The state-selected ion–molecule reaction experiments of Chupka and co-workers were by-products of their spectroscopic studies. Their experiments were carried out in a single chamber in which both photoionization and subsequent reactions took place. The collision energy of the reaction was adjusted by simply changing the repeller or ion extraction voltage of the ion source. Due to a continuous potential drop across the gas chamber, the collision energy was ill-defined. Cross sections thus obtained are phenomenological cross sections representing velocity averages over the microscopic cross sections. In order to compare experimental results with theoretical predictions, appropriate conversion of phenomenological cross sections to microscopic cross sections is necessary. However, there are severe limitations [52] on the accuracy to which microscopic cross sections can be derived by these means.

In the last few years, specially designed ion–molecule reaction apparatuses [34–36, 48, 50, 53–63] that couple photoionization methods with ion beam–gas cell or crossed ion–neutral beam techniques have been reported. The use of the radio frequency (RF) octopole ion guide [53, 57, 60–64] reaction gas cell in the ion beam–gas cell arrangement not only allows absolute state-selected cross sections to be measured with high accuracy but also extends the collision energy range to near thermal energy.

The intensity of state-selected reactant ions prepared by photoionization using laboratory VUV discharge lamps or synchrotron radiation is usually $\lesssim 10^5$ ions/s, making the final product state detection experiment difficult. Recently, Liao et al. [54–56, 61, 65] have developed a crossed ion–neutral beam photoionization apparatus that is equipped with an internal state-sensitive electron transfer detector. By measuring the electron transfer reactivities of product ions with different probing gases, they have been able to perform state-to-state studies on several simple electron transfer reactions. The electron transfer detector is highly sensitive, which makes possible the measurement of the internal state distribution of electron transfer product ions with as few as 2–10 product ions [61] formed per second. Shao et al. [63] have extended the electron transfer detection method and designed a triple-quadrupole double-octopole photoionization apparatus that can be used to measure absolute state-to-state total cross sections for electron transfer, excitation, and relaxation processes involving atomic and simple molecular ions. Under the severe constraint of low reactant ion intensities, these new developments have opened up some new avenues for the experimental study of state-to-state ion–molecule reaction dynamics.

A laser source can be used to prepare reactant ions in highly specific states due to the narrow laser linewidth ($< 1 \text{ cm}^{-1}$). The availability of commercial visible and ultraviolet lasers has prompted the investigation of multiphoton ionization (MPI) processes of gaseous atoms [66–68] and molecules [69–73]. Resonance enhanced multiphoton ionization (REMPI) has emerged as a new and useful spectroscopic technique [74–78]. Multiphoton ionization experiments performed with electron kinetic energy analyses [79–92] give information about the internal state distributions of the photoions. Recently, Zare and co-workers [58, 59, 92] have applied REMPI techniques to examine the vibrational energy effects on several ion–molecule reactions involving NH_3^+ and NO^+.

During the past decade, important strides in the generation of coherent VUV radiation have been made. Tunable VUV laser radiation covering the wavelength (λ) range of 1000–2000 Å with intensities of $\sim 10^{10}$ photons/pulse and narrow linewidths ($< 1 \text{ cm}^{-1}$) can be generated reliably by harmonic generation and nonlinear mixing [93, 94]. The ease of generating coherent VUV radiation with a sufficiently high intensity will have a significant impact on VUV chemistry as well as on state-to-state ion chemistry.

Different VUV light sources have their distinct features. Most of the experimental techniques for state-selected ion–molecule reaction studies have been designed to take advantage of the characteristics of a specific VUV source. Because of the relatively new development of laser ionization methods, the overwhelming majority of state-selected ion–molecule reaction experiments have been performed using laboratory discharge lamps or synchrotron radiation as the photoionization sources. This chapter will review the progress in the application of photoionization methods to the study of ion–molecule reactions from 1974 to 1986. During this period, many ion–molecule reaction photoionization apparatuses have been reported in the literature. The designs of several of these apparatuses will be described. Selected experimental results obtained using these apparatuses will also be described to illustrate their merits and limitations.

2 EXPERIMENTAL TECHNIQUES FOR PHOTOIONIZATION

The unique element of an ion–molecule reaction apparatus designed for state-selected cross-section measurements is the photoionization ion source. The preparation of state-selected photoions requires essentially the same experimental arrangement as that used in PIMS. The recent progress in PIMS has been reviewed by Ng [95]. A photoionization ion source consists of four basic components: light source, monochromator, light detector, and mass spectrometer. Since the description of these components has been given in detail previously, it will not be duplicated here. The monographs of Samson [96], Berkowitz [9], and Kunz [24] are excellent references for the techniques and instrumentation in the VUV. The last reference deals with experimental techniques in working with synchrotron radiation.

A photoionization ion source of an apparatus designed for ionic state selection by the photoelectron–photoion coincidence method [35–51, 97–104] includes a photoelectron energy analyzer. The methods of studying energetic photoelectrons by dispersive energy analyzers have been discussed extensively in numerous publications [9–14]. Dispersive energy analyzers used in previous photoelectron–photoion coincidence studies include hemispherical [33, 37, 38, 42–47, 51, 105], cylindrical mirror [48], retarding field [36, 39–41, 97], time-of-flight (TOF) [33, 34, 49, 50, 102–104] and 127° electrostatic [98, 100, 101] analyzers.

When photoions are formed by a tunable monochromatic VUV light source, a threshold photoelectron (TPE) analyzer [106–114] can be used. The methods of TPE detection have been reviewed by Baer [102] and Berkowitz [9]. The principle of a TPE detector is based on the fact that TPE formed at the ionization threshold have near zero kinetic energies and can be efficiently collected into a small solid angle using a small electric field, whereas energetic photoelectrons are scattered in all directions. The simplest TPE analyzer, a steradiancy analyzer [106–112] is formed by a metal tube. The ratio of the diameter to the length of the tube defines the solid angle in which photoelectrons are detected. The rejection of energetic photoelectrons is greater when a smaller value of the diameter-to-length ratio and a weak electric field are used. The electron transmission of the analyzer also becomes poorer under these conditions. The resolution of a TPE analyzer depends on the wavelength resolution. Since a small fraction of energetic photoelectrons can always reach the detector of a line-of-sight steradiancy analyzer, the observed TPE peaks exhibit long "*hot*" electron tails. This problem can be alleviated by adding a dispersive energy analyzer after the steradiancy analyzer and operating the dispersive analyzer at a relatively low resolution [35, 101]. With a careful design, the collection efficiency for TPE can be as high as 50% compared to an efficiency of $\lesssim 1\%$

for most dispersive analyzers at a similar resolution. Peatman et al. [109–112] have reported an indirect trajectory TPE analyzer that overcomes the "hot" electron problem and has excellent resolution (~ 3 meV). However, this analyzer is inappropriate for coincidence studies because of its geometry and low collection efficiency.

Synchrotron radiation from an electron storage ring provides photon pulses with widths less than 2 ns (full width half-maximum—FWHM) and pulse repetition times of the order of 100 ns. Using an electric field of ~ 1 V/cm and an electron flight path of a few centimeters, the flight times of photoelectrons are expected to be less than 100 ns. Guyon et al. [113] and Baer et al. [114] have used the pulse characteristics of the ACO synchrotron source (Orsay, France) and designed a TPE–TOF analyzer. By gating only the electron signal at TOF corresponding to TPE, a collection efficiency of 50% for TPE and the complete rejection of energetic photoelectrons are achieved at a resolution of 20 meV for the TPE spectrum. This TPE–TOF analyzer is capable of resolving rotational structure in the TPE spectrum of H_2^+ [115].

When photoions are formed by a tunable VUV light source, a greater variety of states can be prepared via the autoionization mechanism. The internal states of an ion accessible by direct photoionization using a line source such as He(I) are limited by the Franck-Condon principle. For example, up to 18 vibrational states of $NO^+(\tilde{X})$ (v = 0 – 18) can be identified in the TPE spectrum [103], whereas only 6 vibrational peaks (v = 0 – 5) are observed in the first electronic band of the He(I) PES for NO^+ [10].

For REMPI, a high pulse energy Nd^{3+}: YAG or excimer-pumped dye laser system is required [58, 59, 78–92]. The laser beam must be focused in order to attain a sufficiently high intensity. The single photon ionization rate from a resonant excited state is equal to the product of the laser intensity (photons·cm^{-2}·s^{-1}) and the photoionization cross section. When laser radiation at 2500 Å with a pulse width of 10 ns and pulse energy of 1 mJ is focused to an area of $\sim 10^{-6}$ cm^2, the photon intensity becomes $\sim 10^{29}$ cm^{-2}·s^{-1}. If the photoionization cross section of the excited molecule is $\sim 10^{-17}$–10^{-18} cm^2, the rate of ionization is estimated to be $\sim 10^{11}$–10^{12} s^{-1} [84]. Under these conditions, an excited molecule with a lifetime of $\gtrsim 10^{-12}$ s can be efficiently ionized.

3 GENERAL THEORY FOR PHOTOIONIZATION

Processes 1–6 represent the basic mechanisms for photoionization of atoms and molecules. The ionization may be a direct process (processes 1 and 3) that involves a direct transition of a bound electron into the ionization continuum or an indirect process (processes 2, 4, 5, and 6) that involves an intermediate state. The theory for MPI will not be discussed in this section. Refer to the monographs by Letokov [116], Lin et al. [117], and Lambropoulous and Smith [118] for theoretical studies of MPI processes.

3.1 Direct Ionization

It has been shown both experimentally [9, 21, 119–125] and theoretically [126, 127] that the threshold for the production of a single quantum state by direct photoionization is approximately a steplike function, i.e., the onset rises sharply at the threshold of each process and then tends to assume a constant value. Since the direct photoionization transition of a molecule is rapid with respect to the vibrational period, it obeys the Franck-Condon principle that predicts that the photoionization cross section is proportional to the Franck-Condon factor between the neutral and ionic vibrational states. For an ionic state having an equilibrium distance different from that in the ground state of the neutral molecule, the maximum vibrational overlap (Franck-Condon region) occurs at a position on the ionic potential energy surface different from that on the neutral ground state. The most commonly observed Franck-Condon overlap conditions in photoionization are shown in Fig. 1 [128]. The transitions are all taken as originating from the ground vibrational level of the neutral electronic state. Three cases are shown:

1. The ionization involves the removal of a nonbonding or "lone pair" electron, which has little effect on the bonding of the molecular ion. In

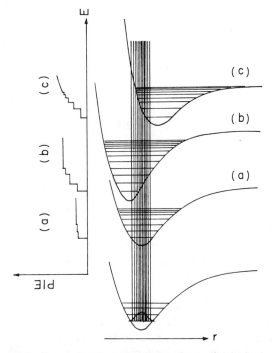

Fig. 1. Application of the Franck-Condon principle to direct photoionization: (*a*) removal of a nonbonding electron, (*b*) removal of an antibonding electron, and (*c*) removal of a bonding electron.

other words, there is little or no change in internuclear distance and vibrational interval from that of the neutral molecule. The photoionization efficiency (PIE) curve, i.e., the ratio of the number of ions produced to the number of transmitted photons plotted as a function of photon energy, will essentially show one big step corresponding to an intense $(0 \leftarrow 0)$ transition (curve a of Fig. 1).

2. The removal of an antibonding electron, such as in the ionization of NO to form NO^+ ($\tilde{X}^2\Pi$) [119–121], results in strengthening the bonding of the molecular ion. This implies a decrease in bond length and a wider vibrational spacing for the ionic state. The PIE curve hence will show a few widely spaced steps (curve b of Fig. 1).

3. The ejection of a bonding electron produces transitions to the steeper slope of the inner ionic potential well. The observed PIE curve (curve c of Fig. 1) is composed of closely spaced steps stemming from an increase in bond length and a reduction in vibrational frequency in the molecular ion compared to those in the neutral molecule. The ionization of C_2H_2 to form $C_2H_2^+$ ($\tilde{X}^2\Pi_u$) [122–124] is an example of this case.

Usually the wavelength resolution used in photoionization experiment does not permit rotational structure to be resolved. Rotational excitations of neutral molecules produce tailing of the onsets and rounding of the steps. The exact shape of the threshold curve depends on the thermal populations of rotational and low-frequency vibrational levels. In addition to NO^+ and $C_2H_2^+$, many other molecular ions, such as $C_2H_4^+$ [129–131], NH_3^+ [21], H_2O^+ [9, 125], and $C_3H_6^+$ [132], have PIE spectra that exhibit a staircase structure near the ionization thresholds. The relative heights of the steps observed in a PIE spectrum such as curves a, b, and c of Fig. 1 can be used to estimate the vibrational distributions of photoions. In many cases the vibrational distributions of photoions determined by the relative vibrational step heights of the PIE spectra and He(I) PES are in agreement [121].

3.2 Autoionization

The removal of an electron from an atom or molecule that does not involve a transition directly into the ionization continuum is known as autoionization. This process arises from the overlapping of discrete energy levels by a continuous range of levels as depicted in Fig. 2. Series 1 represents the Rydberg levels converging to the first IE and series 2 the Rydberg levels leading to a higher ionic state. If an electron is excited to a bound level in series 2 that is higher in energy than the first IE, then the possibility exists for a radiationless transition from the bound molecular Rydberg state to a stable ionic state. The electron ejected in this manner will have a kinetic energy equal to the difference between the energies of the bound state and the lower IE. Unlike direct ionization, autoionization is a resonant process occurring at discrete energies that correspond to members of Rydberg series 2. This process is manifested as a series

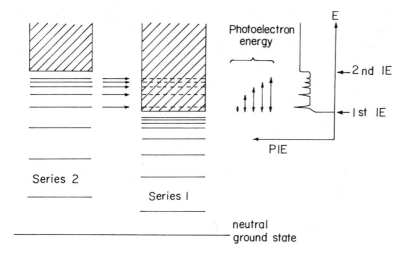

Fig. 2. Autoionization of molecules from Rydberg states above the first IE.

of peaks on the PIE curve shown in Fig. 2. A necessary condition for the decay of a Rydberg state by ionization is that the excitation energy of the ion core be greater than the binding energy between the ion core and the Rydberg electron, which is due primarily to coulombic attraction. Depending on the type of energy transferred from the ionic core to the ejecting Rydberg electron, autoionization processes can be classified as rotational, vibrational, and electronic autoionization. A well-known example of rotational and vibrational autoionization is observed in the PIE spectrum for H_2^+ [133–136]. The autoionization structures found in the photoionization of N_2 [9, 137] involve the full interplay of rotational–vibrational–electronic autoionization processes.

The phenomenon of autoionization was first postulated by Shenstone [138] to explain the broadening and asymmetry of some absorption peaks above the IE of certain species. Fano [139, 140] postulated configuration interaction between the discrete states and the continuum and successfully predicted the position and intensity shifts produced in a Rydberg series due to autoionization [141]. Characteristic asymmetric autoionization structures observed in PIE curves for atomic and molecular ions are now referred to as Fano-Beutler profiles. Since the original work of Fano, there have been extensive studies of autoionization structure in PIE spectra of atoms and molecules [27, 28, 142–148]. In addition to autoionizing profiles, photoionization branching ratios and photoelectron angular distributions have been recognized to be important experimental measurements for the studies of molecular photoionization dynamics [27, 28]. The PIE spectra near the thresholds for diatomic molecular ions such as H_2^+, N_2^+, O_2^+, and N_2O^+ are dominated by autoionization features [9]. The vibrational branching ratios of autoionization peaks in the PIE spectra for H_2^+ [149, 150], N_2^+ [150–152] O_2^+ [32], and

N_2O^+ [153] have been measured. Due to specific selection rules for the autoionization process, the vibrational state distribution of molecular ions formed by autoionization is usually quite different from that expected from the Franck-Cordon principle. The $\Delta v = -1$ selection rule for vibrational autoionization of H_2 [135, 154] makes possible the preparation of $H_2^+(\tilde{X}^2\Sigma_g^+)$ ions in a nearly pure vibrational state via autoionization.

In the absence of autoionization, it is generally assumed that the Franck-Condon approximation holds, and vibrational branching ratios can be calculated from the Franck-Condon factors even when the staircase structure is not observed in the PIE spectrum. However, the effects of shape resonances [27, 28, 152, 155, 156] may cause the vibrational branching ratios to deviate from the Franck-Condon expectations.

4 PRINCIPLES FOR STATE SELECTION BY PHOTOIONIZATION METHODS

4.1 Simple Photoionization Method

An atomic ion in its ground electronic state or a molecular ion in its ground vibrational state can usually be formed with 100% purity by photoionization. Chupka and co-workers [22, 23] have shown that many reactant ions in a pure or a known distribution of internal states can be prepared by the simple photoionization method. This method is based on our understanding of the photoionization mechanisms of atoms and molecules as discussed in Section 3. Reactant ions in a known state can also be produced by processes 5 and 6. However, the cross sections for processes 5 and 6 near the thermochemical thresholds are usually very low. Reactant ions formed at photon energies above the thresholds for processes 5 and 6 will possess finite kinetic energies. Many state-specific anions can be prepared by the ion-pair process [14, 157–160].

4.2 Photoion–Photoelectron Coincidence Method

The photoion–photoelectron coincidence (PIPECO) methods [33–51, 89–96, 100, 105, 106] have been used extensively in the study of ion dynamics since the first such experiment [97] in 1967. The overwhelming majority of the PIPECO experiments have been concerned with the unimolecular decomposition rates and kinetic-energy-release distributions of photofragments [97–105] of state-selected molecular ions. The PIPECO technique is undoubtedly the most powerful and general method for ionic state selection. This subject has been reviewed by Gellender and Baker [99] and Baer [102, 104]. The earlier PIPECO experiments [98–100, 161–163] used a rare gas resonance lamp as the light source and a dispersive analyzer for the energy analysis of photoelectrons. The internal energy states of an ion are then selected by coincidence with the corresponding energy-selected photoelectrons. As pointed out previously, due to the high collection efficiency, the detection of TPE is preferred when a tunable VUV monochromatic light source is used. The basic principle of a PIPECO experiment is the time correlation of an electron–ion pair produced by photoionization. One of the most common coincidence circuits uses a time-to-amplitude converter (TAC) together with a multichannel analyzer (MCA) operated in the pulse height analysis mode [102]. In a PIPECO experiment electrons and ions formed by photoionization are guided toward the electron and ion detectors. Since the flight time for the electron is much shorter than that of the ion, the electronic pulse signifying the detection of an electron is used to start a voltage ramp in the TAC. The ion pulse produced by the arrival of an ion at the ion detector immediately following the electron pulse serves to stop the voltage ramp. After the completion of the start–stop cycle, a pulse with a height proportional to the time interval between the detection of the electron and ion is generated by the TAC and stored in a proper channel of the MCA. When the VUV lamp is a continuous light source and the photoionization occurs continuously in the ion source, the ions that are correlated with the energy-selected electrons will appear in a narrow range of channels in the MCA corresponding to the flight times of the correlated ions. The detection of uncorrelated ions will give rise to a uniform background because these ions arrive randomly at the ion detector at a uniform rate. Since only one ion is detected for one start signal, the TAC arrangement is suitable for experiments with low ionization rates. As a result of false coincidences due to random uncorrelated ions, the probability of observing a true coincidence after an electron start pulse is an exponentially decreasing function of time [164]. At high count rates, the TAC circuitry may be paralyzed [100]. In the latter case, it is necessary to use a start-multistop arrangement. When a TOF mass spectrometer is used in conjunction with the TAC coincidence arrangement, the paralysis effect due to the random background ions is to decrease signals for high mass ions that have longer flight times from the photoionization region to the ion detector.

The problem of false coincidences is more severe in a pulsed experiment, either using a pulsed light source or a pulsed extraction electric field. This is because the flight times for the correlated and uncorrelated ions become similar. In these cases, it is necessary to measure the false coincidence TOF mass spectrum and subtract it from the coincidence spectrum.

The statistics of coincidence counting have been discussed many times in the past [98, 99, 100, 102, 165, 166]. In a PIPECO experiment, the counting rates for energy-selected electrons (E), ions (I), and true coincidence signals (C) can be expressed in terms of the total ionization rate (N), collection efficiencies for the electrons (f_e) and ions (f_i), and the branching ratio (g_e) of the photoionization process:

$$I = Nf_i \tag{7}$$

$$E = Ng_e f_e \tag{8}$$

$$C = Ef_i = Ng_e f_e f_i \tag{9}$$

The actual signals I' and E' observed in the ion and electron detectors, respectively, also include background counts for the ion (I_b) and electron (E_b) detectors:

$$I' = I + I_b \qquad (10)$$

$$E' = E + E_b \qquad (11)$$

The false coincidence rate (C_f) per time interval (Δt) is given by

$$C_f = (E + E_b)(I + I_b)\Delta t \qquad \text{for } I'\Delta t < 1 \qquad (12)$$

Since the net area under the coincidence TOF mass peak and above the uniform false coincidence background level is a measure of the true coincidence signal, Δt can be taken as the full width of the coincidence peak and usually has a value of a few microseconds. The resulting signal-to-noise ratio for coincidence counting is

$$\frac{\text{Signal}}{\text{Noise}} = \frac{C}{\sqrt{(C + 2C_f)}} \qquad (13)$$

$$\approx \left(\frac{Ng_e f_e f_i}{1 + 2N\Delta t}\right)^{1/2} \qquad \text{for } E \gg E_b \text{ and } I \gg I_b \qquad (14)$$

The most effective way of improving the signal-to-noise ratio in a PIPECO experiment is to increase f_e and f_i. Gellender and Baker [98, 99] have shown that under certain experimental conditions, the signal-to-noise ratio can also be increased by increasing N. When the coincidence signal is low, it is also necessary to suppress I_b and E_b.

In the application of PIPECO techniques to measure cross sections for state-selected ion–molecule reactions, we are concerned with two processes, namely, the process for the primary $A^+(j)$ reactant ion formation,

$$A + h\nu \rightarrow A^+(j) + e^-(j) \qquad (15)$$

and that for the secondary ion production,

$$A^+(j) + B \rightarrow C^+ + D \qquad (16)$$

Here, $e^-(j)$ represents the photoelectron ejected with the kinetic energy corresponding to the ionic state j. Neglecting random or false coincidences and assuming thin target conditions, the total state-selected cross section (σ_j) for reaction 16 is given by

$$\sigma_j = \frac{1}{nl}\left(\frac{\text{Rate of coincidences of } [e^-(j) + C^+]}{\text{Rate of coincidences of } [e^-(j) + A^+(j)] \text{ and } [e^-(j) + C^+]}\right) \qquad (17)$$

where n is the density of B and l is the interaction path length. Therefore, the cross-section calculations require the measurements of the PIPECO spectra for the reactant and product ions. The TOF peak of the primary state-selected ions is usually narrow. However, because of the exothermicity of the bimolecular reaction and the spatial scattering pattern of the product ions, the TOF mass peak for the product ions is expected to spread over a wider temporal range. In other words, Δt for the TOF coincidence peak of the secondary ions is greater than that of the primary ion. According to Eq. 12, the false coincidence rate for the energy-selected electron secondary ion coincidence experiment is higher than that for the energy-selected primary ion coincidence counting. This, together with the fact that the intensity for the product ion is usually only $\sim 10\%$ that for the reactant ion, makes the application of the PIPECO technique to bimolecular collisions more difficult.

4.3 Resonant-Enhanced Multiphoton Ionization Method

When a molecule is resonantly excited to a discrete state by n photons and subsequently ionized by m photons, the whole process is abbreviated as $(n + m)$

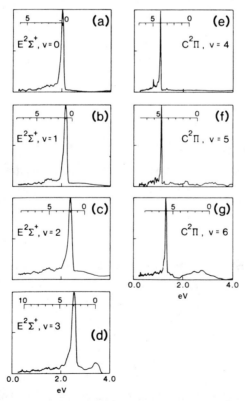

Fig. 3. Photoelectron spectra following the $2 + 1$ REMPI of NO via (a) $E^2\Sigma^+$ $v = 0$, (b) $E^2\Sigma^+$ $v = 1$, (c) $E^2\Sigma^+$ $v = 2$, (d) $E^2\Sigma^+$ $v = 3$, (e) $C^2\Pi$ $v = 4$, (f) $C^2\Pi$ $v = 5$, and (g) $C^2\Pi$ $v = 6$. Reprinted from Ref. [92] with permission of North-Holland Physics Publications.

resonant ionization. In principle, due to specific selection rules for photoionization, a molecular ion in a specific rovibronic state can be prepared by REMPI via an appropriate intermediate state. Since the spectroscopy for each molecule is unique, the scheme for ionic state selection varies from ion to ion.

Photoionization of an excited state is governed by the Franck-Condon principle. Since the removal of a Rydberg electron has little effect on the bonding of the ion core, ionization from a Rydberg state is expected to follow the $\Delta v = 0$ transition. It has been demonstrated in the REMPI of NH_3 [58, 59, 83, 87, 88], NO [83, 89, 92], H_2 [90], and N_2 [91] that $(n + 1)$ resonant ionization via Rydberg states produces mostly molecular ions in the same vibrational states as those of the resonant Rydberg states. Figure 3 shows the photoelectron spectra of $NO^+(\tilde{X}, v)$ following the $(2 + 1)$ resonant ionization of NO via the $E^2\Sigma^+$, $v = 0$–3, and $C^+\Pi$, $v = 4$–6 states [92]. These spectra indicate that the selectivities of the $v = 0$–6 states of $NO^+(\tilde{X})$ are greater than 70%. Deviations from the expected Franck-Condon behavior in $(n + 1)$ resonant ionization can be caused by autoionization and/or configuration mixing of the resonant state.

Since the laser intensity is very high in a MPI experiment, a molecular ion with a high density of states initially formed by MPI may further absorb photons and end up in states different from that indicated by the photoelectron spectrum. This is the major difficulty of using MPI as a general ionic state-selection technique. The photoelectron spectra observed for the $(3 + 1)$ resonant ionization of H_2S via Rydberg states [167] show that HS^+ and S^+ ions are produced mainly by additional photon absorption from the $v = 0$ and 1 states of $H_2S^+(\tilde{X})$, respectively.

5. MEASUREMENTS OF STATE-SELECTED ION–MOLECULE REACTION CROSS SECTIONS

Due to the low intensity of state-selected reactant ions that are prepared by photoionization, all the previous ion–molecule reaction photoionization apparatuses have been designed for total cross-section measurements. In this section the designs of some of these apparatuses, and selected experimental results obtained using them, will be discussed.

5.1 Threshold Electron Secondary Ion Coincidence Apparatus

The threshold electron secondary ion coincidence (TESICO) technique was first successfully applied to bimolecular ion–molecule reactions by Baer and co-workers [33, 34] in a single-chamber arrangement. The apparatus is essentially the same as that used for the study of unimolecular reactions. Their coincidence apparatus and the analytical method to calculate symmetric charge transfer and collision-induced cross sections from coincidence data have been reviewed by Baer [102]. Koyano and Tanaka [35] have extended the TESICO technique to the study of chemical rearrangement reactions. The Okazaki group in Japan and the Orsay group in France [49, 50] are presently the most active in applying this technique to state-selected cross-section measurements.

5.1.1 Okazaki Apparatus [35]

EXPERIMENTAL ARRANGEMENT AND PROCEDURES

The schematic diagram of the TESICO apparatus constructed by Koyano and Tanaka is shown in Fig. 4. It can be divided into four major stages: a photoionization ion source, an electron energy analyzer, a reactant chamber, and a quadrupole mass spectrometer. These stages are connected only through small apertures, and each chamber is evacuated by an independent pumping system.

The reactant ions A^+ and corresponding electrons are formed in the ionization chamber by photoionization of neutral molecules A leaked into the chamber. The tunable VUV monochromatic radiation is produced by a windowless monochromator discharge lamp system. The ions and electrons are repelled in directions perpendicular to the incident photon beam. The electron energy analyzer is a combination of steradiancy and hemispherical analyzers that has a resolution of ~ 20 meV for TPE spectra of Ar^+. In a double-chamber experiment, the primary ions are extracted from the ionization chamber and focused into the reaction chamber in which reactions between A^+ and neutral molecules B take place. The primary ion-beam energy is defined by the potential difference

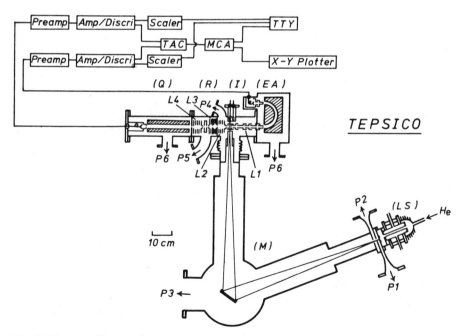

Fig. 4. Schematic diagram of the apparatus TEPSICO. LS; Light source; M; monochromator; I; ionization chamber; R; reaction chamber; Q; quadrupole mass spectrometer; EA; electron energy analyzer; P1–P6; pumping ports. Reprinted from Ref. [35] with permission of American Institute of Physics.

between the photoionization region and the reaction chamber. The product ions, together with unreacted primary ions, are further extracted from the reaction chamber, mass analyzed by the quadrupole mass filter (QMF), and detected by a channeltron multiplier. The ion signals are counted in coincidence with the TPE signals using the TAC–MCA arrangement.

Coincidence experiments are also conducted in a single chamber mode, i.e., both photoionization and reaction take place in the ionization chamber.

SELECTED EXPERIMENTAL RESULTS

Cross Sections for the Reaction $H_2^+(v_0' = 0-3) + H_2(v_0'' = 0) \to H_3^+ + H$

The reaction

$$H_2^+(v_0') + H_2(v_0'' = 0) \to H_3^+ + H \tag{18}$$

has been studied by the single-chamber method because the cross sections for reaction 18 are very high at low center-of-mass collision energies ($E_{c.m}$), and

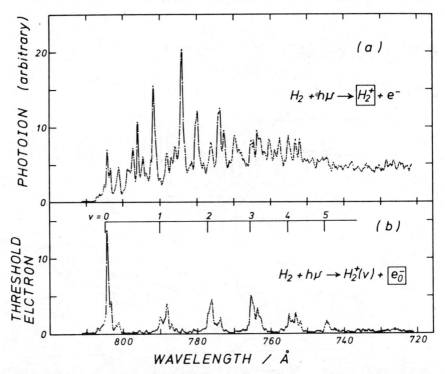

Fig. 5. PIE curve (*a*) and threshold electron spectrum (*b*) of H_2 taken at a pressure of 2×10^{-4} torr and with a wavelength bandwidth of 0.52 Å (FWHM) and threshold electron energy resolution of 20 meV (FWHM). Reprinted from Ref. [35] with permission of American Institute of Physics.

considerable H_3^+ is formed in the ionization chamber even at a fairly low H_2 pressure.

The PIE curve and the TPE spectrum for H_2^+ obtained with a wavelength resolution of 0.52 Å are shown in Fig. 5. The jagged profiles in the TPE spectrum are due to partially resolved rotational structures and to "hot" photoelectrons produced by autoionization near the ionization thresholds.

The coincidence measurements have been made by setting λ at the maxima of the peaks in the TPE spectrum. The typical TOF coincidence spectra for H_2^+ and H_3^+ at an average $E_{c.m.}$ of 0.11 eV and $v_0' = 0$–3 are plotted in Fig. 6. The average value for $E_{c.m.}$ is determined by the TOF of the primary H_2^+ ions. A mass programmer is used to switch repetitively the RF voltage of the QMF corresponding to the masses H_2^+ and H_3^+ throughout the total accumulation time indicated in the figure. The residence time for collection of H_3^+ is three times that for H_2^+ in one period. The use of a QMF has two major advantages in secondary

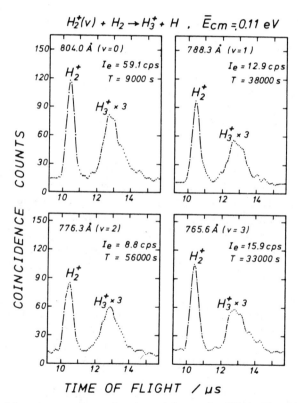

Fig. 6. Signals of the primary and secondary ions of the reaction $H_2^+(v) + H_2 \rightarrow H_3^+ + H$ taken in coincidence with the threshold electrons at four wavelengths indicated. Average collision energy $\bar{E}_{c.m.}$ was 0.11 eV. I_e and T represent the threshold electron count rate and the overall accumulation time, respectively. The labels × 3 on the H_3^+ peaks indicate that the accumulation times for obtaining these peaks were three times those for obtaining corresponding parent ion peaks. Reprinted from Ref. [35] with permission of American Institute of Physics.

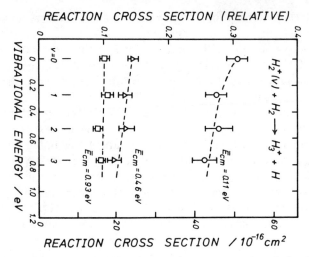

Fig. 7. Cross section for the reaction $H_2^+(v) + H_2 \rightarrow H_3^+ + H$ as a function of vibrational energy of H_2^+, obtained at three different average collision energies indicated. Reprinted from Ref. [35] with permission of American Institute of Physics.

ion coincidence experiments. Since most of the false coincidence counts are due to primary reactant ions, the signal-to-noise ratio for the TOF secondary ion coincidence spectrum is substantially improved by setting the QMF to the mass of the secondary ion. Furthermore, the use of the QMF is found to increase the collection efficiency of the secondary ions.

The relative cross sections of reaction 18 for $v'_0 = 0-3$ as shown in Fig. 7 have been obtained by dividing the area of the H_3^+ coincidence peak by the sum of the areas of the H_2^+ and H_3^+ peaks after correcting for the background. At low collision energies, the experiment reveals that the cross section for reaction 18 decreases with increasing vibrational energy and that the vibrational dependence of the cross section becomes less pronounced as $E_{c.m.}$ is increased. These conclusions are consistent with the previous results of Chupka et al. [19]. The relative cross sections for $v'_0 = 0$ are found to obey the $E_{c.m.}^{-1/2}$ dependence as predicted by the Langevin-Gioumousis-Stevenson (LGS) model [168]. The absolute values for the cross sections have been estimated by using the $E_{c.m.}^{-1/2}$ dependence and the thermal rate constant for $v'_0 = 0$ determined by Theard and Huntress [169]. The comparison of the results of this experiment with those of recent experiments [57, 60] and theory [170, 171] is summarized in a later section.

At the $E_{c.m.}$ of interest in this experiment, the major competing process is the electron transfer reaction

$$H_2^+(v'_0) + H_2(v''_0 = 0) \rightarrow H_2(v') + H_2^+(v'') \tag{19}$$

The electron transfer product $H_2^+(v'')$ ions are expected to be distributed over a wide TOF range on the longer TOF side of the primary ion peak. If these secondary $H_2^+(v'')$ ions have a considerable intensity at the position of the H_3^+ peak, the cross sections deduced would be erroneous. This possibility is excluded stemming from the finding that the coincidence TOF spectrum shows no discernible peak nor background at the position of H_3^+ when the QMF is set at the mass of H_2^+. Nevertheless, since electron transfer product $H_2^+(v'')$ ions may be consumed in the ionization chamber by further reacting with H_2 to form H_3^+, the latter observation cannot be taken as sufficient evidence to exclude the contribution from reaction 19.

Cross Sections for the Reaction $Ar^+(^2P_{3/2,1/2}) + H_2(v'_0 = 0) \rightarrow ArH^+ + H$

The study of the reaction

$$Ar^+(^2P_{3/2,1/2}) + H_2(v = 0) \rightarrow ArH^+ + H \tag{20}$$

has been performed using the ion beam–gas chamber method [37], namely, the double-chamber mode. The double-chamber coincidence technique allows the direct measurements of the cross sections due to $Ar^+(^2P_{3/2})$ and $Ar^+(^2P_{1/2})$ and suppresses the complication of the $H_2^+ + Ar$ reactions. The H_2 pressure in the reaction chamber is quite high (3.2×10^{-3} torr). In determining the absolute cross sections (σ_J) for reaction 20, Tanaka et al. have taken into account the competing electron transfer reaction and secondary reactions involving the product ArH^+ and H_2^+ ions in the reaction chamber, such as

$$Ar^+(^2P_{3/2,1/2}) + H_2(v = 0) \rightarrow Ar + H_2^+ \tag{21}$$

$$\rightarrow \text{scattering} \tag{22}$$

$$ArH^+ + H_2 \rightarrow H_3^+ + Ar \tag{23}$$

$$\rightarrow \text{scattering} \tag{24}$$

Here, Eqs. 22 and 24 represent the losses of reactant and product ions due to scattering from H_2 that are estimated to be $\sim 20\%$. By solving the rate equations for reactions 20–24, they find that σ_J is related to the intensities of ArH^+ (I_{ArH^+}) and Ar^+ (I_{Ar^+}) by the equation

$$\sigma_J = \left(\frac{\eta}{\eta'(m-q)nl}\right) \ln\left[1 + (m-q)\left(\frac{I_{ArH^+}}{I_{Ar^+}}\right)\right] \tag{25}$$

where η and η' are the collection efficiencies for Ar^+ and ArH^+, and m and q are parameters depending on η, η', and the cross sections for reactions 20–24. The values for m and q are determined using a fitting procedure. The value for $\sigma_{1/2}$ thus determined is about 1.5 times greater than that for $\sigma_{3/2}$ (Fig. 8), an

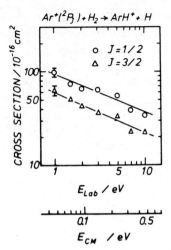

Fig. 8. State-selected cross sections for reaction 20 for the $^2P_{3/2}$ and $^2P_{1/2}$ states of Ar^+ as a function of collision energy. Reprinted from Ref. [37] with permission of American Institute of Physics.

observation consistent with the previous state-selected results obtained using the single-chamber and the simple photoionization method [20].

Despite the difficulties due to secondary reactions and ion collection efficiencies as illustrated in the above examples, the Okazaki apparatus has been one of the most successful ion–molecule reaction apparatuses and has made possible the studies of many state-selected ion–molecule reactions [35, 37, 38, 42–47, 51] Recently, the electron energy analyzer of this apparatus has been replaced by an indirect trajectory TPE analyzer [51] that has a collection efficiency for TPE several times higher than the previous analyzer. The sensitivity of the apparatus is expected to be improved by the same factor.

5.1.2 Orsay Apparatus [50]

EXPERIMENTAL ARRANGEMENT AND PROCEDURES

The TESICO technique used at Orsay is an extension of the threshold electron–photoion coincidence (TEPICO) technique [113, 114] used for the state-selected unimolecular fragmentation study of molecular ions. The schematic diagram of the dual electron–ion TOF apparatus is shown in Fig. 9. Dispersed synchrotron radiation from the electron storage ring ACO is used as the ionization source. Photoelectrons produced in the ionization chamber are accelerated toward the electron detector by a weak electric field (2 V/cm). The TPE detector, a combination of a steradiancy and a TOF analyzer, has a resolution of ~ 20 meV. The nearly total rejection of "hot" electrons is achieved by angular and temporal discrimination. The threshold electron signal triggers an extraction electric pulse for the reactant ions in the ionization chamber. After proper focusing, the reactant ion beam intersects an effusive jet of the

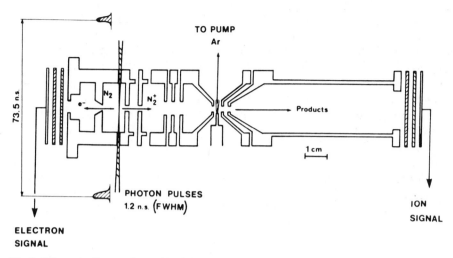

Fig. 9. Schematic diagram (to scale) of the dual electron–ion time of flight (TOF) spectrometer. Reprinted from Ref. [50] with permission of North-Holland Physics Publications.

neutral reactant at the 3-mm reaction region between two conical lenses. The neutral reactant beam effuses from a hypodermic needle placed in the middle of the reaction region. A field of 20 V/cm is maintained in the reaction region to ensure the efficient collection of product ions by the ion TOF detector. The potential of the field-free ion drift region is adjusted according to those in the reaction region in order to satisfy first-order space focusing conditions [172]. State-selected reactant ion and product ion TOF spectra are measured in delayed coincidence with TPE using the TAC–MCA arrangement.

The apparatus is evacuated by two turbomolecular pumps with a combined pumping speed of ~ 1600 L/s. During the experiment, the photoionization region, reaction region, and vacuum chamber pressures are estimated to be 2×10^{-4}, 3×10^{-3}, and $< 4 \times 10^{-5}$ torr, respectively.

The main weakness of the Orsay apparatus is the low resolution of the TOF mass spectrometer at high masses. This makes the studies of many reactions, such as proton transfer reactions, difficult. For example, the TOF mass spectrometer may not be able to distinguish the masses for product ArH$^+$ and Ar$^+$ in the collisions of H$_2^+$ + Ar.

SELECTED EXPERIMENTAL RESULTS

Cross Sections for the Reactions $N_2^+(\tilde{X} \text{ or } \tilde{A}, v') + Ar \rightarrow N_2 + Ar^+$

The state-selected cross-section measurement of the reactions

$$N_2^+(\tilde{X}, v') + Ar(^1S_0) \rightarrow N_2(X, v) + Ar^+(^2P_{3/2,1/2}) \tag{26}$$

$$N_2^+(\tilde{A}, v') + Ar(^1S_0) \rightarrow N_2(X, v) + Ar^+(^2P_{3/2,1/2}) \tag{27}$$

by the Orsay group [50] represents a state-of-the-art experiment in state-selected ion chemistry.

Figure 10 shows the TPE spectrum of N_2 in which the vibrational levels $v' = 0$–4 for $N_2^+(\tilde{X})$ and $v' = 0 - 6$ for $N_2^+(\tilde{A})$ are clearly identified. The He(I) (584 Å) photoelectron spectrum [10] shows a negligible intensity for $\tilde{X}$-state levels $v \geqslant 2$ and also quite a different vibrational distribution for the $\tilde{A}$ state. The modifications observed in the TPE spectrum are due to autoionization from Rydberg levels to lower ionic states. This example illustrates that the TEPICO technique allows the study of the reactivity of higher vibrational levels of the $\tilde{X}$ state.

A typical coincidence TOF spectrum corresponding to the electron transfer reaction of $N_2^+(\tilde{X}, v' = 2) + Ar$ at $E_{c.m.} = 8\,eV$ is shown in Fig. 11. The arrival time of Ar^+ ions is the sum of the N_2^+ TOF from the ionization region to the reaction region and the Ar^+ TOF from the reaction zone to the detector. The

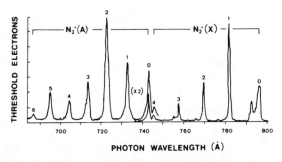

Fig. 10. Threshold photoelectron spectrum of N_2 obtained under the experimental conditions used for the photoelectron–photoion coincidence experiments. The time-averaged count rate on the $N_2^+(X, v' = 1)$ peak is typically 2000 counts/s. Reprinted from Ref. [50] with permission of North-Holland Physics Publications.

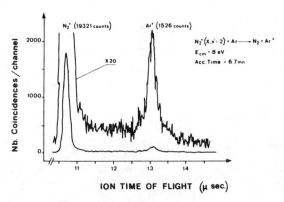

Fig. 11. Typical ion TOF spectrum showing unreacted N_2^+ ions (left-hand peak) and Ar^+-product ions (right-hand peak) resulting from the reaction $N_2^+(X, v' = 2) + Ar \rightarrow N_2 + Ar^+$ at $E_{c.m.} = 8\,eV$. Reprinted from Ref. [50] with permission of North-Holland Physics Publications.

Ar^+ ions produced in the photoionization region arrive at the detector 0.26 μs earlier. The background between the N_2^+ and Ar^+ peaks results from false coincidences, typical of pulsed extraction experiments. As the extraction pulse is applied, there is a finite probability of detecting an ion produced by an earlier, uncorrelated, ionization event. False coincidence spectra are obtained by applying extraction pulses randomly distributed in time with respect to the ionizing photons. They are scaled to the appropriate number of extraction pulses and to the ionization rate and are subtracted from the coincidence spectrum of interest.

The ratios of the Ar^+ to the sum of the Ar^+ and N_2^+ peak areas measured at different vibrational states and $E_{c.m.}$ permit a direct evaluation of the vibrational and kinetic energy dependences of the electron transfer reaction of $N_2^+(\tilde{X}, v' = 0\text{--}4) + Ar$. Since the absolute cross sections for $Ar^+ + Ar$ are known [173], absolute cross sections for reactions 26 and 27 have also been estimated by calibrating the intensity ratios of this experiment against those of the $Ar^+ + Ar$ reaction.

The lifetime for $N_2^+(\tilde{A}, v')$ $[\tau(\tilde{A}, v')]$ varies from 16.6 μs for $v' = 0$ to 8.25μs for $v' = 6$ [174]. The above procedures yield apparent cross sections $[\sigma_{app}(\tilde{A}, v')]$ that have to be corrected for the fact that the $\tilde{A}$ state decays to the $\tilde{X}$ state on a time scale comparable to the flight time t:

$$\sigma_{app}(\tilde{A}, v') = \exp\left(\frac{-t}{\tau_{v'}}\right)\sigma(\tilde{A}, v') + \left[1 - \exp\left(\frac{-t}{\tau_{v'}}\right)\right]\sum_{v''} b_{v'v''}\sigma(\tilde{X}, v'') \quad (28)$$

where $\sigma(\tilde{A}, v')$ and $\sigma(\tilde{X}, v'')$ denote the true cross sections for the $\tilde{A}$ and $\tilde{X}$ states, respectively.

The measured electron transfer cross sections for reaction 26 at $E_{c.m.} = 8$ and 20 eV plotted against the internal energy of the N_2^+ ion are compared to the theoretical cross sections obtained by the recent semiclassical multistate calculation [175] in Fig. 12. The experimental and theoretical results are in reasonable agreement. Vibrational-state-selected cross sections for reaction 26 have also been measured by Kato et al. [43] and most recently by Liao et al. [65] and Shao et al. [62]. These experimental results are consistent with the results of the Orsay group.

Cross Sections for the Symmetric Electron Transfer Reaction $H_2^+(v_0' = 0\text{--}10) + H_2(v_2'' = 0)$

Reaction 19 has been the subject of many theoretical [170, 175–182] and experimental [22, 39, 49, 54, 55, 182, 183] studies. Recently, vibrational-state-selected cross sections have been reported by three groups [39, 49, 54, 55, 182].

Figure 13 shows the coincidence TOF spectrum for the reaction of $H_2^+(v_0' = 6) + H_2$ at $E_{c.m.} = 16$ eV obtained using the Orsay apparatus [49]. Electron transfer product H_2^+ ions and H^+ ions resulting from collision-induced

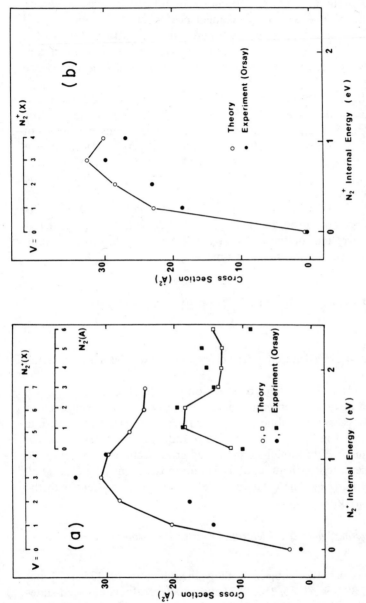

Fig. 12. (a) Electron transfer cross sections for $N_2^+(X, A; v) + Ar \rightarrow N_2 + Ar^+$ at $E_{c.m.} = 20$ eV are plotted against the internal energy of the N_2^+ ion. Theoretical results from Ref. [175] for $v = 0$–7 of the X state and $v = 0$–6 of the A state are shown as open circles and squares, respectively. Experimental results from Ref. [50] are shown as solid circles and squares. The lines drawn through the theoretical values are only a guide for the eye. (b) Electron transfer cross sections for $N_2^+(X; v) + Ar \rightarrow N_2 + Ar^+$ at $E_{c.m.} = 8$ eV are plotted against the internal energy of the N_2^+ ion. Theoretical results from Ref. [175] for $v = 0$–4 are shown as open circles; experimental results from Ref. [50] are shown as solid circles. The line through the theoretical values is simply a guide for the eye. The scales for Figs. 12a and 12b are the same to facilitate comparison. Reprinted from Ref. [175] with permission of North-Holland Physics Publications.

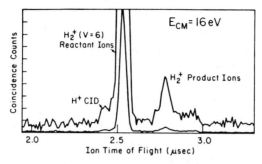

Fig. 13. Example of coincidence TOF distribution for H_2^+ ($v = 6$). The average collection time varied from 5 minutes for the $v = 0$ data to about 40 minutes for the $v = 10$ data. Reprinted from Ref. [49] with permission of North-Holland Physics Publications.

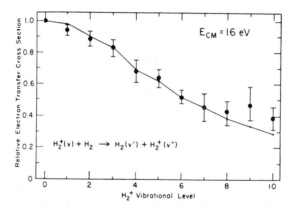

Fig. 14. Relative electron transfer cross sections for $H_2^+ + H_2$ from $v = 0$–10 at a relative translational energy of 16 ± 1.4 eV. The error bars are determined from the standard deviation of several TOF spectra at each reactant ion vibrational level. The solid line segments are the calculated cross sections of Lee and DePristo [181]. Reprinted from Ref. [49] with permission of North-Holland Physics Publications.

dissociation are identified in the spectrum. The measured relative electron transfer cross section for reaction 19 at $E_{c.m.} = 16$ eV decreases monotonically with v_0', an observation in accord with calculated cross sections [181, 182] of Lee and DePristo (Fig. 14). The experimental results at $E_{c.m.} = 16$ eV are also consistent with those obtained by Liao et al. [54, 55]. However, at $E_{c.m.} = 8$ eV the relative electron transfer cross sections measured by the Orsay group [182] differ from both the theoretical predictions and the experimental results of Liao et al. A more detailed comparison of experimental and theoretical results of reaction 19 is given in a later section.

The TOF for H_3^+ formed in the photoionization region is estimated to be 3.1 μs. Cole et al. [49] argued that since the formation of $H_3^+ + H$ is important only at low $E_{c.m.}$, H_3^+ is produced in the ion source rather than in the collision region. This argument is not convincing because slow electron transfer H_2^+ product ions can react further with H_2 to form H_3^+ in the reaction zone. Assuming a reaction cross section of 92 Å^2 at thermal energy [169] for reaction 18, an H_2 background pressure of $\sim 3 \times 10^{-3}$ torr [50] in the collision region, and an interaction length of 0.3 mm, the conversion of electron transfer H_2^+ product ions to H_3^+ is estimated to be 30%. This value probably represents an upper bound because the electron transfer H_2^+ product ions are accelerated out of the reaction region immediately after their formation. The TOF for H_3^+ formed in the collision region is expected to be between 3.1 μs and the TOF of the electron transfer product H_2^+. In order to obtain accurate cross sections for reaction 19, it is necessary to account for the conversion of the electron transfer product H_2^+ to H_3^+ in the collision region.

5.2 Tandem Photoionization Mass Spectrometer [57, 60]

One of the difficulties encountered in the total cross-section measurement of an ion–molecule reaction is to achieve the total collection of product ions. When the ion–molecule reaction of interest is exothermic, a large fraction of the exothermicity can appear as the kinetic energy of a product ion. In principle, all product ions can be collected if a sufficiently high electrostatic field is used at the reaction region to extract the product ions. However, the use of high extraction fields at the reaction zone may lower the kinetic energy resolution.

Teloy and Gerlich [64] have introduced the RF octopole ion-guide reaction gas cell technique that makes possible nearly 100% collection of product ions without degrading the kinetic energy resolution. An RF octopole ion guide is constructed of eight molybdenum rods, symmetrically spaced on a circle of diameter $2r_0$. The RF potential applied to alternate rods is $\pm V_0 \cos(\omega t)$. An RF octopole ion-guide reaction gas cell is formed by placing a collision gas cell at the middle of an RF octopole ion guide. In the limit of very high RF frequency ω, the effective radial potential of the RF octopole is of the form [53]

$$V_{\text{eff}}(r) = \frac{e^2 V_0^2}{4m\omega^2 r_0^2} \left(\frac{r}{r_0}\right)^6 \tag{29}$$

where m is the mass of the ion and r is the radial distance from the octopole axis. This potential is close to a square well, providing a tubular trapping volume with low potential near the center and steep walls at a large radius. In comparison, the effective radial potential of a quadrupole is proportional to $(r/r_0)^2$, indicating that the potential energy spread of ions within a given tubular volume of the quadrupole is greater than that of the octopole. Since the effective potentials for higher-order multipoles have a radial dependence that varies with a higher power of r/r_0, they represent even better approximations to a radial square well than the octopole. Therefore, the kinetic energy resolution achieved in an ion-beam–RF octopole (or higher multipole) ion-guide reaction gas cell arrangement is better than that using an ion-beam–RF quadrupole reaction gas cell combination. Due to the fact that a light ion moves faster than a heavy ion for a given kinetic energy, a higher optimum trapping efficiency for a light ion can be achieved with a higher ω and a sufficiently high V_0. During the experiment, the ion guide and the gas cell are maintained at a direct current (dc) potential that determines the axial ion energy. Product ions formed in the RF octopole ion-guide reaction gas cell are trapped by the radial repulsive wall. The forward-scattered product ions will drift toward the exit of the octopole where they are extracted and detected. The backward-scattered product ions can also be reflected toward the detector by maintaining a sufficiently high repeller voltage at the entrance ion lens. When the reactant ion-beam kinetic energy is low, a high dc repeller voltage applied to the entrance ion lens may prevent the reactant ion beam from entering the ion guide. In this case, the reactant ion beam is pulsed, and a repeller voltage is placed on the entrance lens immediately after the injection of the reactant ions.

Lee and co-workers [53, 184–187] first combined the photoionization and the RF ion-guide technique and measured the vibrational-state-selected cross sections for several simple ion–molecule reactions. Recently, Shao and Ng [57, 60] have developed a tandem photoionization mass spectrometer. The experimental setup is similar to that constructed by Lee and co-workers, except that a quadrupole mass filter is used to reject background ions formed in the photoionization region.

5.2.1 Experimental Arrangement and Procedures

The detailed cross-sectional view of the tandem photoionization mass spectrometer is depicted in Fig. 15 [57, 60]. The apparatus consists of a photoionization source, a reactant quadrupole mass filter (QMF), an RF octopole ion-guide reaction gas cell, a product QMF, and a variant of the Daly-type scintillation ion detector. Tunable monochromatic VUV light is generated by a discharge lamp and a windowless 0.2-m monochromator system. The typical wavelength resolution achieved is ~ 4 Å (FWHM).

The vacuum system is partitioned into four chambers: the photoionization chamber (1), the reactant QMF chamber (6), the RF octopole chamber (7), and the detector chamber (13). These chambers are connected only through small apertures. Each of these chambers is evacuated by a separate pumping system, and the pressures maintained in these chambers during the experiment are shown in Fig. 15.

The reactant ions in a pure state, or a known distribution of states, are prepared by photoionization of neutral precursors in a free jet. Photoions are extracted perpendicular to the free jet. By setting the reactant QMF to the mass of the reactant ion, background ions formed in the photoionization region can be rejected completely, and a pure reactant ion beam is selected to enter the RF octopole ion-guide reaction gas cell. After the reaction between the reactant ions and neutral molecules in the gas cell, the intensities of the reactant and product ions are measured by the product QMF.

Since the sections of the octopole ion guide outside the reaction gas cell are

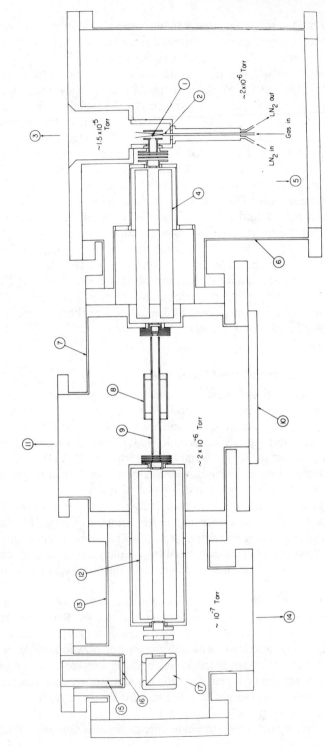

Fig. 15. Cross-sectional view of the tandem photoionization mass spectrometer. 1, Photoionization region; 2, quartz nozzle; 3, to freon-trapped 6-in. diffusion pump (DP); 4, QMF; 5, to liquid-nitrogen-trapped 6-in. DP; 6, QMF chamber; 7, RF octopole chamber; 8, reaction gas cell; 9, RF octopole; 10, auxiliary molecular beam port; 11, to liquid-nitrogen-trapped 4-in. DP; 12, QMF; 13, detector chamber; 14, to liquid-nitrogen-trapped 2-in. DP; 15, photomultiplier; 16, plastic scintillator window; 17, aluminium ion target. Reprinted from Ref. [60] with permission of American Institute of Physics.

fully exposed to the pumping system of the octopole chamber, the density of the neutral reactant in this region is approximately two orders of magnitude lower than that of the gas cell. Thus, the intensity of product ions formed along the octopole ion guide outside the reaction gas cell is negligibly small. This background can be estimated by measuring the intensity of the product ions when the reactant is introduced into the RF octopole chamber instead of the reaction gas cell.

The reactant ion-beam energy, which is defined nominally by the potential difference between the photoionization region and the reaction gas cell, can be determined by the retarding field or TOF method. The absolute cross section σ at a given $E_{c.m.}$ is determined by the relation

$$\sigma = -\left(\frac{1}{nl}\right)\ln\left[1-\left(\frac{i}{I_0}\right)\right] \tag{30}$$

Here, i is the intensity of the product ions after correcting for the intensity of background product ions formed outside the reaction gas cell; I_0 is the measured intensity of the reactant ions when the reaction gas cell is empty. If the condition of total product collection is attained, I_0 is equal to the sum of the intensities for reactant and product ions when the reaction gas cell is filled with the neutral reactant gas.

5.2.2 Selected Experimental Results

Cross Sections for the Reaction $H_2^+(v_0' = 0\text{--}4) + H_2(v_0'' = 0) \rightarrow H_3^+ + H$ [57, 60]

The autoionization of H_2 favors the $|\Delta v| = 1$ process [133, 134], and even when autoionization is only possible by a $|\Delta v| > 1$ process, the ions have been shown to form predominately ($>75\%$) in the highest possible vibrational state [188]. When the wavelength resolution used is not sufficient to resolve the detailed autoionization structure, the vibrational distribution of H_2^+ formed in a given wavelength interval depends on the contributions by autoionization as well as direct ionization. Using the high-resolution PIE spectrum for H_2^+ obtained by Dehmer and Chupka [133], Anderson et al. [53] have estimated the ratio of the H_2^+ produced by direct photoionization to that produced by autoionization. This ratio, together with the assumption that the vibrational distributions of H_2^+ formed by direct photoionization are governed by Franck-Condon factors [189] for transitions from $H_2(X^1\Sigma_g^+, v) \rightarrow H_2^+(\tilde{X}^2\Sigma_g^+, v_0')$, allows Anderson et al. to deduce the vibrational distributions of H_2^+ resulting from photoionization in the wavelength region (745–805 Å) corresponding to the first five vibrational intervals (see Table 1) [53].

Figure 16 compares the PIE spectra for the reactant H_2^+ and product H_3^+ ions obtained at $E_{c.m.} = 0.04\text{--}15\,\text{eV}$ [57, 60]. The PIE spectra of the product H_3^+ and reactant H_2^+ ions in the region of 790–805 Å, which corresponds to the formation of reactant H_2^+ in $v_0' = 0$, have been normalized to have the

Table 1. Estimated Vibrational Distributions for H_2^+ Formed by Photoionization

v_0' (Nominal)	Estimated Vibrational Distribution for $H_2^{+\,a}$				
	0	1	2	3	4
$v_0' = 0$	1	0.11	0.08	0.08	0.07
1		0.89	0.16	0.17	0.14
2			0.76	0.19	0.16
3				0.56	0.15
4					0.48

[a]Reference [53].

same value. These comparisons clearly show that $\sigma(v_0')$ for reaction 18 at $E_{c.m.} = 0.04, 0.25, 0.46, 10,$ and $15\,eV$ are inhibited by vibrational excitation of the reactant H_2^+, whereas those at $E_{c.m.} = 1, 2, 3,$ and $5\,eV$ are enhanced as v_0' is increased. Using the vibrational distributions of H_2^+ shown in Table 1, these spectra allow the calculation of the relative cross sections, $\sigma(v_0')/\sigma(v_0' = 0)$, $v_0' = 0\text{--}4$. By calibrating the values for $\sigma(v_0')/\sigma(v_0' = 0)$, $v_0' = 0\text{--}4$, with the measured values for $\sigma(v_0' = 0)$ at the corresponding $E_{c.m.}$, absolute values for $\sigma(v_0')$, $v_0' = 1\text{--}4$, are also determined.

Table 2 [60] summarizes values for $\sigma(v_0)$, where $v_0' = 0$ and 3, at the $E_{c.m.}$ range of 0.25–5 eV, obtained by previous theoretical [170, 190, 191] and experimental [35] studies and compares them with those determined in this study. The uncertainties for the cross sections obtained in this experiment are due mostly to the high conversion efficiency of electron transfer product H_2^+ to H_3^+ in the reaction gas cell. The experimental and theoretical values for $\sigma(v_0' = 0)$ at $E_{c.m.} = 0.5$ and $1.0\,eV$ are in good agreement. The kinetic energy dependence of $\sigma(v_0')$ over the $E_{c.m.}$ range of 0.25–5 eV is consistent with the prediction of the LGS model, suggesting that the dynamics of reaction 18 at low $E_{c.m.}$ are governed mainly by the charge-induced-dipole interaction. The absolute cross sections measured for $v_0' = 0$ and 3 at $E_{c.m.} = 0.5, 1, 2,$ and $5\,eV$ are in better agreement with the "trajectory surface hopping" calculations of Stine and Muckerman [170] and Eaker and Schatz [191] that include nonadiabatic surface hopping throughout the reaction. The earlier calculation of Eaker and Schatz [190] assumes that the trajectories follow the diabatic surfaces up to a particular intermolecular separation and then the ground adiabatic surfaces thereafter.

Cross Sections for the Electron Transfer Reaction $Ar^+(^2P_{3/2,1/2}) + N_2(X, v = 0)$

The absolute spin-orbit-state-selected cross sections, $\sigma_{3/2}$ and $\sigma_{1/2}$, for the electron transfer reactions

$$Ar^+(^2P_{3/2,1/2}) + N_2(X, v = 0) \to Ar + N_2^+ \qquad (31)$$

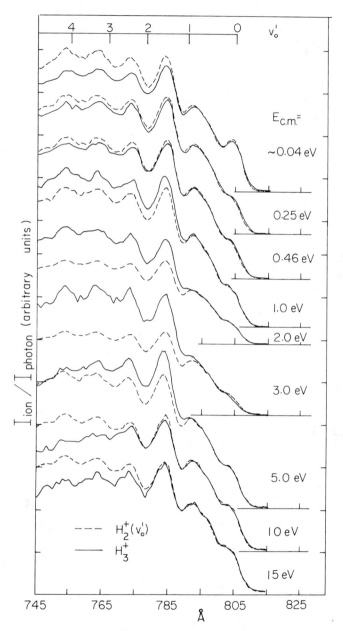

Fig. 16. The comparison of PIE curves for H_3^+ product ions (—) formed at the $E_{c.m.}$ range of ~0.04–15 eV with that for $H_2^+(v_0')$ reactant ions (----). Reprinted from Ref. [60] with permission of American Institute of Physics.

Table 2. Total Cross Sections for the Reaction $H_2^+(v_0' = 0 \text{ or } 3) + H_2(v_0'' = 0) \to H_3^+ + H$

$\sigma(v_0')(\text{Å}^2)$

		Experimental			Theoretical		
$E_{c.m.}$(eV)	v_0'	Ref. [60]	Ref. [35]	Ref. [170]	Ref. [171]	ES[a]	LGS[b]
0.25	0	34.0 ± 4.0	—	47.8 ± 1.4	—	46.4 ± 2.0	30.5
	3	32.3 ± 4.0	—	38.8 ± 2.2	—	37.4 ± 2.0	30.5
0.5	0	26.6 ± 2.0	24.3 ± 1.5^c	30.6 ± 1.7	30.6 ± 1.4^c	30.6 ± 1.4	21.5
	3	26.3 ± 2.0	19.5 ± 2.0^c	26.7 ± 1.7	33.7 ± 1.4^c	26.4 ± 2.0	21.5
1.0	0	$17.4 \pm {}^{1.5}_{2.5}$	16.9 ± 1.5^d	16.0 ± 1.1	18.3 ± 0.8^d	16.6 ± 1.1	15.2
	3	$19.3 \pm {}^{1.5}_{2.5}$	16.2 ± 1.5^d	18.8 ± 1.7	23.0 ± 0.8^d	18.5 ± 1.4	15.2
3.0	0	$7.5 \pm {}^{0.8}_{1.6}$	—	5.8 ± 0.8	—	6.5 ± 0.6	8.8
	3	$11.8 \pm {}^{0.8}_{1.6}$	—	9.3 ± 1.1	—	9.6 ± 0.8	8.8
5.0	0	$6.1 \pm {}^{0.6}_{2.0}$	—	3.9 ± 0.6	—	3.7 ± 0.3	6.8
	3	$6.3 \pm {}^{0.6}_{2.0}$	—	5.3 ± 0.8	—	5.06 ± 0.6	6.8

[a] Results of the TSH calculation obtained by Eaker and Schatz (ES) (private communication).
[b] Values calculated using the LGS model.
[c] Values for $E_{c.m.} = 0.46$ eV.
[d] Values for $E_{c.m.} = 0.93$ eV.

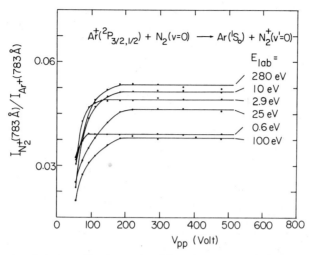

Fig. 17. The ratios of the intensities for product N_2^+ and reactant Ar^+ at $E_{lab} = 0.6$, 2.9, 10, 25, 100, and 280 eV measured at 783 Å, $I_{N_2^+}(783\text{ Å})/I_{Ar^+}(783\text{ Å})$, plotted as a function of the RF peak-to-peak voltage (V_{pp}) applied to the RF octopole ion guide. Reprinted from Ref. [61] with permission of American Institute of Physics.

have been measured by Shao et al. [192] using the tandem photoionization mass spectrometer.

At a given value of the N_2 gas cell pressure, the ratio of the intensities for product N_2^+ and reactant Ar^+ determines the value of the total cross section. The ratios measured at 783 Å and a fixed N_2 gas cell pressure, $I_{N_2^+}(783\text{ Å})/I_{Ar^+}(783\text{ Å})$, in the laboratory collision energy (E_{lab}) range of 0.6–280 eV are plotted as a function of the peak-to-peak (pp) RF voltage (V_{pp}) in Fig. 17. The ω value used in this experiment is ~ 4 MHz. The constant values for the ratios at $V_{pp} \gtrsim 180$ V indicate that total product N_2^+ collection is achieved at $V_{pp} \approx 180$ V. Depending on E_{lab}, when V_{pp} is increased beyond ~ 400–500 V, the transmission for reactant Ar^+ through the RF octopole ion guide is found to decrease significantly, concomitant with a decrease in product N_2^+ ion intensity. The decrease of the ratio for $E_{lab} = 280$ eV in V_{pp} range of ~ 350–500 V is evident in Fig. 17. This observation is consistent with the expectation that the trapping efficiencies for different ions depend on ω, V_{pp}, and ion energy. The value of 280 V for V_{pp} is approximately at the center of the V_{pp} voltage region for which values for $I_{N_2^+}(783\text{ Å})/I_{N_2^+}(783\text{ Å})$ are constant and at the highest values for a given E_{lab}. Therefore, the value for V_{pp} is fixed at 280 V during the experiment.

Figure 18 compares the PIE spectra for the reactant Ar^+ and product N_2^+ ions measured at $E_{c.m.} = 10.3$ eV. The PIE data for the reactant and product ions have been normalized to have the same value at 783 Å. The PIE curve for N_2^+ at energies above the ionization energy (IE) for $Ar^+(^2P_{1/2})$ is less than the corresponding PIE curve for Ar^+, indicating that $\sigma_{1/2}$ is lower than $\sigma_{3/2}$. This observation is in accord with results reported previously [42, 193, 194]. Samson

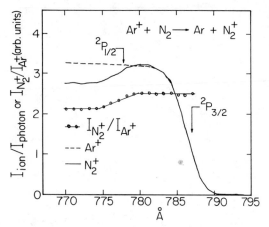

Fig. 18. PIE spectra for reactant Ar^+ (- - -) and product N_2^+ (—) at $E_{c.m.} = 10.3$ eV. ○ The ratio for intensities of N_2^+ and Ar^+ plotted as a function of wavelength. Reprinted from Ref. [61] with permission of American Institute of Physics.

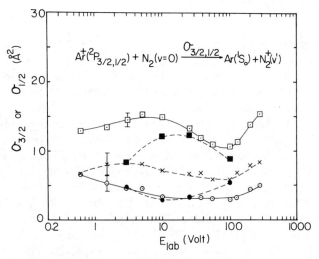

Fig. 19. Total spin-orbit-state-selected cross sections $\sigma_{3/2}$, $\sigma_{1/2}$ for the reactions $Ar^+(^2P_{3/2,1/2}) + N_2(v=0) \rightarrow Ar(^1S_0) + N_2^+(v')$ at the E_{lab} range of 0.6–280 eV. Experimental: □, $\sigma_{3/2}$, x, $\sigma_{1/2}$ measured using the tandem photoionization mass spectrometer (Ref. [61]); ○, $\sigma_{1/2}$ obtained by multiplying $\sigma_{3/2}$ of Ref. [61] and $\sigma_{3/2}/\sigma_{1/2}$ measured using the crossed-ion–neutral beam arrangement. Theoretical (Ref. [196]): ■, $\sigma_{3/2}$; ●, $\sigma_{1/2}$. Reprinted from Ref. [61] with permission of American Institute of Physics.

and Cairns [195] have shown that the spin-orbit-state distributions for Ar^+ produced by photoionization at photon energies above the IE for $Ar^+(^2P_{1/2})$ have the same value of 1.98 throughout the photon energy range of ~ 16–21 eV. The ratio for the intensity of product N_2^+ to that of reactant Ar^+ is also plotted in Fig. 18. The constant values observed for the ratio at wavelengths $< 775\,\text{Å}$ support the conclusion of Samson and Cairns.

The variations of $\sigma_{3/2}$ and $\sigma_{1/2}$ as a function of E_{lab} can be seen in Fig. 19. The theoretical cross sections of Spalburg and Gislason [196] are also included in the figure. The values measured for $\sigma_{3/2}$ and $\sigma_{1/2}$ are in general higher than the theoretical predictions. The experimental and theoretical values are in better agreement at higher E_{lab}. Both $\sigma_{3/2}$ and $\sigma_{1/2}$ increase as E_{lab} is increased from 100 eV. This observation is interpreted as due to the formation of N_2^+ in the $\tilde{A}^2\Pi_u$ state at high E_{lab}. The most recent semiclassical multistate calculation of Parlant and Gislason [197], which uses the ab initio potential of the $[\text{Ar} + N_2]^+$ system, has confirmed this interpretation.

5.3 Crossed Ion–Neutral Beam Photoionization Apparatus

Chantry [198] has shown that the random thermal motion of neutral reactant molecules in a reaction gas cell has the effect of broadening the distribution of collision energies in an ion beam–gas cell experiment. The actual collision energy distribution depends on the nominal $E_{\text{c.m.}}$, the Boltzmann constant k, the target gas temperature T, and γ that is defined to be the ratio of the mass of the reactant ion to the sum of the masses for the ion and neutral reactants. For $E_{\text{c.m.}} > kT$ the distribution peaks approximately at $E_{\text{c.m.}}$ and has a FWHM given approximately by

$$\Delta E = (11.1\gamma k T E_{\text{c.m.}})^{1/2} \tag{32}$$

At very low collision energies, the relative velocity distributions approach the Maxwell-Boltzmann distribution with an effective temperature equal to γT. In other words, the collision energy distributions at low energies are determined primarily by the thermal velocities of the target gas.

In order to attain high kinetic energy resolution, Liao et al. [54–56] have designed an ion–molecule reaction photoionization apparatus that combines the crossed ion–neutral beam method and high-resolution photoionization mass spectrometry. The crossed ion–neutral beam photoionization apparatus is most suitable for state-selected electron transfer cross-section measurements.

5.3.1 Experimental Arrangement and Procedures

The crossed ion–neutral beam photoionization apparatus has been developed from a high-resolution photoionization mass spectrometer [199]. Figure 20 shows a detailed cross-sectional view of the apparatus. The apparatus consists of a windowless 3-m near-normal incidence VUV monochromator, a discharge lamp, two supersonic molecular beam systems, and two QMFs.

The neutral precursors of the reactant ions are introduced into the photoionization region by supersonic expansion through a quartz nozzle (nozzle 1). The reactant ions formed by photoionization are extracted perpendicular to the neutral precursor beam and focused onto the neutral reactant beam at an

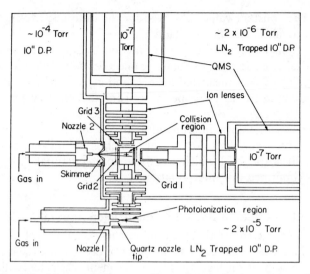

Fig. 20. The schematic of the crossed ion–neutral beam photoionization apparatus. Reprinted from Ref. [54] with permission of American Institute of Physcis.

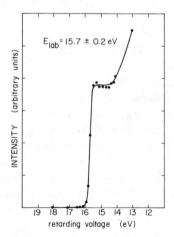

Fig. 21. Retarding potential energy curve for a reactant H_2^+ ion beam with a nominal laboratory collision energy of 16 eV. Reprinted from Ref. [55] with permission of American Institute of Physics.

intersecting angle of 90°. The neutral reactant beam is produced by supersonic expansion through a stainless-steel nozzle (nozzle 2) and then collimated into the scattering chamber by a conical skimmer. The intensity of the reactant ion beam is monitored with the vertical QMF placed along the direction of the reactant ion beam. Slow electron transfer product ions are collected and measured by the horizontal QMF positioned in the direction of the neutral reactant beam. Previous studies [200–202] show that electron transfer product ions are predominantly slow ions, scattered at small laboratory angles with respect to the neutral reactant beam direction.

The collision region is shielded by a small square ion lens and three high-transmission gold grids. Based on the geometric angle subtended by grid 1 at the collision center, the ion lens system of the horizontal QMF accepts product ions scattered 25° away from the neutral reactant ion-beam axis. During the experiment, the square ion lens and grids 2 and 3 are maintained at the same potential, while grid 1 is at a voltage lower than grid 2. Due to the favorable acceptance angle of the horizontal QMF, a high collection efficiency for electron transfer product ions is obtained with a weak extraction field.

In order to achieve high kinetic energy resolution, it is also necessary to minimize the repeller field strength in the photoionization region. For a given repeller field, the resolution in kinetic energy of the reactant ion beam is governed by the height of the photon beam at the photoionization region.

5.3.2 Selected Experimental Results

Cross Sections for the Symmetric Electron Transfer Reaction $H_2^+(v_0', J) + H_2(v_0'' = 0)$ [55, 56].

The retarding potential analysis of the reactant H_2^+ ion beam having a nominal E_{lab} of 16 eV and a repeller field of ~ 2 V/cm at the photoionization region is shown in Fig. 21. The spread in E_{lab} caused by the repeller field is estimated to be $\lesssim 0.2$ eV. The retarding potential analysis involves measuring the reactant H_2^+ ion intensity by the vertical QMF as a function of retarding potential applied to the ion lenses immediately above the square ion lens and grids at the collision region. The actual E_{lab} for the reactant H_2^+ ion beam is found to be 15.7 ± 0.2 eV. Because of the translational cooling and narrow angular spread of the neutral H_2 beam achieved by supersonic expansion, the collision energy spread due to the neutral H_2 beam is negligible. Taking into account the effect of the extraction field at the collision region, a nominal E_{lab} of 16 eV actually corresponds to $E_{c.m.} = 7.9 \pm 0.2$ eV. In an ion-beam–gas cell experiment, ΔE due to the random target motion alone is predicted to be 1.2 eV.

Figure 22a displays the PIE spectrum for the reactant H_2^+ ions measured by the vertical QMF using an optical resolution of 0.3 Å (FWHM). Most of the autoionization features found in this spectrum consist of more than one autoionization state. Nevertheless, the resolution used is sufficient to allow for the correction of photoionization yields arising from direct photoionization. In the wavelength range of ~ 791–807 Å, H_2^+ can be formed only in the $v_0' = 0$ state. The positions of the base lines, which account for photoionization yields due to direct photoionization in wavelength intervals corresponding to the formation of H_2^+ in the $v_0' = 1, 2, 3$, and 4 states, are defined by the lowest PIE measured in each wavelength interval. In arriving at the base lines shown in the figure, Liao et al. [55, 56] have assumed that direct photoionization gives

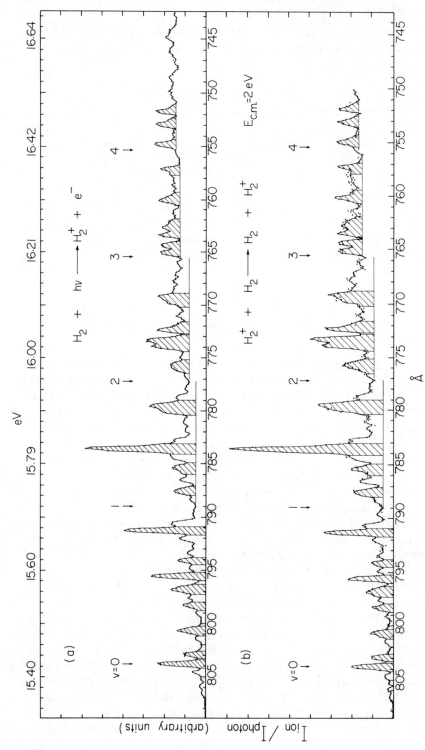

Fig. 22. (a) PIE curve for the H_2^+ reactant ions in the wavelength region of 743–809 Å [wavelength resolution = 0.3 Å (FWHM)]. (b) PIE curve for the H_2^+ product ions in the wavelength region of 750–809 Å observed at $E_{c.m.} = 2$ eV [wavelength resolution = 0.3 Å (FWHM)]. Reprinted from Ref. [55] with permission of American Institute of Physics.

rise to uniform PIE within a vibrational interval. Since the positions of the base lines thus determined necessarily represent the upper limits in photoionization yield due to direct photoionization, photoionization above these base lines is ascribed solely to autoionization. According to the above discussion, the shaded areas of autoionization peaks in a photon energy region higher than the threshold for the formation of $H_2^+(v_0')$ and below that of $H_2^+(v_0'+1)$ are taken as a measure for the intensity of reactant H_2^+ ions formed in the v_0' state. Similar procedures to correct for photoionization yields due to direct photoionization have been applied to a PIE curve for the product H_2^+ ions at $E_{c.m.} = 2\,eV$ (Fig. 22b).

Relative values for $\sigma(v_0')$, $v_0' = 0$–4, at $E_{c.m.} = 2\,eV$ are determined by the ratios of the shaded areas for corresponding autoionization peaks resolved in

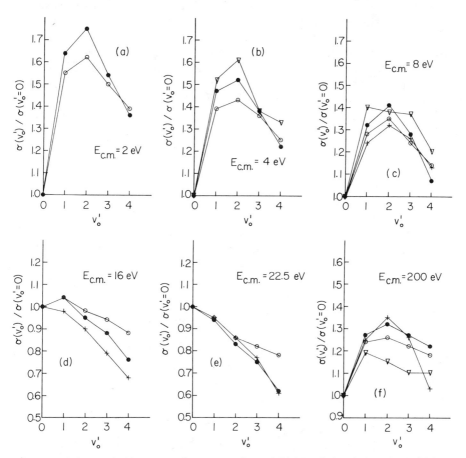

Fig. 23. Relative total electron transfer cross sections $\sigma(v_0')/\sigma(v_0'=0)$ for the reaction $H_2^+(v_0') + H_2(v''=0)$ plotted as a function of v_0'. (a) $E_{c.m.} = 2\,eV$; (b) $E_{c.m.} = 4\,eV$; (c) $E_{c.m.} = 8\,eV$; (d) $E_{c.m.} = 16\,eV$; (e) $E_{c.m.} = 22.5\,eV$; (f) $E_{c.m.} = 200\,eV$. ●, Ref. [55]; ○, Ref. [55], obtained without corrections of direct photoionization; ▽, Ref. [39]; +, theoretical, Ref. [181]. Reprinted from Ref. [55] with permission of American Institute of Physics.

the PIE curves for the product and reactant H_2^+ ions. The shaded areas are arbitrarily chosen for comparison. The measured values for $\sigma(v'_0)/\sigma(v'_0 = 0)$, $v'_0 = 0$–4 at $E_{c.m.} = 2, 4, 8, 16, 22.5$, and 200 eV are plotted in Figs. 23a–23f to compare with the theoretical calculations of Lee and DePristo [181]. The relative cross sections at $E_{c.m.} = 8$ and 200 eV reported by Campbell et al. [39] are also included in Figs. 23c and 23f. The experimental results of Liao et al. [54, 55] and the theoretical predictions are in good agreement, except that the experimental value for $\sigma(v'_0 = 4)/\sigma(v'_0 = 0)$ at $E_{c.m.} = 200$ eV is higher than the theoretical value. This discrepancy can be partly ascribed to the inaccuracy of the coupling potentials at short H_2^+–H_2 distances used in the calculation.

The rotational energy effect on the cross section at $E_{c.m.} = 2$ eV for the symmetric electron transfer reaction $H_2^+(v'_0 = 0, J) + H_2(v''_0 = 0)$ has also been examined by Liao et al. [55]. Figure 24 compares the high resolution [0.14 Å (FWHM)] PIE curves for the product and reactant ions in the region of ~800–808 Å. The rotational states of the reactant H_2^+ ions formed have been inferred by autoionization selection rules [203, 204] and energy constraints [133, 134]. As a result of the preference for ionization to occur with no change in rotational state, H_2^+ ions resulting from autoionizing Rydberg states $Q(1)$ $8p\pi(v = 1)$ and $R(1)$ and $P(1)$ $8p\sigma(v = 1)$ should be in the $J = 1$ rotational state. After normalizing the heights of the autoionization peaks to have the same

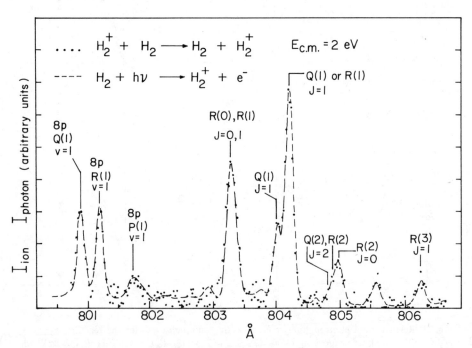

Fig. 24. The comparison of the PIE curve for the product H_2^+ ions formed at $E_{c.m.} = 2$ eV with that for the reactant H_2^+ ions in the region of 800.5–806.5 Å [wavelength resolution = 0.14 Å (FWHM)]. Reprinted from Ref. [55] with permission of American Institute of Physics.

value at 804.12 Å for the reactant and product ions, the PIE spectra for the reactant and product ions are found to be superimposable. This observation indicates that within the experimental uncertainty of $\sim 10\%$, changing the rotational state of $H_2^+(v_0' = 0)$ from $J = 0$ to 2 has no measurable effect on the symmetric electron transfer reaction $H_2^+(v_0' = 0, J) + H_2(v_0'' = 0)$ at $E_{c.m.} = 2$ eV.

Cross Sections for the Electron Transfer Reaction $N_2^+(\tilde{X}, v' = 0\text{--}2) + Ar$ [65]

Figure 25d shows the PIE spectrum of N_2^+ in the region of $\sim 766\text{--}797$ Å obtained using a wavelength resolution of 1.4 Å. The $N_2^+(\tilde{X})$ ions formed at the three strong autoionization peaks, $\lambda = 781$, 776, and 771.5 Å, between the IEs of $N_2^+(\tilde{X}, v' = 1)$ and $N_2^+(\tilde{X}, v' = 2)$, can be in the $v' = 0$ and 1 states. The vibrational distributions of $N_2^+(\tilde{X}, v' = 0$ and 1) produced by photoionization at these autoionization peaks have been measured previously by Tanaka et al. [151] and Berkowitz and Chupka [150]. The former experiment uses the TPE spectrum of N_2^+ and the difference in reactivity between the electron transfer

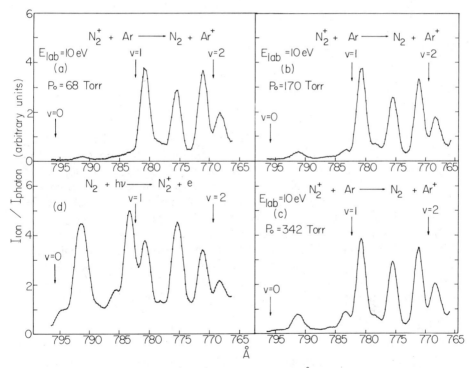

Fig. 25. PIE curve for product Ar^+ in the region of 766–797 Å obtained at $E_{lab} = 10$ eV and a nozzle stagnation pressure of N_2. (a) $P_0 = 68$ torr; (b) $P_0 = 170$ torr; (c) $P_0 = 342$ torr [wavelength resolution = 1.4 Å (FWHM)]; (d) PIE curve for N_2^+ in the region of 766–797 Å [wavelength resolution = 1.4 Å (FWHM)]. Reprinted from Ref. [65] with permission of American Institute of Physics.

Table 3. Vibrational State Distributions of N_2^+ Produced at Selected Autoionization Peaks

		$X_{v'}(\lambda)^a$		
$\lambda(\text{Å})^b$	v'^c	This Work	Ref. [151]	Ref. [150]
781	0	0.49	0.49	0.21
	1	0.51	0.51	0.79
	2	0.00	0.00	0.00
776	0	0.74	0.73	0.72
	1	0.26	0.27	0.28
	2	0.00	0.00	0.00
771.5	0	0.41	0.34	0.27
	1	0.59	0.66	0.73
	2	0.00	0.00	0.00
768.5	0	0.55	0.34	—
	1	0.20	0.31	—
	2	0.25	0.35	—
761	0	0.52	0.31	—
	1	0.21	0.32	—
	2	0.27	0.37	—

^{a}Fraction of $N_2^+(\tilde{X},v')$ formed at λ.
^{b}Positions of autoionization peaks.
^{c}Vibrational state of $N_2^+(\tilde{X})$.

collisions of $N_2^+(\tilde{X}, v'=0) + Ar$ and $N_2^+(\tilde{X}, v'=1) + Ar$, and the latter study uses the retarded field PES to determine the vibrational distributions of N_2^+. The results, which are summarized in Table 3, show that > 50% of N_2^+ formed at $\lambda = 781$ and 771.5 Å are in the $v'=1$ state. The wavelength resolution used in the latter study is higher than that used in the former experiment. Due to the favorable Franck-Condon factor, $N_2^+(\tilde{X})$ ions are produced predominantly in the $v'=0$ state by direct photoionization. Since N_2^+ ions formed at $\lambda = 781$, 776, and 771.5 Å consist of N_2^+ formed by direct ionization and autoionization, and the proportion of N_2^+ arising from direct ionization is greater when a lower wavelength resolution is used, the fractions of $N_2^+(\tilde{X}, v'=0)$ observed by Tanaka et al. are greater than those found by Berkowitz and Chupka. The vibrational distributions of $N_2^+(\tilde{X})$ at the autoionization peaks $\lambda = 768.5$ and 761 Å determined by Tanaka et al. reveal nearly equal distributions of N_2^+ in the $v' = 0$, 1, and 2 states.

The PIE spectra for product Ar^+ at $E_{lab} = 10$ eV observed by the horizontal QMF at N_2 stagnation pressures $P_0 = 68$, 170, and 342 torr are shown in Figs. 25a–25c. These spectra are arbitrarily normalized at $\lambda = 781$ Å. The intensity of product Ar^+ at $\lambda = 791.5$ Å relative to that at $\lambda = 781$ Å increases as P_0 is increased, indicating possible changes of vibrational distributions of reactant

$N_2^+(\tilde{X}, v')$ at high P_0 before the electron transfer collision. The ratio of the intensities for Ar^+ at $\lambda = 791.5$ and 781 Å remains constant at $P_0 \leqslant 68$ torr. This is taken as evidence that changes of vibrational distributions of $N_2^+(\tilde{X}, v')$ ions due to collisions of $N_2^+(\tilde{X}, v')$ with background gas are negligible at $P_0 \leqslant 68$ torr. Thus, the P_0 value of 68 torr is used in the experiment.

Since the wavelength resolution used in the experiment of Liao et al. [65] is only slightly lower than that used by Tanaka et al., the vibrational distributions of $N_2^+(\tilde{X})$ at the autoionization peaks observed in the two studies should be similar. Reactant $N_2^+(\tilde{X})$ in the pure $v' = 0$ state is prepared with high intensity at $\lambda = 791.5$ Å. At a given $E_{c.m.}$ and under thin target conditions, the value for $n'l'\sigma_0(E_{c.m.})$ can be determined from

$$n'l'\sigma_0 = \frac{I_{Ar^+}(E_{c.m.}, 791.5 \text{ Å})}{I_{N_2^+}(791.5 \text{ Å})} \quad (33)$$

where n' is the density of the neutral Ar beam at the collision region and l' is the effective interaction length. The measured total cross section $\sigma_m(E_{c.m.}, \lambda)$ characteristic of $N_2^+(\tilde{X}, v')$ formed at $\lambda = 781, 776$, or 771.5 Å is related to $\sigma_0(E_{c.m.})$, $\sigma_1(E_{c.m.})$, and the fraction $X_{v'}(\lambda)$ of $N_2^+(\tilde{X})$ at $v' = 0$ and 1 by Eqs. 34 and 35:

$$X_0(\lambda) + X_1(\lambda) = 1 \quad (34)$$

$$X_0(\lambda)n'l'\sigma_0(E_{c.m}) + X_1(\lambda)n'l'\sigma_1(E_{c.m.}) = n'l'\sigma_m(E_{c.m.}, \lambda) = \frac{I_{Ar^+}(E_{c.m.}, \lambda)}{I_{N_2^+}(\lambda)} \quad (35)$$

Since $n'l'\sigma_0(E_{c.m.})$ is known, $n'l'\sigma_1(E_{c.m.})$ can be calculated using the known values of $X_0(\lambda)$ and $X_1(\lambda)$, or vice versa. Equations 34 and 35 can be combined to give

$$\frac{n'l'\sigma_m(E_{c.m.}\lambda) - n'l'\sigma_0(E_{c.m.})}{X_1(\lambda)} = n'l'\sigma_1(E_{c.m.}) - n'l'\sigma_0(E_{c.m.}) \quad (36)$$

This equation allows the proportion $X_1(781 \text{ Å}):X_1(776 \text{ Å}):X_1(771.5 \text{ Å})$ to be determined. Using the value of 0.51 for $X_1(781 \text{ Å})$ measured by Tanaka et al., Liao et al. have obtained the vibrational distributions of N_2^+ at $\lambda = 776$ and 771.5 Å listed in Table 3. The values for $X_{v'}(\lambda)$, $v' = 0$ and 1, at 776 Å are essentially identical to those observed previously, whereas the value for $X_0(771.5 \text{ Å})$ is slightly higher than that reported by Tanaka et al.

The $N_2^+(\tilde{X})$ ions formed at $\lambda = 768.5$ and 761 Å are in the $v' = 0, 1$, and 2 states. The values for $n'l'\sigma_2(E_{c.m.})$ can be calculated from Eqs. 37 and 38:

$$X_0(\lambda) + X_1(\lambda) + X_2(\lambda) = 1 \quad (37)$$

$$X_0(\lambda)n'l'\sigma_0(E_{c.m.}) + X_1(\lambda)n'l'\sigma_1(E_{c.m.}) + X_2(\lambda)n'l'\sigma_2(E_{c.m.}) = n'l'\sigma_m(E_{c.m.}, \lambda) \quad (38)$$

when the values for $X_{v'}(\lambda)$, $v' = 0$–2, $n'l'\sigma_0(E_{\text{c.m.}})$, $n'l'\sigma_1(E_{\text{c.m.}})$, and $n'l'\sigma_m(E_{\text{c.m.}}, \lambda)$ are known. Previous experimental studies [43, 50] show that at $E_{\text{c.m.}} \leqslant 10\,\text{eV}$, $\sigma_0(E_{\text{c.m.}})$ is negligible compared to $\sigma_{v'}(E_{\text{c.m.}})$, $v' \geqslant 1$. The TESICO experiments also reveal that $\sigma_1(E_{\text{c.m.}}) < \sigma_2(E_{\text{c.m.}})$ at $E_{\text{c.m.}} \sim 8$–$20\,\text{eV}$ [50], whereas $\sigma_1(E_{\text{c.m.}}) \sim \sigma_2(E_{\text{c.m.}})$ at $E_{\text{c.m.}} \sim 2\,\text{eV}$ [43]. Therefore,

$$X_1(\lambda) + X_2(\lambda) \approx \frac{n'l'\sigma_m(E_{\text{c.m.}} \sim 2\,\text{eV}, \lambda)}{n'l'\sigma_1(E_{\text{c.m.}} \sim 2\,\text{eV})} \tag{39}$$

$$\frac{n'l'\sigma_2(E_{\text{c.m.}} \sim 8\,\text{eV})}{n'l'\sigma_1(E_{\text{c.m.}} \sim 8\,\text{eV})} = \frac{[n'l'\sigma_m(E_{\text{c.m.}} 8\,\text{eV}, \lambda)/n'l'\sigma_1(E_{\text{c.m.}} \sim 8\,\text{eV})] - X_1(\lambda)}{X_2(\lambda)} \tag{40}$$

Using the value of 1.25 for the ratio $n'l'\sigma_2(E_{\text{c.m.}} = 8\,\text{eV})/n'l'\sigma_1(E_{\text{c.m.}} = 8\,\text{eV})$ determined previously, approximate values for $X_{v'}(\lambda)$, $v' = 0$–2, at $\lambda = 768.5$ and 761 Å are obtained by solving Eqs. 39 and 40. The values for the vibrational distributions of N_2^+ at $\lambda = 768$ and 761 Å listed in Table 3 have been adjusted so that values of $n'l'\sigma_2(E_{\text{c.m.}})$ calculated at these wavelengths are in agreement in the $E_{\text{c.m.}}$ range of 1.2–40 eV.

The values for the relative cross sections $\sigma_{v'}(E_{\text{c.m.}})/\sigma_1(E_{\text{c.m.}})$, $v' = 0$–2, for reaction 26 over the $E_{\text{c.m.}}$ range of 1.2–320 eV obtained using Eqs. 33–38 and the values for $X_{v'}$, $v' = 0$–2, deduced in this experiment are plotted as a function of v' in Fig. 26. The experimental results at $E_{\text{c.m.}} = 1.2$–$40\,\text{eV}$ are in reasonable agreement with those obtained in previous experimental [43, 50] and theoretical studies [175, 196], indicating that σ_0 is substantially less than σ_1 and σ_2. As E_{cm} is increased, σ_0 becomes comparable to σ_1 and σ_2 in the $E_{\text{c.m.}}$ range of ~ 140–$200\,\text{eV}$. At $E_{\text{c.m.}} = 260$ and $320\,\text{eV}$, the cross sections are in the order $\sigma_0 > \sigma_1 > \sigma_2$.

6 MEASUREMENTS OF STATE-TO-STATE ION–MOLECULE REACTION CROSS SECTIONS

Many experimental methods such as translational energy measurements [202, 205–211] laser-induced fluorescence [212–225], emission studies [205, 226], and fine structure angular distribution measurements [227, 228] have been used previously to probe the internal energy distributions of product ions formed in simple ion–molecule collisions. Since the product ion intensities observed in a state-selected experiment described in Section 5 are usually between 1000 and 10 counts/s, it is difficult to apply the above-mentioned methods to measure the internal energy distributions of product ions resulting from state-selected ion–molecule collisions.

Previous state-selected electron transfer studies [36–51, 53–56, 61, 62, 65, 92, 186, 187] have revealed dramatic variations in total cross sections with ion internal and translational energies. As a result of the favorable kinematics in crossed ion–neutral beam studies of electron transfer reactions and the large

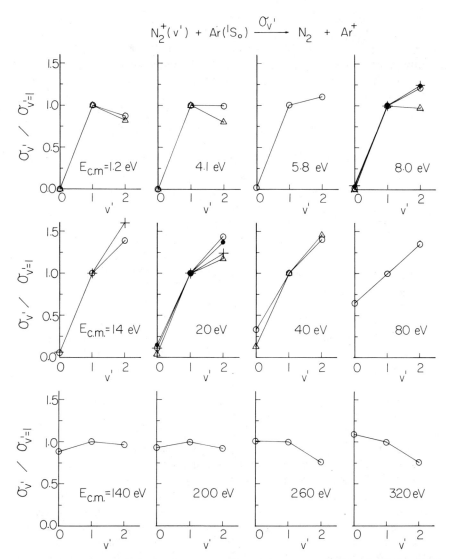

Fig. 26. Relative total cross sections of $\sigma_{v'}/\sigma_{v'=1}$ for reaction 1 at $E_{c.m.}$ = 1.2, 4.1, 5.8, 8.0, 14, 20, 40, 80, 140, 200, 260, and 320 eV plotted as a function of the vibrational quantum number v'. Experiment: ○, Ref. [65]; +, Ref. [50]. Theory: △, Ref. [196]; ●, Ref. [175]. Reprinted from Ref. [65] with permission of American Institute of Physics.

cross sections for electron transfer processes, it is most appropriate to use the electron transfer detection method to determine the internal energy distributions of product ions of electron transfer reactions. Liao et al. [55, 56, 61, 65, 229] have coupled an electron transfer detector to the crossed ion–neutral beam photoionization apparatus and have made state-to-state studies on the electron transfer reactions of $H_2^+(v'=0-1) + H_2(v_0''=0)$ [229], $Ar^+(^2P_{3/2,1/2}) + Ar(^1S_0)$ [56], $Ar^+(^2P_{3/2,1/2}) + N_2(X, v=0)$ [61], $N_2^+(\tilde{X}, v'=0-2) + Ar(^1S_0)$ [65], $Ar^+(^2P_{3/2,1/2}) + H_2(v=0)$ [230], and $H_2^+(v'=0-4) + Ar(^1S_0)$ [230]. Most recently, Shao et al. [63] have developed a triple-quadrupole double-octopole photoionization apparatus and obtained absolute state-to-state excitation cross sections for

$$Ar^+(^2P_{3/2}) + Ar(^1S_0) \rightarrow Ar^+(^2P_{1/2}) + Ar(^1S_0) \quad (41)$$

$$Ar^+(^2P_{3/2}) + N_2(X, v=0) \rightarrow Ar^+(^2P_{1/2}) + N_2(X, v) \quad (42)$$

using an electron transfer detection scheme similar to that of Liao et al. The triple-quadrupole double-octopole photoionization apparatus is capable of determining absolute state-to-state vibrational relaxation cross sections for many simple molecular ions in collisions with various quenching gases.

6.1 Improved Crossed Ion–Neutral Beam Photoionization Apparatus [61, 65]

6.1.1 Experimental Arrangement and Procedures

In order to perform state-to-state studies of electron transfer reactions, the crossed ion–neutral beam apparatus has been modified to add an electron transfer detector. The cross-sectional view of the improved apparatus is shown in Fig. 27. The electron transfer detector, which includes an RF octopole reaction gas cell (14) and a QMF [back horizontal QMF (12)], is placed behind the horizontal QMF [front horizontal QMF (11)]. The RF octopole ion-guide reaction gas cell is formed by enclosing the entire RF octopole ion guide inside a cylindrical cell that consists of a cylindrical wall and front and back ion lenses. The RF octopole ion guide allows the collection of $>80\%$ of the electron transfer product ions formed in the reaction gas cell.

The horizontal detector chamber has been partitioned into two differential pumping chambers. Similar to the previous arrangement, the front horizontal QMF, the vertical QMF, and the RF octopole ion-guide gas cell are evacuated by a liquid-nitrogen trapped 4-in. diffusion pump (DP). The back horizontal QMF is pumped by a water-cooled 2-in. DP. When a pressure of 5×10^{-4} torr of a probing gas (e.g., Ar) is introduced into the RF octopole ion-guide gas cell, the pressures in the front and the back horizontal QMF chambers are $\lesssim 2 \times 10^{-6}$ torr. Maintaining a good vacuum in the horizontal detector chamber is expected to minimize the perturbation of the state distribution of product ions due to collisions of product ions with background gas molecules during the transport of product ions from the collision region to the RF octopole in the guide gas cell.

In a state-to-state experiment, product ions of a state-selected reaction formed at the collision region are collected by the ion lenses and mass selected by the front horizontal QMF before entering the RF octopole ion-guide reaction gas

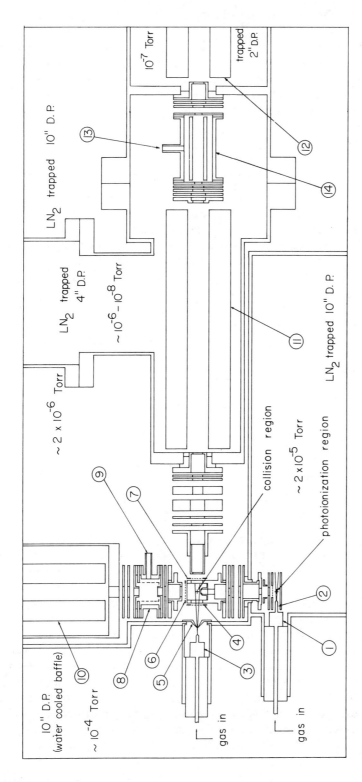

Fig. 27. Cross-sectional view of the improved crossed ion–neutral beam photoionization apparatus. 1, Lower supersonic nozzle; 2, quartz nozzle tip; 3, upper supersonic nozzle; 5, skimmer, 4, 6, 7, 90% transmission gold grids; 8, vertical gas cell; 9, 13, to baratron manometer; 10, vertical QMF; 11, front horizontal QMF; 12, back horizontal QMF; 14, RF octopole reaction gas cell. Reprinted from Refs. [61] and [65] with permission of American Institute of Physics.

cell in which the product ions further react with a probing gas. The reactivity or the cross section for the electron transfer reaction between the product ions and the probing gas measured by the back horizontal QMF contains information about the internal state distribution of the product ions. The detailed calculation of product state distributions will be illustrated in the examples described below.

6.1.2 Selected Experimental Results

State-to-State Cross Sections for the Electron Transfer Reaction $N_2^+(\tilde{X}, v' = 0\text{–}2) + Ar(^1S_0) \to N_2(X, v) + Ar^+(^2P_{3/2,1/2})$

The Ar^+ ions formed in reaction 26 can be in the $^2P_{3/2}$ or $^2P_{1/2}$ state. The principle used here to determine the spin-orbit-state distribution of product $Ar^+(^2P_{3/2,1/2})$ is based on the difference in the total spin-orbit-state-selected electron transfer cross sections $\sigma_{3/2}(H_2^+)$ and $\sigma_{1/2}(H_2^+)$ for reaction 21.

The product $Ar^+(^2P_J)$ ions of reaction 26 formed in the collision region are guided through the RF octopole ion-guide reaction cell where the $Ar^+(^2P_J)$ ions further react with H_2 at $E_{1ab} = 10\,\text{eV}$. The intensity of $H_2^+(i_{H_2^+})$ thus formed is measured by the back horizontal QMF. Since the front horizontal QMF is tuned to the mass of Ar^+, background H_2^+ formed in the collision region due to the electron transfer reaction between the reactant N_2^+ ions and background H_2 is completely suppressed. The total intensity for product $Ar^+(^2P_J)$ is measured by tuning the back horizontal QMF to the mass of Ar^+ when the RF octopole gas cell is empty. For thin target conditions, we have the relations

$$i_{H_2^+} = I_{Ar^+}(E_{c.m.}, \lambda) n l \sigma_m(H_2^+, \lambda) \tag{43}$$

$$X_{3/2}(E_{c.m.}, \lambda) n l \sigma_{3/2}(H_2^+) + X_{1/2}(E_{c.m.}, \lambda) n l \sigma_{1/2}(H_2^+) = n l \sigma_m(H_2^+, \lambda) \tag{44}$$

$$X_{3/2}(E_{c.m.}, \lambda) + X_{1/2}(E_{c.m.}, \lambda) = 1 \tag{45}$$

Here $\sigma_m(H_2^+, \lambda)$ is the measured total electron transfer cross section for reaction 21 at $E_{1ab} = 10\,\text{eV}$ characteristic of product Ar^+ formed at specific values of $E_{c.m.}$ and λ; and $X_{3/2}(E_{c.m.}, \lambda)$ and $X_{1/2}(E_{c.m.}, \lambda)$ are the fractions of the Ar^+ ions in the $^2P_{3/2}$ and $^2P_{1/2}$ states, respectively.

The values for $nl\sigma_{3/2}(H_2^+)$ and $nl\sigma_{1/2}(H_2^+)$ are measured in a separate experiment using the ion beam–RF octopole gas cell arrangement. This involves the formation of Ar^+ ions in the pure $^2P_{3/2}$ state, or in a known distribution of the $^2P_{3/2}$ and $^2P_{1/2}$ states, by photoionization in the photoionization region. The Ar^+ ions are then deflected through the RF octopole ion-guide reaction gas cell by adjusting the voltages of grids 1, 2, and 3 at the collision region. By measuring the intensities of Ar^+ and H_2^+ formed in the RF octopole gas cell using the back horizontal QMF, the values for $nl\sigma_{3/2}(H_2^+)$ and $nl\sigma_{1/2}(H_2^+)$ are calculated. At $E_{1ab} = 10\,\text{eV}$, the value for $nl\sigma_{1/2}(H_2^+)$ is approximately six times that for $nl\sigma_{3/2}(H_2^+)$. The measurements of $nl\sigma_{3/2}(H_2^+)$, $nl\sigma_{1/2}(H_2^+)$, and

$nl\sigma_m(H_2^+, \lambda)$ allow the values for $X_{3/2}(E_{c.m.}, \lambda)$ and $X_{1/2}(E_{c.m.}, \lambda)$ to be determined by solving Eqs. 44 and 45.

At $\lambda = 791.5$ Å, the value for $X_{1/2}(E_{c.m.}, \lambda)$ is equal to the fraction of $Ar^+(^2P_{1/2})$ formed by reaction 26 with $v' = 0$, $X_{0 \to 1/2}(E_{c.m.})$. At $\lambda = 781, 776$, and 771.5 Å, the total cross section $\sigma_{1/2}(E_{c.m.}, \lambda)$ for producing $Ar^+(^2P_{1/2})$ at a given $E_{c.m.}$ can be expressed as

$$\sigma_{1/2}(E_{c.m.}, \lambda) = X_0(\lambda)\sigma_{0 \to 1/2}(E_{c.m.}) + X_1(\lambda)\sigma_{1 \to 1/2}(E_{c.m.}) \quad (46)$$

where $\sigma_{v' \to 1/2}$, $v' = 0$ or 1, are the total state-to-state cross section for reaction 26. By substituting the relations

$$\frac{\sigma_{1/2}(E_{c.m.}, \lambda)}{\sigma_m(E_{c.m.}, \lambda)} = X_{1/2}(E_{c.m.}, \lambda)$$

$$\frac{\sigma_{0 \to 1/2}(E_{c.m.})}{\sigma_0(E_{c.m.})} = X_{0 \to 1/2}(E_{c.m.})$$

$$\frac{\sigma_{1 \to 1/2}(E_{c.m.})}{\sigma_1(E_{c.m.})} = X_{1 \to 1/2}(E_{c.m.}) \quad (47)$$

into Eq. 46, we obtain

$$X_{1/2}(E_{c.m.}, \lambda) = X_0(\lambda)X_{0 \to 1/2}(E_{c.m.})\left(\frac{\sigma_0(E_{c.m.})}{\sigma_m(E_{c.m.}, \lambda)}\right)$$
$$+ X_1(\lambda)X_{1 \to 1/2}(E_{c.m.})\left(\frac{\sigma_1(E_{c.m.})}{\sigma_m(E_{c.m.}, \lambda)}\right) \quad (48)$$

from which the value for $X_{1 \to 1/2}(E_{c.m.})$ can be calculated.

The measurements at $\lambda = 768.5$ and 761 Å involve three vibrational states. The equation used to calculate $X_{2 \to 1/2}$ is

$$X_{1/2}(E_{c.m.}, \lambda) = X_0(\lambda)X_{0 \to 1/2}(E_{c.m.})\left(\frac{\sigma_0(E_{c.m.})}{\sigma_m(E_{c.m.}, \lambda)}\right)$$
$$+ X_1(\lambda)X_{1 \to 1/2}(E_{c.m.})\left(\frac{\sigma_1(E_{c.m.})}{\sigma_m(E_{c.m.}, \lambda)}\right)$$
$$+ X_2(\lambda)X_{2 \to 1/2}(E_{c.m.})\left(\frac{\sigma_2(E_{c.m.})}{\sigma_m(E_{c.m.}, \lambda)}\right) \quad (49)$$

The measured values for $X_{v' \to 1/2}$, $v' = 0$ and 1, plotted as a function of $E_{c.m.}$, are shown in Fig. 28. The values for $X_{v' \to 3/2}$ can be calculated as $1 - X_{v' \to 1/2}$. The values for $X_{0 \to 1/2}$ and $X_{1 \to 1/2}$ approach zero at $E_{c.m.} \leqslant 4$ and 8 eV, respectively. The value for $X_{0 \to 1/2}$ is found to increase as $E_{c.m.}$ is increased. The value for $X_{1 \to 1/2}$ increases from ~ 0.02 at $E_{c.m.} = 4$ eV to ~ 0.14 at $E_{c.m.} = 20$ eV and

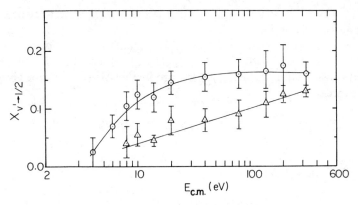

Fig. 28. Values for $X_{v' \to 1/2}$ ($\triangle$, $v' = 0$ and $\bigcirc$, $v' = 1$) plotted as a function of $E_{c.m.}$. Reprinted from Ref. [65] with permission of American Institute of Physics.

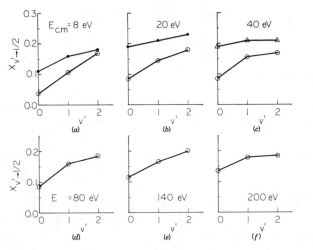

Fig. 29. Values for $X_{v' \to 1/2}$ at $E_{c.m.} = 8, 20, 40, 80, 140,$ and $200\,eV$ plotted as a function of the vibrational quantum number of v'. Experiment: $\bigcirc$, Ref. [65]. Theory: $\triangle$, Ref. [196]; $\bullet$, Ref. [175]. Reprinted Ref. [65] with permission of American Institute of Physics.

remains nearly constant for $E_{c.m.} = 20$–$320\,eV$. In this $E_{c.m.}$ range, the value for $X_{1 \to 1/2}$ is greater than that for $X_{0 \to 1/2}$.

The values for $X_{2 \to 1/2}$ have been determined at selected $E_{c.m.}$. The vibrational dependences for $X_{v' \to 1/2}$, $v' < 0$–2, at $E_{c.m.} = 8, 20, 40, 80, 140,$ and $200\,eV$ are plotted in Fig. 29. For all these collision energies the values for $X_{v' \to 1/2}$ are in the order $X_{0 \to 1/2} < X_{1 \to 1/2} < X_{2 \to 1/2}$. As $E_{c.m.}$ is increased, the differences between the values become smaller.

The theoretical values of Parlant and Gislason [175] for $X_{v' \to 1/2}$, $v' = 0$–2, at $E_{c.m.} = 8$ and $20\,eV$ and those obtained by Spalburg and Gislason [196] at $E_{c.m.} = 41.2\,eV$ are compared to the experimental values in Figs. 29a–29c. The theoretical calculations correctly predict that product Ar$^+$ ions are formed predominantly ($\gtrsim 80\%$) in the $^2P_{3/2}$ state.

This observation supports the theoretical interpretations [175, 196] that energy resonance and vibronic coupling are the important factors that govern the electron transfer product branching ratios for the strongly coupled [N$_2$ + Ar]$^+$ system. Since the Franck-Condon factor consideration is not an important criterion in predicting the final product state distribution, $\Delta v' = 0$ processes are not favored. Instead the electron transfer of the [N$_2$ + Ar]$^+$ system has a strong preference for transitions with $|\Delta v'| = 1$. The predicted variation of the vibrational dependence as a function of $E_{c.m.}$ is also in qualitative agreement with the experimental findings. However, the theoretical values for $X_{v \to 1/2}$ $v' = 0$–2, are higher than the experimental results. For $X_{0 \to 1/2}$, the theoretical values are found to be nearly twice the experimental values.

Most recently, the absolute value for $\sigma_{v'}$, $v' = 0$ and 1, have been measured by Shao et al. [62] using the tandem photoionization mass spectrometer. These measurements, together with the relative values for $\sigma_{v'}$, $v' = 0$–2, obtained by Liao et al. [65] allow the absolute values for σ_2 to be determined. The absolute values for $\sigma_{v'}$, $v' = 0$–2, at $E_{c.m} = 1.2$–140 eV determined by Shao et al. and Liao et al. are listed in Table 4. Since the values for σ_0 at $E_{c.m.} \leqslant 20$ eV have high uncertainties, the values for σ_2 at $E_{c.m.} \leqslant 20$ eV are calculated using only the known values of σ_2/σ_1 and σ_1. Absolute values for σ_2 at $E_{c.m.} = 20$–140 eV determined by multiplying the values of σ_2/σ_0 with the corresponding values for σ_0 are in accord with those calculated using the values for σ_2/σ_1 and σ_1. The TESICO results obtained by Govers et al. [50] and the theoretical cross sections are also included in Table 4.

Using the values of $X_{v' \to J}$ and absolute values for $\sigma_{v'}$ listed in Table 4, absolute partial state-to-state cross sections, $\sigma_{v' \to J}$, $v' = 0$–2, have been determined and are compared to the theoretical results in Table 5. The experimental observations and the theoretical predictions of Parlant and Gislason at $E_{c.m.} = 8$ and 20 eV are in reasonable agreement.

From the consideration of microscopic reversibility [231], we have

$$E_1 g_1 \sigma_{v' \to Jv}(E_1) = E_2 g_2 \sigma_{Jv \to v'}(E_2) \tag{50}$$

Here E_1 is the asymptotic reactant relative kinetic energy, E_2 is the asymptotic product relative kinetic energy, g_1 is the degeneracy for the reactant state, g_2 is the degeneracy for the product state, and $\sigma_{v' \to Jv}(E_1)$ and $\sigma_{Jv \to v'}(E_2)$ are the state-to-state cross sections for reaction 26 and its reverse at E_1 and E_2, respectively. If the degeneracies due to the rotational states of N$_2^+(\tilde{X})$ and N$_2(X)$ are ignored, the values of g_1 and g_2 for reaction 26 are equal to the degeneracy of the electronic state ($^2\Sigma_g^+$) of N$_2^+$ and that of the spin-orbit state (^2P_J) of Ar$^+$, respectively.

Absolute values for $\sigma_{3/2 0 \to 1}$ at selected $E_{c.m.}$ have been determined with accuracy by Liao et al. [61]. According to the theoretical studies, the values for $\sigma_{0 \to 3/2}$ and $\sigma_{1 \to 3/2}$ are overwhelmingly dominated by $\sigma_{0 \to 3/2 0}$ and $\sigma_{1 \to 3/2 0}$,

Table 4. Absolute Vibrational-State-Selected Total Cross Sections, $\sigma_{v'}$, $v' = 0$–2, at $E_{c.m.} = 1.2$–140 eV for the reactions $N_2^+(\tilde{X}, v' = 0$–$2) + Ar(^1S_0) \to N_2 + Ar^+$

			$\sigma_{v'}(\text{Å}^2)$			
		Experiment[a,b]			Theory[c,d]	
$E_{c.m.}$(eV)	$v' = 0$	$v' = 1$	$v' = 2$[e]	$v' = 0$	$v' = 1$	$v' = 2$
1.2	0.28 ± 0.08	26.3 ± 4.2	22.9 ± 4.3	0.00	17.46	14.41
4.1	1.01 ± 0.25	21.9 ± 3.4	21.7 ± 4.0	0.01	24.31	19.90
5.8	1.41 ± 0.35	21.7 ± 3.3	23.9 ± 4.3	—	—	—
8	1.59 ± 0.40	18.8 ± 2.8	22.7 ± 4.1	0.1[g]	26.0[g]	30.8[g]
	(≤0.9)	(18.7 ± 1.6)	(23.1 ± 1.9)	(0.9)	(22.9)	(28.5)
14	1.93 ± 0.40	16.6 ± 2.4	23.1 ± 4.1	—	—	—
	(≤0.9)	(15.6 ± 1.2)	(24.2 ± 2.8)			
20	2.17 ± 0.43	14.3 ± 2.1	20.6 ± 3.6	1.0[g]	25.0[g]	29.6[g]
		19.7 ± 4.4[f]	28.4 ± 6.3[f]	(3.2)	(20.4)	(28.0)
	(1.6 ± 0.2)	(14.3 ± 2.4)	(17.8 ± 2.9)			
40	4.3 ± 0.4	11.6 ± 1.8	16.4 ± 3.0	3.0[h]	21.2[h]	30.8[h]
		13.1 ± 1.8[f]	18.5 ± 2.6[f]			
80	8.8 ± 0.67	11.5 ± 1.5	15.5 ± 2.5	—	—	—
		13.6 ± 1.7[f]	18.3 ± 2.3[f]			
140	14.9 ± 1.1	15.6 ± 1.5	14.3 ± 2.1	—	—	—
		17.0 ± 2.1[f]	16.3 ± 2.0[f]			

[a]Reference [62].
[b]The values in parentheses are obtained from Ref. [50].
[c]Reference [196].
[d]The values in parentheses are obtained from Ref. [175].
[e]Values determined using the absolute values for σ_1 and values for σ_2/σ_1 obtained from Ref. [65].
[f]Values determined using the absolute values for σ_0 and values for σ_1/σ_0 obtained from Ref. [65].
[g]Values quoted in Ref. [175].
[h]Values calculated at $E_{c.m.} = 41.2$ eV.

respectively. Assuming $\sigma_{0 \to 3/2\,0} \simeq \sigma_{0 \to 3/2}$ and $\sigma_{1 \to 3/2\,0} \simeq \sigma_{1 \to 3/2}$, Eq. 50 becomes

$$\sigma_{3/2\,0 \to v'}(E_2) = \left(\frac{E_1}{2E_2}\right)\sigma_{v' \to 3/2}(E_1) \quad v' = 0 \text{ and } 1 \quad (51)$$

Using the values for $\sigma_{0 \to 3/2}$ and $\sigma_{1 \to 3/2}$ listed in Table 5, the values for $\sigma_{3/2\,0 \to v'}$, $v' = 0$ and 1 at $E_{c.m.} = 1$–41.2 eV are calculated. Table 6 compares the calculated and experimental values obtained by Liao et al. The predicted and experimetal values are in good agreement except at $E_{c.m.} = 10.3$ eV where the predicted value for $\sigma_{3/2\,0 \to 1}$ is slightly higher than the experimental value.

The theoretical calculation of Spalburg and Gislason [196] also reveals that at $E_{c.m.} = 8$–41.2 eV, $Ar^+(^2P_{3/2}) + N_2(X, v = 1)$ is the predominant (>95%) product channel in the reaction of $N_2^+(\tilde{X}, v' = 2) + Ar(^1S_0)$. Assuming $\sigma_{2 \to 3/2} \approx$

Table 5. Absolute Partial State-to-State Total Cross Sections, $\sigma_{v'\to J}$, $v' = 0$–2, at $E_{c.m.} = 1.2$–140 eV for the Reactions $N_2^+(\tilde{X}, v' = 0$–2$) + Ar({}^1S_0) \to N_2(X, v) + Ar^+({}^2P_J)$

	Experiment[a,b]						Theory[c,d]					
	$\sigma_{v'\to J}(\text{Å}^2)$											
$E_{c.m.}$(eV)	$\sigma_{0\to 3/2}$	$\sigma_{0\to 1/2}$	$\sigma_{1\to 3/2}$	$\sigma_{1\to 1/2}$	$\sigma_{2\to 3/2}$	$\sigma_{2\to 1/2}$	$\sigma_{0\to 3/2}$	$\sigma_{0\to 1/2}$	$\sigma_{1\to 3/2}$	$\sigma_{1\to 1/2}$	$\sigma_{2\to 3/2}$	$\sigma_{2\to 1/2}$
1.2	0.28 ± 0.08	0.0	26.3 ± 4.2	~0.0	—	—	0.00	0.00	17.37	0.09	9.06	5.35
4.1	1.01 ± 0.25	0.0	21.4 ± 3.4	0.5 ± 0.6	—	—	0.01	0.00	24.31	0.56	16.56	3.34
5.8	1.41 ± 0.35	0.0	20.2 ± 3.1	1.5 ± 0.5	—	—	—	—	—	—	—	—
8	1.53 ± 0.39	0.06 ± 0.04	16.8 ± 2.5	2.0 ± 0.6	18.8 ± 3.6	3.9 ± 1.3	(0.8)	(0.1)	(19.3)	(3.6)	(23.5)	(5.0)
10.3	1.61 ± 0.38[e]	0.09 ± 0.04[e]	15.8 ± 2.4[e]	2.3 ± 0.6[e]	18.6 ± 3.6[e]	4.0 ± 1.3[e]	0.15	0.01	24.86	1.89	25.2	3.94
14	1.84 ± 0.38	0.09 ± 0.04	14.6 ± 2.2	2.0 ± 0.5	19.1 ± 3.6	3.8 ± 1.4	—	—	—	—	—	—
20	2.00 ± 0.40	0.17 ± 0.06	12.2 ± 1.8 (16.8 ± 3.8)	2.1 ± 0.4 (2.9 ± 0.8)	16.9 ± 3.8 (23.3 ± 5.2)	3.7 ± 1.2 (5.1 ± 1.8)	(2.6)	(0.6)	(16.1)	(4.3)	(21.5)	(6.5)
40	4.0 ± 0.4	0.4 ± 0.1	9.8 ± 1.6 (11.2 ± 1.6)	1.8 ± 0.4 (2.0 ± 0.4)	13.6 ± 2.6 (15.3 ± 2.3)	2.8 ± 0.9 (3.1 ± 1.0)	2.40[f]	0.57[f]	16.52[f]	4.63[f]	24.43[f]	6.40[f]
80	8.0 ± 0.6	0.8 ± 0.2	9.7 ± 1.3 (11.4 ± 1.6)	1.8 ± 0.4 (2.2 ± 0.4)	12.6 ± 2.2 (18.4 ± 2.1)	2.9 ± 0.9 (3.4 ± 1.0)	—	—	—	—	—	—
140	13.2 ± 1.0	1.7 ± 0.4	13.0 ± 1.4 (14.2 ± 1.9)	2.6 ± 0.6 (2.8 ± 0.7)	11.4 ± 1.8 (13.0 ± 1.8)	2.9 ± 0.8 (3.3 ± 0.9)	—	—	—	—	—	—

[a]Reference [62].
[b]The values in parentheses are obtained from values of σ_1 and σ_2 determined using the values for σ_0 of Ref. [62] and σ_1/σ_0 and σ_2/σ_0 obtained from Ref. [65].
[c]Reference [196].
[d]The values in parentheses are obtained from Ref. [175].
[e]Values estimated using the estimated values for $\sigma_{v'}$, $v' = 0$–2 at $E_{c.m.} = 10.3$ eV that are interpolated using the $\sigma_{v'}$ values at $E_{c.m.} = 8$ and 14 eV.
[f]Values calculated at $E_{c.m.} = 41.2$ eV.

Table 6. Comparison of Experimental Values and Predicted Values[a] for $\sigma_{3/2+v'}$, $v' = 0$–1, at $E_{c.m.} = 1.2$–41.3 eV from the Consideration of Microscopic Reversibility

$E_{c.m.}$(eV)	$\sigma_{3/2\,0\to 0}(\text{Å}^2)$		$\sigma_{3/2\,0\to 1}(\text{Å}^2)$	
	Predicted	Experimental[b]	Predicted	Experimental[b]
1.2	0.16 ± 0.05[c]	$0.0 + 1.5$ -0.0	12.2 ± 2.0[d]	12.3 ± 2.2
4.1	0.53 ± 0.13[e]	$0.0 + 1.5$ -0.0	10.5 ± 1.7[f]	12.7 ± 2.2
10.3	0.82 ± 0.19	$0.0 + 1.3$ -0.0	7.9 ± 1.2	$12.8 + 0.4$ -2.0
41.2	2.0 ± 0.2	2.4 ± 1.1	4.9 ± 0.8 5.6 ± 0.8	7.2 ± 1.9

[a] Values calculated using Eq. 51 (see text).
[b] Reference [61].
[c] The predicted value at $E_{c.m.} = 1.04$ eV.
[d] The predicted value at $E_{c.m.} = 1.29$ eV.
[e] The predicted value at $E_{c.m.} = 3.94$ eV.
[f] The predicted value at $E_{c.m.} = 4.19$ eV.

$\sigma_{2\to 3/2\,1}$, the values for $\sigma_{3/2\,1\to 2}$ at $E_{c.m.} = 8$, 20, and 40 eV have been calculated and compared to the theoretical predictions in Table 7. The theoretical predictions are found to be higher than the calculated values.

Table 7. Comparison of Theoretical and Predicted Values[a] for $\sigma_{3/2\,1\to 2}$ at $E_{c.m.} = 8$, 20, and 40 eV from the Consideration of Microscopic Reversibility

$E_{c.m.}$(eV)	$\sigma_{3/2\,1\to 2}(\text{Å}^2)$	
	Predicted	Theoretical[b]
8	9.3 ± 1.8	12.2
20	8.4 ± 1.5 11.6 ± 2.7	—
40	6.8 ± 1.3 7.7 ± 1.2	11.6

[a] Values calculated using Eq. 50 (see text).
[b] Reference [196].

State-to-State Cross Sections for the Symmetric Electron Transfer Reaction $H_2^+(v_0') + H_2(v_0'' = 0) \to H_2(v') + H_2^+(v'')$

In this experiment [229], the electron transfer reactions

$$H_2^+(v'') + N_2 \rightarrow H_2 + N_2^+ \tag{52}$$

$$H_2^+(v'') + Ar \rightarrow H_2 + Ar^+ \tag{53}$$

at $E_{1ab} = 20\,eV$ have been used as the probing reactions to deduce the vibrational state distribution of product $H_2^+(v'')$. The theoretical calculation of Lee and DePristo [181] predicts that at $E_{c.m.} \leqslant 16\,eV$, an overwhelming majority of the products H_2^+ formed by reaction 19 are in the $v'' = 0, 1$, and 2 states. Assuming the $H_2^+(v'')$ ions consist only of H_2^+ in the $v'' = 0, 1$, and 2 states, and using similar arguments presented above, we arrive at the following relations:

$$X_0 + X_1 + X_2 = 1 \tag{54}$$

$$X_0 nl\sigma_0(N_2^+) + X_1 nl\sigma_1(N_2^+) + X_2 nl\sigma_2(N_2^+) = nl\sigma_m(N_2^+) \tag{55}$$

$$x_0 nl\sigma_0(Ar^+) + X_1 nl\sigma_1(Ar^+) + X_2 nl\sigma_2(Ar^+) = nl\sigma_m(Ar^+) \tag{56}$$

Here X_0, X_1, and X_2 are the fractions of product H_2^+ ions formed in the $v'' = 0$, 1, and 2 states, respectively; σ_0, σ_1, and σ_2 are the state-selected total cross sections for reactions 52 and 53 when H_2^+ ions are prepared in the 0, 1, and 2 vibrational states, respectively; and $\sigma_m(N_2^+)$ and $\sigma_m(Ar^+)$ represent the total cross sections for reactions 52 and 53 characteristic of the $H_2^+(v'')$ ions formed by reaction 19. The calculations of X_0, X_1, and X_2 need not involve the determination of absolute total cross sections provided $nl\sigma_0$, $nl\sigma_1$, and $nl\sigma_m$ are measured in the same gas cell with the same value of n.

Table 8 lists the values for $nl\sigma_0(N_2^+)$, $nl\sigma_1(N_2^+)$, $nl\sigma_2(N_2^+)$, $nl\sigma_0(Ar^+)$, $nl\sigma_1(Ar^+)$, and $nl\sigma_2(Ar^+)$ determined using the ion-beam–horizontal gas cell

Table 8. Values for $nl\sigma_{v'}(N_2^+)^a$ and $nl\sigma_{v'_0}(Ar^+)$,^{a} $v'_0 = 0$–4, Determined at $E_{1ab} = 20\,eV$

v'_0	$nl\sigma_{v_0}(N_2^+)^{b,c}$	$nl\sigma_{v_0}(Ar^+)^{b,c}$
0	0.01305	0.00780
1	0.03344	0.03010
2	0.02001	0.03566
3	0.01482	0.02580
4	0.01471	0.01838

$^a \sigma_{v_0}(N_2^+)$ and $\sigma_{v_0}(Ar^+)$ represent the total cross sections for the reactions, $H_2^+(v'_0) + N_2 \rightarrow H_2 + N_2^+$ and $H_2^+(v'_0) + Ar \rightarrow H_2 + Ar^+$, respectively.
^{b} The standard deviations due to counting statistics are $\leqslant 3\%$.
$^c nl\sigma_3(N_2^+)$, $nl\sigma_4(N_2^+)$, $nl\sigma_3(Ar^+)$, and $nl\sigma_4(Ar^+)$ are not used in the calculations.

Table 9. Values for $nl\sigma_m(N_2^+)^a$ and $nl\sigma_m(Ar^+)^a$ determined at $E_{1ab} = 20\,eV^b$

$v_0'^c$	$E_{c.m.}(eV)^d$	$nl\sigma_m(N_2^+)^e$	$nl\sigma_m(Ar^+)^e$
0	2	0.01282	0.00860
	4	0.01430	0.01020
	6	0.01509	0.01066
	8	0.01529	0.01100
	12	0.01611	0.01175
	16	0.01659	0.01187
1	2	0.02916	0.02542
	4	0.02711	0.02510
	6	0.02505	0.02403
	8	0.02524	0.02297
	12	0.02434	0.02154
	16	0.02250	0.02032

[a] $\sigma_m(N_2^+)$ and $\sigma_m(Ar^+)$ are the total cross sections for the reactions $H_2^+(v'') + N_2 \to H_2 + N_2^+$ and $H_2^+(v'') + Ar \to H_2 + Ar^+$, respectively.
[b] E_{1ab} is defined to be the difference in potential between the collision region and the horizontal gas cell.
[c] Vibrational state of the reactant H_2^+ ions.
[d] Center-of-mass collisional energies of the reaction $H_2^+(v_0') + H_2 \to H_2(v') + H_2^+(v'')$.
[e] The standard deviations due to counting statistics are $\leq 3\%$.

arrangement. The horizontal gas cell used in this study [229] has a design different from the RF octopole ion-guide reaction gas cell [65]. The large difference between $nl\sigma_0$, $nl\sigma_1$ and $nl\sigma_2$ and the different trends observed in the vibrational dependences of $nl\sigma_{v_0'}(N_2^+)$ and $nl\sigma_{v_0'}(Ar^+)$ provide a sensitive detection for the vibrational state distribution of $H_2^+(v'')$ formed by reaction 19. The measured values for $nl\sigma_m(N_2^+)$ and $nl\sigma_m(Ar^+)$ are listed in Table 9.

Using the values listed in Tables 8 and 9, the fractions of product H_2^+ formed in $v'' = 0$, 1, and 2 have been calculated by solving appropriate 3×3 linear equations. The calculated fractions are listed in Table 10. The theoretical results of Lee and DePristo at $E_{c.m.} = 8$ and $16\,eV$ are also included in the table.

When the reactant H_2^+ ions are in $v_0' = 0$, nearly all the electron transfer product H_2^+ ions formed at $E_{c.m.} = 2\,eV$ are in the $v'' = 0$ state, indicating that resonance electron transfer is the dominant process. As $E_{c.m.}$ changes from 2 to $16\,eV$, the value for X_1 increases steadily from ~ 0.0 to 0.17. The fractions of product H_2^+ observed in the $v'' = 2$ state in this energy range are small. The theoretical values [181] for X_0, X_1, and X_2 at $E_{c.m.} = 8$ and $16\,eV$ with $v_0' = 0$ are in fair agreement with the experimental findings.

For the reactant H_2^+ ions prepared in $v_0' = 1$, the experimental results show that even at $E_{c.m.} = 2\,eV$ the inelastic relaxation channel forming $v'' = 0$ is

Table 10. Vibrational State Distributions of Product $H_2^+(v'')$ formed by the Electron Transfer Reaction $H_2^+(v_0' = 0 \text{ or } 1) + H_2(v_0'' = 0) \to H_2(v') + H_2^+(v'')$ in the center-of-mass ($E_{c.m.}$) collisional energy range of 2–16 eV

$E_{c.m.}$ (eV)	v_0'	$X_0^{a,b}$	$X_1^{a,b}$	$X_2^{a,b}$
2	0	1.00	0.00	0.00
	1	0.21	0.79	0.00
4	0	0.91	0.04	0.05
	1	0.25	0.66	0.09
6	0	0.88	0.09	0.03
	1	0.31	0.54	0.15
8	0	0.86 (0.92)	0.10 (0.07)	0.04 (0.01)
	1	0.34 (0.17)	0.57 (0.76)	0.09 (0.07)
12	0	0.83	0.14	0.03
	1	0.40	0.53	0.07
16	0	0.82 (0.87)	0.17 (0.12)	0.01 (0.01)
	1	0.46 (0.39)	0.43 (0.50)	0.11 (0.11)

[a] X_0, X_1, and X_2 are the fractions of product H_2^+ formed in the $v'' = 0, 1,$ and 2 states. The uncertainties of these values are estimated to be 0.05.
[b] The values in the parentheses are theoretical values (Ref. [181]).

significant. The degree of relaxation also increases as $E_{c.m.}$ increases. At a given $E_{c.m.}$, the value for X_2 observed with $v_0' = 1$ is comparable to that for X_1 with $v_0' = 0$. The most interesting observation is that the extent of the relaxation channel is substantially greater than that of the excitation channel. Although calculations predict such a trend, the theory seems to underestimate the degree of inelastic relaxation. Better agreement between experimental and theoretical results is found at $E_{c.m.} = 16$ eV. At $E_{c.m.} = 8$ eV and $v_0' = 1$, the experimental value for X_0 is twice that of the theoretical value. However, the theoretical study of Lee and DePristo has not taken into account the formation of $H_3^+ + H$, which becomes more important at lower collision energies. This may explain the discrepancy observed between the experimental and theoretical results at $v_0' = 1$.

The efficient vibrational relaxation of $H_2^+(v_0' = 1)$ observed in this study is consistent with the interpretation that electron transfer mainly involves the long-range electron jump mechanism by which the vibrational energy of the reactant H_2^+ ion can be efficiently distributed between the products H_2^+ and H_2. The cross sections for inelastic vibrational excitation channels, which are necessarily governed by short-range collisions, are expected to be smaller than vibrational relaxation channels. This expectation is also in accordance with the experimental observation.

6.2 Triple-Quadrupole Double-Octopole Photoionization Apparatus [62, 63]

6.2.1 Experimental Arrangement and Procedures

The triple-quadrupole double-octopole photoionization apparatus is designed for state-to-state excitation and relaxation studies of atomic and simple molecular ions. It has been developed from the tandem photoionization mass spectrometer [57, 60] by adding an RF octopole reaction gas cell and a QMF in series with the original tandem QMF system. Figure 30 shows the cross-sectional view of the apparatus. It consists essentially of a photoionization ion source, three QMFs, two RF octopole ion-guide reaction gas cells, and an ion detector.

The reactant A^+ ions in a pure internal state, or a known distribution of A_0^+ and A^{+*}, are prepared by photoionization in the ion source. Here A_0^+ and A^{+*} represent the reactant ions in the ground and excited states, respectively. After selection by the first QMF (4), the reactant A_0^+ and/or A^{+*} ions enter the first RF octopole reaction gas cell and react with neutral B molecules according to the reactions

$$A^{+*} + B \xrightarrow{\sigma_r^*} C^+ + D \tag{57}$$

$$A_0^+ + B \xrightarrow{\sigma_r} E^+ + F \tag{58}$$

$$A^{+*} + B \xrightarrow{\sigma_R} A_0^+ + B \tag{59}$$

$$A_0^+ + B \xrightarrow{\sigma_E} A^{+*} + B \tag{60}$$

Reactions 57 and 58 represent the reactive channels for the reactant in the excited and ground states. The reactive channels include the electron transfer reactions. Reactions 59 and 60 are the relaxation and excitation channels. The excitation of A^{+*} is not included in the reaction scheme. In many cases, it is possible to eliminate this process by keeping a sufficiently low value of $E_{c.m.}$. After the reactions, A^{+*} and A_0^+ ions are further selected by the second QMF (11) and react with a probing gas M in the second RF octopole reaction gas cell (13). The probing electron transfer reactions are

$$A_0^+ + M \xrightarrow{\sigma_{pr}} M^+ + A \tag{61}$$

$$A^{+*} + M \xrightarrow{\sigma_{pr}^*} M^+ + A \tag{62}$$

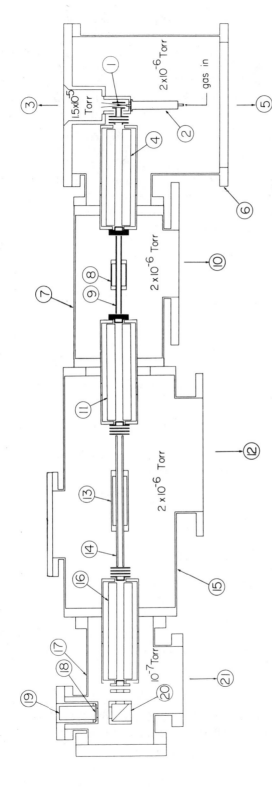

Fig. 30. Cross-sectional view of the triple-quadrupole double-octopole photoionization apparatus. 1 Photoionization region; 2, nozzle; 3, to freon-trapped 6-in. diffusion pump (DP); 4, the first quadrupole mass filter (QMF); 5, to liquid nitrogen (LN$_2$)-trapped 6-in. DP; 6, the first QMF chamber; 7, the first radio frequency (RF) octopole ion-guide chamber; 8, the first reaction gas cell; 9, the first RF octopole ion guide; 10, to LN$_2$-trapped 6-in. DP; 11, the second QMF; 12, to LN$_2$-trapped 4-in. DP; 13, the second reaction gas cell; 14, the second RF octopole ion guide; 15, the second RF octopole ion-guide chamber; 16, the third QMF; 17, detector chamber; 18, plastic scintillator window; 19, photomultiplier tube; 20, aluminum ion target; 21, to LN$_2$-trapped 2-in. DP. Reprinted from Ref. [62] with permission of American Institute of Physics.

The cross sections for the electron transfer reaction (σ_m) characteristic of the mixture of A_0^+ and A^{+*} resulting from the collision of A^+ and B in the first gas cell are measured by the third QMF. Here we assume that the state-selected cross sections, σ_r^*, σ_r, σ_{pr}^*, and σ_{pr}^* have been measured in separate experiments.

The above experimental procedures and proper data analyses [192] can lead to the determination of σ_R and σ_E.

Case 1. Measurement of σ_E with Only Reaction 58 and 60 Involved. In case 1 the reactant A^+ ions are prepared in the pure ground state. After collisions in the first RF octopole ion-guide gas cell, the fractions of A^+ ions in the ground and excited states become X'_g and X'_e, respectively. The probing gas experiment allows the determination of the values for X'_g and X'_e by solving Eqs. 63 and 64:

$$X'_g + X'_e = 1 \tag{63}$$

$$X'_g \sigma_{pr} + X'_e \sigma_{pr}^* = \sigma_m \tag{64}$$

Furthermore, we have the relation

$$I = I_0 \exp(-nl\sigma_r) \tag{65}$$

$$i(A^+) = I_0 \exp[-nl(\sigma_r + \sigma_E)] \tag{66}$$

where I_0 is the intensity of A^+ observed when the first and second gas cells are empty; I is the intensity of A^+ after B is introduced into the first RF octopole gas cell; n is the number density of B in the first gas cell; and l is the length of the first gas cell. Combining Eqs. 65 and 66 gives

$$X'_g = \exp(-nl\sigma_E)$$

or

$$\sigma_E = -\left(\frac{1}{nl}\right) \ln(X'_g) \tag{67}$$

$$\approx \frac{X'_e}{nl} \tag{68}$$

Equations 67 and 68 show that the absolute value for σ_E can be detemined by the calculation of X'_g or X'_e from Eqs. 63 and 64. Equation 68 is valid for thin target conditions.

Case 2. Measurement of σ_R with Only Reactions 57 and 59 Involved. When the reactant ions are prepared purely in the excited state, we only need to consider reactions 57 and 59. Following the same arguments as used in case 1, the absolute value for σ_R can be obtained by the equation

$$\sigma_R = -\left(\frac{1}{nl}\right)\ln(X'_g) \qquad (69)$$

Case 3. Measurement of σ_R with all the Reactions 57–60 Involved. When the reactant A^+ ions are formed in a known distribution of the ground and excited state, we have to consider all the reactions 57–60. Here, we assume the intensities for A_0^+ and A^{+*} before entering the first gas cell are $i_0(A^+)$ and $i_0(A^{+*})$, respectively. After the reactions in the first gas cell, the intensities for A^+ and A^{+*} become $i(A^+)$ and $i(A^{+*})$.

We have

$$i(A^+) = i_0(A^+)\exp[-(\sigma_r + \sigma_E)nl] + i_0(A^{+*})[1 - \exp(-\sigma_R nl)] \qquad (70)$$

Dividing Eq. 70 by I yields

$$X'_g = X_g\left(\frac{I_0}{I}\right)\exp[-(\sigma_r + \sigma_E)nl] + X_e\left(\frac{I_0}{I}\right)[1 - \exp(-\sigma_R nl)] \qquad (71)$$

By definition,

$$I = i(A^+) + i(A^{+*})$$
$$= X_g I_0 \exp(-\sigma_r nl) + X_e I_0 \exp(-\sigma_r^* nl) \qquad (72)$$

where $X_g = i_0(A^+)/I_0$ and $X_e = i_0(A^{+*})/I_0$. Substituting Eq. 72 into 71 results in

$$X'_g = \frac{X_g \exp[-(\sigma_r + \sigma_E)nl] + X_e[1 - \exp(-\sigma_R nl)]}{X_g \exp(-\sigma_r nl) + X_e \exp(-\sigma_r^* nl)} \qquad (73)$$

Equation 73 can be rearranged to give Eq. 74 from which the absolute value for σ_R can be calculated:

$$\sigma_R = -\left(\frac{1}{nl}\right)\ln\Big[1 - \left(\frac{1}{X_e}\right)\{X'_g[X_g\exp(-\sigma_r nl) + X_e\exp(-\sigma_r^* nl)]$$
$$- X_g\exp[-(\sigma_r + \sigma_E)nl]\}\Big] \qquad (74)$$

If the excitation of Ar^{+*} is included in the reaction scheme, it becomes a three-state problem. In order to measure σ_R in a three state case, it is necessary to use two probing gases to probe the final state distribution of A^+.

6.2.2 Experimental Results

The triple-quadrupole double-octopole apparatus has been used to measure the absolute spin-orbit-state excitation cross sections, $\sigma_{3/2 \to 1/2}$, for the reactions 41 and 42. Reaction 21 at $E_{lab} = 10\,eV$ has been used as the probing reaction. The value for E_{lab} of the probing reaction is determined by the potential difference between the photoionization region and the second reaction gas cell. For the direct excitation process, such as reactions 41 and 42, at sufficiently high E_{lab}, the E_{lab} for product $Ar^+(^2P_{1/2})$ ions are expected to be only slightly different from those of $Ar^+(^2P_{3/2})$, whereas product $Ar^+(^2P_J)$ or $N_2^+(\tilde{X}, v')$ ions resulting from the electron transfer processes

$$Ar^+(^2P_{3/2}) + Ar(^1S_0) \to Ar(^1S_0) + Ar^+(^2P_J) \tag{75}$$

$$Ar^+(^2P_{3/2}) + N_2(X, v=0) \to Ar(^1S_0) + N_2^+(\tilde{X}, v') \tag{76}$$

will have low velocities, close to thermal velocities. For an $E_{lab} > 10\,eV$ for reactions 41 and 42, slow electron transfer product ions formed in the first gas cell cannot enter the second gas cell because the second gas cell is biased at a potential positive with respect to the first gas cell. This is an important experimental feature for the isolation of the electron transfer channel from the direct channel in a symmetric system such as $Ar^+ + Ar$.

The most important condition for the success of this experiment is the use of the RF octopole ion guides that make possible the collection of nearly all product $Ar^+(^2P_{3/2, 1/2})$ ions of reactions 41 and 42. When N_2 is introduced into the first gas cell, the attenuation of reactant $Ar^+(^2P_{3/2})$ ions is found to be consistent with the estimate calculated using the known electron transfer cross section for reaction 76.

Table 11 compares the absolute values for $\sigma_{3/2 \to 1/2}$ of reaction 41 measured in this study at $E_{lab} = 60$, 130, and 200 eV with experimental results obtained by Itoh et al. [232]. Both experiments show that $\sigma_{3/2 \to 1/2}$ of reaction 41 increases as E_{lab} is increased, an observation consistent with the theoretical prediction [233]. In the experiment of Itoh et al., the value for $\sigma_{3/2 \to 1/2}$ is deduced by

Table 11. Absolute Spin-Orbit-State Excitation Cross Sections, $\sigma_{3/2 \to 1/2}$, for the Reaction $Ar^+(^2P_{3/2}) + Ar(^1S_0) \to Ar^+(^2P_{1/2}) + Ar(^1S_0)$ at $E_{lab} = 60$, 130, and 200 eV

	E_{lab}(eV)		
	60	130	200
$\sigma_{3/2 \to 1/2}(\text{Å}^2)^{a,b}$	0.61 ± 0.06 (0.31)	1.58 ± 0.28 (1.1)	2.14 ± 0.30 (2.0)

[a] Reference [63].
[b] The values in parentheses are obtained from Ref. [232].

observing the energy loss of Ar^+ along the Ar^+ beam direction with an acceptance angle of 0.45°. At low E_{lab}, where the angular spread of inelastic scattered $Ar^+(^2P_{1/2})$ ions is $> 0.45°$, the values for $\sigma_{3/2 \to 1/2}$ obtained previously are likely to be lower limits. As E_{lab} is increased and the angular spread of $Ar^+(^2P_{1/2})$ ions becomes narrower, the value measured in the energy loss experiment is expected to be closer to the true value for $\sigma_{3/2 \to 1/2}$. This expectation is in accord with the observations that the values for $\sigma_{3/2 \to 1/2}$ determined here are higher than those of Itoh et al. and that the difference between the two measurements decreases as E_{lab} is increased. Although the values for $\sigma_{3/2 \to 1/2}$ obtained in this experiment are higher, they are still lower than the theoretical predictions. As pointed out previously [56], the theoretical predictions will be improved if accurate ab initio potential energy curves for Ar_2^+ are used in the calculation.

Table 12 lists the values for $X_{3/2}$ and $\sigma_{3/2 \to 1/2}$ at $E_{lab} = 25$ eV for reaction 42 obtained in the N_2 gas cell pressure (P) range of $2.38-4.65 \times 10^{-4}$ torr. The average values for $\sigma_{3/2 \to 1/2}$ at $E_{lab} = 25$ eV is 2.18 ± 0.25 Å^2 and is substantially greater than the excitation cross section for reaction 41 at $E_{lab} = 60$ eV. The experimental value is in agreement with the calculated values obtained by Spalburg and Gislason [196] and Parlant and Gislason [175, 197]. The surprisingly large value for $\sigma_{3/2 \to 1/2}$ for reaction 42 has been attributed to the mutual interactions of electron transfer $Ar + N_2^+$ states with the spin-orbit states of Ar^+.

7 CONCLUSIONS AND FUTURE DEVELOPMENTS

Since the pioneering state-selected experiments of Chupka and co-workers, important advances have been made in the application of photoionization

Table 12. Absolute Spin-Orbit-State Excitation Cross Sections, $\sigma_{3/2 \to 1/2}$, for the Reaction $Ar^+(^2P_{3/2}) + N_2(X, v = 0) \to Ar^+(^2P_{1/2}) + N_2(X, v)$ at $E_{lab} = 25$ eV

$P(10^{-4}$ torr$)^a$	$X_{3/2}^b$	$\sigma_{3/2 \to 1/2}$(Å^2)
2.38	0.9894	2.42
2.79	0.9865	2.53
3.31	0.9845	2.09
4.06	0.9877	1.34
4.65	0.9738	2.53
	$\langle \sigma_{3/2 \to 1/2} \rangle^c =$	2.18 ± 0.25
		$(1.98^d; 3^e)$

^{a}The N_2 pressure used in the first reaction gas cell.
^{b}The fraction of Ar^+ in the $^2P_{3/2}$ state.
^{c}Average value.
^{d}The theoretical value obtained from Ref. [196].
^{e}The theoretical value obtained from Refs. [175, 197].

methods to the studies of ion–molecule reaction dynamics. The TESICO technique has been established to be the most general method for state selecting a reactant ion and examining the reactivity of the ion with a neutral reactant. Due to the improvement in sensitivity by careful designs of experimental apparatuses, microscopic state-selected total cross sections can now be measured using the ion beam–gas cell or crossed ion–neutral beam arrangements. The crossed ion–neutral beam arrangement allows the measurement of cross sections at high kinetic energy resolution. The incorporation of the RF octopole ion-guide technique to an ion–molecule reaction photoionization apparatus enables the total collection of product ions formed in a state-selected ion–molecule reaction. An ideal ion–molecule reaction photoionization apparatus for state-selected total cross-section measurements should combine the TESICO, tandem mass spectrometry, crossed ion–neutral beam, and RF octopole ion-guide techniques. Such an apparatus will certainly be developed in the near future.

The electron transfer detection method overcomes the severe constraint of low product ion intensities and allows state-to-state cross sections for several simple electron transfer reactions to be measured with high accuracy. The comparisons of experimental and theoretical cross sections of these systems have provided valuable insights into the dynamics of electron transfer. The novel combination of the electron transfer detection method with tandem photoionization mass spectrometric techniques has been shown to be capable of providing absolute state-to-state relaxation and excitation cross sections for atomic and simple molecular ions. It is logical to incorporate the TESICO method in the crossed ion–neutral beam photoionization apparatus and the triple-quadrupole double-octopole photoionization apparatus. With the inclusion of the TESICO technique, these apparatuses can be used for the studies of many state-to-state ion–molecule reactions involving more complicated molecular ions.

The resolution achieved by the current photoelectron energy analyzers are insufficient for the selection of rotational states except those of H_2^+. State-selected and state-to-state studies involving rotational states will require the use of high-resolution laser techniques. Based on our present understanding of photoionization mechanisms, single photon ionization by VUV lasers holds more promise than MPI as a general technique in state-selected and state-to state ion chemistry. The pulsed laser photoionization source, TOF–PES, and TOF mass spectrometry should be the natural combination in a future ion–molecule reaction laser photoionization apparatus.

The techniques for total cross-section measurements of state-selected ion–molecule reactions have reached a mature state. Techniques for state-to-state total cross-section measurements of rearrangement ion–molecule reactions remain to be developed. The development in this area will rely on the future improvement of high-power VUV lasers. When state-selected reactant ions can be prepared in abundance by VUV lasers in the future, state-selected and state-to-state differential scattering cross section measurements will become possible.

Parallel to the intense activities in VUV photophysics and photochemistry, studies of state-selected and state-to-state ion–molecule reactions by photoionization methods will continue to develop and become an exciting field in the next decade.

ACKNOWLEDGMENT

The author is grateful to Professor H. J. Svec for reading the manuscript.

REFERENCES

1. R. D. Levine and R. B. Bernstein, *Molecular Reaction Dynamics*, Oxford, New York, 1974.
2. P. R. Brooks and E. F. Hayes, Eds., *State-to-State Chemistry*, American Chemical Society, Washington, D.C., 1977.
3. R. B. Bernstein, *Chemical Dynamics Via Molecular Beam and Laser Techniques*, Oxford, New York, 1982.
4. R. N. Zare and P. J. Dagdigian, *Science*, **185**, 739 (1974).
5. J. L. Kinsey, *Ann. Rev. Phys. Chem.* **28**, 349 (1977).
6. M. A. A. Clyne and I. S. McDermid, *Adv. Chem. Phys.* **50**, 1 (1982).
7. R. Altkorn and R. N. Zare, *Ann. Rev. Phys. Chem.* **35**, 265 (1984).
8. F. F. Crim, *Ann. Rev. Phys. Chem.*, **35**, 657 (1984).
9. J. Berkowitz, *Photoabsorption, Photoionization, and Photoelectron Spectroscopy*, Academic Press, New York, 1979.
10. D. W. Turner, C. Baker, A. D. Baker, and C. R. Brundle, *Molecular Photoelectron Spectroscopy*, Wiley, New York, 1970.
11. J. W. Rabalais, *Principles of Ultraviolet Photoelectron Spectroscopy*, Wiley, New York, 1977.
12. T. A. Carlson, *Photoelectron and Auger Spectroscopy*, Plenum, New York; 1975.
13. C. R. Bundle and A. D. Baker, Eds., *Electron Spectroscopy*, Academic Press, New York, 1977.
14. D. A. Shirley, Ed., *Electron Spectroscopy*, North-Holland, Amsterdam, 1972.
15. H. M. Rosenstock, K. Draxl, B. W. Steiner, and J. T. Herron, *J. Phys. Chem. Ref. Data*, **6**, Suppl. 1 (1977).
16. R. D. Levin and S. G. Lias, *Ionization Potential and Appearance Potential Measurements, 1971–1981*, Natl. Stand. Ref. Data Ser., Natl. Bur. Stand., U.S., 1982, vol. 71.
17. P. M. Dehmer and W. A. Chupka, *J. Chem. Phys.*, **62**, 4525 (1976).
18. P. M. Dehmer and W. A. Chupka, *J. Chem. Phys.*, **67**, 2740 (1977).
19. W. A. Chupka, M. E. Russell, and K. Refaey, *J. Chem. Phys.*, **48**, 1518 (1968).
20. W. A. Chupka and M. E. Russell, *J. Chem. Phys.*, **49**, 5426 (1968).
21. W. A. Chupka and M. E. Russell, *J. Chem. Phys.*, **48**, 1527 (1968).
22. W. A. Chupka, in J. L. Franklin, Ed., *Ion–Molecule Reactions*, Plenum, New York, 1972, p. 33.
23. W. A. Chupka, in C. Sandorfy, P. J. Ausloos, and M. B. Robin, Eds., *Chemical Spectroscopy and Photochemistry in the Vacuum Ultraviolet*, NATO-Advanced Study Institutes Series C, Reidel, Boston, 1973, vol. 8, p. 433.
24. C. Kunz, Ed., *Synchrotron Radiaton: Techniques and Applications*, Springer-Verlag, Berlin, 1979.
25. F. Lamahni, Ed., *Photophysics and Photochemistry Above 6 eV*, Elsevier, Amsterdam, 1985.

26. S. Leach, in S. McGlynn, G. Findley, and R. Hueber, Eds., *Photophysics and Photochemistry in the Vacuum Ultraviolet* Reidel, Boston, 1985, p. 293.
27. J. L. Dehmer, D. Dill, and A. C. Parr, in S. McGlynn, G. Findley, and R. Hueber, Eds., *Photophysics and Photochemistry in the Vacuum Ultraviolet*, Reidel, Boston, 1985, p. 341.
28. J. L. Dehmer, A. C. Parr, and S. H. Southworth, in G. V. Marr, Ed., *Handbook on Synchrotron Radiation*, North-Holland, Amsterdam, vol. II 1987, p. 241.
29. J. B. West, A. C. Parr, B. E. Cole, D. L. Ederer, R. Stockbauer, and J. L. Dehmer, *J. Phys. B*, **13**, L105 (1980).
30. A. C. Parr, D. L. Ederer, J. B. West, D. M. P. Holland, and J. L. Dehmer, *J. Chem. Phys.*, **76**, 4349 (1982).
31. D. M. P. Holland and J. B. West, *Z. Phys. D*, **4**, 367 (1987).
32. D. M. P. Holland and J. B. West, *J. Phys. B*, **20**, 1479 (1987).
33. T. Baer, L. Squires, and A. S. Werner, *Chem. Phys.*, **6**, 325 (1974).
34. L. Squires and T. Baer, *J. Chem. Phys.*, **65**, 4001 (1976).
35. I. Koyano and K. Tanaka, *J. Chem. Phys.*, **72**, 4858 (1980).
36. F. M. Campbell, R. Browning, and C. J. Latimer, *J. Phys. B*, **13**, 4257 (1980).
37. K. Tanaka, J. Durup, T. Kato, and I. Koyano, *J. Chem. Phys.*, **74**, 5561 (1981).
38. K. Tanaka, T. Kato, and I. Koyano, *J. Chem. Phys.*, **75**, 4941 (1981).
39. F. M. Campbell, R. Browning, and C. J. Latimer, *J. Phys. B*, **14**, 3493 (1981).
40. F. M. Campbell, R. Browning, and C. J. Latimer, *J. Phys. B*, **14**, 1183 (1981).
41. C. J. Latimer and F. M. Campbell, *J. Phys. B*, **15**, 1765 (1982).
42. T. Kato, K. Tanaka, and I. Koyano, *J. Chem. Phys.*, **77**, 337 (1982).
43. T. Kato, K. Tanaka, and I. Koyano, *J. Chem. Phys.*, **77**, 837 (1982).
44. K. Tanaka, T. Kato, P. M. Guyon, and I. Koyano, *J. Chem. Phys.*, **79**, 4302 (1983).
45. T. Kato, K. Tanaka, and I. Koyano, *J. Chem. Phys.*, **79**, 5969 (1983).
46. T. Kato, *J. Chem. Phys.*, **80**, 6105 (1984).
47. T. Kato, K. Tanaka, and I. Koyano, *J. Chem. Phys.*, **81**, 5666 (1984).
48. D. van Pijkeren, E. Bottjes, J. van Eck, and A. Niehaus, *Chem. Phys.*, **91**, 293 (1984).
49. S. K. Cole, T. Baer, P.-M. Guyon, and T. R. Govers, *Chem. Phys. Lett.*, **109**, 285 (1984).
50. T. R. Govers, P.-M. Guyon, T. Baer, K. Cole, H. Fröhlich, and M. Lavollée, *Chem. Phys.*, **87**, 373 (1984).
51. K. Tanaka, T. Kato, and I. Koyano, *J. Chem. Phys.*, **84**, 750 (1986).
52. J. C. Light, *J. Chem. Phys.*, **41**, 586 (1964).
53. S. L. Anderson, F. A. Houle, D. Gerlich, and Y. T. Lee, *J. Chem. Phys.*, **75**, 2153 (1981).
54. C.-L. Liao, C.-X. Liao, and C. Y. Ng, *Chem. Phys. Lett.*, **103**, 418 (1984).
55. C.-L. Liao, C.-X. Liao, and C. Y. Ng, *J. Chem. Phys.*, **81**, 5672 (1984).
56. C.-L. Liao, C.-X. Liao, and C. Y. Ng, *J. Chem. Phys. Lett.*, **82**, 5489 (1985).
57. J.-D. Shao and C. Y. Ng, *Chem. Phys. Lett.*, **118**, 481 (1985).
58. R. J. S. Morrison, W. E. Conaway, and R. N. Zare, *Chem. Phys. Lett.*, **113**, 435 (1985).
59. R. J. S. Morrison, W. E. Conaway, T. Ebata, and R. N. Zare, *J. Chem. Phys.*, **84**, 5527 (1986).
60. J.-D. Shao and C. Y. Ng, *J. Chem. Phys.*, **84**, 4317 (1986).
61. C.-L. Liao, J.-D. Shao, R. Xu, G. D. Flesch, Y.-G. Li, and C. Y. Ng, *J. Chem. Phys.*, **85**, 3874 (1986).
62. J.-D. Shao, Y.-G. Li, G. D. Flesch, and C. Y. Ng, *J. Chem. Phys.*, **86**, 170 (1987).
63. J. D. Shao, Y.-G. Li, G. D. Flesch, and C. Y. Ng, *Chem. Phys. Lett.*, **132**, 58 (1986).
64. E. Teloy and D. Gerlich, *Chem. Phys.*, **4**, 417 (1974).

65. C.-L. Liao, R. Xu, and C. Y. Ng, *J. Chem. Phys.*, **85**, 7136 (1986).
66. P. Lambropoulous, *Adv. At. Mol. Phys.*, **12**, 87 (1976).
67. V. S. Letokhov, *Comments At. Mol. Phys.*, **8**, 39 (1978).
68. G. Mainfray, *Comments At. Mol. Phys.*, **9**, 87 (1980).
69. L. Zandee, R. B. Bernstein, and D. A. Lichtin, *J. Chem. Phys.*, **69**, 3427 (1978).
70. L. Zandee and R. B. Bernstein, *J. Chem. Phys.*, **70**, 2574 (1979).
71. L. Zandee and R. B. Bernstein, *J. Chem. Phys.*, **71**, 1359 (1979).
72. P. M. Johnson, *Acc. Chem. Res.*, **13**, 20 (1980).
73. P. M. Johnson, *Appl. Opt.*, **19**, 3920 (1980).
74. T. G. DiGuiseppe, J. W. Hudgens, and M. C. Lin, *J. Chem. Phys.*, **76**, 1982 (1982).
75. M. N. R. Ashfold, J. M. Bayley, and R. N. Dixon, *J. Chem. Phys.*, **79**, 4080 (1983).
76. P. Chen, W. A. Chupka, and S. D. Colson, *Chem. Phys. Lett.*, **121**, 405 (1985).
77. P. Chen, S. D. Colson, W. A. Chupka, and J. A. Berson, *J. Phys. Chem.*, **90**, 2319 (1986).
78. H. Reisler and C. Wittig, *Adv. Chem. Phys.*, **60**, 1 (1985).
79. S. L. Anderson, D. M. Rider, and R. N. Zare, *Chem. Phys. Lett.*, **93**, 11 (1982).
80. J. L. Durant, D. M. Rider, S. L. Anderson, F. D. Proch, and R. N. Zare, *J. Chem. Phys.*, **80**, 1817 (1984).
81. S. L. Anderson, G. Kubiak, and R. N. Zare, *Chem. Phys. Lett.*, **105**, 22 (1984).
82. S. L. Anderson, L. Goodman, K. Krogh-Jespersen, A. G. Ozkabak, R. N. Zare, and C. Zheng, *J. Chem. Phys.*, **82**, 5329 (1985).
83. A. Achiba, K. Sato, K. Shobatake, and K. Kimura, *J. Chem. Phys.*, **78**, 5474 (1983).
84. K. Kimura, *Adv. Chem. Phys.*, **60**, 161 (1985); *Int. Rev. Phys. Chem.*, **6**, 195 (1987).
85. J. T. Meek, S. R. Long, and J. P. Reilly, *J. Phys. Chem.*, **86**, 2809 (1982).
86. S. R. Long, J. T. Meek, and J. P. Reilly, *J. Chem. Phys.*, **79**, 3206 (1983).
87. J. A. Glownia, S. J. Riley, S. D. Colson, J. C. Miller, and R. N. Compton, *J. Chem. Phys.*, **77**, 68 (1982).
88. W. E. Conway, R. J. S. Morrison, and R. N. Zare, *Chem. Phys. Lett.*, **113**, 429 (1985).
89. J. C. Miller and R. N. Compton, *J. Chem. Phys.*, **75**, 22 (1981).
90. S. T. Pratt, J. L. Dehmer, and P. M. Dehmer, *Chem. Phys. Lett.*, **105**, 28 (1984).
91. S. T. Pratt, J. L. Dehmer, and P. M. Dehmer, *J. Chem. Phys.*, **80**, 1706 (1984).
92. T. Ebata and R. N. Zare, *Chem. Phys. Lett.*, **130**, 467 (1986).
93. S. C. Wallace, *Adv. Chem. Phys.*, **47**, 153 (1981) and references therein.
94. J. C. Miller and R. N. Compton, in S. McGlynn, G. Findley, and R. Hueber, Eds., *Photophysics and Photochemistry in the Vacuum Ultraviolet*, Reidel, Boston, 1985, p. 133; S. C. Wallace, *ibid.*, p.105; and references therein.
95. C. Y. Ng, *Adv. Chem. Phys.*, **52**, 263 (1983).
96. J. A. R. Samson, *Techniques of Vacuum Ultraviolet Spectroscopy*, Wiley, New York, 1967.
97. B. Brehm and E. von Puttkamer, *Z. Naturforsch. A*, **22**, 8 (1967).
98. M. E. Gellender and A. D. Baker, *Int. J. Mass Spectrom. Ion Phys.*, **17**, 1 (1975).
99. M. E. Gellender and A. D. Baker, in C. R. Brundle and A. D. Baker, Eds., *Electron Spectroscopy*, Academic Press, New York, 1977, vol. 1, p. 435.
100. J. H. D. Eland. *Int. J. Mass Spectrom. Ion Phys.*, **8**, 143 (1972).
101. R. Stockbauer, *J. Chem. Phys.*, **58**, 3800 (1973).
102. T. Baer, in M. T. Bowers, Ed., *Gas Phase Ion Chemistry*, Academic Press, New York, 1979. vol. 1, p. 153.
103. T. P. Murray and T. Baer, *Int. J. Mass Spectrom. Ion Phys.*, **30**, 165 (1979).

104. T. Baer, *Adv. Chem. Phys.*, **69**, 111 (1986).
105. J. Dannacher, H. M. Rosenstock, R. Buff, A. C. Parr, R. L. Stockbauer, R. Bombach, and J. P. Stadelmann, *Chem. Phys.*, **75**, 23 (1983).
106. D. Villarejo, R. R. Herm, and M. G. Inghram, *J. Chem. Phys.*, **46**, 4995 (1967).
107. T. Baer, W. B. Peatman, and E. W. Schlag, *Chem. Phys. Lett.*, **4**, 243 (1969).
108. R. Spohr, P. M. Guyon, W. A. Chupka, and J. Berkowitz, *Rev. Sci. Instrum.*, **42**, 1872 (1971).
109. W. B. Peatman, G. B. Kasting, and D. J. Wilson, *J. Electron Spectrosc. Relat. Phenom.*, **7**, 233 (1975).
110. W. B. Peatman, *Chem. Phys. Lett.*, **36**, 495 (1975).
111. W. B. Peatman, *J. Chem. Phys.*, **64**, 4093 (1976).
112. W. B. Peatman, *J. Chem. Phys.*, **64**, 4368 (1976).
113. P. M. Guyon, T. Baer, L. F. A. Ferreira, I. Nenner, A. Tabché-Fouhailé, R. Botter, and T. R. Govers, *J. Phys. B.*, **11**, L141 (1978).
114. T. Baer, P. M. Guyon, I. Nenner, T. R. Govers, A. Tabché-Fouhailé, R. Botter, L. F. A. Ferreira, *J. Chem. Phys.*, **70**, 1585 (1979).
115. P.-M. Guyon and T. R. Govers, to be published.
116. V. S. Letokov, *Nonlinear Laser Chemistry*, Springer, Berlin, 1983.
117. S. H. Lin, Y. Fujimura, H. J. Neusser, and E. W. Schlag, *Multiphoton Spectroscopy of Molecules*, Academic Press, New York, 1984.
118. P. Lambropoulos and S. J. Smith, *Multiphoton Processes*, Springer, Berlin, 1984.
119. K. Watanabe, *J. Chem. Phys.*, **22**, 1564 (1954).
120. P. C. Killgoar, Jr., G. E. Leroi, J. Berkowitz, and W. A. Chupka, *J. Chem. Phys.*, **58**, 803 (1973).
121. Y. Ono, S. H. Linn, H. F. Prest, C. Y. Ng, and E. Miescher, *J. Chem. Phys.*, **73**, 4855 (1980).
122. V. H. Dibeler and R. M. Reese, *J. Chem. Phys.*, **40**, 2034 (1964).
123. V. H. Dibeler and J. A. Walker, *Int. J. Mass Spectrom. Ion Phys.*, **11**, 49 (1973).
124. Y. Ono, E. A. Osuch, and C. Y. Ng, *J. Chem. Phys.*, **76**, 3905 (1982).
125. V. H. Dibeler, J. A. Walker, and H. M. Rosenstock, *J. Res. Nat. Bur. Stand*, **70A**, 459 (1966).
126. E. P. Wigner, *Phys. Rev.*, **73**, 1002 (1948).
127. S. Geltman, *Phys. Rev.*, **102**, 171 (1956).
128. C. R. Brundle and M. B. Robin, in F. Nachod and G. Zuckerman, Eds., *Determination of Organic Structures by Physical Methods*, Academic Press, New York, 1971, vol. 3, p. 1.
129. D. Reinke, R. Kraessig, and H. Baumgärtel, *Z. Naturforsch.*, **28A**, 1021 (1973).
130. K. V. Wood and J. W. Taylor, *Int. J. Mass Spectrom. Ion Phys.*, **30**, 307 (1979).
131. Y. Ono, S. H. Linn, W.-B. Tzeng, and C. Y. Ng, *J. Chem. Phys.*, **80**, 1482 (1984).
132. W.-B. Tzeng, Y. Ono, S. H. Linn, and C. Y. Ng, *J. Chem. Phys.*, **83**, 2803 (1985).
133. P. M. Dehmer and W. A. Chupka, *J. Chem. Phys.*, **65**, 2243 (1976).
134. W. A. Chupka and J. Berkowitz, *J. Chem. Phys.*, **51**, 4244 (1969).
135. M. Raoult and Ch. Jungen, *J. Chem. Phys.*, **74**, 3388 (1981).
136. Ch. Jungen and D. Dill, *J. Chem. Phys.*, **73**, 3338 (1980).
137. P. M. Dehmer and W. A. Chupka, Argonne National Laboratory Report ANL-77-65, 1977, p. 28.
138. A. G. Shenstone, *Phys. Rev.*, **38**, 873 (1931).
139. U. Fano, *Nuovo Cimento*, **12**, 156 (1935).
140. U. Fano, *Phys. Rev.*, **124**, 1866 (1961).
141. H. Beutler, *Z. Phys.*, **93**, 177 (1935).
142. U. Fano and J. W. Cooper, *Phys. Rev. A*, **137**, 1364 (1965).

143. F. H. Mies, *Phys. Rev.*, **175**, 164 (1968).
144. J. N. Bardsley, *Chem. Phys. Lett.*, **2**, 329 (1968).
145. A. L. Smith, *Philos. Trans. R. Soc. London*, **A268**, 169 (1970).
146. U. Fano and J. W. Cooper, *Rev. Mod. Phys.*, **40**, 441 (1968).
147. G. V. Marr, *Photoionization Processes in Gases*, Academic Press, New York, 1967.
148. C. H. Greene and Ch. Jungen, *Adv. At. Mol. Phys.*, **21**, 51 (1985).
149. P. M. Dehmer and W. A. Chupka, *J. Chem. Phys.*, **66**, 1972 (1977).
150. J. Berkowitz and W. A. Chupka, *J. Chem. Phys.*, **51**, 2341 (1969).
151. K. Tanaka, T. Kato, and I. Koyano, *Chem. Phys. Lett.*, **97**, 562 (1983).
152. D. M. P. Holland and J. B. West, *J. Phys. B*, submitted.
153. E. D. Poliakoff, M.-H. Ho, M. G. White, and G. E. Leroi, *Chem. Phys. Lett.*, **130**, 91 (1986).
154. R. S. Berry and S. E. Nielsen, *Phys. Rev. A*, **1**, 395 (1970).
155. J. B. West, A. C. Parr, B. E. Cole, D. L. Ederer, R. Stockbauer, and J. L. Dehmer, *J. Phys. B*, **13**, L105 (1980).
156. R. R. Lucchese and B. V. McKoy, *J. Phys. B*, **14**, L629 (1981).
157. V. H. Dibeler, J. A. Walker, K. E. McCulloh, and H. M. Rosenstock, *Int. J. Mass Spectrom. Ion Phys.*, **7**, 209 (1971).
158. V. H. Dibeler, J. A. Walker, and K. E. McCulloh, *J. Chem. Phys.*, **53**, 4414 (1970).
159. V. H. Dibeler and J. A. Walker, *J. Chem. Phys.*, **57**, 1007 (1967).
160. H. Oertel, H. Schenk, and H. Baumgärtel, *Chem. Phys.*, **46**, 251 (1980).
161. J. Dannacher and J. P. Stadelmann, *Chem. Phys.*, **48**, 79 (1980).
162. C. S. T. Cant, C. J. Danby, and J. H. D. Eland, *J. Chem. Soc. Faraday Trans. 2*, **71**, 1015 (1975).
163. I. Powis, P. I. Mansell, and C. J. Danby, *Int. J. Mass Spectrom. Ion Phys.*, **32**, 15 (1979).
164. C. Holzapfel, *Rev. Sci. Instrum.*, **45**, 894 (1974).
165. E. von Puttkamer, *Z. Naturforsch.*, **25A**, 1062 (1970).
166. A. Harvey, M-de-L. F. Monteiro, and R. I. Reed, *Int. J. Mass Spectrom. Ion Phys.*, **4**, 365 (1970).
167. Y. Achiba, K. Sato, K. Shobatake, and K. Kimura, *J. Chem. Phys.*, **77**, 2709 (1982).
168. G. Gioumousis and D. P. Stevenson, *J. Chem. Phys.*, **29**, 294 (1958).
169. L. P. Theard and W. T. Huntress, Jr., *J. Chem. Phys.*, **60**, 2840 (1974).
170. Data of Stine and Muckerman (1980). See J. T. Muckerman, in *Theoretical Chemistry*, D. Henderson, Ed., Academic, New York, 1981.
171. C. W. Eaker and G. C. Schatz, *J. Phys. Chem.*, **89**, 2612 (1985).
172. W. C. Wiley and I. H. McLaren, *Rev. Sci. Instrum.*, **26**, 1156 (1955).
173. P. Mahadevan and G. D. Magnuson, *Phys. Rev.*, **171**, 103 (1968).
174. D. C. Cartwright, *J. Chem. Phys.*, **58**, 178 (1973).
175. G. Parlant and E. A. Gislason, *Chem. Phys.*, **101**, 227 (1986).
176. E. F. Gurnee and J. L. Magee, *J. Chem. Phys.*, **26**, 1237 (1957).
177. J. J. Leventhal, T. F. Moran, and L. Friedman, *J. Chem. Phys.*, **46**, 4666 (1967).
178. D. R. Bates and R. H. G. Reid, *Proc. R. Soc. London Ser. A*, **310**, 1 (1969).
179. R. N. Stock and H. Newman, *J. Chem. Phys.*, **61**, 3852 (1974).
180. K. J. McCann, M. R. Flannery, J. V. Hornstein, and T. F. Moran, *J. Chem. Phys.*, **63**, 4998 (1975).
181. C.-Y. Lee and A. E. DePristo, *J. Chem. Phys.*, **80**, 1116 (1984).
182. S. K. Cole and A. E. DePristo, *J. Chem. Phys.*, **85**, 1389 (1986).
183. C. F. Barnett, J. A. Ray, E. Ricci, M. I. Wilker, E. W. McDaniel, E. W. Thomas, and H. B. Gilbody, Oak Ridge National Laboratory Report 5206, 1977, and references therein.

184. T. Turner, O. Dutuit, and Y. T. Lee, *J. Chem. Phys.*, **81**, 3475 (1984).
185. T. Turner and Y. T. Lee, *J. Chem. Phys.*, **81**, 5638 (1984).
186. S. L. Anderson, T. Turner, B. H. Mahan, and Y. T. Lee, *J. Chem. Phys.*, **77**, 1842 (1982).
187. F. A. Houle, S. L. Anderson, D. Gerlich, T. Turner, and Y. T. Lee, *Chem. Phys. Lett.*, **82**, 392 (1981).
188. P. M. Dehmer and W. A. Chupka, *J. Chem. Phys.*, **66**, 1972 (1977).
189. J. A. R. Samson, *J. Electron. Spectrosc. Relat. Phenom.*, **8**, 123 (1976).
190. C. W. Eaker and G. C. Schatz, *J. Phys. Chem.*, **89**, 2612 (1985).
191. C. W. Eaker and G. C. Schatz (private communication).
192. J.-D. Shao, Ph.D. thesis, Iowa State University, Ames, Iowa, 1986.
193. P.-M. Guyon and T. R. Govers, to be published.
194. C.-L. Liao, R. Xu, and C. Y. Ng, *J. Chem. Phys.*, **84**, 1948 (1986).
195. J. A. R. Samson and R. B. Cairns, *Phys. Rev.*, **173**, 80 (1968).
196. M. R. Spalburg and E. A. Gislason, *Chem. Phys.*, **94**, 339 (1985).
197. G. Parlant and E. A. Gislason, *J. Chem. Phys.*, **88**, 1622 (1988).
198. P. J. Chantry, *J. Chem. Phys.*, **55**, 2746 (1971).
199. Y. Ono, S. H. Linn, H. F. Prest, M. E. Gress, and C. Y. Ng, *J. Chem. Phys.*, **73**, 2523 (1980).
200. K. B. McAfee, Jr., R. S. Hozack, and R. E. Johnson, *Phys. Rev. Lett.*, **44**, 1247 (1980).
201. E. W. Kaiser, A. Crowe, and W. E. Falconer, *J. Chem. Phys.*, **61**, 2720 (1974).
202. B. Friedrich, W. Trafton, A. Rockwood, S. L. Howard, and J. H. Futrell, *J. Chem. Phys.*, **80**, 2537 (1984).
203. R. de L. Dronig, *Z. Phys.*, **50**, 347 (1928).
204. H. Beutler and H. O. Jünger, *Z. Phys.*, **100**, 80 (1936); **101**, 285, 304 (1936).
205. R. Derai, S. Fenistein, M. Gerard-Ain, T. R. Govers, R. Marx, G. Mauclaire, C. Z. Profous, and C. Sourisseau, *Chem. Phys.*, **44**, 65 (1979).
206. G. Mauclaire, R. Derai, S. Feistein, R. Marx, and R. Johnson, *J. Chem. Phys.*, **70**, 4017 (1979).
207. R. Derai, G. Mauclaire, and R. Marx, *Chem. Phys. Lett.*, **86**, 275 (1982).
208. P. M. Hierl, V. Pacak, and Z. Herman, *J. Chem. Phys.*, **67**, 2678 (1977).
209. Z. Herman, V. Pacak, A. J. Yencha, and J. H. Futrell, *Chem. Phys. Lett.*, **37**, 329 (1976).
210. J. Glosik, B. Friedrich, and Z. Herman, *Chem. Phys.*, **60**, 369 (1981).
211. T. Matsuo, N. Kobayashi, and Y. Kaneko, *J. Phys. Soc. Jpn.*, **51**, 1558 (1982).
212. P. J. Dagdigian and J. P. Doering, *Chem. Phys. Lett.*, **64**, 200 (1979).
213. J. Danon and R. Marx, *Chem. Phys.*, **68**, 255 (1982).
214. D. R. Guyer, L. Hüwel, and S. R. Leone, *J. Chem. Phys.*, **79**, 1259 (1983).
215. A. O. Langford, V. M. Bierbaum, and S. R. Leone, *J. Chem. Phys.*, **84**, 2158 (1986).
216. A. O. Langford, V. M. Bierbaum, and S. R. Leone, *J. Chem. Phys.*, **83**, 3913 (1985).
217. G.-H. Lin, Jürgen Maier, and S. R. Leone, *J. Chem. Phys.*, **84**, 2180 (1986).
218. G.-H. Lin, Jürgen Maier, and S. R. Leone, *J. Chem. Phys.*, **82**, 5527 (1985).
219. C. E. Hamilton, V. M. Bierbaum, and S. R. Leone, *J. Chem. Phys.*, **83**, 601 (1985).
220. C. E. Hamilton, V. M. Bierbaum, and S. R. Leone, *J. Chem. Phys.*, **83**, 2284 (1985).
221. C. E. Hamilton, V. M. Bierbaum, and S. R. Leone, *J. Chem. Phys.*, **80**, 1831 (1984).
222. L. Hüwel, D. R. Guyer, G.-H. Lin, and S. R. Leone, *J. Chem. Phys.*, **81**, 3520 (1984).
223. V. E. Bondybey and T. A. Miller, *J. Chem. Phys.*, **69**, 3597 (1978).
224. D. H. Katayama, T. A. Miller, and V. E. Bondybey, *J. Chem. Phys.*, **72**, 5469 (1980).
225. B. H. Mahan, C. Martner, and A. O'Keefe, *J. Chem. Phys.*, **76**, 4433 (1982).

226. T. R. Govers, M. Gerad, and R. Marx, *Chem. Phys.*, **15**, 185 (1976).
227. K. B. McAfee, Jr., W. E. Falconer, R. S. Hozack, and D. J. McClure, *Phys. Rev. A*, **21**, 827 (1980).
228. K. B. McAfee, Jr., and R. S. Hozack, *J. Chem. Phys.*, **83**, 5690 (1986).
229. C.-L. Liao and C. Y. Ng, *J. Chem. Phys.*, **84**, 197 (1986).
230. R. Xu, C.-L. Liao, and C. Y. Ng, to be published.
231. J. Ross, J. C. Light, and K. E. Schuler, in A. R. Hochstim, Ed., *Kinetic Processes in Gases and Plasma*, Academic, New York, 1969.
232. Y. Itoh, N. Kobayashi, and Y. Kaneko, *J. Phys. Soc. Jpn.*, **50**, 3541 (1981).
233. R. E. Johnson, *J. Phys. B*, **3**, 539 (1970).

Chapter **IX**

SPECTROSCOPIC PROBES

Masaharu Tsuji

---

1 **Introduction**
2 **Ion-Beam Methods**
   2.1 Ion Beam Coupled with UV and Visible Emission Detection
   2.2 Ion Beam Coupled with Infrared Chemiluminescence Detection
   2.3 Ion Beam Coupled with Laser-Induced Fluorescence Detection
3 **Ion-Trap Methods**
   3.1 Ion Cyclotron Resonance Coupled with UV and Visible Emission Detection
      3.1.1 ICR Cell
      3.1.2 Trapping and Ejection of the Ions
      3.1.3 Pulsing and Counting Sequences
      3.1.4 Emission Spectra Excited in the ICR Cell
   3.2 ICR Coupled with LIF Detection
   3.3 Quadrupole Ion Trap Coupled with LIF Detection
4 **Flowing Afterglow Methods**
   4.1 FA Coupled with UV and Visible Emission Detection
   4.2 FA Coupled with IRCL Detection
   4.3 FA Coupled with LIF Detection
   4.4 FA Ion Source Coupled with a Supersonic Nozzle Expansion
      4.4.1 FA Coupled with a Low-Pressure Chamber for UV and Visible Emission Detection
      4.4.2 FA Coupled with a Low-Pressure Chamber for LIF Detection
5 **Drift Tube Methods**
   5.1 DT Coupled with UV and Visible Emission Detection
   5.2 DT Coupled with LIF Detection
References

---

1 INTRODUCTION

Emission and laser-induced fluorescence (LIF) spectroscopies are valuable techniques for the determination of product state internal energy distributions and rate constants of ion–molecule reactions. First, measurements of nascent product state distributions provide state-to-state information on ion–molecule reactions. Second, these techniques can detect short-lived intermediate species (e.g., excited radical ions), giving useful information about molecular structure of unstable species. Third, as with other spectroscopic methods, these optical detection methods can be used to monitor the concentration of specific products.

Figure 1 illustrates the complementary nature of emission and LIF spectroscopies. Measurements of ultraviolet (UV) and visible emissions provide information about relative vibrational and rotational state distributions of an electronic excited state and branching ratio of each emitting electronic state, whereas the vibrational state distribution of an electronic ground state for $v'' \geq 1$ can be determined from the observation of infrared (IR) emission. The LIF method can be used to monitor vibrational and rotational state distributions

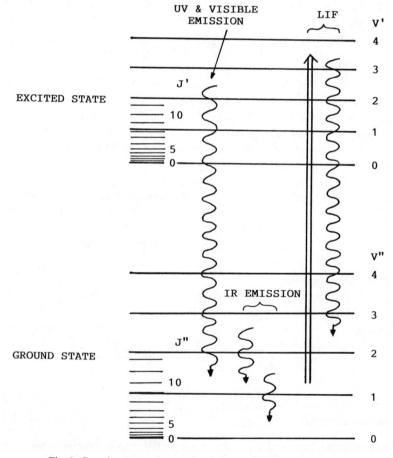

Fig. 1. Complementary nature of emission and LIF spectroscopies.

of nonemitting products in electronic ground states and long-lived metastable states. An advantage of LIF detection is that the $v''=0$ population can be directly obtained.

The optical detection methods described above have been applied to the product state analysis of the following types of ion–molecule reactions:

$X^+ + AB$	$\rightarrow AB^+ + X$	(Charge transfer)	(1)
	$\rightarrow A^+ + B + X$	(Dissociative charge transfer)	(2)
	$\rightarrow XA^+ + B + X$	(Heavy atom transfer)	(3)
	$\rightarrow X^+ + AB^*$	(Collisional excitation)	(4)
	$\rightarrow X^{+*} + AB$	(Projectile excitation)	(5)
$X^- + AB$	$\rightarrow XA + B^-$	(Heavy atom transfer)	(6)
	$\rightarrow XAB + e^-$	(Associative detachment)	(7)
$X^{2+} + AB$	$\rightarrow AB^+ + X^+$	(Charge transfer)	(8)
	$\rightarrow A^+ + B^+ + X$	(Dissociative charge transfer)	(9)
	$\rightarrow A^+ + X^+ + B$	(Dissociative charge transfer)	(10)

Here X^+, X^- and X^{2+} stand for reactant ions, and AB represents a target molecule. Optical emissions from XA^+ and XA in reactions 3 and 6 are called chemiluminescence because luminescence originates as a result of new chemical bond formation. Among the devices used to investigate the above ion–molecule reactions are ion beams, trapped-ion cells, flowing afterglow instruments, and drift tubes. This chapter will describe experimental details of each device coupled with optical detection methods by using selected examples. The conventional ion-beam, ion cyclotron resonance (ICR), flowing afterglow (FA), and drift tube (DT) apparatuses have been reviewed in Chapters 7, 1, and 4, respectively. Therefore, the emphasis in this chapter will be on the modifications of each device required for optical studies, with the techniques, advantages and constraints.

We will assume the reader has fundamental knowledge of UV/visible and IR spectroscopy. For basic principles of emission and absorption spectroscopy from the UV to IR regions, refer to the standard textbooks of Herzberg [1–4]. Because of recent extensive publications on laser spectroscopy, fundamentals of lasers, and applications of laser spectroscopy, refer to representative reviews and textbooks [5–17] for additional information. Most of the standard instruments and techniques used for optical studies, such as monochromators, optics, and tunable dye lasers, are omitted here. In places where we omitted details of a given technique because of space limitations, we have given appropriate references to the literature.

Although there seems to be no up-to-date, comprehensive review of spectroscopic probes of ionic chemistry, several review articles recently appeared for each technique. The article by Tiernan and Lifshitz [18] summarizes the application of UV and visible emission spectroscopy to ion–molecule reactions up to 1981. Reviews by Leone and co-workers [19, 20] give a recent development of FA and DT apparatuses for ion–molecule reactions by IR chemiluminescence and LIF techniques. Reviews by Ottinger [21, 22] cover chemiluminescent probes of ion–molecule reactions, and Leventhal's article [23] provides information obtained from UV and visible emissions of charge-transfer reactions. These articles demonstrate the use of optical techniques in extending our understanding of gas-phase ionic processes.

2 ION-BEAM METHODS

2.1 Ion Beam Coupled with UV and Visible Emission Detection

Many of the earlier ion-beam experiments coupled with UV and visible emission spectroscopy were conducted at relatively high collisional energies above a few hundred eV [18]. Beam-target gas-luminescence apparatuses have recently been developed to study ion–molecule reactions at low energies down to a few electron volts [21–26]. Figure 2 shows a schematic representation of such an apparatus. The apparatus consists basically of the following components:

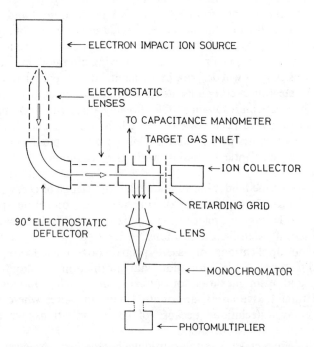

Fig. 2. Schematic diagram of an ion-beam apparatus for studying UV and visible emissions from ion–molecule reactions.

1. *Ion Source.* Reactant ions are formed by either electron impact or an electrical discharge source (e.g., hollow cathode). As reactant ions for beam-luminescence studies, the following atomic and diatomic ions have been used, as summarized in tables of Tiernan and Lifshitz [18]: H^+, D^+, Li^+, Na^+, K^+, Be^+, Mg^+, Ca^+, B^+, Al^+, C^+, N^+, O^+, F^+, He^+, Ne^+, Ar^+, Kr^+, Xe^+, H_2^+, N_2^+, He_2^+, Ne_2^+, CO^{2+}, and OD^+. Recently, P^+ [27], He^{2+}, Ne^{2+}, Ar^{2+} [28] and C^{2+} [29] have also been used.

2. *Reactant-Ion Separator.* Mass analysis of a desired ion is performed with a magnetic sector field (or in some cases with quadrupole mass filters).

3. *Reactant-Ion Accelerator.* A series of cylindrical lenses is usually used to focus and accelerate an ion beam at a desired collision energy.

4. *Collision Chamber.* A target gas is introduced into a collision chamber from a needle valve in the form of either an effusive beam or a nozzle beam. In the former case, the velocity of the target beam is characterized by a Boltzmann distribution, whereas in the latter case, the beam has a much better defined velocity and angular width.

As target gases, not only stable molecules but also unstable species such as hydrogen atoms have been used [30]. Hydrogen atoms were produced by a 2450-MHz microwave discharge of H_2 in a quartz tube at the optimum pressure of 0.4 torr and introduced into a low-pressure collision chamber through a Pyrex disk with a small hole. Recently, excited Na atoms, the state of which has been selected by means of a tunable dye laser, have been used as a target species [31].

5. *Monochromator.* In general, the optimum slit width B of a monochromator at wavelength λ is given by

$$B = \lambda \cdot \frac{f}{D} \qquad (11)$$

where f is the focal length of a collimator lens or mirror and D is its effective diameter. For example, the B value at 400 nm is calculated to be 4 μm for a 1-m Czerny-Turner type of Jarrell Ash monochromator equipped with two concave mirrors ($f = 1000$ mm) and a grating of 1180 groove·mm^{-1} in an area of 100×102 mm^2. Since the reciprocal dispersion of this monochromator is 0.82 nm·mm^{-1} in first order, the slit width of 4 μm corresponds to 0.00328 nm. On the other hand, from the theoretical resolution of this monochromator in first order (118000), the difference of two lines that can be separated at 400 nm is estimated to be

$$\delta\lambda = \frac{400 \text{ nm}}{118000} = 0.00339 \text{ nm} \qquad (12)$$

This value is close to the spectral resolution at a slit width of 4 μm. According to Eq. 11, the optimum slit width is about 7–15 μm for a large monochromator

($\geqslant 1$ m) and about 2–7 μm for medium and small ones ($\leqslant 0.5$ m). Such a high resolution is unnecessary in most measurements except for studies of vibrational and rotational structures so that larger slit widths are recommended to obtain higher intensity. Since light signals are very weak in ion–beam experiments at low collisional energies, small and bright monochromators (focal length 25–50 cm) are mostly used with large entrance and exit slit widths. Principles and detailed techniques for use of monochromators have been given elsewhere [32–34].

6. *Optical Detection System.* Photoemission in the collision region is observed through a fused silica or MgF_2 window and focused onto the entrance slit of a monochromator. Light signals from the monochromator are detected by using a cooled photomultiplier in the counting mode; techniques of photon counting will be described in Section 2.3. A multichannel analyzer (e.g., 1024 channels) is used for data acquisition. Both the multichannel analyzer and a stepping motor that increments the monochromator in predetermined steps are controlled by appropriate circuitry. The spectral range of interest is scanned repetitively in order to achieve signal averaging; typically, accumulation of the data for 2–100 successive scans over several hours is required to achieve a sufficient signal-to-noise (S/N) ratio. For these measurements monochromators whose wavelength scan is controlled by a pulse stepping motor must be used.

7. *Ion Collection.* The ion beam is collected in a Faraday cup, and the beam intensity is monitored with an electrometer. The beam intensity is attenuated passing through the collision chamber. The correction for beam attenuation is approximately given by the equation

$$i_{\text{eff}}(p) = i(p) \cdot \sqrt{\frac{i_0}{i(p)}} \qquad (13)$$

where $i_{\text{eff}}(p)$ is an effective ion-beam intensity and $i(p)$ and i_0 represent the beam currents at target gas pressure p and without target gas, respectively.

Advantages of beam apparatuses are as follows: First, various kinds of atomic and molecular ions can be chosen as a reactant ion, including multicharged species and metastable ions with lifetimes longer than about 10^{-5} s. Second, since the reactant ion is usually mass analyzed, interference from other unwanted ions and active species (e.g., metastable atoms) is removed. Third, the collisional energy of reactant ions can be tuned over a wide range above $\sim 1\,\text{eV}_{\text{lab}}$. Therefore, endoergic product channels are open at high collisional energies. Fourth, beam experiments are performed at lower gas pressures in a collision chamber than FA experiments described in Section 4, permitting us to measure photoemissions primarily from excited states in most cases. Fifth, new ion fluorescences, not observed in conventional discharge and electron-impact methods, can be obtained (e.g., $BH^+(B'^2\Sigma^+ - X^2\Sigma^+)$ from $B^+(^1S) + H_2$ [35] and $AlH^+(B^2\Sigma^+ - X^2\Sigma^+)$ from $Al^+ + H_2$ [36]. Sixth, using photon counting and multichannel analyzer systems, low detection limits of about 10^{-3} Å^2 (cross section) are achieved.

The following types of measurements are usually performed.

1. *Spectral Identification.* Scanning the wavelength of a monochromator at constant experimental conditions, UV and visible emissions are observed with optical resolution as high as possible. Each band is assigned with reference to reported tables. Tables of Harrison [37], Striganov and Sventitskii [38], Zaidel' et al. [39], and Reader and Corliss [40] are useful for the assignment of atomic spectra, whereas those of Rosen [41], Pearse and Gaydon [42], and Huber and Herzberg [43] are helpful for the identification of molecular spectra. If a new emission system is observed, vibrational and rotational analyses on the basis of standard procedures as described in Herzberg's textbooks [2–4] provide information on the type of electronic transition such as $^2\Sigma - {}^2\Pi$ and $^1B_1 - {}^1A_1$ from which emitting excited species can be identified in most cases. In general, a band degrading to violet implies that a rotational constant of the upper state is larger than that of the lower state, whereas the inverse relation holds for a blue-shaded band.

2. *Population Analysis.* Vibrational and rotational state distributions of an excited product are obtained from corrected photon emission intensities. The sensitivity of an optical detection system is usually calibrated by using a standard tungsten or halogen lamp in the UV and visible region above ~ 250 nm and a deuterium lamp in the UV region of 200–300 nm. Intensity calibration is also possible by using known relative intensities of molecular emission bands in a $(v', v'' + \Delta v'')$ progression where only a few levels are populated [44]. The relative intensity (photon·s^{-1}) of a transition from the v', N' level to the v'', N'' level is given by

$$I_{v'N',v''N''} \propto N_{v'N'} R_e^2(\bar{r}_{v'v''}) q_{v'v''} v_{v'N',v''N''}^3 S_{N'N''} g_{N'}^{-1} \tag{14}$$

where $N_{v'N'}$ is the rotational population in a given vibrational level, $R_e(\bar{r}_{v'v''})$ is the electronic transition moment in the r-centroid approximation, $q_{v'v''}$ is the Franck-Condon (FC) factor, $v_{v'N',v''N''}$ is the transition frequency, $S_{N'N''}$ is the rotational line strength, and $g_{N'} = 2N' + 1$ is the degree of degeneracy. FC factors are usually calculated by using Morse or Rydberg-Klein-Rees (RKR) potentials. The dependence of FC factors on the rotational quantum number due to the effect of vibrational-rotational interaction should be taken into account for molecules with $\gamma > 0.001$ [45]:

$$\gamma = \frac{2B_e}{\omega_e} \tag{15}$$

where B_e and ω_e, respectively, are equilibrium rotational and vibrational constants of a molecule. Equation 14 is modified to include only rotational terms by assuming that the other terms are constant for a particular vibrational band:

$$I_{N'N''} \propto N_{N'} v_{N'N''}^3 S_{N'N''} g_{N'}^{-1} \tag{16}$$

For a Boltzmann distribution of rotational states with an effective rotational temperature T_R, the rotational population $N_{N'}$ is expressed by

$$N_{N'} \propto g_{N'} \exp\left(\frac{-hcB_{v'}N'(N'+1)}{kT_R}\right) \tag{17}$$

Then, Eq. 16 is reduced to

$$\ln\left(\frac{I_{N'N''}}{v_{N'N''}^3 S_{N'N''}}\right) = \text{const} - \frac{hcB_{v'}N'(N'+1)}{kT_R} \tag{18}$$

and a plot of $\ln(I_{N'N''}/v_{N'N''}^3 S_{N'N''})$ vs. $N'(N'+1)$ yields a straight line with a slope $-hcB_{v'}/kT_R$.

When an observed spectrum is well resolved, vibrational and rotational distributions can be determined directly from the above equations. On the other hand, when the spectrum is not sufficiently resolved, the population analysis is perfomed on the basis of computer simulation of the observed band envelope. The band envelope is computed by superimposing the intensity $I_{v'N',v''N''}$ multiplied by a slit function. The slit function is approximated by a normalized Gaussian function (19) estimated from the observed shape of a heavy atomic line like a mercury line with the identical slit width:

$$G(v-v_0;\gamma_D) = \frac{\sqrt{\ln 2}}{\gamma_D\sqrt{\pi}} \exp\left(-\frac{(\ln 2)(v-v_0)^2}{\gamma_D^2}\right) \tag{19}$$

where G is the intensity in the center of a band with a wavenumber v_0, and γ_D is a half-width at half-maximum (HWHM) of the band. The vibrational and rotational distributions are adjusted by trial and error in such a way that the calculated band envelope reproduces the observed spectrum. The rotational distribution is usually approximated assuming a single or double Boltzmann function with effective rotational temperature(s). Ottinger and co-workers have estimated vibrational and rotational distributions of excited products on the basis of spectral simulation of emission systems such as the following: $CO^+(A^2\Pi - X^2\Sigma^+)$ from $C^+(^2P) + O_2$ [46] and $N^+ + CO$ [47], $CH(D)^{+\cdot}(A^1\Pi - X^1\Sigma^+)$ from $C^+(^2P) + H(D)_2$ [48], $CH(D)^+(b^3\Sigma^- - a^3\Pi)$ from $C^+(^4P) + H(D)_2$ [49], $NH(D)^+(A^2\Delta - X^2\Pi)$ from $N^+(^1D) + H(D)_2$ [50], $NH(A^3\Pi - X^3\Sigma^-)$ from $N^+(^3P) + RH$ (R=CH_3, C_2H_3, C_2H_5, C_3H_7) [51], $OH(A^2\Sigma^+ - X^2\Pi)$ from $O^+ + RH$ (R=CH_3, C_2H_3, C_2H_5, C_3H_7) [52], $AlH^+(A^2\Pi - X^2\Sigma^+)$ from $Al^+ + H_2$ [36], and $PH(D)^+(A^2\Delta - X^2\Pi)$ from $P^+(^1D) + H(D)_2$ [27].

For low-resolution spectra, relative vibrational populations are obtained from

$$I_{v'v''} \propto N_{v'} R_e^2(\bar{r}_{v'v''}) q_{v'v''} v_{v'v''}^3 \tag{20}$$

The dependence of R_e on $\bar{r}$ is estimated by plotting values of $[I_{v'v''}/q_{v'v''}v_{v'v''}^3]^{1/2}$

as a function of r-centroid and rescaling data from individual v'' progression on the basis of Fraser's method [53] to account for the different populations among the upper state vibrational levels. In most cases, $R_e(\bar{r})$ functions can be expressed by either a linear function or a parabolic one. If the $R_e(\bar{r})$ function cannot be obtained from the above method because of the absence of an appropriate v'' progression, $N_{v'}$ values are usually estimated assuming $R_e(\bar{r})$ to be constant.

When FC factors are unknown but all branches of an emission system are measured within the observed wavelengths, the vibrational population can be obtained by summing up intensity from the same emitting v' level:

$$N_{v'} \propto \sum_{v''} \frac{I_{v'v''}}{R_e^2(\bar{r}_{v'v''})v_{v'v''}^3} \tag{21}$$

3. *Emission Cross Section.* Absolute emission cross sections σ_{em} are determined from the equation

$$\sigma_{em} = \frac{4\pi P(\omega)}{\omega(I^+/q)nLk(\lambda)} \tag{22}$$

where $P(\omega)$ is the number of photons emitted into the solid angle ω, (I^+/q) is the number of singly positively charged ions passing per second through the collision chamber (I^+ = ion current, q = elementary charge), n is the number density of target molecules per cubic centimeter determined from gas pressure measurements, L is the observation length (cm), and $k(\lambda)$ is the calibration factor for the quantum yield of the optical equipment as determined using a calibrated standard tungsten lamp. With n, L held constant, $k(\lambda)$ known, and ω fixed in an experimental arrangement, Eq. 22 is simplified to the following expression:

$$\sigma_{em} = \frac{A(\lambda)P(\omega)}{I^+} \tag{23}$$

$A(\lambda)$ being a known quantity. A detailed procedure to estimate detection efficiency and geometric factors by using a standard light source is described in Ref. [54].

Absolute emission cross sections can also be determined by reference to known emission cross sections which can be studied in an identical apparatus; e.g., $He^+ + Ar \rightarrow Ar^{+*} + He$ charge-transfer reaction [26, 55, 56]. If an electron beam is passed through the ion-beam apparatus, emission spectra excited by electron impact can also be used; e.g., $e^- + N_2 \rightarrow N_2^+(B) + 2e^-$ [57]. This reference reaction method is easier because it avoids the measurement of absolute emission intensities. The best way to evaluate emission cross sections from the reference reaction method is to compare total intensities between an emission and a reference band in prepared mixtures of a reagent and a reference sample. If mixtures are not used, then consecutive experiments should be performed with the reagent and the reference sample, with no changes in experimental conditions.

Emission cross sections obtained from the steady-state photoemissions of the excited product equal excitation cross sections, provided there is no quenching, nonradiative decay, or pumping of the excited product state from the observation zone and the radiative lifetime is shorter than the residence time in the observation zone. These conditions are normally satisfied for molecules with radiative lifetimes shorter than about 1 μs. If the radiative lifetime is longer than 1 μs, loss of emitting products from the observation zone should be taken into account. An approximate correction for this effect can be estimated from the equation

$$\frac{N'}{N_0} = 1 - \frac{\lambda_0}{l_2 - l_1}\left[\exp\left(\frac{-l_1}{\lambda_0}\right) - \exp\left(\frac{-l_2}{\lambda_0}\right)\right] \quad (24)$$

where N_0 and N' are the total number of excited species in the state of interest and the total number observed, respectively, λ_0 is the mean flight distance of an emitting species calculated under the assumption that the excited product has roughly the laboratory energy of the primary ion, and l_1 and l_2 are the distances from the observation region, including a section of the ion beam from l_1 to l_2, measured from a collision chamber entrance slit. Assuming that $l_2 - l_1$ is constant in each case, N'/N_0 is larger if $l_1 \sim \lambda_0$ than if $l_1 = 0$. This shows that a geometry with the observation region well behind the entrance slit is preferred. For example, when the $OH^+(A-X)$ emission with a radiative lifetime of 1 μs is detected at $l_1 = 0$ and $l_2 = 2$ cm in the 100 eV $O^+ + H_2$ CT reaction, about 70% of $OH^+(A)$ escapes the field of view [24].

Metastable ions as well as ground-state ions are often involved in a reactant beam and take part in the formation of excited species. Since quenching cross sections of metastable ions are generally much larger than those of ground-state ions, the fractions of metastable and ground-state ions can be determined by the differential attenuation technique described in Chapter 7. If an electron-impact ion source is used and the electron energy is adjustable below and above the appearance potentials of metastable ions, the effects of two active species for a reaction can be distinguished. For example, Rothwell et al. [58] have used this technique to examine the formation of excited Ar atoms in the $Ar^+ + H_2$ and $Ar^{+*} + H_2$ reactions.

It is possible to estimate relative emission cross sections of an excited species due to reactions of metastable ions (σ_m) and ground-state ions (σ_g), when a direct current (dc) discharge ion source is used. The relative emission intensity normalized to the ion current at an anode-to-cathode voltage U of the ion source to that at some reference voltage U_0 is given by

$$\frac{Z(U)}{Z(U_0)} = \frac{\sigma_m f(U) + \sigma_g[1 - f(U)]}{\sigma_m f(U_0) + \sigma_g[1 - f(U_0)]}$$

$$= \frac{1 + (\gamma - 1)f(U)}{1 + (\gamma - 1)f(U_0)}$$

$$= F(U; \gamma), \quad (25)$$

where f is the metastable fraction and $\gamma = \sigma_m/\sigma_g$. Since the fractions of metastable ions at various voltages are known from the attenuation technique, the $F(U;\gamma)$ functions are described assuming various γ values. By comparing the calculated $F(U;\gamma)$ function at various voltages with the observed one, the γ value is obtained. For example, the cross-section ratio between the $B^+(^1S) + D_2$ and $B^+(^3P) + D_2$ reactions for the formation of excited boron atoms has been determined using this method [59].

If the radiative lifetime of one component of an excited state is sufficiently different from that of the other component, the relative contribution of each component to the observed emission can be determined from the measurement of the emission intensity as a function of distance into the target cell. Van Zyl et al. [60] have used this technique for studying the formation of specific angular momentum states in one-electron atoms produced by charge transfer in a beam–gas experiment. Figure 3 shows the basic photon detector to measure hydrogen Balmer emissions formed in $H^+ + Ar$ collisions. Photons produced along the

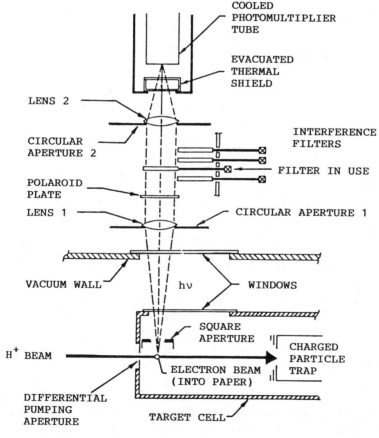

Fig. 3. The target cell and photon detector used by Van Zyl et al. [60].

beam axis pass through a target cell and vacuum-wall windows, are focused into a parallel beam by lens 1, pass through an appropriate interference filter, and are imaged onto a cooled photomultiplier by lens 2. The detector's field is restricted by a square aperture inside the target cell and circular apertures fronting lenses 1 and 2. A Polaroid plate can be inserted into the photon path when the polarization of the emitted light is measured. The entire photon detector and the aperture inside the target cell can be moved along the proton beam axis to permit observation of Balmer lines at various penetration depths into the collision cell.

For example, since the lifetime of hydrogen atoms in the $4s$ state is longer than those in the $4p$ and $4d$ states, the former component in H_β emission decays more slowly than the latter component. Assuming that $l_1 = 0$ and that the target-cell pressure has a step-function onset at $l_2 = 0$, the point of entry into the cell, the fraction of equilibrium between photoemission and collisional excitation per unit length F_{4l} is calculated from the following equation under single collision conditions:

$$F_{4l} = 1 - \exp\left(\frac{-l_2}{v\tau_{4l}}\right), \tag{26}$$

where l_2 corresponds to the distance into the cell, v is the hydrogen atom velocity, which is assumed to be the same as that of the incident ion beam, and τ_{4l} is the known radiative lifetime. First, the dependence of H_β signal intensity on l_2 was measured at various target-gas pressures, and the zero-pressure extrapolated data were evaluated to remove the effects of secondary collisions. Then the data obtained were fitted to the equation

$$C(l_2) = C_{4s}(F_{4s} + K_p) + (C_{4p} + C_{4d})F_{4p+4d} \tag{27}$$

where C_{4l} are coefficients giving the fractions of the total H_β signal resulting from decay of the $4l$ state, and K_p is a small correction applied to F_{4s} to account for the fact that the pressure profile near the cell entrance does not have a true step-function onset and that the pressure before the cell entrance is finite. By using this technique, the emission cross section of each component has been estimated for H^+ energies above 300 eV.

4. *Kinetic Energy Dependence.* Spectra are recorded as a function of the reactant ion kinetic energy. The variation of internal energy distributions of products gives information on the reaction mechanism. As a typical example, vibrational distributions of $N_2^+(B)$ produced from reactions of various atomic ions with N_2 deviate significantly from the FC distribution at low projectile-ion velocities ($< 10^8$ cm·s^{-1}), but they approach the FC-like distribution with increasing reactant-ion velocity [61]. The non-FC distribution at low energies has been explained by a distorted FC model presented by Lipeles [62]. According to this model, the target N_2 molecules in the ground state are distorted as the

projectile ion approaches. Vertical transitions then occur from the perturbed N_2 molecules into the $N_2^+(B)$ state.

The dependence of the emission cross section of a band on the ion kinetic energy, the so-called excitation function, provides information about the excitation process. Figure 4 shows the excitation function of an emission produced from an endothermic process. By linearly extrapolating the excitation function to low ion kinetic energies, the appearance potential (threshold energy) of the emitting product, E_0, can be determined. In some cases, a few additional shoulders appear in the excitation function at higher energies, as shown in Fig. 4, due to openings of new exit product channels. From the measurements of the appearance potentials of such shoulders, the second and the third threshold energies, E_1 and E_2, are determined. By comparing the observed value with the calculated one obtained from known thermochemical and spectroscopic data, dominant excitation process(es) near each threshold can be determined. For the precise determination of threshold energies, the kinetic energy spread of the reactant ion must be minimized. In a beam–gas experiment, the thermal target motion introduces a spread into the relative energy. Detailed analytical formulas taking account of this effect have been presented by Chantry [63].

In the case of exothermic processes, the excitation function need not exhibit such a threshold. In fact, thresholds have not been observed in most cases. However, the formation of Mg^{+*} from the $He^+ + Mg$ reaction is an exceptional case in which a threshold was observed [64]. It was expected as a result of the infinite number of continuum $(He + Mg^{2+} + e^-)$ states available to the system.

The overall shape of an excitation function also provides information on the reaction dynamics. For example, Kusunoki and Ottinger [65] have measured

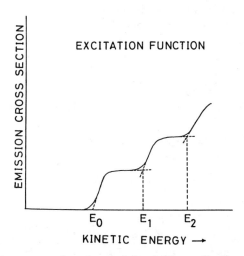

Fig. 4. Optical emission cross section of an emission band as a function of ion kinetic energy (excitation function).

excitation functions for NH($A^3\Pi$) produced from the reactions

$$N^+ + RH \rightarrow NH(A) + R^+$$
$$R = CH_3, C_2H_3, C_2H_5, C_3H_7 \tag{28}$$

On the basis of the fact that there is a certain maximum kinetic energy in a low-energy range, whereas the excitation functions fall off nearly exponentially at high energy, a low-energy strippinglike and high-energy knockout mechanism could be distinguished with the four hydrocarbons.

The measurements of absolute excitation cross sections in ion–molecule reactions have been extended to a state-selected excited reactant by Kushawaha et al. [31]. The absolute cross sections for Lyman-α (H_α) production as a function of kinetic energy have been compared for these processes:

$$H^+ + Na(3s) \rightarrow H(n=2) + Na^+ - 1.74\,eV \tag{29}$$

$$H^+ + Na(3p) \rightarrow H(n=2) + Na^+ + 0.37\,eV \tag{30}$$

In process 30, sodium atoms in the excited $3p$ level were prepared by the $3s \rightarrow 3p$ laser excitation. Experiments were carried out in a beam–gas condition; the number density of sodium vapor was about $10^{13}\,cm^{-3}$. Light from a single-frequency continuous wave (cw) dye laser was introduced into a collision cell and interpenetrated the ion beam. The laser power density was about $10\,W \cdot cm^{-2}$, which was sufficient to saturate the $3s \rightarrow 3p$ transition in sodium. L_α radiation was detected at $90°$ to (both) the ion and laser beams through a narrow bandpass filter. A solar-blind photomultiplier was used in the counting mode. The laser beam was mechanically chopped, and the photon signal was detected in synchronization with the chopper, yielding comparative data for processes 29 and 30. An enhancement of the H_α production less than a factor of 2 has been observed when the sodium was irradiated with laser light tuned to a D line for reactant ion energies below 35 eV. General techniques used in laser excitation and fluorescence detection will be described in Section 2.3.

5. *Ion-Beam Current and Pressure Dependence.* The number of reactant ions (n) and target molecules (m) that take part in the formation of an excited emitting state can be determined from the dependence of the emission intensity on the ion-beam current and the target gas pressure, respectively,

$$nX^+ + mAB \xrightarrow{k_{31}} AB^{+*} + X \tag{31}$$

$$I \propto \frac{d[AB^{+*}]}{dt} = k_{31}[X^+]^n[AB]^m \tag{32}$$

Although most emitting species excited in beam experiments have been demonstrated to be primary products on the basis of a linear pressure dependence of the emission intensity, a few exceptional cases have been reported. For example, the CN(B − X) emission in the $C^+(^2P) + NO$ reaction was found to depend quadratically on the NO gas pressure [66, 67]. It was interpreted as a result of the reaction sequence

$$C^+(^2P) + NO \rightarrow C + NO^+ \tag{33}$$

$$C + NO \rightarrow CN(B) + O \tag{34}$$

The formation of the CH(B − X) emission in the $Ar^+ + C_2H_2$ reaction was proportional to noninteger powers of the target gas pressure between 1 and 2 [68]. These results have been analyzed by assuming simultaneous excitation through a primary process and a secondary one.

6. *Radiation polarization.* In general, the polarization of radiation is defined as

$$P = \frac{I_\parallel - I_\perp}{I_\parallel + I_\perp} \tag{35}$$

where $I_\parallel$ and $I_\perp$ are the observed emission intensities with the planes of polarization parallel and perpendicular to an ion-beam axis. $I_\parallel$ and $I_\perp$ can be measured using a polarizer. When an observed emission is polarized, the emission intensity exhibits an angular distribution. The emission intensity observed at an observation angle θ from the reactant ion beam is given by

$$I(\theta) = \frac{3I_0(1 - P\cos^2\theta)}{4\pi(3 - P)} \tag{36}$$

where I_0 is the total emission intensity. According to Eq. 36, the emission intensity is independent of the polarization at an observation angle of 54.5° ($\cos^2\theta = 1/3$).

As a polarizer, Polaroid plates are generally used. Polaroid, which is the most popular dichroic polarizer, is made of polymer films in which long-chain molecules with appropriate absorbing side groups are oriented by stretching. The stretched film is then sandwiched between glass or plastic sheets. Raman lines of CCl_4 excited by a linearly polarized He—Cd laser can be used as a method to determine the values of maximum and minimum transmission of the polarizer to linearly polarized light. General techniques required for the polarization measurements of light emission have been published [69–71].

As an example, Van Zyl et al. [60] measured the polarization of hydrogen Balmer lines in $H^+ + Ar$ collisions. For the case of H_α, contributions from the 3s, 3p, and 3d states are possible. Among them, the radiation from the spherically symmetric 3s state should not be polarized, whereas the radiation from the other two states should be polarized. Thus, the measured polarization reflects the relative contribution of the latter excited states. In the 100-eV proton-energy region, the P value was nearly zero, indicating that radiation from the $3s$ state may be important in this region. The polarizations of the radiations from decays of the $m_l = 0$, 1, and 2 magnetic sublevels of the $3d$ state of hydrogen should be $+0.48$, $+0.26$, and -0.70, respectively. The measured H_α polarization value of about -0.3 in the 1-keV region where the contribution of the $3d$ state is dominant suggests that a substantial part of the total radiation from decay of the $3d$ state comes from the $m_l = 2$ magnetic substate. As shown in the above example, the measurements of polarization give detailed information about excitation mechanisms in ion–molecule reactions.

2.2 Ion Beam Coupled with Infrared Chemiluminescence Detection

Although few beam experiments have been detected by IR emission spectroscopy, Ding [72] has studied the following ion–molecule reaction by measuring the infrared chemiluminescence (IRCL) from the H_3^+ product:

$$H_2^+ + H_2 \rightarrow H_3^+ + H \tag{37}$$

The reaction was examined in a large reaction chamber filled with low-pressure H_2 (about 5×10^{-5} torr), where the reactant H_2^+ ions were generated by electron impact. The experiments in the large chamber at the low pressure were necessary to reduce effects of secondary collisions of vibrationally excited H_3^+ ions with H_2 and walls because IR emissions are generally long-lived, ranging from 10^{-3} to 1 s. The IRCL was measured in the 2.5–5.5-μm region using a high-resolution Fourier transform (FT) IR apparatus. The vibrational population of the third vibrational level of the asymmetric stretch mode was high, and the rotational distribution showed maxima for levels with $J \sim K$. These results led Ding to conclude that reaction 37 predominantly proceeds via an in-plane collision encounter. To our knowledge, this is the only example in which the ion-beam apparatus coupled with IRCL detection system has been applied to research on ion–molecule reactions. Detail techniques used for the IRCL detection will be described in Section 4.2.

2.3 Ion Beam Coupled with Laser-Induced Fluorescence Detection

The laser-induced fluorescence (LIF) method is a powerful technique for product state analysis under high resolution. It has been successfully applied to ion–molecule reactions to probe ro-vibrational distributions of reactant ions [73, 74] and products [75, 76] in the ground states. Figure 5 shows a typical arrangement of the LIF apparatus used by Ding and Richter [76]. It consists of a pumping laser, a tunable dye laser, a light baffle, a reaction chamber, a concave mirror, photodetectors, a photon counter, and an on-line computer to calibrate and control the experiments. As a probe laser, they used a pulsed oscillator-amplifier dye laser combination (Lambda Physik FL2000) pumped by either a nitrogen laser (Lambda Physik M1000) or an XeCl excimer laser

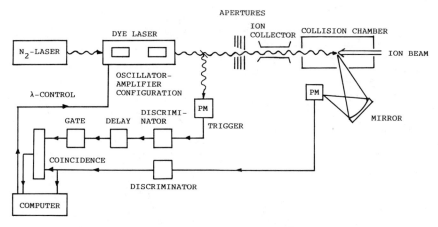

Fig. 5. Experimental arrangement for observing LIF of an ion beam and products in ion–molecule reactions [76].

(Lambda Physik EMG102). The tuning range of the dye laser is between above 340 nm and the near-infrared region with proper dye selection. The laser beam is directed into the reaction chamber through the light baffle fitted with several apertures to reduce scattered light. Details of tunable dye lasers and pumping N_2 and excimer lasers have been published [9–17].

Distinct from neutral beam experiments, a coaxial arrangement of laser and ion beams is superior to the perpendicular one in obtaining strong LIF signal because of the maximum overlap between the two beams. An additional merit is that the Doppler width caused by energy spread of the ion beam is reduced if experiments are carried out at sufficiently high ion energies. This is due to the fact that the velocity spread Δv of an ion beam of constant energy spread ΔE is diminished with increasing energy E:

$$\Delta v \propto \frac{\Delta E}{\sqrt{E}} \qquad (38)$$

An advantage of beam experiments is that essentially Doppler-free spectra can be obtained at high ion energies.

The LIF signal is usually observed through a slit along the ion path focused onto a photodetector by a mirror or lens through either a cutoff filter or a bandpass filter. Figure 6 shows a general method for appropriate filter selection in the LIF measurement of a (v', v'') band. When a cutoff filter is used, all LIF signals from the $(v', v'' - n: n = 1, 2, ...)$ transitions are observed. A cutoff wavelength is usually set between the (v', v'') and $(v', v'' - 1)$ bands as shown by curve 1. On the other hand a bandpass filter isolates LIF signal from a specific $(v', v'' - n)$ transition with a favorable FC factor as shown by curve 2, 3, or 4. A bandpass filter with a high transmittance within the $(v', v'' - n)$ band should

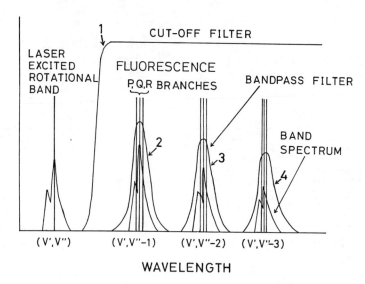

Fig. 6. Cutoff and bandpass filters used for LIF measurements.

be selected. When the transmittance is not constant within the wavelength region of the $(v', v'' - n)$ band, the observed LIF spectrum must be corrected; the transmittance curve can be measured by using a conventional UV spectrometer. Although the intensity of the LIF signal measured with a cutoff filter is much stronger than that obtained with a bandpass filter, the S/N ratio is usually lower because of stronger background signal. General references about optical filters have been given elsewhere [77–79].

Photodiodes that are made of silicon, PIN, and GaAsP are usually used to detect a small portion of dye laser. They are used as a power monitor and to provide a trigger pulse for a photon counter or a boxcar integrator. The photomultiplier type of photodetector is commonly used to detect LIF in the UV, visible, and near-infrared region. In general, photomultipliers consist of a photoemissive surface (photocathode), secondary electron multipliers called dynodes, and an anode. If photoelectrons emitted from the photocathode are accelerated in vacuo and allowed to strike a series of secondary electron-emitting surfaces, a considerable electron multiplication can be achieved and a substantial current can be collected at the anode. The choice of photomultiplier is governed by factors such as the following:

1. The wavelength region of the emission to be detected
2. The intensity of the radiation to be detected
3. The time response required to resolve high speed events.
4. The environmental conditions under which the detector is to be used (e.g., size and temperature)
5. The cost

When a desired photomultiplier is selected, the following characteristics of photomultipliers tabulated in commercial guide books should be compared.

1. *Quantum Efficiency* $[Q(\lambda)]$. The quantum efficiency of a photoemissive surface is defined as

$$Q(\lambda) = \text{electrons emitted/photon absorbed} \tag{39}$$

Practical quantum efficiencies for photoemissive materials range up to about 0.4.

2. *Radiant Sensitivity* $[E(\lambda)]$. The radiant sensitivity is the current per unit area of photoemissive surface produced by unit irradiance:

$$E(\lambda) = \frac{i_k}{P} = \frac{\lambda Q(\lambda)}{1.2395} \times 10^{-5} \tag{40}$$

where i_k is the cathode current in amperes and P the total radiant power in watts.

3. *Cathode Sensitivity* S_k. The cathode sensitivity is represented by

$$S_k = \frac{i_k}{L} \tag{41}$$

where L is the luminous flux of a standard tungsten lamp at 2857 K in lumens

4. *Anode Sensitivity* S_p. The anode sensitivity is given by

$$S_p = \frac{i_p}{L} \tag{42}$$

where i_p is an output current at the anode. When a secondary electron emission coefficient at each dynode is given as δ_i, S_p is obtained from

$$S_p = S_k \prod_{i=1}^{n} \delta_i \tag{43}$$

where n is the number of dynodes.

5. *Spectral Response*. The dependence of quantum efficiencies on the wavelength is shown in Fig. 7 for representative EMI photomultipliers. Such spectral response curves usually provide the most important information in the selection of multipliers. Since the spectral response depends significantly on the wavelengths, observed spectra must be corrected for quantitative measurements of spectral intensities.

6. *Secondary Emission Ratio* [gain: μ]. Gain is defined by the ratio

$$\mu = \frac{S_p}{S_k} = \prod_{i=1}^{n} \delta_i \tag{44}$$

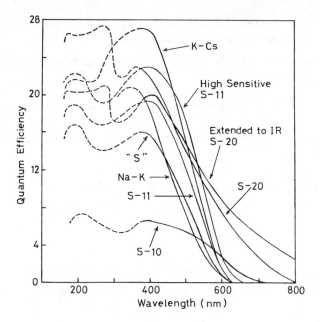

Fig. 7. Wavelength dependence of quantum efficiencies of typical EMI photomultipliers.

Although μ increases with increasing n, sources of noise are multiplied simultaneously. Therefore, the maximum n is limited to about 13.

There are various kinds of photomultipliers. Among them, end-on photomultipliers are most widely used. In such multipliers, photoemissive materials are deposited on the window. Practical photoemissive materials fall into two main categories: classical photoemitters and negative-electron-affinity (NEA) materials. The former photoemitters generally involve an alkali metal or metals, a group V element such as phosphorus, arsenic, antimony, or bismuth and sometimes silver and/or oxygen. The latter photoemitters use a photoconductive single-crystal semiconductor substrate with a very thin surface coating of cesium and usually a small amount of oxygen. Since these surfaces are easily damaged by irradiation by a strong light, careful treatment is required.

The accelerating voltages are supplied to dynodes by a resistive voltage divider called a dynode chain. Typical electronic circuits of dynode chains for alternating current (ac) and dc amplification are shown in Fig. 8. The optimum supply voltage for each dynode is given in guide books. In order to ensure linearity between the intensity of incident radiation and anode current, the following relation should hold between anode current i_p and dynode current i_d:

$$i_d > 10 i_p \tag{45}$$

In general, the output signal of a photomultiplier is either accumulated by

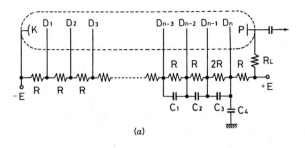

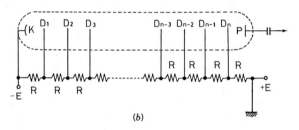

Fig. 8. Dynode chains for (a) ac and (b) dc amplifications: $R = 100\,k\Omega$, $C_1 = C_2 = 0.001\,\mu F$, $C_3 = 0.01\,\mu F$, $C_4 = 0.05\,\mu F$.

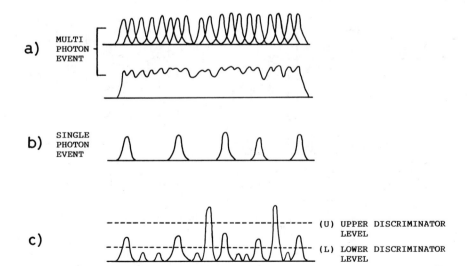

Fig. 9. (a) Multiphoton event and (b) single-photon event in the output signal of a photomultiplier and (c) discrimination between photoelectron pulses and noise pulses by upper and lower discriminator levels.

a photon counter or averaged by a boxcar integrator. The principle of the photon counting technique is as follows. There are two types of responses of a photomultiplier for incident photons as shown Fig. 9.

1. *Multiphotoelectron Event (MPE)*. When a number of photoelectrons are ejected simultaneously from a photocathode due to strong incident radiation, intervals of two signals pulses are shorter than the pulse width. In such a case, each photoelectron pulse cannot be detected separately (Fig. 9a), and spectral measurements are usually carried out by using either an electrometer (picoammeter) to monitor anode current or an operational amplifier to convert anode current to voltage. Figure 10 displays an electronic circuit of the dc amplifier used in our laboratory. It is composed of a dc amplifier in the left part and a smoothing circuit in the right part. In this circuit, the S/N ratio can be improved by using a larger capacity condenser in either the left or center part because of a large time constant, although a slower wavelength scan of monochromator is necessary to follow the long time constant.

2. *Single Photoelectron Event (SPE)*. When photoelectrons ejected from a photocathode are observed discontinuously due to weak incident radiation, each photoelectron pulse can be distinguished (Fig. 9b). The number of photoelectron pulses in this case is proportional to the intensity of the incident light. This allows us to count photons for a constant time to obtain a good

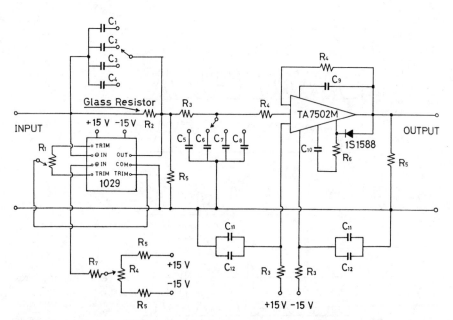

Fig. 10. A typical dc amplifier with Teledyne Philbrick 1029 and Toshiba TA 7502M operational amplifiers: $R_1 = 50\,k\Omega$, $R_2 = 100\,M\Omega$, $R_3 = 100\,k\Omega$, $R_4 = 30\,k\Omega$, $R_5 = 10\,k\Omega$, $R_6 = 15\,k\Omega$, $C_1 = 47\,\mu F$, $C_2 = 10\,\mu F$, $C_3 = 1\,\mu F$, $C_4 = 0.1\,\mu F$, $C_5 = 1\,\mu F$, $C_6 = 3\,\mu F$, $C_7 = 4.7\,\mu F$, $C_8 = 10\,\mu F$, $C_9 = 200\,pF$, $C_{10} = 5000\,pF$, $C_{11} = 22\,\mu F$, and $C_{12} = 0.1\,\mu F$.

S/N ratio. The S/N ratio is given by

$$\frac{S}{N} = \frac{N_s t}{\sqrt{(N_s + N_d)t}} \tag{46}$$

where N_s and N_d are the average numbers of signal pulses and noise pulses per unit time, respectively, and t is the count time. $N_s t$ is obtained by subtracting $N_d t$ measured without incident radiation from the total pulses $(N_s + N_d)t$. According to this equation, the S/N ratio is improved by increasing the count time t.

Since various kinds of noise pulses are overlapped with the photoelectron pulses, they must be removed effectively. Noise pulses of a photomultiplier are classified into the following three types by the scale of pulse height: (1) lower amplitude noise pulses than photoelectron pulses come from thermal electrons from dynodes, (2) higher amplitude noise pulses than photoelectron pulses originate from emission from glass and afterpulses, and (3) noises having comparable pulse heights with photoelectron pulses arise from thermal electrons from the photocathode. Dominant noise from (1) and low-level noise (3) can be removed by using two discriminators (Fig. 9c). The upper and lower discriminator levels should be set like U and L in Fig. 9c, such that higher pulses from (2) and lower pulses from (1) are removed, respectively. Noise pulses (3) can be reduced by using a small cathode photomultiplier or by cooling the cathode to about $-20°C$. In most cases, a favorable pulse height distribution of noise pulses is achieved for a low-noise photomultiplier made especially for photon-counting experiments. Under circumstances in which photomultipliers must be operated in close proximity to magnetic fields, magnetic shields to enclose the tube are available from various tube suppliers.

Various boxcar integrators have been used in which a high time resolution of 100 ps is involved. Figure 11 shows the principle of the boxcar integrator. It consists of an analog switch S and an RC filter to improve the S/N ratio. A part of the input signal is sampled within a time interval ΔT after a trigger pulse T_D and the output voltage is added to the RC integrator with a time constant $\tau = RC$. The signal intensity is averaged by repeating such sampling N times. The effective number of integrating cycles to improve the S/N ratio is given by

$$N_{\text{eff}} = \frac{2\tau}{\Delta T} \tag{47}$$

The S/N ratio is improved by $\sqrt{N_{\text{eff}}}$ because the signal intensity increases by a factor of N_{eff}, whereas the noise does by a factor of $\sqrt{N_{\text{eff}}}$. Boxcar integrators can be used not only for signal averaging but also for reproduction of a time profile of a signal by scanning T_D in coincidence with the signal.

An advantage of the boxcar integrator for LIF measurements is that it can

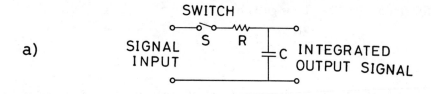

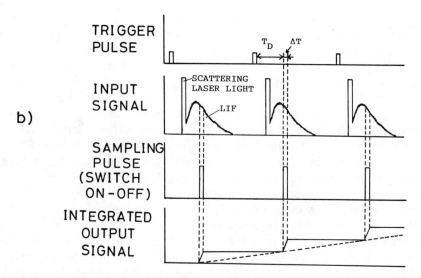

Fig. 11. The principle of the boxcar integrator: (*a*) fundamental electronic circuit and (*b*) operating waveforms.

be used even though a strong background emission is present, as in the cases of flames and discharges. Although MPE occurs in such conditions, a weak LIF signal superimposed upon the strong background emission can be detected if the fluctuation of the background signal is small. On the other hand, the photon counting technique is more effective when the signal intensity is very weak but the background emission is either very weak or absent. A transient digitizer (e.g., Tektronix 7912AD) is also useful to monitor LIF signals, especially for measurements of the time profile of the fluorescence signal. As in the cases of other experimental techniques, an on-line microcomputer is powerful for overall control, data acquisition, and processing.

Here, the use of the ion-beam/LIF technique is demonstrated by a beam–gas study of Ding and Richter [76] on the $N_2^+ + N_2$ reaction at a high projectile ion energy of 800 eV. The LIF spectra of $N_2^+(B^2\Sigma_u^+ - X^2\Sigma_g^+)$ have been measured using a gated photon counter. The signal from a photomultiplier consists of the LIF signal, the scattered laser signal, and emission from metastable ions. The signal pulses were gated for approximately 100 ns to reduce the continuous background mainly from the $A^2\Pi_u - X^2\Sigma_g^+$ emission of N_2^+. Delaying the gate by approximately 50 ns from the laser trigger pulse reduced the influence of the scattered laser light significantly. A high ratio of 10^{16} between the number of incoming photons and the number of signal photons has been obtained.

There are three contributions to the observed LIF spectrum:

$$\begin{array}{c}\text{The unscattered primary } N_2^+ \text{ ions} \\ \text{The inelastically scattered } N_2^+(X)\end{array} \quad (48)$$

$$N_2^+(X:v, N) + N_2 \rightarrow N_2^+(X:v', N') + N_2 \quad (49)$$

and the charge-exchange process

$$N_2^+(X:v, N) + N_2 \rightarrow N_2 + N_2^+(X:v', N') \quad (50)$$

The upper two processes produce ions having roughly the same velocity, which is determined by the energy of incident ion beam. A Doppler shift of about 0.94 Å is expected at 800 eV. Meanwhile, the great majority of product ions from process 50 have roughly thermal velocity because of no significant momentum transfer. The Doppler shift thus enabled Ding and Richter to distinguish between processes 48 and 49 and process 50. The contribution from process 48 is negligible when N_2 gas of about 10^{-4} torr filled the gas cell.

Simulation techniques were applied to obtain ro-vibrational distributions for both the CT products and the inelastic fast ions. The signal intensity for a LIF line (v', N', v'', N'') obtained by monitoring the (v', v) fluorescence is related to the population of a lower ro-vibrational level $N_{v''N''}$:

$$L_{v'N', v''N''} = kN_{v''N''} q_{v'v''} v_{v'N', v''N''} S_{N'N''} g_{N''}^{-1} \sum_v (q_{v'v} v_{v'v}^3) \quad (51)$$

where k is a proportionality constant that includes the electronic transition moment assumed to be constant in this case. The vibrational population ratio between a lower vibrational level $(v'' = l)$ and an upper one $(v' = u)$ is obtained from the relation

$$\frac{N_u}{N_l} = \left(\frac{QN_{uN''}}{N_{N''}}\right)_u \bigg/ \left(\frac{QN_{lN''}}{N_{N''}}\right)_l \quad (52)$$

where Q is the rotational state sum and $N_{N''}$ is a rotational distribution function given by Eq. 17. In the present example, the rotational distributions were expressed by a temperature of 1000 K for the inelastic products and 300 K for the CT products. With background pressures in the range of 10^{-7} torr, the rotational temperature of the primary ions (process 48) was also evaluated to be 650 K [76]. No vibrational excitation was found for inelastic collisions, whereas the CT process exhibited a significant vibrational excitation.

The power of the combination of ion-beam and LIF techniques in ion–molecule reactions is that initial ro-vibrational distribution of products in ground and metastable states can be detected under high resolution. A disadvantage is that the effect of radiative cascade from upper states often complicates the analysis of the data [75]. The application of the LIF method to ion-beam experiments has been severely restricted due to low densities obtainable in free ion beams. Because of space charge effects, these densities hardly surmount 10^6–$10^7\,\text{cm}^{-3}$ for ion energies of several hundreds electron volts. Low LIF intensity makes it difficult to use normal laser techniques such as frequency doubling that extends the tuning range down to about 200 nm and an intracavity etalon that significantly improves spectral resolution (about 0.01 Å), because of considerable reduction of the laser power. It has also restricted the application of new laser techniques such as various laser–laser double-resonance experiments, providing new and valuable insight into the reaction dynamics.

3 ION-TRAP METHODS

In the preceding paragraphs we have presented various optical techniques used for studying ion–molecule reactions in beam experiments. The most significant advantage of beam experiments is that the reactant ion energy is variable over a wide range. However, their disadvantage is that experiments at lower energies below a few electron volts are difficult because of a great reduction of the beam intensity and a large energy spread of the reactant beam. The ion-trap method is a powerful technique for investigating ion–molecules reactions at near thermal energy and low pressures. The ion-trap apparatus is based on the ability to confine a sufficient number of ions in an electric field of suitable geometry [80]. Many types of ion traps have been developed and applied to ion–molecule chemistry. Among them, ICR spectrometers described in Chapter 1 have been used most widely. Although extensive ICR studies have been carried out to determine rate constants and product distributions of ion–molecule reactions, ion-trap experiments coupled with optical techniques have been performed by only two groups. Marx and co-workers have modified a conventional ICR drift cell to incorporate UV/visible emission and LIF spectroscopies [81–92]. Mahan and co-workers have performed a series of LIF studies in a three-dimensional radiofrequency (*RF*) quadrupole trap [93–100]. This paragraph describes techniques used in these trapped ion-optical experiments by using selected examples.

3.1 Ion Cyclotron Resonance Coupled with UV and Visible Emission Detection

3.1.1 ICR Cell

Figure 12 shows a cross-sectional drawing of a drift type of ICR cell fitted with an optical detection system; the cell is inserted in the pole gap of an electro-

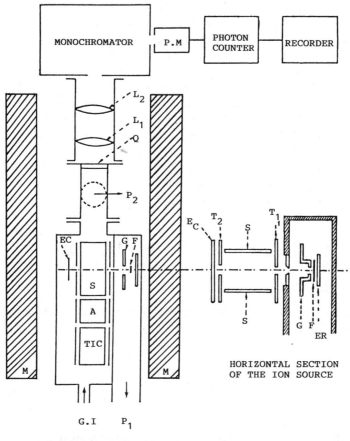

Fig. 12. ICR cell for the study of ion–molecule reaction by UV and visible emission spectroscopy [44]: F, filament; ER, electron reflector; G, grid; EC, electron collector; T_1 and T_2, trapping plates; S, source drift plate; A, analyzer drift plates; TIC, total ion current drift plates; M, magnetic pole caps, G.I, gas introduction; P_1, to electron source pumping; P_2, to main chamber pumping; Q, quartz window; L_1 and L_2, quartz lenses.

magnet. The cell consists of a source region in the middle of which all optical events take place, a short ion analyzer section, a total ion current collector, and a separated electron source. The electron source consists of a tungsten or rhenium ribbon filament, a repeller, and a grid. Except for the gold electron collector and the filament, all the electrodes are made of electrolytically polished Inconel alloy. The geometry of the cell is rectangular due to a compromise between two contradictory requirements. A square cell is favorable for good trapping, whereas a flat one is best suited quick ejection. There are two separate pumping systems: P_1 for the electron source and P_2 for the main chamber. These were necessary to reduce stray light emitted from the filament and to maintain a low enough pressure ($< 10^{-3}$ torr) in the electron source for the optical experiments at relatively high pressures (10^{-3}–10^{-1} torr). A conventional ionization gauge directly connected to the cell could not be used for the pressure measurement because of the high operating pressures and the presence of a strong magnetic field. Values measured by ionization gauges located on P_1 and P_2 were used after correction for pressure drop along the pipes. The emission is observed through a quartz window. Two fixed quartz lenses focus the light emitted around the beam axis on the entrance slit of a 60-cm monochromator. The slit is parallel to the electron beam for an effective collection of light emission.

The favorable features of this apparatus are as follows: (1) an open structure allowing easy light collection; (2) the possibility of trapping the ions long enough in the observation region for a significant fraction of them to react; (3) the possibility of using ion ejection to select the processes producing the observed excited species; and (4) lower operating pressure range than in usual FA (see Section 4), allowing the observation of long-lived emitting species that would otherwise be quenched.

3.1.2 Trapping and Ejection of the Ions

Ions generated by the impact of an electron beam on source gas molecules are trapped by applying an electrostatic potential on the plates perpendicular to the magnetic field. In order to achieve high trapping efficiency, relatively high magnetic fields (9500 G) and trapping voltages (3–5 eV) were used in comparison with conventional ICR experiments. In these conditions, up to 3×10^4 ions were trapped for 3×10^{-5} s. For positive ions this trapping time could be increased up to 10^{-4} s if the electron beam was present all the time to reduce the ion losses due to space charge.

Different types of ion ejection may be applicable to an ICR cell.

1. *Selective Ion Cyclotron Ejection.* This ejection is performed by using an oscillator connected to one of the drift plates. The ejection time is given by

$$\tau = \frac{Bd}{\varepsilon_0} \tag{53}$$

where B, d, and ε_0 are the magnetic field, the distance between drift plates, and the electric field amplitude, respectively. This ejection is rather slow ($\tau \geqslant 5 \times 10^{-5}$ s) under such operating conditions, although it has the advantage of being selective and the velocity is independent of the mass of the ion.

2. *Selective Ion Ejection at the Trapping Electrodes.* This ejection results from the excitation of the oscillatory motion of the ions between trapping electrodes, with an audiofrequency field of frequency

$$\omega_T = \left(\frac{4eV_T}{mD^2}\right)^{1/2} \tag{54}$$

where V_T, m, and D are the trapping potential, the mass of the ion, and the distance between trapping plates, respectively. This method is applicable to a square cell, although resolution is quite poor, particularly if one tries to increase speed by increasing the ejection field amplitude.

3. *Nonselective Detrapping.* This ejection is carried out by reversing the sign of the potential applied to one of the trapping plates. The ejection time for an ion starting from the the center of the cell is given by

$$\tau' = D\left(\frac{m}{2eV_T}\right)^{1/2} \qquad (55)$$

Detrapping ejection is always more rapid, especially for light ions. As an example, He$^+$ ions are ejected in about 1 μs with $V_T = 5$ V. This time is sufficiently short to obtain significant difference signals even with rate constants as high as about 10^{-9} cm^3 s^{-1}. Only this ejection technique could be used in the optical experiments of Marx and collaborators. The effect of kinetic energy of ions accelerated by the trapping potential must be considered when a high trapping potential is used. However, the contribution of fast ions has been estimated to be small under their operating conditions.

3.1.3 Pulsing and Counting Sequences

In the ICR cell, the following processes participate in the formation of excited species (C*):

1. Primary excitation of a target molecule B by the electron beam

$$e^- + B \rightarrow C^* \qquad (56)$$

2. Secondary reactions of metastable neutrals produced by the electrons

$$A^* + B \rightarrow C^* \qquad (57)$$

3. Reactions of primary ions in stable or metastable states

$$A^+ + B \xrightarrow{k_{58}} C^* \qquad (58)$$

In order to observe photons from ion–molecule reaction 58 selectively, a modulation system shown in Fig. 13 was used. The electron beam is modulated on and off by pulsing the potential applied to the grid of the electron gun. Ions are alternatively trapped or ejected from the observation region, and the corresponding photons are accumulated in channels A and B of a synchronous photon counter, respectively. After a sufficiently large number N of trapping/ejection cycles (typically $N = 10^6$), the difference signal of interest $\Delta S = S_A - S_B$ is registered. For example, delay and modulation times used for the study of the He$^+$ + N$_2$ → N$_2^+$(C) + He system are as follows: $t_1 = 15$–$30\,\mu$s

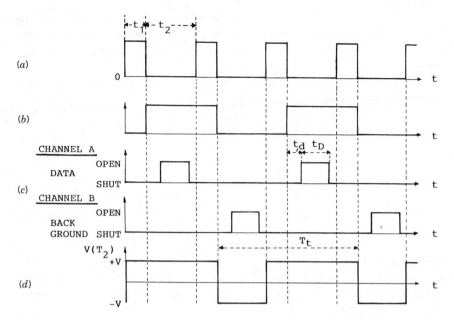

Fig. 13. Pulsing sequences: (*a*) electron-beam pulse; (*b*) triggering of a dual photon counter: data channel activated by leading edge, background channel activated by the trailing edge; (*c*) opening of the two channels gates during a time t_D after a delay t_d; (*d*) signal applied to the trapping plate T_2: T_1 is maintained at potential $+V$. T_t denotes one period of the pulsing sequence.

with t_d about 5 μs and $t_D = 10$–20 μs [44]. Care must be taken when long-lived excited species are produced in the ICR cell. Various methods to check whether the observed emitting species certainly result from a primary ion–molecule reaction have been described in detail [44].

If the decay of primary ions is first order, ΔS is given by

$$\Delta S = \left(\frac{N\beta k_{58}\sigma_+ il[A]}{k_+^2 [B]} \right) \{1 - \exp(-k_+[B]t_1)\}$$
$$\times \{1 - \exp(-k_+[B]t_D)\}\{\exp(-k_+[B]t_d)\} \qquad (59)$$

where β is an apparatus function depending on the photon collection efficiency of the optical arrangement and on the spatial distribution of the primary species, σ_+ is the ionization cross section of primary ions A^+, i is the electron beam intensity, l is the path length of the electrons, and k_+ is the overall decay rate constant of A^+. The following points have been noted to maximize ΔS [44]: First, for a fixed total accumulation time $T = NT_t$, ΔS becomes a maximum when $t_1 = t_2$ and there is no benefit to have t_2 longer than the overall decay time of A^+. Second, ΔS is linearly proportional to [A] (Eq. 59). However, when the mean free path of A^+ ions becomes shorter than their distance to the detrapping plate at high [A], the ejection time increases so that the delay time t_d

has to be increased. Since ΔS is a decreasing function of t_d, this gives a higher limit to [A]. Third, there is a rather broad maximum corresponding to the most favorable pressure range for [B] because ΔS increases linearly with [B] at low [B], whereas it decreases rapidly as a function of $\{\exp(-k_+[B]t_d)\}/[B]$ at high [B]. The limits on [B] depend mainly on k_+. Fourth, the electron beam intensity i is limited dominantly by space charge to about $5\,\mu A$. When the ion density is too high, ΔS becomes small because of an increase in diffusion rate out of the reaction zone.

3.1.4 Emission Spectra Excited in the ICR Cell

Emission spectra are recorded by stepping the grating of the monochromator. Each step is followed by a starting pulse that resets the counters and initiates a new sequence at the new wavelength. This automatic step-to-step process can also be used at a fixed wavelength to vary the electron beam energy for the measurement of an excitation function or to vary the delay t_d for the measurement of a rate constant. The overall rate constant k_+ is obtained from a slope of a logarithmic plot of ΔS vs. t_d.

For example, the emission spectrum produced from the $He^+ + H_2O$ reaction at thermal energy has been studied in the ICR cell [84]. The spectrum was observed in a (He, H_2O) mixture, the partial pressure of which was typically about 3×10^{-3} torr for He and 3×10^{-4} torr H_2O. The dominant band was $OH^+(A^3\Pi - X^3\Sigma^-)$ excited by the dissociative CT reaction,

$$He^+ + H_2O \rightarrow OH^+(A^3\Pi) + H + He \tag{60}$$

The measurement of the excitation function, using the He* 491.1-nm line as a reference, gave a threshold at $24.5 \pm 0.5\,eV$ for process 60, which is in good agreement with the recombination energy of $He^+(24.58\,eV)$. The rate constant of process 60 was estimated by comparing the total emission intensity with the known value for $N_2^+(B)$ in the thermal $He^+ + N_2$ CT reaction in a prepared mixture of (He, N_2, H_2O). In addition to the He^+/H_2O [84] and He^+/N_2 [44, 85] systems, the following systems have been investigated in ICR-optical experiments: He^+/D_2O [83, 84], He^+/N_2O [87], He^+/C_2H_2 [86], and Ar^+/H_2O [88]. The most valuable feature of this technique is that excited products of ion–molecule reactions can be detected under single collision conditions because of the low pressure in the ICR cell. A disadvantage of this technique is that emission intensity is much weaker than in the conventional FA apparatus and in the FA ion source coupled with a low-pressure cell, as will be described in Section 4. The optical resolution is therefore too poor to analyze rotational distributions of excited products.

3.2 ICR Cell Coupled with LIF Detection

Danon et al. [90] modified their ICR apparatus to allow for laser excitation and subsequent observation of time-resolved fluorescence spectra of trapped ions. Their main purpose was to probe internal energy distribution of ground and optically metastable products of ion–molecule reactions. A schematic drawing of the ICR–LIF apparatus is shown in Fig. 14. The experimental setup is essentially identical with that described in the previous paragraph except for the addition of a laser excitation and fluorescence detection system. These workers used a N_2 laser (Molectron UV14) to pump a Hänsch-type tunable dye laser (Sopra) with an oscillator–amplifier configuration. Typically, 8-ns pulses with 0.6-mJ peak energy were obtained at 50-Hz maximum repetition rate, the half-width being 0.02 nm at 460 nm. The ICR cell is a rectangular trapped-ion-type single-section cell. End plates are removed so that the laser beam can cross the cell freely. The laser beam is focused into the center of the ICR cell by using two prisms and two quartz lenses. The resulting LIF emission is observed through a side arm at right angles to both the laser and electron beams; the electron beam is perpendicular to the plane of the drawing. Two apertures cut into each drift plate and covered with gold wire mesh allow the detection of LIF emission in the center of the ICR cell (see Fig. 15). Two quartz lenses are used for the LIF detection, one for an effective collection with a large solid angle and the other for focusing light on the photocathode of a photomultiplier. Optical filters are inserted between these lenses to reduce the remaining scattered light from the laser and the filament of the electron gun. The electron source and the reaction chamber are pumped by one turbomolecular pump through two arms.

Various background emissions and LIF emissions occur simultaneously in the ICR cell. In order to detect the LIF spectrum selectively, experiments are

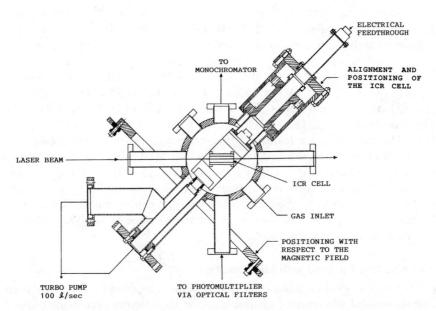

Fig. 14. ICR apparatus for observing LIF from products of ion–molecule reactions [90].

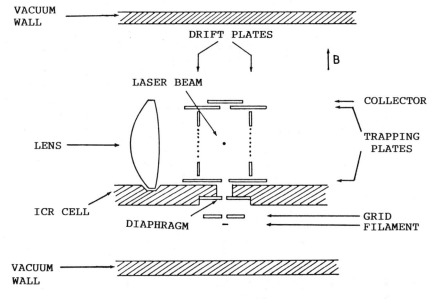

Fig. 15. ICR cell for observing LIF spectra [90].

performed in a pulsed mode shown in Fig. 16. The experimental cycle consists of four sequences; A, B, C, and D. The photons detected in these sequences correspond to the following processes:

$$C_A = I_{LIF} + IM_{LIF} + N_{LIF} + I_L + N_L + SL + SF \tag{61}$$

$$C_B = I_L + N_L + SF \tag{62}$$

$$C_C = N_{LIF} + N_L + SL + SF \tag{63}$$

$$C_D = N_L + SF \tag{64}$$

Here, I_{LIF}, IM_{LIF}, and N_{LIF} are LIF signals of ions resulting from electron impact, ion–molecule reactions, and neutrals, whereas I_L and N_L are the emission signals of long-lived states of ions and neutrals produced by electron impact or ion–molecule reactions, respectively. SL and SF are the scattered light from the laser and the filament of the electron gun, respectively. It can be deduced that

$$(C_A - C_B) - (C_C - C_D) = I_{LIF} + IM_{LIF} \tag{65}$$

A photon-counting system fitted with a multichannel time-to-digital converter was used to perform time-resolved detection. In this system, time-resolved LIF signals, between 10 ns and 5 μs could be measured with 10-ns resolution. A fast

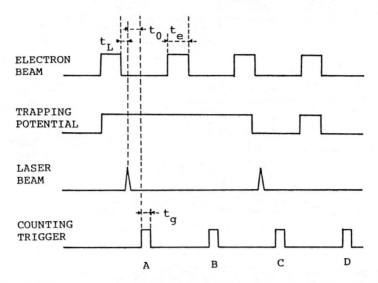

Fig. 16. Pulsing sequences used for LIF detection of trapped ions in the ICR cell [90].

microprocessor was used for overall control, data acquisition, and processing. A block diagram of the experimental system is shown in Fig. 17.

The ICR–LIF technique has only been applied to the measurement of the $CO^+(A-X)$ LIF spectra in pure CO and (CO, Ar) mixtures [90–92]. The main purpose was the study of internal state distributions of $CO^+(X)$ from the CT reaction:

$$Ar^+(^2P_{1/2}, {}^2P_{3/2}) + CO(X^1\Sigma^+ : v=0) \xrightarrow{k_{66}} Ar(^1S_0) + CO^+(X^2\Sigma^+ : v'') \quad (66)$$

Since no signal from neutral species was present, the photon counts due to ions from $C_A - C_B$ were observed at a constant trapping voltage. A relatively large delay time between ionization and laser excitation was necessary to reduce the background signal from $CO^+(A-X, B-X)$ emissions excited by electron impact and to avoid an overflow of the counting system. Typical delay and modulation times were as follows: $t_e = 40\,\mu s$, $t_L = 13\,\mu s$, $t_0 = 10$–$20\,\mu s$, and $t_g = 3.5\,\mu s$. Photon counts were accumulated typically during 4000 laser pulses.

The trapping efficiency of $CO^+(X)$ was estimated from the intensity change in the LIF signal when the delay time t_L was increased from 13 to 21 μs. A first-order exponential decay with a rate constant inversely proportional to CO pressure was observed at all CO pressures studied, indicating that the main decay is due to diffusion along the magnetic field. The diffusion rate constant was $k_D = 6 \times 10^4\,s^{-1}$ in pure CO at 2.5 mtorr. This value is much smaller than a free diffusion constant of $k_{D'} = 1 \times 10^6\,s^{-1}$, reflecting efficient ion trapping. When the laser beam was turned on, 1×10^6 ions·cm^{-3} were trapped in the center of

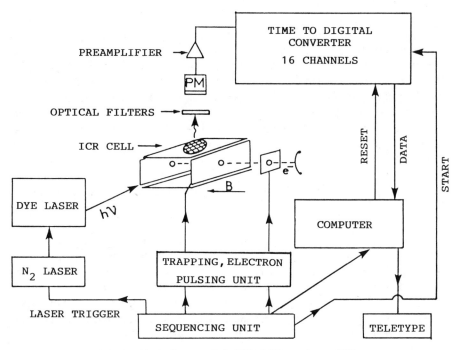

Fig. 17. Block diagram of the ICR–LIF experiment [90].

the cell 13 μs after the end of the electron pulse. The limiting sensitivity of $CO^+(X)$ was evaluated to be 2×10^5 ions·cm^{-3}.

The following two types of LIF experiments have been attempted:

1. *Radiative Lifetime and Quenching of* $CO^+(A:v' = 1)$. In general, when an excited species C* decay both radiatively and nonradiatively by collisions with a quencher M

$$C^* \xrightarrow{k_r} C + hv \tag{67}$$

$$C^* + M \xrightarrow{k_n} C + M^* \tag{68}$$

the measured fluorescence decay rate $k(=1/\tau)$ is expressed by

$$k = k_r + k_n[M] \tag{69}$$

The radiative lifetime $\tau_r(=1/k_r)$ and the quenching rate constant k_n can be obtained from a Stern-Volmer plot of $k(=1/\tau)$ against [M]. By using this method, radiative lifetimes and collisional quenching rate constants of $CO^+(A:v' = 1)$ for CO and Ar have been evaluated.

2. *LIF from the* $Ar^+ + CO$ *CT Reaction.* A fairly high CO pressure was necessary to obtain a detectable amount of $CO^+(X)$ because of the low detection sensitivity and the slow CT reaction rate ($k_{66} = 5 \times 10^{-11}$ cm^3 s^{-1}). Under such conditions the other processes, mainly electron-impact ionization of CO and the $CO^+ + CO$ CT reaction, were significant and completely smeared out the initial population. A kinetic simulation of the observed dependence of LIF intensity on the CO or Ar pressure provided the following conclusion: The thermal energy $Ar^+ + CO$ CT reaction yields vibrationally excited levels other than the $CO^+(X^2\Sigma^+ : v'' = 0)$ level but they are effectively quenched to $v'' = 0$ through fast quasiresonant CT processes such as

$$CO^+(X^2\Sigma^+ : v'') + CO(X^1\Sigma^+ : v = 0) \to CO(X^1\Sigma^+ : v = v'') + CO^+(X^2\Sigma^+ : v'' = 0) \tag{70}$$

In order to obtain direct and quantitative information on the vibrational and rotational distributions of $CO^+(X)$ in process 66, a higher sensitivity apparatus at low pressure was required by reducing the scattered light from the electron source and increasing the trapping efficiency of ions. The competition of fast deactivation processes of product ions is a general problem in the LIF experiments, especially at high pressure, since most vibrationally excited ions are more reactive than those in the vibrational ground state. It was concluded that the production rate should be at least of the order of their deactivation rate to determine nascent interal state distributions of ion–molecule reactions in the ICR–LIF apparatus. Successful LIF studies on the nascent internal state distribution of $CO^+(X)$ in process 66 have been performed in FA (see Section 4.3) and FA coupled with a low-pressure reaction cell (see Section 4.4.2).

3.3 Quadrupole Ion Trap Coupled with LIF Detection

Figure 18 shows a schematic representation of the three-dimensional quadrupole trap that is capable of storing 10^6–10^8 ions·cm^{-3} for tens of millisecond at a pressure of 10^{-4}–10^{-6} torr [95]. The trap is of cylindrical geometry rather than a general hyperbolic design primarily to minimize electric-field distortion introduced by the presence of laser-beam entrance and exit holes in the center electrode. The trap cell consists of a hollow, cylindrical center electrode and two end caps that are electrically grounded. The geometry is determined by the cap-to-cap spacing ($2Z_1$) and the radius of the cylindrical center electrode (r_1). In this design, $r_1 = 1.43$ cm, and cap electrode cones were available to permit $Z_1 = r_1$ or $2Z_1^2 = r_1^2$. The latter geometry creates an electric field closely approximated by the ideal hyperbolic case.

Trapping is achieved by applying an RF voltage at some dc bias level. Variation of the peak RF voltage determines the center position of a charge-to-mass window such that all ions within the window are trapped, whereas the dc-to-rf ratio determines the range of mass window. The maximum mass was ~ 100 amu and the maximum useable resolution $m/\Delta m$ was ~ 40 in this design.

Given the general design of the trap, the depth of the pseudopotential well

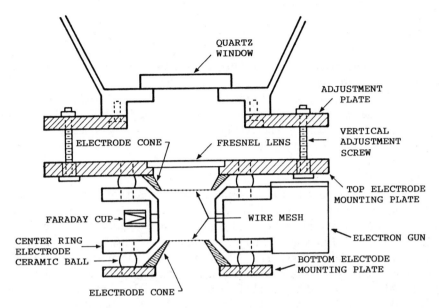

Fig. 18. Schematic representation depicting a vertical slice of the quadrupole ion trap used for the LIF study [95].

can be expressed as

$$\bar{D} = \frac{eV^2}{4Z_1^2 M f^2} = 6.109 \times 10^{-3} \frac{V^2}{Z_1^2 M f^2} \quad \text{V cm}^{-1}\text{amu}^{-1}\text{MHz}^{-1} \quad (71)$$

where V is the maximum RF voltage between the electrodes, M is the mass in amu, and f is the applied field frequency in megahertz. Typical values for these parameters were $Z_1 = 1$, $V = 200$, $M = 13$, and $f = 1$, and the corresponding well depth was 19 V. The well depth determines the number of ions that can be trapped. The maximum concentration of ions is calculated by

$$n_{\max}(\text{ions} \cdot \text{cm}^{-3}) = 1.66 \times 10^6 \frac{\bar{D}}{Z_1^2} \quad (72)$$

For a well depth of 19 eV, 3×10^7 ions·cm^{-3} could be trapped. According to these equations, the number of trapped ions is proportional to f^{-2}. Thus, in order to trap large numbers of ions, a relatively low RF frequency (0.5–1.0 MHz) should be used. This situation results in a driven RF motion of ions that, while of small amplitude, significantly broadens the Doppler profile of the ions. This effect makes high-resolution LIF studies of polyatomic ions difficult.

Ions are created within the trap by electron impact ionization of a selected source gas. The pressure ranged from 10^{-6} to 10^{-5} torr. Under these conditions, the trapped ions experience collisions with background gas molecules in a millisecond time scale. When ion–molecule reactions are studied, an additional buffer gas is added that varies the pressure over a wide range. The electron gun consists of a resistively heated thoriated tungsten wire mounted on a ceramic base and enclosed within a metal shield that is floated at the center electrode potential. This shielding serves to reduce scattering light emitted from the hot filament. A series of lenses focus the electrons through a hole in the center electrode and into the trap. The average electron current is typically 50 μA as monitored by a small Faraday cup. One of the focusing lenses also serves as an electron shutter through the application of a high-voltage pulse to the lens element.

As in the case of the ICR–LIF experiment, the RF ion trap–LIF measurement was performed in a pulse mode with a repetition rate of 10–40 Hz. Each pulsing sequence consists of ion creation and confinement, excitation of the ions, and LIF detection. The ions are formed by a short ($\sim$ 1–2 ms) pulse of electrons and after a short (0.1–1.0 ms) delay, are probed by a 10-ns pulse from a tunable dye laser. During the delay time any necessary mass selection of the ions is allowed to occur, and any excited electronic state relaxes radiatively. LIF is observed at right angles to the laser beam (Fig. 19). A lens-and-mirror system directs some of the fluorescence through wire mesh end electrodes to a cooled photomultiplier tube. In order to minimize scattered laser light, the laser beam is collimated to ca. 0.5 cm diameter with two lenses and directed through 0.5 m arms fitted with light baffles. In addition, the use of a gated photon counting system reduces the effects of scattered laser light. After the LIF detection, the

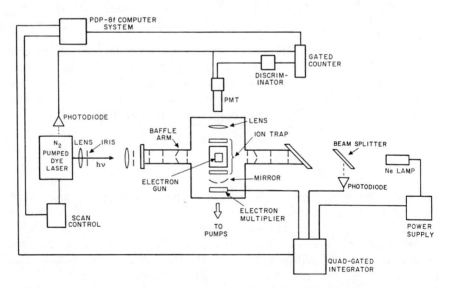

Fig. 19. Block diagram of the quadrupole ion trap–LIF apparatus used in the frequency scanning experiments [95].

experimental cycle is completed by pulsing the ions out of the trap to a magnetic strip electron multiplier. To obtain a typical spectral scan, LIF from 500 such cycles is accumulated at each 0.1-Å wavelength increment. This LIF signal is normalized to both the laser power monitored by a photodiode and the total density of ions in the trap measured by the electron multiplier. A hollow cathode discharge in Ne is mounted past the exit window to monitor the laser wavelength optogalvanically. The entire experiment is controlled by a computer system.

The following types of measurements have been carried out.

1. *LIF Spectra and Radiative Lifetimes.* LIF spectra of parent cations and fragment ions have been observed: $N_2^+(B^2\Sigma_u^+ - X^2\Sigma_g^+)$, $CO^+(A^2\Pi - X^2\Sigma^+)$, $CH^+(A^1\Pi - X^1\Sigma^+)$, $CD^+(A^1\Pi - X^1\Sigma^+)$, $HCl^+(A^2\Sigma^+ - X^2\Pi)$, $HBr^+(A^2\Sigma^+ - X^2\Pi)$, $BrCN^+(\tilde{B}^2\Pi - \tilde{X}^2\Pi)$, and 1,3,5-trifluorobenzene cation $(\tilde{B}^2A_2'' - \tilde{X}^2E'')$ [93–100]. In general, the trapping of fragment ions with large kinetic energies is more difficult than that of parent cations. Although the rather high velocity and chaotic motion characteristic of trapped ions restricts the spectral resolution, this technique has successfully been applied to spectroscopic studies of $CH^+(A^1\Pi - X^1\Sigma^+)$ and $CD^+(A^1\Pi - X^1\Sigma^+)$ with relatively large rotational constants [93, 96, 97]. New lines at high J that could not be fit by the existing molecular constants were identified due to a high rotational excitation. A new set of constants was obtained by refitting observed CH^+ and CD^+ lines. For $BrCN^+$, the vibrational frequency of the v_1 mode (primarily C—Br stretch) of the $\tilde{B}^2\Pi$ state and the difference of spin-orbit constants $A'' - A'$ were determined from the vibrational analysis of the $\tilde{B}^2\Pi - \tilde{X}^2\Pi$ system. However, the complicated structure due to overlapping of many sequence bands and Fermi resonances made difficult the detailed analysis.

When radiative lifetimes were measured, the laser wavelength was fixed at a given transition, and the resulting LIF signal from a photomultiplier was fed into a discriminator and then into a transient digitizer (Tracor Northern NS 575) with a channel width of 10 ns. The LIF signal as a function of time was then accumulated in the digitizer for a period of several thousand experimental cycles. The dye laser was then detuned from the transition, typically several angstrom, and a background signal was subtracted for an equivalent time period. Data from the digitizer were analyzed by an on-line computer. The radiative lifetimes of the following ions have been measured: $N_2^+(B^2\Sigma_2^+)$ [94, 98], $CO^+(A^2\Pi)$ [96, 100], $CH^+(A^1\Pi)$ [96, 97], $CD^+(A^1\Pi)$ [97], $HCl^+(A^2\Sigma^+)$ [99], and $HBr^+(A^2\Sigma^+)$ [99]. An advantage of this technique in comparison with the pulsed electron-impact method [101] is the absence of rapid spatial dissipation of ions from the viewing region due to ion–ion repulsive forces. For example, the measured lifetime of $CH^+(A^1\Pi)$ was 815 ± 25 ns, in good agreement with the upper range of the theoretical value (660–800 ns [102]), and was larger than the value of 630 ns obtained under electron-impact excitation [101].

2. *Ion State Distributions.* The RF trap can be operated at sufficiently low pressures so that ions can be confined in a collision-free environment for long times compared to fluorescence decay times and, generally, for times that are

limited by the collision rate between an ion and a reagent gas. Under such conditions, nascent internal state distributions of ionized molecules and fragments have been obtained. For example, hot $CH^+(X)$ and $CD^+(X)$ fragment ions have been observed from electron impact on CH_4 and C_2H_2 [97]. The vibrational and rotational temperatures were estimated at about 5500 and 3000 K, respectively, for both molecules.

When a radiative cascade contributes to the formation of a ground or optically metastable state, the effect must be taken into account. For example, the (4, 5) and (7, 7) LIF bands of $CO^+(A-X)$ have been observed under electron impact on CO at roughly 150 eV, although the ionization into these high vibrational levels is unfavorable owing to small FC factors for ionization [100]. The formation of these high vibrational levels was explained as a result of the $A-X$ radiative cascade.

3. *Product State Analysis and Collisional Energy Transfer.* The RF trap coupled with LIF detection has been used for the determination of ro-vibrational state distributions of $N_2^+(X)$ and $CO^+(X)$ and for the study of energy transfer of $N_2^+(X)$ ions by collisions with N_2 and inert gases [94, 98, 100]. For example, the $N_2^+(X:v=0)$ rotational distribution was characterized by a rotational temperature of 300 ± 25 K at an N_2 pressure of 5×10^{-6} torr. From the observed Doppler width of well-resolved LIF bands, the effective translational energy of $N_2^+(X:v=0)$ was estimated to be 0.3–0.5 eV. When a buffer gas was mixed to study the effect of collisions between $N_2^+(X:v=0)$ and the buffer gas, this translational energy caused a rapid heating of the rotational distribution as a result of translational to rotational energy transfer:

$$N_2^+(X^2\Sigma_g^+:v=0, \text{low } N) + Ar \xrightarrow{T-R} N_2^+(X^2\Sigma_g^+:v=0, \text{high } N) + Ar \qquad (73)$$

The degree of such $T-R$ transfer can be controlled by changing either the Ar gas pressure or the interval between the ion formation and LIF detection. After 20 collisions by Ar atoms, $N_2^+(X)$ is excited up to $T_R = 4000$–6000 K. Using this technique, it may be possible to populate collisionally high rotational levels in selected ions and to observe, by suitable means, rotational predissociation, thus providing accurate estimates of dissociation energies.

Under collision-free conditions, the R head of the (1, 1) band of $N_2^+(B-X)$ was observed. The band became weak with increasing Ar pressure and almost disappeared after 20 collisions due to a fast CT reaction:

$$N_2^+(X^2\Sigma_g^+:v=1) + Ar \rightarrow N_2(X^1\Sigma_g^+:v=0, 1) + Ar^+ \qquad (74)$$

A similar CT channel is energetically closed for the $N_2^+(X^2\Sigma_g^+:v=0)$ state. In the N_2^+–N_2 system, $N_2^+(X^2\Sigma_g^+:v=1)$ was efficiently quenched by the CT process:

$$N_2^+(X^2\Sigma_g^+:v=1) + N_2(X^1\Sigma_g^+:v=0) \rightarrow N_2(X^1\Sigma_g^+:v=0, 1) + N_2^+(X^2\Sigma_g^+:v=0) \qquad (75)$$

The quenching cross section of $N_2^+(X^2\Sigma_g^+)$ by process 75 was estimated to be 20 times larger than that by process 74. As shown in the above example, the RF trap–LIF technique provides valuable information about the dependence of CT reaction rates on vibrational energy of reactant ions.

4 FLOWING AFTERGLOW METHODS

Techniques used for studying ion–molecule reactions at thermal energy are ion-trap methods described in the preceding section and FA methods discussed in this section. Although ion-trap experiments coupled with optical spectroscopy have been quite restricted, extensive optical studies have been performed with FA by observing UV and visible emission, IRCL, and LIF. Here, the modifications of the conventional FA method required for optical spectroscopic studies are described.

4.1 FA Coupled with UV and Visible Emission Detection

The FA apparatus was originally coupled with emission spectroscopy in the UV and visible region for product state analysis in energy-transfer reactions by long-lived metastable species such as $Ar(^3P_{0,2})$, $He(2^3S)$, $N_2(A^3\Sigma_u^+)$, and $CO(a^3\Pi)$ [103–107]. In most cases, ionic active species involved in the discharge flow have been treated as undesired impurities. For example, although He^+ and He_2^+ ions have been known to take part in the formation of some excited species in the He afterglow, considerable care has been taken to reduce their contribution [108–110]. In such FA optical studies for neutral reactions, a hollow cathode discharge source and a relatively low-capacity pumping system ($\sim 1000\,L\cdot min^{-1}$) have usually been used. Combining a microwave discharge with a high-capacity pumping system (7000–30,000 $L\cdot min^{-1}$), FA optical

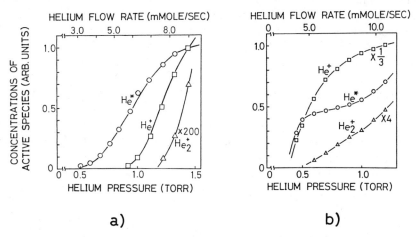

Fig. 20. Dependence of relative concentration of He^+, He_2^+, and $He(2^3S)$ on the He pressure or the He flow rate when the flow system was maintained by (a) 7000 $L\cdot min^{-1}$ and (b) 10,000 $L\cdot min^{-1}$ pumps. Distance between the discharge and the mixing region 11 cm; microwave power 70 W.

spectroscopic studies have recently been extended to ion–molecule reactions [111–144]. For generation of ionic active species, the microwave discharge is preferred over the hollow cathode discharge for the following reasons: (1) Based upon our experience, ionic active species are formed in He, Ne, and Ar afterglows by the microwave discharge, whereas the hollow cathode discharge produces ions only in the He afterglow; (2) in the He afterglow, relative concentrations of ionic active species to those of metastable atoms are high; and (3) a stable discharge is maintained at a wider pressure range for a long time without contamination of the discharge section by electrode materials. The high- and low-pressure limits of the microwave discharge used for generating ionic active species are governed mainly by the loss of the ions due to collisions with the walls of the discharge tube. In order to obtain a high ion density ($> 10^9$ cm^{-3}), a fast linear flow velocity ($\sim 10^4$ cm·s^{-1}) should be established with a fast pumping system. As one example, Fig. 20 shows the dependence of relative concentrations of active species in the He afterglow, He$^+$, He$_2^+$, and He(2^3S), on the He pressure, when 7000 and 10,000 L·min^{-1} pumps are used in our laboratory [119, 144]. The relative concentration of each active species is estimated by comparing the total emission intensity of the N$_2^+$ $(B - X)$ system produced in reactions of He$^+$, He$_2^+$, and He(2^3S) with N$_2$ [119]. A comparison of Figs. 20a and b demonstrates that by using a higher capacity-pumping system the concentrations of ionic active species relative to those of metastable atoms are significantly enhanced and experiments at lower rare gas pressures become possible. When a 30,000 L·min^{-1} booster pump is used, experiments with He$^+$ and He$_2^+$ can be extended to lower pressure (about 0.1 torr) [138]. To achieve a high concentration of ionic active species, not only an increase in the pumping speed but also a short distance between the discharge and mixing zone and a large discharge tube are generally recommended. As shown in Fig. 20, [He$_2^+$] increases more rapidly than [He$^+$] with increasing the He pressure, because He$_2^+$ ions are formed through secondary reactions such as $2\text{He}(2^3S) \rightarrow \text{He}_2^+ + e^-$ ($k = 1.1 \pm 0.2 \times 10^{-9}$ cm^3 s^{-1} [145] and He$^+ + 2\text{He} \rightarrow \text{He}_2^+ +$ He($k = 1.1 \pm 0.1 \times 10^{-31}$ cm^6 s^{-1} [146]. The [He$_2^+$]/[He$^+$] ratio can also be enhanced by extending the distance between the discharge and mixing zone. For example, the [He$_2^+$]/[He$^+$] ratio at 30 cm downstream from the discharge region is enhanced by a factor of ~ 5 relative to that 11 cm downstream. A similar method is effective for the enhancement of Ar$_2^+$ ions in the Ar flow and Ne$_2^+$ ions in the Ne flow. Table 1 summarizes rare gas ions that can be studied by the FA method, their available energies, and emission systems used for the diagnosis of each active species. The recombination energies of molecular ions, He$_2^+$, Ne$_2^+$, and Ar$_2^+$, have ranges because the ground states of He$_2$, Ne$_2$, and Ar$_2$ are repulsive. Energies for metastable Ar ions are shown for both the Ar$^{+*}(3p^43d) \rightarrow$ Ar$^*(3p^53d)$ CT process and the Ar$^{+*}(3p^43d) \rightarrow$ Ar$^+(^2P_{3/2}:3p^5)$ energy transfer process, because the two electron processes, Ar$^{+*}(3p^43d) \rightarrow$ Ar($^1S:3p^6$), are generally unfavorable [147].

The FA optical apparatus fitted with a microwave discharge and a fast pumping system has originally been used by Collins and Robertson [148, 149]

Table 1. Emission Systems used for Diagnosis of Rare Gas Ions in the FA

Rare-Gas Ions	Available Energies (eV)			Emission Systems	Reagent	Ref.
He^+	24.58			$N_2^+(B^2\Sigma_u^+ - X^2\Sigma_g^+)$	N_2	[119, 150]
				$CO^+(A^2\Pi - X^2\Sigma^+)$	CO_2, OCS	[117, 135]
				$CS^+(B^2\Sigma^+ - X^2\Sigma^+)$	CS_2, OCS	[117, 119]
He_2^+	18.3–20.3			$N_2^+(B^2\Sigma_u^+ - X^2\Sigma_g^+)$	N_2	[119, 150]
				$CO_2^+(\tilde{B}^2\Sigma_u^+ - \tilde{X}^2\Sigma_g^+)$	CO_2	[127]
Ne^+	$\begin{cases} 21.56(^2P_{3/2}) \\ 21.66(^2P_{1/2}) \end{cases}$			$CO^+(A^2\Pi - X^2\Sigma^+)$	OCS	[128, 137]
				$SiBr^+(a^3\Pi_{0^+,1} - X^1\Sigma^+)$	$SiBr_4$	[132]
Ne_2^+	18.9–19.3			$N_2^+(B^2\Sigma_u^+ - X^2\Sigma_g^+)$	N_2	[131]
Ar^+	$\begin{cases} 15.76(^2P_{3/2}) \\ 15.93(^2P_{1/2}) \end{cases}$			$OCS^+(\tilde{A}^2\Pi - \tilde{X}^2\Pi)$	OCS	[128]
				$HBr(A^2\Sigma^+ - X^2\Pi)$	HBr	[139]
Ar_2^+	$12.6–13.5^a$, 14.4^b			$CS_2^+(\tilde{A}^2\Pi_u - \tilde{X}^2\Pi_g)$	CS_2	[134]
				$C_4H_2^+(\tilde{A}^2\Pi_u - \tilde{x}^2\Pi_g)$	C_4H_2	[138]
Ar^{+*}	$Ar^{+*} \to Ar^+$,	$Ar^{+*} \to Ar^*$		$CH(A^2\Delta - X^2\Pi)$	CH_3CN, CH_4	[111, 112]
	16.41	17.87–18.33	$(^4D_{7/2})$			
	17.63	19.09–19.55	$(^4F_{9/2})$			
	17.69	19.15–19.61	$(^4F_{7/2})$			
	18.49	19.95–20.41	$(^2F_{7/2})$			
	19.12	20.58–21.04	$(^2G_{9/2, 7/2})$			
	20.27	21.73–22.19	$(^2F_{7/2})$			

^{a} Reference [134].
^{b} Reference [171].

who have found that He^+ and He_2^+ take part in some He afterglow reactions of simple molecules. A similar apparatus has also been used for investigating reactions due to metastable argon ions by Kuchitsu and co-workers [111–115] and those due to Ar^+ [128, 139], Ar_2^+ [134], Ne^+ [128, 132, 137], and Ne_2^+ [131] by Tsuji and co-workers. As an example, Fig. 21 shows the FA apparatus used in our laboratory for the study of the $Ar^+ + CS \rightarrow CS^+(B) + Ar$ CT reaction by UV and visible emission spectroscopy. The flow reactor, which consists of two quartz discharge tubes (i.d. 11 mm) and a stainless-steel main flow tube (i.d. 60 mm), is evacuated continuously by a 10,000 L·min^{-1} booster pump connected with a 1600 L·min^{-1} oil rotary pump. A buffer rare gas is admitted to a discharge tube after purification by passage through cooled, activated molecular sieve traps. Liquid-nitrogen-cooled sieve traps can be used with He at atmospheric pressure, whereas they must be used with Ar at low pressure to prevent condensation of liquid Ar. The buffer-gas pressure in the flow tube is typically 0.1–5 torr, which can be varied by controlling the gas flow or the pumping

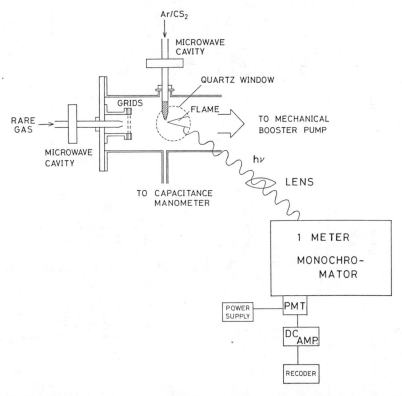

Fig. 21. The flowing afterglow apparatus used for investigating the charge-transfer reaction of Ar^+ with CS radicals by UV and visible emission spectroscopy. The outside of a side flow tube for generating CS radicals (hatched region) is coated with black paint in order to reduce stray light from the discharge.

speed. Reactant ions and metastable atoms are produced through a 2450-MHz microwave discharge (power 40–70 W). These active species undergo many collisions with the buffer gas before reaching the reaction zone and are thus effectively thermalized. The FA experiment is, therefore, generally applicable for investigating thermal-energy reactions.

In order to determine whether ionic species or metastable atoms contribute to the production of emitting excited species, ion collector grids are placed between the discharge and the reaction zone. A pair of nickel electrodes (20 mesh, transparency $\sim 90\%$) are spaced at intervals of 2 mm. By applying a suitable positive or negative electrostatic potential (typically $+20$ or -15 V) to one grid, more than 99% of the ions are trapped. In our experience, a longer distance between the discharge section and the grids (>15 cm) is recommended in the Ar and Ne afterglows to trap ionic active species sufficiently. For ion collection in the He afterglow generated by a hollow cathode discharge, Piper et al. [150] used a pair of parallel wire electrodes inserted at the right angle bend (light trap) and extended to the edge of the mixing zone.

A small amount of a reagent gas is injected through an entry port 10–30 cm downstream from the center of the discharge. The entry port is usually a simple needle (i.d. 0.4–1.0 mm) or a perforated ring inlet for uniform mixing. For non-volatile reagents, the sample was preheated by a ribbon heater up to about 150°C [125]. A crucible that can be heated up to 400°C was used for metal halides [151–153]. As reagent gases, unstable radicals such as CS and PN have also been used [133]. They were generated in microwave discharges of Ar/CS_2 and $Ar/PCl_3/N_2$ mixtures in a side quartz tube as shown in Fig. 21. Vibrationally excited N_2 molecules generated by a microwave discharge of a mixture of N_2 and a rare gas have been used as a reagent to examine the effect of vibrational excitation [154].

By the addition of a reagent gas, a conical reaction flame usually appears around the gas inlet as shown in Fig. 21. The emission from the flame is collected by a quartz lens through a quartz window (i.d. 50 mm) and focused onto the entrance slit of a monochromator. A dc amplifier (Fig. 10) or photon-counting equipment is used for photoelectric measurements. When a strong UV emission is present, the second-order emission often appears in the visible region. Such an undesired emission can be removed using a suitable cutoff filter (e.g., Pyrex glass filter and color filters). Other techniques required for spectral measurements and data analysis are essentially the same as those described in beam experiments.

The overlap of emissions due to ionic reactions with those due to metastable reactions has often made detailed FA optical studies on ion–molecule reactions difficult. To exclusively detect fluorescence from ionic reactions, Tsuji et al. [117, 127, 129, 135] used a pulse modulation technique shown in Fig. 22. Square pulses (typically $+20$ or -15 V, 100–150 Hz) are applied to one of the ion-collector grids resulting in modulation of the flux of ionic species. The up-and-down counting modes of a photon counter are synchronized to photon signals due to ionic plus neutral reactions and neutral reactions, respectively. The

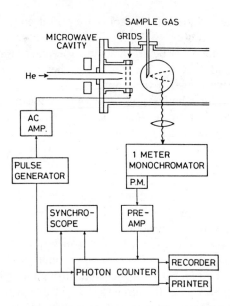

Fig. 22. Block diagram of a pulse modulation system used for studying ion–molecule reactions in the flowing afterglow.

difference signal of interest is accumulated for 200–500 cycles for each step and is registered. Here, the use of this technique is demonstrated by the He afterglow reactions of CO_2 [127, 135]. Although a weak $CO^+(A^2\Pi - X^2\Sigma^+)$ emission due to the $He^+ + CO_2$ dissociative CT reaction is superimposed upon strong $CO_2^+(\tilde{A}^2\Pi_u - \tilde{X}^2\Sigma_g^+, \tilde{B}^2\Sigma_u^+ - \tilde{X}^2\Sigma_g^+)$ emissions due to the $He(2^3S) + CO_2$ Penning ionization in Fig. 23a, the former emission is selectively observed in Fig. 23b by using the modulation technique. By increasing the $[He_2^+]/[He^+]$ ratio significantly, $CO_2^+(A - X, B - X)$ emissions due to the $He_2^+ + CO$ CT reaction can be detected without interference of the $CO^+(A - X)$ emission from the $He^+ + CO_2$ CT reaction [127].

Tsuji et al. have recently extended FA optical spectroscopic studies to secondary ion–molecule reactions such as $C^+(^2P) + O_2$, CO_2, NO_2, and N_2O, leading to $CO^+(A^2\Pi - X^2\Sigma^+)$ chemiluminescence, and $C^+(^2P) + CS_2$ and OCS leading to $CS^+(A^2\Pi - X^2\Sigma^+)$ chemiluminescence [155–159]. As an example, Fig. 24 shows the FA apparatus designed for the chemiluminescent study of the $C^+(^2P) + O_2 \rightarrow CO^+(A) + O$ reaction, $C^+(^2P)$ ions being produced from the thermal-energy dissociative CT reaction between He^+ and CO 10 cm upstream of the O_2 gas inlet [155]. Modulation of C^+ ions was necessary because a strong $CO^+(A - X)$ emission due to $He(2^3S) + CO$ Penning ionization, which occurs upstream of the reaction zone, caused a weak background emission. Modulated $C^+(^2P)$ ions were obtained indirectly by pulsing the He^+ ions. A similar apparatus has been used for studies of molecular ion–molecule reactions such as $CO^+(X^2\Sigma^+) + CS_2 \rightarrow CS_2^+(\tilde{A}^2\Pi_u) + CO$ [160] and $N_2^+(X^2\Sigma_g^+) + OCS \rightarrow$

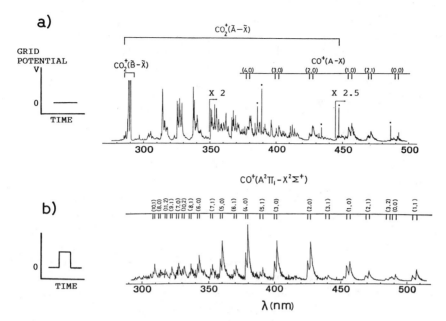

Fig. 23. Emission spectra of CO_2 in the He flowing afterglow due to (a) He^+, $He(2^3S) + CO_2$ reactions and (b) $He^+ + CO_2$ reaction.

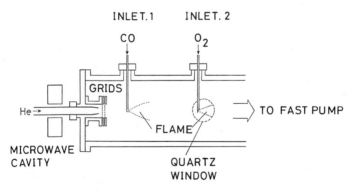

Fig. 24. The flowing afterglow reactor used for the chemiluminescent study of the $C^+(^2P) + O_2 \rightarrow CO^+(A) + O$ reaction.

$OCS^+(A^2\Pi) + N_2$ [161]. $CO^+(X)$ and $N_2^+(X)$ ions in the vibrational ground states were generated upstream through the $He(2^3S) + CO$ and N_2 Penning ionization, respectively.

The following studies have been carried out by using the FA optical techniques.

1. *Spectroscopic Studies of New Diatomic Ions.* Thermal-energy CT reaction, especially by He^+, is a promising way of generating new ion fluorescence in

the UV and visible region. Molecular constants of the following diatomic ions have been determined on the basis of spectrosopic analyses in the He afterglow: $SO^+(A^2\Pi - X^2\Pi)$ [118, 124, 162, 163], $SeO^+(A^2\Pi - X^2\Pi)$ [164], $CS^+(B^2\Sigma^+ - A^2\Pi_i)$ [119], $CCl^+(a^3\Pi_1 - X^1\Sigma^+)$ [126], $SiCl^+(a^3\Pi_{0^+,1} - X^1\Sigma^+)$ [120], $GeCl^+(a^3\Sigma^+ - X^1\Sigma^+)$ [122], $SnCl^+(a^3\Sigma^+ - X^1\Sigma^+)$ [121], $CBr^+(A^1\Pi - X^1\Sigma^+, a^3\Pi_1 - X^1\Sigma^+)$ [125, 140], $SiBr^+(a^3\Pi_1 - X^1\Sigma^+)$ [132], $PCl^+(A^2\Pi - X^2\Pi)$ [165], $PBr^+(A^2\Pi - X^2\Pi)$ [166], $SbCl^+(A^2\Pi - X^2\Pi)$ [166], $AsCl^+(A^2\Pi - X^2\Pi)$ [167], $GeH(D)^+(a^3\Pi_{0^+,1} - X^1\Sigma^+)$ [123, 130, 143], and $SnH(D)^+(a^3\Pi_{0^+,1} - {}^1\Sigma^+)$ [141, 142].

2. *Product state distributions.* Electronic branching ratios of emitting species have been determined by comparing total intensities of each emission system. A similar method has been used to estimate branching ratios of excited atoms. For example, Langford et al. [168] measured absolute branching ratios for production of $O(^1D)$ and $O(^1S)$ in the $N^+ + O_2$ reaction at 300 K. The $O(^1S)$ atoms were monitored by the $O(^1S) \rightarrow O(^1D)$ emission at 557.7 nm, whereas the $O(^1D)$ atoms were detected via sensitized fluorescence at 760 nm from $O_2(b^1\Sigma_g^+)$ formed by energy transfer from $O(^1D)$ to $O_2(X^3\Sigma_g^-)$. Relative vibrational and rotational populations of an excited state have been estimated from the measurements of intensity distributions. These data give detailed information on energy disposal in ion–molecule reactions. Since the operating pressure in FA experiments is rather high, the effect of collisional relaxation should be checked by measuring the dependence of electronic, vibrational, and rotational distributions on buffer gas pressure. Nascent distributions are estimated by a smooth extrapolation of data to zero pressure. If cascading processes contribute to the formation of an emitting species, their effects must be considered. For example, the vibrational population of $CO^+(A)$ produced from the $He_2^+ + CO$ CT reaction was determined by taking account of radiative cascade from the $CO^+(B - A)$ transition [127]. The dependence of relative vibrational distributions on the rare gas pressure has also been used to estimate relaxation rates in molecular ions. Marcoux et al. [169] have deduced relaxation rates in $CO^+(A)$ by steady-state emission intensities in a flow reactor as a function of pressure. More direct and accurate evaluation of relaxation rates is possible by using LIF techniques described in Section 4.3.

3. *Rate Constant for the Formation of an Excited State.* Emission rate constants in CT reactions of He^+ and He_2^+ with some molecules have been determined with reference to the emission intensity of $N_2^+(B - X)$ produced from the $He^+ + N_2$ and $He_2^+ + N_2$ reactions [117, 127, 135, 136]. An emission rate constant of 4×10^{-12} cm^3 s^{-1} was used for the former reference reaction, and that of the latter reference reaction was assumed to be 1.3×10^{-9} cm^3 s^{-1}. When the fluorescence quantum yield of an emitting species is less than unity, the emission rate constant does not correspond to the formation rate constant. In such a case, the formation rate constant can be obtained by dividing the emission rate constant by the known fluorescence quantum yield. The emission cross section can be calculated from the observed rate constant by using the

relation

$$\sigma_{em} = \frac{k_{em}}{\langle v \rangle} = \frac{k_{em}}{(8kT/\pi\mu)^{1/2}} \tag{76}$$

where $\langle v \rangle$, k, T, and μ are the relative average velocity of reactant particles, Boltzmann constant, the absolute temperature, and the reduced mass, respectively. Since no reference reaction is known for reactions by the other rare gas ions and $C^+(^2P)$ ions, absolute emission rate constants cannot be determined at the present time.

When a reagent such as Cd atoms is used, the reduction of density along the flow tube (z axis) due to diffusion through a buffer rare gas and deposition on a wall must be considered for the determination of emission rate constants. Baltayan et al. [153] have determined rate constants for excitation of individual Cd^{+*} levels in the $He^+ + Cd$ CT reaction. The (3, 9) band of the N_2^+ ($C-X$) emission from the $He^+ + N_2$ reactant with an emission rate constant of 2.2×10^{-11} cm^3 s^{-1} has been used as a reference. The decay of the intensity of the (3, 9) band at 191.4 nm was first recorded with a small flow of N_2 ($f_0 \sim 0.2$ μmol·s^{-1}) and a measured amount of Cd ($f_1 \sim 0.1$–1 mg·min^{-1}). After that, the heating of Cd crucible was stopped and enough N_2 (flow rate f_0') was added to obtain exactly the same decay of the (3, 9) band intensity as in the first case. The rate constant k_m for reaction of He^+ with Cd was determined from the relation

$$k_m = \frac{f_0' k_0 - f_0 k_0}{f_1 (1 - z/2z_m)} \tag{77}$$

where k_m and k_0 denote reaction rate constants with Cd and N_2, respectively, and z_m can be estimated from the relation

$$z_m = \frac{P\Lambda^2 v}{D_m} \tag{78}$$

Here, P is the working pressure, Λ the characteristic diffusion length [104], v the flow velocity, and D_m the diffusion coefficient of Cd in 1 torr of He.

The FA apparatus coupled with UV and visible emission spectroscopy is a powerful technique for spectroscopic studies of molecular ions and kinetic and dynamical studies of ion–molecule reactions. Advantages of this technique are as follows: First, because of a high density and a large volume of ions, emission intensity is strong. Therefore, the detection limit of an emission system is low ($\sim 10^{-14}$ cm^3 s^{-1}), and high resolution measurements that can separate vibrational and rotational structures effectively are possible in most cases. Second, since the FA optical apparatus is very simple, it is easy to handle in comparison with beam and ion-trap experiments. Third, product state analysis in ion–molecule reactions is possible at thermal energy, where the application of beam experiments is difficult. Kinetic energies of reactant ions have been varied over the range 80–900 K (0.01–0.11 eV) by changing the temperature of a flow tube in mass spectroscopic studies [170]. Richardson and Setser [108] used a liquid N_2 trap at 77 K in their FA study on Penning ionization to examine the effect of kinetic energy. However, this technique has not been applied to optical studies on ion–molecule reactions. Fourth, a combination of various reactant ions and reagents offers great versatility in reaction system that can be studied.

Disadvantages are as follows: First, since reactant ions are not mass selected, more than one ion–molecule reaction is often capable of exciting the same emission. This complicates the identification of the responsible reaction. In this respect, the use of a quadrupole mass spectrometer provides complementary information on the responsible reactant ions [171]. The presence of metastable atoms in the flow tube often interferes with the observation and analysis of ion–molecule reactions. Second, since the operating pressure is rather high, initial internal state distributions are often modified or lost by collisional relaxation. This effect is especially severe for molecules with long radiative lifetimes (> about 1 μs) and large quenching rate constants. Third, a large flow rate is necessary to obtain a high ion density. This requires high consumption of expensive rare gases and reagents.

4.2 FA Coupled with IRCL Detection

The FA apparatus coupled with IRCL detection has been well established as a powerful technique to probe final state distributions of neutral–neutral reactions (A + BC → AB + C) [172, 173]. Leone and co-workers [174–183] have extended this technique to studies of ion–molecule reactions. Since the maximum achievable ion densities in FA are smaller than neutral radical concentrations by many orders of magnitudes, resulting IRCL signals in ion–molecule reactions are much weaker than those in neutral reactions. Leone et al. overcome this difficulty by using continuously variable bandpass filters that comprise a low-resolution monochromator: Because initial rotational distributions are completely lost by a fast collisional relaxation in the He buffer gas, high resolution detection is generally unnecessary for obtaining vibrational distributions.

Figure 25 shows a schematic diagram of the FA apparatus for the detection of IRCL produced from ion–molecule reactions. The flow tube with a large internal diameter (73 mm) consists of an electron impact ion source, fixed and movable inlets, an observation window for IRCL, and a quadrupole mass filter used to detect all positive and negative ion products through a sampling orifice. The flow system is evacuated by using a large-capacity Roots pump (39,600 L·min^{-1}) backed by a 4800 L·min^{-1} mechanical pump. Ions are produced in a fast flow ($\sim 8 \times 10^3$ cm s^{-1}) of helium at about 0.5 torr by the addition of a small amount of source gas into the ionizer. F_2, CCl_4, BrCN, and O_2 are used as efficient sources of F^-, Cl^-, CN^-, and O^-,

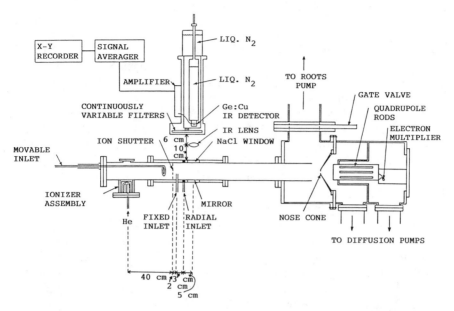

Fig. 25. Schematic diagram of the flowing afterglow apparatus used for IR chemiluminescent studies of ion–molecule reactions [19, 178].

respectively, and N^+ and N_2^+ are generated by the addition of N_2. The ionizer source is composed of a thorium-oxide coated iridium filament, or a tantalum wire filament, with about 16-mm active length mounted 2 mm from a repeller plate. A negative potential of 100 V is applied to both the filament and the repeller plate. Regulated emission currents of 0.25–100 mA are used to obtain high ion densities of 10^8–10^9 cm^{-3} in the reaction zone. Reactant ions travel along the flow tube about 40 cm downstream from the ion source before introducing a reagent gas. In order to minimize the effects of radiative cascading, vibrational relaxation, and diffusion to the walls on the product state distributions, it is desirable to minimize the partial pressure of a reagent gas and the distance between the point of the reagent addition and the viewing region (typically 5 cm). The IRCL is observed through a NaCl window. Larger window-to-detector separations are used to minimize collection of off-normal light, because the wavelength transmission depends on the angle of incidence. An aluminum mirror is placed opposite the window to increase the collection efficiency of IRCL. A CaF_2 lens is used to collect and focus the IRCL onto the detector.

IRCL signals are detected by using either a 1.27-cm-diameter 77 K InSb photovoltaic detector with a wavelength response of 1.0–5.5 μm and a detectivity of $D^* \sim 1 \times 10^{11}$ cm Hz$^{1/2}$ W^{-1} at 5 μm or a copper-doped germanium detector with a wavelength response of 2–30 μm and a detectivity of $D^* \sim 1 \times 10^{12}$ cm Hz$^{1/2}$ W^{-1} at 3 μm. The former InSb detector has been used for the reactions of Cl^- and CN^-. A set of room temperature fixed frequency interference filters mounted below the detector is used to resolve IRCL. For the precise determination of N_1, a cold gas filter cell of HCl, HF, or HCN is used to absorb the IRCL selectively from $v' = 1$ to $v' = 0$. This gas filter is a 1 by 2.86 cm diameter cell containing ~ 100 torr gas that is capable of blocking the fundamental emission to greater than 99%. The output from the InSb detector is amplified and summed in a signal averager. The latter Cu:Ge detector is applied to studies involving F^- and N^+. In front of the detector, a rotable circular variable interference filter (CVF) and a 1.5×12 mm slit are placed to provide wavelength resolution of the transmitted light: Both are cooled at 77 K to reduce the blackbody noise flux and improve the S/N ratio. The CVF is a 90° segment of a quartz annulus coated with continuously varying thicknesses of multiple dielectrics. Each radial slice through the annulus is a narrow bandpass interference filter whose peak transmission wavelength varies linearly in the 2.5–14.5 μm range around the circumference of the annulus. The resolving power of the CVF–slit combination is low; most spectra have been measured with a resolution of 0.06–0.11 μm. Liquid-He-cooled MgF_2 and sapphire windows mounted directly on the detector heat sink serve as a long wavelength ($\lambda > 5.5$ μm) blackbody cutoff filter. The output of the detector is amplified by current feedback circuitry and accumulated in the signal averager. The relative wavelength response of the detector and CVF combination is calibrated by using a standard blackbody souce.

The reactant ions must be modulated to distinguish the ion–molecule product signal from other interfering signals. Ion modulation is achieved by applying a retarding wave potential to a 95% transmission tungsten mesh stretched across the flow tube. For negative ions, the square wave operates between ground and from -10 to -50 V for 0.3 ms and returned to ground for the remainder of the 5 or 10 ms period. In order to obtain a good S/N ratio, modulated IRCL signals are averaged for up to 100,000 pulses at each CVF wavelength setting.

Total spectra are represented by 40–95 points at wavelength intervals of 0.008–0.03 μm. If the vibrational structure is well resolved, relative vibrational populations $(N_{v'})$ are obtained by dividing the intensity of each vibrational band by its Einstein coefficient A_v and the detector response. When spectral resolution is insufficient to separate each vibrational band, a curve-fitting procedure is used [178]. Since rotational levels are rapidly thermalized by collisions with a buffer gas, a Boltzmann rotational distribution can be assumed. For each $(v', v' - 1)$ band, a stick spectrum consisting of the wavelength and relative emission intensity of the various rotational lines is calculated from the relation

$$I_{em}(v'J' - v''J'') \propto |R_{v'J',v''J''}|^2 v^3_{v'J',v''J''}(J' + J'' + 1) \times (Q^{rot}_{v'})^{-1} \exp\left(\frac{-E^{rot}_{v'J'}}{kT_R}\right) \tag{79}$$

where $R_{v'J',v''J''}$ is the transition moment, $Q^{rot}_{v'}$ is the rotational partition function of a vibrational level v' at a temperature T_R, and $E^{rot}_{v'J'}$ is the rotational energy of the $v'J'$ level. At each CVF wavelength λ_i, the stick spectrum for each $(v', v' - 1)$ band is convolved with an experimentally determined filter transmission Gaussian function to yield the calculated $(v', v' - 1)$ emission intensity per unit population at λ_i, denoted by $I_{v'}(\lambda_i)$. The relative vibrational distribution is derived by a linear least-squares fit of the calculated spectrum

$$I_{cal}(\lambda_i) = \sum_{v'} N_{v'} I_{v'}(\lambda_i) \qquad (80)$$

to the observed spectrum with $A_{v'} N_{v'}$ as the adjustable parameters in the fit. Here, the $A_{v'}$ factor is hidden in $I_{v'}$ of the above equation. For the precise determination of vibrational distributions from the IRCL technique, collisional relaxation by a buffer gas, wall deactivation, radiative cascade, and reactive loss {e.g., HF($v \geqslant 3$) + H → H$_2$ + F [180]} must be considered. These effects can be accounted for, and the corrections are generally small, typically smaller than about 10% of the raw population [175].

The IRCL technique provides information about nascent vibrational distributions in ground-state products, which are major products in many ion–molecule reactions at thermal energy. Diatomic molecules that can be studied by the IRCL method are HX and DX (X = F, Cl, and Br), CO, NeH$^+$, ArH$^+$, NO$^+$, whereas a number of polyatomic molecules may also detected, including CO$_2$, HCN, OCS, N$_2$O, NO$_2$, H$_2$O, and CH$_4$. For polyatomic molecules, vibrational relaxation by the helium buffer gas is considerably more rapid than for diatomic molecules, making it difficult to measure initial vibrational distributions. So far, the following types of ion–molecules reactions have been investigated by this method:

1. Proton transfer reactions [175, 177, 178, 183]; e.g.,

$$Y^- + HX \rightarrow HY(v) + X^- (Y = F, Cl, \text{ and } CN; X = Cl, Br, \text{ and } I) \qquad (81)$$

2. Associative detachment reactions [174, 176, 179, 180]; e.g.,

$$O^- + CO \rightarrow CO_2(v) + e^- \qquad (82)$$

$$F^- + H \rightarrow HF(v) + e^- \qquad (83)$$

3. Heavy atom transfer [181, 182]; e.g.,

$$N^+ + O_2 \rightarrow NO^+(v) + O \qquad (84)$$

$$SF_6^- + H \rightarrow SF_5^- + HF(v) \qquad (85)$$

Further detailed techniques required for IRCL studies of the above ionic reactions have been reported by Zwier et al. [175] and Weisshaar et al. [178].

4.3 FA Coupled with LIF Detection

The FA apparatus incorporated with the LIF detection system is an ideal and complementary technique to probe ionic and neutral products in the ground and metastable states with often higher sensitivity and resolution than afforded by the IRCL technique. This method had been dominantly used to probe ionic products resulting from energy transfer reactions by rare gas metastable atoms [e.g., He(2^3S), Ar($^3P_{0,2}$)]. Engelking and Smith [184] were the first to use a FA–LIF reactor to detect $N_2^+(X)$ ions produced from He(2^3S) + N_2 Penning ionization. Subsequently, Miller, Bondybey, Katayama, and Welsh [185–193] and Maier and co-workers [194–198] have performed extensive spectroscopic studies on a number of diatomic and polyatomic ions by using this method. The former investigators have also carried out a series of detailed studies on collision-induced energy transfer of N_2^+ and CO^+ between the $A^2\Pi$ and $X^2\Sigma^+$ states [188–193].

The FA–LIF study has recently been extended to ion–molecule reactions by Tsuji and co-workers [136, 199–201] and Leone and co-workers [202, 203]. As an example, Fig. 26 shows a sketch of the reactor used in our laboratory for a LIF study of $N_2^+(X)$ formed in the dissociative CT reaction [136]

$$\text{He}^+ + \text{N}_2\text{O} \rightarrow \text{N}_2^+(X^2\Sigma_g^+ : v'' = 0\text{--}3) + \text{O} + \text{He} \tag{86}$$

Incorporation of the LIF excitation system required the following flow tube modification to allow entry of laser light and reduction of background and scattered lights:

1. A hollow cathode discharge is used as a source of He$^+$ ions. The discharge

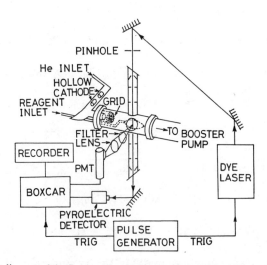

Fig. 26. Schematic diagram of the flowing afterglow apparatus used for LIF studies of ion–molecule reaction.

operates at 400–500 V and 20–30 mA for the optimum production of the reaction ion. A right angle bend with a light trap is inserted between the discharge and mixing zone to reduce background emission from the discharge source. A hollow cathode discharge can be replaced by a microwave discharge, which can operate at a lower He gas pressure [200]. Since the S/N ratio is dominantly limited by the continuous background light, a stable discharge must be maintained to reduce the fluctuation of the baseline of the LIF spectrum.

2. Two baffle arms (30 or 70 cm) equipped with several apertures and Brewster angle quartz windows are attached perpendicularly to the flow tube. All surfaces within the flow tube are blackened with a flat coating. These changes are effective to reduce stray and scattered light.

3. The reagent gas of N_2 is introduced from a movable inlet (i.d. 1 mm). To avoid collisional vibrational relaxation of $N_2^+(X:v''>0)$, the N_2 pressure must be kept low and the distance between the reagent gas inlet and the observation zone should be minimized (5–10 mm).

The N_2^+ ions in the ground $X^2\Sigma_g^+$ state were excited into the $B^2\Sigma_u^+$ state by using an N_2 laser pumped dye laser system (Molectron UV 22, DL 14P system). The repetition rate was typically 20 Hz and the optical resolution (full width half-maximum—FWHM) was about 0.01 nm as measured by an He line at 388.9 nm. The N_2 laser was triggered by a pulse generator (Hewlett Packward 8013B). To reduce background signals, the photomultiplier view is restricted to the small volume of gas intersected by the laser beam by a combination lens and slit arrangement. The LIF was observed through a quartz window fitted with a narrow blackened slit (5 mm), passed through an interference filter and a $f=5$ cm focusing lens, and observed by a photomultiplier (Hamamatsu Photonics R955). The signal was averaged by a digital boxcar integrator (NF Circuit Block Co: BX-531) that was gated on just after any scattered laser light. The digital output of the boxcar integrator was recorded on a strip chart recorder. The relative laser power was simultaneously monitored by a pyroelectric detector (Molectron P1-15H) and averaged in the second channel of the boxcar. The relative contribution between the $He^+ + N_2O$ reaction and the $He(2^3S) + N_2O$ reaction has been examined by using an ion collector grid placed at the exit opening of the discharge tube.

The rotational distribution of $N_2^+(X)$ was completely thermalized by collisions with the buffer He gas during the time between formation and detection, providing no information about initial rotational distribution as in the cases of IRCL experiments. The relative vibrational populations of $N_2^+(X:v''=0-3)$ have been determined from Eqs. 51 and 52. A recommended method for the determination of a precise vibrational distribution $N_{v''}$ is to measure intensities of the LIF spectra resulting from various $v'' \to v'$ excitations and the observation of a common $v' \to v$ transition. This method has the advantage that it eliminates the need to correct for detection efficiency or loss processes of the fluorescing state. In the present case, the laser was tuned to the $N_2^+(X:v'') \to N_2^+(B:v'=0)$ transition, and the resulting fluorescence from the $N_2^+(B:v'=0) \to N_2^+(X:v=0)$

transition was recorded. Since FA experiments are performed at collision-dominant conditions, the effect of collisional relaxation must be examined by observing the dependence of vibrational distributions on the sample-gas and buffer-gas pressures. In general, the vibrational relaxation by collisions with rare gas is slow, but it is fast for sample gas. Therefore, the LIF spectra should be measured at sample-gas pressures as low as possible. The initial vibrational populations are estimated by a smooth extrapolation of pressure-dependence data to zero pressure. The analyses of both $N_2^+(B-X)$ LIF spectra through such a procedure and the $N_2^+(B-X)$ emission spectrum from the $He^+ + N_2O$ reaction have shown that the contribution of the $B-X$ radiative cascade for the formation of $N_2^+(X)$ is very small (about 2%), and initial vibrational populations of N_2^+ are inverted for $v'' = 0$–3 in reaction 86.

Tsuji et al. applied a similar FA–LIF technique to studying the following ion–molecule reactions [199–201]:

$$He^+(He_2^+) + N_2 \rightarrow N_2^+(X:v'' = 0\text{–}6) + He(2He) \tag{87}$$

$$C^+ + RH(R=C_2H_3, C_2H_5, C_3H_5, \text{ and } C_3H_7) \rightarrow CH(X:v'' = 0, 1) + R^+ \tag{88}$$

$$C^+ + CH_3OH \rightarrow CH(X:v'' = 0, 1) + CH_2OH^+(CH_3O^+) \tag{89}$$

In reactions 88 and 89, C^+ ions in the ground ^2P state were formed by the $He^+ + CO$ dissociative CT reaction, and the contribution of metastable $C(^1D, ^1S)$ atoms produced from the $He(2^3S) + CO$ dissociative excitation was examined by inserting an ion collector grid between the CO inlet and the reaction zone. Isotope labeling experiments are valuable to obtain information on reaction processes. For example, it was shown that the $CH(X)$ radicals produced in reaction 89 arise from the selective attack on the methyl site by using CH_3OD.

The FA–LIF apparatus of Leone et al. is similar to that described above. The flow reactor and operating conditions are essentially identical with those used for their IRCL experiments (Fig. 25). As an ion shutter, at a distance of 8 cm upstream of the reagent gas inlet is a series of electrically insulated guard rings. The tunable laser source in the 380–700 nm range is a pulsed Nd-YAG laser pumped dye laser system (5 ns pulses, 10 Hz, 1 cm^{-1} linewidth). Frequency doubling of the output laser with an autotracking, angle-tuned KD*P crystal extended the tuning range to 217 nm. Most experiments have been performed with 0.1–1 mJ·cm^{-2} to avoid saturation of a molecular transition. LIF spectra were observed by using either a boxcar integrator or a photon-counting system. The use of the photon-counting detection system greatly improved the sensitivity: The detection limit was estimated to be $\sim 10^4$ cm^{-3} for molecules in a particular state when the signal from 200 laser pulses was collected.

The apparatus of Leone et al. has been successfully used for the following types of ion–molecule reactions [202, 203]:

$$O^- + HF \rightarrow OH(X:v'' = 0, 1) + F^- \tag{90}$$

$$Ar^+ + CO \rightarrow CO^+(X:v'' = 0-6) + Ar \tag{91}$$

Here, the use of their apparatus is demonstrated by the study on CT reaction 91 by Hamilton et al. [203]. The Ar^+ ions were produced from the $He(2^3S) + Ar$ Penning ionization. The density in the reaction zone was typically $2-4 \times 10^7 \, cm^{-3}$, which was of the same order as the ICR method described in Section 3.2. Although no information on the nascent vibrational distribution can be extracted from the ICR–LIF measurements because of poor S/N ratio, it was estimated in the FA–LIF method. The use of a high sensitivity detection system made it possible to lower the CO pressure to about 10^{-5} torr, where secondary quenching processes such as 70 were greatly reduced. The relative populations have been estimated from the following procedure: First, the nascent population ratio, $N_{v''=4}/N_{v''=0}$, was determined by extrapolating the pressure dependence of the ratio between 10^{-5} and 10^{-4} torr to zero CO pressure. The data indicated that there was a $68 \pm 18\%$ reduction in the $N_{v''=4}/N_{v''=0}$ ratio as the CO pressure was lowered from the value 3.75×10^{-5} torr, which was the CO pressure used in the measurements of the population ratios, to zero. The other ratios, $N_{v''=1-3,5,6}/N_{v''=0}$ were obtained by assuming that the ratios are reduced by the same percentage as the CO pressure is lowered to zero. Finally, the $v'' = 0$ population was corrected for the relaxation of all $v'' > 0$ states into $v'' = 0$. It was shown that $CO^+(X:v'' = 0-6)$ vibrational populations have a broad distribution with a peak at $v'' = 5$.

The data for $v'' = 4$ vs. CO pressure were used to extract a rate constant for the relaxation of $v'' = 4$:

$$Ar^+ + CO \xrightarrow{k_{92}} CO^+(X:v'' = 4) + Ar \tag{92}$$

$$CO^+(X:v'' = 4) + CO \xrightarrow{k_{93}} CO + CO^+(v = 0) \tag{93}$$

The k_{93} value was estimated from the kinetic relation

$$[CO^+(X:v'' = 4)]$$
$$\propto \frac{k_{92}[CO](\exp\{-(k_d^{Ar^+} + k_{92}[CO])t\} - \exp\{-(k_d^{CO^+} + k_{93}[CO])t\})}{(k_d^{CO^+} + k_{93}[CO]) - (k_d^{Ar^+} + k_{92}[CO])} \tag{94}$$

Here $k_d^{Ar^+}$ and $k_d^{CO^+}$ are rate constants for the diffusive losses of Ar^+ and CO^+, and t is the average time required for an ion to travel from the CO inlet to the LIF probe. By substituting known parameters for Eq. 94, the k_{93} value was evaluated to be $6.0 \pm 2.5 \times 10^{-10} \, cm^3 \, s^{-1}$. An advantage of the FA–LIF technique is a high detection limit, making a number of spectroscopic and dynamic studies possible. Disadvantages are occurrence of fast vibrational relaxation and secondary reaction of products and interference due to reactions by metastable atoms. For example, ion vibrational relaxation and an obscuring Penning ionization reaction precluded accurate measurements for the $N^+ + CO \rightarrow CO^+(v'' = 0 - 2) + N$ system [202].

A great advantage of the LIF technique is the capability of the monochromatic excitation source to populate selectively a desired ro-vibrational state of the species under study. Such a selective excitation in a short pulse mode provides a convenient and direct method of studying energy disposal in the excited molecules. The FA–LIF technique is an ideal method for investigating the energy flow and dynamics of relaxation in vibrationally or electronically excited species. The FA cell used for such a dynamical study of ion–molecule collisions was similar to that described in Fig. 26. $He(2^3S)$ atoms were generated by a hollow cathode discharge in a flow system that was maintained by a small capacity pump (650 L·min^{-1}). Vibrationally excited $CO^+(X)$ and $N_2^+(X)$ ions were prepared by the $He(2^3S)$ Penning ionization of CO and N_2. The reagent gas entered through a concentric nozzle. Mixing between the metastable atoms and the reagent gas occurred in a conical region downstream from the discharge-tube inlet into the reaction chamber. The tip of this conical mixing zone was centered in the viewing windows and was crossed by the laser beam. The entire discharge and flow tube section could be surrounded by a liquid N_2 trap. It was not used for dynamical studies but was used for spectroscopic studies: Low rotational temperature spectra without overlapping with a hot band often reduce congestion so that spectral interpretations become easier.

In a typical experiment, a desired vibronic level was excited by a 10-ns pulse of an N_2 laser-pumped dye laser, and the resulting fluorescence was monitored as a function of time by resolving the signal in a fast waveform recorder with 10-ns resolution (Biomation 8100). The output was averaged in a signal averager (Nicolet 1072). Such measurements to determine the radiative decay time for each vibronic level were carried out over a wide range of pressure. A Stern-Volmer treatment of data gives both the radiative lifetimes of the excited levels and their quenching rate constants. This method has been used for the $^{12}CO^+(A)$–He [189, 191, 192], $^{13}CO^+(A)$–He [192], $N_2^+(A)$–He [190, 193], $N_2^+(B)$–N_2 [204], and $N_2^+(B)$–Ne [204] systems. Dynamical information has also been obtained from the study of the time-resolved behavior of the vibrationally relaxed fluorescence. This was accomplished by exciting a given vibrational level of an emitting state and using an interference filter to isolate emission from a desired lower level. The application of kinetic models for the vibrational deactivation of $CO^+(A)$ in He required the postulation of a three-level system to fit the decay profile [189]. It was proposed that $CO^+(A)$ vibrational levels relax via intermediate, highly excited, vibrational levels of the ground $X^2\Sigma^+$ state of CO^+. A similar mechanism is proposed for $N_2^+(A)$ in He.

In subsequent experiments, the populations of the ground-state $N_2^+(X)$ levels were directly monitored by using an optical–optical double resonance (OODR) technique. Figure 27 shows the relevant potential curves of N_2^+ and the pumped and probed transitions. The population arriving into high vibrational levels of the X state following the initial A state excitation was probed by a second laser, tuned to a suitable $X - B$ vibronic transition. In the first OODR experiments of Katayama et al. [190], the N_2-pumped dye laser was supplemented by a flashlamp-pumped dye laser (Chromatix CMX-4). The CMX-4 laser was free running at 15 Hz. Scattered light from the CMX-4 laser pulse optically triggered, within adjustable time delay in the range of 0.1–10 μs, the second N_2-pumped dye laser. The pulse width of the N_2-pumped dye laser was ~ 10 ns and that of the flashlamp pumped laser was 1 μs. The following three OODR experiments were performed for the $N_2^+(A)$–He system, and the results obtained provided direct evidence of the predicted relaxation process. First, the blue laser was tuned to the band head of either the $2' - 4''$, $3' - 5''$, or $4' - 6''$ transition of the

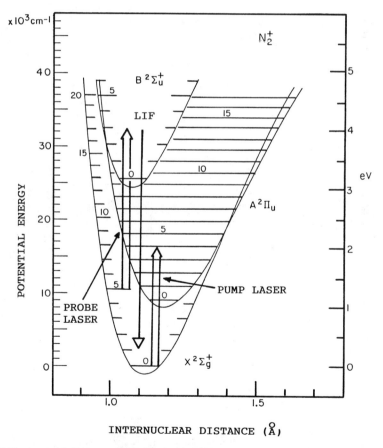

Fig. 27. Energy level diagram of N_2^+ showing transitions excited by both the red and blue lasers and the fluorescence transition upon which the double resonance is observed.

$B - X$ system and the corresponding LIF signal for the $2' - 0''$, $3' - 1''$, and $4' - 2''$ $B - X$ bands was monitored through a bandpass filter. The red laser was then switched on and off centered on the $4' - 0''$ $A - X$ band head. When the red laser was blocked, a clear decrease in the LIF signal was found. Second, the blue laser's frequency was set at the $3' - 5''$ $B - X$ band head. The red laser was then scanned, and an excitation spectrum of the $4' - 0''$ $A - X$ band system was recorded while observing the $3' - 1''$ $B - X$ emission. When the wavelength of the blue laser was detuned from the $3' - 5''$ $B - X$ band, no excitation spectrum was observed. Third, the red laser was centered on the $4' - 0''$ $A - X$ band head. The blue laser was then scanned over the $4' - 6''$ $B - X$ transition detecting the $4' - 2''$ emission band. An enhancement of the intensity of the $4' - 6''$ $B - X$ excitation spectrum was observed when the red laser was switched on.

Katayama [193] carried out a further detailed OODR study on collision-induced electronic energy transfer between the $A^2\Pi_{ui}(v = 4)$ and $X^2\Sigma_g^+(v = 8)$ rotational manifolds of N_2^+. The block diagram of the OODR experiments is shown in Fig. 28. Two dye lasers were pumped almost simultaneously by the 532-nm (green) and 355-nm (UV) harmonics of the Nd:YAG laser. The light beams from these dye lasers form the pump and probe beams, respectively. The pump and probe beams were coincident in time at the region of N_2^+ ions, and the pulse duration was approximately 20 ns, in which collision-induced electronic transfer indeed occurs. On the basis of high-resolution OODR experiments, collisional selection or propensity rules have been obtained for a homonuclear molecule that has no perturbations between the states involved.

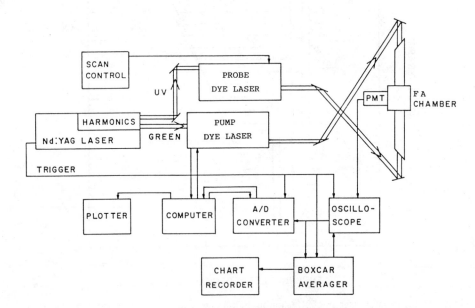

Fig. 28. Schematic diagram of the optical–optical double resonance experiment [193].

4.4 FA Ion Source Coupled with a Supersonic Nozzle Expansion

In order to obtain a complete picture of energy disposal in ion–molecule reactions, nascent ro-vibrational state distributions must be determined. A serious problem with the FA method is that information about initial ro-vibrational distributions is often lost by collision with a buffer gas before detection. The rotational relaxation is especially fast, although nascent rotational distributions provide important information about the reaction dynamics. The FA apparatus has recently been combined with a sampling orifice to obtain a supersonic expansion of reactant ions in a low-pressure reaction chamber. This apparatus preserves the advantageous features of the high ion density and thermalized ions available in the FA but is one in which ion–molecule reactions can be studied under single-collision conditions. This technique was developed by Leone's group [205–208] to probe ground-state products by using the saturated LIF detection method. Very recently, Tsuji's group [209] coupled this technique with a UV and visible emission detection system to probe excited products. This section describes these promising devices.

4.4.1 FA Coupled with a Low-Pressure Chamber for UV and Visible Emission Detection

Figure 29 shows an overview of our apparatus that consists of an FA source for the generation of thermalized ions and a high vacuum chamber [209]. The FA reactor and operating conditions are essentially identical with those used in previous FA optical studies (see Section 4.1). Reactant ions and metastable atoms [e.g., He^+ and $He(2^3S)$] are produced by a microwave discharge in the FA, and a small portion of them is expanded through a 1.5–3.0-mm-diameter orifice centered on the flow tube axis into the separately pumped reaction chamber. This chamber is evacuated by a $3700 \, L \cdot s^{-1}$ diffusion pump. The distance between the end of a quartz discharge tube and the sampling orifice is variable between 3–10 cm by shifting the discharge tube. A flow profile of reactant ions from the exist of the discharge tube to the orifice can be checked by adding a small amount of N_2 to the FA and observing the blue $N_2^+(B-X)$ emission through the first quartz window in Fig. 29.

A reagent gas is introduced to the reaction chamber through a movable stainless-steel needle (i.d. 0.5 mm). A reaction flame due to $N_2^+(B-X)$ from the $He(2^3S)$, He^+, $He_2^+ + N_2$ reactions can be seen easily by eye along an expanded discharge flow, and it can be used to check the flow profile of expanded reactant species and to determine relative concentrations of the He active species. Total pressure in the reaction chamber is $< 1 \times 10^{-3}$ torr, and the sample gas pressure is $< 5 \times 10^{-4}$ torr. An ion collector grid is placed either at the exit opening of the discharge tube or in front of the orifice to distinguish a desired emission of an ion–molecule reaction from all other interfering signals. The emission is observed through the second quartz window and focused on the entrance slit of a monochromator. Since the observed emission is generally strong (only weaker than the conventional FA by a factor of 10–20), the signal from a photomultiplier is amplified by a dc amplifier in most cases. When a

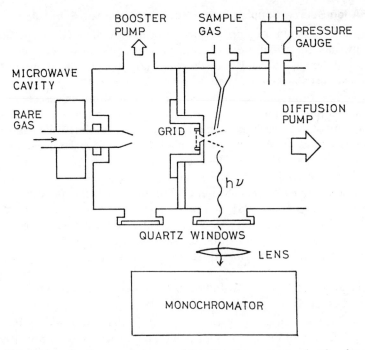

Fig. 29. The flowing afterglow apparatus coupled with a low pressure chamber for studying ion–molecule reactions at near thermal energy under single collision conditions by UV and visible emission spectroscopy.

high resolution spectrum is necessary or a band intensity is very weak, a photon-counting system (Hamamatsu Photonics C767) is used. This apparatus has been applied to studying thermal energy ion–molecule reactions {e.g., $He^+ + SiH_4 \rightarrow SiH^+(A) + 3H + He$ [209]}. A great advantage of this technique in comparison with the ICR method (Section 3.1) is that emission intensities are much stronger because of the higher density of reactant ions ($> 10^8 \, cm^3 \, s^{-1}$). Therefore, the detection limit is lower, and measurements of higher resolution spectra are possible.

4.4.2 FA Coupled with a Low-Pressure Chamber for LIF Detection

Leone and co-workers [205–208] modified their FA reactor used in IRCL and LIF experiments to perform ion–molecule LIF studies in a low-pressure chamber. An overview of the apparatus is shown in Fig. 30. It consists of an FA source for the production of thermalized ions, a high-vacuum chamber in which reaction and product detection take place, and a quadrupole mass spectrometer to monitor ionic reactants and products. A large capacity diffusion pump ($10000 \, L \cdot s^{-1}$) is used to evacuate the reaction chamber. Densities of thermalized ions are estimated to be about $10^6 \, cm^{-3}$ with a background pressure of 3×10^{-4} torr. All measurements have been performed at pressures less than

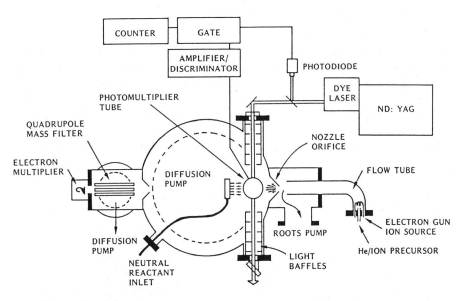

Fig. 30. The flowing afterglow apparatus coupled with a low-pressure chamber for studying ion–molecule reactions at near thermal energy under single collision conditions by LIF spectroscopy [205].

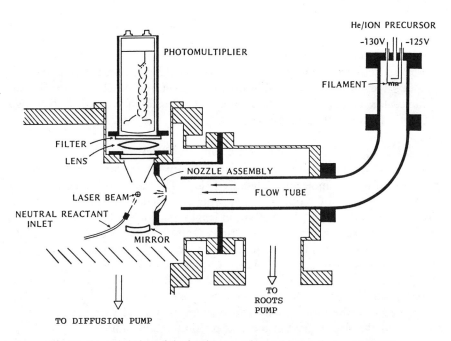

Fig. 31. Expanded view of the free jet expansion and interaction region [205].

6×10^{-4} torr. Figure 31 shows an expanded view of the free jet expansion and interaction region. In order to obtain a sufficient S/N ratio of LIF signal, the pulsed excitation laser is operated in the saturating regime. This apparatus has been successfully applied to probe nascent ro-vibrational state distributions of products in near thermal energy (typically $0.1\,\text{eV}_{\text{cm}}$) CT reactions: $N^+ + CO \rightarrow CO^+(X:v'') + N$ [205, 208], $Ar^+ + N_2 \rightarrow N_2^+(X:v'') + Ar$ [206], and $Ar^+ + CO \rightarrow CO^+(X:v'') + Ar$ [207]. All experiments have been performed on a CW nozzle beam. Detailed procedures used for the determination of ro-vibrational distributions by the saturated LIF method, including the technique for estimating angle-averaged velocities, have been given by Hüwel et al. [206].

5 DRIFT TUBE METHODS

Energies of reactant ions are variable over the temperature range of 80–900 K (0.01–0.11 eV) in FA experiments [170], whereas beam experiments are usually performed at energies above a few electron volts. A DT apparatus provides an ideal way of studying ion–molecule reactions as a function of reactant ion energy from ambient temperature to about 3 eV, in which the application of the above two techniques is difficult. Here, the energy of a reactant ion is controlled by varying an electric field in the drift region. The energy dependence of ion–molecule reaction rate constants and product distributions at energies below a few electron volts gives an important determination of the dynamical characteristics of the reaction. Extensive DT mass spectroscopic studies on low-energy ion–molecule reactions have been carried out since the first flow DT experiment of McFarland et al. [210]. Relatively little optical work has been carried out in the DT apparatus. Tsuji et al. [211] modified their FA apparatus to perform the first flow DT study coupled with UV and visible emission spectroscopy. Duncan and co-workers [212] used a DT–LIF apparatus as a new technique to study collisional excitation and deactivation processes of molecular ions in a drift field. In this section, techniques used in these two DT optical studies are presented.

5.1 DT Coupled with UV and Visible Emission Detection

Figure 32 shows the DT apparatus used in our laboratory to probe excited products from ion–molecule reactions [211]. The flow DT experiments by emission spectroscopy require the following two changes to the conventional FA: First, a drift section about 20 cm in length is incorporated. It is composed of 13 stainless-steel cylindrical guard rings (0.50 cm wide × 9.0 cm o.d. × 7.0 cm i.d.) separated by 2-mm ceramic spacers and electrically connected in series by 10-kΩ resistors. A continuous voltage is applied to guard rings to establish a uniform drift field within the guard rings. Second, for the optical experiment, a hole approximately 5-mm square is cut in a guard ring to allow optical viewing of the inside of the drift region. This opening is crossed with small wires to ensure that the electric field in the drift region remains uniform.

A reagent gas was introduced through a glass nozzle, and the resulting flame

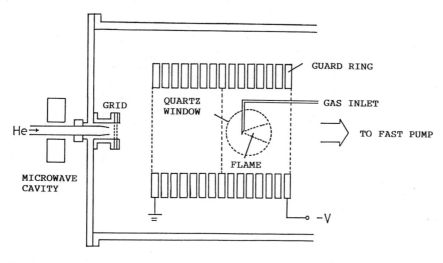

Fig. 32. The DT apparatus used for the study of low-energy ion–molecule reactions by UV and visible emission spectroscopy.

spectrum was observed in the same manner as used for previous FA studies in our laboratory.

When the variation of mobility with the ratio of the electric field to the buffer gas number density, E/N, is expressed in units of townsends (Td = 10^{-17} V·cm^2), the average ion drift velocity is calculated by

$$v_d = \mu_0 N_0 \left(\frac{E}{N}\right) \quad (95)$$

where μ_0 is the ion mobility at 1 atm and 273 K and $N_0 = 2.69 \times 10^{19}$ cm^{-3}. The mobility in various gases measured until 1978 has been compiled in tables of Ellis et al. [213, 214]. For a given drift velocity, the average ion kinetic energy (KE) in the center-of-mass system is calculated from the Wannier formula [215]:

$$\text{KE}_{\text{ion}} = \tfrac{1}{2} M_b v_d^2 + \tfrac{3}{2} kT \quad (96)$$

where m, M_b, and T are the ion mass, the buffer gas mass, and the thermal temperature of the buffer gas, respectively.

When the mobility is unknown, the average ion flow velocity can be determined experimentally by measuring the time of flight (TOF) spectrum of a reactant ion between an ion shutter and the reaction zone. The TOF spectrum is obtained by applying a 10–50-μs square pulse to an ion collector grid and observing the arrival time to the reaction zone by using a photon counter with a time-resolving sampler system (Hamamatsu Photonics C767–C767TR). The measured velocity corresponds to the sum of the ion flow drift velocity and the flow velocity of a buffer gas. As an example, Figs. 33a and b show TOF spectra

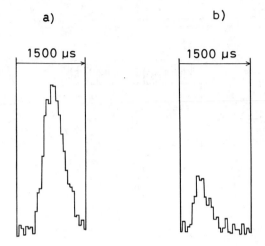

Fig. 33. TOF spectra of He$^+$ ions for (a) E/N = 0 and (b) E/N = 32 Td.

of He$^+$ ions measured by detecting $N_2^+(B-X)$ emission resulting from the He$^+$ + N$_2$ CT reaction for E/N = 0 and 32 Td, respectively; the radiative lifetime of $N_2^+(B:\tau =$ about 60 ns [95]) is sufficiently short in comparison with the arrival times of ion pulses. The upper limit of the kinetic energy is dominantly determined by the initiation of a spark discharge when a high drift field is applied to the guard rings. In order avoid the spark discharge, surfaces of reaction chamber and guard rings must be kept clean. By using this apparatus, the dependence of the relative vibrational populations of $N_2^+(C)$ on the He$^+$ ion energy is examined, and the reaction dynamics at near thermal energies are discussed [211].

5.2 DT Coupled with LIF Detection

Figure 34 shows the DT apparatus designed by Duncan et al. [212] for the LIF study of the translation-to-rotation energy transfer in the N_2^+ + He system. The apparatus is essentially the same as that used in their FA–LIF studies, with the exception of the addition of a drift section and change in the direction of the laser beam. The drift region consists of 54 steel guard rings (0.95 cm wide × 9.53 cm o.d. × 7.30 cm i.d.) separated by 0.77-mm Mylar spacers. A uniform drift field is created by applying a dc potential between each guard ring that is connected in series by 499 Ω, 0.1% precision resistors. The translational energy of N_2^+ ions is varied from thermal energy to 0.054 eV$_{cm}$. In order to obtain well-defined operating conditions of separated charges, low ion density, and well-characterized ion mobility, the ranges of conditions at low enough ion density and long enough drift length for well-characterized collision energies are limited by S/N levels to ⩽ 16 Td.

A Nd-YAG laser-pumped dye laser (∼5 ns duration; 10 Hz, ∼0.5 cm^{-1} bandwidth) is directed down the axis of the flow tube, rather than across it, to

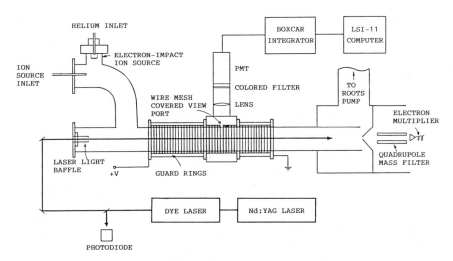

Fig. 34. Schematic diagram of DT apparatus for detection of LIF spectra of ions [212].

minimize perturbations in the electric field. In order to minimize laser-scattered light, the guard rings are blackened with an electrically conductive nickel coating. LIF is used to probe the degree of rotational excitation of N_2^+ ions within the drift field. To avoid saturation, the absolute dye laser power is kept below $1.0\,\mathrm{mJ \cdot cm^{-2}}$. LIF signal is observed through a 2×1 cm opening of the center guard ring. The opening is covered with a fine tungsten wire mesh to maintain a uniform electric field. The LIF detection system is essentially the same as that used in their FA–LIF studies described in Section 4.3.

When a drift field of 30.9 Td was applied, $N_2^+(X)$ ions with a rotational temperature of about 700 K were obtained due to translation–rotation energy transfer. The dependence of the rotational temperature on the length of the drift section before laser irradiation and on the time delay between application of the field and laser probing was measured to examine the rate of the rotational excitation. The latter measurements have been performed under the following operating conditions. In order to separate positive and negative charge carriers, a dc electric field ($1\,\mathrm{V \cdot cm^{-1}}$) was applied to the first seven guard rings. The remaining downstream guard rings were held at ground potential except for 1-ms intervals of time beginning a few microseconds before each probe laser pulse. A square pulse was applied to the drift field during this time. The overall rise time of the field was about $10\,\mu s$. The rotational distribution of $N_2^+(X)$ was measured as a function of time in the drift field by changing the time delay between the beginning of the pulsed drift field and the probe laser pulse in a time resolution of a microsecond. An oscilloscope was used to measure the time delay. Since an upper limit on the time required for the rotational excitation was on the order of a microsecond, the characteristic time of the pulsed experiment was too slow to detect time-resolved data. However, it was demonstrated that this technique was effective for the study of rapid translation-to-rotation energy transfer processes in ion–molecule reactions as in the RF ion trap–LIF technique described in Section 3.3.

REFERENCES

1. G. Herzberg, *Atomic Spectra and Atomic Structure*, Dover, New York, 1944.
2. G. Herzberg, *Molecular Spectra and Molecular Structure, Vol. I, Spectra of Diatomic Molecules*, Van Nostrand Reinhold, New York, 1950, 2nd ed.
3. G. Herzberg, *Molecular Spectra and Molecular Structure, Vol. II, Infrared and Raman Spectra of Polyatomic Molecules*, Van Nostrand Reinhold, New York, 1945.
4. G. Herzberg, *Molecular Spectra and Molecular Structure, Vol III, Electronic Spectra and Electronic Structure of Polyatomic Molecules*, Van Nostrand Reinhold, New York, 1966.
5. B. A. Lengyel, *Lasers*, Wiley-Interscience, New York, 1971, 2nd ed.
6. A. E. Siegman, *An Introduction to Lasers and Masers*, McGraw-Hill, New York, 1971.
7. D. C. O'Shea, W. R. Callen, and W. T. Rhodes, *Introduction to Lasers and Their Applications*, Addison-Wesley, Reading, Mass., 1977.
8. J. T. Verdeyen, *Laser Electronics*, Prentice-Hall, Englewood Cliffs, N. J., 1981.
9. *Handbook of Laser Science and Technology, Vol. I: Lasers in all Media*, M. Weber Ed., CRC Press, Boca Raton, Fla., 1982.
10. *Laser Applications in Chemistry*, K. L. Kompa and J. Wanner Eds., NATO ASI Series, Series B: Phys., Vol. 105, Plenum, New York, 1984.
11. Springer Series in Chemical Physics, Springer-Verlag, Berlin–Heidelberg-New York: *Advances in Laser Chemistry*, A. H. Zewail, Ed., 1978, Vol. 3; *Laser Spectroscopy: Basic Concepts and Instrumentation*, W. Demtröder, 1981, Vol. 5; *Laser-Induced Processes in Molecules: Physics and Chemistry*, K. L. Kompa and S. D. Smith, Eds., 1979, Vol. 6; *Lasers and Chemical Change*, A. Ben-Shaul, Y. Haas, K. L. Kompa, and R. D. Levine, Eds., 1981, Vol. 10; *Gasdynamics Lasers*, S. A. Losev, 1981, vol. 12; *Nonlinear Laser Chemistry: Multiple-Photon Excitation*, V. S. Letokhov, 1983, vol. 22.
12. *Chemical and Biochemical Applications of Lasers*, C. B. Moore Ed., Academic Press, New York, 1974–80, Vols. I–V.
13. *Laser Handbook*, North Holland, Amsterdam: vols. 1–2, F. T. Arecchi and E. O. Schulz-Dubois, Eds., 1972; vol. 3, M. L. Stitch, ed., 1979.
14. *Handbook of Chemical Lasers*, R. W. F. Gross and J. F. Bott, Eds., Wiley, New York, 1976.
15. *High-Power Lasers and Applications*, K. L. Kompa and H. Walther, Eds., Springer-Verlag, Berlin-Heidelberg-New York, 1978.
16. *Dye Lasers*, in F. P. Schäfer, Ed., *Topics in Applied Physics*, Springer-Verlag, Berlin-Heidelberg-New York, 1973, vol. 1, 2nd ed.
17. *Excimer Lasers*, C. K. Rhodes, Ed., Springer-Verlag, Berlin-Heidelberg-New York, 1979.
18. T. O. Tiernan and C. Lifshitz, *Adv. Chem. Phys.*, **45**, 81 (1981).
19. V. M. Bierbaum, G. B. Ellison, and S. R. Leone, in M. T. Bowers, Ed., *Gas Phase Ion Chemistry, vol. 3, Ions and Light*, Academic Press, New York, 1984, p. 1.
20. C. E. Hamilton and S. R. Leone, in A. Fontijn, Ed., *Gas-Phase Chemiluminescence and Chemi-Ionization*, North-Holland, Amsterdam, 1985, p. 139.
21. Ch. Ottinger, in M. T. Bowers, Ed., *Gas Phase Ion Chemistry, vol. 3, Ions and Light*, Academic Press, New York, 1984, p. 249.
22. Ch. Ottinger, in A. Fontijn, Ed., *Gas-Phase Chemiluminescence and Chemi-Ionization*, North-Holland, Amsterdam, 1985, p. 117.
23. J. J. Leventhal, in M. T. Bowers, Ed., *Gas Phase Ion Chemistry, vol. 3, Ions and Light*, Academic Press, New York, 1984, p. 309.

24. H. H. Harris and J. J. Leventhal, *J. Chem. Phys.*, **64**, 3185 (1976).
25. Ch. Ottinger and J. Simonis, *Chem. Phys.*, **28**, 97 (1978).
26. F. Howorka, I. Kuen, H. Villinger, and W. Lindinger, *Phys. Rev. A*, **26**, 93 (1982).
27. B. Müller and Ch. Ottinger, *J. Chem. Phys.*, **85**, 243 (1986).
28. I. Kuen, H. Störi, and F. Howorka, *Phys. Rev. A*, **28**, 119 (1983).
29. T. Ishikawa, H. Sasaki, and H. Inouye, *Chem. Lett.*, 975 (1981).
30. B. Müller, Ch. Ottinger, and M. Yang, *Z. Phys. A*, **320**, 61 (1985).
31. V. S. Kushawaha, C. E. Burkhardt, and J.J. Leventhal, *Phys. Rev. Lett.*, **45**, 1686 (1980).
32. R. A. Sawyer, *Experimental Spectroscopy*, Prentice-Hall, New York, 1951.
33. J. F. James and R. S. Sternberg, *The Design of Optical Spectrometers*, Chapman and Hall, London, 1969.
34. H. S. Strobel, *Chemical Instrumentation*, Addison-Wesley, Reading, Mass., 1973.
35. Ch. Ottinger and J. Reichmuth, *J. Chem. Phys.*, **74**, 928 (1981).
36. B. Müller and Ch. Ottinger, *J. Chem. Phys.*, **85**, 232 (1986).
37. G. R. Harrison, MIT Wavelength Tables, MIT Press, Cambridge, Mass., 1969.
38. A. R. Striganov and N. S. Sventitskii, *Tables of Spectral Lines of Neutral and Ionized Atoms*, IFI/Plenum, New York, 1968.
39. A. N. Zaidel', V. K. Prokof'év, S. M. Raiskii, V. A. Slavnyi, and E. Ya. Shreider, *Tables of Spectral Lines*. IFI/Plenum. New York. 1970.
40. J. Reader and C. H. Corliss, *Wavelengths and Transition Probabilities for Atoms and Atomic Ions, Part I. Wavelengths*, Natl. Stand. Ref. Data Ser., Nat. Bur. Stand. (U.S.), **68**, 1980.
41. B. Rosen, *Spectroscopic Data Relative to Diatomic Molecules*, Pergamon, Oxford, 1970.
42. R. W. B. Pearse and A. G. Gaydon, *The Identification of Molecular Spectra*, Chapman and Hall, London, 1976, 4th ed.
43. K. P. Huber and G. Herzberg, *Molecular Spectra and Molecular Structure, Vol. IV, Constants of Diatomic Molecules*, Van Nostrand Reinhold, New York, 1979.
44. G. Mauclaire, R. Marx, C. Sourisseau, C. Van de Runstraat, and S. Fenistein, *Int. J. Mass Spectrom. Ion Phys.*, **22**, 339 (1976).
45. L. A. Kuznetsova, N. E. Kuz'menko, Yu. Ya. Kuzyakov, and Yu. A. Plastinin, *Sov. Phys. Usp.*, **17**, 405 (1974).
46. Ch. Ottinger and J. Simonis, *Phys. Rev. Lett.*, **35**, 924 (1975).
47. D. Neuschäfer, Ch. Ottinger, and S. Zimmermann, *Chem. Phys.*, **55**, 313 (1981).
48. I. Kusunoki and Ch. Ottinger, *J. Chem. Phys.*, **71**, 4227 (1979).
49. I. Kusunoki and Ch. Ottinger, *J. Chem. Phys.*, **76**, 1845 (1982).
50. I. Kusunoki and Ch. Ottinger, *J. Chem. Phys.*, **80**, 1872 (1984).
51. I. Kusunoki and Ch. Ottinger, *J. Chem. Phys.*, **70**, 699 (1979).
52. I. Kusunoki and Ch. Ottinger, *J. Chem. Phys.*, **71**, 894 (1979).
53. P. A. Fraser, *Can J. Phys.*, **32**, 515 (1954).
54. E. W. Thomas, *Excitation in Heavy Particle Collisions*, Wiley-Interscience, New York, 1972.
55. M. Lipeles, R. D. Swift, M. S. Longmire, and M. P. Weinreb, *Phys. Rev. Lett.*, **24**, 799 (1970).
56. J. Simonis, Ph.D. Thesis, Bonn, 1976, published as Bericht 3/1977, Max-Planck-Institut für Strömungsforschung, Göttingen.
57. J. M. Ajello, *J. Chem. Phys.*, **55**, 3158 (1971).
58. H. L. Rothwell, R. C. Amme, and B. Van Zyl, *J. Chem. Phys.*, **68**, 4326 (1978).
59. Ch. Ottinger, J. Reichmuth, and M. Yang, *Chem. Phys.*, **76**, 61 (1983).
60. (a) B. Van Zyl, H. Neumann, H. L. Rothwell, Jr., and R. C. Amme, *Phys. Rev. A*, **21**, 716 (1980); (b) B. Van Zyl, H. L. Rothwell, Jr., and H. Neumann, *Phys. Rev. A.*, **21**, 730 (1980).

61. J. H. Moore, Jr., and J. P. Doering, *Phys. Rev.*, **177**, 218 (1969).
62. M. Lipeles, *J. Chem. Phys.*, **51**, 1252 (1969).
63. P. J. Chantry, *J. Chem. Phys.*, **55**, 2746 (1971).
64. G. D. Myers and J. J. Leventhal, *Phys. Rev. A*, **18**, 434 (1978).
65. I. Kusunoki and Ch. Ottinger, *J. Chem. Phys.*, **70**, 710 (1979).
66. D. Brandt, Ch. Ottinger, and J. Simonis, *Ber Bunsenges. Phys. Chem.*, **77**, 648 (1973).
67. D. Brandt and Ch. Ottinger, *Chem. Phys. Lett.*, **23**, 257 (1973).
68. H. Inouye, N. Shimakura, and H. Sasaki, *J. Phys. B*, **15**, 2423 (1982).
69. T. R. Gilson and P. J. Hendra, *Laser Raman Spectroscopy*, Wiley-Interscience, London, 1970.
70. W. A. Shurchuff, *Polarized Light*, Harvard, Cambridge, Mass., 1962.
71. R. M. A. Azzam and N. M. Basham, *Ellipsometry and Polarized Light*, North-Holland, Amsterdam, 1977.
72. A. Ding, *Faraday Discuss. Chem. Soc.*, **67**, 353 (1979).
73. R. D. Brown, P. D. Godfrey, J. G. Crofts, Z. Ninkov, and S. Vaccani, *Chem. Phys. Lett.*, **62**, 195 (1979).
74. A. Ding, K. Richter, and M. Menzinger, *Chem. Phys. Lett.*, **77**, 523 (1981).
75. P. J. Dagdigian and J. P. Doering, *Chem. Phys. Lett.*, **64**, 200 (1979).
76. A. Ding and K. Richter, *Z. Phys. A*, **307**, 31 (1982).
77. *Handbook of Chemistry and Physics*, R. C. Weast, Ed., CRC Press, Boca Raton, Fla., 1986, 67th ed.
78. *Handbook of Lasers*, R. J. Pressley, Ed., CRC Press, Cleveland, Ohio, 1971.
79. H. A. Macleod, *Thin-Film Optical Filters*, American Elsevier, New York, 1969.
80. P. H. Dawson and N. R. Whetten, *Adv. Electron Electron. Phys.*, **27**, 59 (1969).
81. R. Marx, G. Mauclaire, M. Wallart, and A. Deroulede, in A. R. West, Ed., *Advances in Mass Spectrometry*, Applied Science, England, 1974, p. 735.
82. R. Marx, in P. Ausloos, Ed., *Interactions between Ions and Molecules*, Plenum, New York, 1975, p. 563.
83. M. Gérard, T. R. Govers, C. A. Van de Runstraat, and R. Marx, *Chem. Phys. Lett.*, **44**, 154 (1976).
84. T. R. Govers, M. Gérard, and R. Marx, *Chem. Phys.*, **15**, 185 (1976).
85. T. R. Govers, M. Gérard, G. Mauclaire, and R. Marx, *Chem. Phys.*, **23**, 411 (1977).
86. M. Gérard, T. R. Govers, and R. Marx, *Chem. Phys.*, **30**, 75 (1978).
87. M. Gérard, T. R. Govers, and R. Marx, *Chem. Phys.*, **36**, 247 (1979).
88. R. Derai, S. Fenistein, M. Gérard-Aïn, T. R. Govers, R. Marx, G. Mauclaire, C. Z. Profous, and C. Sourisseau, *Chem. Phys.*, **44**, 65 (1979).
89. R. Marx, in P. Ausloos, Ed., *Kinetics of Ion-Molecule Reactions*, Plenum, New York, 1979, p. 103.
90. J. Danon, G. Mauclaire, T. R. Govers, and R. Marx, *J. Chem. Phys.*, **76**, 1255 (1982).
91. J. Danon and R. Marx, *Chem. Phys.*, **68**, 255 (1982).
92. R. Marx, in M. A. A. Ferreira, Ed., *Ionic Processes in the Gas Phase*, NATO ASI Series, Series C: Math. and Phys. Sci., D. Reidel, Dordrecht, Holland, 1984, vol. 118, p. 67.
93. F. J. Grieman, B. H. Mahan, and A. O'Keefe, *J. Chem. Phys.*, **72**, 4246 (1980).
94. F. J. Grieman, B. H. Mahan, and A. O'Keefe, *J. Chem. Phys.*, **74**, 857 (1981).
95. B. H. Mahan and A. O'Keefe, *J. Chem. Phys.*, **74**, 5606 (1981).
96. B. H. Mahan and A. O'Keefe, *Astrophys. J.*, **248**, 1209 (1981).
97. F. J. Grieman, B. H. Mahan, A. O'Keefe, and J. S. Winn, *Discuss. Faraday Soc.*, **71**, 191 (1981).

98. B. H. Mahan, C. Martner, and A. O'Keefe, *J. Chem. Phys.*, **76**, 4433 (1982).
99. C. C. Martner, J. Pfaff, N. H. Rosenbaum, A. O'Keefe, and R. J. Saykally, *J. Chem. Phys.*, **78**, 7073 (1983).
100. J. S. Winn, in J. Berkowitz and K.-O. Groeneveld, Ed., *Molecular Ions*, NATO ASI Series, Series. B: Phys., 1983, vol. 90, p.53.
101. P. Erman, *Astrophys. J.*, **213**, L89 (1977).
102. M. Yoshimine, S. Green, and P. Thaddeus, *Astrophys. J.*, **183**, 899 (1973).
103. D. H. Stedman and D. W. Setser, *Prog. React. Kinet.*, **6**, 1 (1971).
104. J. H. Kolts and D. W. Setser, in D. W. Setser, Ed., *Reactive Intermediates in the Gas Phase*, Academic Press, New York, 1979, p. 152.
105. M. F. Golde, Special Periodic Report: *Gas Kinetics and Energy Transfer*, The Chemical Society, London, 1977, vol. 2, p. 123.
106. M. Tsuji and Y. Nishimura, *J. Spectrosc. Soc. Japan (Bunko Kenkyu)*, **32**, 77 (1983).
107. A. J. Yencha, in C. R. Brundle and A. D. Baker, Eds., *Electron Spectroscopy: Theory, Techniques, and Applications*, Academic Press, New York, 1984, vol. 5.
108. W. C. Richardson and D. W. Setser, *J. Chem. Phys.*, **58**, 1809 (1973).
109. D. W. Setser, *Int. J. Mass Spectrom. Ion Phys.*, **11**, 301 (1973).
110. R. S. F. Chang, G. W. Taylor, and D. W. Setser, *Chem. Phys.*, **35**, 201 (1978).
111. K. Suzuki and K. Kuchitsu, *Bull. Chem. Soc. Jpn.*, **50**, 1905 (1977).
112. K. Suzuki and K. Kuchitsu, *Chem. Phys. Lett.*, **56**, 50 (1978).
113. K. Suzuki, I. Nishiyama, Y. Ozaki, and K. Kuchitsu, *Chem. Phys. Lett.*, **58**, 145 (1978).
114. I. Nishiyama, Y. Ozaki, K. Suzuki, and K. Kuchitsu, *Chem. Phys. Lett.*, **67**, 258 (1979).
115. K. Suzuki and K. Kuchitsu, *J. Photochem.*, **10**, 401 (1979).
116. M. Tsuji, K. Tsuji, and Y. Nishimura, *Int. J. Mass Spectrom. Ion Phys.*, **30**, 175 (1979).
117. M. Tsuji, M. Matsuo, and Y. Nishimura, *Int. J. Mass Spectrom. Ion Phys.*, **34**, 273 (1980).
118. M. Tsuji, C. Yamagiwa, M. Endoh, and Y. Nishimura, *Chem. Phys. Lett.*, **73**, 407 (1980).
119. M. Tsuji, H. Obase, M. Matsuo, M. Endoh, and Y. Nishimura, *Chem. Phys.*, **50**, 195 (1980).
120. M. Tsuji, T. Mizuguchi, and Y. Nishimura, *Can. J. Phys.*, **59**, 985 (1981).
121. M. Tsuji, Y. Nishimura, and T. Mizuguchi, *Chem. Phys. Lett.*, **83**, 483 (1981).
122. M. Tsuji, T. Mizuguchi, and Y. Nishimura, *Chem. Phys. Lett.*, **84**, 318 (1981).
123. M. Tsuji, S. Shimada, and Y. Nishimura, *Chem. Phys. Lett.*, **89**, 75 (1982).
124. I. Murakami, M. Tsuji, and Y. Nishimura, *Chem. Phys. Lett.*, **92**, 131 (1982).
125. M. Tsuji, K. Shinohara, T. Mizuguchi, and Y. Nishimura, *Can. J. Phys.*, **61**, 251 (1983).
126. M. Tsuji, T. Mizuguchi, K. Shinohara, and Y. Nishimura, *Can. J. Phys.*, **61**, 838 (1983).
127. M. Endoh, M. Tsuji, and Y. Nishimura, *Chem. Phys.*, **82**, 67 (1983).
128. H. Sekiya, M. Tsuji, and Y. Nishimura, *Chem. Phys. Lett.*, **100**, 494 (1983).
129. M. Endoh, M. Tsuji, and Y. Nishimura, *J. Chem. Phys.*, **79**, 5368 (1983).
130. M. Tsuji, S. Yamaguchi, S. Shimada, and Y. Nishimura, *J. Mol. Spectrosc.*, **103**, 498 (1984).
131. H. Obase, M. Tsuji, and Y. Nishimura, *Chem. Phys. Lett.*, **105**, 214 (1984).
132. M. Tsuji, K. Shinohara, S. Nishitani, T. Mizuguchi, and Y. Nishimura, *Can J. Phys.*, **62**, 353 (1984).
133. H. Obase, M. Tsuji, Y. Nishimura, *Chem. Phys.*, **87**, 93 (1984).
134. M. Endoh, M. Tsuji, and Y. Nishimura, *Chem. Phys. Lett.*, **109**, 35 (1984).
135. M. Tsuji, M. Endoh, T. Susuki, K. Mizukami, and Y. Nishimura, *J. Chem. Phys.*, **81**, 3559 (1984).
136. H. Sekiya, M. Endoh, M. Tsuji, and Y. Nishimura, *Nippon Kagaku Kaishi*, 1498 (1984).
137. H. Sekiya, M. Tsuji, and Y. Nishimura, *Bull. Chem. Soc. Jpn*, **57**, 3329 (1984).

138. M. Tsuji and J. P. Maier, *Chem. Phys.*, **97**, 397 (1985).
139. H. Obase, M. Tsuji, and Y. Nishimura, *Chem. Phys.*, **99**, 111 (1985).
140. M. Tsuji, R. Kuhn, J. P. Maier, S. Nishitani, K. Shinohara, H. Obase, and Y. Nishimura, *Chem. Phys. Lett.*, **119**, 473 (1985).
141. S. Yamaguchi, H. Obase, M. Tsuji, and Y. Nishimura, *Chem. Phys. Lett.*, **119**, 477 (1985).
142. S. Yamaguchi, H. Obase, M. Tsuji, and Y. Nishimura, *Can. J. Phys.*, **64**, 700 (1986).
143. S. Yamaguchi, M. Tsuji, and Y. Nishimura, *Can J. Phys.*, **64**, 1374 (1986).
144. M. Endoh, Ph.D. Thesis, Department of Molecular Science and Technology, Graduate School of Engineering Sciences, Kyushu University, 1984.
145. R. Deloche, P. Monchicourt, M. Cheret, and F. Lambert, *Phys. Rev. A*, **13**, 1140 (1976).
146. R. Johnsen, A. Chen, and M. A. Biondi, *J. Chem. Phys.*, **73**, 1717 (1980).
147. T. Matsuo, N. Kobayashi, and Y. Kaneko, *J. Phys. Soc. Jpn*, **50**, 3482 (1981); **51**, 1558 (1982).
148. C. B. Collins and W. W. Robertson, *J. Chem. Phys.*, **40**, 701 (1964).
149. W. W. Robertson, *J. Chem. Phys.*, **44**, 2456 (1966).
150. L. G. Piper, L. Gundel, J. E. Velazco, and D. W. Setser, *J. Chem. Phys.*, **62**, 3883 (1975).
151. P. Baltayan, J. C. Pebay-Peyroula, and N. Sadeghi, *J. Chem. Phys.*, **78**, 2942 (1983).
152. P. Baltayan, J. C. Pebay-Peyroula, and N. Sadeghi, *Chem. Phys. Lett.*, **104**, 168 (1984).
153. P. Baltayan, J. C. Pebay-Peyroula, and N. Sadeghi, *J. Phys. B*, **18**, 3615 (1985).
154. A. L. Schmeltekopf, E. E. Ferguson, and F. C. Fehsenfeld, *J. Chem. Phys.*, **48**, 2966 (1968).
155. M. Tsuji, T. Susuki, M. Endoh, and Y. Nishimura, *Chem. Phys. Lett.*, **86**, 411 (1982).
156. M. Tsuji, T. Susuki, K. Mizukami, and Y. Nishimura, *J. Chem. Phys.*, **83**, 1677 (1985).
157. K. Mizukami, H. Obase, M. Tsuji, and Y. Nishimura, *Chem. Phys. Lett.*, **116**, 510 (1985).
158. M. Tsuji, K. Mizukami, H. Obase, and Y. Nishimura, *J. Phys. Chem.*, **92**, 1163 (1988).
159. M. Tsuji, I. Nagano, T. Susuki, K. Mizukami, H. Obase, and Y. Nishimura, *J. Phys. Chem.*, **90**, 3998 (1986).
160. M. Tsuji, K. Mizukami, H. Sekiya, H. Obase, S. Shimada, and Y. Nishimura, *Chem. Phys. Lett.*, **107**, 389 (1984).
161. H. Obase, M. Tsuji, and Y. Nishimura, *Chem. Phys. Lett.*, **141**, 133 (1987).
162. J. A. Coxon and S. C. Foster, *J. Mol. Spectrosc.*, **103**, 281 (1984).
163. J. L. Hardwick, Y. Luo, D. H. Winicur, and J. A. Coxon, *Can. J. Phys.*, **62**, 1792 (1984).
164. J. A. Coxon, S. Naxakis, and A. B. Yamashita, *Chem. Phys. Lett.*, **117**, 235 (1985).
165. J. A. Coxon and S. Naxakis, *Chem. Phys. Lett.*, **119**, 223 (1985).
166. A. B. Yamashita, private communication.
167. J. A. Coxon, S. Naxakis, and A. B. Yamashita, *Spectrochim. Acta*, **41A**, 1409 (1985).
168. A. O. Langford, V. M. Bierbaum, and S. R. Leone, *J. Chem. Phys.*, **84**, 2158 (1986).
169. P. J. Marcoux, M. Van Swaay, and D. W. Setser, *J. Phys. Chem.*, **83**, 3168 (1979).
170. D. B. Dunkin, F. C. Fehsenfeld, A. L. Schmeltekopf, E. E. Ferguson, *J. Chem. Phys.*, **49**, 1365 (1968).
171. B. L. Upschulte, R. J. Shul, R. Passarella, R. E. Leuchtner, R. G. Keesee, and A. W. Castleman, Jr., *J. Phys. Chem.*, **90**, 100 (1986).
172. B. E. Holmes and D. W. Setser, in I. W. M. Smith Ed., *Physical Chemistry of Fast Reactions, vol. 2, Reaction Dynamics*, Plenum, New York, 1980, p. 83.
173. S. R. Leone, *Acc. Chem. Res.*, **16**, 88 (1983).
174. V. M. Bierbaum, G. B. Ellison, J. H. Futrell, and S. R. Leone, *J. Chem. Phys.*, **67**, 2375 (1977).
175. T. S. Zwier, V. M. Bierbaum, G. B. Ellison, and S. R. Leone, *J. Chem. Phys.*, **72**, 5426 (1980).

176. T. S. Zwier, M. M. Maricq, C. J. S. M. Simpson, V. M. Bierbaum, G. B. Ellison, and S. R. Leone, *Phys. Rev. Lett.*, **44**, 1050 (1980).
177. M. M. Maricq, M. A. Smith, C. J. S. M. Simpson, G. G. Ellison, *J. Chem. Phys.*, **74**, 6154 (1981).
178. J. C. Weisshaar, T. S. Zwier, and S. R. Leone, *J. Chem. Phys.*, **75**, 4873 (1981).
179. T. S. Zwier, J. C. Weisshaar, and S. R. Leone, *J. Chem. Phys.*, **75**, 4885 (1981).
180. M. A. Smith and S. R. Leone, *J. Chem. Phys.*, **78**, 1325 (1983).
181. M. A. Smith, V. M. Bierbaum, and S. R. Leone, *Chem. Phys. Lett.*, **94**, 398 (1983).
182. C. E. Hamilton, V. M. Bierbaum, and S. R. Leone, *J. Chem. Phys.*, **80**, 1831 (1984).
183. A. O. Langford, V. M. Bierbaum, and S. R. Leone, *J. Chem. Phys.*, **83**, 3913 (1985).
184. P. C. Engelking and A. L. Smith, *Chem. Phys. Lett.*, **36**, 21 (1975).
185. T. A. Miller and V. E. Bondybey, *J. Chim. Phys. Phys.-Chim. Biol.*, **77**, 695 (1980).
186. T. A. Miller and V. E. Bondybey, *Appl. Spectrosc. Rev.*, **18**, 105 (1982); *Philos. Trans. R. Soc. London A*, **307**, 617 (1982).
187. T. A. Miller, *Ann. Rev. Phys. Chem.*, **33**, 257 (1982).
188. T. A. Miller and V. E. Bondybey, in T. A. Miller and V. E. Bondybey, Eds., *Molecular Ions: Spectroscopy, Structure and Chemistry*, North-Holland, Amsterdam, 1983, p. 201.
189. V. E. Bondybey and T. A. Miller, *J. Chem. Phys.*, **69**, 3597 (1978).
190. D. H. Katayama, T. A. Miller, and V. E. Bondybey, *J. Chem. Phys.*, **72**, 5469 (1980).
191. D. H. Katayama and J. A. Welsh, *J. Chem. Phys.*, **75**, 4224 (1981).
192. D. H. Katayama and J. A. Welsh, *J. Chem. Phys.*, **79**, 3627 (1983).
193. D. H. Katayama, *J. Chem. Phys.*, **81**, 3495 (1984).
194. J. P. Maier, *Angew. Chem. Int. Ed. Engl.*, **20**, 638 (1981); *Acc. Chem. Res.*, **15**, 18 (1982).
195. J. P. Maier, O. Marthaler, L. Misev, and F. Thommen in J. Berkowitz and K.-O. Groeneveld, Eds., *Molecular Ions*, Plenum, London, 1983. p. 125.
196. D. Klapstein, J. P. Maier, and L. Misev, in T. A. Miller and V. E. Bondybey, Eds., *Molecular Ions: Spectroscopy, Structure and Chemistry*, North-Holland, Amsterdam, 1983, p. 175.
197. J. P. Maier, D. Klapstein, S. Leutwyler, L. Misev, and F. Thommen, in M. A. A. Ferreira, Ed., *Ionic Processes in the Gas Phase*, NATO ASI Series, Series C: Math. and Phys. Sci., Reidel, Dordrecht, Holland, 1984, vol. 118, p. 159.
198. J. P. Maier, *J. Electron Spectrosc. Relat. Phenom.*, **40**, 203 (1986).
199. H. Sekiya, T. Hirayama, M. Endoh, M. Tsuji, and Y. Nishimura, *Chem. Phys.*, **101**, 291 (1986).
200. M. Tsuji, I. Nagano, N. Nishiyama, H. Obase, H. Sekiya, and Y. Nishimura, *J. Phys. Chem.*, **90**, 3106 (1986).
201. M. Tsuji, I. Nagano, N. Nishiyama, H. Sekiya, and Y. Nishimura, *Laser Chem.*, **7**, 333 (1987).
202. C. E. Hamilton, M. A. Duncan, T. S. Zwier, J. C. Weisshaar, G. B. Ellison, V. M. Bierbaum, and S. R. Leone, *Chem. Phys. Lett.*, **94**, 4 (1983).
203. C. E. Hamilton, V. M. Bierbaum, and S. R. Leone, *J. Chem. Phys.*, **83**, 2284 (1985).
204. A. Plain and J. Jolly, *Chem. Phys. Lett.*, **111**, 133 (1984).
205. D. R. Guyer, L. Hüwel, and S. R. Leone, *J. Chem. Phys.*, **79**, 1259 (1983).
206. L. Hüwel, D. R. Guyer, G. H. Lin, and S. R. Leone, *J. Chem. Phys.*, **81**, 3520 (1984).
207. G. H. Lin, J. Maier, and S. R. Leone, *J. Chem. Phys.*, **82**, 5527 (1985).
208. G. H. Lin, J. Maier, and S. R. Leone, *J. Chem. Phys.*, **84**, 2180 (1986).
209. S. Yamaguchi, M. Tsuji, H. Obase, H. Sekiya, and Y. Nishimura, *J. Chem. Phys.*, **86**, 4952 (1987).
210. M. McFarland, D. L. Albritton, F. C. Fehsenfeld, E. E. Ferguson, and A. L. Schemeltekopf, *J. Chem. Phys.*, **59**, 6610 (1973).
211. M. Tsuji, I. Nagano, and Y. Nishimura, to be published.

212. M. A. Duncan, V. M. Bierbaum, G. B. Ellison, and S. R. Leone, *J. Chem. Phys.*, **79**, 5448 (1983).
213. H. W. Ellis, R. Y. Pai, E. W. McDaniel, E. A. Mason, and L. A. Viehland, *At. Data Nucl. Data Tables*, **17**, 177 (1976).
214. H. W. Ellis, E. W. McDaniel, D. L. Albritton, L. A. Viehland, S. L. Lin, and E. A. Mason, *At. Data Nucl. Data Tables*, **22**, 179 (1978).
215. G. H. Wannier, *Bell. Syst. Tech. J.*, **32**, 170 (1953).

Chapter **X**

INFRARED LASER PHOTOLYSIS

Cris E. Johnson
and
John I. Brauman

---

1 Introduction
2 Infrared Source—The CO$_2$ Laser
   2.1 Description
   2.2 Pulsed Laser
   2.3 CW Laser
   2.4 Optics
   2.5 Measurement of Laser Parameters
   2.6 Other Infrared Sources
3 Theory of IRMP Dissociation
4 Experimental Configurations
   4.1 Ion Cyclotron Resonance Spectrometer
   4.2 Ion-Beam Spectrometer
   4.3 Other Techniques
5 IR Photolysis Experiments
   5.1 Photophysical Studies
      5.1.1 Intensity Comparison
      5.1.2 Vibrational Relaxation
      5.1.3 Fluence Saturation
      5.1.4 Wavelength Dependence in the Quasi-continuum
   5.2 Photochemical Studies
      5.2.1 Isotopic and Isomeric Selectivity
      5.2.2 Unimolecular and Bimolecular Reaction Dynamics
      5.2.3 Vibrationally Induced Electron Detachment
6 Conclusion
References

---

1 INTRODUCTION

Photolysis is a "loosening, dissolution, or decomposition caused by the action of radiant energy." The definition has been properly expanded to include all types of chemical change induced by the interaction of matter with electromagnetic radiation. This includes bond dissociations, elimination and rearrangement reactions, the removal of an electron, and the formation of an excited state. The current discussion will focus on these types of change induced by the infrared (IR) laser.

It is well known that the absorption of infrared radiation causes vibrational transitions within a molecule. It is also well understood that a molecule with sufficient vibrational excitation can undergo bond dissociation or rearrangement reactions. However, the energy content of a single infrared photon is typically about 3 kcal mol^{-1}, while the bond dissociation energies are on the order of 50–100 kcal mol^{-1}. Even lower energy isomerization reactions typically require 10–30 kcal mol^{-1}. Therefore, it is clear that these reactions require the absorption of many infrared photons. The development of high-powered laser sources in the infrared was quickly followed by the demonstration that bond dissociation reactions could be induced, as was first suggested by Isenor and Richardson [1]. It was subsequently demonstrated that the sequential absorption of many infrared photons by a single molecule can occur, even in the absence of collisions, and that this process occurs in a large number of molecules and ions [2]. This novel form of photochemical activation and dissociation has been termed infrared multiple-photon (IRMP) dissociation. Furthermore, great interest was generated by the prospects of selectivity in laser-induced reactions. Such selectivity has been unequivocally demonstrated with laser isotope separation (LIS), the activation and separation of a single isotope from a mixture, and with the separation of geometric isomers. Selectivity within a molecule, the so-called mode-selective chemistry by which different reactions occur due to the excitation of different vibrational modes within the molecule, has been reported on several occasions, but such claims have generally been refuted [2c]. Although such mode-selective chemistry has yet to be demonstrated, the field of IRMP photochemistry has met with tremendous growth and has been proven useful in a variety of photophysical and photochemical studies.

The following types of information have been obtained using the technique of infrared laser photolysis:

1. Photophysical information about IRMP excitation
2. Photodissociation products including identification of the lowest energy product channels
3. Rates of vibrational relaxation
4. Unimolecular and bimolecular reaction dynamics
5. Observation of vibrationally induced electron detachment of anions and information about vibrational-electronic (V-E) coupling

2 INFRARED LASER SOURCE—THE CO_2 LASER

The most widely used source of infrared radiation is the CO_2 laser; the great majority of photolysis reactions has been carried out using this source. For this reason, the CO_2 laser will be discussed in some detail, whereas alternate sources will be mentioned only briefly.

2.1 Description

Laser action involves the rotational lines of two vibrational transitions within the CO_2 molecule. CO_2 is a linear, triatomic molecule containing three normal vibrational modes: the symmetric stretch (v_1), the doubly degenerate bending mode (v_2), and the antisymmetric stretch (v_3). A vibrational state of CO_2 can be specified by the notation $n_1 n_2^l n_3$, where n_i is the number of vibrational quanta in normal mode v_i, and l specifies the vibrational angular momentum of the degenerate v_2 mode. The value of l depends upon the relative magnitudes and phases of the orthogonal bending modes [3]. A partial vibrational level energy diagram is given in Fig. 1. The diagram indicates the laser transitions present between the $00^01 \rightarrow 02^00$ and $00^01 \rightarrow 10^00$ transitions that result in a series of lines centered around 9.4 and 10.4 μm, respectively.

Antisymmetrization of the total wave function of $^{12}C^{16}O_2$ requires that state 00^01 contain only rotational states of odd values of j, whereas states 10^00 and 02^00 contain rotational states with only even values of j. Therefore, the laser output will consist of a series of lines corresponding to the P and R branches of a ro-vibrational transition with alternate lines missing. This corresponds to usable radiation in the range 920–992 cm^{-1} for the branch centered at 10.4 μm and 1030–1085 cm^{-1} for the branch centered at 9.4 μm, with a separation between lines of 1–2 cm^{-1}.

The normal mode approximation assumes the vibrational modes are separable harmonic oscillators and does not account for anharmonic interactions between the modes. One important consequence of such interactions is that the 10^00 and 02^00 states are significantly mixed due to the presence of a Fermi resonance between these states [3, 4]. For simplicity, however, these states will continue to be referred to as 10^00 and 02^00.

The lasing medium usually consists of a mixture of CO_2, N_2, and He, although in the pulsed CO_2 laser the nitrogen is sometimes removed. Many other gases have been mixed with CO_2, but no mixtures have approached the power output available in the CO_2/N_2/He mixture [5]. A typical ratio of gases in the mix is 1:1:8 for $CO_2:N_2$:He. The population inversion in the medium is significantly increased by the presence of N_2 and He. The most significant excitation mechanism appears to be the direct excitation of N_2 by electron impact from electrons in the plasma, although the direct excitation of CO_2 is also possible. The cross section for this process in N_2 is very large compared to many other gases and is approximately 4×10^{-16} cm^2 [6]. The excited states of nitrogen lie very close in energy to the excited asymmetric stretch mode in CO_2 (see Fig. 1). Thus, the energy can be transferred from N_2 to ground-state CO_2 molecules by collisions. This leads to a large population present in the 00^01 state. Furthermore, the 10^00 and 02^00 states are nearly twice the energy of the 01^10 state. This allows a collision between CO_2 in the 10^00 or 02^00 state and a ground-state molecule to produce two molecules in the 01^10 state, thus

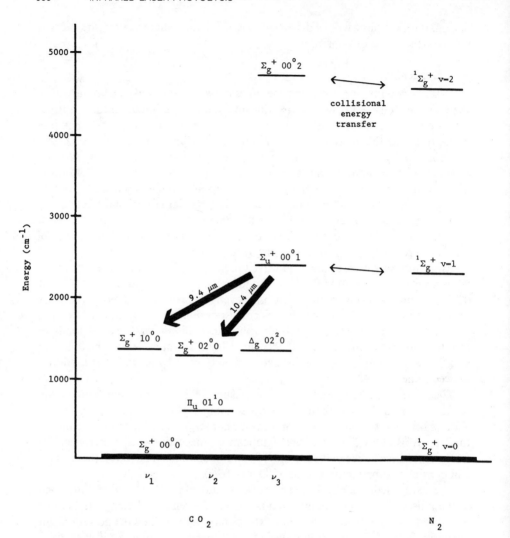

Fig. 1. Partial vibrational level energy diagram for CO_2 and N_2. Heavy arrows indicate laser transitions.

depopulating the lower states of the laser transition. Helium significantly increases this depopulation by increasing the rate of energy transfer from the 01^10 state to the ground state [7]. Therefore, the $CO_2/N_2/He$ mixture is an efficient medium for obtaining the necessary population inversion for lasing to occur. The laser action generated in this mixture can either be pulsed or continuous wave.

To understand the distinctions between the output characteristics of the pulsed and cw lasers, the parameters photon flux, intensity, and fluence need to be clearly understood. Therefore, these parameters are defined below:

Photon Flux (Φ) The number of photons per unit area per unit time; given in units of photons cm^{-2} s^{-1}

Intensity (I) The power per unit area; given in W cm^{-2}

Fluence (F) The energy per unit area; given in J cm^{-2}

Note that the photon flux is equal to the intensity divided by the energy per photon (Eq. 2.1), and the fluence is the time integral of intensity (Eq. 2.2):

$$\Phi = \frac{I}{h\nu} \tag{2.1}$$

$$F = \int_0^\tau I(t)\,dt \tag{2.2}$$

For a constant intensity pulse, Eq. 2.2 reduces to the product of the intensity and the laser pulse length, τ (Eq. 2.3):

$$F = I\tau \quad \text{for constant } I \tag{2.3}$$

2.2 Pulsed Laser

The pulsed CO_2 laser is capable of generating extremely high-powered pulses (megawatts) of short duration (microseconds or less). TEA is an abbreviation for transverse excitation at atmospheric pressure; the gas mixture is present at atmospheric pressure, which is much higher than in the cw laser. Most lasers use a flow design so that dissociation products are removed from the laser tube. A high-voltage power supply (typically 30–40 kV, 20 mA) is used to charge energy storage capacitors (0.075 μF). A discharge is initiated by the means of a thyratron switch or a spark gap that discharges the capacitors onto a pair of electrodes. The electrodes are present along the length of the cavity and have a precise separation. The discharge occurs between the electrodes transverse to the axis of the laser tube. This results in a plasma that initiates the excitation processes indicated above. Lasing occurs until the energy from the discharge is consumed and the population inversion can no longer be maintained. The capacitors are then recharged, and the process is repeated. The repetition rate depends upon the power supply and the charging characteristics of the capacitors and is typically 0.5–5 Hz. For a given power supply, a higher power pulse requires a slower repetition rate.

The resulting laser pulse consists of a high-intensity spike of about 150–200 ns in duration, followed by a lower-intensity tail extending to approximately 3 μs. Each portion contains roughly half of the pulse energy. The pulse consists of a collection of equally spaced peaks separated by approximately 20 ns, which represents the round-trip transit time through the laser cavity and is due to longitudinal mode beating. The laser can also be operated without the nitrogen gas, in which case the tail is no longer present but the energy per pulse is decreased. This indicates that the nitrogen provides energy storage for the discharge and increases the overall lasing efficiency by collisional energy transfer through the tail region. The energy output of the laser is typically several joules per pulse with a beam diameter of 2–5 cm^2. The laser fluence is therefore 1–2 J/cm^2. Given the pulse width indicated above, the laser intensity is therefore in the megawatt range. The wavelength is tunable over the rotational lines mentioned above if a diffraction grating is incorporated in the cavity (replacing the rear reflector); the bandwidth (full width half-maximum—FWHM) is typically 0.1 cm^{-1}.

2.3 CW Laser

The cw laser produces continuous output at lower power than the pulsed laser. The power output of a cw laser is proportional to the length of the laser tube (up to a few meters), so the cost of high-powered cw lasers can be extremely large. However, 30–50-W (total output) cw CO_2 lasers are readily obtainable, and even 100–1000-W cw lasers are commercially available. It should be noted that the design and construction of a fairly low power cw CO_2 laser is simple enough that it can be "home built." In our laboratory, a 30-W laser was assembled using a commercial power supply, output coupler, diffraction grating, and optical mounts. The laser tube specifications were calculated, and the tube was blown in a glass shop. The parts were assembled, and the operating characteristics were experimentally optimized. The result was a functional CO_2 laser at about one-half the cost of a similar commercially available product.

The important variables in optimizing the design of a cw CO_2 laser are the tube diameter and length, the current in the discharge, and the rate of gas flow. A current-limited high-voltage direct current (dc) power supply (20–30 kV) is connected to a pair of electrodes located at each end of the laser tube. The electrode materials and design are relatively noncritical, although design of the cavity containing the electrode, especially the cathode, is important [8]. One common arrangement is a pin anode and a ring cathode, the cathode being placed in an enlarged cavity with the inner diameter of the ring matching the tube diameter. A glow discharge is generated along the length of the tube, and a steady state is established between excitation and deexcitation (i.e., stimulated emission, collisional relaxation, and spontaneous emission) processes. The current, gas pressure, and flow rate can be varied to optimize the power output for each value of the tube diameter. Typical values are a 10–30-mm tube diameter with several torr of gas, 20–50 mA of current, and a flow rate of 100–200 volume changes per minute. The presence of a continuous discharge necessitates that the discharge tube be kept cool. The lasing efficiency increases with a decrease in the surface temperature of the tube, so that the lowest temperature obtainable is best. However, water circulating around the outside of the discharge tube is used almost exclusively because of cost considerations.

The cw laser is diffraction grating tunable over the same rotational lines as the pulsed laser, and, similarly, the power output at each individual line is considerably less than the total power output. The intensity of a laser that gives

30 W of total output is typically 1–15 W cm^{-2} on a single line and is fairly constant during operation of the laser. The beam diameter is dependent upon the diameter of the discharge tube and the type of resonator used, but it is usually 1–2 cm^2. Fluences similar to those obtained with the pulsed CO_2 laser can, therefore, be obtained with long irradiance times, on the order of 1 s.

2.4 Optics

Laser optics consist of internal optics, devices to produce a resonating cavity, and external optics, devices to control and manipulate the laser beam outside the cavity. It is important to realize that the classical properties of the optical resonator, not the quantum properties of the laser medium itself, provide the desirable characteristics of laser radiation such as high monochromaticity, small bandwidth, and directionality. Although many designs in optical resonators are possible, all result in establishing standing waves within the cavity called resonator modes. There are two types of modes in the cavity–longitudinal modes (along the laser axis) and transverse modes (perpendicular to the axis).

Longitudinal modes result because light waves of any integral number of half wavelengths will constructively interfere upon successive trips through the cavity. In general, the frequency separation of these longitudinal modes is much smaller than the natural linewidth of the laser gain profile because of Doppler broadening of the transition. Thus, a number of equally spaced longitudinal modes will be amplified by the resonator and will be simultaneously present in the cavity. In the pulsed CO_2 laser, these longitudinal modes tend to couple to one another or "beat" against each other in a process called "self-mode-locking." This results in a constant phase ("locked") oscillation, which produces a modulation in the laser output. The output consists of a series of equally spaced high-intensity spikes, as mentioned above. The pulsed laser has been constrained to a single longitudinal mode by incorporating a low-pressure cw gain section into the laser cavity [9]. This results in a non-mode-locked pulse that is much smoother than the multimode pulse and allows the laser output to be more easily characterized.

Transverse modes determine the spatial properties of the beam perpendicular to the laser axis. They are designated TEM_{mn} for transverse electric and magnetic, where m and n are the number of nodes along a pair of orthogonal axes. The lowest-order modes (small values of m and n) have the smallest diffraction losses and beam divergence and so are the most desirable for laser oscillation. The most common transverse modes are TEM_{00}, which has a Gaussian beam profile, and the combination of TEM_{01} and TEM_{10}, which results in a "doughnut"—a ring of intensity with a dark center [10]. The output can be constrained to TEM_{00} by placing an aperture in the laser cavity. One advantage of the cw laser is that the small discharge tube helps to restrain oscillation to the lower-order modes. With small adjustments (usually adjusting the current below that for optimum power output), TEM_{00} can be obtained.

The external optics, or optical train, consists of devices to guide the beam to the desired location within the experimental apparatus or to change certain properties of the beam. A description of the most significant items of optical equipment is given below:

1. **Mirrors**—Plane, spherical, and parabolic reflectors are available with a gold coating to provide high reflectivity in the IR. For a spherical mirror, a short radius of curvature specifies a more highly focusing mirror.
2. **Beam splitter**—An optical flat or cube designed to reflect a known percentage of the beam (e.g., to a power meter) while transmitting the rest. The transmitted beam is displaced slightly due to refraction in the medium. Various ratios of reflected beam to transmitted beam are obtainable.
3. **Attenuator**—CaF_2 flats or polyethylene sheets that reduce the intensity of the beam.
4. **Aperture**—A "hole" or an opening in an absorbing surface (blackened aluminum) that is usually adjustable. The aperture is used to remove the low-intensity outer edges of the beam or to restrict the beam diameter.
5. **Beam stop**—A highly adsorptive material used to block the beam. Fire brick serves the purpose well, as does a blackened metal surface.
6. **Lens**—Refractive material that changes direction of incident radiation; used in pairs for beam expansion and contraction and singly for focusing.
7. **Shutter**—An aperture that can be triggered open and closed.
8. **Beam chopper**—A rotating disk with slits, which divides a continuous beam into a series of pulses.

Alignment of the optical setup can be a significant task. The optical components are placed upon an optical rail colinear with the laser beam, and a HeNe laser is used to facilitate the alignment. The use of a "breadboard," a small optics table with precisely spaced threaded holes, allows the optical rails and other optics to be mounted in a reasonably stable arrangement. Alignment of the beam can be difficult compared to visible wavelengths because the beam is invisible. For the pulsed laser, a graphite paddle or thermal imaging paper is used to "image" the beam during alignment. Fluorescent paddles are used with the cw laser. Finally, it should be noted that significant hazard exists to the eyes while aligning the infrared laser. Eye protection must be worn at all times while operating a CO_2 laser.

2.5 Measurement of Laser Parameters

The high-energy densities and short pulse widths obtainable with laser sources make the measurement of laser parameters a significant task. The parameters that need to be determined are laser power or energy (including a temporal profile), wavelength, and beam diameter. Laser power meters can be characterized as either calorimetric, in which the absorption of radiant energy causes heat, or photoelectric, in which absorption of radiant energy causes an electron to be emitted. All power measuring devices have a maximum power above which the absorption surface is damaged and the calibration (at least) is ruined. A brief description of the most common IR detectors is given below.

Calorimetric Detectors

1. Power probe—An absorbing surface with calibrated dial thermometer. The probe is placed in the laser beam (cw) for a specified time interval (several seconds).
2. Thermopile detector—Calibrated detector head with absorbing surface; heating produces an electrical output to a meter.
3. Pyro-electric detector—High-precision, pyro-electric crystal (such as $LiTaO_3$) produces a voltage when radiatively heated.

Photo-Electric Detectors

1. Photodiode—Material such as p-type semiconductor produces a free electron upon absorption of a photon; current collects at cathode.
2. Photon drag detector—absorption of an infrared photon by doped germanium crystal causes a potential drop that is used to measure intensity. The fast response allows temporal profile to be determined.

The wavelength of the laser beam is determined using a spectrum analyzer. The beam is passed through dispersive elements inside the analyzer and impinges upon a fluorescent strip with calibrated markings of wavelength. The incident beam must be nearly colinear with the axis of the analyzer. For cw measurement a shutter is used or a fraction of the beam is diverted with a beam splitter. The beam diameter is determined using an imaging plate or burn patterns from thermal imaging paper.

2.6 Other Infrared Sources

The CO_2 laser has been used almost exclusively in infrared laser photolysis because of its high output power, low cost, and ease of use. The CO_2 laser is not, however, the only infrared source available. Infrared source technology is rapidly improving, although available output power remains the chief limitation. Table 1 lists several infrared lasers that are commercially available and their output wavelength ranges.

Table 1. Some Commercially Available Infrared Lasers

Laser	Wavelength (μm)
CO_2	9–11
CO	5–6
HF/DF	2.5–3/3.5–4
F-center	1.4–1.6
	2.3–3.3
Diode (semiconductor)	0.8–1.5

In addition to traditional laser sources, infrared radiation has been obtained with the optical parameteric oscillator (OPO) [11]. This source takes advantage of a nonlinear effect in a lithium niobate crystal, $LiNbO_3$, to produce tunable infrared radiation from the difference frequency of a visible laser "pump," such as a Nd:YAG laser, and a tunable visible source, such as a dye laser. These parametric oscillators have been used successfully in the spectroscopy of ions, although the damage threshold of the crystal limits their use in photolysis reactions.

3 THEORY OF IRMP DISSOCIATION

The ab initio theoretical description of the time evolution of a collection of polyatomic molecules or ions interacting with an intense, monochromatic radiation field is far too complex to solve at the present time. The usual approach is to reduce the time-dependent Schrödinger equation for a multiple-level system to a set of coupled differential equations known as the master equation [12–15]. Out of these theoretical treatments a simple model has emerged that provides significant agreement with experiments. Multiple-photon excitation involves the sequential absorption of photons through three distinct regimes: the discrete level regime, the quasicontinuum, and the dissociation threshold above which a true continuum of product states exists (see Fig. 2).

The discrete level regime consists of the individual spectroscopic states of the molecule at low energies. Excitation requires that the laser frequency be in resonance with the energy separation of any two states. However, the vibrational states are not equally spaced because of vibrational anharmonicity. Therefore, consecutive absorption of photons would quickly cause the laser to be out of resonance. For this reason, some additional mechanism is necessary to explain the ease of excitation through this regime. When rotational states are included there may exist ro-vibrational transitions that will continue to be resonant with the laser, the rotational energy compensating for the anharmonicity. Also, the phenomenon of power broadening, by which high-powered lasers can induce transitions that are significantly off resonance, is commonly cited as a possible explanation of excitation through the discrete regime. For these reasons, the discrete regime is expected to exhibit dissociation yields that may be highly intensity and wavelength dependent. In small molecules with a low density of states, the discrete regime may prevail for the absorption of the first few photons, whereas in larger molecules the absorption of a single photon may place the molecule into the quasicontinuum.

As the internal energy of the molecule increases, the density of states increases until eventually the separation of states is less than the laser bandwidth. This condition defines the quasicontinuum, so called because some pair of energy levels is always connected by a photon of arbitrary frequency. Thus, radiation of any frequency can be absorbed, although some modes absorb more strongly than others (vide infra). This vibrational energy is quickly transferred to other ro-vibrational states (intramolecular V-V energy transfer). This process occurs

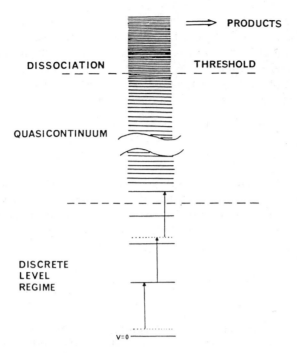

Fig. 2. Schematic energy diagram for IRMP absorption and dissociation. Dotted lines indicate rotational sublevels.

much faster than the absorption of a photon [16] so that the system is best described in terms of energy levels rather than particular states.

Eventually, the molecule has absorbed sufficient energy to dissociate. Above the quasicontinuum, a dissociation threshold exists. Below the threshold, the rate constant for dissociation is zero, and dissociation can not occur. Above the threshold, the dissociation rate constant is expected to increase with increasing internal energy, and the molecule can either react or absorb another photon. Eventually, the dissociation will compete effectively with the rate of photon absorption (pumping rate), and the molecule will dissociate. The rate of intramolecular energy redistribution is sufficiently fast to allow a statistical rate theory, such as RRKM theory [17], to characterize the dissociation.

At each level of internal energy a deactivation process, namely, stimulated emission, also occurs. The master equation (Eq. 3.1) specifying the time dependence of the population of each level therefore involves the rate constants for absorption, k_n^a, for stimulated emission, k_n^e, and for dissociation, k_n^{diss}. In Eq. 3.1, P_i is the population of level i, and I is the intensity of radiation:

$$\frac{dP_n}{dt} = k_{n-1}^a I P_{n-1} + k_{n+1}^e I P_{n+1} - (k_n^a + k_n^e) I P_n - k_n^{diss} P_n \qquad (3.1)$$

Additionally, terms involving spontaneous emission, collisional relaxation, or other processes can be included. The solution of the master equation involves the specification of all rate constants indicated above and then solving $n-1$ linear differential equations.

For a single photon absorption process leading to the dissociation of species N, the dissociation rate is given by a linear rate law (Eq. 3.2):

$$\frac{-d[N]}{dt} = \sigma\Phi[N] \qquad (3.2)$$

σ is the cross section for absorption in units of square centimeters, Φ is the photon flux, and $[N]$ is the concentration of species N. Quack has shown [18] that when a master equation involving a series of incoherent, single-step processes is a good description, the unimolecular dissociation rate constant, $k_{uni}(t)$, is given by Eq. 3.3:

$$\frac{-d\ln(1-F_D)}{dt} = k_{uni}(t) \qquad (3.3)$$

F_D is the fraction of reactant that has decomposed, or the product yield, induced by a constant intensity laser pulse of length τ. The rate constant, $k_{uni}(t)$ is phenomenological; that is, it describes the overall absorption and dissociation process. After a short period of time, called the induction period, the level populations reach a steady state and k_{uni} reaches a steady-state value, k_{uni}^{ss}. A bimolecular rate constant can be defined by Eq. 3.4:

$$k_I^{ss} = \frac{k_{uni}^{ss}}{I} \qquad (3.4)$$

Substitution into Eq. 3.3 and integration yields Eq. 3.5:

$$-\ln(1-F_D) = \int_0^\tau k_I^{ss} I\, dt = k_I^{ss} I\tau \qquad (3.5)$$

Equation 3.5 indicates that it is the laser fluence (the product of the intensity and the irradiation time, τ) that governs the dissociation yield under these conditions, even though the dissociation rate is intensity dependent. The k_I^{ss} can be obtained from the plot of $-\ln(1-F_D)$ vs. fluence, called an LRF plot for log reactant vs. fluence. After the induction period, the system reaches steady state and the plot becomes linear. The slope of this linear portion of the LRF plot is the bimolecular rate constant k_I^{ss}. A phenomenological cross section can be defined by Eq. 3.6:

$$k_I^{ss} = \sigma_{ss}\Phi \qquad (3.6)$$

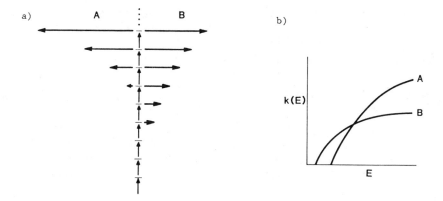

Fig. 3. (*a*) Schematic diagram of a two-channel reaction. Vertical arrows represent pumping process (photon absorption), and horizontal arrows represent the rate constants for reaction A and B. The product ratios vary with internal energy. (*b*) Variation of dissociation rate constants with internal energy.

The cross section is obtained from the division of k_I^{ss} by the photon flux. Although Eqs. 3.3–3.6 apply to constant intensity pulses, they have been used to characterize radiation from a TEA pulsed laser with good agreement. In the pulsed laser, the pulse length is held fixed and the fluence (and intensity) is varied by placing an attenuator in the beam path.

It may be possible to observe multiple products when the rate of photon absorption is sufficiently fast so that a second product channel is reached before reaction via the lowest energy channel. This pumping rate is directly proportional to the laser intensity, so that a high-intensity TEA laser is expected to pump significantly above the reaction threshold, whereas the cw laser produces dissociation products at or near the threshold. The ideal situation to observe such photochemical branching (production of multiple products from a single photochemically activated transition state) would occur when the rate constant for the lower channel increases slowly with increasing internal energy, while the rate constant for the upper channel increases more rapidly with internal energy. This situation is depicted in Fig. 3, where the length of the arrows represent increasing magnitude of the rate constant. Under these conditions, it may be possible to observe a change in the product branching ratios with a change in laser intensity. It is also possible to imagine multiple products resulting from the formation of a chemical intermediate that subsequently branches or from secondary photochemical reactions of the primary product. Such situations must, therefore, be carefully differentiated.

4 EXPERIMENTAL CONFIGURATIONS

Infrared photolysis of gas phase ions has been mainly studied using ion cyclotron resonance (ICR) and ion-beam techniques. Although these techniques

4.1 Ion Cyclotron Resonance Spectrometer

Ion cyclotron resonance spectrometry provides the ability to trap ions for extended periods of time. The entire ion population present in the ICR cell can easily be irradiated with the use of a collimated, unfocused beam. The typical operating pressures (10^{-8}–10^{-5} torr) provide sufficient time between ion–molecule collisions to allow pulsed dissociation experiments to be carried out under collisionless conditions. Furthermore, the technique of double resonance allows ionic photoproducts to be associated unambiguously with a particular ionic precursor.

The typical trapped-ion cell is easily modified to accommodate IR photolysis experiments. Figure 4 indicates the configuration of an ICR spectrometer in our laboratory. Access to the cell is afforded the laser with the use of a zinc selenide or potassium chloride window mounted on a Conflat flange and sealed with Viton O-rings. The salt window allows transmission in the visible region of the spectrum for two-color experiments, but it is easily damaged with the pulsed laser at power densities around 5–10 MW cm^{-2}. The ZnSe is only transparent to the red of 580 nm (yellow), but it has a much higher damage

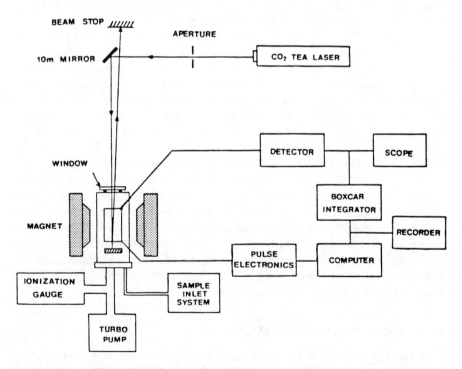

Fig. 4. Block diagram of a pulsed ICR photolysis apparatus.

threshold and is suitable for use with high-powered lasers. It does scratch very easily, however. The pulsed laser is operated multimode, although no evidence of hot spots was found by scanning across the beam with a 2-mm aperture. The spatial profile of the beam was smooth and was nearly Gaussian. The cw laser is operated in TEM_{00} and has a Gaussian beam profile. The pulsed laser beam is passed through an aperture to eliminate the low-intensity edges of the beam, and it is then directed into the cell with a 10 m radius of curvature gold-coated spherical mirror. The spherical mirror focuses the beam slightly to increase the intensity of the laser beam admitted to the cell. However, the focal length is sufficiently long that the beam remains essentially collimated along its path through the cell. The cw laser beam is directed into the cell with a gold-coated planar mirror.

Figure 5 shows a detailed view of the cell configuration. The end cell plate was removed to admit the laser beam, without affecting the trapping characteristics of the cell adversely. The laser is reflected from the cell with the use of an optically flat gold-coated OFHC (oxygen free high conductivity) copper mirror. Gold-coated OFHC copper was also used for the remaining cell plates to avoid pressure surges of the laser caused by desorption from the surfaces during the alignment of the laser, although molybdenum is sufficient for this

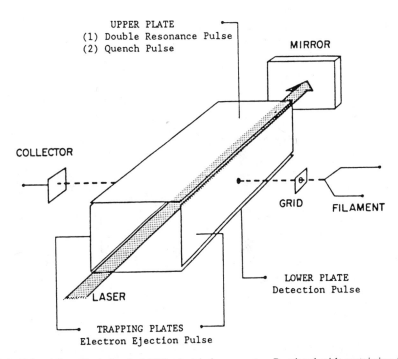

Fig. 5. Cell configuration of pulsed ICR photolysis apparatus. Reprinted with permission from J. M. Jasinski, R. N. Rosenfeld, F. K. Meyer, and J. I. Brauman, *J. Am. Chem. Soc.*, **104**, 652 (1982). Copyright 1982 American Chemical Society.

purpose as well. The spherical mirror is adjusted so that there is a separation between the beams sufficient to direct the reflected beam below the spherical mirror and into a beam stop. Within the cell, the angle between the incoming and the reflected beams is small enough so that the beams overlap spatially. The path length of the laser beam through the cell is sufficiently short so that the reflected beam overlaps temporally as well. Therefore, the total fluence within the cell is very nearly twice the fluence of the incoming beam. The fluence (and intensity) is controlled by placing an attenuator in the optical train of the pulsed laser or by changing the irradiation time in the cw laser.

It should be noted that the cell configuration of Beauchamp and co-workers' apparatus is slightly different. This cell configuration is shown in Fig. 6. The laser is directed into the source region of the continuous-mode (drift) ICR cell through a wire mesh of high transmittance (0.92). The fluence within the cell is also twice the fluence of the transmitted incoming beam (1.84 times the fluence measured outside the cell).

One minor complication in the ICR detection occurs as a result of photolysis. The product yield from a photolysis experiment is determined by measuring the ratio of ion signal with and without radiation. As the ions are irradiated, a sufficient number may dissociate and cause a change in the ion population present in the cell. This change in the distribution of charge in the cell affects the electric field experienced by the non-reactant ions, which in turn affects the detection frequency of the ions. For this reason, it is possible that photolysis can cause a shift in the frequency of the observed ion signal maximum called a resonance shift. Since the signal maximum is no longer being detected, such resonance shifts can lead to incorrect peak heights, which would be interpreted

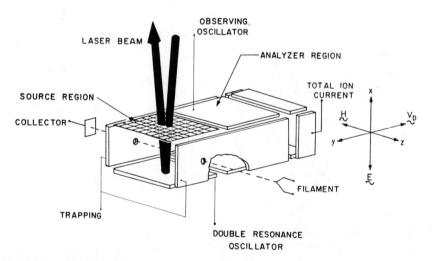

Fig. 6. Cell configuration of continuous-mode (drift) ICR photolysis apparatus. Reprinted with permission from D. S. Bomse, R. L. Woodin, and J. L. Beauchamp, *J. Am. Chem. Soc.*, **101**, 5503 (1979). Copyright 1979 American Chemical Society.

as too large a decrease in product signal upon irradiation. Therefore, it is necessary to scan across the ion signal peaks during the data aquisition of a photolysis experiment so that the correct signal maxima can be determined. This "peakscan" is easily incorporated into a data aquisition program of an ICR suitably interfaced to a microcomputer. Of course, the Fourier transform (FT) ICR detection does not suffer from this liability because the entire mass spectrum is obtained.

4.2 Ion-Beam Spectrometer

Ion-beam spectrometers allow the measurement of photofragment spectra and the kinetic energy release of the dissociation products. Infrared decompositions have been studied with the laser in either a crossed-beam or a coaxial configuration [19–22]. The coaxial configuration provides a significant increase in resolution, provides increased sensitivity due to a larger interaction region with the laser beam, and allows the use of the Doppler shift to bring the laser into resonance ("Doppler tuning").

The configuration of a laser–ion coaxial beam spectrometer is given in Fig. 7. The primary difficulty of effecting IR photolysis in a coaxial configuration is

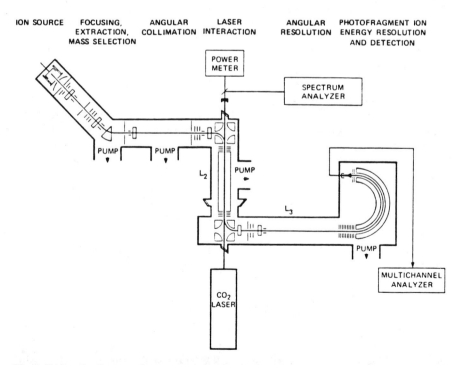

Fig. 7. Schematic diagram of a coaxial laser–ion-beam spectrometer. Reprinted with permission from M. J. Coggiola, P. C. Cosby, H. Helm, J. R. Peterson, and R. C. Dunbar, *J. Phys. Chem.*, **91**, 2794 (1987). Copyright 1987 American Chemical Society.

deflecting the ion beam so that it merges with the laser beam. This is elegantly accomplished with the use of a quadrupole deflector [23] in which the quadrupole is oriented perpendicularly to the bending plane. The quadrupole deflector eliminates the necessity of using a magnetic deflector or of putting a hole in an electrostatic deflector to admit the laser beam, which would cause a change in the electric field. Regardless of the deflector type used, the deflector can be tuned to allow either the photofragment or the parent ion beam to be sent to the analyzer region. This allows the photofragment to be distinguished from the decomposition of metastable ions formed in the ion source. The interaction time between the laser and the ion beam is not long enough for the ions to undergo a multiple-photon absorption process. However, some of the ions formed in the source have sufficient internal energy for a single photon to effect decomposition. Therefore, the photodissociation process in the ion-beam spectrometer corresponds to the last step in a multiple-photon process.

4.3 Other Techniques

Infrared photolysis has also been studied using a quadrupole ion store (QUISTOR) [24] and in a time-of-flight (TOF) mass spectrometer [25].

5 IR PHOTOLYSIS EXPERIMENTS

A brief overview of experimental results will be given in order to provide an indication of the types of information obtainable with the technique of IR laser photolysis. This is not intended to be an exhaustive review of the photolysis reactions of gas-phase ions, only a representative sample. The experiments can be roughly divided into photophysical studies and photochemical studies. Photophysical experiments provide information about the IRMP activation process itself, whereas photochemical studies affect a variety of chemical transformations.

5.1 Photophysical Studies

Photophysical experiments have mostly consisted of examining the photodissociation yield while varying the laser fluence, wavelength, and intensity, as well as sample pressure and temperature. The results have helped to formulate the simple model of IRMP activation mentioned above and are generally in agreement with it.

5.1.1 Intensity Comparison

Except for the contributions of the discrete regime for small molecules, IRMP photodissociation (IRMPD) yields depend only upon laser fluence. This suggested the possibility that low-intensity cw radiation would provide the same reactivity observed with pulsed lasers. IRMPD was first demonstrated using cw CO_2 radiation by Beauchamp and co-workers for the decomposition of the proton-bound dimer of diethyl ether $[(C_2H_5)_2O]_2H^+$ and $C_3F_6^+$ using a power of only 4 W cm^{-2} [26]. Subsequent comparison of the low-intensity results with results using pulsed CO_2 radiation by Jasinski et al. [27] for the same ions allows examination of photophysical and photochemical processes over a wide range of irradiation intensity.

The phenomenological cross section, σ_{ss}, was obtained from the slope of the linear portion of an LRF plot as discussed above. Fluence thresholds were also determined. Both experiments exhibited an induction period followed by a linear portion, indicating that the use of a master equation provided a good photophysical description of each system. However, the cross sections obtained were slightly smaller and the fluence thresholds were approximately a factor of 10 larger for the cw experiments. Since no intensity effects were observed in the range accessible to each experiment and the fluence plots were linear in each case, the magnitude discrepancy was thought to be due to relaxation processes, collisional and radiative, present during the long cw irradiation. The wavelength dependence was similar for both systems, with the pulsed laser giving a somewhat broader spectrum in $C_3F_6^+$. The value of maximum photodissociation yield was blue-shifted in the pulsed spectrum as compared to the cw spectrum, although it is reasonable that both are red-shifted from the maximum 0–1 low-power absorption of $C_3F_6^+$, since this is an expected consequence of anharmonicity.

5.1.2 Vibrational Relaxation

Infrared multiple-photon dissociation has been used to probe vibrational relaxation in ions containing excess internal energy from either electron impact or chemical activation. The experiments have been performed using ICR spectrometry. The photodissociation yield is monitored while the delay time between the formation of the ion and irradiation with the laser is increased. Ions containing excess internal energy are dissociated more readily since they are initially closer to the dissociation threshold. Thus as the ion relaxes the photodissociation yield decreases.

Jasinski and Brauman studied the relaxation of CF_3O^- with various bath gases [28] and found collisional efficiencies to be surprisingly small. The collisional efficiency is the experimentally determined collisional relaxation rate constant divided by the theoretical ion–molecule collision rate constant as calculated using, for example, Average Dipole Orientation (ADO) theory [29]. Furthermore, radiative relaxation via spontaneous emission was found to be significant due to the low pressures in the ICR cell. The relaxation rates were determined by plotting $-\ln(1 - F_d)$ vs. delay time for the laser and obtaining an apparent first-order rate constant from the slope. This was done for a variety of pressures, and then a "quenching plot" was obtained by plotting the k_{app} vs. pressure (Stern-Volmer plot). The slope of these plots was a straight line. The slope was interpreted as a collisional relaxation rate constant and the intercept as the radiative relaxation rate constant. The collisional efficiencies determined in this manner were found to range from 0.03 to 0.16 depending upon bath gas. The radiative relaxation rate was found to be on the order of $20 \, \text{s}^{-1}$.

Rosenfeld et al. studied the collisional relaxation of CH_3OHF^- [30] and found that between 40 and 80 collisions were required to remove the 15 kcal/mol of excess energy present due to the exothermicity of the reaction that generates the ion. Thorne and Beauchamp determined the relaxation rate constant for CF_3^+ [31] using cw radiation and found it to be of similar magnitude. Barfknecht and Brauman studied the IRMP dissociation of vibrationally excited iodobenzene and bromo-3-(trifluoromethyl)-benzene cations prepared by the internal conversion of a visible photon [32]. Absolute relaxation rates were obtained for a variety of bath gases. These results are in agreement with those obtained by Dunbar using two-photon visible photodissociation in bromobenzene cation [33] and iodobenzene cation [34]. Such results are surprising because it was thought that the long-range interactions during an ion–molecule collision would result in a long-lived collision complex in which energy transfer would readily take place. Therefore, a statistical distribution of vibrational energy in the complex would be expected. However, these studies indicate that vibrational energy transfer is somewhat slower than that predicted by a statistical distribution and that the excitation can be present for a significantly large number of collisions.

5.1.3 Fluence Saturation

Although the master equation model for IRMP excitation predicts linear LRF plots, deviations from linearity have been observed in a number of cases. This nonlinear behavior is referred to as "saturation" of the product yield with increasing fluence, or "turnover" of the LRF plot. Among gas-phase ions, saturation effects have been observed in photodissociation yields of CH_3OHF^- [35] and photodetachment yields of benzyl anion, $C_6H_5CH_2^-$ [36]. In CH_3OHF^-, the effect of saturation is to produce an LRF plot containing two straight line segments with different slopes when the ions contain excess internal energy (i.e., at short delay times following their formation). When the ions are allowed to relax sufficiently, however, a linear plot results. In benzyl anion, bent plots occur whether the ions are allowed to relax or not. Relaxation does have an effect, however; the amount of photodetachment product is decreased at each value of fluence relative to the excited ions.

Within the framework of the linear rate law model, there are two possible explanations for such bent fluence plots. One possible cause is the presence of two distinct subpopulations with different cross sections of absorption. Thus, one subpopulation is resonant with the laser and is easily pumped whereas the remaining ions are nonresonant and are less easily pumped. The presence of a resonant subpopulation could be the result of the population of particular rotational levels in the discrete regime. If these rotational levels cause the molecules to be in resonance whereas those occupying other levels are not, then the subpopulation could be depleted by a sufficient number of photons, thus leaving a "hole." Collisions would repopulate these levels–a process referred to as "rotational hole filling." A second cause of fluence saturation could be the presence of a "bottleneck"—a slow step in the pumping process. Molecules that contain sufficient internal energy to lie above the bottleneck would be unaffected by it, whereas those below the bottleneck would react with a different phenomenological cross section. Therefore, as the fluence is increased a sufficiently large value of fluence would be reached where the entire excited population is being pumped; further increase results in less product than expected. If all molecules are below the bottleneck, then they all are pumped at the same rate, and a linear plot results. CH_3OHF^- appears to be an example of saturation due to some sort of bottleneck, whereas the lack of a linear plot in relaxed benzyl anion indicates some true inhomogeneity in the ion population.

5.1.4 Wavelength Dependence in Quasicontinuum

Variations in photodissociation yield with changes in laser wavelength are commonly observed. Such variations are expected in the discrete regime because only certain wavelengths are resonant with the laser given the limited number of ro-vibrational transitions present in this regime. The quasicontinuum, however, is defined by the condition that any laser wavelength will be in resonance with some transition. Therefore, photodissociation yields for molecules already present in the quasicontinuum are not expected to be quite as strongly wavelength dependent. However, there is increasing evidence that the quasicontinuum is itself "structured" and that some vibrational modes are more readily excited. In fact, most of the excitations involve combination bands in which the strong absorption at $\sim 10\,\mu m$ plays a dominant role. The dissociation of excited CF_3X^+ cations (where $X = Br, I, Cl$) in a molecular beam by a single infrared photon has shown a strong wavelength dependence [19b, 37]. These highly excited molecules are clearly in the quasicontinuum; such dissociations are equivalent to the last step in a multiple-photon dissociation.

5.2 Photochemical Studies

5.2.1 Isotopic and Isomeric Selectivity

As noted earlier, the discovery that infrared photolysis could be used to excite one isotope selectively and thereby cause separation of that isotope from a mixture provided much of the impetus for early investigation in IR photochemistry. Two distinct mechanisms for laser isotope separation are possible. The first is a significant shift in the wavelength of absorption upon isotopic substitution allowing separation by merely changing the laser wavelength. The second involves an increase in the density of states of the molecule that changes the rate of intramolecular energy transfer and may provide a more efficient pathway through the discrete regime. This mechanism may be especially important for molecules having a low density of states. In either case, one isotope has an enhanced rate of excitation relative to the others and can therefore be separated. Bomse and Beauchamp have demonstrated such a separation in a mixture of $(CD_3)_2Cl^+$, $(CH_3)_2Cl^+$, and $CH_3CD_3Cl^+$ using cw CO_2 radiation

[38]. $(CD_3)_2Cl^+$ was removed from the mixture because it was the only isotope dissociated as shown in Eqs. 5.1–5.3:

$$(CD_3)_2Cl^+ \rightarrow CD_2Cl^+ + CD_4 \qquad (5.1)$$

$$(CH_3)_2Cl^+ \rightarrow \text{no reaction} \qquad (5.2)$$

$$CD_3CH_3Cl^+ \rightarrow \text{no reaction} \qquad (5.3)$$

Isomeric selectivity has been demonstrated for a mixture of $C_7H_7^-$ isomers by Wight and Beauchamp [39]. A single isomer is distinguishable from a mixture by the production of different photoproducts or by a different wavelength dependence of product yield (photodissociation or photodetachment spectrum). In the case of $C_7H_7^-$ isomers, benzyl anion and cycloheptatrienyl anion produce different photodetachment spectra, whereas norbornadienyl anion does not detach at all under cw irradiation.

5.2.2 Unimolecular and Bimolecular Reaction Dynamics

IRMP activation has been used to elucidate the dissociation reaction mechanism of a highly energized ion. The *t*-butoxide anion can be dissociated with a pulsed CO_2 laser to give acetone enolate and methane. Two possible mechanisms for this reaction are a concerted process involving a four-center elimination reaction (Eq. 5.4) and a stepwise mechanism involving a proton transfer from an intermediate ion–molecule complex (Eq. 5.5). Note that the stepwise mechanism can occur by either a homolytic cleavage (producing $\cdot CH_3$) or a heterolytic cleavage (producing CH_3^-) of the C—C bond:

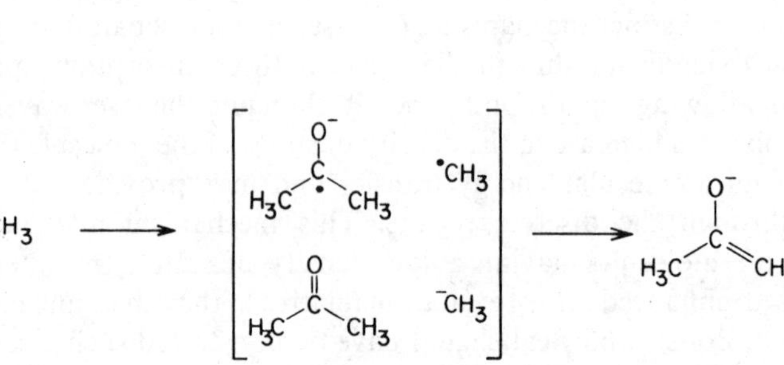

(5.4)

(5.5)

The mechanism of this reaction was elucidated by comparing the intramolecular kinetic hydrogen isotope effects of various deuterated t-butylalkoxides with the theoretical effects expected on the basis of the above mechanisms [40]. The photolysis of 2-methyl-2-propoxide-1,1,1-d_3 produced more than one photoproduct (see Eq. 5.6):

$$H_3C-\underset{\underset{CD_3}{|}}{\overset{\overset{O^-}{|}}{C}}-CH_3 \longrightarrow \begin{cases} CD_3-\overset{O}{\overset{||}{C}}-CH_2^- + CH_4 \\ CH_3-\overset{O}{\overset{||}{C}}-CD_2^- + CH_3D \\ CH_3-\overset{O}{\overset{||}{C}}-CH_2^- + CD_3H \end{cases}$$

(5.6)

Measurement of the relative intensities of the various acetone enolates (only the ionic product is detected) allows primary (H vs. D) and secondary (CH_3 vs. CD_3) isotope effects to be determined. Small primary effects and a relatively large secondary effect (loss of CH_3 favored over loss of CD_3) were observed. Furthermore, irradiation with the cw laser showed an amplification of the secondary effect but not the primary effect. This indicates that the primary and secondary isotope effects have different energy dependences, because decomposition with the cw laser takes place near the threshold for decomposition, whereas the pulsed laser imparts additional internal energy because of the increased pumping rate. The results appear to be consistent only with a stepwise mechanism.

It is possible to probe bimolecular reaction dynamics using IRMP activation. Intermediates of gas-phase ion–molecule reactions are more stable than the reactants or products because of dipole induced-dipole interactions in the intermediate and because the isolated ions are not solvated in the gas phase. These intermediates can often be generated and then irradiated with the CO_2 laser, activating them above the dissociation thresholds that represent product or reactant formation in the bimolecular reaction. Bomse and Beauchamp have examined proton-bound alcohol dimers of the form $(ROH)H^+(R'OH)$ and has shown that irradiation with the cw laser can be used to identify the lowest energy channel [41]. In addition, the pulsed laser has been used to examine the proton-bound fluoride–alkoxide ion complexes of the form $ROHF^-$ [42].

These ions are the presumed intermediates of proton transfer between the appropriate alcohols in the former case and between the alkoxide ion and fluoride ion in the latter. Similarly, the mechanism of the proton transfer reaction between alkyl carbonium ions and alkyl amines was studied by irradiating ethyl(i-propyl) ammonium ion, $C_2H_5(i\text{-}C_3H_7)NH_2^+$ [43]. This ion is the intermediate of an addition–elimination mechanism and also lies on the hypersurface of a proton transfer mechanism involving an ion–molecule complex. Comparison of the photochemical results, product distributions for the proton transfer reactions, and high-pressure product distributions allowed the conclusion that the available addition–elimination pathway for proton transfer was not followed. Detection of the photoproducts of such intermediates provides information about the potential surfaces and thermochemistry of the appropriate bimolecular reaction.

Multiple products were obtained in $(CH_3)_3CCH_2OHF^-$ (Eq. 5.7) and $CF_3COCH_2^-$ (Eq. 5.8) [42]:

$$(CH_3)_3CCH_2OHF^- \begin{cases} \rightarrow F^- + (CH_3)_3CCH_2OH \\ \rightarrow (CH_3)_3CCH_2O^- + HF \end{cases} \quad (5.7)$$

$$CF_3COCH_2^- \begin{cases} \rightarrow CF_3^- + H_2CCO \\ \rightarrow HCCO^- + CF_3H \end{cases} \quad (5.8)$$

In the proton-bound fluoride-alkoxide ion complex, the proton transfer was determined to be approximately thermoneutral, so that both product channels are approximately equal in energy.

In the trifluoroacetone enolate anion, the branching is thought to arise from the competing reactions of the photochemically generated intermediate ion–molecule complex of $(CF_3^- \cdot H_2CCO)$.

5.2.4 Vibrationally Induced Electron Detachment

For many negative ions the energy required to remove an electron is less than the energy required for bond dissociation [44]. Upon irradiation with an infrared laser these ions undergo photodetachment, producing the neutral and a free electron, in a process referred to as vibrationally induced electron detachment (VED). This process was first demonstrated in benzyl anion (Eq. 5.9) using a pulsed laser [45]:

$$C_6H_5CH_2^- \rightarrow C_6H_5CH_2{\cdot} + e^- \quad (5.9)$$

VED has subsequently been demonstrated in a variety of anions, such as SF_6^- [46], allyl, 2,4-hexadienyl, 1,3-cycloheptadienyl, and anilide anions [47], and in allyl anions using cw radiation [48]. VED involves a bound-free rather than a bound–bound transition. The vibrational energy deposited in the anion from

IRMP activation must be transferred to an electronic degree of freedom. Similar V-E transfer processes occur in electron scattering and vibrational autoionization of neutral molecules and associative detachment reactions of anions.

An estimate of the rate of VED has been obtained in the detachment of acetone enolate anion [49]. Upon irradiation from a pulsed CO_2 laser, acetone enolate was found to yield both the photodetachment product and a photodissociation product, HC_2O^- (Eq. 5.10):

$$\underset{CH_3}{\overset{O}{\underset{\|}{C}}}-CH_2^- \longrightarrow \begin{cases} \underset{CH_3}{\overset{O^\cdot}{\underset{|}{C}}}-CH_2 + e^- \\ H-C\equiv C-O^- + CH_4 \end{cases} \quad (5.10)$$

The photodetachment channel was lower in energy than the dissociation channel, but the detachment rate increased only slowly with increasing internal energy so that the dissociation channel could be accessed before complete detachment. The dissociation rate constant could be calculated using RRKM theory, and the rate of photon absorption was known from the laser intensity; this allowed the detachment rate to be limited to about 10^7 s^{-1} at approximately 1 eV above the detachment threshold. Thus, the pumping rate was used as an internal "clock" to determine the rate of VED.

6 CONCLUSION

Infrared laser photolysis is a novel activation technique that allows a variety of photophysical and photochemical information to be obtained. This technique has been demonstrated for many gas-phase ions using both low- and high-intensity radiation sources under collisionless and collisional regimes. Technological advances in laser sources should broaden the applicability and use of the technique by extending it to different wavelength regions in the infrared.

ACKNOWLEDGMENT

We are grateful to the National Science Foundation for support of this research.

REFERENCES

1. N. R. Isenor and M. C. Richardson, *Appl. Phys. Lett.*, **18**, 224 (1971).
2. For reviews see (a) R. V. Ambartzumanian and V. S. Letokhov, "Multiple Photon Infrared Laser Photochemistry," in C. B. Moore, Ed., *Chemical and Biochemical Applications of Lasers*, Academic Press, New York, 1977, vol. 3. (b) W. C. Danen and J. C. Jang, "Multiphoton Infrared Excitation and Reaction of Organic Compounds," in J. I. Steinfeld, Ed., *Laser-Induced Chemical Processes*, Plenum, New York, 1981. (c) P. A. Schulz, Aa. S. Subdø, D. J. Krajnovich, H. S. Kwok, Y. R. Shen, and Y. T. Lee, *Ann. Rev. Phys. Chem.*, **30**, 379 (1979).
3. G. Herzberg, *Molecular Spectra and Molecular Structure, II. Infrared and Raman Spectra of Polyatomic Molecules*, Von Nostrand Reinhold, New York, 1945.
4. G. Amat and M. Pimbert, *J. Mol. Spec.*, **16**, 278 (1965).
5. D. C. Tyte, "Carbon Dioxide Lasers," in D. W. Goodwin, Ed., *Advances in Quantum Electronics*, Academic Press, New York, 1970, vol. I.
6. G. J. Schulz, *Phys. Rev.*, **A135**, 988 (1964).
7. C. K. Rhodes and A. Szöke, "Gaseous Lasers: Atomic, Molecular, and Ionic," in F. T. Arrechi and E. O. Schulz-Dubois, Eds., *Laser Handbook*, North-Holland, Amsterdam, 1972, vol. 1.
8. J. P. Goldsborough, "Design of Gas Lasers," in F. T. Arrechi and E. O. Schulz-Dubois, Eds., *Laser Handbook*, North-Holland, Amsterdam, 1972, vol. 1.
9. (a) A. Gondhalekar, E. Holzhauer, and N. R. Heckenberg, *Phys. Lett.* 1973, 46A, 229. (b) A. Gondhalekar, E. Holzhauer, and N. R. Heckenberg, *IEEE J. Quantum Electron.*, **QE-11**, 103 (1975).
10. O. Svelto, *Principles of the Laser*, Plenum, New York, 1982.
11. R. G. Smith, "Optical Parametric Oscillators," in F. T. Arrechi and E. O. Schulz-Dubois, Eds., *Laser Handbook*, North-Holland, Amsterdam, 1972, vol. 1.
12. (a) M. Quack, *J. Chem. Phys.*, **69**, 1282 (1978); (b) M. Quack, *Ber. Bunsenges. Phys. Chem.* **83**, 757 (1979); (c) M. Quack, *Ber. Bunsenges. Phys. Chem.* **83**, 1287 (1979).
13. J. C. Stephenson, D. S. King, M. F. Goodman, and J. Stone, *J. Chem. Phys.*, **70**, 4496 (1979).
14. J. Troe, *J. Chem. Phys.*, **73**, 3205 (1980).
15. A. C. Baldwin, J. R. Barker, D. M. Golden, R. Duperrex, and H. van den Bergh, *Chem. Phys. Lett.*, **62**, 178 (1979).
16. R. E. Smalley, *Ann. Rev. Phys. Chem.*, **34**, 129 (1983).
17. (a) W. Forst, *Theory of Unimolecular Reactions*, Academic Press, New York, 1973. (b) P. J. Robinson and K. A. Holbrook, *Unimolecular Reactions*, Wiley-Interscience, London, 1972.
18. (a) M. Quack, *J. Chem. Phys.*, **70**, 1069 (1979). (b) M. Quack, P. Humbert, and H. van den Bergh, *J. Chem. Phys.*, **73**, 247 (1980).
19. (a) B. A. Huber, T. M. Miller, P. C. Cosby, H. D. Zeman, R. L. Leon, J. T. Moseley, and J. R. Peterson, *Rev. Sci. Instrum.*, **48**, 1306 (1977). (b) M. J. Coggiola, P. C. Cosby, and J. R. Peterson, *J. Chem. Phys.*, **72**, 6507 (1980).
20. D. E. Tolliver, G. A. Kyrala, and W. H. Wing, *Phys. Rev. Lett.*, **43**, 1719 (1979).
21. W. H. Wing, G. A. Ruff, W. E. Lamb, Jr., and J. J. Spezeski, *Phys. Rev. Lett.*, **36**, 1488 (1976).
22. A. Von Hellfeld, D. Feldman, K. H. Welge, and A. P. Fournier, *Opt. Commun.*, **30**, 193 (1979).
23. H. D. Zeman, *Rev. Sci. Instrum.*, **48**, 1079 (1977).
24. (a) R. J. Hughes, R. E. March, and A. B. Young, *Can. J. Chem.*, **61**, 824 (1983). (b) A. B. Young, R. E. March, and R. J. Hughes, *Can. J. Chem.*, **63**, 2324 (1985).
25. J. S. Chou, D. Sumida, and C. Wittig, *Chem. Phys. Lett.*, **100**, 209 (1983).
26. (a) R. L. Woodin, D. S. Bomse, and J. L. Beauchamp, *J. Am. Chem. Soc.*, **100**, 3248 (1978). (b) R. L. Woodin, D. S. Bomse, and J. L. Beauchamp, *Chem. Phys. Lett.*, **63**, 630 (1979). (c) D. S. Bomse, R. L. Woodin, and J. L. Beauchamp, *J. Am. Chem. Soc.*, **101**, 5503 (1979).

27. J. M. Jasinski, R. N. Rosenfeld, F. K. Meyer, and J. I. Brauman, *J. Am. Chem. Soc.*, **104**, 652 (1982).
28. J. M. Jasinski and J. I. Brauman, *J. Chem. Phys.*, **73**, 6191 (1980).
29. T. Su and M. T. Bowers, *Int. J. Mass Spectrom. Ion Phys.*, **12**, 347 (1973).
30. R. N. Rosenfeld, J. M. Jasinski, and J. I. Brauman, *J. Am. Chem. Soc.*, **104**, 658 (1982).
31. L. R. Thorne and J. L. Beauchamp, *Chem. Phys. Lett.*, **74**, 5100 (1981).
32. A. T. Barfknecht and J. I. Brauman, *J. Chem. Phys.*, **84**, 3870 (1986).
33. M. S. Kim and R. C. Dunbar, *J. Phys. Chem.*, **60**, 247 (1979).
34. N. B. Lev and R. C. Dunbar, *J. Phys. Chem.*, **87**, 1924 (1983).
35. R. N. Rosenfeld, J. M. Jasinski, and J. I. Brauman, *Chem. Phys. Lett.*, **71**, 400 (1980).
36. K. E. Salomon, Ph.D. thesis, Stanford University, 1986.
37. M. F. Jarrold et al. *J. Phys. Chem.*, **87**, 2213 (1983).
38. D. S. Bomse and J. L. Beauchamp, *Chem. Phys. Lett.*, **77**, 25 (1981).
39. C. A. Wight and J. L. Beauchamp, *J. Am. Chem. Soc.*, **103**, 6499 (1981).
40. (a) W. Tumas, R. F. Foster, M. J. Pellerite, and J. I. Brauman, *J. Am. Chem. Soc.*, **105**, 7464 (1983). (b) W. Tumas, R. F. Foster, M. J. Pellerite, and J. I. Brauman, *J. Am. Chem. Soc.*, **109**, 961 (1987).
41. D. S. Bomse and J. L. Beauchamp, *J. Am. Chem. Soc.*, **103**, 3292 (1981).
42. (a) C. R. Moylan, J. M. Jasinski, and J. I. Brauman, *Chem. Phys. Lett.*, **98**, 1 (1983). (b) C. R. Moylan, J. M. Jasinski, and J. I. Brauman, *J. Am. Chem. Soc.*, **107**, 1934 (1985).
43. C. R. Moylan and J. I. Brauman, *J. Am. Chem. Soc.*, **107**, 761 (1985).
44. B. K. Janousek and J. I. Brauman, "Electron Affinities," in M. T. Bowers, Eds., *Gas Phase Ion Chemistry*, Academic Press, New York, 1979, vol. 2.
45. R. N. Rosenfeld, J. M. Jasinski, and J. I. Brauman, *J. Chem. Phys.*, **71**, 1030 (1979).
46. P. S. Drzaic and J. I. Brauman, *Chem. Phys. Lett.*, **83**, 508 (1981).
47. F. K. Meyer *et al.*, *J. Am. Chem. Soc.*, **104**, 663 (1982).
48. C. A. Wight and J. L. Beauchamp, *J. Phys. Chem.*, **88**, 4426 (1984).
49. R. F. Foster, W. Tumas, and J. I. Brauman, *J. Chem. Phys.*, **79**, 4644 (1983).

Chapter **XI**

PULSED METHODS FOR CLUSTER ION SPECTROSCOPY

Mark A. Johnson
and
W. Carl Lineberger

---

1 **Introduction**
2 **A Brief Retrospective: "Traditional" Methods of Ion Spectroscopy**
3 **Pulsed Ion Sources**
   3.1 Positively Charged Clusters
       3.1.1 Molecular or "van der Waals" Ionic Clusters
       3.1.2 Cluster Ions of Metals and Semiconductors
   3.2 Negatively Charged Clusters
       3.2.1 Molecular Ion Clusters
       3.2.2 Cluster Ions of Metals and Semiconductors
2 **Mass Selection**
   4.1 Primary Ion Beam
   4.2 Interaction of the Ion Beam with a Pulsed Laser
5 **Determination of Charged Photofragments Using Tandem Time-of-Flight Mass Spectrometry**
   5.1 Tandem TOF Using a Reflectron
   5.2 Tandem TOF Using Two Pulsed Fields
6 **Pulsed Photoelectron Spectroscopy Using Time-of-Flight Electron Energy Analysis**
7 **A Few Results**
   7.1 Mechanism of Optical Absorption in "van der Waals" Cluster Ions: Localization of Charge within Large Clusters
   7.2 Statistical Photodissociation Dynamics of Large Molecular Ion Clusters
   7.3 Formation of "Magic Numbers" in Cluster Ion Mass Spectra
   7.4 Size Dependence of the Br_2^- Recombination Quantum Yield upon Photodissociation of Br_2^- Within $Br_2^- \cdot (CO_2)_n$ Clusters: Observation of "Caging" of the Atomic Fragments
8 **Outlook**
References

1 INTRODUCTION

A central goal of physical chemistry is the explanation, on a molecular level, of chemical and physical processes occurring on a macroscopic scale. Traditionally, researchers have been divided into two groups: one primarily interested in the exact details of very small systems and one interested in understanding very complex systems as they occur in nature. The themes of each endeavor are correspondingly different, with small molecule research being directed toward observation of single eigenstates in order to determine empirically the molecular Hamiltonian and with condensed phase research relying heavily on ensemble averaging in modeling "real" physical processes. Over the past few years, these two approaches have been converging as experimentalists are increasingly able to carry out a variety of laser spectroscopy experiments on very selectively prepared ensembles of molecular aggregates or clusters containing from 2 to about 100 molecules. At the same time, theoretical techniques are becoming available that can handle small aggregates at Hückel [1], ab initio [2-8], and molecular dynamics [9] levels, and the forthcoming decade promises to be rich in interaction between theory and experiment.

Ionic clusters play an important role in this field since mass spectrometry allows ionic clusters of precisely determined composition to be selected for study. Indicative of this importance is the virtual explosion of experiments over the past 5 years, all aimed at elucidating the properties of ionic clusters. [10,11] In a recent review article on ionic clusters, for example, Castleman and Keesee [10] catalog over 450 recent publications in this field. In fact, whereas in 1983 only one report appeared [12] on the photodissociation of a cluster ion containing more than two monomer units, in 1987 at least 12 research groups [12-23] have been directing major efforts toward the photodissociation spectroscopy of mass selected ion clusters. The topics of these studies range from ion solvation using van der Waals [14] type clusters such as $(CO_2)_n^+$ to the elucidation of semiconductor [24] and metal [25] band structures as a function of cluster size and composition. A number of novel experimental approaches have contributed to the rapid development of this field, and the purpose of this chapter is to review several of these techniques using pulsed ion beams in conjunction with pulsed lasers. Indeed, a primary dividend arising from these methods has been a simplification of ion-beam experiments to the extent that they are now generally applicable to a wide range of chemical species.

One of the most important recent developments is the generation of pulsed beams of clusters ions based on the ionized free jet. This advance has been paralleled by recent tandem time-of-flight (TOF) techniques allowing the efficient overlap of ion beams with low duty cycle pulsed lasers. The combination of these methods makes it relatively easy to mass select a specific size ionic cluster of either positive or negative polarity, to interact nearly 100% of the selected ions with a pulsed laser, and, finally, to detect all of the photofragments–daughter ions, photodetached electrons, and fast neutrals.

This flexibility is especially important in the investigation of cluster ions since these species are likely too complex for application of high-resolution spectroscopic techniques common in molecular physics, yet they are also small enough that statistical averaging and periodicity, so useful in understanding the behavior of bulk systems, are not directly applicable. Therefore, a large effort is being made to characterize the behavior of systems as the size and composition is systematically varied. The goal of this effort is to extract regularities in behavior from the database in order to gain an empirical understanding upon which to build more sophisticated models and more detailed experiments. In this review, we catalog some recent developments in experimental techniques that are currently making a major impact and summarize some representative trends that have been observed using these techniques. We also mention some important remaining challenges for experimentalists.

2 A BRIEF RETROSPECTIVE: "TRADITIONAL" METHODS OF ION SPECTROSCOPY

The *pulsed* instruments outlined here are fundamentally different in character from the continuous ion-beam experiments discussed in a previous review [26]. Traditionally, ions are formed in a discharge environment and are continuously extracted and mass selected, typically by a Wien filter, quadrupole mass filter, or magnetic sector and steered into a long arm for collinear interaction with a continuous laser. An example of such an experiment is shown in Fig. 1, which depicts a fast ion-beam apparatus developed by Huber et al. [27] and used for very high-resolution photofragmentation spectroscopy of molecular ions. Such instruments have also been productive in studies of negative ion threshold photodetachment, where Doppler broadening can be reduced to ~ 20 MHz using the kinematic compression in a fast beam. In addition, there have been some reports on cluster ion spectroscopy using these techniques. Fayet and Wöste [18] have studied Ag_n^- photodissociation in a continuous ion beam where photodissociation was carried out in an octupole trap; Okimura et al. [20] have obtained infrared (IR) photodissociation spectra of H_n^+ and $H_3O^+ \cdot H_2$ also using an ion trap, and Castleman et al. [12] have studied the photodissociation of $CO_3^- \cdot (H_2O)_n$ clusters with continuous beams and a photofragment energy analyzer. Helm et al. [17] used a photofragment energy analyzer to study the spectroscopy of Cs_n^+ clusters. Much of the existing spectral data on binary and ternary molecular ion complexes was obtained by Peterson and co-workers [28] at SRI using mass-spectroscometric sampling of a high-pressure drift tube ion source in which the ions are irradiated at the entrance aperture of the mass spectrometer chamber. Bowers et al. [29] have used a continuous apparatus to obtain the photofragment angular distributions from a number of cation dimers and trimers [e.g., $(NO)_n^+$]. Using a triple quadrupole instrument equipped with a sputtering ion source, Magnera et al. [19] have obtained a structured electronic spectrum for the $N_2^+ \cdot Ne_n$ system. Ion cyclotron resonance (ICR) experiments have also been useful in obtaining data on binary ion-solvent systems [30], metal ion complexes [31], and the photochemistry of

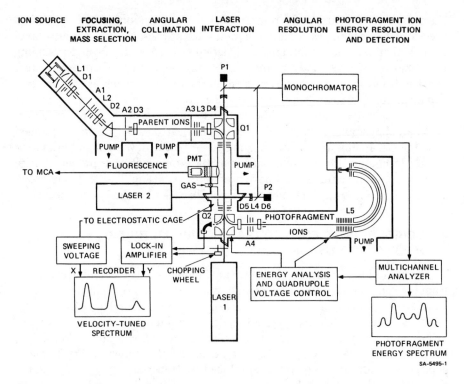

Fig. 1. Schematic diagram of the continuous wave (cw) coaxial ion–laser beam photofragmentation spectrometer developed at SRI International. Reproduced with permission from Huber et al. [27].

organic cations [32]. The experimental methods enumerated above each offer special advantages that complement the pulsed beam methods discussed in this review. ICR experiments, for instance, afford an unparalleled capability of measuring decay rate constants and can also obtain extremely high mass resolution. This review presents a new, complementary method, which we expect to join these proven techniques in the field of cluster ion research.

3 PULSED ION SOURCES

3.1. Positively Charged Clusters

For 20 years, the ionized free jet has been recognized as the method of choice for forming cluster ions. Until recently, however, most cluster ion experiments have used the jet as a source of neutral clusters that were then converted to cationic clusters either by electron impact or photoionization. The ionic mass spectrum has typically been used as an indication of the neutral cluster distribution formed in the jet. Such experiments have been very successful in determining the bimolecular reaction kinetics of clusters with small molecules

[33–35]. These experiments all depend upon the ability to relate the intensity of a given ion observed in the mass spectrometer to the concentration of a neutral precursor(s). As a consequence, there is always concern over the extent to which fragmentation occurring in the ionization process may complicate or even invalidate the analysis. Very few studies have yielded definitive results regarding fragmentation branching ratios for the nascent ions. Buck and Meyer [36] have used elegant momentum transfer experiments to establish the fragmentation of argon clusters upon ionization. They report that clusters at least as large as $n = 6$ form primarily dimers upon electron impact ionization. This extensive fragmentation is actually responsible for the appearance of intensity anomalies (magic numbers [37]) in the ion cluster distribution, as we elaborate below. A second problem with neutral ionization methods (more important from the perspective of the techniques described here) arises from the internal energy content of the ionic cluster formed after ionization and fragmentation of a neutral. The sequence of events that likely occurs is vertical ionization of the neutral cluster to an internuclear configuration of the ionic cluster far from its equilibrium geometry [38]. The ion then relaxes toward the equilibrium geometry with concomitant vibrational excitation of the cluster. If the vibrational excitation exceeds the binding energy of neutrals, monomers may be ejected by the process

$$M_n + e^- \rightarrow [M_n]^{+*} \rightarrow (n-p)M + M_p^+ + 2e^-$$

until the internal energy of M_p^+ is below the bond dissociation energy $(M_{p-1}^+) - M$. We therefore expect that the product ionic cluster will be in general internally excited to an extent roughly comparable to the bond energy of one neutral monomer unit.

In order to overcome difficulties associated with the internal excitation inherent in direct ionization sources, a significant effort has been devoted to develop alternative preparation schemes yielding controlled internal energy ion clusters. The technical demands on the source are quite different for forming clusters of volatile species as compared to the case of refractory precursors, so we review the methodology of each type of source separately.

3.1.1 Molecular or "van der Waals" Ionic Clusters

The solution to the internal energy problem was demonstrated as early as 1973, when Searcy and Fenn [39] showed that large, cationic water clusters were formed in an ionized free-jet by sequential association of neutral monomers onto an ionic core. The dependence of the cluster size distribution on jet conditions clearly showed that the three-body association reactions,

$$H_3O^+ \cdot (H_2O)_n + H_2O + Ar \rightarrow H_3O^+ \cdot (H_2O)_{n+1} + Ar$$

were occuring in the high-density region of the expansion where small ions were forming nucleation centers for the synthesis of the larger clusters. Although there is at present only indirect evidence regarding the internal temperature of the cluster ions, it is likely that the more probable two-body collisions will relax the nascent cluster before another monomer is added to the cluster and releases another bond energy of internal excitation.

In Fenn's work, as well as in more recent studies by Beuhler and Friedman [40] on large water cation clusters and by Harris et al. [41] on rare gas clusters, ionization was effected by a high-voltage pin located on the high-pressure side of the nozzle. Since the ions were formed by a corona discharge and entrained in the jet, it may seem surprising that the clusters are not broken apart due to high energy collisions as the ions are driven by the high electric field. Not only are the clusters intact, but Carrick and Engelking [42] have recently dispersed the fluorescence from molecular ions and neutral radicals created in a discharged nozzle and found that the electronically excited states are rotationally cooled to about 20 K. It is apparent that the expansion is relatively unaffected by the voltage creating the discharge because a high free electron density forms an overall electrically neutral plasma that shields the core of the jet.

Although the methodology of producing internally cold cluster ions has been in hand for some time, as of 1984 these ion sources were intrinsically continuous and as such were best used in conjunction with cw lasers. The broad tunability of pulsed lasers and the vast reduction in the scale of molecular beam machines using pulsed expansions provided a powerful incentive to develop ways of coupling pulsed expansions to the cold ion-beam techniques based on nucleation.

This coupling was initially carried out by spectroscopists primarily interested in isolating cold monomer ions for high-resolution experiments. The approach envisioned by several spectroscopy groups involved first cooling neutral species and then ionizing them with high energy electron impact to create ions to be studied. High energy electrons (> 100 eV) were used so that high ionization potential species could be studied and so that little angular momentum would be imparted to the cation provided the electron molecule interaction is fast compared to nuclear motion (i.e., within the Born approximation) [43]. Using this method, Miller et al. [44, 45] at Bell Laboratories, Carrington and Tuckett [46] at Cambridge, Klapstein, et al. [47] at Basel, Alexander et al. [48] at JILA and Johnson et al. [49] at Stanford all reported cold spectra of cations in the years 1979–1983. In carrying out these experiments, care was taken to minimize external electric and magnetic fields in order to avoid changes in the ion trajectories resulting in collisional heating of the ions by collision with the neutral expansion. In virtually all of the experiments, a molecular beam was ionized by an electron beam that crossed about 1 cm downstream from a continuous nozzle, in an arrangement in which the filament of the electron gun was floated negative and the electrons accelerated to ground when they crossed the molecular beam, so that cations are created in a field-free environment. Typical electron beam currents are on the order 0.1–1 mA. Moreover, since the ionization is carried outside and independent of the nozzle, the technique is completely compatible with pulsed valves and hence forms a high-intensity pulsed cluster ion source.

Early experiments concentrated on determining the rotational temperature of excited states formed directly by electron impact ionization [45, 46], and the rotational temperature determined for the $N_2^+ B^2\Sigma_u^+$ state is typically 20 K. In 1982, Johnson et al. [49] and Miller [50] obtained laser-induced fluorescence (LIF) spectra of CO_2^+ and $C_6F_5H^+$, respectively, and showed that the ground electronic states of the ions, which drifted for tens of microseconds in the field-free expansion, were also rotationally cooled to about 20 K. In the CO_2^+ study, optical–optical double resonance experiments were also carried out that labeled a specific lower state rotational level by selective removal of its population and detection of the "hole" in the rotational distribution with a second laser about 500 ns later. No collisional satellite double resonance transitions were observed. In 1983, Johnson et al. [51] incorporated the pulsed valve into a cold ion apparatus to enhance signal to noise (S/N) and reported the observation of microwave transitions of the CO^+ ion. In their microwave-optical double resonance experiment, the ions drift over 10 cm between rotational excitation and detection of the population shift caused by the microwave absorption. No collisional features were observed in the double resonance spectra. These results demonstrate that not only are ions cooled in the low temperature of the jet but that downstream from the nozzle they exist in a nearly collisionless environment. A crude estimate of the ion density based on the LIF signal intensity reveals that the cation density in the pulsed expansion is quite large, $\sim 10^9/\text{cm}^3$ at a distance of 15 cm downstream from the orifice, far in excess of the space-charge limited density.

In addition to establishing that the jet is indeed a collisionless environment for ions, the microwave experiments also showed that *small electric and magnetic fields had no effect on the rotational distribution*. The ability of the jet to support such large ion densities, the insensitivity of the system to external fields, and the efficiency with which ions are rotationally cooled are explained by the high electron density appearing in the ionized jet. The free electron density was directly measured using the Fabry-Perot cavity that introduced the microwave field into the vacuum chamber [51]. Because the microwave frequency resonant with the cavity depends on the index of refraction of the medium in the cavity, the resonant frequency was observed to shift when the electron beam was turned on. The refractive index of free electrons is known to depend on the free electron density [52], and the cavity shift of several Megahertz at 120 GHz reveals a free electron density of about $10^9/\text{cm}^3$, roughly the same concentration as the cations indicated by LIF. It is also clear that these electrons must have energies much less than 1 eV or the electron–neutral collisions would cause rotational excitation [45]. The free jet is thus a *cold, overall electrically neutral plasma* with a sufficiently high free electron density to have a short penetration depth (Debye wavelength) for external electric fields.

The pulsed jet cold cation source can be regarded as a partially controlled ionization analog of the high-voltage corona discharge type of apparatus. It is now clear that significant cluster formation must have been occurring in the monomer ion spectroscopy studies but went unnoticed since the cluster ions usually have completely different electronic structure than the monomer ions and therefore do not absorb near the monomer bands. The external electron beam ionization scheme is ideal for use with a pulsed valve and has the additional advantage of creating the ions in a well-defined environment. The scale of the pulsed ion source apparatus is very much reduced from that of the continuous ion sources and is easily configured for laser spectroscopy experiments.

As an indication of the ease with which clusters are formed with the ionized pulsed jet, Fig. 2 presents mass spectra [14] from the Boulder cooled ion source as a function of the distance Z between the 1-keV electron beam and the orifice (see Fig. 10 for definition of Z). The top trace shows ionization at 2.8 cm, and the cation spectrum contains only the CO_2^+ monomer ion. This corresponds to the condition used for the monomer ion spectroscopy experiments. As the electron beam is moved closer to the orifice, the lower traces in Fig. 2 show a gradual shifting of the charged species to larger and larger clusters. The mechanism of the cluster growth is seen to be sequential addition of neutrals, as expected from the nucleation experiments [39–41]. Indeed, we can "fine tune" the cluster distribution with the electron beam position, giving this apparatus great flexibility. In contrast with the corona discharge sources, in which the cluster distribution is altered by varying beam conditions (e.g., backing pressure, stagnation temperature, nozzle diameter, and seed ratio), each requiring a significant change in the experimental configuration, the electron beam effects changes immediately observable when detected with TOF mass spectrometry.

The fact that cations seeded into the expansion form nucleation sites for cluster growth results from the enhanced bimolecular collision rates of the ion–neutral encounters compared with those between neutrals in the low ambient temperature of the jet. For neutrals, the rate of bimolecular collisions suffered by a molecule is given by

$$R = N \cdot V_{rel} \sigma(V_{rel}) \tag{1}$$

where V_{rel} is the center-of-mass collision velocity, $\sigma(V_{rel})$ is the velocity-dependent cross section, and N is the collision gas density. For hard sphere collisions, $\sigma(V_{rel}) \approx \pi D^2$ so that $R \propto \sqrt{T}$ and rate of neutral–neutral collisions falls off rapidly with decreasing temperature. On the other hand, at least in the Langevin approximation [53], the rate of ion–molecule collisions is governed by the long-range ion-induced dipole interaction that yields a cross section proportional to $1/\sqrt{T}$ and a nearly temperature-independent collision rate given by $k_L \cdot N$ where k_L is the Langevin rate constant, usually on the order 10^{-9} cm^3/molecule-s. An ion can therefore be introduced into the jet in a regime in which the neutrals are essentially collisionless, while an ion still suffers a high rate of collisions.

As an illustration of the above statements, we estimate the behavior expected for an expansion such as that used in generating the $(CO_2)_n^+$ clusters in Fig. 2. The expansion conditions are 2 atm stagnation pressure and a nozzle diameter of 1 mm. We can calculate the density $n(x)$ and local translational temperature $T(x)$

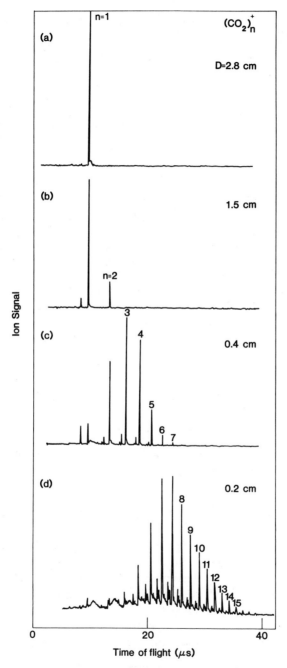

Fig. 2. Time-of-flight mass spectra using a 1-keV electron beam to ionize a CO_2 pulsed free jet expansion, where the distance from the electron beam to the nozzle is varied. Clusters result from successive attachment of CO_2 monomers not from ionization of neutral clusters. Reproduced with permission from Johnson et al. [14].

at a distance x downstream on the axis of the expansion according to the well-known formulas [54]:

$$n(x) = n_0 \left[1 + \frac{\gamma - 1}{2} A^2 \left(\frac{x}{D} \right)^{2(\gamma - 1)} \right]^{-1/(\gamma - 1)} \qquad (2)$$

$$T(x) = T_0 \left[1 + \frac{\gamma - 1}{2} A^2 \left(\frac{x}{D} \right)^{2(\gamma - 1)} \right]^{-1} \qquad (3)$$

where $\gamma = C_p/C_v$, D = nozzle diameter, A is related to the Mach number M by

$$M = A \left(\frac{x}{D} \right)^{\gamma - 1} \qquad (4)$$

which is tabulated for various species. Using these expressions, we estimate the density and temperature 5 mm downstream to be 2.3×10^{16} molecules/cm^3 and 15 K. If we take the hard sphere collision cross section for CO_2 to be about 50 Å^2, Eq. 1 implies that the rate of neutral collisions is $3 \times 10^5 \, \text{s}^{-1}$, whereas ionic collisions occur at the Langevin rate of $2 \times 10^7 \, \text{s}^{-1}$. Thus ion–molecule collisions are about 100 times more probable than collisions between neutrals. For efficient nucleation to take place, these bimolecular collisions must occur during the lifetime T_c of the ion–molecule collision complex $[CO_2-CO_2^+]^*$

$$CO_2^+ + CO_2 \rightarrow [CO_2-CO_2^+]^*$$

$$[CO_2-CO_2^+]^* + CO_2 \rightarrow (CO_2)_2^+ + CO_2.$$

If the stabilizing collision is assumed to proceed with the bimolecular ion–molecule rate, efficient clustering occurs when

$$\frac{1}{T_c} = n(x) \cdot k_L$$

For the CO_2 cluster spectrum shown in Fig. 2, clustering is seen to be almost complete at $x = 3$ mm from the nozzle, which implies that the complex lifetime must be on the order of tens of nanoseconds. Thus there are two effects that can allow ionic species to dominate clustering: the enhanced bimolecular collision rate and the enhanced complex lifetimes.

Van Koppen et al. [55] have recently investigated the temperature dependence of the three-body association rates for the system

$$N_2^+ + N_2 + M \rightarrow N_4^+ + M$$

and found that the third-order rate coefficient increases continuously from 450 down to 80 K. Their experimentally determined expression for the rate constant k is

$$k = \frac{k_c \beta k_s}{k_b + \beta k_s [M]}$$

where k_c is the bimolecular collision rate, k_s is the rate of stabilizing collisions, each with efficiency β, and k_b is the unimolecular dissociation rate of the $[N_4^+]^*$ complex. Using Langevin rates for k_s and k_c, these authors find that the temperature dependence is due to the increase in RRKM lifetime of the complex

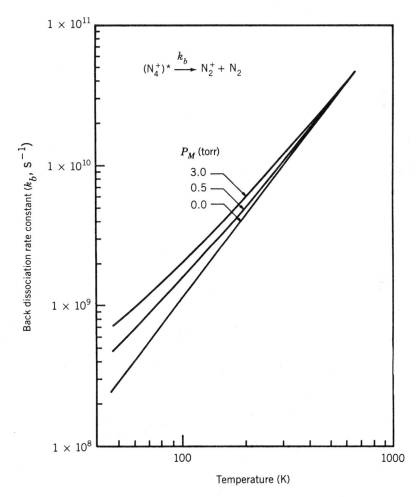

Fig. 3. Temperature dependence of the $(N_4^+)^*$ unimolecular decomposition rate constant. Calculated for the nitrogen association complex at several values of the total pressure P_M. Reproduced, with permission from van Koppen et al. [55].

as the temperature is decreased. The derived values of the rate constant k_b for dissociation of the $[N_4^+]^*$ complex,

$$[N_4^+]^* \xrightarrow{k_b} N_2^+ + N_2$$

are plotted in Fig. 3, where the complex lifetimes $1/k_b$ are indeed found to be in the range of nanoseconds. Interestingly, extrapolation of their data to 15 K (the temperature of the CO_2 jet) yields an estimate of about 20 ns for the $[N_4^+]^*$ collision complex. This estimate is actually quite consistent with our estimate of 10–50 ns for the $(CO_2)_2^+$ complex lifetime based on calculated jet conditions and the observed loss of the monomer ion.

By moving the electron beam very close to the nozzle, the ion collision frequency continues to rise, and quite large clusters can be formed with the nucleation technique. Figure 4 displays the cluster distribution generated from a pure carbon dioxide expansion, showing cations containing up to 60 CO_2 monomers. At least a portion of the falloff at large clusters is due to decreasing sensitivity of the ion detector. The simple addition of a "conversion dynode" [40] would likely extend considerably the maximum useful cluster size. In comparison, the continuous corona discharge used by Harris et al. [41] was able to generate larger clusters of rare gases containing up to 200 atoms, and Beuhler and Friedman [40] have made water cation clusters with $m/e \sim 59{,}000$! Presently, the pulsed jet techniques appear most useful for cluster sizes below 100, but the use of more energetic ionizing electrons and higher stagnation pressures would likely increase this limit substantially.

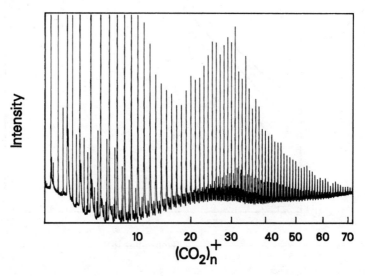

Fig. 4. Large cation clusters of CO_2 obtained from 1-keV electron beam ionization of a CO_2 pulsed expansion. Levinger et al., unpublished results.

3.1.2 Cluster Ions of Metals and Semiconductors

The development of metal and semiconductor cluster ion beams parallels that of the molecular ion clusters. In 1972, Takagi et al. [56] reported an apparatus for generating metal cluster ions based on a heated oven to vaporize the metal, followed by expansion of the metal through a nozzle and ionization of the expansion by electron bombardment. The approach is similar to that reported by Searcy and Fenn [39] in their 1973 study of water cation clusters, except that ionization is achieved by passing the beam through a high current ($\sim 1/2$ A) electron shower instead of a discharge. The metal ion cluster beam technique has grown into a large area of activity in materials research, since the beam can be used to prepare many kinds of semiconductor and metallic surfaces otherwise difficult to generate [57]. Although these "hot oven" methods are very productive, they are not readily transferred to the pulsed jet technique, since the valve would be required to operate under extremely corrosive high temperature conditions. They also require large amounts of material and probably do not achieve the cooling available from the hard pulsed expansion. Fortunately, recent results show that the ions of refractory materials may also be introduced into the pulsed-free jet [58, 59].

Laser methods for preparing cations of refractory materials actually date back to 1963 when Honig and Woolston [60] at RCA observed ions emanating from various surfaces (Cu, Mo, Ta, W, C, Ge, SiC, Al_2O_3) when exposed to a focused 0.4-J pulse from a ruby laser. These authors found that up to 3×10^{16} electrons could be ejected per pulse, yielding peak positive ion currents of up to 500 mA. Later in that year, Howe [61] determined the temperature of the plasma to be in the range of 3000–10,000 K by dispersing the CN optical emission. He graphically describes [61] the effect: "It is known that the output beam from a pulsed ruby laser can be focused upon the surface of a suitable target to produce high temperatures and spectacular pyrotechnic displays. When the laser beam was focused on this material [graphite], a mushroom-shaped jet, blue white in color and over 1 cm long, resulted." In the following year, Berkowitz and Chupka [62] used mass spectrometry to isolate the *neutral* component of the laser plume in the vaporization of graphite and found a C_n distribution which for $n < 11$ shows significant even–odd intensity alternations. Somewhat prophetically, they warned that this distribution probably did not reflect the primary cluster distribution created by the laser because the local density in the focal region was sufficiently high to create a miniature supersonic expansion, altering the distribution via three-body association reactions! Over the past 30 years, the use of laser vaporization to characterize surfaces has become a standard technique, laser desorption mass spectrometry (LDMS) [63].

Although the ability of pulsed lasers to create ions and cluster ions was known from the earliest days of the ruby laser, the development of effective laser vaporization sources required another dozen years. One obstacle was that the surface is destroyed by the laser after a few pulses, requiring that a fresh surface be presented to the focal point on every few shots. Velghe and Leach [64] overcame this difficulty in 1976 with the apparatus shown in Fig. 5, where a rod of metal is

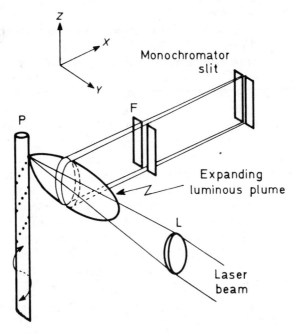

Fig. 5. Schematic diagram used by Velghe and Leach [64] of the laser vaporization apparatus that incorporated a rotating target. Reproduced with permission from Ref. [62].

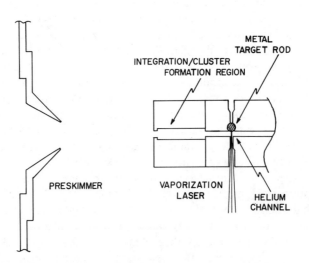

Fig. 6. Cluster ion source used by Brucat et al. [58] to generate cold metal clusters by carrying out the vaporization before expansion. Reproduced with permission.

continuously rotated and advanced. The primary interest of these authors was the spectroscopic characterization of small radicals formed in the neat vapor or by subsequent reactions. Cody et al. [65a] have also used the laser vaporization technique as a pulsed ion source for kinetic and spectroscopic studies in the pulsed ICR trap, and Weisshaar and co-workers [65b] have created metal ions using laser vaporization for study in a fast flow reactor.

From the extensive literature on the generation of ions by laser vaporization, it is clear that it is a viable pulsed alternative to the ionized metal oven for cluster ion formation. To use it effectively, however, we must reckon with the difficulty that ions generated in the plume are very hot and as such are not ideal targets for spectroscopic studies. In 1981, Smalley et al. [66] and Bondybey and English [67] demonstrated a method for cooling the metal atoms formed in laser vaporization. The Smalley ion source is depicted in Fig. 6. In this approach, the metal is vaporized inside a channel about 2 mm in diameter, and the vapor is entrained in a high-pressure burst of helium from a pulsed valve. The rotating rod is incorporated into the channel of the jet. The laser is timed to overlap with the peak density of the helium flow, and the atoms are entrained in the channel and cooled to room temperature and then expanded in a free jet. Large, cold neutral clusters are formed after expansion; they are then photoionized with an excimer

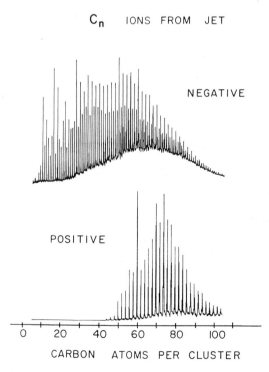

Fig. 7. C_n^+ and C^- cluster distribution obtained by O'Brien et al. [68] using the laser vaporization ion source. Reproduced with permission.

laser to detect the cluster distribution. High-resolution studies using resonant two photon ionization (R2PI) [66] and laser-induced fluorescence [67] have been carried out on many dimers (e.g., Pb_2, Mo_2, Sn_2) and establish that the clusters are cold. Unfortunately, ionization of cold neutral clusters does not guarantee that the ions will be vibrationally cold, since the ionization step is a Franck-Condon process and the internal vibrational energy of the ion is determined by changes in structure between ion and neutral. Devices based upon the initial Smalley design and its further refinements are now used in many laboratories.

Brucat et al. [58] and Bloomfield et al. [59] have recently shown that the plasma initiated by the vaporization laser is itself an efficient ion source. Thus, with the laser vaporization cluster sources, cluster ions are *directly* produced so that photoionization of neutral clusters is not required. High-resolution studies of the Nb_2^+ system confirm that cationic clusters are indeed rotationally cold. In addition, the shower of free electrons generated by the vaporization laser via the photoelectric effect allows efficient production of a neutral plasma described in the earlier discussion of van der Waals clusters. Because the electrons move much more rapidly than the clusters, vaporization in the high molecular density region as shown in Fig. 6 seems essential to confine the electrons before they fly away from the plume. O'Brien et al. [68] have used this source to obtain the mass spectra of ionic clusters of carbon shown in Fig. 7, indicating that both positive and negative ions are efficiently produced. As we will see in the next section, the presence of a high density of low energy electrons in the jet plasma generally creates a very fertile medium for synthesizing negative ions and cluster ions. The formation mechanism for negative cluster ions is significantly different than that operative for cations, however, as we discuss in the next section.

3.2 Negatively Charged Clusters

3.2.1 Molecular Ion Clusters

Negatively charged clusters have generally been considered more difficult to prepare, since they cannot be created by simple ionization of a parent neutral as can positive ions. Two processes commonly exploited [69] for negative ion production are three-body stabilization

$$RA + e^- + M \rightarrow RA^- + M \tag{5}$$

and dissociative attachment

$$RA + e^- \rightarrow A^- + R \tag{6}$$

An example of the dissociative attachment mechanism for molecular species is the formation of O^- from N_2O [70]:

$$e^- + N_2O \rightarrow O^- + N_2$$

which is often used as a starting reagent in negative ion flowing afterglow reactors.

In general, we might not expect 6 to be very efficient since the time scale for electron motion is much faster than that for the nuclei, and this mechanism requires "catching" the electron while it is transiently attached to the molecule. Therefore, it may be surprising that Klots and Compton [71] have demonstrated a cluster variation of dissociative electron attachment, "evaporative attachment," where the "third-body M" in 5 is actually a part of a cluster. Thus the process is

$$e^- + M_n \rightarrow M_m^- + (n-m)M \tag{7}$$

where the negative ion is stabilized by ejection of neutral monomers. This evaporative attachment mechanism 7 can be considered to be a combination of 5 and 6, where the incipient M^- species is stabilized by collision with monomers already attached to the cluster.

The formation of oxygen negative ion clusters affords a good example of how the electron attachment occurs. First, consider the inelastic collision of a low energy electron with an isolated O_2 molecule

$$e^- + O_2(v=0) \rightarrow O_2(v=1) + e^-$$

Spence and Schulz [72] have determined the electron energy dependence of this process and found that the cross section for vibrational excitation shows several peaks occurring at collision energies corresponding to resonant formation of autodetaching vibrational states of O_2^- with $v > 3$,

$$e^- + O_2 \rightarrow O_2^-{}^*(v > 3) \rightarrow O_2(v=1) + e^-$$

Now these transient $O_2^-(v > 3)$ negative ion states are estimated [73] to be rather long lived (about 100 ps) so that at resonant energies we might expect an enhancement of the three-body association reaction,

$$e^- + O_2 + O_2 \rightarrow O_2^- + O_2$$

Such an enhancement is indeed found, as shown by the experimental results of Spence and Schulz [74] where the electron energy dependence of the inelastic scattering process is seen to be nearly identical with that of electron attachment. In 1978, Klots and Compton [71] realized the implication of these data on the possibility of forming negative ion clusters. They reasoned that attachment to neutral clusters such as $O_2 \cdot O_2$ should be very efficient since the capture cross section for the electrons is likely to be similar to that of the monomer, whereas the stabilization step is very efficient, owing to the long lifetime of the resonance and the expected short vibrational predissociation lifetimes that stabilize the cluster ion before the electron detaches. Klots and Compton [71] demonstrated that negative cluster ions of light molecular gases such as CO_2 and N_2O are efficiently formed when low energy electrons are injected into a supersonic expansion of pure CO_2 or N_2O. Shimamori and Fessenden [75] find that the oxygen dimer has a significant effect on the thermal electron attachment rate to oxygen as evidenced by the anomalous temperature dependence of the three-body rate constant.

In light of the above discussion, it is clear that the efficiency of negative cluster ion formation will depend on the electron energy. Märk and co-workers [76–78] have recently reported the electron energy dependence of cluster formation for the CO_2, N_2O, and O_2 systems. For all three systems, it is clear that large negative ion formation is optimized with low energy (effectively zero energy) electrons. Thus the problem of negative ion formation lies in finding efficient ways to inject low energy electrons into the jet.

Many ingenious approaches have met with varying degrees of success with the problem of injecting the low energy electrons. The inherent problem with producing the ions with an electron gun is, of course, that the coulombic repulsion limits the electron density and hence low energy electron beams typically have very low flux. Bowen and co-workers [79] have used electron transfer from alkali atoms to species such as $(Cl_2)_n$ to create a variety of negative molecular ion clusters. Haberland et al. [80] have been particularly successful in forming clusters of $(NH_3)_n^-$ and $(H_2O)_n^-$ using both the photoelectric effect to inject electrons in the high density region of the expansion and discharging the expansion with a heated filament. More recently, Bowen and his students [81] have used a variation on Haberland's filament approach to generate a variety of clusters of both metals and nonmetals.

Perhaps surprisingly, the high energy (~ 1 keV) electron impact ionized jets described in the previous section are excellent sources of negative ion clusters. This could have been anticipated based on the description of the characteristics of the ionized jets as a dense free electron plasma. Moreover, since these jets have been observed to preserve the rotational cooling achieved by the expansion, the average electron energies must be low (< 1 eV) or the electron neutral collisions would result in rotational excitation. The mechanism for production of these low energy electrons starting with keV electrons must involve the secondary electron formed in the initial ionization step:

$$e^- + O_2 \rightarrow O_2^+ + e^-(<10\,\text{eV}) + e^-(\sim 990\,\text{eV})$$

That is, one electron comes off slowly, and the high energy electron loses only a small fraction of its original energy and continues on through the jet. For ionization in the dense part of the jet, the high energy electron beam penetrates through the gas, leaving a trail of low energy electrons. Since the jet is observed to be electrically neutral, the high energy electron must eventually scatter out of the jet, whereas the low energy electron cools in the jet through inelastic scattering events.

For example, the electron impact ionization cross section [82] of N_2 for a 1-keV electron is about 10^{-16} cm^2 so that at a density just outside the throat of a 1-

mm nozzle of about $10^{17}/cm^3$, each primary fast electron suffers an average of about one ionizing collision, producing one cation and one low energy electron. At higher molecular densities, the low energy electron yield can, in principle, *exceed* that obtained from the 1-keV gun. For oxygen, the electron thermalization rate has been determined [83] to be 1.7 torr-μs so that just outside the nozzle where the density is equivalent to about 10 torr pressure we expect rapid thermalization of the electrons resulting from ionization. Note that rare gases are about two orders of magnitude less efficient [83] at electron thermalization, so that helium expansions may well be less favorable negative ion sources.

To illustrate the performance of this source, the top trace in Fig. 8 presents the negative ion mass spectrum obtained from a CO_2 expansion [84] ionized under similar conditions as the cation cluster spectrum shown in Fig. 2. The peak intensities of the anion spectra are about a factor of 10 lower than the cations. *Ionization at 2.5 cm from the nozzle creates very different size cluster ions in the positive and negative channels.* Large negative ion clusters (up to 17 monomers) are produced in a regime in which only the monomer cation is evident in the mass spectrum. The implication is that the production mechanisms

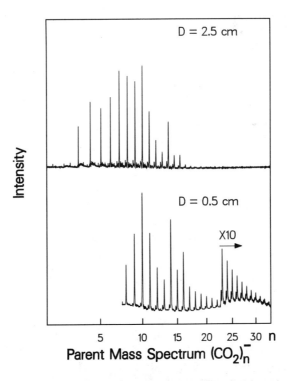

Fig. 8. Dependence of negative cluster ion distributions on the distance Z from the electron beam to the nozzle taken from Alexander et al. [84] Unlike the positive ion clusters (Fig. 2), the negative cluster ions are relatively insensitive to the ionization distance. Reproduced with permission.

are different since the negative ion clusters can be formed by evaporative attachment to neutral clusters, whereas cations are formed by nucleation of neutrals onto monomer ions. This is particularly evident in the CO_2 case since the monomer ion CO_2^- is not observed in the mass spectrum and in fact is not a stable species [85]! It has been observed as a long-lived species, however, in both ion-molecule reactions and charge exchange with Cs [85] and by low energy electron attachment to neutral carbon dioxide clusters [77]. The anion cluster spectrum in Fig. 8 can also be used to establish a bound on the electron energy. Knapp et al. [77] have shown that the energy threshold for production of CO_3^- is about 3 eV. This ion is absent from the 1-keV electron impact ionized jet mass spectrum, indicating that most of the electron in the expansion are less than 1 eV. The anion spectrum can be affected by the position of ionization, however, and the bottom trace in Fig. 8 shows the $(CO_2)_n^-$ distribution for ionization at 0.5 cm, indicating that larger clusters can be formed by nucleation reactions. *Negative ions appear to be different from cations in that the nucleation ions are themselves clusters.*

This asymmetry between positive and negative ion formation is caused by the kinetics of the ion formation step. For positive ions, the association rates are likely to be similar for the monomer ion and small cluster ions, and the rate of formation of monomer ion relative to cluster ions by direct electron impact is governed primarily by relative abundances in the jet. Since under typical conditions (1-atm backing pressure, 300 K, nonhydrogen bonded species) neutral clusters are only about 0.1% of the monomer density [86], most cation clusters are formed from the monomer ion. Note also that ionization usually takes place close to one nozzle diameter away from the orifice, implying that only mild cooling has occurred and clustering of the neutral is not complete. For the negative ions, there is no direct monomer channel leading to ion production, so the competing processes are binary and ternary electron–molecule collisions:

$$e^- + O_2 + O_2 \xrightarrow{k_1} O_2^- + O_2$$

$$e^- + (O_2)_n \xrightarrow{k_2} (O_2)_m^- + (n-m)O_2$$

These processes lead to an expression for the relative rates R_1 and R_2 for O_2^- formation:

$$\frac{R_1}{R_2} = \frac{[O_2]^2 k_1}{[(O_2)_2] k_2}$$

If the density of O_2 monomers is about $10^{17}/cm^3$ just outside the nozzle and the percentage of dimers in the beam is 0.1%, $k_1 \sim 10^{-31} cm^3/s$ [75] and $k_2 \sim 10^{-9} cm^3/s$, the rate for anion formation by three-body addition to a monomer is one-tenth that for evaporative attachment to a neutral cluster: Most negative ions are thus created from neutral precursor clusters in the jet. The suggestion that the negative cluster ions are created by evaporative attachment in the pulsed expansion is confirmed by the cluster size distributions shown in Fig. 8. This figure shows that the size distribution is relatively insensitive to changes of over an order of magnitude in the neutral concentration at the point of ionization.

The role of neutral clusters in negative ion formation is also supported by comparing the electron beam generated cluster size distribution with that obtained using controlled low energy electron impact [76] and electron transfer from Rydberg states in collision with rare gas atoms [87]. The $(CO_2)_n^-$ distribution obtained from the low energy attachment,

$$e^- + (CO_2)_n \rightarrow (CO_2)_m^- + (n-m)CO_2$$

was recently determined by Stamatovic et al. [76], and their spectrum is quite similar to the CO_2 negative cluster ion formed by the high energy electron impact source shown in Fig. 8, including the intensity anomalies described as "magic numbers." This similarity is also observed for water clusters, where the low energy electron injection distribution is observed [88] to create clusters beginning at about $n = 13$, a pattern that is also observed for the 1 keV electron beam source [89].

3.2.2 Cluster Ions of Metals and Semiconductors

The essential features of the electron beam ionized jet that lead to large negative cluster ion formation and extensive cooling are duplicated in the pulsed jet-expansion chamber version of the laser vaporization experiments. Zheng et al. [90] at Rice and Bloomfield et al. [59] at AT&T Bell Laboratories have recently shown that indeed the laser vaporization technique is capable of generating intense, pulsed negative cluster beams of refractory materials. The group at AT&T Bell [59] has reported negative ions of carbon and silicon and the Rice group [15, 58, 90] has applied the technique to Ag, Cu, Nb, Ni, and C. Based upon the results of these experiments, the evaporative attachment mechanism involved for negative ion formation from neutral van der Waals clusters may not account for cluster negative ion formation in metals. This fact is due to the much larger binding energy of metal atoms in the cluster ion compared to that of the van der Waals case. In the Cu_n^- system, for example, the binding energy [15, 58] is on the same order as the electron affinity of the cluster. Therefore, the incipient ion cannot eject multiple neutral atoms but rather must be stabilized by collisions with helium.

Finally, an advantage of either version of the negative ion jet ion sources is that the cluster ions are likely to be cooled internally to an extent that high-resolution spectroscopy experiments may be attempted. Certainly, significant cooling occurs since clusters of weakly bound species such as argon and neon atoms can be formed with both metals and nonmetals. In the case of cations, spectroscopic experiments [58] indicate that the clusters are vibrationally and rotationally cold. However, we know of only one direct determination [22] (on O_4^-) of the internal energy distribution of anion clusters and therefore caution against presuming that these species are cooled to the translational temperature of the jet. Indeed, such experiments would be extremely valuable and represent a significant challenge for experimentalists. It is likely that internal energy measurements will use spectroscopic methods. With this in mind, the next section summarizes the pulsed spectroscopic techniques currently in use to characterize the cluster ions.

4 MASS SELECTION

4.1 Primary Ion Beam

Although the use of pulsed jet cluster ion sources is well established, there is an intrinsic disadvantage when using such sources in conjunction with mass spectrometry to select a specific cluster size for study. This disadvantage results from the necessity of allowing the ions to continue to expand in the jet until the neutral density is sufficiently low that the ionic clusters are not collisionally activated upon extraction from the expansion. In addition, the plasma that acts to shield the ions in the expansion from external fields must be broken down by the extraction voltage as well. A free drift region of about 15–20 cm is sufficient to allow extraction, but because of the $1/r^2$ fall off of the density, the initially compact ion cloud is now spread out over about 5 cm on either side of the flow axis. Thus, although TOF mass spectrometry is the natural choice for mass selection, we are faced with two problems before it can be effectively implemented for laser experiments: first, the inherent mass resolution limit caused by the large spatial spread, and, second, the necessity to obtain spatial compression of the mass selected ions in order to achieve efficient overlap with a pulsed laser beam.

These problems are simultaneously overcome using a space-focusing TOF mass spectrometer first demonstrated by Wiley and McLaren [91] in 1955. These authors show that the resolution loss due to the spatial extent of the ions in the source can be compensated, to the order of the first moment of the spatial distribution, simply by accelerating the ions through *two* regions with appropriately chosen, different electric fields, followed by the usual drift region. Wiley and McLaren [91] were interested in improving the resolution of an electron bombardment ion source, where the spatial extent of the ions was on the order of a few millimeters, the diameter of the electron beam. Since the ions in the jet source are spread over several centimeters, the focusing conditions must be altered somewhat from the Wiley-McLaren condition. We consider the spectrometer to be optimized when the arrival time T of an ion is nearly independent of its initial transverse position in the source. Figure 9 presents the arrival times as a function of transverse position in the source for an optimized configuration, such as exists (Fig. 10) on the Boulder apparatus. It is clear that the optimum condition only exists for ions distant from the extraction hole. In this mass focusing technique, mass peaks are quite narrow at the half-width, but the line shape is generally asymmetric. There is a sharp edge on the long time side, but the peak has a significant tail toward shorter times extending over several microseconds, and we must guard against the possibility that a low-level

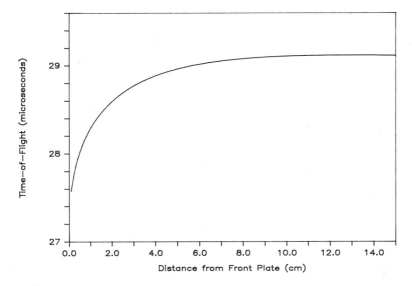

Fig. 9. Calculated arrival times as a function of the initial transverse position of ions in the source for an optimized Wiley-McLaren [91] TOF mass spectrometer, From M. J. DeLuca and M. A. Johnson, unpublished results.

background of ions from nearby masses might contribute to the observed signal in photofragmentation experiments.

To couple this two-field TOF spectrometer to the cluster ion source, the ions prepared in the high density region close to the nozzle are allowed to drift 15–20 cm in a field-free region in the flow of the free jet. The expanding plasma passes between two plates separated by about 15 cm to avoid substantial interruption of the flow, since this could cause collisional dissociation or fragmentation. When the ion cloud passing between the plates reaches maximum intensity, the potential of one of the plates is rapidly switched with respect to the other, causing ions to accelerate toward the plates. Most ions simply collide with the plates; however, a cylindrical volume of ions passes through a 4-mm-diameter aperture, is accelerated in a second region about 5 cm long, and is then allowed to drift 1.5 m where the beam is focused in a plane perpendicular to the drift axis.

The apparatus used by Johnson et al. [14] is shown schematically in Fig. 10, and typical mass spectra are shown in Fig. 2, 4, and 8. The time evolution of ions of a given mass is indicated by the small dots in Fig. 10, and the compression the ion beam as it nears the focus is illustrated by the density of the dots. A mass resolution $m/\Delta m$ of 200–300 can be obtained for the parent ion whose energy spread is about 500 eV (resulting from the differing formation positions in the first transverse field) with a mean energy of order 3 keV. For mass spectral characterization of the beam, an electron multiplier is placed at the focus, whereas the laser experiments are carried out by timing the laser to intersect the ion beam at this space focus at the precise time that the ion of interest is *transiently* focused.

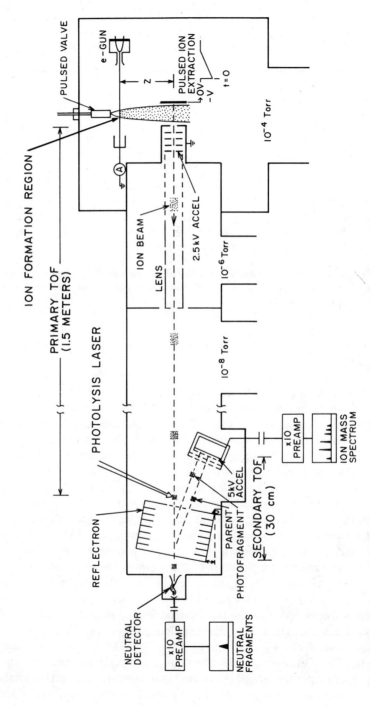

Fig. 10. Schematic diagram of the tandem TOF photofragmentation mass spectrometer used by Alexander et al. [107]. Reproduced with permission.

This does not require physically removing the multiplier, however, since the position of the focus can be varied in a calculable fashion with the voltages used in the two-stage acceleration region.

Note that there is no necessity to pulse the ion formation (e.g., with a pulsed electron beam) since the leading edge of the draw out pulse initiates the TOF process, and the extraction voltage remains on until well after the pulsed valve closes. As long as this voltage is on, subsequent ions entering the region between the plates are accelerated into the plates well before the aperture and are not injected into the mass spectrometer.

4.2 Interaction of the Ion Beam with a Pulsed Laser

The space focusing described above that improves the resolution of mass spectrometry also creates an ideal ion target cloud for interaction with pulsed lasers. A 50-ns wide mass peak for a 2-keV ion beam implies that the ion cloud is less than 0.5 cm long. If a simple Einzel lens is used to focus the beam, about 10^5 ions can be compressed into the focal volume of the laser beam with an area of $0.5 \, cm^2$. *Essentially 100% overlap can be achieved between the mass selected ion cloud and a pulsed laser.* Since the pulse width of most common pulsed lasers (Nd:YAG and excimer, pulse width $< 20 \, ns$) is less than the time an ion is focused, mass selection of the parent ion can be achieved by timing the laser to pick off the desired mass at the focus during the 20-ns pulse width of the laser. Thus, with this relatively simple technology, the most difficult problem of using pulsed lasers in an ion-beam apparatus—poor beam overlap owing to the duty cycle of the pulsed laser—is substantially overcome. The TOF method in essence compresses about $20 \, \mu s$ of ion output from the jet down to 50 ns for optimal compatibility with the pulsed laser.

This new compatibility has significantly opened up the possibilities for laser studies on cluster ion beams. We now review two areas that have already benefited from application of these pulsed techniques: photofragmentation and photoelectron spectroscopies.

5 DETERMINATION OF CHARGED PHOTOFRAGMENTS USING TANDEM TIME-OF-FLIGHT MASS SPECTROMETRY

Whereas the resolution of the parent mass selection can be quite good, as described in the previous section, obtaining the mass spectrum of the photofragments is again complicated by the large energy spread of the parent ions. Two approaches have been used to overcome this difficulty:

1. Laser interaction with the high energy beam followed by fragment analysis using a reflecting field or "reflectron"
2. Deceleration before laser interaction followed by a second pulsed Wiley-McLaren mass spectrometer to mass analyze photofragments

Each approach has advantages, and we discuss then in turn.

5.1 Tandem TOF Using a Reflectron

In 1984, Johnson et al. [14] demonstrated the reflectron method on photofragmentation of $(CO_2)_n^+$ clusters with the apparatus shown in Fig. 10. Photofragments are separated from parent ions by reflecting the ions back at a 170° angle using a weak electric field so that the ions penetrate far into the reflecting field region before turning around. The first use of the reflectron technique was reported by Mamyrin et al. [92] in 1973, who demonstrated a resolution improvement of about a factor of 10 in TOF mass spectrometry by incorporating the reflecting region. The reason for this improvement is that, like the Wiley-McLaren scheme, the arrival times of ions are essentially independent of the energy of the ion. Boesl et al. [93] used this device for high-resolution multiphoton ionization experiments, and in 1981 they pointed out that the reflectron is also useful for determining the decomposition products from metastable decay of ions in the drift region before the reflecting field. Echt et al. [94] used a reflectron to observe collisional and metastable decay of ammonia cation clusters formed by multiphoton ionization of neutral ammonia clusters. Thus, based on the success of these workers in determining fragmentation patterns, the reflectron was a logical choice for the photofragmentation experiments. In addition, the reflectron does not require decelerating the ion beam, so that the focusing characteristics of the high energy cluster beam could be maintained.

The principle of the single field reflectron is that higher energy ions of a given mass spend longer in the reflectron field than lower energy ions of the same mass, but the higher energy ion traverses a drift region more quickly. Therefore, at some distance down the drift tube the faster ions delayed in the reflectron will catch up to the slower ions, at which point the ions at this m/e will be focused. If E_r is the reflectron electric field tilted at angle ϕ with respect to the axis of the ion beam, L the length of the drift region, E_B the ion beam energy, the arrival time of an ion with mass m_p is

$$T(m_p, E_B) = \frac{2m_p \sec \phi}{eE_r} \sqrt{\frac{2E_B}{m_p}} + L \sec 2\phi \bigg/ \sqrt{\frac{2E_B}{m_p}} \qquad (8)$$

If we impose the condition that $dT/dE_B = 0$, i.e., the arrival time of a fragment ion with mass m_f is independent of initial energy, we find

$$T(m_f, E_B) = L\left(\frac{m_f \sec \phi}{2\langle E_f \rangle} \sqrt{\frac{2E_B}{m_p}} + \sec 2\phi \bigg/ \sqrt{\frac{2E_B}{m_p}}\right) \qquad (9)$$

where $\langle E_f \rangle$ is the average energy of the fragment ions which penetrate a distance $L/4$ into the reflectron. Equation 9 shows different mass ions with identical initial velocities are dispersed so that the photofragmented ions are indeed mass resolved. A resolution of about 150 has been demonstrated for the daughter ions in the reflectron experiments.

5.2 Tandem TOF Using Two Pulsed Fields

Bloomfield et al. [13] first carried out pulsed photofragmentation experiments with a tandem TOF instrument, investigating the Si_2^+–Si_{12}^+ systems. Brucat et al. [58] use a similar technique, and their apparatus is shown in Fig. 11. In their approach, the parent beam is decelerated before injection into the second mass spectrometer, which allows similar drift lengths and extraction voltages to be used on each spectrometer. Both spectrometers use the Wiley-McLaren focusing, and the second stage has been oriented either parallel and perpendicular to the first. Since the ions are decelerated before laser interaction, the laser timing cannot be used to achieve high mass resolution, since the focus of the first spectrometer occurs when the beam is at high energy. Therefore, a "pulsed mass isolator" or "mass gate" was used for parent mass selection, a device that consists of a short region that is electrically isolated and houses a set of pulsed deflector plates. This region is placed at the focus with the plates normally set to deflect the ion beam but when the desired mass passes through the plates are pulsed off so that this mass continues undeflected into the deceleration region. The resolution $m/\Delta m$ reported using this technique, is greater than 100 for parent and photofragment ions.

The Smalley group [15, 58] has conducted several careful power dependences of the photofragmentation and photodetachment process for the metal cluster ions and has found multiphoton phenomena are prevalent at moderate laser powers. Studies of the fluence dependence of Fe_6^+ photodissociation [58] show that only Fe_5^+ is a one-photon product, with all of the smaller cation fragments resulting from two- and three-photon absorption. Strong nonlinear behavior is apparent at fluences below $10\,mJ/cm^2$, demonstrating that extreme caution must be exercised in carrying out the pulsed experiments to avoid multiphoton effects.

Smalley et al. [15] have used careful determination of the onset of one-photon

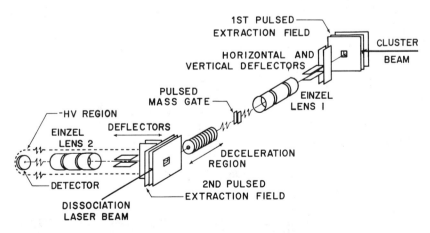

Fig. 11. Schematic of the tandem TOF photofragmentation mass spectrometer used by Brucat et al. [58] that uses two pulsed acceleration regions. Reproduced with permission.

induced photodissociation to measure the sequential bond dissociation energies and electron affinities of copper clusters and are presently pursuing this method on several transition metal systems. This group has also carried out photofragmentation studies of negative metal cluster ions of copper and silver. Interestingly, $(Ag)_{10}^-$ only photodetaches, whereas the Ag_7^- photodissociates exclusively to Ag_3^- and Ag_5^-. Both photodetachment and photodissociation channels are competing in the trimer anion.

6 PULSED PHOTOELECTRON SPECTROSCOPY USING TIME-OF-FLIGHT ELECTRON ENERGY ANALYSIS

The marriage of the pulsed laser and ion source has created new possibilities for photoelectron spectroscopy. Photoelectron spectroscopy (PES) of negative molecular ions was demonstrated first on O_2^- and NO^- in 1972 by Celotta et al. [95] using a continuous ion beam and the intracavity 488-nm radiation of an argon ion laser. Until 1986, all negative ion PES machines used concepts derived from this first device. Much of the work on photodetachment of molecular ions has recently been reviewed [97]. The groups have demonstrated that the continuous methods can also be applied to cluster ions. Bowen [21, 81] has reported spectra of $NO^- \cdot (N_2O)_n$, $(N_2O)_n^-$, $(H_2O)_n^-$, $H^- \cdot (NH_3)_n$, and $NO^- \cdot (RG)_n$ (RG = rare gas and $n \leqslant 3$), and Leopold and Lineberger [25] have recently obtained photoelectron spectra of Cu_n^- with $n < 11$. The desirability of ultraviolet (UV) photons for PES studies of clusters derives from the typical solvation energies of ions. An ion with an electron affinity of 1 or 2 eV is readily studied with the 488-nm line from an argon laser, but each solvent molecule clustered to the ion will increase the electron affinity roughly by the solvation energy, usually in the range of 0.25–0.7 eV/solvent molecule [98]. Thus, after the first few solvent molecules the electron affinity approaches the photon energy and the clusters do not photodetach. The photon energy available from continuous wave (cw) sources is also not an intrinsic limitation since ion lasers can be operated with (marginally) adequate flux on several cw ion laser lines between 350 and 370 nm. Moreover, new materials, such as β-$BaBO_3$, hold the promise of frequency doubling with adequate intensity, provided that the doubled light is injected into a buildup cavity located totally inside the vacuum container. These approaches aimed at improving cw PES spectroscopy for use on cluster ions are, however, rather substantial undertakings, whereas the pulsed methods described below allow PES to be easily carried out with conventional pulsed UV lasers and intense pulsed nozzle sources.

Posey et al. have constructed the apparatus shown in Fig. 12, a machine that first demonstrated [22] negative ion pulsed photoelectron spectroscopy. The principle of this instrument is quite simple—the ions are photodetached at the space focus from the extraction TOF mass spectrometer. The electron spectrometer on this apparatus is very similar to TOF photoelectron spectrometers used in multiphoton ionization experiments on neutrals [99] except that the photodetachment spectrum of the ions is usually a one-photon event. The

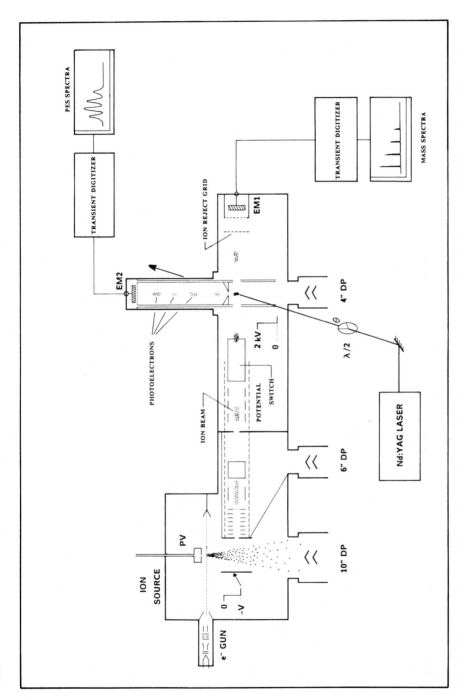

Fig. 12. Schematic diagram of the pulsed negative ion photoelectron spectrometer developed by Posey et al. [22]. Reproduced with permission.

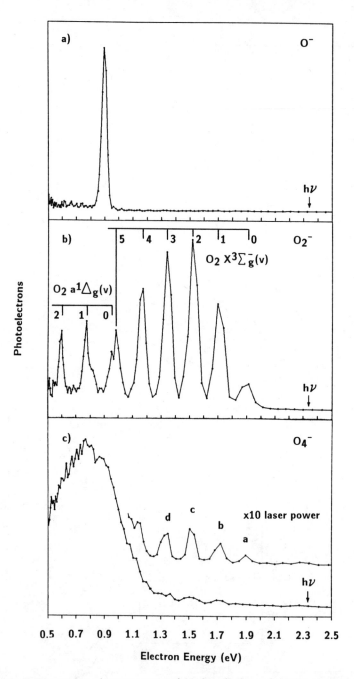

Fig. 13. Pulsed 532-nm photoelectron spectra of (a) O^-, (b) O_2^- and (c) O_4^- obtained with the spectrometer shown in Fig. 12. The peaks labeled a, b, c, d in the inset of the bottom trace were recorded at 10 times the laser power (40 mJ/pulse) as the full trace and result from photodetachment of an O_2^- photofragment.

electron drift region is a 60 cm long, magnetically isolated drift tube. The arrival times are recorded in a transient digitizer to generate the photoelectron spectrum. Typical spectra of O^-, O_2^-, and O_4^- are shown in Fig. 13. The resolution of this spectrometer is about 35 meV at 1 V, compared to state-of-the-art cw PES spectrometers with resolution of 3–8 meV [100]. The collection efficiency of the TOF spectrometer is about 10^{-3}, comparable to that of the cw instruments. The PES spectrum also yields an independent estimate of the ion density at the laser crossing region, since the charge density in the focused ion beam is sufficiently large to cause a shift in the position of electron peaks owing to the coulombic repulsion between the departing electron and the ion cloud [22]. Shifts on the order 100 meV at 1 eV are routine for the O_2^- system, indicating that the charge density is on the order $10^5/cm^3$. Using the pulsed apparatus, Johnson and co-workers have studied three isomers of the $N_2O_2^-$ system, $(O_2)_n^-$ with $1 \leqslant n \leqslant 10$ and $(CO_2)_n^-$ with $2 < n < 16$. A powerful feature of the pulsed apparatus is illustrated by the O_4^- data, shown on the bottom trace of Fig. 13. It shows that O_2^- is formed by photodissociation of O_4^- and then photodetached by a second photon. It is clear that with a second stage of mass selection, photoelectron spectra of photofragment ions may be recorded and used to evaluate the product vibrational energy disposal in cluster ion fragmentation studies.

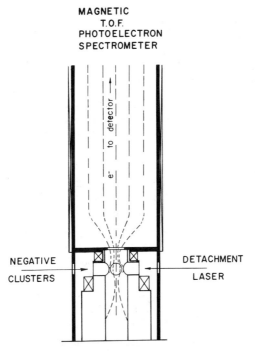

Fig. 14. Magnetic parallelizer used in the pulsed photoelectron spectrometer described by Cheshnovsky et al. [101]. Reproduced with permission.

Recently, Cheshnovsky et al. [101] at Rice University have also demonstrated pulsed PES of a $(Cu_n)^-$ negative cluster ion beam. The copper clusters were formed by laser vaporization, and TOF was again used to energy analyze the electrons. In order to increase the collection efficiency of this spectrometer, however, a magnetic parallelizer was used to achieve an efficiency of about 30%. This device, shown schematically in Fig. 14, is an adaptation of a PES spectrometer used by Kruit and Read [102] and operates by adiabatically turning the electron flight direction without changing its energy. By photodetaching the electron in a region of high magnetic field that expands rapidly into a weak solenoidal field, the electrons are turned and directed along the flight tube axis without changing their energies. Provided the turning radius is small compared to the flight length, good resolution can be maintained with high (up to 98%) collection efficiency. The resolution reported by Cheshnovsky et al. [101] is about 100 meV, but much higher resolution has been obtained by Kruit and Read, and it should be realizable in this context with several obvious improvements.

7 A FEW RESULTS

7.1 Mechanism of Optical Absorption in "van der Waals" Cluster Ions: Localization of Charge within Large Clusters

Lineberger and co-workers [14, 103] have studied the visible and near UV photodestruction cross sections for $(CO_2)_n^+$ and Ar_n^+, finding strong, continuous absorption for each species extending over several hundred nanometers. This finding at first appears curious in light of the fact that none of the species CO_2, Ar, CO_2^+ or Ar^+ possesses an optically allowed transition above 350 nm. The $(CO_2)_2^+$ cation is known [28] to have a broad visible absorption band. To understand the origin of this absorption for the larger clusters, relative cross sections were measured for $(CO_2)_n^+$, $2 < n < 11$, with the results displayed in Fig. 15. Interestingly, the cross section for all the clusters is quantitatively similar at 532 nm but is strongly dependent on cluster size for excitation at 1064 nm. The overall picture is that the higher clusters generally absorb in the same spectral region as the dimer cation $(CO_2)_2^+$. Smith and Lee [28] have studied the photodestruction spectra of $(CO_2)_2^+$ and find broad, continuous photodissociation extending from 400 to 1000 nm. They suggest that it arises from excitation to a potential curve strongly repulsive between the CO_2 centers and involves binding electrons of the dimer ($D_0 = 11.8$ kcal/mol) [104]. In this model, the nearly Gaussian shape of the absorption spectrum is due to the ground-state vibrational wavefunction corresponding to the relative motion of the two CO_2 molecules. Since the shape of the absorption for the dimer is controlled by the shape of the repulsive surface via the "reflection principle" for dissociative transitions [105], the blue edge of the absorption cross section "probes" the inner turning point of the dimer. Kim et al. [29] have obtained much more complete photodissociation spectra of the trimer $(CO_2)_3^+$ and find a spectrum remarkably similar to the dimer cation in the range

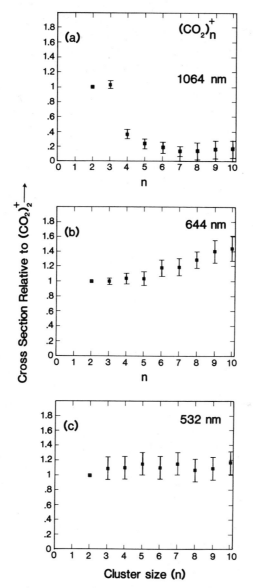

Fig. 15. Photodestruction cross sections of $(CO_2)_n^+$ at (a) 532 nm, (b) 650 nm and (c) 1064 nm, for $2 \leq n \leq 11$. Larger ionic clusters up to at least $n = 26$ appear to absorb in the spectral region as the dimer. Reproduced with permission from Johnson et al. [14].

400–700 nm. The similarity of the photodestruction cross section for the higher clusters with that for the dimer over most of the spectrum leads to the suggestion that the dimer carries the charge within the larger clusters.

Similar studies [103] on photoabsorption of Ar_2^+–Ar_{60}^+ show a broad absorption band but indicate the complexity to be expected. The species Ar_3^+ through Ar_{19}^+ all have a similar photoabsorption cross section, but it is larger than Ar_2^+ by several orders of magnitude throughout the visible spectrum. Similarly, a new, stronger, red absorption first appears at Ar_{20}^+ and persists at least to Ar_{40}^+. These results indicate the danger inherent in drawing structural conclusions from a limited set of observations. These data are likely not completely consistent with a model that views Ar_n^+ as having a dimer core for all n. This difference in core structure would alter existing theories [37] of the ionization dynamics of neutral argon clusters.

Thus, using the characteristic dimer photodestruction spectrum as a diagnostic of the charge distribution, a model for the larger clusters is emerging where the charge is generally contained on a smaller cluster, with the additional monomers adding on to this charged cluster remaining essentially neutral. Substantial changes may occur at "shell closings" in the larger cluster ions, as indicated by the Ar_n^+ studies. This result was actually anticipated by workers concerned with the mechanism of "magic number" formation in cluster ion mass spectra, as described in the next section.

7.2 Statistical Photodissociation Dynamics of Large Molecular Ion Clusters

Photofragmentation studies have been carried out on the homogeneous systems $(CO_2)_n^+$, $(CO_2)_n^-$, and Ar_n^+ by Lineberger and co-workers [14, 84, 103, 106], and a general trend is observed for these systems whereby large clusters (with about 15 or more monomer units) absorb a photon and eject a constant number of neutral monomers independent of parent cluster size. The fragmentation spectra for the $(CO_2)_n^+$ system with $2 < n < 26$ have been reported [106], and an expanded region including $2 < n < 46$ recently obtained by Levinger et al. is displayed in Fig. 16. The fragment ions are recorded on the horizontal axis, and the parent ion size is displayed on the depth axis. Above about $n = 15$, it is seen that the CO_2 systems all eject about six neutrals for excitation with 1064-nm light, which appears as a "ridge" in the fragment ion patterns. Note that there are no energy constraints limiting the fragment size, so that the entire lined region represents possible fragment channels. Figure 17 shows the dependence of the number of neutrals ejected in the constant loss regime on photon energy. A linear dependence is found, and the slope yields an upper bound for the binding energy for the monomer to the cluster ion of 5.3 kcal/mol, nearly the heat of sublimation of dry ice (5.7 kcal/mol)! The fact that the number of neutrals ejected appears to be governed only by the binding energy points to a model for the photodissociation in which the absorbed photon energy is degraded into thermal energy and finally to evaporation of the CO_2 molecules on the surface of the cluster. Presumably the absorber is photodissociated within the cluster, and the recoiling fragments are

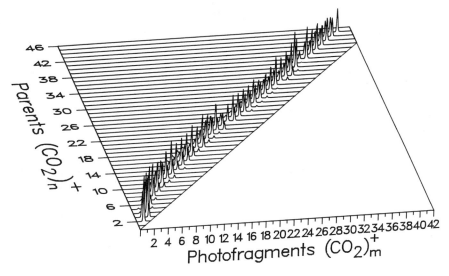

Fig. 16. Photofragmentation spectra of $(CO_2)_n^+$ taken at 1064 nm, obtained by Levinger, Ray, and Lineberger (unpublished results).

captured by the cluster to effect the energy degradation. This suggestion leads to a series of experiments investigating the "cage effect", experiments that are discussed in Section 7.4.

Interestingly, the negative cluster ions of CO_2 are also observed to photodissociate (see Section 7.3), and the number of neutrals ejected is included in Fig. 17. The asymptotic number ejected falls nicely on the line established for the cation data, indicating that the evaporation step is insensitive to charge on the cluster. Further evidence for the evaporation mechanism can be found [107] in the $Br_2^- \cdot (CO_2)_n$ system, where the Br_2^- component can be photodissociated within the cluster. (The Br_2^- ion, isoelectronic with Xe_2^+, will have a continuous absorption similar to that of Ar_2^+.) The asymptotic number of CO_2 monomers ejected from the $Br_2^- \cdot (CO_2)_n$ system at 700 nm appears to fall on the same line as $(CO_2)_n^-$ and $(CO_2)_n^+$, suggesting that the effect is even independent of interior composition in addition to charge. Actually, the bromine system has an additional feature of interest in that the photodissociation dynamics of small clusters contains information on the "cage effect" on the atomic recombination, as discussed more fully below.

The widths of the fragment ion distributions depicted in Fig. 16 are quite narrow, about 2.5 monomers, and the shape of these distributions contains information on the dynamics of the photodissociation process. Engelking [108] has used a statistical (RRK) model to reproduce these fragmentation patterns and finds qualitative agreement with the experimental results. The photodissociation behavior for all $(CO_2)_n^+$ clusters can be reproduced with an average bond energy of 3.6 ± 0.6 kcal/mol for clusters larger than the dimer. The difference between

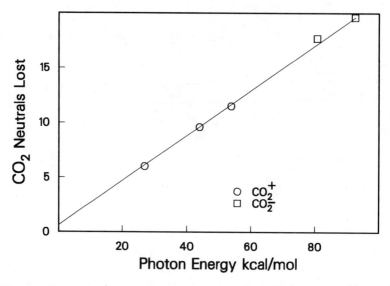

Fig. 17. Asymptotic number of neutral CO_2 molecules ejected from carbon dioxide positive and negative ion clusters as a function of photon energy. Reproduced with permission from Alexander et al. [107].

this value and the slope of the line in Fig. 17 is suggested [108] to result from the translational energy disposal. Another general conclusion of these studies is that the width of the fragment distribution is controlled by the difference between the monomer binding energy and the thermal energy carried off per monomer at a temperature characteristic of the photoexcited cluster. A good illustration of this point is the photofragmentation spectrum [103] of the Ar_n^+ system. The binding energy in the argon system is only about 1 kcal/mol so that, assuming an equivalent heat capacity for the argon and carbon dioxide systems, the fragment ion distribution would be expected to be much larger for the argon clusters, as is observed [about 2 monomers wide for CO_2 vs. 6 for Ar (see Fig. 20)].

7.3 Formation of "Magic Numbers" in Cluster Ion Mass Spectra

The anion $(CO_2)_n^-$ mass spectrum shown in Fig. 8 contains an example of what is called a "magic number" [41], essentially an anomalously abundant cluster in the mass spectrum with respect to the intensity of neighboring clusters. In the CO_2^- anion case, the $n = 14$ cluster is more abundant than 13 or 15 by about a factor of 4. Such anomalies are common, especially in rare gas cation systems [41], and have been the subject of much recent study to determine if they reveal a property of the parent *neutral* cluster or if they result from kinetic or thermodynamic characteristics of the *ionic* clusters. Recent work by Haberland [38] and Saenz et al. [37] suggests that the magic numbers in the rare gas cluster ions result from ionization of neutral clusters creating a nascent ion cluster far from its equilibrium internuclear geometry. This displacement is due to the propensity of the ionic cluster to compress the charge into a small, relatively tightly bound core, as we discussed in the previous section. The energized cluster then relaxes by ejecting neutral molecules until the internal energy of the ion cluster approaches the binding energy of one monomer. If, as the ionic cluster is sequentially ejecting neutrals, an especially stable cluster ion is formed with a substantially higher binding energy, then the activation energy for this cluster to dissociate will be proportionally higher, and a larger portion of the ensemble of ions will be trapped at this cluster size than at adjacent masses. In this view, the magic numbers in the mass spectrum are associated with properties of the ions, not the parent neutrals. Such a model is certainly proven in the case of the formation of the $H_3O^+\cdot(H_2O)_{20}$ cluster, which has been shown [110] to develop in real time after ionization. It is also possible to test this theory further by photodissociating a large cluster ion with a wavelength that creates fragment masses in the size range near a magic number. Since the model does not depend on how the cluster is energized, photolysis should also result in magic numbers in the photofragment distribution and have the advantage that the starting cluster size and initial internal energy are well defined. In addition, there is no possibility of the result being convoluted with the initial distribution of neutral clusters. Alexander et al. [84] have recently carried out such experiments on the $(CO_2)_n^-$ clusters, with the result shown in Fig. 18. The $n = 14$ cluster is seen to indeed occur with increased intensity compared to neighboring peaks, indicating that an enhanced stability is associated with this cluster.

Saenz et al. [37] have proposed a similar model to account for the magic numbers observed in rare gas cation clusters. Levinger et al. [103] have obtained the fragmentation spectrum of Ar_{46}^+ presented in Fig. 19. The $n = 20$ cluster, which appears weak in the parent ion mass spectrum, is also a minor fraction of

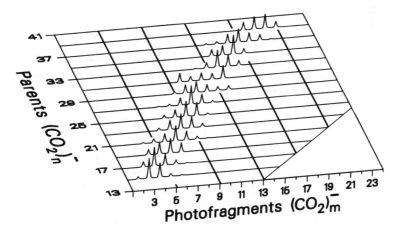

Fig. 18. Photofragmentation spectra of $(CO_2)_n^-$ cluster ions following 308-nm excitation. These data show that the intensity anomalies present in the negative ion mass spectra (Fig. 8) are also present in the photofragmentation distributions. Reproduced with permission from Alexander et al. [84].

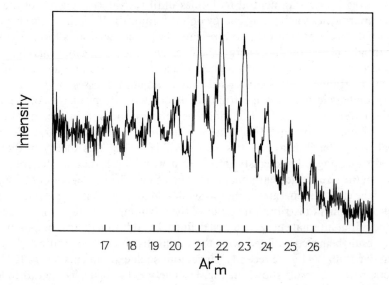

Fig. 19. Photofragmentation spectra of Ar_{48}^+ at 700 nm, where fragments in the neighborhood of $n = 20$ are created. Reproduced with permission from Levinger et al. [103].

the photofragment distribution, showing that the effect exists in a single photofragment spectrum. As in the $(CO_2)_n^-$ case, the effect is even more striking when all processes leading to a given fragment are summed. Bloomfield et al. [13] have also observed a magic number phenomenon in the photodissociation of Si_n^+ clusters, where in this case a propensity is observed to eject the Si_6^+ fragment, indicating that this ion is particularly stable.

7.4 Size Dependence of the Br_2^- Recombination Quantum Yield upon Photodissociation of Br_2^- within $Br_2^- \cdot (CO_2)_n$ Clusters: Observation of "Caging" of the Atomic Fragments

Several experiments have recently been carried out to examine the dynamics of atom recombination upon photodissociation of diatomic molecules in a cluster [111] or condensed phase [112] medium. The solution experiments [112] have been primarily concerned with direct measurement of the time scale of the recombination, whereas cluster experiments [111] have concentrated on the state distribution of the nascent diatomic via dispersed fluorescence. Cluster ion experiments afford the possibility of measuring the complementary process, where the atoms are ejected before recombination can occur. Alexander et al. [107] have carried out such experiments on the $Br_2^- \cdot (CO_2)_n$ system. The ion is grown by successive CO_2 nucleation onto the initially formed Br_2^- ion; the Br_2^- moiety is then photodissociated via a charge resonance band similar to that found for Ar_2^+. The branching ratio for the processes

$$Br_2^- \cdot (CO_2)_n + hv \rightarrow Br^- \cdot (CO_2)_m + (n-m)CO_2 + Br \quad \text{Uncaged products}$$
$$\rightarrow Br_2^- \cdot (CO_2)_p + (n-p)CO_2 \quad \text{Caged products}$$

was measured to determine the cluster size n at which the bromine atoms are effectively caged at various excitation energies. The fraction of caged photoproducts as a function of cluster size is shown in Fig. 20 for several photolysis wavelengths. Excitation at 800 and 720 nm corresponds to roughly 0.5 and 0.7-eV bromine atom recoil energies, respectively, whereas excitation at 355 nm provides 2.3-eV recoil energy. The long wavelength data show a sudden onset for caging between $n = 11$ and 12, and uncaged products are not observed from any larger cluster ions. This sharp transition is relatively insensitive to the recoil energy of the bromine atoms between 0.3 and 0.6 eV, demonstrating that it is not due to a peculiarity in the absorption spectrum around 700 nm. The onset of strong caging at $n = 12$ seems likely to be caused by a structural effect, where carbon dioxide monomers surround the bromine molecular ion. This conclusion is reinforced by the observation that the $n = 12$ cluster appears as a magic number in the $Br_2^- \cdot (CO_2)_n^-$ parent cluster ion mass spectrum. Interestingly, a magic number also occurs in the negative ion spectrum of carbon dioxide at $n = 14$. Van de Waal [113] has noted that 13 cluster CO_2 neutrals ought to form a stable geometrical structure with 12 CO_2 molecules arranged in an icosahedron

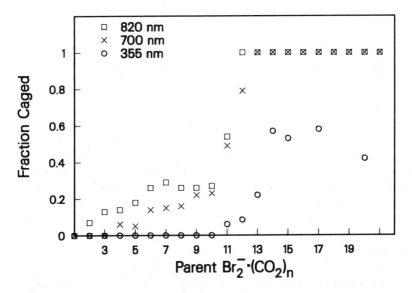

Fig. 20. Quantum yield for recombination of the bromine atoms upon photodissociation of $Br_2^- \cdot (CO_2)_n$ clusters as a function of cluster size and photolysis wavelength. Reproduced with permission from Alexander et al. [107].

around a central monomer. Recent experiments [107] by Alexander et al. at 355 nm increase the recoil energy to ~ 2.3 eV; a sharp increase in caging, to 50%, is observed near $n = 12-14$, but caging is not complete up to at least $n = 20$. The reason for the behavior of the ions is by no means clear, and it would be very useful to determine the infrared spectra of these systems to test these structural conclusions. Amar [114] has carried out calculations on the analogous $Br_2^- \cdot Ar_n$ system and finds evidence for two different types of caging processes, one dominated by attractive forces and another essentially structural in origin.

8 OUTLOOK

Pulsed methods in ion spectroscopy are making a profound impact on the field of cluster ion laser spectroscopy. The intensity and flexibility of the pulsed beam ion sources have been ideally matched by the versatility of pulsed laser sources to produce a scenario in which experimenters have greatly reduced constraints on the types of cluster ionic systems amenable to study. Consequently, the barrier between theory and experiment in this field should erode as experimentalists become able to test subtle theoretical predictions. For instance, it is now plausible to consider experiments in which a chemical reaction is triggered within a cluster in a fashion such that a laser energizes the reactants at a known geometry in a specific chemical environment. The reaction dynamics can then be studied as a unimolecular decomposition problem, and the role of a specific number and kind of solvent molecules can be investigated at a level thought to be out of the question only a decade ago. However, the challenges are commensurate with the goal. A few important unresolved issues that are currently impeding the progress of this field involve finding ways to determine the following characteristics of intermediate size cluster ions:

Intermolecular geometry and rigidity of molecular clusters
Internal energy content
Interatomic geometry of metal clusters
Electronic structure of both molecular and metallic clusters

Significant progress has been made in some of these areas, such as electronic structure where photoelectron spectroscopy appears to be a useful tool in identifying the molecular orbital or band structure of the aggregate. Other areas, such as the general determination of cluster geometry, have barely been approached and offer substantial opportunities for innovation. A host of interesting questions present themselves as targets of opportunity as we progress in defining the physical attributes of the clusters. Among these are the question of the existence of phase transitions [114] in the microscopic cluster environment and the nature of the free electron in condensed media [115]. High-resolution spectroscopy is on the verge of providing the first glimpse of structural information. Infrared spectroscopy via vibrational photofragmentation appears to be an obvious candidate, as well as the electronic transition excited through known chromophores in both the ion and ligands. Non-gas-phase techniques such as electron spin resonance (ESR) and nuclear magnetic resonance (NMR) carried out on matrix-deposited, mass-selected ion beams seem very exciting and may represent powerful avenues to establish symmetry and structure.

In the meantime, mapping out the photochemistry of these new species is certain to provide an immediate return from existing experiments. Other directions that presently appear as opportune targets include subpicosecond pump–probe experiments that record the time evolution of reactions occurring inside clusters of specific composition and high-resolution spectroscopy experiments that probe how the product mix of vibrationally induced reactions is changed by the character of the motion excited. Another area that appears on the verge of delivering a large body of information to be correlated and digested is the photoelectron spectroscopy studies of semiconductors and metals, where the evolution of the band gaps can be directly traced as a function of size and composition. In conclusion, we can look forward to the exploitation of these advances in pulsed technology to move the realm of gas-phase oriented chemical physics research further into the arena of condensed phases, with hope of it converging toward a fundamental understanding of how reactions occur in bulk media.

ACKNOWLEDGMENTS

The authors extend thanks to Professors R. E. Smalley, M. T. Bowers, and P. J. Brucat and to Drs. S. Leach and J. Peterson for providing us with original figures and enlightening discussions regarding the content of this article. Special thanks are due to Meryl Mantione whose tireless attention to detail in the preparation of this manuscript salvaged the effort. W. C. L. acknowledges support from the NSF under grants CHE 83-16628 and PHY 86-04504, and M. A. J. wishes to thank the NSF under CHE 86-02195 and the Office of Naval Research and to acknowledge the Donors of the Petroleum Research Fund, administered by the American Chemical Society, for partial support of his research. Professor Johnson's research is also supported by a grant from Research Corporation and has benefited from his designation as a Shell Faculty Fellow.

REFERENCES

1. D. M. Lindsay, Y. Wang, and T. George, *J. Chem. Phys.*, **86**, 3493 (1987).
2. M. J. Frisch, J. E. Del Bene, J. S. Binkley, and H. F. Schaefer III, *J. Chem. Phys.*, **84**, 2279 (1986).
3. J. Kalcher, P. Rosmus, and M. Quack, *Can. J. Phys.*, **62**, 1323 (1984).
4. J. N. Allison and W. A. Goddard, III, *Surf. Sci.*, **115**, 553 (1982); C. F. Melius, T. H. Upton and W. A. Goddard, III, *Solid State Commun.*, **28**, 501 (1978).
5. W. L. Jorgensen and J. K. Buckner, *J. Phys. Chem.*, **90**, 4651 (1986).
6. K. Ohta and K. Morokuma, *J. Phys. Chem.*, **91**, 401 (1987).
7. R. P. Messmer, S. H. Lamson, and D. E. Salahub, *Phys. Rev. B*, **5**, 3576 (1982).
8. C. W. Bauschlicher, *J. Chem. Phys.*, **84**, 250 (1986).

9. G. Natanson, F. G. Amar, and R. S. Berry, *J. Chem. Phys.*, **78**, 399 (1983); F. G. Amar and R. S. Berry, *J. Chem. Phys.*, **85**, 5943 (1986).
10. A. W. Castleman, Jr. and R. G. Keesee, *Chem. Rev.*, **86**, 589 (1986).
11. T. D. Märk and A. W. Castleman, Jr., *Adv. At. Mol. Phys.*, **20**, 65 (1985); A. W. Castleman, Jr., and R. G. Keesee, *Ann. Rev. Phys. Chem.*, **37**, 525 (1986).
12. A. W. Castleman, Jr., D. E. Hunton, T. G. Lindeman, and D. N. Lindsay, *Int. J. Mass Spectrom. Ion Phys.*, **47**, 199 (1983).
13. L. A. Bloomfield, R. R. Freeman, and W. L. Brown, *Phys. Rev. Lett.*, **54**, 2246 (1985).
14. M. A. Johnson, M. L. Alexander, and W. C. Lineberger, *Chem. Phys. Lett.*, **112**, 285 (1984).
15. L.-S. Zheng, C. M. Karner, P. J. Brucat, S. Yang, C. L. Pettiette, M. J. Craycraft, and R. E. Smalley, *J. Chem. Phys.*, **85**, 1681 (1986).
16. H.-S. Kim, M. F. Jarrold, and M. T. Bowers, *J. Chem. Phys.*, **84**, 4882 (1986).
17. H. Helm, P. C. Cosby, and D. L. Heustis, *J. Chem. Phys.*, **78**, 6451 (1983).
18. P. Fayet and L. Wöste, *Surf. Sci.*, **156**, 134 (1985).
19. T. F. Magnera, D. E. David, and J. Michl, *Chem. Phys. Lett.*, **123**, 327 (1986).
20. M. Okimura, L. I. Yeh, and Y. T. Lee, *J. Chem. Phys.*, **83**, 3705 (1985).
21. J. V. Coe, J. T. Snodgrass, C. B. Freidhoff, K. M. McHugh, and K. H. Bowen, *Chem. Phys. Lett.*, **124**, 274 (1986).
22. L. A. Posey, M. J. DeLuca, and M. A. Johnson, *Chem. Phys. Lett.*, **131**, 170 (1986).
23. M. H. Shen, J. Winniczek, and J. M. Farrar, *J. Phys. Chem.* **91**, 6447 (1987); S. C. Ostrander, L. Sanders, and J. C. Weisshaar, *J. Chem. Phys.*, **84**, 529 (1986).
24. S. C. O'Brien, Y. Liu, Q. Zhang, J. R. Heath, F. K. Tittel, R. F. Curl, and R. E. Smalley, *J. Chem. Phys.*, **84**, 4074 (1986); Y. Liu, Q.-L. Zhang, F. K. Tittel, R. F. Curl, and R. E. Smalley, *J. Chem. Phys., J. Chem. Phys.* **85**, 7434 (1986).
25. D. G. Leopold and W. C. Lineberger, *J. Chem. Phys.*, **85**, 51 (1986).
26. J. Moseley and J. Durup, *Ann. Rev. Phys. Chem.*, **32**, 53 (1981).
27. B. A. Huber, T. M. Miller, P. C. Cosby, H. D. Zeman, R. L. Leon, J. T. Moseley, and J. R. Peterson, *Rev. Sci. Instrum.*, **48**, 1307 (1977).
28. G. P. Smith and L. C. Lee, *J. Chem. Phys.*, **69**, 5393 (1978); G. P. Smith and L. C. Lee, *J. Chem. Phys.*, **69**, 5393 (1978); P. C. Cosby, R. A. Bennett, J. R. Peterson, and J. T. Moseley, *J. Chem. Phys.*, **63**, 1612 (1975); P. C. Cosby, J. H. Ling, J. R. Peterson, and J. T. Moseley, *J. Chem. Phys.*, **65**, 5267 (1976); G. P. Smith, L. C. Lee, P. C. Cosby, J. R. Peterson, and J. T. Moseley, *J. Chem. Phys.*, **68**, 3818 (1978); L. C. Lee and G. P. Smith, *J. Chem. Phys.*, **70**, 1727 (1979); G. P. Smith, L. C. Lee, and J. T. Moseley, *J. Chem. Phys.*, **71**, 4034 (1979).
29. H.-S. Kim and M. T. Bowers, *J. Chem. Phys.*, **85**, 2718 (1986); M. F. Jarrold, A. J. Illies, and M. T. Bowers, *J. Chem. Phys.*, **81**, 222 (1984); A. J. Illies, M. F. Jarrold, W. Wagner-Redeker, and M. T. Bowers, *J. Phys. Chem.*, **88**, 5204 (1984); H.-S. Kim, M. F. Jarrold, and M. T. Bowers, *J. Chem. Phys.*, **84**, 4882 (1986); M. F. Jarrold, L. Misev, and M. T. Bowers, *J. Chem. Phys.*, **81**, 4369 (1984); M. F. Jarrold, A. J. Illies, W. Wagner-Redeker, and M. T. Bowers, *Chem. Phys.*, **82**, 1832 (1985).
30. C. R. Moylan, J. A. Dodd, and J. I. Brauman, *Chem. Phys. Lett.*, **118**, 38 (1985).
31. L. R. Thorne and J. L. Beauchamp, in M. T. Bowers, Ed., *Gas Phase Ion Chemistry*, Academic Press, New York, 1984, vol. 3, p. 42.
32. R. C. Dunbar, in M. T. Bowers, Ed., *Gas Phase Ion Chemistry*, Academic Press, New York, 1984, vol. 3, p. 42.
33. S. C. Richtsmeier, E. K. Parks, K. Liu, L. G. Pobo, and S. J. Riley, *J. Chem. Phys.*, **82**, 3659 (1985).
34. M. E. Geusic, M. D. Morse, S. C. O'Brien, and R. E. Smalley, *J. Chem. Phys.*, **82**, 590 (1985).
35. R. L. Whetten, D. M. Cox, D. J. Trevor, and A. Kaldor, *Phys. Rev. Lett.*, **54**, 494 (1985).

36. U. Buck and H. Meyer, *J. Chem. Phys.*, **84**, 4854 (1986).
37. J. Saenz, J. M. Soler, and N. Garcia, *Chem. Phys. Lett.*, **114**, 15 (1985).
38. H. Haberland, in J. Eichler, I. V. Hertel and N. Stolterfoht, Eds., *Electronic and Atomic Collisions*, Elsevier, B. V., 1984, p. 597.
39. J. Q. Searcy and J. B. Fenn, *J. Chem. Phys.*, **61**, 5282 (1973).
40. R. J. Beuhler and L. Friedman, *J. Chem. Phys.*, **77**, 2549 (1982).
41. I. A. Harris, R. S. Kidwell, and J. A. Northby, *Phys. Rev. Lett.*, **53**, 2390 (1984).
42. P. G. Carrick and P. C. Engelking, *Chem. Phys. Lett.*, **108**, 505 (1984); A. T. Droege and P. C. Engelking, *Chem. Phys. Lett.*, **96**, 316 (1983); P. G. Carrick, S. D. Brossard, and P. C. Engelking, *J. Chem. Phys.*, **83**, 1995 (1985).
43. L. I. Schiff, *Quantum Mechanics*, McGraw Hill, New York, 1955, 3rd ed., p. 324.
44. T. A. Miller, B. R. Zegarski, T. J. Sears, and V. E. Bondybey, *J. Phys. Chem.*, **84**, 3154 (1980).
45. B. M. DeKoven, D. H. Levy, H. H. Harris, B. R. Zegarski, and T. A. Miller, *J. Chem. Phys.*, **74**, 5659 (1981).
46. A. Carrington and R. P. Tuckett, *Chem. Phys. Lett.*, **74**, 19 (1980).
47. D. Klapstein, J. P. Maier, and L. Misev, *J. Chem. Phys.*, **78**, 5393 (1983).
48. M. L. Alexander, K. R. Lykke, and W. C. Lineberger, in *Proceedings*, 1983 Conference on the Dynamics of Molecular Collisions, Gull Lake, MN (unpublished).
49. M. A. Johnson, J. Rostas, and R. N. Zare, *Chem. Phys. Lett.*, **92**, 225 (1982).
50. M. Heaven, T. A. Miller, and V. E. Bondybey, *J. Chem. Phys.*, **76**, 3831 (1982).
51. M. A. Johnson, M. L. Alexander, I. Hertel, and W. C. Lineberger, *Chem. Phys. Lett.*, **105**, 374 (1984).
52. M. A. Heald and C. B. Wharton, *Plasma Diagnostics with Microwaves*, Wiley, New York, 1965, p. 155.
53. T. Su and M. T. Bowers, in M. T. Bowers, Ed., *Gas Phase Ion Chemistry*, Academic Press, New York, 1979, vol. 3, p. 84.
54. H. Ashkenas and F. Sherman, in J. H. de Leeuw, Ed., *Rarefied Gas Dynamics*, Academic Press, New York, 1968; B. D. Kay, *Dynamics, Energetics, and Structure of Hydrogen Bonded Clusters: Elucidating the Transition from the Gaseous to the Condensed State*, Ph.D. Thesis, University of Colorado, Boulder, Col., 1982.
55. P. A. M. van Koppen, M. F. Jarrold, and M. T. Bowers, *J. Chem. Phys.*, **81**, 288 (1984).
56. T. Takagi, I. Yamada, M. Kuroni, and S. Kobiyama, in *Proceedings, 2nd International Conference on Ion Sources*, Vienna, 1972, p. 790.
57. T. Takagi, in T. Takagi, Ed., *Proceedings of the International Workshop on Ionized Cluster Beam Technique*, 1986.
58. P. J. Brucat, L.-S. Zheng, C. L. Pettiette, S. Yang, and R. E. Smalley, *J. Chem. Phys.*, **84**, 3078 (1986).
59. L. A. Bloomfield, M. E. Geusic, R. R. Freeman, and W. L. Brown, *Chem. Phys. Lett.*, **131**, 33 (1985).
60. R. E. Honig and J. R. Woolston, *Appl. Phys. Lett.*, **2**, 138 (1963).
61. J. A. Howe, *J. Chem. Phys.*, **39**, 1362 (1963).
62. J. Berkowitz and W. A. Chupka, *J. Chem. Phys.*, **40**, 2735 (1964).
63. R. E. Honig, in A. Benninghoven, R. J. Colton, D. S. Simsons, and H. W. Werner, Eds., *Secondary Ion Mass Spectrometry SIMS V*, Springer-Verlag, 1985, p. 2.
64. M. Velghe and S. Leach, *J. de Phys.*, **34**, C2–111 (1973); S. Leach and M. Velghe, *J. Quant. Spectrosc. Radiat. Transfer*, **16**, 861 (1976); A. Frad and S. Leach, *Chem. Phys. Lett.*, **12**, 599 (1972).
65a. R. B. Cody, R. C. Burnier, W. D. Reents, Jr., T. J. Carlin, D. A. McCrery, R. K. Lengel, and

B. S. Freiser, *Int. J. Mass Spectrom. Ion Phys.*, **33**, 37 (1980); B. R. Tonkin and J. C. Weisshaar *J. Am. Chem. Soc.* **108**, 7128 (1986)

66. J. B. Hopkins, P. R. R. Langridge-Smith, M. D. Morse, and R. E. Smalley, *J. Chem. Phys.*, **78**, 1627 (1983); D. E. Powers, S. G. Hansen, M. E. Geusic, D. L. Michalopoulos, and R. E. Smalley, *J. Chem. Phys.*, **78**, 2866 (1983).
67. V. E. Bondybey and J. H. English, *J. Chem. Phys.*, **76**, 2165 (1982).
68. S. C. O'Brien, J. R. Heath, H. W. Kroto, R. F. Curl, and R. E. Smalley, *Chem. Phys. Lett.*, **132**, 99 (1986).
69. H. S. W. Massey, *Negative Ions*, Cambridge University Press, London and New York, 1976, 3rd Ed.
70. D. A. Parkes, *J. Chem. Soc. (Lond.)*, **68**, 2103 (1972).
71. C. E. Klots and R. N. Compton, *J. Chem. Phys.*, **69**, 1636 (1978).
72. D. Spence and G. J. Schulz, *Phys. Rev. A*, **2**, 1802 (1970).
73. D. Spence and G. Schulz, *Phys. Rev. A*, **3**, 1968 (1971).
74. D. Spence and G. J. Schulz, *Phys. Rev. A*, **5**, 724 (1972).
75. H. Shimamori and R. W. Fessenden, *J. Chem. Phys.*, **74**, 453 (1981).
76. A. Stamatovic, K. Leiter, W. Ritter, K. Stephan, and T. D. Märk, *J. Chem. Phys.*, **83**, 2942 (1985).
77. M. Knapp, O. Echt, D. Kreisle, T. D. Märk, and E. Recknagel, *Chem. Phys. Lett.*, **126**, 225 (1986).
78. T. D. Märk, K. Leiter, W. Ritter, and A. Stamatovic, *Phys. Rev. Lett.*, **55**, 2559 (1985).
79. K. H. Bowen, G. W. Liesegang, R. A. Sanders, and D. R. Herschbach, *J. Phys. Chem.*, **87**, 557 (1983).
80. H. Haberland, H.-G. Schindler, and D. R. Worsnop, *J. Chem. Phys.*, **81**, 3742 (1984); H. Haberland, H.-G. Schindler, and D. R. Worsnop, *Ber. Bunsenges Phys. Chem.*, **88**, 270 (1984); H. Haberland and M. Winterer, *Rev. Sci. Instrum.*, **54**, 764 (1983).
81. J. V. Coe, J. T. Snodgrass, C. B. Freidhoff, K. M. McHugh, and K. H. Bowen, *J. Chem. Phys.*, **83**, 3169 (1985); J. T. Snodgrass, J. V. Coe, C. B. Freidhoff, K. M. McHugh, and K. H. Bowen, *Chem. Phys. Lett.*, **122**, 352 (1985).
82. G. Culp and A. T. Stair, *J. Chim. Phys.*, **64**, 57 (1967).
83. J. M. Warman and M. C. Sauer, Jr., *J. Chem. Phys.*, **62**, 1971 (1975).
84. M. L. Alexander, M. A. Johnson, N. E. Levinger, and W. C. Lineberger, *Phys. Rev. Lett.*, **57**, 976 (1986).
85. T. O. Tiernan and R. L. C. Wu, *Adv. Mass Spectrom.*, **7a**, 136 (1974); R. N. Compton, P. W. Reinhardt, and C. D. Cooper, *J. Chem. Phys.*, **63**, 3821 (1975); J. F. Paulson, *J. Chem. Phys.*, **52**, 963 (1970); S. Y. Tang, E. W. Rothe, and G. P. Reck, *J. Chem. Phys.*, **61**, 2592 (1974); S. V. Krishna and V. S. Venkatasubramanian, *J. Chem. Phys.*, **79**, 6423 (1984).
86. D. E. Stogryn and J. O. Hirschfelder, *J. Chem. Phys.*, **31**, 1531 (1959).
87. T. Kondow and K. Mitsuke, *J. Chem. Phys.*, **83**, 2612 (1985); T. Kondow, *J. Phys. Chem.*, **91**, 1307 (1987).
88. M. Knapp, O. Echt, D. Kreisle, and E. Recknagel, *J. Chem. Phys.*, **85**, 636 (1986).
89. M. DeLuca, L. A. Posey, and M. A. Johnson, unpublished results; $(H_2O)_n^-$ clusters were observed in the electron gun ionized pulsed jet with $n > 15$, similar to the distribution found in Ref. [88] where low-energy electrons were selectively injected into the jet.
90. L.-S. Zheng, P. J. Brucat, C. L. Pettiette, S. Yang, and R. E. Smalley, *J. Chem. Phys.*, **83**, 4273 (1985).
91. W. C. Wiley and I. H. McLaren, *Rev. Sci. Instrum.*, **26**, 1150 (1955).
92. B. A. Mamyrin, V. I. Karataev, D. V. Schmikk, and V. A. Zagulin, *Sov. Phys. JETP*, **37**, 45 (1973).

93. U. Boesl, H. J. Neusser, R. Weinkauf, and E. W. Schlag, *J. Phys. Chem.*, **86**, 4857 (1982).
94. O. Echt, P. D. Dao, S. Morgan, and A. W. Castleman, Jr., *J. Chem. Phys.*, **82**, 4076 (1985).
95. R. J. Celotta, R. A. Bennett, J. L. Hall, M. W. Siegel, and J. Levine, *Phys. Rev. A*, **6**, 631 (1972).
96. M. W. Siegel, R. J. Celotta, J. L. Hall, J. Levine, and R. A. Bennett, *Phys. Rev. A*, **26**, 607 (1972).
97. R. R. Corderman and W. C. Lineberger, *Ann. Rev. Phys. Chem.*, **30**, 347 (1979); R. D. Mead, A. E. Stevens, and W. C. Lineberger, in M. T. Bowers, Ed., *Gas Phase Ion Chemistry*, Academic Press, New York, 1984, vol. 3, pp. 214–248; P. S. Drzaic, J. Marks, and J. I. Brauman, in M. T. Bowers, Ed., *Gas Phase Ion Chemistry*, Academic Press, New York, 1984, vol. 3, pp. 167–211.
98. R. G. Keesee and A. W. Castleman, Jr., *J. Phys. Chem. Ref. Data*, **15**, 1011 (1986).
99. S. R. Long, J. T. Meek, and J. P. Reilley, *J. Chem. Phys.*, **79**, 3206 (1983); K. Kimura, *Adv. Chem. Phys.*, **60**, 161 (1985).
100. C. S. Feigerle, Ph.D. Thesis, University of Colorado, Boulder, Colo., 1982, unpublished.
101. O. Cheshnovsky, P. J. Brucat, S. Yang, C. L. Pettiette, M. J. Craycraft, and R. E. Smalley, in P. Jena, S. Khanna, and B. Rao, Eds., *NATO ASI Series*, Plenum, New York, 1987.
102. P. Kruit and F. H. Read, *J. Phys.*, **E16**, 313 (1983).
103. N. E. Levinger, D. Ray, M. L. Alexander, and W. C. Lineberger, *J. Chem. Phys.*, submitted.
104. S. H. Lin and C. Y. Ng, *J. Chem. Phys.*, **75**, 4921 (1981).
105. H. Herzberg, *Spectra of Diatomic Molecules*, Van Nostrand, New York, 1966, 2nd ed., p. 393.
106. M. L. Alexander, M. A. Johnson, and W. C. Lineberger, *J. Chem. Phys.*, **82**, 5288 (1985).
107. M. L. Alexander, N. E. Levinger, M. A. Johnson, D. Ray, and W. C. Lineberger, *J. Chem. Phys.*, accepted.
108. P. C. Engelking, *J. Chem. Phys.*, **85**, 3103 (1986).
109. M. L. Alexander, M. A. Johnson, and N. E. Levinger, *Phys. Rev. Lett.*, **57**, 976 (1986).
110. O. Echt, D. Kreisle, M. Lnapp, and E. Recknagel, *Chem. Phys. Lett.*, **108**, 401 (1984).
111. G. Kubiak, P. S. H. Fitch, L. Wharton, and D. H. Levy, *J. Chem. Phys.*, **68**, 4477 (1978); J. J. Valentini and J. B. Cross, *J. Chem. Phys.*, **77**, 572 (1982); K. L. Saenger, G. M. McClelland, and D. R. Herschbach, *J. Phys. Chem.*, **85**, 3333 (1981).
112. T. J. Chuang, G. W. Hoffman, and K. B. Eisenthal, *Chem. Phys. Lett.*, **25**, 201 (1974); D. F. Kelley, N. A. Abul-Haj, P. Bado, and K. R. Wilson, *J. Phys. Chem.*, **88**, 655 (1984); M. Berg, A. L. Harris, and C. B. Harris, *Phys. Rev. Lett.*, **54**, 951 (1985).
113. B. W. van de Waal, *J. Chem. Phys.*, **79**, 3948 (1983).
114. F. G. Amar, in P. Jena, S. Khanna, and B. Rao, Eds., *The Chemistry and Physics of small Clusters*, NATO ASI Series, Plenum, New York, 1987.
115. E. J. Hart and J. W. Boag, *J. Am. Chem. Soc.*, **84**, 4090 (1962); J. W. Boag and E. J. Hart, *Nature*, **197**, 45 (1965); J. P. Keene, *Nature*, **197**, 47 (1963); E. G. Hart and M. Anbar, *The Hydrated Electron*, Wiley, New York, 1970; J. Schnitker, P. J. Rossky, and G. A. Kenney-Wallace, *J. Chem. Phys.*, **85**, 2996 (1986).

INDEX

Abstraction reaction, 373, 378
Acetylene, 148
Acid-base reactions, 127
Activation barrier, 397, 400, 403
Activation of ions:
  by collisions, see Collision-induced dissociation
  by electrons (EIEIO), 110
  by photons, see Photodissociation
Addition, radical, to ethylene, 157
Adiabatic collision, 396
Alcohol dimers, proton-bound, 585
ω-Alkenyl phenyl ethers, 149
Alkyl bromides, 121
Alkyl cations, 153
Allene, 123
Ambipolar diffusion, 172
Ambipolar diffusion coefficient, 172
Ammonia, 126, 141, 154
  deuterated, 156
tert-Amyl cation, 149, 153, 154, 157
Analytical applications, in FTMS:
  biological samples, 71, 79, 99–100
  detection limits, 94, 98
  dynamic range, 74–75, 105, 106
  exact mass, 70–72
    calibration, 70–72
    errors in, 70–72, 100
  high molecular weight samples, 71–72, 100–103, 109
  isomer distinction, 63, 98. See Ion structures
  library search, 96
  mixtures:
    by GC, 71
    by MS/MS, 78
  nonvolatile, thermally stable, 71, 79, 99–100
  pattern recognition, 97
  quantitation, 100
  sensitivity, 66, 94, 108
Angular distribution:
  for photon emission, 503
  for reactive scattering, 330, 332, 367, 393
Angular momentum conservation, dynamical constraints, 399
Angular resolution, 362
Antibonding molecular orbital, 395, 399

Appearance energy, 418, 498, 501
Ar^+:
  charge transfer reactions:
    in kinetic energy ICR, 47, 48
    in tandem ICR, 42
    with state-selected ions, 446–451, 462, 478–480
  reaction with H_2, 435–436
  spin-orbit distribution created by ionization, 450
Argon resonance lamp, 121
Ar_n^+:
  magic numbers, 626–628
  photodestruction cross section, 624
Associative detachment reactions, 175, 176, 192
Atom abstraction, from NH_3, 196
Attenuation:
  primary beam, 327, 352, 494, 498
  probe of excited states in ion beam, 340
Autoionization, 418, 422
  electronic, 425
  rotational, 425
  vibrational, 425
Averages over initial states, final states, 327

Backward scattering, 382
Bandwidth, of pulsed CO_2 laser, 568
Beam modulation, 378
Beam-gas cell geometry, 330, 370–376, 394, 419, 442–445
Beer-Lambert law, 327
Benzyl anion, 586
Beta particle, 289
Bond energies, determination from cross section thresholds, 402. See also Thermochemistry, bond energies
Bottleneck, 379, 583
Branching ratio, 380, 393
Bridge detectors, see Capacitance bridge detectors in ICR
1-Bromobutane, 121, 128
Bromocyclohexane, 126, 139, 140, 153
Bromopentane, 148
1-Bromopentane, 121, 122, 125
$Br_2^-(CO_2)_n$:
  cage effect in, 627–629

637

$Br_2^-(CO_2)_n$ (*Continued*):
   photofragmentation of, 627–629
1,3-Butadiene, 123
Butane, 147, 148
   1,4-T_2, 300
tert-Butanol, 158
2-Butanol, 128, 158
cis-2-Butene, 144
trans-2-Butene, 144
1-Butene, 122, 144, 147, 148
2-Butene, *cis* and *trans*, 122
2-Butene, 123, 147
Butenes, deuterated, 146
tert-Butoxide ion, 584
Butyl cations, 125
Butyl phenyl ether, 124, 143, 146, 147, 148
sec-Butyl ether, 128
tert-Butyl fluoride, 159, 160
2-Butylcyclopentanone, 147

$(CD_3)_2Cl^+$, 583
$(CD_3)_3CF$, 161
$(CH_3)_2CF_2$, 160
$(CH_3)_2Cl^+$, 583
$(CH_3)_3CCH_2OHF^-$, 586
$(CH_2H_5)_3CF$, 161
Calibration, in FTMS, *see* Analytical applications, exact mass
Capacitance bridge detectors in ICR, 29–30, 30–31, 32, 39, 44, 47, 52–56
   comparison, with marginal oscillator, 55
   design, 52–54
   sensitivity, 52, 54, 55
Capacitance manometer, 139
Carbenium ions, 293, 295
Cascade, radiative, 528, 536, 544
Cavity modes:
   longitudinal, 569
   transverse (TEM*mn*), 569
CD_2=$CFCH_3$, 161
$CD_3CF_2CH_3$, 161
Center of mass system, 362
Centrifugal barrier, 392
Centroid, 334, 363
$CF_2(CH_3)_2$, 161
$CF_3COCH_2^-$, 586
CF_3O^-, 581
CF_3X^+, 583
Chain reactions, radiation-induced, 298
Charge-induced dipole interaction, 446
Charge stripping, 120
Charge transfer reactions in ICR, 334, 362, 382, 384, 386, 393, 396, 437–442, 446–451, 453–460

   dissociative, 339
   doubly-charged ions, 385
   reagent preparation by, 335, 388
   studied in ICR, 42–44, 44–45, 46, 48
Chemical kinetics, *see* Reaction kinetics
Chemical shift differences, deuterium induced, 125
Chemiluminescence, 333, 491
   infrared, 491, 504, 538–541
   population, 540–541
   ultraviolet and visible emission, 534
Chromatography:
   gas, 93–94, 98. *See also* Analytical applications, mixtures
      data handling, *see* Data manipulation, clipped representation
      detection limits, 98
      mass accuracy, 71
      resolution, 82, 94
      sensitivity, 94
   liquid, 108
CH_2=$CFCD_3$, 161
CH_3^+, 293
   association reactions, 196, 211
CH_3CBr=CH_2, 161
$CH_3CD_3Cl^+$, 583
CH_3CF=CH_2, 159, 161
CH_3CH=$CHBr$, 161
CH_3OHF^-, 582
CH_4, formation of, by charge transfer reactions in ICR, 42, 44
CH_5^+, 302
CH_n^+, from decay of CH_3T, 295
Cluster ions, 126
   anion formation kinetics, 606
   anionic formation mechanisms, 606
   formation by association, 128, 595, 598
   formation by nucleation, 596, 598
   kinetics of formation, 599, 607
   kinetics in supersonic expansion, 600
   metal, 344, 403–408
   reactions, 405–408, 627–629
   reaction size dependence, 406–408
CO^+:
   charge transfer reactions, tandem ICR, 42, 44
   collision frequency, with CO, 23
Coincidence counting, statistics, 427–428
Collision:
   collinear, 399
   ion-surface, 128
   molecular, 133
   nonionizing, 132, 146
Collisional damping, 67
Collisional quenching, 200, 306

Collisional rate coefficient, 174
Collisional relaxation, 581
Collisional stabilization, 197
Collision cell, 389. *See also* Beam–gas cell geometry
Collision complex, 378, 399, 403
    charge transfer, 382
    ion–molecule, 149, 153
    lifetime, 369
    osculation, 369
Collision frequencies in ICR:
    definition, 8, 22
    determination:
        by frequency heterodyning, 23
        by pressure broadening, 22–23
        by relaxation of capacitance bridge signal, 33
        by transient power absorption, 23
    effect, on ion motion, 7–8, 22
    Langevin, 8, 22
    and power absorption, 9–11, 17, 31
Collision-induced dissociation, *see also* MS/MS
    bond energies from, 404, 406
    comparison of FTMS to other methods, 73–76, 81–82
    in ICR spectrometry, 76
    for ion structure. *See* Ion structures in FTMS
    for ion synthesis, 84–86
    for laser desorbed ions, 79, 100
    multiple collisions in, 81–82
    resolution in, 78, 82–84
    translational ion energy, 75–76, 86–87
Collision-limited power absorption in ICR, 9–10
Colloidal graphite, use in ion optics, 360
Contour flux diagram, 367–369, 377–378, 380–381, 382–384
    direct *vs.* complex collisions, 369
    symmetry about 90 degrees, 367
Conversion dynode, 361
Coriolis coupling, 397
Correlation diagram, 390
    diabatic, 403
    electronic state, 395
    orbital, 395, 397
Cost, in FTMS, 110
CO_2 radiation, cw, 580
CO_2-TEA laser, 567
$(CO_2)_n^+$:
    formation by association, 598
    neutral bonding energetics, 624
    photodestruction cross sections, 622–625
    photofragmentation at 1.06 microns, 624
$(CO_2)_n^-$:
    formation by association in supersonic jet, 611
    magic numbers in fragments, 626–627
    neutral bonding energetics, 625
Cross section, 121
    absolute, 335, 389, 393, 445
    absorption, 574
    collision-induced, 430
    differential, 335
        state-to-state, 474–480
    electron transfer, 441
    emission, 498–501
    excitation, 498, 502
    ionization, 123
    microscopic, 419
    phenomenological, 419
    photoionization, 422
    quenching, 529
    reference reaction method, 497
    scattering, 132
    state-selected, total, 328
    state-to-state:
        definition, 329
        measurement, absolute, 460–480
        total, 420
        vibrational relaxation, 462
    total, 327, 330, 336, 351, 373, 374, 397
Crossed beam method, 327, 333–334, 376–388, 419, 451–453, 462–464
CT_4, 299
Curve crossing, 387, 395
    avoided, 403
    role of reagent vibration, 397
Cyclization, 127
Cycloalkene, 152
Cycloalkyl cation, 152
Cyclobutane, 122
Cyclobutyl cation, proof of existence, 309
Cyclohexane, multitritiated, 310
Cyclohexene, 126, 152
Cyclohexyl cation:
    isomerization, 310
    proof of existence, 310
    RRKM calculations on, 311
Cyclopentene, 152
Cyclopropane, corner-protonated, 157
Cyclopropylcarbinyl cation, 309
Cyclotron frequency, 4
    effect of ICR trapping field on, 6, 29
Cyclotron heating, 24, 29. *See also* Double resonance in ICR; Ions(s)
    kinetic energy, in ICR, 44–48
Cyclotron motion, 130, 132
    frequency, 62, 65
$C_2H_3^+$, excess internal energy, 293
C_2T_2, 299

$C_3F_6^+$, 580
$C_4H_9^+$, 122
$C_6H_5CH_2^-$, 582, 586
C_6T_6, 299

D/H exchange, 193, 194, 195
Daly detector, 361, 377, 389, 394
  negative ion operation, 361
Damping, see Collisional damping
Data analysis:
  in FTMS:
    apodization, 67–68
    clipped representation, 94
    fast Fourier transform (FFT), 68
    maximum entropy method (MEM), 68
    peak shapes, 68
    transients, 67, 72
    zero-filling, 72
  in ICR:
    in drift ICR, 18–21
    in equilibrium measurements, 25
    in kinetic energy ICR, 45–46, 47
    in tandem ICR, 40–41
    in trapped ICR, 30–31
Daughter ion, 289
  excess internal energy of polyatomics, 293
  kinetic energy, 291
Debye shielding length, 130, 597
Decarbonylation reaction, 381–382
Decay, chemical effects, 295
Decay products, detection and characterization, 296
Decay technique, see Nuclear decay technique
Decelerator, 338
  exponential field, 377
Deconvolution of fluxes in beam experiments, 332, 367
Deconvolution of intensities in ICR, 18–21, 30. See also Data analysis, in ICR
Dempster ICR spectrometer, 63
Dempster mass filter, 37, 38, 39
Deprotonation:
  of methylcyclopentyl cation, 156
  of protonated sec-butyl ether, 158
Desorption ionization, see Ionization methods, desorption
Detection, in FTMS:
  heterodyne, 68, 72
  image current, 66, 71
  limits, 98
  marginal oscillator, 63
  Nyquist criterion, 63
  pulsed, 63
  tailored excitation (SWIFT), 104–106

  theory, 64–69
Detection sensitivity, 329
Detection solid angle, 331
Detector resolution, 331
Detectors for ICR, see Ion detectors for ICR
Deuteron transfer, 389, 393
Dibutyl ether, 148, 155
Diethyl ether, 122
  proton-bound dimer, 580
Diffusion, of ions in PHPMS, 242
Dimer, proton-bound, 158
Dimethylcyclopropanes, 123
Dimethyl ether, 122
Directed ion flow tube (DRIFT), 17, 40
Direct reaction dynamics, 397
Discharge:
  electrical, 124, 147, 343–344
  microwave, 333
Discrete level regime, 572
Dissociative attachment reactions, 180, 341
Dissociative detachment, 192
Dissociative recombination, 179
Distribution:
  centroid, 334
  internal energy, 337, 339
  kinetic energy, 345
  product energy, 332, 367, 393
  product kinetic energy, 334, 379
Doppler shift, 513
Doppler width, 505, 525, 528
Double resonance in FTMS, 73–75. See also MS/MS
Double resonance in ICR:
  calculation of ion kinetic energy in:
    effect of collisions on, 17
    effect of phase on, 17
    errors in, 16–17
    at high pressure, 17
    at low pressure, 15, 31
    in tandem ICR, 41–42
  for determining reaction sequences, 15, 16
  interference, with detector, 31, 39
  ion ejection, 15–16, 516–517
    calculation of ejection time for, 15–16
    due to trapping field, see Trapping ejection in ICR
    in trapped ICR, 28, 31
    in tricyclotron spectrometer, 44
    use, with electrometer detection, 47, 56
    method of applying, 39, 64
    use, in tandem ICR ion source, 37
    use to determine photoproducts, 576
Doubly charged ions, 200
DRIFT, see Directed ion flow tube (DRIFT)

Drift ICR, 2, 5–8, 12–25
  cells for, 12–14
  collision frequencies by, 21–23
  data analysis in, 17–21, 25
  double resonance in, 15–17
  electronics for, 14–15
  equilibrium measurements by, 24–25
  hydrogen reactions in, 25
  inelastic excitation in, 25
  ion ejection in, 15–16
  ion energy in, 24
  metastable molecules in, 25
  pressure range in, 12, 14
  reaction kinetics by, 17–21
  resolution in, 12
  sensitivity, 21
  temperature-variable, 23–24
  trapping ejection in, 25
  vacuum system for, 14
Drift potential, *see* Ion cyclotron resonance cell
Drift tube, 552–556
  laser-induced fluorescence, 554–556
  ultraviolet and visible emission, 552–556
Drift velocity, ion, 553. *See also* Ion motion in ICR
Dual-cell FTMS, 109–110. *See also* Instrumentation for FTMS, dual-cell FTMS
Dye laser, 493, 502–506, 543–544, 547, 548
Dynamic range, 74–75, 105

EBFlow, *See* Electron bombardment flow reactor
Ejection in FTMS, *see* Double resonance in FTMS
Ejection in ICR, *see* Double resonance in ICR; Trapping ejection in ICR
Elastic scattering, 387
Electrical conditions, in PHPMS ion source, 242
Electric fields in ICR, 4, 5, 28. *See also* Ion motion in ICR
Electrometer detectors in ICR:
  in drift ICR, 56
  equipment for, 56
  in kinetic energy ICR, 46
  sensitivity, 56
  in trapped ICR, 30, 56
Electron(s), trapping ejection, 25
Electron affinity, 178
  negative, 342
Electron attachment, 179, 607
Electron attachment coefficients, 180
Electron beam, 120
Electron beam current, 156
Electron bombardment flow reactor, 119–164
Electron gun, 134
  in PHPMS, 232
Electronic states:
  excited, production in ionization, 333
  metastable, 342
  selection, 398
  specificity of reactions, 398–399
Electron-ion recombination, 179
Electron jump mechanism, 382, 384
Electron multiplier:
  continuous dynode, 361
  discrete dynode, 361
  dynamic range, 361
Electron shaking, 289
Electron transfer detector, 420. *See also* Charge transfer
Electron transfer reaction, 420. *See also* Charge transfer
Emitter, thermionic, 342
Encounter complex, 378
Endothermic reactions, 208
Endothermic trapping, 197
Energy:
  disposal, in elementary reactions, 326
  laboratory kinetic, 332
  primary ion beam, 350
  relative kinetic, 331, 334
  total available, 364
Energy analyzer:
  cylindrical, 357–358
  deflection, 357
  pass energy, 359
  resolution, 359
  retarding field, 356
  spherical, 358
  steradiancy, 421
  threshold photoelectron, 421
  toroidal, 357
  virtual slits, 359
  zoom lens, 359, 370
Energy resolution, 362
Energy resonance, in charge transfer, 388
Energy transfer:
  collisional, 385–387, 528, 542, 548, 554–556
  intermolecular, V-E, 587
  intramolecular, V-E, 572
  intramolecular, V-V, 572
  vibrational, 386
Entropies by ICR, 24
Entropy bottleneck, 379, 583
Equilibria:
  ion-molecule:
    association, 275
    clustering, 275
    examples, 222

Equilibria (*Continued*):
  ion-molecular (*Continued*):
    instrumental artifacts, 279
    by PHPMS, 265
Equilibrium constant determination, 24–25, 33, 271
Ethane, 148
  1,2-T_2, 300
Ethyl bromide, multitritiated, 301
Ethyl(i-propyl)ammonium ion, 586
Ethylcyclopropane, 122
Ethylene, 148, 157
Evaporative attachment, 607
Exchange reaction, 373, 378
Excitation:
  function, 502, 519
  inelastic, by ICR, 25
  internal, 336, 338
  of ions in ICR, *see* Double resonance in ICR; Ion(s), kinetic energy, in ICR; Ion motion in ICR, power absorption by methods, 87–88, 106. *See also* Collision-induced dissociation; Detection, in FTMS; Double resonance in FTMS
Excited electronic state reactivity, 382, 398–400, 403, 437–438
External source, *see* Instrumentation for FTMS, external source

FA, *see* Flowing afterglow
FALP, *see* Flowing afterglow/Langmuir probe
FALP apparatus, 170
FALP technique, 179, 181
False coincidence, 427
Fano-Beutler profile, 425
Filaments, LaB$_6$-coated rhenium, 143
Filaments for ICR, 14, 23
Filter gas technique, 199, 200
Filter lens, 356
Filters, light, 502, 505, 533, 539–540
Flow dynamics, 171
Flowing afterglow, 17, 33, 120, 165, 168, 181, 529–552
  infrared chemiluminescence, 538, 541
  laser-induced fluorescence, 542–552
  supersonic nozzle, 549–550
  ultraviolet and visible emission, 529–538
Flowing afterglow/Langmuir probe, 167, 170, 179, 181
Fluence, 567
  saturation, 582
Fluorocarbon, 161
3-Fluoro-2-pentene, 161
Flux conservation, intensity transformation, 366

Forced oscillator, 387
Fourier transform ion cyclotron resonance spectrometry (FTICR), 3, 26, 30, 64
Fourier transform mass spectrometry (FTMS), 61–118
  advantages, 64
Franck-Condon factor, 339, 387, 423
Franck-Condon principle, 422
Franck-Condon transition, 339
Fraser's method, 497
Free energies by ICR, 24–25
Frequency sweeping in ICR, 50–51. *See also* Ion detectors for ICR, scanning
Frequency synthesizer, 73
Fringing fields, effect, on ICR double resonance, 8, 16–17
FTICR, *see* Fourier transform ion cyclotron resonance spectrometry
FTMS, *see* Fourier Transform Mass Spectrometry
Furan, reaction with ^{3}HeT$^+$, 315

Gamma-Radiolysis, 120
Gas chromatography, 124. *See also* Chromotography, gas
Gas handling, in PHPMS, 240
GC-MS, 124, 153, 159
GLPC, 124, 143, 153, 154, 158, 159
Guided beam technique, 336, 393–408
  potentials, 336, 350

Halide transfer, 123
Harmonic generation in UV production, 420
^{3}HeX$^+$, 295
He$^+$, in kinetic energy ICR, 47, 48
Heats of reaction by ICR, 24–25
Helium, as buffer gas in tandem ICR, 40
Helium-3, 289
Heteroaromatic rings, gas phase protonation, 315
Heterodyne ICR, for collision frequency determination, 23
Heterodyne mode, *see* Detection, in FTMS, heterodyne
Hexadienes, 152
5-Hexenyl phenyl ether, 152
H$_n$CO$^+$, n = 0–3, 189
HPMS (high pressure mass spectrometry), 222–283
H$_n$S$^+$, n = 0–3, 189
Hydrocarbon ions, 189
Hydrocarbons, 121, 126, 143, 144
Hydrogen, reactivity of atomic and ionic, in ICR, 25

Hydrogen atom transfer, 388, 390
1,2-Hydrogen shifts, 127
Hydronium ions, hydrated, 194
H_2^+:
  charge transfer reactions of vibrationally state selected ions, 439–442, 453–457
  chemical reaction with H_2 with state-selected ions, 432–435, 445–446
  state-to-state charge transfer reactions, 470–473
  H_3^+ ion, 180, 433–434, 441–442, 445–447

ICR, see Ion cyclotron resonance spectrometry
"Ideal" beam experiment, 329
Image current, see Detection, image current
Inelastic excitation by electrons in ICR, 25
Inelastic scattering, 362, 382, 388
  probe of collision complex, 380
  vibrational, 387
Infrared detectors:
  calorimetric:
    power probe, 571
    pyroelectric, 571
    thermopile, 571
  photoelectric:
    photodiode, 571
    photon drag, 571
Infrared emission, from nascent product molecules, 176. See also Chemiluminescence, infrared
Infrared laser photolysis, 563
Infrared laser sources, 571
Infrared multiple-photon dissociation (IRMPD), 564, 581
Inorganic negative ions, 177
Insertion complex, 403
Insertion mechanism, 395, 399, 402
Instrumentation for FTMS:
  dual-cell FTMS, 109–110
  external source, 98, 108–110
  gas chromatography, see Chromatography, gas
  liquid chromatography, 109
  quadrupole, FTMS, 108–109
  supersonic metal cluster source, 109
Intensities in ICR, see Ion cyclotron resonance spectrometry, line shape in; Resonant intensity in ICR
Internal energy of ions, in ICR, 42–44, 44–45
Interstellar chemistry, 190
Interstellar clouds, ion chemistry, 183
Interstellar molecules, isotope fractionation in, 193
Intramolecular energy transfer, 380

Ion(s):
  cyclotron frequency, see Ion motion in ICR
  drift velocity, in ICR, 4–8
  internal energy, tandem ICR study, 42–44
  kinetic energy, in ICR
    in double resonance, 15, 16–17, 31, 41–42
    in drift ICR, 24
    in kinetic energy ICR, 44–48
    relaxation, due to collisions, 24, 40
    relaxation, due to image current heating, 24
    in tandem ICR, 40, 41–42
  odd-electron, 123
  reactions, in recognizable series, 189
  structural isomers, 203
  thermalization, in ICR, 33, 40
  trapping ejection, see Trapping ejection in ICR
Ion acceleration and focusing, in PHPMS, 237
Ion beam method, 325–416, 417–487, 492–514
  collision cell, 330, 370–376, 394, 419, 442–445, 493
  infrared chemiluminescence, 504
  laser-induced fluorescence, 504–506, 512–514
  ultraviolet and visible emission, 529–538
Ion capture radius, 130, 134
Ion collisions in ICR, see Collision frequencies in ICR; Reaction kinetics
Ion current in ICR:
  in double resonance, 16
  in drift cell, 13
  in electrometer detection, 56
  in tandem ICR, 34–37, 39–40
Ion cyclotron resonance cell:
  configuration for laser photolysis, 576
  construction, 12–14, 27–28, 37, 39, 46, 47
  cubic, 28
  drift, 12–14, 514–515
  drift potential in, 5, 13. See also Electric fields in ICR
  drift velocity in, 6. See also Ion motion in ICR
  electronics for, 14–15, 28–30.
  electron space charge in, 7
  filaments in, 14, 23
  flat, 5, 7–8, 12–13
    excitation field in, 16
  kinetic energy, 46–47
  materials for, 12, 27, 37, 39
  square, 5, 12–13
  tandem, 34–39
  as tandem ICR source, 37

Ion cyclotron resonance cell (*Continued*):
    temperature-controlled, 23–24, 34
    trapped, 5, 26–28
        cubic, 63–65
        cylindrical, 28, 65
        hyperbolic, 31–32, 65, 71
        storage capacity, 32–33
        store/drift, 29
        trapped/drift, 46–47
    trapping ejection from, *see* Trapping ejection in ICR
    trapping potential in, 4–5. *See also* Electric fields in ICR
Ion cyclotron resonance spectrometry, 1–56, 120, 514–524, 576–579
    charge transfer reactions in, *see* Charge-transfer reactions in ICR
    data analysis in, *see* Data analysis in ICR
    and decay-induced fragmentation, 294
    detectors for, *see* Ion detection in ICR; Ion detectors for ICR; Sensitivity in ICR
    double resonance, *see* Double resonance in ICR
    drift, *see* Drift ICR
    ejection in, *see* Double resonance in ICR; Trapping ejection in ICR
    Fourier transform, *see* Fourier transform ion cyclotron resonance spectrometry (FTICR)
    history, 2
    and infrared photolysis, 575
    intensities in, *see* Ion cyclotron resonance spectrometry, line shape in; Resonant intensity in ICR
    ion detection in, *see* Ion detection in ICR
    ion energy in, *see* Ion(s), kinetic energy
    ion motion in, *see* Ion motion in ICR
    ion-neutral collision frequencies in, *see* Collision frequencies in ICR
    ion optics in, 37–39
    kinetic energy, *see* Kinetic energy ICR; Ion(s), kinetic energy, in ICR
    laser-induced fluorescence in, 519–524
    line shape in, 10, 11. *See also* Resolution in ICR; Resonant intensity in ICR
        collisional broadening, 22–23
    magnet for, *see* Magnet(s) for ICR
    modulation and pulsing techniques, 14–15, 29, 38, 39, 46–47
    power absorption in, *see* Ion motion in ICR
    pressure measurement in, 14, 34
    reaction kinetics by, *see* Reaction kinetics
    resolution in, *see* Resolution in ICR
    reviews, 2
    tandem, *see* Tandem ICR
    temperature-variable, 24–25
    thermodynamics measurements by, 24–25
    transient, *see* Transient ICR
    trapped, *see* Trapped ICR
    ultraviolet and visible emission in, 514–519
Ion detection in ICR, *see* Data analysis in ICR; Ion detectors for ICR
    in kinetic energy ICR, 46–47
    in tandem ICR, 39
    in trapped ICR, 29–30
    in tricyclotron spectrometer, 44
Ion detectors for ICR, 48–56
    calibration, 49–51
    capacitance bridge, *see* Capacitance bridge detectors in ICR
    electrometer, *see* Electrometer detectors in ICR
    marginal oscillator, *see* Marginal oscillator detectors in ICR
    Q-meter, *see* Q-meter detector for ICR
    "rapid scan", 30
    relative sensitivity, 55, 56
    scanning, 49, 52, 56
Ion ejection in ICR, *see* Double resonance in ICR; Trapping ejection in ICR
Ionic species, effect of solvation on, 306
Ion–ion neutralization, 179
Ion–ion recombination coefficients, 181
Ionization energy, 418
Ionization methods, 337–347
    chemical (CI), 341
        advantages, 96–98
        metal ions for, 97
        pattern recognition, *see* Analytical applications, in FTMS, pattern recognition
        reagentless, 97
        resolution in, *see* Resolution, in chemical ionization
        self-CI, 97
    desorption:
        cesium ion, 72, 79, 101, 110
        laser, 72–79, 98–100, 101, 109
        plasma, 103
        secondary ion mass spectrometry (SIMS), 101, 109. *See also* Desorption, cesium ion
    direct, 423
    electron impact, 96, 109, 124, 338–341, 343, 608
    multiphoton, 94–98, 373–374, 420
    negative surface, 342
Ion–molecule association, 125

INDEX 645

Ion–molecule collisions, in dense gases, 319
Ion–molecule complex, 142
Ion–molecule equilibria, by PHPMS, 265
Ion–molecule reactions, in FTMS, 63–64
  acid–base, 63
  anions, 86
  cations, 84
  clusters, 88–89, 98, 109
  endothermic, 88
  kinetics, 63–64
  metal ions, 79–86, 88–89, 97–98
  organic, 80–84
  organometallic, 63, 79–80, 84–86, 97–98
Ion motion in ICR, 3–11, 62–66
  cyclotron frequency, 4–7, 8
    in double resonance, 15
    perturbations, 6, 28, 29, 31–32
  drift motion, 4
    effect of collisions on, 7
  equations of motion, 3–11
  power absorption by, 8–9. See also Resonant intensity in ICR
    collision-limited, 9
    effect of phase on, 17
    frequency dependence of, 9. See also Ion cyclotron resonance spectrometry, line shape in
    general equations for, 8–9
    in transient ICR, 23
    zero-collision, 9–10
  in static fields, 3–8
  in trapping field, 11. See also Trapping
  ejection in ICR
Ion–neutral collisions, see Collision frequencies in ICR
Ion–neutral reactions, see Reaction kinetics
Ionogenic process, 288
  time scale of events in, 290
Ion optics, 346
  astigmatism, 347
  decelerator, 350
  in tandem ICR, 38–39
Ionospheric reactions, 174
Ion-pair formation, 418
Ion plasma conditions, in PHPMS, 242
Ion production, see Ionization
Ion source:
  cluster, 594–606
  continuous, 593
  discharge, 343–344
  drift cell, 394, 399
  in ICR, electron space charge in, 7
  in-line electron impact, 340–341
  laser vaporization, 345. See also Laser vaporization
  photoionization, 344, 421–422
  in PHPMS, 235
  pulsed cluster ion, 594–606
  refractory species, 603
  "soft", 341
  surface, 341–342
  thermal, 341–342
  van der Waals, 595
Ion spectroscopy, continuous beams, 593
Ion structures in FTMS:
  by collision-induced dissociation, 79–83
  by ion–molecule reactions, 63
  by photodissociation, 89–93. See also Photodissociation
  organic ions, 80–86
  organometallic ions, 79–86
Ion transmission, in quadrupole PHPMS, 239
Ion trap, 514–529
Ion trapping efficiency, see Trapped ICR; Trapping, efficiency, in ICR
IRMP dissociation, 564, 580
  theory, 572
IRMP reaction dynamics, unimolecular and bimolecular, 584
IRMPD wavelength dependence, 581, 583
IRMPD, see IRMP dissociation
Isoamyl bromide, 153, 154
Isobutane, 148
Isobutene, 122, 123, 148
Isomeric selectivity:
  in $C_7H_7^-$ photolysis, 584
  in photochemical studies, 583
Isomerization barrier, 379–381
Isomers:
  HCO^+ and COH^+, 203
  linear and cyclic, $C_3H_3^+$, 205
Isotope effect:
  intramolecular, 339, 397, 585
Isotope exchange, 177, 193, 195
Isotope scrambling, minimization, 302
Isotopic label, 125, 128, 196
Isotopic refrigeration, 197
Isotopic selectivity, in photochemical studies, 583

Jacobian, 366

Kinematic analysis, 362–369
  deconvolution, 367
  intensity transformation, 365
  limiting cases, 367

646 INDEX

Kinematic analysis (*Continued*):
  velocity transformation, 363
Kinetic energy ICR, 2, 3, 44–48
  cells for, 46–47
  data analysis in, 45–46, 47
  detectors for, 46–47
  trapping in, 45–46
Kinetic integral plot, in PHPMS, 261
Kinetics, *see* Reaction kinetics
Kinetic treatment, complex reactions, in PHPMS, 258
Knudsen number, 354
Kr^+, charge transfer reactions of, in tandem ICR, 42–44

Lab - center of mass (c. m.) transformation, 362
Langevin collision frequency, 7–8, 22
Langevin-Gioumousis-Stevenson rate, 133, 174, 378, 392, 395, 434
Langevin potential, 7, 16
Langmuir probe, 179
Langmuir-Kingdon equation, 342
Lanthanum hexaboride (LaB_6), 134
Laser:
  CO_2, description, 564
  cw, 568
  pulsed, 567
Laser-induced fluorescence, 177, 489, 505–514, 519–529, 542–556
Laser isotope separation, 564, 583
Laser optics:
  alignment, 570
  description, 570
  external, 569
  internal, 569
Laser parameters, definition and measurement, 570
Lasers, in FTMS:
  desorption, *see* Ionization methods, desorption, laser
  multiphoton ionization, *see* Ionization methods, multiphoton
  photodissociation, *see* Photodissociation
  relaxation studies, radiative and collisional, of vibrationally excited ions, 92–93
Laser vaporization, 344–346, 405
  anionic clusters from, 611
  clusters from, 603–605
  cw ion source, 345
  jet cooling, 605
  metal anion clusters, 611
  pulsed ion source, 345
  refractory materials, 611

LGS, *see* Langevin-Gioumousis-Stevenson model
LIF, *see* Laser-induced fluorescence
Ligand switching reactions, 175
Light source, vacuum ultraviolet, 420
Linear rate law, 574, 582
Line shape in ICR, *see* Ion cyclotron resonance spectrometry, line shape in; Resolution in ICR; Resonant intensity in ICR
LIS, *see* Laser isotope separation
Log reactant *vs.* fluence (LRF), 574
Long-range force, 335
Lorentzian line shape in ICR, 10, 11, 22–23

Macor, in ICR cells, 12, 14, 27, 39
Magic numbers:
  in cluster ion mass spectra, 626–628
  formation mechanism, 626–628
Magnetic field:
  homogeneity, in ICR, 31
  ion motion in, 3–8
Magnet(s) for ICR, 13, 28
  field homogeneity, 31
  field modulation, 15
  superconducting, 27, 28
Magnetic sector, resolution, 346
Magnetron motion, 130
  in trapped ICR, 28
Marginal oscillator detectors in ICR, 29, 30, 31, 32, 39, 48–52, 63
  calibration, 51–52
  comparison, with capacitance bridge, 55
  design, 48–49
  frequency sweeping in, 50–51
  pulsed, 50
  sensitivity, 50, 55
Mass accuracy, *see* Analytical applications, in FTMS, exact mass
Mass analysis, in PHPMS, 238
Mass defect, 70
Mass spectrometer:
  Fourier transform, 345
  magnetic, 346
  photoionization, 451–453
  pulsed electron high pressure, 221
  quadrupole, 140, 349, 432
  tandem photoionization, 442–443
  time-of-flight, 427, 580
  Wien filter, 38–40, 347–348, 388
Mass spectrometry, 127, 128, 346–350, 417–487
  by ICR, 33, 37
  photoionization, 417–487
Master equation, 572

Maximum entropy method (MEM), see Data analysis in FTMS, maximum entropy method
Mean free path, 121, 129, 133
Merged-beam technique, 334–336, 388–393
Metal ion chemistry, gas phase, 382, 398–408
Metastable ions, 494, 498, 512, 517
Metastable molecules, reactions of, in ICR, 25
Metastable states, 342, 382, 398–400, 403, 437–438
Methane, tritiated, GC analysis of, 303
Methylallene, 146
Methyl bromide, multitritiated, 301
2-Methyl-1-butene, 154
3-Methyl-1-butene, 157
2-Methyl-2-butene, 123, 154
2-Methylbutene, 149, 154, 157
Methylcyclobutane, 122
1-Methylcyclopentene, 152, 154, 156
Methylcyclopropane, 122, 146, 148
Methylenecyclopentane, 154, 156
N-Methylpyrrole, reaction with ^{3}HeT$^+$, 315
Microchannel plate, angular distributions from, 388
Microscopic reversibility, 467
MOde-selective chemistry, 564
Modulation techniques in ICR, see Ion cyclotron resonance spectrometry, modulation and pulsing techniques
Molecular beam, supersonic, 451–453
Momentum analysis, see Mass spectrometer, magnetic
Momentum transfer, 291. See also Collision frequencies in ICR
Monitor gas/monitor ion technique, 199
Monitor gas technique, 199
Monochromator, light, 493–494
MPI, see Multiphoton ionization
MS/MS, 72–89, 104, 106. See also Collision-induced dissociation
Multichannel analyzer, 494
Multiphoton ionization, 373–374, 420
reagent preparation, 371
Multiple labeling, 296
Multitritiated compounds:
dilution, 301
purification and characterization, 302
synthesis, 299

N atoms, reactions, 191
N$^+$, reaction, with O$_2$, 40
Narrow band excitation, see Detection, in FTMS, heterodyne
Ne$^+$, in kinetic energy ICR, 47, 48

Ne, as buffer gas in tandem ICR, 40
Negative ions, 341
cluster formation, 608–610
reactions, 376, 378
Neopentane, 147
Neopentyl phenyl ether, 149
Neutral source:
choice, 305
effusive, 352
gas cell, 352
speed distribution, 353
supersonic, 345, 353–355, 451–453, 549–552
Neutralization of ions, at surfaces, 123
Newton diagram, 363
Newtonian lens equation, 346
NH$_3^+$, formation, by charge transfer reactions in ICR, 42, 47, 48
NH$_n^+$, n = 0-4, 189
NMR, 124, 141
fluorine, 128, 159, 161
Fourier-transform, 159
NO$^+$, vibrationally excited, quenching, 202
Normalization to total ion intensity, in PHPMS, 255
Nozzle beam, supersonic expansion, 345, 353–355, 451–453, 549–552
laser-induced fluorescence, 550
ultraviolet and visible emission, 549–550
Nuclear decay technique, 287, 296
Nucleogenic ions, decay, 289
Nucleophilic displacement, 381
N$_2^+$:
charge transfer reactions of, in tandem ICR, 42
charge transfer reactions of vibrationally state-selected ions, 437–439, 457–460
collision frequency of, with N$_2$, 23
lifetime of A-state, 439
state-to-state charge transfer reactions, 464–470
N$_2$:
excitation of vibronic states, 25
metastable, ionizing reactions, 25

O atoms, reactions, 191
Octupole ion guide:
effective electrostatic potential, 442. See also Guided beam technique
electrostatic, 336, 350–351, 442–443
Omegatron, 2, 56, 62
Optical detection, 494–500
boxcar integrator, 506, 510, 543–544
dc amplifier, 510, 533, 549

Optical detection (*Continued*):
  photon counting, 494, 505–514, 521–522, 526–527, 533, 544, 550, 553
  sensitivity, 495
Optical–optical double resonance, 546–547
Organometallic chemistry, 374
Oxidative addition, to C-C, C-H bonds, 375
O_2^+:
  charge transfer reactions, in tandem ICR, 42
  in kinetic energy ICR, 47, 48
  reaction with CH_4, 209
  vibrationally excited, quenching of, 202
$(O_2)_n^-$:
  kinetics of formation in ionized jet, 607
  photoelectron spectroscopy, 620–621

Particle detection, 356, 360–361
1-Pentene, 157
2-Pentene, *cis* and *trans*, 123, 157
4-Pentenyl phenyl ether, 152
Pentyl ether, 156
Phenylium ion:
  automerization, 311
  degenerate rearrangement, 314
  singlet *vs.* triplet, 311
Photochemical branching, 575
Photodetector, 505–512
  infrared, 539–540
  photodiode, 506, 527
  photomultiplier, 507–511
Photodissociation, 63, 89–93
  bond energies, 90–91
  clusters, 624–631
  comparison to collision-induced dissociation, 93
  criteria for observing, 89, 91
  information in, 91–92
  ion structure, 91–92
  multiphoton process, 92–93
Photoelectron spectroscopy, 418
  pulsed magnetic mirror, 622
  pulsed TOF method, 618–620
Photoionization, 120, 344, 351, 417–481
  crossed ion–neutral beam apparatus, 451–453
  dissociative, 418
  selection rules, 429
  state-selection by, 426–430
  triple-quadrupole double-octupole, 474–480
Photoionization efficiency curve, 424
Photoionization mass spectrometry, *see* Mass Spectrometry, photoionization
Photoion–photoelectron coincidence, 426–429
  apparatus, 421
  kinetic energy release in, 426

  method, 421
Photolysis, 121, 563–589
Photon flux, 567
PHPMS, *see* Pulsed electron high pressure mass spectrometry
PIE, *see* Photoionization efficiency curve
PIMS, *see* Mass spectrometry, photoionization
PIPECO, *see* Photoion–photoelectron coincidence
Plasma, 129, 134, 141
  Debye wavelength, 130, 597
  inductively coupled, 124
  neutral, 596
  sheath, 344
  shielding from external fields, 596–597
  in sputtering, 344
Poisson distribution, 387
Polarization, light, 500, 503–504
Population analysis, internal state distribution
  infrared chemiluminescence, 541
  laser induced fluorescence, 513, 527, 543–545, 550, 552
  ultraviolet and visible emission, 495–500, 536
Positive column, 344
Positive ion–electron recombination coefficients, 179
Potential:
  effective electrostatic, 351
  ionization, 342
  quadrupolar, 4–5
Potential energy surface, 326
  diabatic, 446
  double minimum, 376, 378
  long range, 390
  potential wells, 327
Power absorption in ICR, *see* Ion motion in ICR
Power broadening, 572
Predissociation, vibrational, 203, 339
Pressure saturation, 213
Products from decay ions, analysis, 307
Propane, 148
  1,2-T_2, 300
Propene, 123, 148
2-Propenyl cation, 160
Propyl ether, 157
n-Propyl radicals, 157
Proton affinity, of O and N atoms, 192
Proton transfer reaction, 175, 177, 381, 391
  from NH_3^+, 196
Pulsed electron circuits, in PHPMS, 241
Pulsed electron high pressure mass spectrometer (PHPMS), 221–286. *See also* HPMS
Pulsed valve:
  for chemical ionization, 82, 97

for collision-induced dissociation, 82
for gas chromatography, 82, 94
for ion-molecule reactions, 82
Pulse sequence, for MS/MS, 73–74
Pulsing and modulation techniques in beam experiments:
  detection technique, 511, 517–518, 521–522, 526
  electron beam, 517, 526
  ion modulation, 533–534, 540, 553, 555
Pumping systems in PHPMS, 231
Pyrrole, reaction with $^3HeT^+$, 315

Q (quality factor), 49–50, 51
Q-meter detector for ICR, 55
Q-spoiler standard signal for ICR, 51
Quadrupole:
  doublet, 347
  electrostatic lens, 351
  mass analyzer, in PHPMS, 239. See also Mass spectrometer, quadruple
Quadrupole FTMS, see Instrumentation for FTMS, quadrupole-FTMS
Quasi-Equilibrium Theory (QET), 339
Quasicontinuum, 572
  definition, 572
Quenching:
  chemical, 339
  collisional, 340, 351
  in ion trap, 406
Quench pulse, in trapped ICR, 28–29
QUISTOR, 580

Radiative association rate coefficients, 198
Radiative lifetime, 498–500, 523, 527, 546
Radiative relaxation, 581
Radicals, 120, 355, 493, 533, 544
  n-Butyl, 122, 127
  1-pentyl, 123
  2-pentyl, 123
Radio-chromatographic techniques, 299
  GC, 299, 308
  GLC, 299
  HPLC, 299, 308
Radio-frequency excitation, 65–72, 103–106
Radiolytic artifacts, prevention, 297
Radiolytic ions, 296
Radiolytic products, 297
Rapid-scan detector for ICR, 30
Rate constant, see also Reaction kinetics
  determination by PHPMS, 254
  diffusion, 522, 545
  quenching, 523, 546
  reference reaction, 536–637

thermal, 329
Rate of product formation, 329
Reaction dynamics, 326, 370–408, 418, 430–481
Reaction energy by ICR, see Kinetic energy ICR
Reaction kinetics, determination:
  by double resonance ICR, 16–17
  by drift ICR, 17–21, 25
  pressure sweeping in, 21
  by tandem ICR, 40, 41, 42–44
  by trapped ICR, 33
Reaction sequences, determination, see Double resonance in ICR
Reaction thermodynamics by ICR, equilibrium measurements in, 24–25, 33. See also Kinetic energy ICR
Rearrangement, 127
  of carbon skeleton, 123
  cationic, 153
  of cyclohexyl cation, 152
  gamma-hydrogen, 147
Rebound reaction, 378
Recoil energy, 291
Recoil speed, center of mass, 364
Recombination energy, 387
Reflectron, 616
Relaxation processes:
  collisional, 581
  of equilibria, 266
  radiative, 581
REMPI, see Resonance enhanced multiphoton ionization
Resolution in FTMS, 66–67, 69–70
  in collision-induced dissociation, 78, 82, 108
  digital, 72
  external source, 109
  factors affecting, 66, 67
  heterodyne mode, 68, 72
  high, 67, 69–70
  maximum entropy method (MEM), 68
Resolution in ICR:
  in drift ICR, 12
  SWIFT excitation, 105–106
  in tandem ICR, 39
  theoretical, 11
  in trapped ICR, 31–32
Resonance enhanced multiphoton ionization (REMPI), 373–374, 420, 429–430
Resonance lamp, rare gas, 426
Resonances, scattering:
  in $Ar^+ + N_2$, 384
  in $F + H_2$, 384
Resonance shifts, in photolysis experiments, 578

INDEX

Resonant intensity in ICR:
  in equilibrium measurements, 25
  of nonreactive ions, 10
  of reactive ions, 10–11, 18–20, 21
  in trapped ICR, 30–31
Retention times in trapped ICR, see Trapping, efficiency
Rhenium, 136
  in ICR filaments, 14
Rotational hole filling, 582
RRKM calculations, 153, 157
RRKM theory, 573
Rydberg state, as precursor to ion, 373, 424

Scanning detectors, see Ion detectors for ICR
Scattering angle:
  c.m., 363, 364
  lab, 363
Scintillation detector, see Daly detector
Secondary electron emission, 361, 386
Selected ion flow drift tube (SIFDT), 205, 208
  Birmingham apparatus, 183
  variable-temperature (VT-SIFDT), 209
Selected ion flow tube (SIFT), 33, 40, 165, 183
  double, 186
  injection system, 185
  variable temperature (VT-SIFT), 183
Sensitivity in ICR, see Capacitance bridge detectors in ICR; Electrometer detectors in ICR; Ion detectors for ICR; Marginal oscillator detectors in ICR; Q-meter detector for ICR
  calibration, 51–52
  in drift ICR, 21
  in tandem ICR, 39, 40
  in trapped ICR, 32–33
Shaking effect, 291
Short range forces, 327
SIFDT, see Selected ion flow drift tube
SIFT, see Selected ion flow tube
Signal generation, see also Detection, in FTMS
  sources of signal and noise, 64–65, 67–68
Signal levels, in beam experiments, 327
Signal to noise ratio, 494, 506, 511, 540, 543, 552
Single-collision conditions, 326
Skimmer, 353
Slit function, 496
S_N^2 reaction, 388
Space charge limit, 131, 140, 337
  beam spreading, 337
  effect in FTMS, 71
Specific activity, choice, 304
Spectator Stripping mechanism, 370, 373, 393

Spin-orbit coupling, 390
Spin-orbit state, 382, 396, 435–436, 446–451, 462, 478–480
Sputtering, 405
  cluster ions from, 345
  ion-impact, 344
  negative ions, 344
Statistical models:
  cluster fragmentation, 625–626
  in unimolecular decay, 379
Stereochemistry, 158
Stereoisomers, 125
Stern-Volmer plot, 523, 546
Store/drift cell, in trapped ICR, 29
Straight line trajectory, 387
Structural isomers, 2–5
Superacids, 180
Supersonic expansion:
  cooling of internal degrees of freedom, 345, 353–354
  intensity gain over effusion, 354–355
  Mach number, 353
  neutral beam production from, 353
  pulsed valves, 82, 94, 97, 355
  seeded beam, 355
  speed distribution, 353
Surface charge, 360
Surface hopping, nonadiabatic, 446
Surface ionization, electronic states produced, 398
Surface studies, in FTMS, 100
Synchrotron radiation, 419

Tandem ICR, 2, 3, 33–44, 62
  cell for, 34, 39
  charge transfer reactions in, 42–44
  data analysis in, 40–41
  ion detection in, 39, 40–41
  ion kinetic energy in, 41–42
  ion optics for, 37–39
  ion sources for, 34–37
  pressure range in, 34–37, 40
  product distributions by, 40
  reaction kinetics in, 40, 41, 42–44
  resolution in, 39
  sensitivity, 39, 40
  temperature-variable, 34
  tricyclotron, 44
  vacuum system for, 33–44
Tandem mass spectrometer, definition, 331
Temperature control:
  in ICR, 23–24, 34
  in PHPMS, 236
Ternary association reactions, 178, 196, 210

INDEX 651

Terrestrial atmosphere, reactions in, 181
Terrestrial ionosphere, reactions in, 174
TESICO, see Threshold electron secondary ion coincidence method
Thermalization of ions, 33, 40, 44
Thermochemical data, see also Thermochemistry, bond energies
  from association and clustering equilibria, 282
  from equilibria, 270
  from reaction thresholds, 402
Thermochemistry, 88
  bond energies, 88–90
Thiophene, reaction with $^3HeT^+$, 315
Thoria, 134
Threshold electron secondary ion coincidence method, 430–436
  thermochemistry from, 402
TICR, see Trapped ICR
Time of flight (TOF), 362
  laser overlap, 615
  mass selection, 612–615
  parent cluster selection, 612–615
  photofragment analysis, 612–615
  tandem mass spectrometer, 612–615
  two pulsed fields, 617
TOF, see Time of flight
Trajectory surface hopping calculation, 446
Transient digitizer, 512, 527
Transient ICR, for collision frequency determination, 23
Transition state:
  loose, 402
  tight, 379, 401
Translational spectroscopy, 371, 382, 384
Transverse excitation at atmospheric pressure (TEA), 567
Trapped ICR, 2, 5, 25–33
  cells for, 5, 26–28, 63–65, 71
  data analysis in, 30–31
  double resonance, 28, 31
  electronics for, 28–30
  ion trapping efficiency, 32
  for mass spectrometry, 33
  power absorption in, 30–31
  pressure range in, 26
  reaction kinetics by, 25–26, 33
  resolution in, 29, 30, 31–32
  sensitivity, 26, 30, 32–33
Trapping:
  efficiency, in ICR, 32
  in FTMS, 63–65, 71
  in kinetic energy ICR, 45–46
Trapping ejection in ICR, 11, 13, 15, 25. See also Double resonance in ICR

ejection time for, 11
Trapping potential, see Electric fields in ICR; Ion cyclotron resonance cell
Triatomic ions, isomeric forms of, 204
Tricyclotron spectrometer, 44
Trimethylamine, 154
Tritiated molecules:
  decay, 288, 292
  decay-induced fragmentation, 294
  hydrocarbons, 295
Tritium atoms:
  decay, 289
  decay of chemically bound, 291
Tritium decay, molecular consequences of, 291

Ultraviolet and visible emission, 489–504, 529–538, 546–556
  spectral identification, 494, 535–536
Unimolecular decay:
  metastable reaction products, 373
  rate from PIPECO, 426
  statistical models, 379

Vacuum systems in ICR, 14, 33–34
Vacuum ultraviolet chemistry, 419
Vacuum ultraviolet radiation, 418
  coherent, 420
Variable-temperature selected ion flow drift tube (VT-SIFDT), 168
Vector:
  centroid, 332
  relative velocity, 332, 335
Velocity, relative, 332, 335
Vespel, 134
  in ICR cells, 12, 27
Vibrational energy:
  in kinetic energy ICR, 48
  selection, 371–374, 377–378, 417–487
  in tandem ICR, 33, 42
Vibrational energy transfer, 362
Vibrationally excited ions, quenching rate coefficient of, 208
Vibrationally induced electron detachment (VED), 586
Vibrational relaxation, 202, 581
  state-to-state cross section, 462
Vibrational state resolution, in translation, 365, 370–371
Vibrational state selection, 373, 417–487
Vinyl cation, addition to methane, 316
Vinylic cations, 161
VUV, see Vacuum ultraviolet radiation

Wannier formula, 553

Wien filter, 347, 388
  stigmatic operation, 348
  in tandem ICR, 38–39, 40
Work function, 342

Xe^+, charge transfer reactions:
  in kinetic energy ICR, 47, 48
  in tandem ICR, 42
XT molecules:
  decay, 291
  hydrogen-like, decay, 295

Zero-collision power absorption in ICR, 9–10
Zero-point energy, of ^{12}CO and ^{13}CO, 193